Berger Automating with STEP 7 in STL and SCL

Automating with STEP 7 in STL and SCL

Programmable Controllers
SIMATIC S7-300/400

by Hans Berger

6th revised and enlarged edition, 2012

Publicis Publishing

Bibliographic information published by the Deutsche Nationalbibliothek
The Deutsche Nationalbibliothek lists this publication in the Deutsche Nationalbibliografie; detailed bibliographic data are available in the Internet at http://dnb.d-nb.de.

This book contains one Trial DVD. **"SIMATIC STEP 7 Professional, Edition 2010 SR1, Trial License"** encompasses: SIMATIC STEP 7 V5.5 SP1, S7-GRAPH V5.3 SP7, S7-SCL V5.3 SP6, S7-PLCSIM V5.4 SP5 and can be used for trial purposes for 14 days.

This Software can only be used with the Microsoft Windows XP 32 Bit Professional Edition SP3 or Microsoft Windows 7 32/64 Bit Professional Edition SP1 or Microsoft Windows 7 32/64 Bit Ultimate Edition SP1 operating systems.

Additional information can be found in the Internet at:
http://www.siemens.com/sce/contact
http://www.siemens.com/sce/modules
http://www.siemens.com/sce/tp

The programming examples concentrate on describing the STL and SCL functions and providing SIMATIC S7 users with programming tips for solving specific tasks with this controller.

The programming examples given in the book do not pretend to be complete solutions or to be executable on future STEP 7 releases or S7-300/400 versions. Additional care must be taken in order to comply with the relevant safety regulations.

The author and publisher are always grateful to hear your responses to the contents of the book.

Publicis Publishing
P.O. Box 3240
91050 Erlangen
E-mail: publishing-distribution@publicis.de

Internet: www.publicis-books.de

ISBN 978-3-89578-412-5

6th edition, 2012

Editor: Siemens Aktiengesellschaft, Berlin and Munich
Publisher: Publicis Publishing, Erlangen

Printed in Germany

Preface

The SIMATIC automation system unites all the subsystems of an automation solution under a uniform system architecture to form a homogeneous whole from the field level right up to process control. This Totally Integrated Automation (TIA) enables integrated configuring and programming, data management and communications throughout the complete automation system.

As the basic tool for SIMATIC, STEP 7 plays an integrating role in Totally Integrated Automation. STEP 7 is used to configure and program the SIMATIC S7, SIMATIC C7 and SIMATIC WinAC automation systems. Microsoft Windows has been chosen as the operating system to take advantage of the familiar user interface of standard PCs as also used in office environments.

For block programming STEP 7 provides programming languages that comply with DIN EN 6.1131-3: STL (statement list; an Assembler-like language), LAD (ladder logic; a representation similar to relay logic diagrams), FBD (function block diagram) and the S7-SCL optional package (Structured Control Language, a Pascal-like high-level language). Several optional packages supplement these languages: S7-GRAPH (sequential control), S7-HiGraph (programming with state-transition diagrams) and CFC (connecting blocks; similar to function block diagram). The various methods of representation allow every user to select the suitable control function description. This broad adaptability in representing the control task to be solved significantly simplifies working with STEP 7.

This book describes the STL and SCL programming languages for S7-300/400. As a valuable supplement to the description of the languages, and following an introduction to the S7-300/400 automation system, it provides valuable, practice-oriented information on the basic handling of STEP 7 when configuring, networking and programming SIMATIC PLCs. The description of the "Basic Functions" of a binary control, such as logic operations or latching/unlatching functions, makes it particularly easy for first-time users or users changing from relay contactor controls to become acquainted with STEP 7. The digital functions explain how digital values are combined; for example, basic calculations, comparisons or data type conversion.

The book shows how you can control program processing (program flow) and design structured programs. In addition to the cyclically processed main program, you can also incorporate event-driven program sections as well as influence the behavior of the controller at startup and in the event of errors/faults.

One section of the book is dedicated to the description of the SCL programming language. SCL is especially suitable for programming complex algorithms or for tasks in the data management area, and it supplements STL towards higher-level programming languages. The book concludes with the description of a program for converting STEP 5 programs to STEP 7 programs, and a general overview of the system functions and the function set for STL and SCL.

The contents of this book describe Version 5.5 of the STEP 7 programming software and Version 5.3 SP5 of the S7-SCL optional package.

Nuremberg, May 2012

Hans Berger

The Contents of the Book at a Glance

Processing the user program

Working with complex variables, indirect addressing

Description of the Programming Language SCL

S5/S7 Converter, block libraries, overviews

Program Processing

20 Main Program
Program Structure; Scan Cycle Control (Response Time, Start Information, Background Scanning); Program Functions; Communications via Distributed I/O and Global Data; S7 and S7-Basic Communications

21 Interrupt Handling
Time-of-Day Interrupts; Time-Delay Interrupts; Watchdog Interrupts; Hardware Interrupts; DPV1 Interrupts; Multiprocessor Interrupt; Handling Interrupts

22 Restart Characteristics
Cold Restart, Hot Restart, Warm Restart; STOP, HOLD, Memory Reset; Parameterizing Modules

23 Error Handling
Synchronous Errors; Asynchronous Errors; System Diagnostics

Variable Handling

24 Data Types
Structure of the Data Types, Declaration and Use of Elementary and Complex Data Types; Programming of User Defined Data Types UDT

25 Indirect Addressing
Area Pointer, DB Pointer, ANY Pointer; Indirect Addressing via Memory and Register (Area-internal and Area-crossing); Working with Address Registers

26 Direct Variable Access
Load Variable Address Data Storage of Variables in the Memory; Data Storage when Transferring Parameters; "Variable" ANY Pointer; Brief Description of the "Message Frame Example"

Structured Control Language SCL

27 Introduction, Language Elements
Addressing, Operators, Expressions, Value Assignments

28 Control Statements
IF, CASE, FOR, WHILE, REPEAT, CONTINUE, EXIT, GOTO, RETURN

29 SCL Block Calls
Function Value; OK Variable, EN/ENO Mechanism, Description of Examples

30 SCL Functions
Timer Functions; Counter Functions; Conversion and Math Functions; Shifting and Rotating

31 IEC Functions
Conversion and Comparison Functions; STRING Functions; Date/Time-of-Day Functions; Numerical Functions

Appendix

32 S5/S7 Converter
Preparations for Conversion; Converting STEP 5 Programs; Postprocessing

33 Block Libraries
Organization Blocks; System Function Blocks; IEC Function Blocks; S5-S7 Converting Blocks; TI-S7 Converting Blocks; PID Control Blocks; DP Functions

34 STL Operation Overview
Basic Functions; Digital Functions; Program Flow Control; Indirect Addressing

35 SCL Statement and Function Overview
Operators; Control Statements; Block Calls; Standard Functions

The Programming Examples

The present book provides many figures representing the use of the STL and SCL programming languages. All programming examples can be downloaded from the publisher's website www.publicis-books.de. There are two libraries, one for STL examples (STL_Book) and one for SCL examples (SCL_Book). When dearchived with the Retrieve function, these libraries occupy approximately 2.9 or 1.7 MB (dependent on the PG/PC file system used).

The library STL_Book contains eight programs that are essentially illustrations of the STL method of representation. Two extensive examples show the programming of functions, function blocks and local instances (Conveyor Example) and the handling of data (Message Frame Example). All the examples exist as source files and contain symbols and comments.

The library SCL_Book contains five programs with representations of the SCL statements and

Library STL_Book

Basic Functions Examples of STL representation	**Program Processing** Examples of SFC Calls
FB 104 Chapter 4: Binary Logic Operations	FB 120 Chapter 20: Main Program
FB 105 Chapter 5: Memory Functions	FB 121 Chapter 21: Interrupt Handling
FB 106 Chapter 6: Transfer Functions	FB 122 Chapter 22: Restart Characteristics
FB 107 Chapter 7: Timer Functions	FB 123 Chapter 23: Error Handling
FB 108 Chapter 8: Counter Functions	
Digital Functions Examples of STL representation	**Variable Handling** Examples of Data Types and Variable Processing
FB 109 Chapter 9: Comparison Functions	FB 124 Chapter 24: Data Types
FB 110 Chapter 10: Arithmetic Functions	FB 125 Chapter 25: Indirect Addressing
FB 111 Chapter 11: Math Functions	FB 126 Chapter 26: Direct Variable Access
FB 112 Chapter 12: Conversion Functions	FB 101 Elementary Data Types
FB 113 Chapter 13: Shift Functions	FB 102 Complex Data Types
FB 114 Chapter 14: Word Logic	FB 103 Parameter Types
Program Flow Control Examples of STL representation	**Conveyor Example** Examples of Basic Functions and Local Instances
FB 115 Chapter 15: Status Bits	FC 11 Conveyor Belt Controller
FB 116 Chapter 16: Jump Functions	FC 12 Counter Control
FB 117 Chapter 17: Master Control Relay	FB 20 Feed
FB 118 Chapter 18: Block Functions	FB 21 Conveyor Belt
FB 119 Chapter 19: Block Parameters	FB 22 Parts Counter
Source File Block Programming (Chapter 3)	
Message Frame Example Handling Data examples	**General Examples**
UDT 51 Data Structure Header	FC 41 Range Monitor
UDT 52 Data Structure Message Frame	FC 42 Limit Value Detection
FB 51 Generate Message Frame	FC 43 Compound Interest Calculation
FB 52 Save Message Frame	FC 44 Double-Word-Wise Edge Evaluation
FC 61 Clock Check	FC 45 Converting S5 Floating-Point to S7 REAL
FC 62 Generate Checksum	FC 46 Converting S7 REAL to S5 Floating-Point
FC 63 Convert Date	FC 47 Copy Data Area (ANY Pointer)

the SCL functions. The programs "Conveyor Example" The library SCL_Book contains five programs with representations of the SCL statements and the SCL functions. The programs "Conveyor Example" and "Message Frame Example" and "Message Frame Example" show the same functions as the STL examples of the same name. The program "General Examples" contains SCL functions for processing complex data types, data storage and – for SCL programmers – a statement for programming simple STL functions for SCL programs.

To try the programs out, set up a project corresponding to your hardware configuration and then copy the program, including the symbol table from the library to the project. Now you can call the example programs, adapt them for your own purposes and test them online.

Library SCL_Book

27 Language Elements Examples of SCL Representation (Chapter 27)	**30 SCL Functions** Examples of SCL Representation (Chapter 30)
FC 271 Delimiter Example OB 1 Main Program for the Delimiter Example FB 271 Operators, Expressions, Assignments FB 272 Indirect Addressing	FB 301 Timer Functions FB 302 Counter Functions FB 303 Conversion Functions FB 304 Math Functions FB 305 Shifting and Rotating
28 Control Statements Examples of SCL Representation (Chapter 28)	**31 IEC Functions** Examples of SCL Representation (Chapter 31)
FB 281 IF Statement FB 282 CASE Statement FB 283 FOR Statement FB 284 WHILE Statement FB 285 REPEAT Statement	FB 311 Conversion Functions FB 312 Comparison Functions FB 313 String Functions FB 314 Date/Time-of-day Functions FB 315 Numerical Functions
29 SCL Block Calls Examples of SCL Representation (Chapter 29)	**General Examples**
FC 291 FC Block with Function Value FC 292 FC Block without Function Value FB 291 FB Block FB 292 Example Calls for FC and FB Blocks FC 293 FC Block for EN/ENO Example FB 293 FB Block for EN/ENO Example FB 294 Calls for EN/ENO Examples	FC 61 DT_TO_STRING FC 62 DT_TO_DATE FC 63 DT_TO_TOD FB 61 Variable Length FB 62 Checksum FB 63 Ring Buffer FB 64 FIFO Register STL Functions for SCL Programming
Conveyor Example Examples of Basic Functions and Local Instances	**Message Frame Example** Handling Data examples
FC 11 Conveyor Belt Controller FC 12 Counter Control FB 20 Feed FB 21 Conveyor Belt FB 22 Parts Counter	UDT 51 Data Structure Header UDT 52 Data Structure Message Frame FB 51 Generate Message Frame FB 52 Save Message Frame FC 61 Clock Check

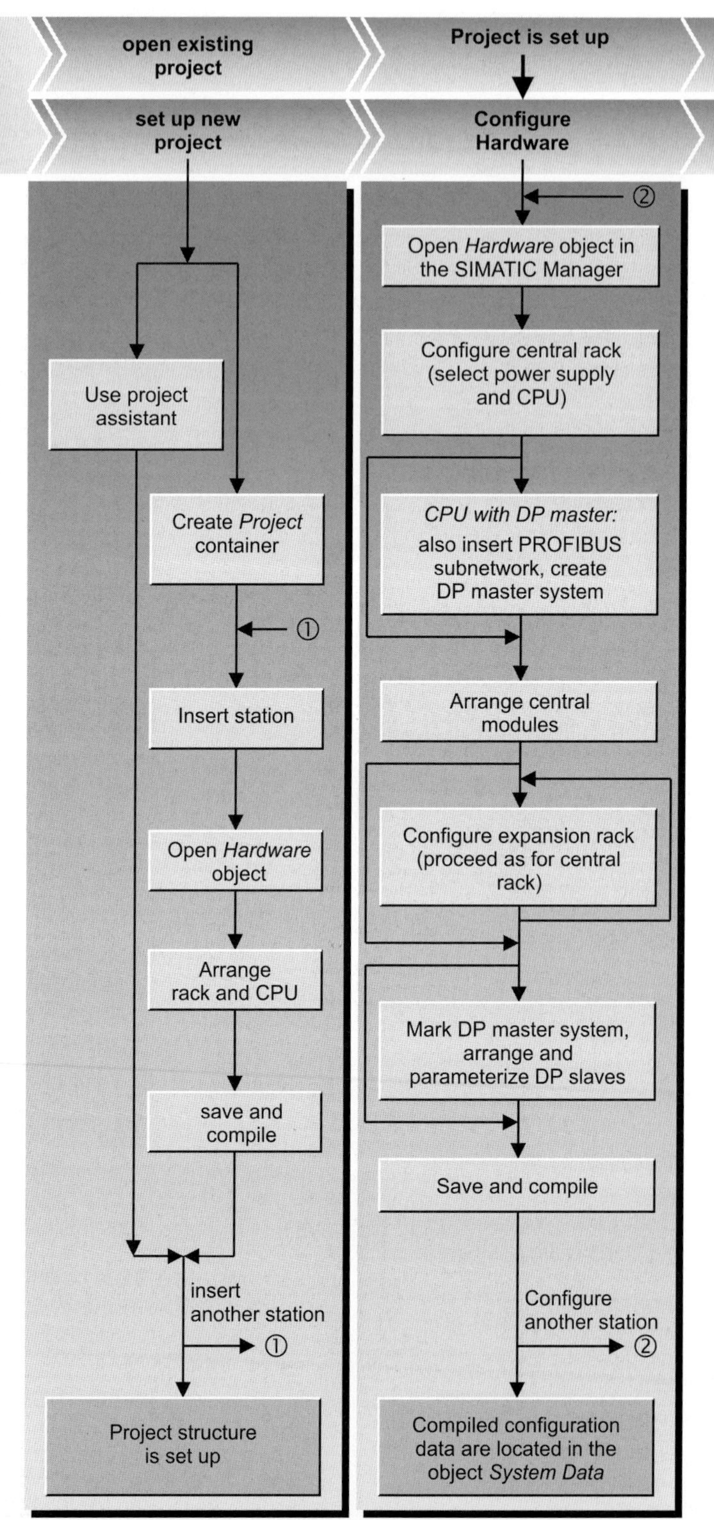

Automating with STEP 7

This double page shows the basic procedure for using the STEP 7 programming software.

Start the SIMATIC Manager and set up a new project or open an existing project. All the data for an automation task are stored in the form of objects in a project. When you set up a project, you create containers for the accumulated data by setting up the required stations with at least the CPUs; then the containers for the user programs are also created. You can also create a program container direct in the project.

In the next steps, you configure the hardware and, if applicable, the communications connections. Following this, you create and test the program.

The order for creating the automation data is not fixed. Only the following general regulation applies: if you want to process objects (data), they must exist; if you want to insert objects, the relevant containers must be available.

You can interrupt processing in a project at any time and continue again from any location the next time you start the SIMATIC Manager.

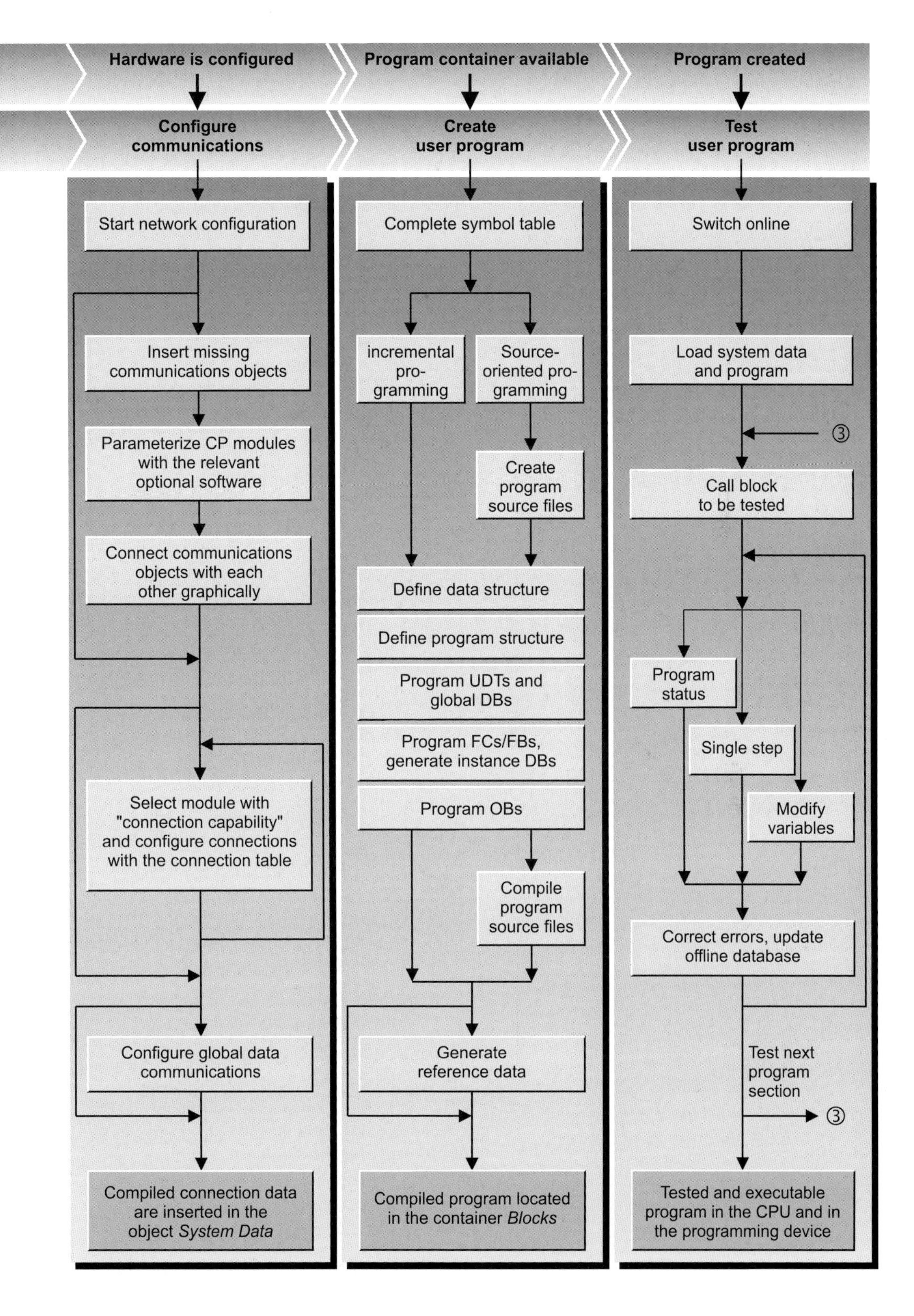
Hardware is configured
Program container available
Program created
Configure communications
Create user program
Test user program
Start network configuration
Insert missing communications objects
Parameterize CP modules with the relevant optional software
Connect communications objects with each other graphically
Select module with "connection capability" and configure connections with the connection table
Configure global data communications
Compiled connection data are inserted in the object System Data
Complete symbol table
incremental pro-gramming
Source-oriented pro-gramming
Create program source files
Define data structure
Define program structure
Program UDTs and global DBs
Program FCs/FBs, generate instance DBs
Program OBs
Compile program source files
Generate reference data
Compiled program located in the container Blocks
Switch online
Load system data and program
③
Call block to be tested
Program status
Single step
Modify variables
Correct errors, update offline database
Test next program section
③
Tested and executable program in the CPU and in the programming device

Table of Contents

Introduction

This section of the book provides an overview of the SIMATIC S7-300/400.

The **S7-300/400 programmable controller** is of modular design. The modules with which it is configured can be central (in the vicinity of the CPU) or distributed without any special settings or parameter assignments having to be made. In SIMATIC S7 systems, distributed I/O is an integral part of the system. The CPU, with its various memory areas, forms the hardware basis for processing of the user programs. A load memory contains the complete user program: the parts of the program relevant to its execution at any given time are in a work memory whose short access times are the prerequisite for fast program processing.

STEP 7 is the programming software for S7-300/400 and the automation tool is the SIMATIC Manager. The SIMATIC Manager is an application for the Windows operating systems from Microsoft and contains all functions needed to set up a project. When necessary, the SIMATIC Manager starts additional tools, for example to configure stations, initialize modules, and to write and test programs.

You formulate your automation solution in the STEP 7 programming languages. The **SIMATIC S7 program** is structured, that is to say, it consists of blocks with defined functions that are composed of networks or rungs. Different priority classes allow a graduated interruptibility of the user program currently executing. STEP 7 works with variables of various data types starting with binary variables (data type BOOL) through digital variables (e.g. data type INT or REAL for computing tasks) up to complex data types such as arrays or structures (combinations of variables of different types to form a single variable).

The first chapter contains an overview of the hardware of the S7-300/400 automation system and the second chapter contains the same overview of the STEP 7 programming software. The basis of the description is the functional scope for STEP 7 Version 5.5.

Chapter 3 “SIMATIC S7 Program” serves as an introduction to the most important elements of an S7 program and shows the programming of individual blocks in the programming languages STL and SCL. The functions and statements of STL and SCL are then described in the subsequent chapters of the book. All the descriptions are explained using brief examples.

1 SIMATIC S7-300/400 Programmable Controller
Structure of the programmable controller; distributed I/O; communications; module addresses; address areas

2 STEP 7 Programming Software
SIMATIC Manager; processing a project; configuring a station; configuring a network; writing programs (symbol table, program editor); switching online; testing programs

3 SIMATIC S7 Program
Program processing with priority classes; program blocks; addressing variables; programming blocks with STL and SCL; variables and constants; data types (overview)

1 SIMATIC S7-300/400 Programmable Controller

1.1 Structure of the Programmable Controller

1.1.1 Components

The SIMATIC S7-300/400 is a modular programmable controller comprising the following components:

- ▷ Racks;
  Accommodate the modules and connect them to each other
- ▷ Power supply (PS);
  Provides the internal supply voltages
- ▷ Central processing unit (CPU);
  Stores and processes the user program
- ▷ Interface modules (IMs);
  Connect the racks to one another
- ▷ Signal modules (SMs);
  Adapt the signals from the system to the internal signal level or control actuators via digital and analog signals
- ▷ Function modules (FMs);
  Execute complex or time-critical processes independently of the CPU
- ▷ Communications processors (CPs)
  Establish the connection to subsidiary networks (subnets)
- ▷ Subnets
  Connect programmable controllers to each other or to other devices

A programmable controller (or station) may consist of several racks, which are linked to one another via bus cables. The power supply, CPU and I/O modules (SMs, FMs and CPs) are plugged into the central rack. If there is not enough room in the central rack for the I/O modules or if you want some or all I/O modules to be separate from the central rack, expansion racks are available which are connected to the central rack via interface modules (Figure 1.1).

It is also possible to connect distributed I/O to a station (see Chapter 1.2.1 “PROFIBUS DP”).

The racks connect the modules with two buses: the I/O bus (or P bus) and the communication bus (or K bus). The I/O bus is designed for high-speed exchange of input and output signals, the communication bus for the exchange of large amounts of data. The communication bus connects the CPU and the programming device interface (MPI) with function modules and communications processors.

1.1.2 S7-300 Station

Centralized configuration

In an S7-300 controller, as many as 8 I/O modules can be plugged into the central rack. Should this single-tier configuration prove insufficient, you have two options for controllers equipped with a CPU 313 or a more advanced CPU:

- ▷ Either choose a two-tier configuration (with IM 365 up to 1 meter between racks)
- ▷ or choose a configuration of up to four tiers (with IM 360 and IM 361 up to 10 meters between racks)

You can operate a maximum of 8 modules in a rack. The number of modules may be limited by the maximum permissible current per rack, which is 1.2 A.

The modules are linked to one another via a backplane bus, which combines the functions of the P and K buses.

Local bus segment

A special feature regarding configuration is the use of the FM 356 application module. An FM 356 is able to “split” a module's backplane bus and to take over control of the remaining modules in the split-off “local bus segment”

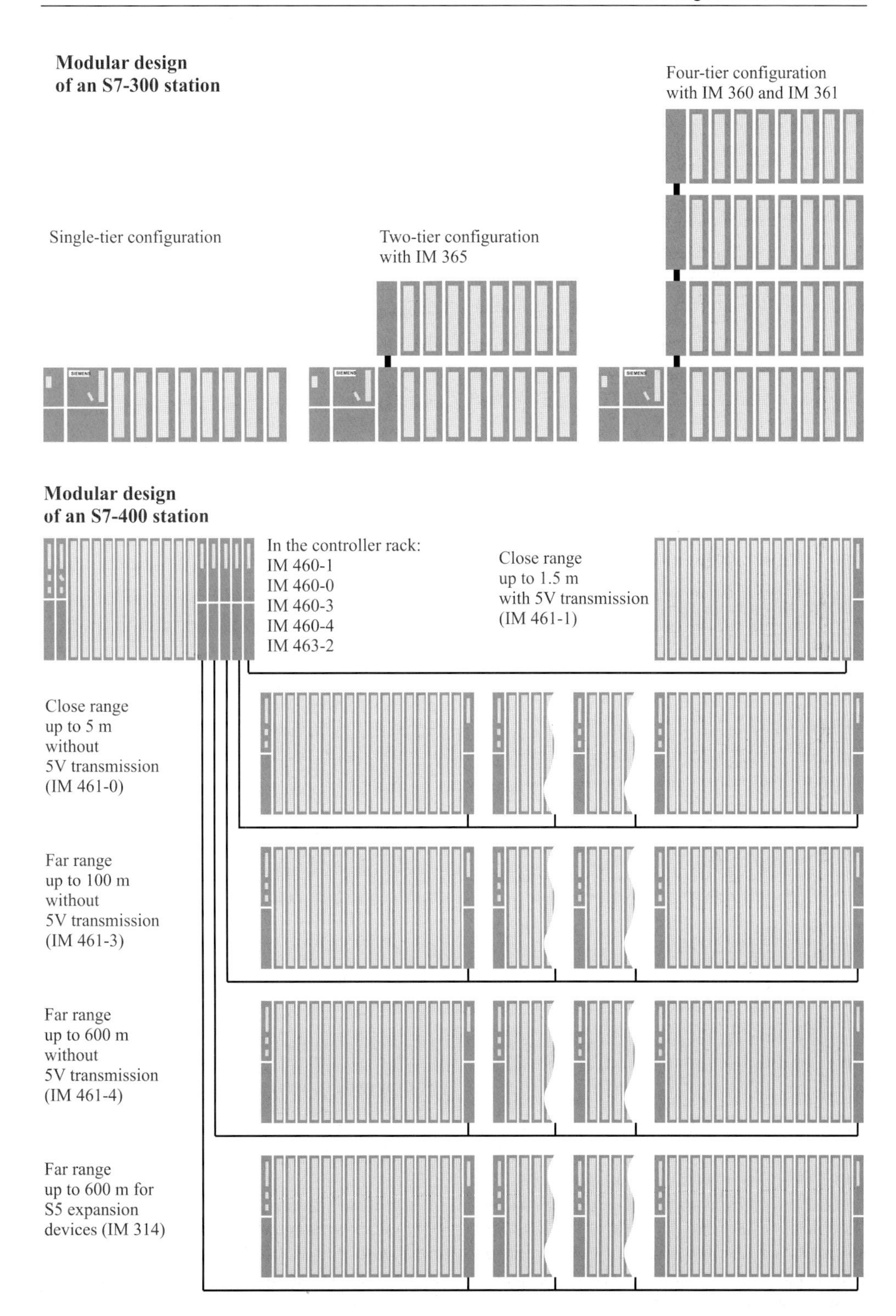

Figure 1.1 Hardware Configuration for S7-300/400

itself. The limitations mentioned above regarding the number of modules and the power consumption also apply in this case.

Standard CPUs

The standard CPUs are available in different versions with regard to memory size and processing speed. They range from the "smallest" CPU 312 for smaller applications with moderate processing speed requirements up to the CPU 319-3 PN/DP with a large program memory and fast program execution for cross-sector automation tasks. Equipped with the corresponding interfaces, some CPUs can be used as the central controller for the distributed I/O over PROFIBUS and PROFINET.

A micro memory card (MMC) is required for using the standard CPUs – as is the case with all innovated S7-300 CPUs. This memory medium opens up new application possibilities compared to the previously used memory card (see Chapter 1.1.6 "CPU Memory Areas").

The now discontinued CPU 318 can be replaced by the CPUs 317 or 319.

Compact CPUs

The 3xxC CPUs permit a compact design for mini programmable controllers. Depending on the version, they already contain:

▷ Integral I/O
Digital and analog inputs/outputs

▷ Integral technological functions
Counting, measurement, control, positioning

▷ Integral communications interfaces
PROFIBUS DP master or slave, point-to-point coupling (PtP)

The technological functions are system blocks which use the onboard I/Os of the CPU.

Technology CPUs

The CPUs 3xxT combine open-loop control functions with simple motion control functions. The control section is designed as with a standard CPU. It is configured, parameterized and programmed using STEP 7. The technology objects and the motion control section require the S7-Technology option package that is integrated in the SIMATIC Manager following the installation.

The technology CPUs have a PROFIBUS DP interface which permits use as a DP master or DP slave. The CPUs are used for cross-sector automation tasks in series machine construction, special machine construction, and plant construction.

Fail-safe CPUs

The CPUs 3xxF are used in production plants with increased safety requirements. Corresponding PROFIBUS and PROFINET interfaces allow use of safety-related distributed I/O with the PROFIsafe bus profile (see "S7 Distributed Safety" under 1.1.5 "Safety-related SIMATIC"). Standard modules for normal applications can be used parallel to safety-related operation.

SIPLUS

The SIPLUS product family offers modules that can be used in harsh environments. The SIPLUS components are based on standard devices which have been specially converted for the respective application, for example for an extended temperature range, increased resistance to vibration and shock, or voltage ranges differing from the standard. Please therefore note the technical data for the respective SIPLUS module. In order to carry out the configuration with STEP 7, use the equivalent type (the standard module on which it is based); this is specified, for example, on the module's nameplate.

1.1.3 S7-400 Station

Centralized configuration

The controller rack of the S7-400 is available in the versions UR1 (18 slots), UR2 (9 slots) and CR3 (4 slots). UR1 and UR2 can also be used as expansion racks. The power supply and the CPU also occupy slots in the racks, possibly even two or more per module. If necessary, the number of available slots can be increased using expansion racks: UR1 and ER1 have 18 slots each, UR2 and ER2 have 9 slots each.

Using the IM 460-1 and IM 461-1 interface modules, one expansion rack per interface can be located up to 1.5 m away from the controller rack, and the 5 V supply is also transmitted. Up to 4 expansion ranks can also be operated via IM 460-0 and 461-0 in the local range up to 5 m. For longer distances, the IM 460-3 and IM 461-3 or the IM 460-4 and 461-4 enable up to 4 expansion racks to be operated up to 100 m or 600 m away.

A maximum of 21 expansion racks can be connected to a central rack. To distinguish between racks, you set the number of the rack on the coding switch of the receiving IM.

The backplane bus consists of a parallel P bus and a serial K bus. Expansion racks ER1 and ER2 are designed for "simple" signal modules which generate no process interrupts, do not have to be supplied with 24 V voltage via the P bus, require no back-up voltage, and have no K bus connection. The K bus is in racks UR1, UR2 and CR2 either when these racks are used as central racks or expansion racks with the numbers 1 to 6.

Connecting segmented rack

A special feature is the segmented rack CR2. The rack can accommodate two CPUs with a shared power supply while keeping them functionally separate. The two CPUs can exchange data with one another via the K bus, but have completely separate P buses for their own signal modules.

Multiprocessor mode

In an S7-400, as many as four specially designed CPUs in a suitable rack UR can take part in multiprocessor mode. Each module in this station is assigned to only one CPU, both with its address and its interrupts. See Chapters 20.3.6 "Multiprocessing Mode" and 21.7 "Multiprocessor Interrupt" for more details.

Connection of SIMATIC S5 modules

The IM 463-2 interface module allows you to connect S5 expansion units (EG 183U, EG 185U, EG 186U as well as ER 701-2 and ER 701-3) to an S7-400, and also allows centralized expansion of the expansion units. An IM 314 in the S5 expansion unit handles the link. You can operate all analog and digital modules allowed in these expansion units. An S7-400 can accommodate as many as four IM 463-2 interface modules; as many as four S5 expansion units can be connected in a distributed configuration to each of an IM 463-2's interfaces.

1.1.4 Fault-Tolerant SIMATIC

For applications with high fault tolerance demands for machines and processes, there are two versions of SIMATIC S7 fault-tolerant programmable controllers with a redundant design: software redundancy and S7-400H/FH.

Software redundancy

Using SIMATIC S7-300/400 standard components, you can establish a software-based redundant system with a master station controlling the process and a standby station assuming control in the event of the master failing.

Fault tolerance through software redundancy is suitable for slow processes because transfer to the standby station can require several seconds depending on the configuration of the programmable controllers. The process signals are "frozen" during this time. The standby station then continues operation with the data last valid in the master station.

Redundancy of the input/output modules is implemented with distributed I/O (ET 200M with IM 153-2 interface module for redundant PROFIBUS DP). The software redundancy can be configured with STEP 7 Version 5.2 and higher.

Fault-tolerant SIMATIC S7-400H

The SIMATIC S7-400H is a fault-tolerant programmable controller with redundant configuration comprising two central racks, each with an H CPU and a synchronization module for data comparison via fiber optic cable. Both controllers operate in "hot standby" mode; in the event of a fault, the intact controller assumes operation alone via automatic bumpless transfer. The UR2-H rack with 2x9 slots offers the possibility for also installing a fault-tolerant system in one single rack.

The I/O can have normal availability (single-channel, single-sided configuration) or enhanced availability (single-channel, switched configuration with ET 200M). Communication takes place with a single or redundant bus.

The user program is the same as for a non-redundant device; the redundancy function is provided exclusively by the hardware that is used and is kept hidden from the user. The software package required for configuration is included in STEP 7 from V5.3. The provided standard libraries *Redundant IO* contain blocks for supporting the redundant I/O.

1.1.5 Safety-related SIMATIC

Fail-safe programmable controllers control processes in which the safe state can be achieved by direct switching-off. They are used in plants with increased safety requirements.

The safety functions are mainly located in the safety-related user program of a correspondingly designed CPU and in the failsafe input and output modules. An F-CPU complies with the safety requirements up to AK 6 in accordance with DIN V 19250/DIN V VDE 0801, up to SIL 3 in accordance with IEC 61508, and up to Category 4 in accordance with EN 954-1. Safety functions can be executed parallel to a non-safety-related user program in the same CPU.

Safety-related communication over PROFIBUS DP – also over PROFINET IO with S7 Distributed Safety – uses the PROFIsafe bus profile. This permits transmission of safety-related and non-safety-related data on a single bus cable.

Safety Integrated for the manufacturing industry

S7 Distributed Safety is a failsafe automation system for the protection of machines and personnel mainly for applications with machine controls and in the process industry.

Controllers from the SIMATIC S7-300, S7-400, and ET 200S ranges are available as F-CPUs. The safety-related I/O modules are connected to S7-400 over PROFIBUS DP or PROFINET IO using the safety-related PROFIsafe bus profile. With S7-300 and ET 200S, use of safety-related I/O modules is additionally possible in the central rack.

The hardware configuration and programming of the non-safety-related user program are carried out using the standard applications of STEP 7.

The *SIMATIC S7 Distributed Safety* option package is required to program the safety-related parts of the program. With this option package you can use the F-LAD or F-FBD programming languages to create the blocks which contain the safety-related program. Interfacing to the I/O is carried out using the process image as with the standard program. S7 Distributed Safety also includes a library with TÜV-certified safety blocks. There is an additional library available with F-blocks for press and burner controls.

The safety-related user program can be executed parallel to the standard user program. If an error is detected in the safety-related part of the program, the CPU enters the STOP state.

Safety Integrated for the process industry

S7 F/FH Systems is a failsafe automation system based on S7-400 mainly for applications in the process industry. The safety-related I/O modules are connected over PROFIBUS DP using the safety-related PROFIsafe bus profile.

An S7-400 F-CPU is provided with the safety-related control functions by application of an *S7 F Systems Runtime license*. A non-safety-related user program can be executed parallel to the safety-related plant unit.

In addition to fail-safety, the S7-400FH also provides increased availability. If a detected fault results in a STOP of the master CPU, a reaction-free switch is made to the CPU running in hot standby mode. The *S7 H Systems* option package is additionally required for operation as S7-400FH.

The hardware configuration and programming of the non-safety-related user program are carried out using the standard applications of STEP 7.

The *S7 F Systems* option package is additionally required for programming the safety-related program parts, and additionally the *CFC* option package V5.0 SP3 and higher and the *S7-SCL* option package V5.0 and higher.

The safety-related program is programmed using CFC (Continuous Function Chart). Programmed, safety-related function blocks from the supplied F-library can be called and interconnected in this manner. Alongside functions for programming safety functions, they also include fault detection and fault reaction functions. This ensures that if there are failures or errors, the F-system can be stopped in or transferred to a safe mode. If a fault is detected in the safety program, the safety-related part of the plant is switched off, whereas the remaining part can continue to operate.

Fail-safe I/O

Failsafe signal modules (F-modules or F-submodules) are required for safety operation. The fail-safety is achieved through the integrated safety functions and the corresponding wiring of sensors and actuators.

The F-modules can also be used in standard applications with enhanced diagnostics requirements. Redundant F-modules can be used with S7 F/FH systems to increase the availability both in standard and safety-related operation.

The failsafe I/O is available in various versions:

▷ The fail-safe signal modules in S7-300 design are used in the ET 200M distributed I/O device or – with S7-Distributed Safety – also centrally.

▷ Failsafe I/O modules are available for the distributed I/O devices in the designs ET 200S, ET 200pro, and ET 200eco.

▷ Failsafe interface modules are also available as F-CPUs for the ET 200S and ET 200pro distributed I/O devices.

▷ Failsafe DP standard slaves and – with S7-Distributed Safety also IO standard devices – can be used which can handle the PROFIsafe bus profile.

Failsafe CPUs and signal modules are also available in SIPLUS design.

1.1.6 CPU Memory Areas

Figure 1.2 shows the memory areas in the programming device, in the CPU, and in the signal modules which are important for your program.

The programming device contains the *off-line data*. These consist of the user program (program code and user data), the system data (e.g. hardware configuration, network and connection configuration) and further project-specific data such as e.g. the symbol table and comments.

The *online data* consists of the user program and the system data on the CPU, stored in two storage areas: the load memory and the work memory. In addition, the system memory is also present here.

Finally, the I/O modules contain memories for the signal statuses of the inputs and outputs.

The central processing units have a slot for a plug-in *memory module*. This memory module also contains the load memory, or parts thereof (see "Physical design of CPU memory", further below). The memory module is designed as memory card (S7-400-CPUs) or as micro memory card (S7-300 CPUs and derived ET 200 CPUs). A firmware update for the CPU operating system can also be performed via the memory module.

Memory card

The memory submodule for the S7-400 CPUs is the memory card (MC). There are two types of memory card: RAM cards and flash EPROM cards.

If you only want to expand the load memory, use a RAM card. A RAM card allows you to modify the entire user program online. This is necessary, for example, for larger programs when testing and during commissioning. RAM memory cards lose their contents when unplugged.

If you want to protect your user program against power failure following testing and commissioning, including configuration data and module parameters, use a flash EPROM card. In this case, load the entire program offline onto the flash EPROM card with the card plugged into the programming device. With the relevant CPUs, you can also load the program online with the memory card plugged into the CPU.

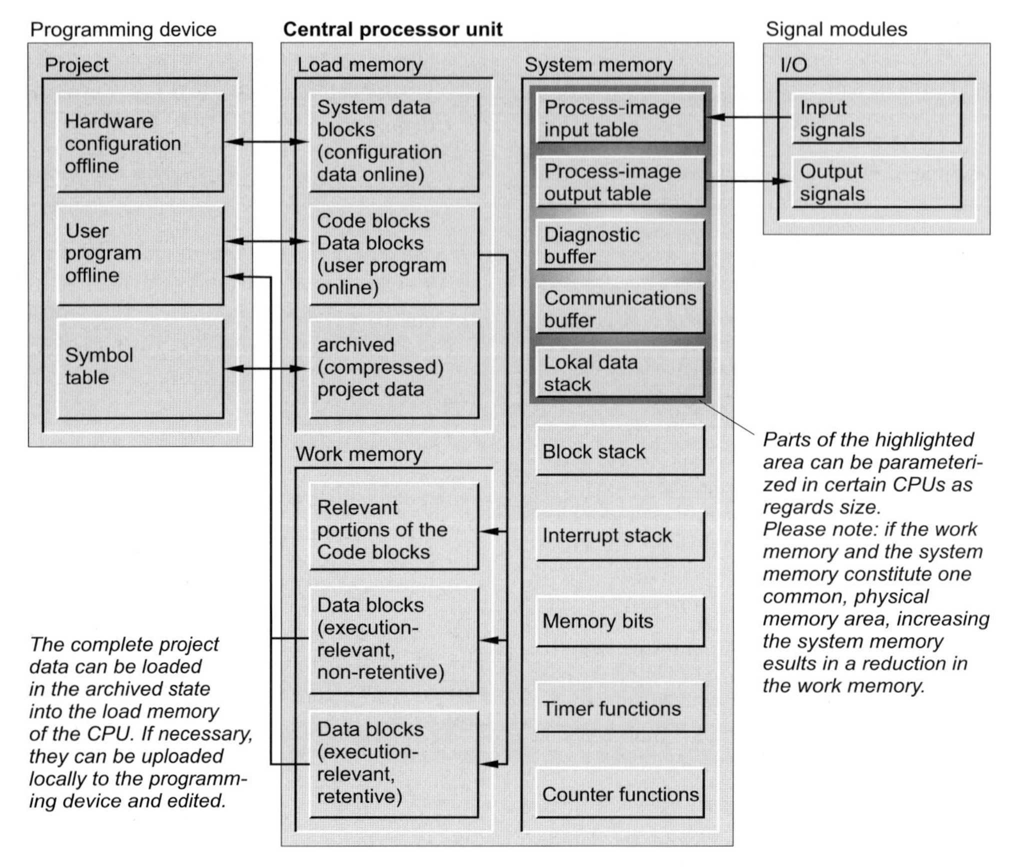

Figure 1.2 Memory areas on the CPU

Micro memory card

The memory submodule for the newer S7-300 CPUs is a micro memory card (MMC). The data on the MMC are non-volatile, but can be read, written and deleted just like with a RAM. This response permits data backup without a battery.

The MMC contains the complete load memory, so that an MMC is always required for operation. The MMC can be used as a portable storage medium for user programs or firmware updates. Using special system functions, you can read or write data blocks on the MMC from the user program, e.g. read recipes from the MMC, or create a measured-value archive on the MMC and provide it with data.

Load memory

The entire user program, including configuration data, is in the load memory (system data). From the programming device, the user program is always initially loaded into the load memory and from there into the work memory. The program in the load memory is not executed as the control program.

With a CPU 300 and a CPU ET200, the load memory is present completely on the micro memory card. Thus the contents of the load memory are retained even if the CPU is de-energized.

If the load memory with a CPU 400 consists of an integrated RAM or RAM memory card, a backup battery is required in order to keep the user program retentive. With an integrated

EEPROM or a plug-in flash EPROM memory card as the load memory, the CPU can be operated without battery backup.

From STEP 7 V5.1 and with appropriately equipped CPUs, you can save all the project data as a compressed archive file in the load memory (see Chapter 2.2.2 "Managing, Rearranging and Archiving").

Work memory

Work memory is designed in the form of high-speed RAM fully integrated in the CPU. The operating system of the CPU copies the "execution-relevant" program code and the user data into the work memory. "Relevant" is a characteristic of the existing objects and does not mean that a particular code block will necessarily be called and executed. The "actual" control program is executed in the work memory.

Specific to the product, the work memory can be either a coherent area or divided according to program and data memories, where the latter is also divided into retentive and non-retentive parts.

When writing back the user program into the programming device, the blocks are fetched from the load memory, supplemented by the current values of the data addresses from the work memory (further information available in Chapters 2.6.4 "Loading the User Program into the CPU" and 2.6.5 "Block Handling").

System memory

System memory contains the addresses (variables) that you access in your program. The addresses are combined into areas (address areas) containing a CPU-specific number of addresses. Addresses may be, for example, inputs used to scan the signal states of momentary-contact switches and limit switches, and outputs that you can use to control contactors and lamps.

The system memory on a CPU contains the following address areas:

- ▷ Inputs (I)
  Inputs are an image ("process image") of the digital input modules.
- ▷ Outputs (Q)
  Outputs are an image ("process image") of the digital output modules.
- ▷ Bit memories (M)
  are information stores which are directly accessible from any point in the user program.
- ▷ Timers (T)
  Timers are locations used to implement waiting and monitoring times.
- ▷ Counters (C)
  Counters are software-level locations, which can be used for up and down counting.
- ▷ Temporary local data (L)
  Locations used as dynamic intermediate buffers during block processing. The temporary local data are located in the L stack, which the CPU occupies dynamically during program execution.

The letters enclosed in parentheses represent the abbreviations to be used for the different addresses when writing programs. You may also assign a symbol to each variable and then use the symbol in place of the address identifier.

The system memory also contains buffers for communication jobs and system messages (diagnostics buffer). The size of these data buffers, as well as the size of the process input image, the process output image and the L stack, are parameterizable on certain CPUs.

Physical design of CPU memory

The physical design of the load memory differs according to the type of CPU (Figure 1.3).

A CPU 300 or CPU ET 200 does not have an integrated load memory. A micro memory card containing the load memory must always be inserted to permit operation. The load memory can be written and read like a RAM. The physical design means that the number of write operations is limited (no cyclic writing by user program). You can use the menu command COPY RAM TO ROM to transfer the current values of the data operands from the work memory to the load memory.

With a CPU 300 with firmware version V2.0.12 or later, the work memory for the user data consists of a retentive part and a non-retentive part.

S7-300 and ET CPUs without adjustable data retentivity

S7-300 and ET CPUs with adjustable data retentivity

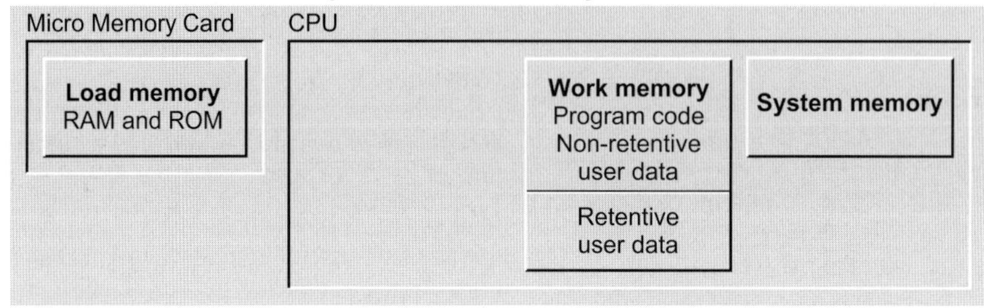

S7-400 CPU

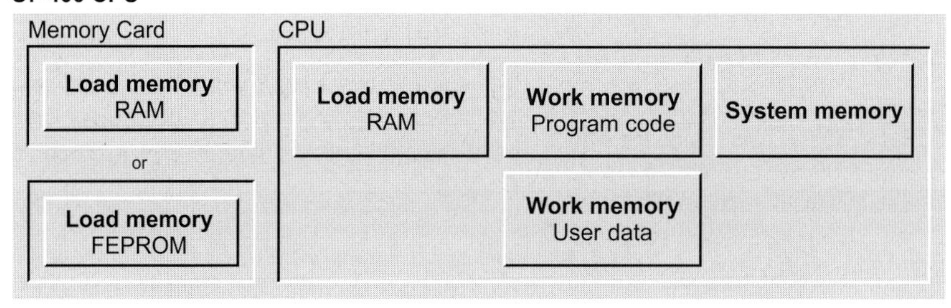

Figure 1.3 Physical Design of the CPU Memory

The control program is also present in the non-retentive part.

The integrated RAM load memory in a CPU 400 is designed for small programs or for modification of individual blocks if the load memory is a flash EPROM memory card. If the complete control program is larger than the integrated load memory, you will need a RAM memory card for testing. The tested program is then copied with the programming device to a flash EPROM memory card, which you insert into the CPU for operation.

The work memory of a CPU 400 is divided into two parts: One part saves the program code, the other the user data. The system and work memories in a CPU 400 constitute one (physical) unit. If the process image changes in size, this affects the size of the work memory.

1.2 Distributed I/O

Distributed I/O refers to modules connected via PROFIBUS DP or PROFINET IO. PROFIBUS DP uses the PROFIBUS subnet for data transmission, PROFINET IO the Industrial Ethernet subnet (for further information, see Chapter 1.3.2 “Subnets”).

1.2.1 PROFIBUS DP

PROFIBUS DP provides a standardized interface for transferring predominantly binary process data between an “interface module” in the (central) programmable controller and the field devices. This “interface module” is called the DP master and the field devices are the DP slaves.

The DP master and all the slaves it controls form a DP master system. There can be up to 32 stations in one segment and up to 127 stations in the entire network. A DP master can control a number of DP slaves specific to itself. You can also connect programming devices to the PROFIBUS DP network as well as, for example, devices for human machine interface, ET 200 devices or SIMATIC S5 DP slaves.

DP master system

PROFIBUS DP is usually operated as a "mono master system", that is, one DP master controls several DP slaves. The DP master is the only master on the bus, with the exception of a temporarily available programming device (diagnostics and service device). The DP master and the DP slaves assigned to it form a DP master system (Figure 1.4).

You can also install several DP master systems on one PROFIBUS subnet (multi master system). However, this increases the response time in individual cases because when a DP master has initialized "its" DP slaves, the access rights fall to the next DP master that in turn initializes "its" DP slaves, etc.

You can reduce the response time if a DP master system contains only a few DP slaves. Since it is possible to operate several DP masters in one S7 station, you can distribute the DP slaves of a station over several DP master systems. In multiprocessor mode, every CPU has its own DP master systems.

DP master

The DP master is the active node on the PROFIBUS network. It exchanges cyclic data with "its" DP slaves. A DP master can be

▷ A CPU with integral DP master interface or plug-in interface submodule (e.g. CPU 315-2DP, CPU 417)

▷ An interface module in conjunction with a CPU (e.g. IM 467)

▷ A CP in conjunction with a CPU (e.g. CP 342-5, CP 443-5)

There are "Class 1 masters" for data exchange in process operation and "Class 2 masters" for service and diagnostics (e.g. a programming device).

DP slaves

The DP slaves are the passive nodes on PROFIBUS. In SIMATIC S7, a distinction is made between

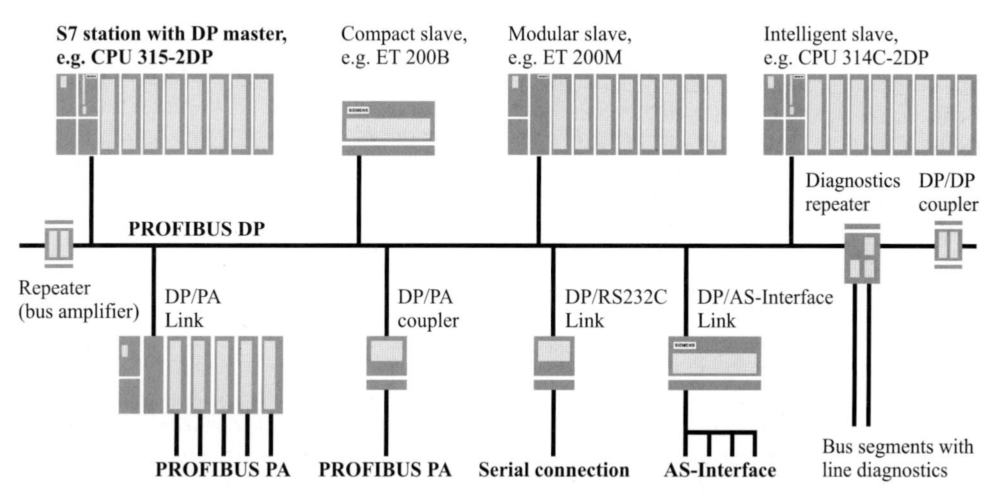

Figure 1.4 Components of a PROFIBUS DP Master System in an RS485 Segment

- ▷ Compact DP slaves
  They behave like a single module towards the DP master
- ▷ Modular DP slaves
  They comprise several modules (submodules
- ▷ Intelligent DP slaves
  They contain a control program that controls the lower-level (own) modules

Compact PROFIBUS DP slaves

Examples of compact DP slaves are the ET 200L, the ET 200R, and the ET 200eco. Bus gateways such as the DP/AS-i Link also behave like a compact slave on the PROFIBUS DP.

Modular PROFIBUS DP slaves

Examples of modular DP slaves are the ET 200iSP, the ET 200M, the ET 200S, and the ET 200pro.

Intelligent PROFIBUS DP slaves

Intelligent DP slaves are, for example, CPUs with integrated DP (slave) interface, or an S7-300 station with the CP 342-5 communications processor. Equally, an ET 200pro station with the IM 154-8 PN/DP CPU interface module or an ET 200S station with the IM 151-7 CPU interface module can be operated as intelligent DP slaves.

RS 485 repeater

The RS 485 repeater combines two bus segments in a PROFIBUS subnet. The number of bus stations and the size of the subnet can then be increased.

The repeater provides signal regeneration and electrical isolation. It can be operated at transmission rates up to 12 Mbit/s – including 45.45 kbit/s for PROFIBUS PA.

The RS 485 repeater is not configured; it need only be considered when calculating the bus parameters.

Diagnostics repeater

You can use a diagnostics repeater to determine the topology in a PROFIBUS segment (RS 485 copper cable) during operation, and to carry out line diagnostics. The diagnostics repeater provides signal regeneration and electrical isolation for the connected segments. The maximum segment length is 100 m in each case; the transmission rate can be between 9.6 kbit/s and 12 Mbit/s.

The diagnostics repeater has connections for 3 bus segments. The line from the DP master is connected to the feed terminals of the bus segment DP1. The two other connections DP2 and DP3 contain the measuring circuits for determination of the topology and line diagnostics on the bus segments connected here. Up to 9 further diagnostics repeaters can be connected in series.

The diagnostics repeater is handled in the master system like a DP slave. In the event of a fault, it transmits the determined diagnostics data to the DP master. These data are the topology of the bus segment (bus station and cable lengths), the contents of the segment diagnostics buffer (last 10 events with fault information, location and cause) and the statistics data (comments on the quality of the bus system). In addition, the diagnostics repeater provides monitoring functions for isochronous mode.

The diagnostics data can be fetched using a programming device with STEP 7 V5.2 or higher, and also displayed graphically. From the user program, the line diagnostics is triggered using the system function SFC 103 DP_TOPOL, and read using SFC 59 RD_REC or SFB 52 RDREC. To adjust the clock on the diagnostics repeater, read the CPU time using the system function SFC 1 READ_CLK and transmit it using SFC 58 WR_REC or SFB 53 WRREC.

The diagnostics repeater is configured and parameterized using STEP 7. A GSD file is available for operation on non-Siemens masters.

1.2.2 PROFINET IO

PROFINET IO offers a standardized interface for transmitting mainly binary process data over Industrial Ethernet between an “interface module” in the (central) programmable controller and the field devices. This “interface module” is referred to as the IO controller, and the field devices as the IO devices. The IO controller and all IO devices controlled by it constitute a PROFINET IO system.

PROFINET IO System

A PROFINET IO system comprises the IO controller in the central station and the IO devices (field devices) assigned to it. The connecting Industrial Ethernet subnet can be shared with other nodes and applications (Figure 1.5).

IO controller

The IO controller is the active node on the PROFINET. It exchanges data cyclically with "its" IO devices. An IO controller can be:

- A CPU with integrated PROFINET interface (e.g. CPU 317-2PN/DP)
- A CP module in association with a CPU (e.g. CP 343-1)

IO device

IO devices are the passive stations on the PROFINET. In SIMATIC S7, a distinction is made between

- Compact IO devices
  These behave like a single module with regard to the IO controller
- Modular IO devices
  These comprise several modules (submodules)
- Intelligent IO devices
  These contain a control program that controls the lower-level (own) modules

Compact PROFINET IO devices

An example of a compact IO device is the ET 200eco. Bus gateways such as the IE/AS-i Link PN IO also behave like a compact slave on the PROFINET IO.

Modular PROFINET IO devices

Examples of modular IO devices are the ET 200M, the ET 200S, and the ET 200pro.

Intelligent PROFINET IO devices

Intelligent IO devices are, for example, CPUs with integrated PN interface. Equally, an ET 200pro station with the IM 154-8 PN/DP CPU interface module or an ET 200S station with the IM 151-8 PN/DP CPU interface module can be operated as intelligent IO devices.

IO supervisor

Devices for parameterization, commissioning, diagnostics, and operator control and monitoring, e.g. programming or HMI devices, are referred to as IO supervisors.

1.2.3 Actuator/Sensor Interface

The actuator/sensor interface (AS-i) is a networking system for the lowest process level in automation plants in accordance with the open

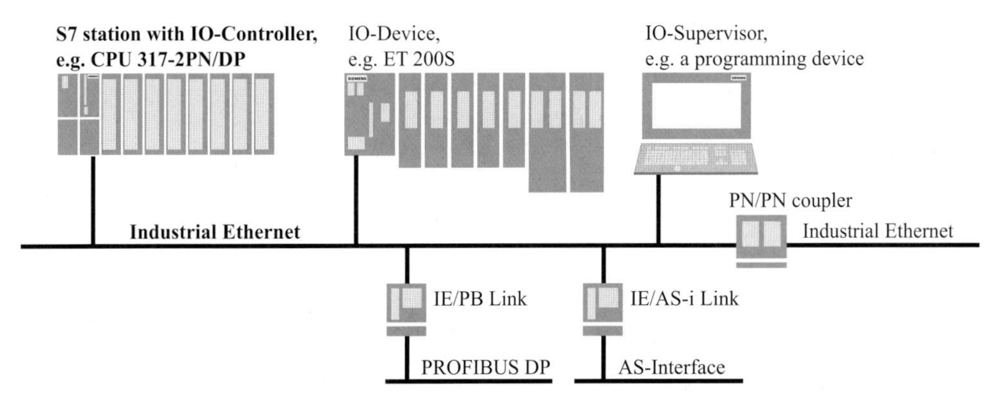

Figure 1.5 Components of a PROFINET IO system

international standard EN 50295. An AS-i master controls up to 62 AS-i slaves over a two-wire AS-i cable which transmits both the control signals and the supply voltage (Figure 1.6).

An AS-i segment can have a maximum length of 100 m; when combined with repeaters and extension plugs, a maximum length of 600 m can be achieved.

With the *ASIsafe* safety concept, you can connect safety sensors such as emergency-off switches, door contact switches, or safety light arrays directly to the AS-i network up to Category 4 in accordance with EN 954-1 or SIL3 in accordance with IEC 61508. This requires safe AS-i slaves for connecting the safety sensors and a safety monitor that combines the safe inputs with parameterizable logic and ensures safe shutdown.

AS-i master

Standard AS-i masters can control up to 31 standard AS-i slaves with a cycle time of max. 5 ms. With enhanced AS-i masters, the quantity framework is increased to a maximum of 62 AS-i slaves with extended addressing range and a cycle time of max. 10 ms. Pairs of slaves with extended addressing range occupy one address; if standard slaves are connected to an enhanced master, they each occupy one address.

The **AS-i master CP 343-2** is used in an S7-300 station or in an ET-200M station. It supports the following AS-i slaves:

- ▷ Standard slaves
- ▷ Slaves with extended addressing range, A/B slaves)
- ▷ Analog slaves in accordance with slave profile 7.3 or 7.4

In *standard mode,* the CP 343-2 responds like an I/O module: it occupies 16 input bytes and 16 output bytes in the analog address space (128 and upwards). Up to 31 standard slaves or up to 62 A/B slaves (slaves with extended addressing range) can be used with a CP 343-2. The AS-i slaves are parameterized with default values stored in the CP.

In *extended mode,* the complete scope of functions is available in accordance with the AS-i master specification. When using the provided FC block, master calls from the user program (transmission of parameters during ongoing operation, checking of the reference/actual configuration, testing and diagnostics) can be carried out in addition to standard operation.

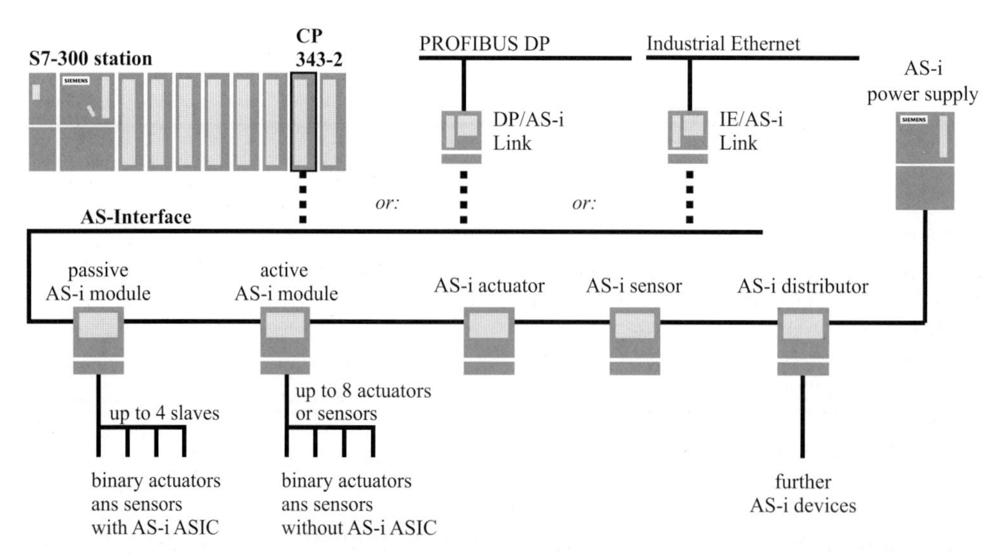

Figure 1.6 Connecting the AS-i Bus System to SIMATIC S7

AS-i slaves

AS-i slaves can be bus-capable actuators or sensors with AS-i ASIC, or also AS-i modules. Actuators and sensors with AS-i ASIC are connected to a passive module. Conventional actuators and sensors can be connected to an active module.

AS-i slaves are available in the standard version where a standard slave occupies one of the max. 31 possible addresses. The standard slaves are addressed by the user program like binary inputs and outputs.

AS-i slaves with extended addressing range (A/B slaves) occupy one address in pairs, so that max. 62 slaves can be operated on one master. "A slaves" are handled like standard slaves, "B slaves" are addressed using data records. AS-i A/B slaves can also be used to record and transmit analog values.

1.2.4 Routers

Routers permit data exchange between devices on different subnets as well as the passing on of configuration and parameterization information between different subnets (Figure 1.7).

Connection of two PROFIBUS subnets

You can use the **DP/DP coupler** (revision level 2) to connect two PROFIBUS subnets together, and can then exchange data between the DP masters. The two subnets are electrically isolated, and can be operated at different transmission rates up to max. 12 Mbit/s. In both subnets, the DP/DP coupler is assigned to the respective DP master as a DP slave with a freely selectable station address in each case.

The maximum size of the transfer memory is 244 byte input data and 244 byte output data, divided in up to 16 areas. Input areas in one subnet must correspond to output areas in the other. Up to 128 byte can be consistently transmitted. If the page with input data fails, the corresponding output data on the other page are held at their last value.

The DP/DP coupler is configured and parameterized using STEP 7. A GSD file is available for operation on non-Siemens masters.

Connection of PROFIBUS DP to PROFIBUS PA

PROFIBUS PA (Process Automation) is a bus system for process engineering in intrinsically-safe areas (Ex area Zone 1) e.g. in the chemical industry, as well as in non-intrinsically-safe areas such as the food and drinks industry.

The protocol for PROFIBUS PA is based on the EN 50170 standard, Volume 2 (PROFIBUS DP), and the transmission method is based on IEC 1158-2.

There are two possible methods of connecting PROFIBUS DP to PROFIBUS PA:

- ▷ DP/PA coupler, if the PROFIBUS DP network can be operated at 45.45 kbits/s
- ▷ DP/PA link that converts the data transfer rates of PROFIBUS DP to the transfer rate of PROFIBUS PA

The **DP/PA coupler** enables connection of PA field devices to PROFIBUS DP. On PROFIBUS DP, the DP/PA coupler is a DP slave operated at 45.45 kbits/s. Up to 31 PA field devices can be connected to one DP/PA coupler. These field devices form a PROFIBUS PA segment with a data transfer rate of 31.25 kbits/s. Taken together, all PROFIBUS PA segments form a shared PROFIBUS PA bus system.

The DP/PA coupler is available in two variants: a non-Ex version with up to 400 mA output current and an Ex version with up to 100 mA output current.

The **DP/PA link** enables connection of PA field devices to PROFIBUS DP at a data transfer rate of 9.6 kbits/s to 12 Mbits/s. A DP/PA link consists of an IM 157 interface module and up to 5 DP/PA couplers linked together via SIMATIC S7 bus connectors. It takes the bus system comprising all the PROFIBUS PA segments and maps it to a PROFIBUS DP slave. You can connect up to 31 PA field devices per DP/PA link.

SIMATIC PDM (Process Device Manager, previously SIPROM) is a vendor-independent tool for parameterizing, startup and diagnostics of intelligent field devices with PROFIBUS PA or HART functionality. The DDL (Device Description Language) is available for parameterizing HART transducers (Highway Addressable Remote Transducers).

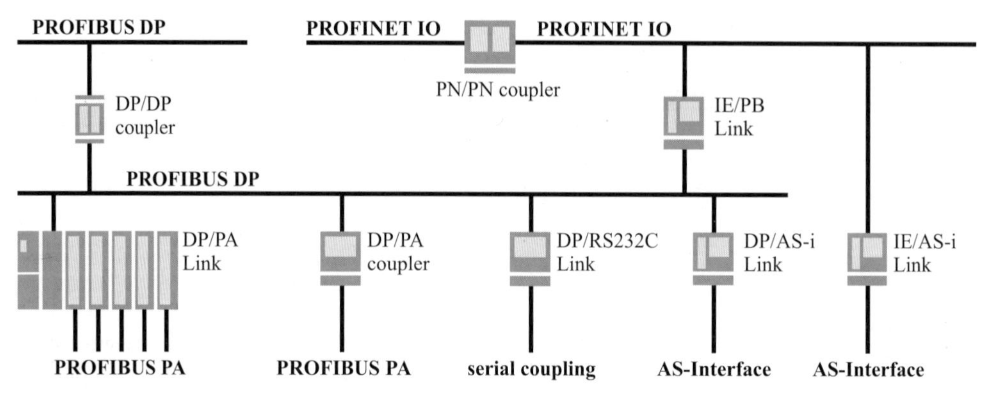

Figure 1.7 Routers

With STEP 7 V5.1 SP3 and higher, the I&C modules are parameterized with the hardware configuration; you must not use SIMATIC PDM any more.

Connection of PROFIBUS DP to AS-Interface

A **DP/AS-Interface link** enables connection of the PROFIBUS DP to the AS-Interface. On PROFIBUS DP, the link is a modular DP slave with a transmission rate of up to 12 Mbit/s and IP 20 degree of protection. On the AS-Interface, it is an AS-i master which controls the AS-i slaves. The link is available in the versions *DP/AS-i Link 20E* and *DP/AS-i Link Advanced.* The following AS-i slaves can be controlled:

- ▷ Standard slaves, AS-i analog slaves
- ▷ Slaves with extended addressing mode (A/B slaves)
- ▷ Slaves with data transfer mechanisms according to AS-i specification V3.0 (DP/AS-i Link Advanced)

Connection of PROFIBUS DP to a Serial Interface

The **PROFIBUS DP/RS 232C** link is a converter between an RS 232C (V.24) interface and PROFIBUS DP. Devices with an RS 232C interface can be connected to PROFIBUS DP with the DP/RS 232C link. The DP/RS 232C link supports the 3964R and free ASCII protocol procedures.

The PROFIBUS DP/RS 232C link is connected to the device via a point-to-point connection. Conversion to the PROFIBUS DP protocol takes place in the PROFIBUS DP/RS 232C link. The data are transferred consistently in both directions. Up to 224 bytes of user data are transmitted per frame.

The data transfer rate on PROFIBUS DP can be up to 12 Mbits/s; RS 232C can be operated at up to 38.4 kbits/s with no parity, even or odd parity, 8 data bits and 1 stop bit.

Connection of two PROFINET subnets

With the **PN/PN coupler,** you interconnect two Ethernet subnets in order to exchange data between the IO controllers of both subnets. There is galvanic isolation between the subnets.

The PN/PN coupler is a 120-mm-wide module that is installed on a DIN rail. The subnets are connected using RJ45 connectors. Two connections with internal switch function are available for each subnet.

From the viewpoint of the relevant IO controller, the PN/PN coupler is an IO device in its own PROFINET IO system. Both IO devices are linked by a data transfer area with 256 input bytes and 256 output bytes, divisible into a maximum of 16 areas. Input areas in one subnet must correspond to output areas in the other.

The PN/PN coupler is configured and parameterized with STEP 7. A GSDML file is available for other configuring tools.

Connection of PROFINET IO to PROFIBUS DP

You can use the **IE/PB Link PNIO** to connect the Industrial Ethernet and PROFIBUS subnets. If you are using PROFINET IO, the IE/PB Link PNIO takes over the role of a proxy for the DP slaves on the PROFIBUS. An IO controller can access both DP slaves and IO devices over the IE/PB Link. In standard mode, the IE/PB Link allows PG/OP communication and S7 routing between subnets.

The IE/PB Link PNIO is a double-width module of S7-300 design. You can connect the IE/PB Link to Industrial Ethernet using an 8-pole RJ45 female connector and to PROFIBUS using a 9-pole SUB-D female connector.

The IE/PB Link is configured using STEP 7 as the IO device, to which a DP master system is connected. When switching on, the IO controller also provides the subordinate DP slaves with the configuration data.

Please note that limitations exist on the PROFIBUS DP following an IE/PB Link. For example, you cannot connect a DP/PA Link, the DP segment has no CiR capability, and isochronous mode cannot be configured.

Connection of PROFINET IO to AS-Interface

An **IE/AS-i Link** enables the connection of PROFINET IO to the AS-Interface. On PROFINET IO, the link is an IO Device. On the AS-Interface, it is the AS-i master that controls the AS-i slaves. The IO controller can access the individual binary and analog values of the AS-i slaves directly.

Connection to PROFINET is made via two RJ45 connectors with internal switch function. The AS-Interface bus is connected to 4-pin plug-in screw-type contacts.

The link is available in single or double master versions (in accordance with AS-Interface specification V3.0) for connection of 62 AS-i slaves each plus integral analog value transmission. The following AS-i slaves can be controlled:

- ▷ Standard slaves, AS-i analog slaves
- ▷ Slaves with extended addressing range (A/B slaves)
- ▷ Slaves with data transfer mechanisms in accordance with AS-i specification V3.0

The IE/AS-i link is configured and parameterized using STEP 7. A GSDML file is available for other configuration tools.

1.3 Communications

Communications – data exchange between programmable modules – is an integral component of SIMATIC S7. Almost all communications functions are handled via the operating system. You can exchange data without any additional hardware and with just one connecting cable between the two CPUs. If you use CP modules, you can achieve powerful network links and the facility of linking to non-Siemens systems.

SIMATIC NET is the umbrella term for SIMATIC communications. It represents information exchange between programmable controllers and between programmable controllers and human machine interface devices. There are various communications paths available depending on performance requirements.

1.3.1 Introduction

The most significant communications objects are initially SIMATIC stations or non-Siemens devices between which you want to exchange data. You require modules with communications capability here. With SIMATIC S7, all CPUs have an MPI interface over which they can handle communications.

In addition, there are communications processors (CPs) available that enable data exchange at higher throughput rates and with different protocols. You must link these modules via networks. A network is the hardware connection between communication nodes.

Data is exchanged via a “connection” in accordance with a specific execution plan (“communications service”) which is based, among other things, on a specific coordination procedure (“protocol”). S7 connection is the standard between S7 modules with communications capability, for example.

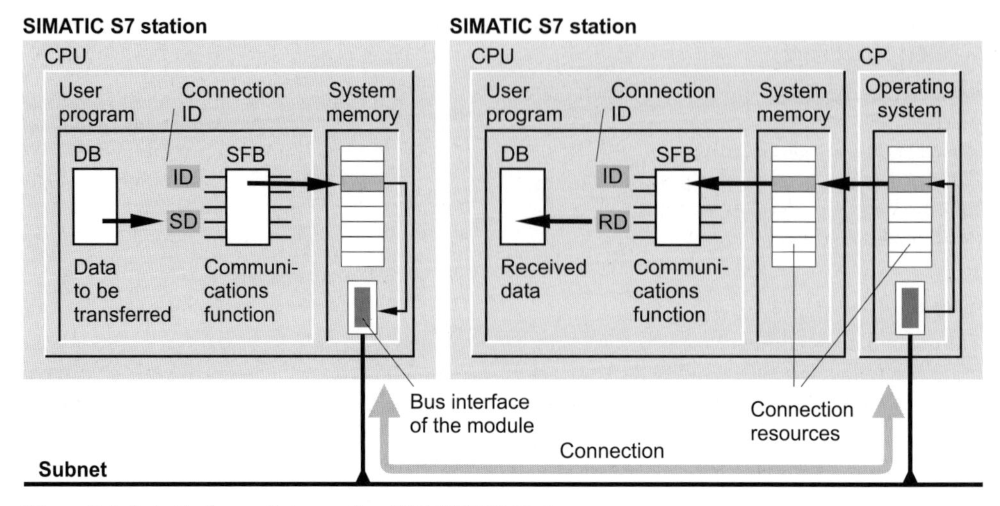

Figure 1.8 Data Exchange Between Two SIMATIC S7 Stations

Figure 1.8 uses the example of an S7 connection to show the objects involved in communication between two stations. In the user program of the left station, the data to be transmitted are present in a data block (DB). In the example, the communications function is a system function block (SFB). You assign a pointer to the RD parameter for the data to be transmitted, and trigger the transmission from the program. The communications function is additionally assigned a connection ID with which the used connection is specified. The connection occupies one connection resource in the CPU's system memory. The data are sent via the module's bus interface, e.g. to a CP module in another station. Connection resources are occupied in both the CP module and the CPU. As a result of the connection ID (and the configured connection path), the communications function "recognizes" the data addressed to it in the receiver station, and writes these by means of the pointer in the RD parameter to the data block of the user program.

Network

A network is a connection between several devices for the purpose of communication. It comprises one or more identical or different subnets linked together.

Subnet

In a subnet, all the communications nodes are linked via a hardware connection with uniform physical characteristics and transmission parameters, such as the data transfer rate, and they exchange data via a shared transmission procedure. SIMATIC recognizes MPI, PROFIBUS, Industrial Ethernet and point-to-point connection (PTP) as subnets.

Communications service

A communications service determines how the data are exchanged between communications nodes and how the data are to be handled. It is based on a protocol that describes, amongst other things, the coordination procedure between the communications nodes.

The services most frequently used with SIMATIC are: PG communication, OP communication, S7 basic communication, S7 communication, global data communication, PtP communication, S5-compatible communication (SEND/RECEIVE interface).

Connection

A connection defines the communications relationships between two communications nodes. It is the logical assignment of two nodes for the execution of a specific communications service and also contains special characteristics such as the type of connection (dynamic, static) and how it is established.

SIMATIC recognizes the following connection types: S7 connection, S7 connection (fault-tolerant), point-to-point connection, FMS and FDL connection, ISO transport connection, ISO-on-TCP and TCP connection, UDP connection and e-mail connection.

Communications functions

The communications functions are the user program's interface to the communications service. For SIMATIC S7-internal communications, the communications functions are integrated in the operating system of the CPU and they are called via system blocks. Loadable blocks are available for communication with non-Siemens devices via communications processors.

Overview of communications objects

Table 1.1 shows the relationships between subnets, modules with communications capability and communications services. PG/OP communication over the MPI, PROFIBUS and Industrial Ethernet subnets is possible in addition to the communications services shown.

1.3.2 Subnets

Subnets are communications paths with the same physical characteristics and the same communications procedure. Subnets are the central objects for communication in the SIMATIC Manager.

The subnets differ in their performance capability:

▷ MPI
Low-cost method of networking a few SIMATIC devices with small data volumes.

▷ PROFIBUS
High-speed exchange of small and mid-range volumes of data, used primarily with distributed I/O

▷ Industrial Ethernet
Communications between computers and programmable controllers for high-speed exchange of large volumes of data, also used with distributed I/O (PROFINET IO)

▷ Point-to-point (PTP)
Serial link between two communications partners with special protocols

From STEP 7 V5, you can use a programming device to reach SIMATIC S7 stations via subnets, for the purposes of, say, programming or parameterizing. The gateways between the subnets must be located in an S7 station with "routing capability".

MPI

Every CPU has an "interface with multipoint capability" (multipoint interface, or MPI). It enables establishment of subnets in which CPUs, human machine interface devices and programming devices can exchange data with each other. Data exchange is handled via a Siemens proprietary protocol.

The maximum number of nodes on the MPI network is 32. Each node has access to the bus for a specific length of time and may send data frames. After this time, it passes the access rights to the next node ("token passing" access procedure).

As transmission medium, MPI uses either a shielded twisted-pair cable or a glass or plastic fiber-optic cable. The maximum cable length in a bus segment with non-electrically isolated interfaces can be up to 50 m depending on the transmission rate, and up to 1000 m with electrically isolated interfaces. This can be increased by inserting RS-485 repeaters (up to 1100 m) or optical link modules (up to > 100 km). The data transfer rate is usually 187.5 kbits/s.

Over the MPI network, you can exchange data between CPUs with global data communications, station-external S7 basic communications or S7 communications. No additional modules are required.

PROFIBUS

PROFIBUS stands for "Process Fieldbus" and is a vendor-independent standard complying with IEC 61158/EN 50170 for universal automation (PROFIBUS DP and PROFIBUS FMS) and IEC 61158-2 for process automation (PROFIBUS PA).

The maximum number of nodes in a PROFIBUS network is 127, where the network is

Table 1.1 Communications Objects

Subnet	Modules	Communications Service, Connection	Configuring, Interface
MPI	All CPUs	Global data communication	GD table
		Station-external S7 basic communications	SFC calls
		S7 communications	Connection table, FB/SFB calls
PROFI-BUS	CPUs with DP interface	PROFIBUS DP (DP master or DP slave)	Hardware configuration, SFB/SFC calls, inputs/outputs
		Station-internal S7 basic communications	SFC calls
	IM 467	PROFIBUS DP (DP master)	Hardware configuration, SFB/SFC calls, inputs/outputs
		Station-internal S7 basic communications	SFC calls
	CP 342-5 CP 443-5 Extended	CP 342-5: PROFIBUS DP-V0 CP 433-5 Ext.: PROFIBUS DP-V1 (DP master or DP slave)	Hardware configuration, SFB/SFC calls, inputs/outputs
		Station-internal S7 basic communications	SFC calls
		S7 communications	Connection table, FB/SFB calls
		S5-compatible communications	NCM, connection table, SEND/RECEIVE
	CP 343-5 CP 443-5 Basic	Station-internal S7 basic communications	SFC calls
		S7 communications	Connection table, FB/SFB calls
		S5-compatible communications	NCM, connection table, SEND/RECEIVE
		PROFIBUS FMS	NCM, connection table, FMS interface
Industrial Ethernet	CPUs with PN interface	PROFINET IO (IO controller)	Hardware configuration, SFB/SFC calls, inputs/outputs
		IE communications	FB calls
	CP 343-1 Lean CP 343-1 CP 443-1	S7 communications	Connection table, FB/SFB calls
		S5-compatible communications Transport protocols TCP/IP and UDP, also ISO with CP 443-1	NCM, connection table, SEND/RECEIVE
	CP 343-1 IT CP 443-1 Advanced CP 443-1 IT	S7 communications	Connection table, FB/SFB calls
		S5-compatible communications Transport protocols TCP/IP and UDP also ISO with CP 443-1	NCM, connection table, SEND/RECEIVE
		IT communication (HTTP, FTP, e-mail)	NCM, connection table, SEND/RECEIVE
	CP 343-1 PN	S7 communications	Connection table, FB/SFB calls
		S5-compatible communications Transport protocols TCP and UDP	NCM, connection table, SEND/RECEIVE

NCM is the configuring software for the CP modules (integrated in STEP 7 from V5.2 and higher)

divided into segments with up to 32 nodes. A distinction is made between active and passive nodes. An active node receives access rights to the bus for a specific length of time and may send data frames. After this time, it passes the access rights to the next node ("token passing" access procedure). If passive nodes (slaves) are assigned to an active node (master), the master executes data exchange with the slaves assigned to it while it is in possession of the access rights. A passive node does not receive access rights.

The physical connection of the PROFIBUS network can be electrical, optical or wireless with various data transmission rates. The length of a segment depends on the transmission rate. The electrical network can have a bus or tree topology. It uses a shielded, twisted two-wire cable (RS485 interface). The transmission rate can be adjusted in steps from 9.6 kbit/s to 12 Mbit/s (31.25 kbit/s with PROFIBUS PA).

The optical network uses either plastic, PCF or glass fiber-optic cables. This is suitable for large distances, provides electrical isolation, and is insensitive to electromagnetic influences. The transmission rate can be adjusted in steps from 9.6 kbit/s to 12 Mbit/s. With optical link modules (OLMs), it is possible to produce a line, ring or star topology. An OLM also enables connection between electrical and optical networks with a mixed design. A cost-optimized version is the design as a line topology with integral interface and optical bus terminal (OBT).

Single or multiple PROFIBUS slaves or segments with PROFIBUS slaves can be linked by a wireless connection when using the PROFIBUS infrared link module (ILM). With a maximum transmission rate of 1.5 Mbit/s and a maximum range of 15 m, communication is possible with moving parts.

You implement connection of distributed I/O via a PROFIBUS network; the relevant PROFIBUS DP communications service is implicit. You can use either CPUs with integral or plug-in DP master, or the relevant CPs. You can also operate station-internal S7 basic communications or S7 communications via this network.

You can transfer data with PROFIBUS-FMS and PROFIBUS-FDL using the relevant CPs. There are loadable blocks (FMS interface or SEND/RECEIVE interface) available as the interface to the user program.

Industrial Ethernet

Industrial Ethernet is the subnet for connecting computers and programmable controllers, with focus on the industrial area, defined by the international standard IEEE 802.3/802.3u. The standards IEEE 802.11 a/b/g/h define the connection to wireless local area networks (WLAN) and Industrial Wireless LANs (IWLAN).

The number of nodes which can be networked with Industrial Ethernet is unlimited; up to 1024 nodes are permitted per segment. Before accessing, each node checks to see if another node is currently transmitting. If this is the case, the node waits for a random time before attempting another access (CSMA/CD access procedure). All nodes have equal access rights.

The physical connections with Industrial Ethernet consist of point-to-point connections between the communication nodes: each node is connected to exactly one peer. To enable several nodes to communicate with one another, they are connected to a "distributor" (switch or hub) which has several connections.

A *switch* is an active bus element which regenerates the received signals, assigns them priorities, and only distributes them to the devices connected to it. A *hub* adjusts itself to the lowest transmission rate at the connections and passes on all signals without priority to all connected devices.

The network can be configured as a line, star, tree, or ring topology. The transmission rates are 10 Mbit/s, 100 Mbit/s (Fast Ethernet) or 1000 Mbit/s (Gigabit Ethernet, not with PROFINET).

Industrial Ethernet can be designed physically as an electrical, optical, or wireless network. FastConnect Twisted Pair cables (FC TP) with RJ45 connections or Industrial Twisted Pair cables (ITP) with sub-D-connections are available for electrical cabling. Fiber-optic cables (FOC) can be glass-fiber, PCF, or POF. These offer galvanic isolation, are insensitive to electromagnetic interferences, and are suitable for long distances. Wireless transmission uses frequencies of 2.4 GHz and 5 GHz, with transmission rates up to 54 Mbit/s (depending on country approval).

You can exchange data with S7 and IE communications via Industrial Ethernet and you can use the S7 functions. With appropriately designed modules, you can also establish ISO transport connections, ISO-on-TCP connections, TCP, UDP and e-mail connections.

PROFINET

PROFINET is the open Industrial Ethernet standard of PROFIBUS International (PNO). PROFINET uses the Industrial Ethernet subnet as the physical medium for data transmission, and takes into account the requirements of industrial automation. For example, PROFINET provides real-time (RT) communication with field devices, and isochronous real-time (IRT) properties for motion control. Compatibility with TCP/IP and the IT standards of Industrial Ethernet is retained.

Siemens applies PROFINET in two automation concepts:

▷ *Component Based Automation* (CBA) uses PROFINET for communication between control devices as components in distributed systems. The configuration tool is SIMATIC iMap.

▷ *PROFINET IO* uses PROFINET to transmit data to and from field devices (distributed I/O. The configuration tool is SIMATIC STEP 7.

Point-to-point connection

A point-to-point connection (PTP) enables data exchange via a serial link. A point-to-point connection is handled by the SIMATIC Manager as a subnet and configured similarly.

The transmission medium is an electrical cable with interface-dependent assignment. RS 232C (V.24), 20 mA (TTY) and RS 422/485 are available as interfaces. The data transfer rate is in the range 300 bits/s to 19.2 kbits/s with a 20 mA interface or 76.8 kbits/s with RS 232C and RS 422/485. The cable length depends on the physical interface and the data transfer rate; it is 10 m with RS 232C, 1000 m with a 20 mA interface at 9.6 kbits/s and 1200 with RS 422/485 at 19.2 kbits/s.

3964 (R), RK 512, printer drivers and an ASCII driver are available as protocols (procedures), the latter enabling definition of user-specific procedures.

AS-Interface

The AS-Interface (actuator/sensor interface, AS-i) networks the appropriately designed binary sensors and actuators in accordance with the AS-Interface specification IEC TG 178. The AS-Interface does not appear in the SIMATIC Manager as a subnet; only the AS-I master is configured with the hardware configuration or with the network configuration.

The transmission medium is an unshielded twisted-pair cable that supplies the actuators and sensors with both data and power (power supply required). Network range can be up to 600 m with repeaters and extension plugs. The data transfer rate is set at 167 kbits/s.

A master controls up to 62 slaves through cyclic scanning and so guarantees a defined response time.

1.3.3 Communications Services

Data exchange over the subnets is controlled by different communications services – depending on the connection selected. The services are provided by the CPU or CP modules. In addition to communication with field devices (PROFIBUS DP, PROFIBUS PA and PROFINET IO, see Chapters 1.2.1 "PROFIBUS DP" and 1.2.2 "PROFINET IO"), the services described below are available depending on the module used.

PG communication

PG communication is used to exchange data between an engineering station and a SIMATIC station. It is used, for example, by a programming device in online mode to execute the "Monitor variables" or "Read diagnostics buffer" functions or to download user programs. The communications functions required for PG communication are integrated in the operating system of the SIMATIC modules. PG communication can be executed over the MPI, PROFIBUS and Industrial Ethernet subnets. Using S7 routing, PG communication can also be used cross-subnet.

OP communication

OP communication is used to exchange data between an operator station and a SIMATIC station. It is used, for example, by an HMI device for operation and monitoring or for reading and writing variables. The communications functions required for OP communication are integrated in the operating system of the SIMATIC modules. OP communication can be executed over the MPI, PROFIBUS and Industrial Ethernet subnets.

S7 basic communication

S7 basic communication is an event-driven service for exchanging smaller quantities of data between a CPU and a module in the same SIMATIC station ("station-internal") or between a CPU and a module in a different SIMATIC station ("station-external"). The connections are established dynamically as required. The communications functions required for S7 basic communication are integrated in the CPU's operating system. They can trigger data transmission, for example using system functions SFC in the user program. Station-internal S7 basic communication is executed over PROFIBUS, station-external over MPI.

S7 communication

S7 communication is an event-driven service for exchanging larger quantities of data between CPU modules with control and monitoring functions. The connections are static, and are configured using STEP 7. The communications functions required for S7 communication are either integrated in the CPU's operating system (system function blocks SFB) or are loadable function blocks (FB). S7 communication can be executed over the MPI, PROFIBUS and Industrial Ethernet subnets.

IE communication

By means of "Open communication over Industrial Ethernet" (in short IE communication), you can transmit data between two devices connected to the Ethernet subnet. Communication can be implemented using the protocols TCP native in accordance with RFC 793, ISO-on-TCP in accordance with RFC 1006, or UDP in accordance with RFC 768. The communication functions are loadable function blocks (FB) which are available in STEP 7 in the *Standard Library* under *Communication Blocks*. The function blocks are called in the main program and control the establishment and clearance of connections as well as data transmission.

Global data communication

Global data communication enables exchange of small volumes of data between several CPUs without additional programming overhead in the user program. Transfer can be cyclic or event-driven. The communications functions required are integrated in the CPU's operating system. Global data communication is possible over the MPI bus or C bus.

PTP communication

PTP communication (point-to-point) transmits data over a serial interface, e.g. between a SIMATIC station and a printer. The communications functions required are integrated in the operating system, e.g. as system function blocks SFB. Data exchange is possible using various transmission procedures.

S5-compatible communication

S5-compatible communication is an event-driven service for data transmission between SIMATIC stations and third-party stations. The connections are static, and are configured using STEP 7. The communications functions are usually loadable functions FC with which you can control the transmission from the user program. Data are sent and received over the SEND/RECEIVE interface, and can be fetched and written over the FETCH/WRITE interface (S7 is the passive partner). S5-compatible communication with Industrial Ethernet can take place over the TCP, ISO-on-TCP, ISO-Transport and UDP connections, and with PROFIBUS over FDL.

Standard communication

Standard communication is carried out with standardized, cross-vendor protocols for data transmission.

PROFIBUS FMS (Fieldbus Message Specification) provides services for program-driven, device-independent transfer of structured variables (FMS variables) in accordance with EN 50170 Volume 2. Data exchange takes place with static FMS connections over a PROFIBUS subnet. The communications functions are loadable function blocks FB with which you can control the transmission from the user program.

An IT communications processor provides a SIMATIC station with interfacing to the **IT communication**. Transmission over Industrial Ethernet comprises PG/OP/S7 communication and S5-compatible communication (SEND/RECEIVE) with the ISO, TCP/IP and UPD transport protocols. SMTP (Simple Mail Transfer Protocol) for e-mail, HTTP (Hyper Text Transfer Protocol) for access with Web browsers, and FTP (File Transfer Protocol) can additionally be used for program-driven data exchange with devices with different operating systems.

1.3.4 Connections

A connection is either dynamic or static depending on the communications service selected. Dynamic connections are not configured; their buildup or cleardown is event-driven ("Communications via non-configured connections"). There can only ever be one non-configured connection to a communications partner.

Static connections are configured in the connection table; they are built up at startup and remain throughout the entire program execution ("communications via configured connections"). Several connections can be established in parallel to one communications partner. You use a "Connection type" to select the desired communications service in the network configuration (see Chapter 2.4 "Configuring the Network").

You do not need to configure connections with the network configuration for global data communications and PROFIBUS DP or for SFC communications in the case of S7 functions. You define the communications partners for global data communications in the global data table; in the case of PROFIBUS DP and S7 basic communications, you define the partners via the node addresses.

Connection resources

Each connection requires connection resources on the participating communications partner for the end point of the connection or the transition point in a CP module. If, for example, S7 functions are executed via a bus interface of the CPU, a connection is assigned in the CPU; the same functions via the MPI interface of the CP occupy one connection in the CP and one connection in the CPU.

Each CPU has a specific number of possible connections. Limitations and rules exist with respect to the usability of the connection resources. Not every connection resource can be used e.g. for every type of connection. One connection is reserved for a programming device and one connection for an OP (these cannot be used for any other purpose).

Connection resources are also required temporarily for the "non-configured connections" in S7 basic communications.

1.4 Module Addresses

1.4.1 Signal Path

When you wire your machine or plant, you determine which signals are connected where on the programmable controller (Figure 1.9).

An input signal, for example the signal from momentary-contact switch +HP01-S10, the one for "Switch motor on", is run to an input module, where it is connected to a specific terminal. This terminal has an "address" called the I/O address (for instance byte 5, bit 2).

Before every program execution start, the CPU then automatically copies the signal to the process input image, where it is then accessed as an "input" address (I 6.2, for example). The expression "I 5.2" is the absolute address.

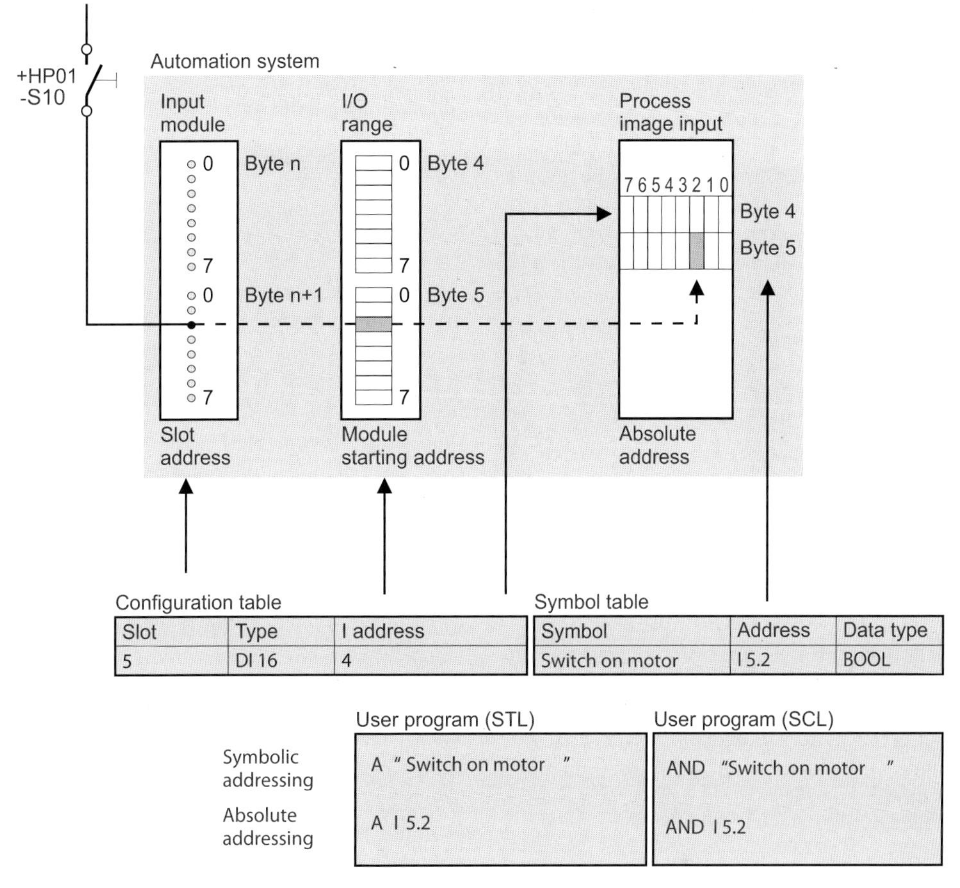

Figure 1.9 Correlation between Module Address, Absolute Address and Symbolic Address (Path of a Signal from Sensor to Scanning in the Program)

You can now give this input a name by assigning an alphanumeric symbol corresponding to this input signal (such as "Switch motor on") to the absolute address in the symbol table. The expression "Switch motor on" is the symbolic address.

1.4.2 Slot Address

Every slot has a fixed address in the programmable controller (an S7 station). This slot address consists of the number of the mounting rack and the number of the slot. A module is uniquely described using the slot address ("geographical address").

If the module contains interface cards, each of these cards is also assigned a submodule address. In this way, each binary and analog signal and each serial connection in the system has its own unique address.

Correspondingly, distributed I/O modules also have a "geographical address". In this case, the number of the DP master system or the PROFINET IO system and the station number replace the rack number.

You use STEP 7's "Hardware Configuration" tool to plan the hardware configuration of an S7 station as per the physical location of the modules. This tool also makes it possible to set the module start addresses and parameterize the modules (see Chapter 2.3 "Configuring Stations").

1.4.3 Logical Address

The logical address corresponds to the absolute address. It is also referred to as the user data address, since you can use it to address the user data of the input/output modules in the user program, either using the process image (inputs I and outputs Q) or directly on the modules (peripheral inputs PI and peripheral outputs PQ). The range of logical addresses starts at zero and ends at a CPU-specific upper limit.

In the case of digital modules, the individual signals (bits) are bundled into groups of eight called "bytes". There are modules with one, two or four bytes. These bytes have the relative addresses 0, 1, 2 and 3; addressing of the bytes begins at the module start address. Example: In the case of a digital module with four bytes and the start address 8, the individual bytes are accessed by addresses 8, 9, 10 and 11. In the case of analog modules, the individual analog signals (voltages, currents) are called "channels", each of which occupies two bytes. Analog modules are available, depending on design, with 2, 4, 8 and 16 channels, corresponding to 4, 8, 16 or 32 bytes of address.

By means of the hardware configuration, you assign a logical address to each byte of a used module. As standard, addresses are assigned starting with zero; however, you can change the proposed address. The logical addresses of the individual modules must not overlap. The logical addresses are assigned separately for the input and output modules, meaning that an input byte can have the same number as an output byte.

The user data of the distributed I/O can also be addressed byte-by-byte using a logical address. In order to guarantee unambiguous assignment of all user data of a CPU (or more exactly: all user data on a P bus), the logical addresses of the distributed I/O must not overlap with the logical addresses of the central modules.

The digital modules are usually arranged according to address in the process image so that their signal states can be automatically updated and they can be accessed with the address areas "Input" and "Output". Analog modules, FMs and CPs receive an address that is not in the process image.

1.4.4 Module Start Address

The module start address is the smallest logical (user data) address of a module; it identifies the relative byte zero of the module. The subsequent module bytes are then assigned consecutive addresses.

In the case of mixed modules having input and output areas, the lower area start address is defined as the module start address. If the input and output areas have the same start address, use the input address.

By means of the hardware configuration, you define the position of the user data addresses in the CPU's address volume through specification of the module start address. The module start address is also the lowest logical address in the modules of the distributed I/O and even for the virtual slots in the transfer memory of an intelligent DP slave.

The module start address serves in many cases to identify a module. It has no particular significance in addition to this.

1.4.5 Diagnostics Address

Appropriately equipped modules can supply diagnostics data that you can evaluate in your program. If centralized modules have a user data address (module start address), you access the module via this address when reading the diagnostics data. If the modules have no user data address (e.g. power supplies), or if they are part of the distributed I/O, there is a diagnostics address for this purpose.

The diagnostics address is always an address in the I/O input area and occupies one byte. The user data length of this address is zero; if it is located in the process image, as is permitted, it is not taken into account by the CPU when updating the process image.

STEP 7 automatically assigns the diagnostics address counting down from the highest possible I/O address. You can change the diagnostics address with the Hardware Configuration function.

The diagnostics data can only be read with special system functions; accessing this address with load statements has no effect (see also Chapter 20.4 "Communication via Distributed I/O").

1.4.6 Addresses for Bus Nodes

MPI

Modules that are nodes on an MPI network (CPUs, FMs and CPs) also have an **MPI address**. This address is decisive for the link to programming devices, human machine interface devices and for global data communications.

Please note that with older revision levels of the S7-300 CPUs, the FM and CP modules operated in the same station receive an MPI address derived from the MPI address of the CPU. In the case of newer S7-300 CPUs, the MPI addresses of FM and CP modules in the same station can be determined independently of the MPI address of the CPU.

PROFIBUS DP

Every DP station (e.g. DP master, DP slave, programming device) on the PROFIBUS also has a **node address** (station number) with which it can be unambiguously addressed on the bus.

PROFINET IO

Nodes on Industrial Ethernet have a factory-set **MAC address** which is unambiguous worldwide. An **IP address** is additionally necessary for identification on the bus, and is configured for the IO controller. The IP addresses for the IO devices are derived from the IP address of the IO controller. The IO controller (the interface) and each IO device is additionally assigned a **device name.** The IO device is addressed from the user program by means of a **device number** (station number).

1.5 Address Areas

The address areas available in every programmable controller are

▷ the peripheral inputs and outputs

▷ the process input image and the process output image

▷ the bit memory area

▷ the timer and counter functions (see Chapters 7 "Timer Functions" and 8 "Counter Functions")

▷ the L stack (see Chapter 18.1.5 "Temporary Local Data")

To this are added the code and data blocks with the block-local variables, depending on the user program.

1.5.1 User Data Area

In SIMATIC S7, each module can have two address areas: a user data area, which can be directly addressed with Load and Transfer statements, and a system data area for transferring data records.

When modules are accessed, it makes no difference whether they are in racks with centralized configuration or used as distributed I/O. All modules occupy the same (logical) address space.

A module's user data properties depend on the module type. In the case of signal modules, they are either digital or analog input/output signals, and in the case of function modules and communications processors, they might, for example, be control or status information. The volume of user data is module-specific. There are modules that occupy one, two, four or more bytes in this area. Addressing always begins at relative byte 0. The address of byte 0 is the module start address; it is stipulated in the configuration table.

The user data represent the I/O address area, subdivided, depending on the direction of transfer, into peripheral inputs (PIs) and peripheral outputs (PQs). If the user data are in the area of the process images, the CPU automatically handles the transfers when updating the process images.

Peripheral inputs

You use the peripheral input (PI) address area when you read from the user data area on input modules. Part of the PI address area leads to the process image. This part always begins at I/O address 0; the length of the area is CPU-specific.

With a Direct I/O Read operation, you can access the modules whose interfaces do not lead to the process input image (for instance analog input modules). The signal states of modules that lead to the process input image can also be read with a Direct Read operation. The momentary signal states of the input bits are then scanned. Please note that this signal state may differ from the relevant inputs in the process image since the process input image is updated at the beginning of the program scan.

Peripheral inputs may occupy the same absolute addresses as peripheral outputs.

Peripheral outputs

You use the peripheral output (PQ) address area when you write values to the user data area on an output module. Part of the PQ address area leads to the process image. This part always begins at I/O address 0; the length of the area is CPU-specific.

With a Direct I/O Write operation, you can access modules whose interfaces do not lead to the process output image (such as analog output modules). The signal states of modules controlled by the process output image can also be directly affected. The signal states of the output bits then change immediately. Please note that a Direct I/O Write operation also updates the signal states of the relevant modules in the process output image! Thus, there is no difference between the process output image and the signal states on the output modules.

Peripheral outputs can reserve the same absolute addresses as peripheral inputs.

1.5.2 Process Image

The process image contains the image of the digital input and output modules and is therefore organized into a process image input and process image output. You address the process image input via the operand range I inputs and the process image output via the operand range Q outputs. In general, the machine or process is controlled via the inputs and the outputs.

The process image can be subdivided into subsidiary process images that can be updated either automatically or via the user program. Please refer to Chapter 20.2.1 "Process Image Updating".

On the S7-300 CPUs and, from 10/98, also on S7-400 CPUs, you can use the addresses of the process image not occupied by modules as additional memory area similar to the bit memory area. This applies both for the process input image and the process output image.

On suitably equipped CPUs, say, the CPU 417, the size of the process image can be parameterized. If you enlarge the process image, you reduce the size of the work memory accordingly. Following a change to the size of the process image, the CPU executes initialization of the work memory, with the same effect as a cold restart.

Inputs

An input is an image of the corresponding bit on a digital input module. Scanning an input is the same as scanning the bit on the module itself. Prior to program execution in every program cycle, the CPU's operating system copies the signal state from the module to the process input image.

The use of a process input image has many advantages:

▷ Inputs can be scanned and linked bit by bit (I/O bits cannot be directly addressed).

▷ Scanning an input is much faster than accessing an input module (for example, you avoid the transient recovery time on the I/O bus, and the system memory response times are shorter than the module's response times). The program is therefore executed that much more quickly.

▷ The signal state of an input is the same throughout the entire program cycle (there is data consistency throughout a program cycle). When a bit on an input module changes, the change in the signal state is transferred to the input at the start of the next program cycle.

▷ Inputs can also be set and reset because they are located in random access memory. Digital input modules can only be read. Inputs can be set during debugging or startup to simulate sensor states, thus simplifying program testing.

These advantages are offset by an increased program response time (please also refer to Chapter 20.2.4 "Response Time".

Outputs

An output is an image of the corresponding bit on a digital output module. Setting an output is the same as setting the bit on the output module itself. The CPU's operating system copies the signal state from the process output image to the module.

The use of a process output image has many advantages:

▷ Outputs can be set and reset bit by bit (direct addressing of I/O bits is not possible).

▷ Setting an output is much faster than accessing an output module (for example, you avoid the transient recovery time on the I/O bus, and the system memory response times are shorter than the module response times). The program is therefore executed that much more quickly.

▷ A multiple signal state change at an output during a program cycle does not affect the bit on the output module. It is the signal state of the output at the end of the program cycle that is transferred to the module.

▷ Outputs can also be scanned because they are located in random access memory. While it is possible to write to digital output modules, it is not possible to read them. The scanning and linking of the outputs makes additional storage of the output bit to be scanned unnecessary.

These advantages are offset by an increased program response time. Chapter 20.2.4 "Response Time" describes how a programmable controller's response time comes about.

1.5.3 Consistent User Data

Data consistency means that data must be handled as an entity. Transmission of the data field must not be interrupted, and the data source and target must not be changed from the other side during the transmission. For example, if you transmit four bytes individually, the transmitted program can be interrupted between each byte by a higher-priority program which modifies the data in the source or target area.

With direct access to the user data (loading and transferring), the data are read and written as byte, word or doubleword. The load and transfer statements, upon which the MOVE box with LAD/CSF and the assignment of variables with elementary data types with SCL are based, are executed without interruption. If you wish to transmit a data field with more than four bytes between the system and work memories without interruption, use the system function SFC 81 UBLKMOV.

Data transmission between DP slave and DP master is consistent for a complete slave, even if the transfer area is divided into several consistent blocks, e.g. as with an intelligent DP slave. The data consistency with direct slave-to-slave traffic is the same as with direct access (1-byte, 2-byte and 4-byte consistency). The same applies to data transmission between IO controller and IO devices on the PROFINET IO.

When configuring stations of the distributed I/O with three or more than four bytes of user data, you can specify the consistent user data areas. These areas are transmitted with the system functions SFC 14 DPRD_DAT and SFC 15 DPWR_DAT consistent to the parameterized target area (e.g. data area in the work memory or process image).

Note that the "normal" updating of the process images can be interrupted following each transmitted doubleword. An exception for newer CPUs is the transmission of user data blocks with distributed I/O by means of a partial process image if the user data blocks have been configured in the hardware as consistent. You can also influence these data blocks in the process image by means of a direct access, but it could be the case that you destroy the data consistency.

CPU-specific data apply to the maximum size of a consistent area in the case of data transmission for global data communication, S7 basic communication and S7 communication by the operating system (said technical specifications in the CP manual).

Diagnostics and parameter data are always transmitted consistent in data sets (e.g. diagnostics data with the SFC 13 DPMRM_DG or SFB

54 RALRM or parameter data transmitted to and from modules using the SFB 52 RDREC and SFB 53 WRREC).

1.5.4 Bit Memories

Bit memories can be regarded as the controller's "auxiliary contactors". Bit memories are used primarily for storing binary signal states. They can be treated as outputs, but are not "externalized". Bit memories are located in the CPU's system memory area, and are therefore available at all times. The number of bit memories is CPU-specific.

Bit memories are used to store intermediate results that are valid beyond block boundaries and are processed in more than one block. Besides the data in global data blocks, the following are also available for storing intermediate results

- ▷ Temporary local data, which are available in all blocks but valid for the current block call only, and
- ▷ Static local data, which are available only in function blocks but valid over multiple block calls.

Retentive bit memories

Some bit memories may be designated "retentive", which means that these bit memories retain their signal states even under off-circuit conditions. Retentivity always begins with memory byte 0 and ends at the designated location. Retentivity is set when the CPU is parameterized. Please refer to Chapter 22.2.4 "Retentivity".

Clock memories

Many procedures in the controller require a periodic signal. Such a signal can be implemented using timers (clock pulse generator), watchdog interrupts (time-controlled program execution), or simply by using clock memory.

Clock memories are bits whose signal states change periodically with a mark-to-space ratio of 1:1. The bits are combined into a byte, and correspond to fixed frequencies (Figure 1.10). You specify the number of clock memory bits when you parameterize the CPU. Please note that the updating of clock memories is asynchronous to execution of the main program.

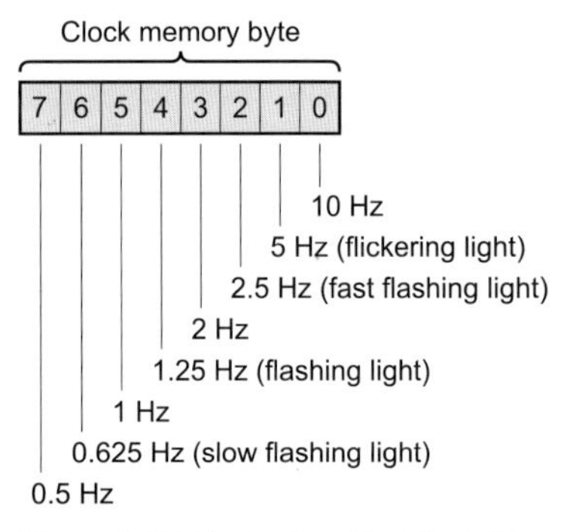

Figure 1.10 Contents of the Clock Memory Byte

2 STEP 7 Programming Software

2.1 STEP 7 Basic Package

This chapter describes the STEP 7 basic package, Version 5.5. After you have been given an overview of the properties of the automation system in the first chapter, you can read here how these properties are set.

The basic package contains the statement list (STL), ladder logic (LAD) and function block diagram (FBD) programming languages. In addition to the basic package, option packages such as S7-SCL (Structured Control Language), S7-GRAPH (sequence planning) and S7-HiGraph (state-transition diagram) are also available.

2.1.1 Installation

STEP 7 V5.5 is a 32-bit application which executes with MS Windows XP Professional and MS Windows 7 Professional, Ultimate and Enterprise. MS Internet Explorer V6.0 or higher is required under all operating systems. You require administration rights in order to install STEP 7, and to work with STEP 7 you must at least be logged-on as a main user.

If you wish to work rapidly with STEP 7, or if you are working on large projects with perhaps several hundred modules, you should use a programming device or PC with up-to-date processing power.

STEP 7 V5.5 occupies approximately 650 to 900 MB on the hard disk depending on the scope of installation and the number of installed languages. A swap file is also necessary and its size must be at least twice that of the main memory.

You should ensure there is sufficient memory on the drive containing your project data. The memory requirements may increase for certain operations, such as copying a project. If there is insufficient space for the swap-out file, errors such as program crashes may occur. You are recommended not to store the project data on the drive containing the Windows swap-out file.

The SETUP program on the DVD is used for the installation, or STEP 7 is already factory-installed on the programming device. In addition to STEP 7, the DVD also includes, inter alia, the Automation License Manager (see Chapter 2.1.2 "Automation License Manager") and the STEP 7 electronic manuals with Acrobat Reader.

A bus interface is required for the online connection to a programmable controller. This can be a multi-point interface, a PROFIBUS interface, or an Ethernet interface.

If you want to use PC memory cards or micro memory cards, you will need a prommer.

STEP 7 V5 has multi-user capability, that is, a project that is stored, say, on a central server can be edited simultaneously from several workstations. You make the necessary settings in the Windows Control Panel with the "SIMATIC Workstation" program. In the dialog box that appears, you can parameterize the workstation as a single-user system or a multi-user system with the protocols used.

You can deinstall STEP 7 using the Setup program or in the usual manner for MS Windows using the "Software" program in the Windows control panel.

2.1.2 Automation License Manager

You require a license (right of use) in order to use STEP 7. This consists of the Certificate of License and the electronic License Key. The License Key is provided on the License Key disk or on a USB flash drive.

A License Key can be present on the License Key Disk, on a USB flash drive, and on local or networked hard disk drives. A License Key only functions if it is present on a hard disk drive which is not write-protected. You can transmit and manage the License Keys using the *Automation License Manager*. Installation of the Automation License Manager is a prerequisite for operation of STEP 7. You can install the Automation License Manager together with STEP 7 or individually.

The Certificate of License defines the type of License Keys:

▷ Single License
This privilege of use is for an unlimited time and for any computer.

▷ Floating License
This privilege of use is for an unlimited time and for access via a network.

▷ Trial License
This privilege of use is limited to max. 14 days or a particular number of days starting from the first use. It can be used for test and validation purposes.

▷ Upgrade License
This privilege of use permits upgrading of a License Key from a previous version to the current one.

Following installation of STEP 7, you will be prompted for your licensing, if the hard disk does not already contain a License Key. You can also provide your licensing later.

The License Key is saved on the hard disk in particularly identified blocks. Please observe the notes for handling License Keys provided in the help text Automation License Manager to prevent unintentional destruction of the License Key.

2.1.3 SIMATIC Manager

The SIMATIC Manager is the main tool in STEP 7; you will find its icon in Windows.

The SIMATIC Manager is started by double-clicking on its icon.

When first started, the project wizard is displayed. This can be used for simple creation of new projects. You can deactivate it with the check box "Display Wizard on starting the SIMATIC Manager" since it can also be called, if required, via the menu command FILE → 'NEW PROJECT' WIZARD.

Programming begins with opening or creating a 'project'. The example projects supplied are a good basis for familiarization.

When you open example project ZEn01_09_S7_Zebra with FILE → OPEN, you will see the split project window: on the left is the structure of the open object (the object hierarchy), and on the right is the selected object. Clicking on the box containing a plus sign in the left window displays additional levels of the structure; selecting an object in the left half of the window displays its contents in the right half of the window (Figure 2.1).

Under the SIMATIC Manager, you work with the objects in the STEP 7 world. These "logical" objects correspond to 'real' objects in your plant. A project contains the entire plant, a station corresponds to a programmable controller. A project may contain several stations connected to one another, for example, via an MPI subnet. A station contains a CPU, and the CPU contains a program, in our case an S7 program. This program, in turn, is a 'container' for other objects, such as the object *Blocks*, which contains, among other things, the compiled blocks.

The STEP 7 objects are connected to one another via a tree structure. Figure 2.2 shows the most important parts of the tree structure (the "main branch", as it were) when you are working with the STEP 7 basic package for S7 applications in offline view. The objects shown in bold type are containers for other objects.

All objects in the Figure are available to you in the offline view. These are the objects that are on the programming device's hard disk. If your programming device is online on a CPU (normally a PLC target system), you can switch to the online view by selecting VIEW → ONLINE. This option displays yet another project window containing the objects on the destination device; the objects shown underlined in the Figure are then no longer included.

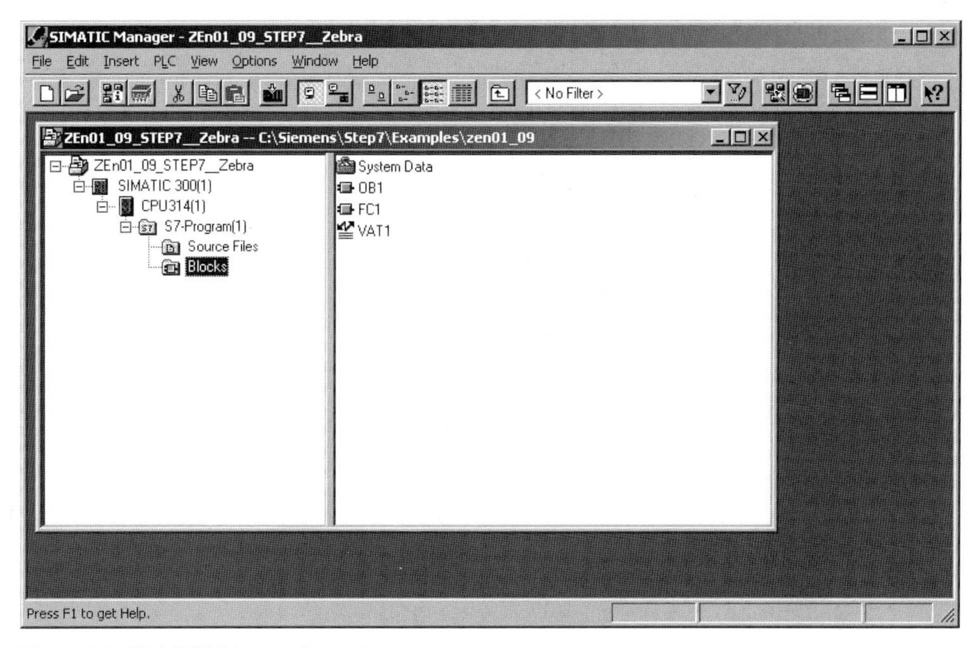

Figure 2.1 SIMATIC Manager Example

You can see from the title bar of the active project window whether you are working offline or online. For clearer differentiation, the title bar and the window title can be set to a different color than the offline window. For this purpose, select OPTIONS → CUSTOMIZE and modify the entries in the "View" tab.

Select OPTIONS → CUSTOMIZE to change the SIMATIC Manager's basic settings, such as the session language, the archive program and the storage location for projects and libraries, and configuring the archive program.

Editing sequences

The following applies for the general editing of objects:

To *select an object* means to click on it once with the mouse so that it is highlighted (this is possible in both halves of the project window).

To *name an object* means to click on the name of the selected object (a frame will appear around the name and you can change the name in the window) or select the menu item EDIT → OBJECT PROPERTIES and change the name in the dialog box. With some objects such as a CPU, you can only change the name with the relevant tool (application), in this case with the Hardware Configuration.

To *open an object*, double-click on that object. If the object is a container for other objects, the SIMATIC Manager displays the contents of the object in the right half of the window. If the object is on the lowest hierarchical level, the SIMATIC Manager starts the appropriate tool for editing the object (for instance, double-clicking on a block starts the editor, allowing the block to be edited).

In this book, the menu items in the standard menu bar at the top of the window are described as operator sequences. Programmers experienced in the use of the operator interface use the icons from the toolbar. The use of the right mouse button is very effective. Clicking on an object once with the right mouse button screens a menu showing the current editing options.

2.1.4 Projects and Libraries

In STEP 7, the 'main objects' at the top of the object hierarchy are projects and libraries.

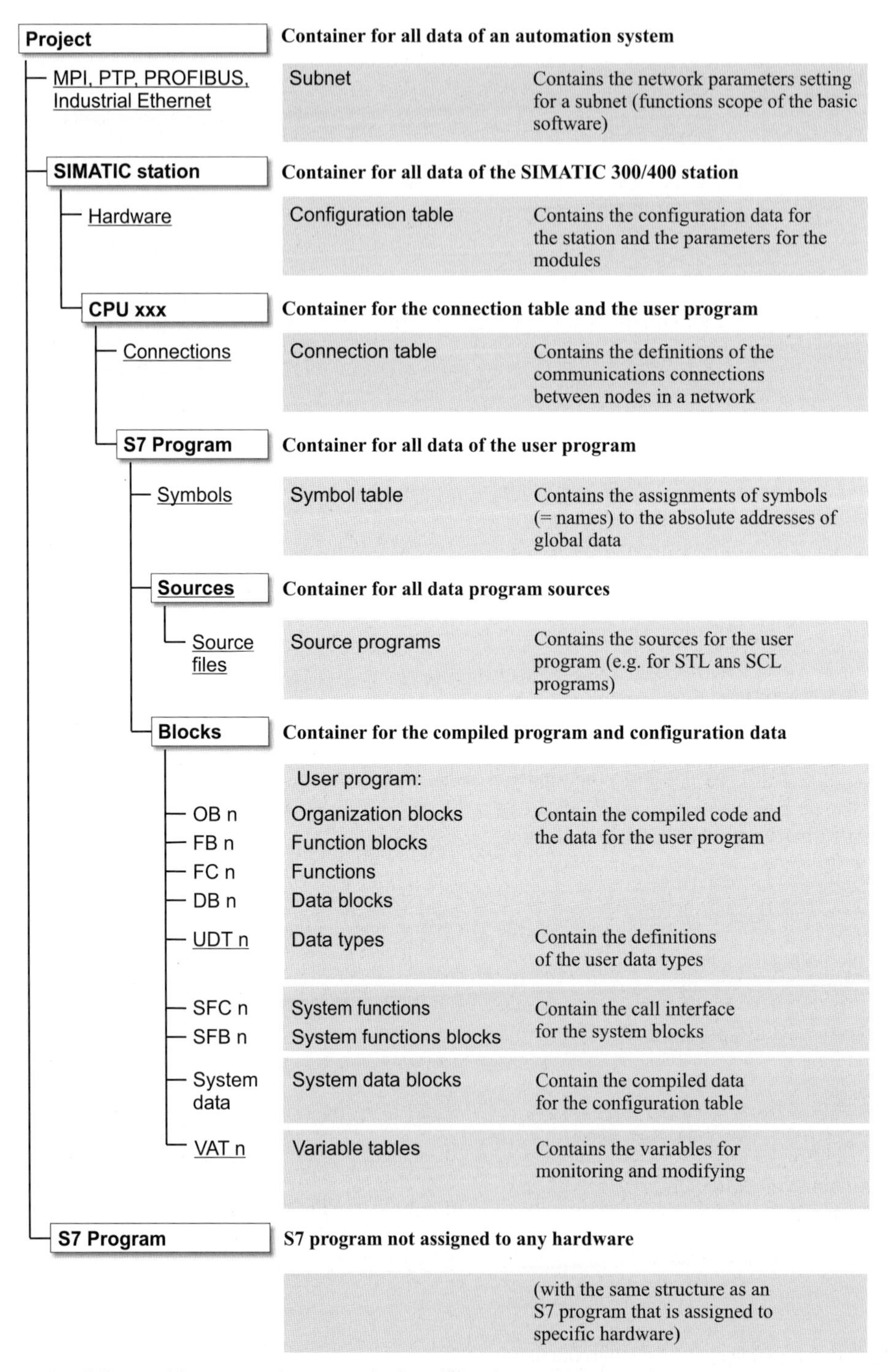

(The different objects are only present in the offline data management.)

Figure 2.2 Object Hierarchy in a STEP 7 Project

Starting with STEP 7 V5.2, you can combine projects and libraries into multiprojects (see Chapter 2.1.5 "Multiprojects").

Projects are used for the systematic storing of data and programs needed for solving an automation task. Essentially, these are

- ▷ the hardware configuration data,
- ▷ the parameterization data for the modules,
- ▷ the configuring data for communication via networks,
- ▷ the programs (code and data, symbols, sources).

The objects in a project are arranged hierarchically. The opening of a project is the first step in editing all (subordinate) objects which that object contains. The following sections discuss how to edit these objects.

Libraries are used for storing reusable program components. Libraries are organized hierarchically. They may contain STEP 7 programs which in turn may contain a user program (a container for compiled blocks), a container for source programs, and a symbol table. With the exception of online connections (no debugging possible), the creation of a program or program section in a library provides the same functionality as in an object.

As supplied, STEP 7 V5 includes the *Standard Library* containing the following programs:

- ▷ System Function Blocks
  Contains the call interfaces of the system blocks for offline programming integrated in the CPU
- ▷ S5-S7 Converting Blocks
  Contains loadable functions for the S5/S7 converter (replacement of S5 standard function blocks in conjunction with program conversion)
- ▷ TI-S7 Converting Blocks
  Contains additional loadable functions and function blocks for the T1-S7 converter
- ▷ IEC Function Blocks
  Contains loadable functions for editing variables of the complex data types DATE_AND_TIME and STRING
- ▷ Communication Blocks
  Contains loadable functions for controlling CP modules
- ▷ Miscellaneous Blocks
  Contains blocks for time tagging and time synchronization
- ▷ PID Control Blocks
  Contains loadable function blocks for closed-loop control
- ▷ Organization Blocks
  Contains the templates for the organization blocks (essentially the variable declaration for the start information)

You will find an overview of the contents of these libraries in Chapter 33 "Block Libraries". Should you, for example, purchase an S7 module with standard blocks, the associated installation program installs the standard blocks as a library on the hard disk. You can then copy these blocks from the library to your project. A library is opened with FILE → OPEN, and can then be edited in the same way as a project. You can also create your own libraries.

The menu item FILE → NEW generates a new object at the top of the object hierarchy (project, library). The location in the directory structure where the SIMATIC Manager is to create a project or library must be specified under the menu item OPTIONS → CUSTOMIZE or in the "New dialog" box.

The INSERT menu is used to add new objects to existing ones (such as adding a new block to a program). Before doing so, however, you must first select the object container in which you want to insert the new object from the left half of the SIMATIC Manager window.

You copy object containers and objects with EDIT → COPY and EDIT → PASTE or, as is usual with Windows, by dragging the selected object with the mouse from one window and dropping it in another. Please note that you cannot undo deletion of an object or an object container in the SIMATIC Manager.

2.1.5 Multiprojects

Projects and libraries are combined into an entity in a multiproject. The multiproject permits editing of communications connections such as S7 connections between the projects. This means that a multiproject can be handled almost like an individual project. Limitations: stations connected together by means of direct data exchange ("internode communication") or global data communication must be present in the same project.

With a multiproject, it is possible to carry out independent, parallel editing of individual projects by different employees. The individual projects can be in different directories in a networked environment. Cross-project functions, e.g. the balancing of subnets and connections, are then carried out centrally by editing the multiproject.

The generation of a multiproject is also an advantage if one wishes to make the individual projects smaller and clearer.

You can archive and retrieve a multiproject just like a project or library.

2.1.6 Online Help

The SIMATIC Manager's online help provides information you need during your programming session without the need to refer to hardcopy manuals. You can select the topics you need information on by selecting the HELP menu. The online help option GETTING STARTED, for instance, provides a brief summary on how to use the SIMATIC Manager.

HELP → CONTENTS starts the central STEP 7 Help function from any application. This contains all the basic knowledge. If you click the "Home" symbol in the menu bar, you will be provided with an introduction to the central topics of STEP 7: Starting with STEP 7, Configuration & programming, Testing & troubleshooting, as well as SIMATIC on the Internet.

HELP → CONTEXT-SENSITIVE HELP F1 provides context-sensitive help, i.e. if you press F1, you get information concerning an object selected by the mouse or concerning the current error message.

In the symbol bar, there is a button with an arrow and a question mark. If you click on this button, a question mark is added to the mouse pointer. With this "Help" mouse pointer, you can now click on an object on the screen, e.g. a symbol or a menu command, and you will get the associated online help.

2.2 Editing Projects

When you set up a project, you create "containers" for the resulting data, then you generate the data and fill these containers. Normally, you create a project with the relevant hardware, configure the hardware, or at least the CPU, and receive in return containers for the user program. However, you can also put an S7 program directly into the project container without involving any hardware at all. Note that initializing of the modules (address modifications, CPU settings, configuring connections) is possible only with the Hardware Configuration tool.

We strongly recommend that the entire project editing process be carried out using the SIMATIC Manager. Creating, copying or deleting directories or files as well as changing names (!) with the Windows Explorer within the structure of a project can cause problems with the SIMATIC Manager.

2.2.1 Creating Projects

Project wizard

The STEP 7 Wizard helps you in creating a new project. You specify the CPU used and the wizard creates for you a project with an S7 station and the selected CPU as well as an S7 program container, a source container and a block container with the selected organization blocks. You can start the project wizard using FILE → 'NEW PROJECT' WIZARD.

Creating a project with the S7 station

If you want to create a project "manually", this section outlines the necessary actions for you. You will find general information on operator entries for object editing in Chapter 2.1.3 "SIMATIC Manager".

Creating a new project

Select FILE → NEW, enter a name in the dialog box "Name", change the type and storage location if necessary, and confirm with "OK" or RETURN.

Inserting a new station in the project

Select the project and insert a station with INSERT → STATION → SIMATIC 300 STATION (in this case an S7-300).

Configuring a station

Click on the plus box next to the project in the left half of the project window and select the station; the SIMATIC Manager displays the Hardware object in the right half of the window. Double-clicking on *Hardware* starts the Hardware Configuration tool, with which you edit the configuration tables.

If the module catalog is not on the screen, call it up with VIEW → CATALOG.

You begin configuring by selecting the rail with the mouse, for instance under "SIMATIC 300" and "RACK 300", "holding" it, dragging it to the free portion in the upper half of the station window, and "letting it go" (drag & drop). You then see a table representing the slots on the rail.

Next, select the required modules from the module catalog and, using the procedure described above, drag and drop them in the appropriate slots. To enable further editing of the project structure, a station requires at least one CPU, for instance the CPU 314 in slot 2. You can add all other modules later. Editing of the hardware configuration is discussed in detail in Chapter 2.3 "Configuring Stations".

Store and compile the station, then close and return to the SIMATIC Manager. In addition to the hardware configuration, the open station now also shows the CPU.

When it configures the CPU, the SIMATIC Manager also creates an S7 program with all objects. The project structure is now complete.

Viewing the contents of the S7 program

Open the CPU; in the right half of the project window you will see the symbols for the *S7 program* and for the connection table.

Open the S7 program; the SIMATIC Manager displays the symbols for the compiled user program (the compiled blocks), the container for the source programs, and the symbol table in the right half of the window.

Open the user program (*Blocks*); the SIMATIC Manager displays the symbols for the compiled configuration data *(System data)* and an empty organization block for the main program (OB 1) in the right half of the window).

Editing user program objects

We have now arrived at the lowest level of the object hierarchy. The first time OB 1 is opened, the window with the object properties is displayed and the editor needed to edit the program in the organization block is opened. You add another empty block for incremental editing by opening INSERT → S7 BLOCK → ... (*Blocks* must be highlighted) and selecting the required block type from the list provided.

When opened, the *System data* object shows a list of available system data blocks. You receive the compiled configuration data. These system data blocks are edited via the *Hardware* object in the container *Station*. You can transfer *System data* to the CPU with PLC → DOWNLOAD and parameterize the CPU in this way.

The object container *Source Files* is empty. With *Source Files* selected, you can select INSERT → S7 SOFTWARE → STL SOURCE to insert an empty source text file or you can select INSERT → EXTERNAL SOURCE to transfer a source text file created, say, with another editor in ASCII format to the *Source Files* container.

Creating a project without an S7 station

If you wish, you can create a program without first having to configure a station. To do so, generate the container for your program yourself. Select the project and generate an S7 program with INSERT → PROGRAM → S7 PROGRAM. Under this S7 program, the SIMATIC Manager creates the object *Symbols* and the object containers *Sources* and *Blocks*. *Blocks* contains an empty OB 1.

Creating a library

You can also create a program under a library, for instance if you want to use it more than once. In this way, the standard program is always available and you can copy it entirely or in part into your current program.

Please note that you cannot establish online connections in a library, which means that you can debug a STEP 7 program only within a project.

2.2.2 Managing, Rearranging and Archiving

The SIMATIC Manager maintains a list of all known "main objects", arranged according to user projects, libraries, example projects and multiprojects. You install the example projects and the standard libraries in conjunction with STEP 7 and you install the user projects, the multiprojects and your own libraries yourself.

When it executes FILE → MANAGE, the SIMATIC Manager shows you all the projects and libraries known to it with name and path. You can now delete ("Hide") projects or libraries from the list which you no longer wish to display, or incorporate new projects and libraries into the list ("Display").

When it executes FILE → REORGANIZE, the SIMATIC Manager eliminates the gaps created by deletions and optimizes data memory similarly to the way a defragmentation program optimizes the data memory on the hard disk. The reorganization can take some time, depending on the data movements involved.

You can also archive a project or library (FILE → ARCHIVE). In this case, the SIMATIC Manager stores the selected object (the project or library directory with all subdirectories and files) in compressed form in an archive file.

Projects and libraries cannot be edited in the archived (compressed) state. You can unpack an archived object with FILE → RETRIEVE and then you can edit it further. The retrieved objects are automatically accepted into the project or library management system.

You make the settings for archiving and retrieving on the "Archive" tab under OPTIONS → CUSTOMIZE. You select the archiving program from a drop-down list: Arj (*.arj), PKZip 12.4 (*.zip), or WinZip (*.zip).

You then set the options for archiving on this tab, for example the target directory for archiving and retrieving or "Automatically create archive path" (no further inputs are then required when archiving, since the name of the archive file is generated from the project name).

Archiving a project in the CPU

With the appropriately designed CPUs, you can store a project in archived (compressed) form in the load memory of the CPU, that is, on the memory card or micro memory card. In this way, you can save all project data required for complete execution of the user program, such as symbols or source files, direct at the machine or plant. If it becomes necessary to modify or supplement the program, you load the locally stored data onto the programming device, correct the user program and save the up-to-date project data again to the CPU.

When loading the project data onto a memory card or micro memory card plugged into the CPU, open the project, mark the CPU and select PLC → SAVE TO MEMORY CARD. In the reverse direction, transfer the stored data back to the programming device with PLC → RETRIEVE FROM MEMORY CARD. Please note that when you write to a memory card plugged into the CPU, the entire contents of the load memory are written to the CPU, including the system data and the user programs.

If you want to fetch back the project data stored on the CPU without creating a project on the programming device, select the relevant CPU with PLC → DISPLAY ACCESSIBLE NODES. If the memory card is plugged into the module receptacle of the programming device, select the memory card with FILE → S7 MEMORY CARD → OPEN before transferring.

2.2.3 Project Versions

Since STEP 7 V5 has become available, there are three different versions of SIMATIC projects. STEP 7 V1 creates version 1 projects, STEP 7 V2 version 2 projects, and STEP 7

V3/V4/V5.0 can be used to create and edit both version 2 and version 3 projects. With STEP 7 from version V5.1, you can create and edit V3 projects and V3 libraries.

If you have a version 1 project, you can convert it into a version 2 project with FILE → OPEN VERSION 1 PROJECT. The project structure with the programs, the compiled version 1 blocks, the STL source programs, the symbol table and the hardware configuration remain unchanged.

You can create and edit version 2 projects with STEP 7 versions V2, V3, V4 and V5.0 (Figure 2.3). STEP 7 version V5.1 and higher works exclusively with version 3 projects.

Up to STEP 7 version 5.3 you can convert a V1 project to a V2 project with FILE → OPEN VERSION 1 PROJECT. With FILE → OPEN you can open a V2 project and convert it to a V3 project. It is not possible to create a V2 project or save a project as a V2 project.

2.2.4 Creating and Editing Multiprojects

Use FILE → NEW to create a new multiproject in the SIMATIC Manager by selecting "Multiproject" as the type in the displayed dialog box. With the multiproject selected, use FILE → MULTIPROJECT → CREATE IN MULTIPROJECT to create a new project or a new library in the multiproject. The newly created project or the newly created library can be edited as described in the previous chapters. Use FILE → MULTIPROJECT → INSERT INTO MULTIPROJECT to incorporate projects and libraries which already exist into the multiproject.

You can also delete projects and libraries again from the multiproject: Mark the project or library, and then select FILE → MULTIPROJECT → REMOVE FROM MULTIPROJECT. This does not delete the project or the library itself.

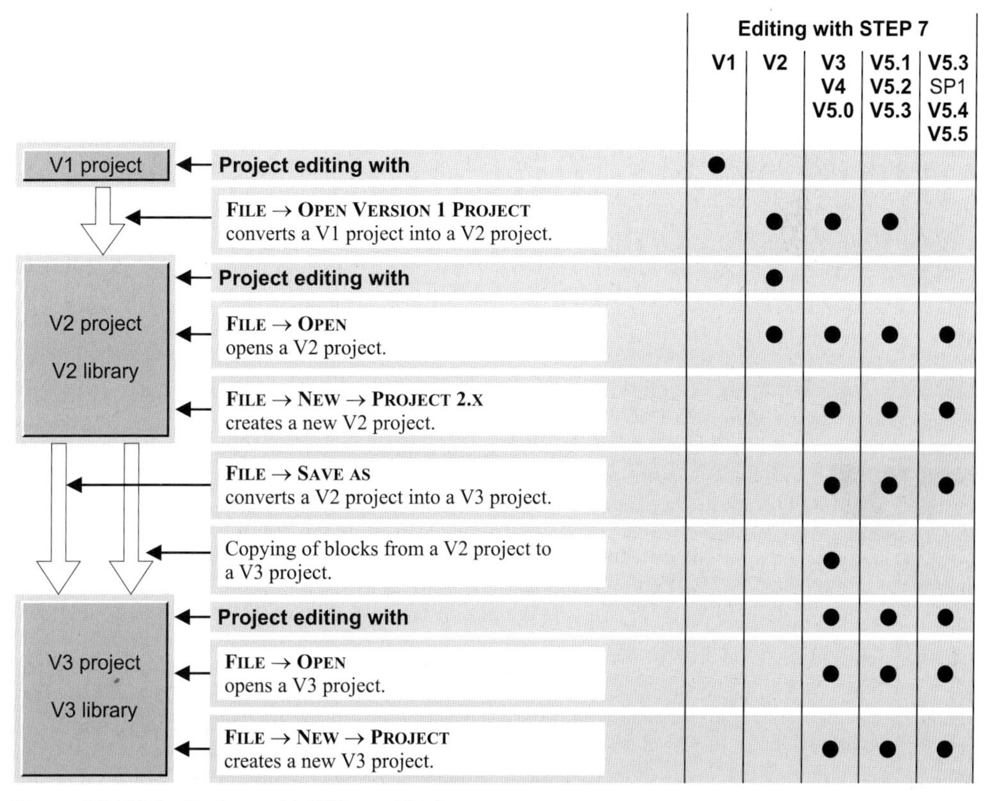

Figure 2.3 Editing Projects with Different Versions

Use FILE → MULTIPROJECT → ADJUST PROJECTS to start a wizard which supports you when adjusting cross-project connections and when combining subnets (Chapter 2.4.6 "Adjusting Projects in the Multiproject").

Using FILE → MULTIPROJECT → DEFINE AS MASTER DATA LIBRARY you can identify one of the libraries in a multiproject as the "Master data library" which e.g. is to accommodate the common blocks of the projects in this multiproject. This library must then only contain one single S7 program.

The menu commands FILE → SAVE AS, FILE → REARRANGE, FILE → MANAGE and FILE → ARCHIVE can also be used on a multiproject, and function just like with a project (see Chapter 2.2.2 "Managing, Rearranging and Archiving"). In addition, archived multiprojects can be transferred to the load memory of a correspondingly designed CPU. Limitations exist when archiving a multiproject whose components are distributed among network drives.

2.3 Configuring Stations

You use the Hardware Configuration tool to plan your programmable controller's configuration. Configuring is carried out offline without connection to the CPU. You can also use this tool to address and parameterize the modules. You can create the hardware configuration at the planning stage or you can wait until the hardware has already been installed.

You start the Hardware Configuration by selecting the station and then EDIT → OPEN OBJECT or by double-clicking on the Hardware object in the opened container *SIMATIC 300/400-STATION*. You make the basic settings of the Hardware Configuration with OPTIONS → CUSTOMIZE.

When configuring has been completed, STATION → CONSISTENCY CHECK will show you whether your entries were free of errors. STATION → SAVE stores the configuration tables with all parameter assignment data in your project on the hard disk.

STATION → SAVE AND COMPILE not only saves but also compiles the configuration tables and stores the compiled data in the *System data* object in the offline container *Blocks*. After compiling, you can transfer the configuration data to a CPU with PLC → DOWNLOAD. The object *System data* in the online container *Blocks* represents the current configuration data on the CPU. You can 'return' these data to the hard disk with PLC → UPLOAD.

You export the data of the hardware configuration with STATION → EXPORT. STEP 7 then creates a file in ASCII format that contains the configuration data and parameterization data of the modules. You can choose between a text format that contains the data in "readable" English characters, or a compact format with hexadecimal data. You can also import a correspondingly structured ASCII file.

Checksum

The Hardware Configuration generates a checksum via a correctly compiled station and stores it in the system data. Identical system configurations have the same checksum so that you can, for example, easily compare an online configuration with an offline configuration.

The checksum is a property of the *System data* object. To read the checksum, open the *Blocks* container in the S7 program, select the *System data* object and open it with EDIT → OPEN OBJECT.

The user program also has an appropriate checksum. You can find this along with the checksum of the system data in the properties of *Blocks*: select the *Blocks* container and then EDIT → OBJECT PROPERTIES on the "Checksums" tab.

Station window

When opened, the Hardware Configuration displays the station window and the hardware catalog (Figure 2.4). Enlarge or maximize the station window to facilitate editing. In the upper section, it shows the S7 stations in the form of tables (one per rack) which are connected together via the interface modules if several racks are used. If distributed I/O is connected the structure of the DP master system or of the PROFINET IO system is shown. The DP stations and IO devices are displayed as symbols.

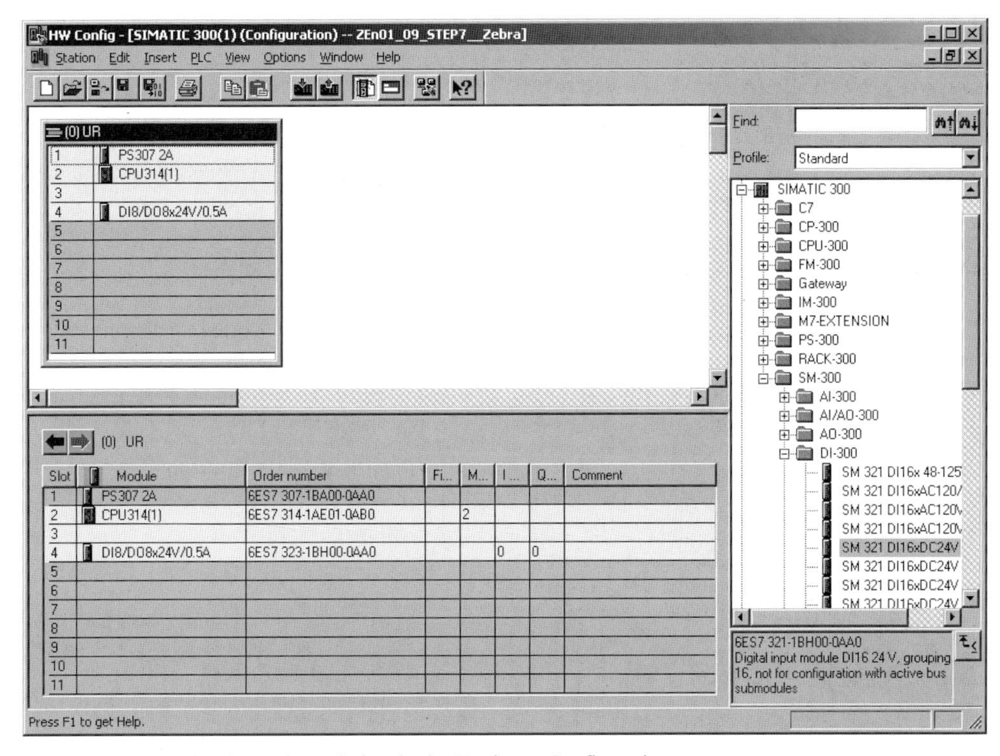

Figure 2.4 Example of a Station Window in the Hardware Configuration

The lower section of the station window shows the configuration table that gives a detailed view of the rack, DP slave or IO device selected in the upper section.

Hardware catalog

You can fade the hardware catalog in and out with VIEW → CATALOG. It contains all available mounting racks, modules and interface submodules known to STEP 7. With OPTIONS → EDIT CATALOG PROFILE, you can compile your own hardware catalog that shows only the modules you want to work with – in the structure you select. By double-clicking on the title bar, you can "dock" the hardware catalog onto the right edge of the station window or release it again.

Installing a hardware update

You can update components for the hardware catalog using OPTIONS → INSTALL HW UPDATES. In the following dialogs, select whether you wish to download the update from the Internet or copy it from a CD. Enter the Internet address and the storage path. When you click the "Install" button, the Hardware Configuration tool transfers the data to the hardware catalog.

Product support information

Use HELP → PRODUCT SUPPORT INFORMATION to display information from the Internet on the marked module. You must first enable this function using OPTIONS → CUSTOMIZE and by setting a valid Internet address. The marked module can be present in the hardware catalog or already in the configured rack.

Configuration table

The Hardware Configuration tool works with tables that each represent an S7 station (a mounting rack), a DP station or an IO-Device. A configuration table shows the slots with the modules arranged in the slots or the properties of the modules such as their addresses and order numbers. A double-click on a module line opens the properties window of the module and allows parameterization of the module.

2.3.1 Arranging Modules

You have created for example, a SIMATIC 300/400 station using the SIMATIC Manager and wish to equip this station with an S7 CPU and the associated modules. To do this, open the station (select the station in the SIMATIC Manager followed by EDIT → OPEN OBJECT or double-click on the *Hardware* object in the opened folder *SIMATIC 300/400 STATION*).

First define the rack. You can find it for S7-400 stations in the hardware catalog under “SIMATIC 400” and “RACK-400” and for S7-300 stations under “SIMATIC 300” and “RACK-300”. Select and “hold” the rack or DIN rail using the mouse, drag it into the top part of the station window, and drop it at any position (drag & drop). An empty configuration table is displayed for the central rack.

To create a station with an ET 200 CPU, select and open the SIMATIC 300 station in the SIMATIC Manager. In the Hardware Catalog under "PROFIBUS DP" or "PROFINET IO" and "I/O", you can then drag the desired CPU, e.g. IM154-8 CPU under ET 200pro, into the top part of the station window using the mouse or select it using a double-click. The configuration table is then equipped with the CPU.

Next, select the required modules from the module catalog and, in the manner described above, drag and drop them in the appropriate slots. The permissible slots have a green background. A “No Parking” symbol tells you cannot drop the selected module at the intended slot.

Please note for stations with an ET 200 CPU that you may only use the modules which are present under the respective CPU in the Hardware Catalog.

You can also mark the slots to be fitted and select INSERT → INSERT OBJECT. The Hardware Configuration then shows you in a popup window all modules permissible for this slot, of which you can select one.

In the case of single-tier S7-300 stations, slot 3 remains empty; it is reserved for the interface module to the expansion rack.

You can generate the configuration table for another rack by dragging the selected rack from the catalog and dropping it in the station window. In S7-400 systems, a non-interconnected rack (or more precisely: the relevant receive interface module) is assigned an interface via the “Link” tab in the Properties window of a Send IM (select module and EDIT → OBJECT PROPERTIES).

The arrangement of distributed I/O stations is described in Chapter 20.4 “Communication via Distributed I/O”.

2.3.2 Addressing Modules

When arranging modules, the Hardware Configuration tool automatically assigns a module start address. You can view this address in the lower half of the station window or in the object properties for the relevant modules in the tab “Addresses”. If you deselect the option “System selection” for S7-300 modules in this tab, you can change the module addresses. When doing so, please observe the addressing rules for S7-300 and S7-400 systems as well as the addressing capacity of the individual modules.

There are modules that have both inputs and outputs for which you can (theoretically) reserve different start addresses. However, please note carefully the special information provided in the product manuals; the large majority of function and communications modules require the same start address for inputs and outputs.

When assigning the module start address on correspondingly designed CPUs, you can also make the assignment to a subsidiary process image. If there is more than one CPU in the central rack of an S7-400, multiprocessor mode is automatically set and you must assign the module to a CPU.

With VIEW → ADDRESS OVERVIEW, you get a window containing all the module addresses currently in use for the CPU selected.

Modules on the MPI bus or communications bus have an MPI address. You may also change this address. Note, however, that the new MPI address becomes effective as soon as the configuration data are transferred to the CPU.

Symbols for user data addresses

In the Hardware Configuration tool, you can assign to the inputs and outputs symbols (names) that are transferred to the Symbol Table.

After you have arranged and addressed the digital and analog modules, you save the station data. Then you select the module (line) and EDIT → SYMBOLS. In the window that then opens, you can assign a symbol, a data type and a comment to the absolute address for each channel (bit-by-bit for digital modules and word-by-word for analog modules.

The "Add Symbol" button enters the absolute addresses as symbols in place of the absolute addresses without symbols. The "Apply" button transfers the symbols into the Symbol Table. "OK" also closes the dialog box.

2.3.3 Parameterizing Modules

When you parameterize a module, you define its properties. It is necessary to parameterize a module only when you want to change the default parameters. A requirement for parameterization is that the module is located in a configuration table.

Double-click on the module in the configuration table or select the module and then EDIT → OBJECT PROPERTIES. Several tabs with the specifiable parameters for this module are displayed in the dialog box. When you use this method to parameterize a CPU, you are specifying the run characteristics of your user program.

Some modules allow you to set their parameters at runtime via the user program with system functions (see Chapter 22.5.2 "System Blocks for Module Parameterization").

Module identification

Innovated S7 CPUs, PROFIBUS DPV1 slaves, and PROFINET IO devices can support functions for identification and maintenance (I&M functions). For example, you can assign a plant identifier and a location identifier to a station for subsequent evaluation in the program. Using the plant identifier, you can assign a name to parts of the plant in accordance with functional aspects. The location identifier is part of the equipment identifier and describes e.g. the exact position of a SIMATIC device in the process engineering plant.

To enter the I&M data, select the module in the Hardware Configuration and select EDIT → OBJECT PROPERTIES. You can then – with a correspondingly equipped module – enter the plant identifier and the location identifier in the "General" tab or in the "Identification" tab. In online mode, select the module, and you can then exchange the I&M data between offline data management and module using PLC → DOWNLOAD MODULE IDENTIFICATION or PLC → DOWNLOAD MODULE IDENTIFICATION IN PG.

To evaluate the I&M data, use SFC 51 RDSYSST to read the system state list with the ID 16#011C index 16#0003 for the plant identifier and index 16#000B for the location identifier.

2.3.4 Networking Modules with MPI

You define the nodes for the MPI subsidiary (subnet) with the Module Properties. Select the CPU, or the MPI interface card, if the CPU is equipped with one, in the configuration table and open it with EDIT → OBJECT PROPERTIES. The dialog box that then appears contains the "Properties" button in the "Interface" box of the "General" tab. If you click on this button you are taken to another dialog box with a "Parameter" tab where you can find the suitable subnet.

This is also an opportunity to set the MPI address that you have provided for this CPU. Please note that on older S7-300 CPUs, FMs or CPs with MPI connection automatically receive an MPI address derived from the CPU.

The highest MPI address must be greater than or equal to the highest MPI address assigned in the subnet (take account of automatic assignment of FMs and CPs!). It must have the same value for all nodes in the subnet.

Tip: if you have several stations with the same type of CPUs, assign different names (identifiers) to the CPUs in the different stations. They all have the name "CPUxxx(1)" as default so in the subnet they can only be differentiated by their MPI addresses. If you do not want to assign a name yourself, you can, for example, change the default identifier from "CPUxxx(1)" to "CPUxxx(n)" where "n" is equal to the MPI address.

When assigning the MPI address, please also take into account the possibility of connecting a programming device or operator panel (OP) to the MPI network at a later date for service or maintenance purposes. You should connect permanently installed programming devices or OPs direct to the MPI network; for plug-in devices via a spur line, there is an MPI connector with a heavy-gauge threaded-joint socket. Tip: reserve address 0 for a service programming device, address 1 for a service OP and address 2 for a replacement CPU (corresponds to the default addresses).

2.3.5 Monitoring and Modifying Modules

With the Hardware Configuration, you can carry out a wiring check of the machine or plant without the user program. A requirement for this is that the programming device is connected to a station (online) and the configuration has been saved, compiled and loaded into the CPU. Now you can address every digital and analog module. Select a module and then PLC → MONITOR/MODIFY, and set the Monitor and Modify operating modes and the trigger conditions.

With the "Status Value" button, the Hardware Configuration shows you the signal states or the values of the module channels. The "Modify Value" button writes the value specified in the Modify Value column to the module.

If the "I/O Display" checkbox is active, the peripheral inputs/outputs (module memory) are displayed instead of the inputs/outputs (process image). The "Enable Periph. Outputs" checkbox revokes the output disable of the output modules if the CPU is in STOP mode (see Chapter 2.7.5 "Enabling Peripheral Outputs").

You can find other methods of monitoring and modifying inputs and outputs in Chapters 2.7.3 "Monitoring and Modifying Variables" and 2.7.4 "Forcing Variables".

2.4 Configuring the Network

The basis for communications with SIMATIC is the networking of the S7 stations. The required objects are the subnets and the modules with communications capability in the stations. You can create new subnets and stations with the SIMATIC Manager within the project hierarchy. You then add the modules with communications capability (CPUs and CPs) using the Hardware Configuration tool; at the same time, you assign the communications interfaces of these modules to a subnet. You then define the communications relationships between these modules – the connections – with the Network Configuration tool in the connection table.

The Network Configuration tool allows graphical representation and documentation of the configured networks and their nodes. You can also create all necessary subnets and stations with the Network Configuration tool; then you assign the stations to the subnets and parameterize the node properties of the modules with communications capability.

You can proceed as follows to define the communications relationships via the networking configuration tool:

- ▷ Open the MPI subnet created as standard in the project container (if it is no longer available, simply create a new subnet with INSERT → SUBNET).
- ▷ Use the Network Configuration tool to create the necessary stations and – if required – further subnets.
- ▷ Open the stations and provide them with the modules with communications capability.
- ▷ Connect the modules with the relevant subnets.
- ▷ Adapt the network parameters, if necessary.
- ▷ Define the communication connections in the connection table, if required.

You can also configure global data communications within the Network Configuration: select the MPI subnet and then select OPTIONS → DEFINE GLOBAL DATA (see Chapter 20.5 "Global Data Communication").

NETWORK → SAVE saves an incomplete Network Configuration. You can check the consistency of a Network Configuration with NETWORK → CHECK CONSISTENCY. You close the Network Configuration with NETWORK → SAVE AND COMPILE.

Network window

To start the Network Configuration, you must have created a project. Together with the project, the SIMATIC Manager automatically creates an MPI subnet.

A double-click on this or any other subnet starts the Network Configuration. You can also reach the Network Configuration if you open the object Connections in the CPU container.

In the upper section, the Network Configuration window shows all previously created subnets and stations (nodes) in the project with the configured connections (Figure 2.5).

The connection table is displayed in the lower section of the window if a module with "communications capability", e.g. an S7-400 CPU, is selected in the upper section of the window.

A second window displays the network object catalog with a selection of the available SIMATIC stations, subnets and DP stations. You can fade the catalog in and out with VIEW → CATALOG and you can "dock" it onto the right edge of the network window (double-click on the title bar). With VIEW → ZOOM IN, VIEW → ZOOM OUT and VIEW → ZOOM FACTOR..., you can adjust the clarity of the graphical representation.

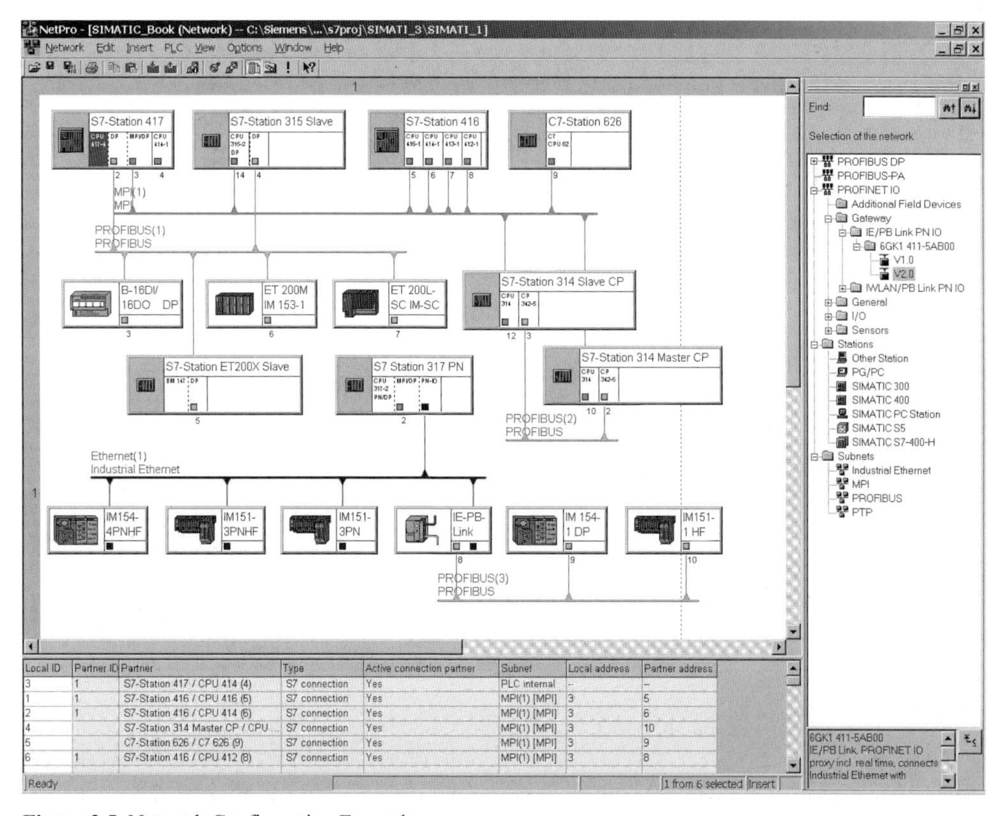

Figure 2.5 Network Configuration Example

2.4.1 Configuring the Network View

Selecting and arranging the components

You begin the Network Configuration by selecting a subnet that you select in the catalog with the mouse, hold and drag to the network window. The subnet is represented in the window as a horizontal line. Impermissible positions are indicated with a 'prohibited' sign on the mouse pointer.

You proceed in the same way for the desired stations, at first without connection to the subnet. The stations are still "empty". A double-click on a station opens the Hardware Configuration tool allowing you to configure the station or at least the module(s) with network connection. Save the station and return to the Network Configuration.

The interface of a module with communications capability is represented in the Network Configuration as a small box under the module view. Click on this box, hold and drag it to the relevant subnet. The connection to the subnet is represented as a vertical line.

Proceed in exactly the same way with all other nodes.

You can move created subnets and stations in the network window. In this way, you can also represent your hardware configuration visually. You may obtain a clearer and more compact arrangement if you reduce the displayed lengths of the subnets using VIEW → REDUCED SUBNET LENGTHS.

Setting communications properties

After creating the graphical view, you parameterize the subnets: select the subnets and then EDIT → OBJECT PROPERTIES. The properties window that then appears includes the S7 subnet ID in the "General" tab. The ID consists of two hexadecimal numbers, the project number and the subnet number. You require this S7 subnet ID if you want to go online with the programming device without a suitable project in order to reach other nodes via the subnet. You set the network properties in the "Network Settings" tab, e.g. the data transfer rate or the highest node address.

When you select the network connection of a node, you can define the network properties of the node with EDIT → OBJECT PROPERTIES, e.g., the node address and the subnet it is connected to, or you can create a new subnet.

On the "Interfaces" tab of the station properties, you can see an overview of all modules with communications capability, with the node addresses and the subnet types used.

You define the module properties of the nodes in a similar way (with the same operator inputs as in the Hardware Configuration tool).

2.4.2 Configuring Distributed I/O with the Network Configuration

You can also use the Network Configuration to configure the distributed I/O with PROFIBUS DP or PROFINET IO. Select VIEW → WITH DP SLAVES/IO DEVICES to display or fade out DP slaves and IO devices in the network view.

PROFIBUS DP

You require the following in order to configure a DP master system:

▷ A PROFIBUS subnet (if not already available, drag the PROFIBUS subnet from the network object catalog to the network window),

▷ A DP master in a station (if not already available, drag the station from the network object catalog to the network window, open the station and select a DP master with the Hardware Configuration tool, either integrated in the CPU or as an autonomous module),

▷ The connection from the DP master to the PROFIBUS subnet (either select the subnet in the Hardware Configuration tool or click on the network connection to the DP master in the Network Configuration, "hold" and drag to the PROFIBUS network).

In the network window, select the DP master to which the slave is to be assigned. Find the DP slave in the network object catalog under "PROFIBUS DP" and the relevant sub-catalog, drag it to the network window and fill out the properties window that appears.

You parameterize the DP slave by selecting it and then selecting EDIT → OPEN OBJECT. The Hardware Configuration is started. Now you can set the user data addresses or, in the case of modular slaves, select the I/O modules (see Chapter 2.3 "Configuring Stations").

You can only connect an intelligent DP slave to a subnet if you have previously created it (see Chapter 20.4.2 "Configuring PROFIBUS DP"). In the network object catalog, you can find the type of intelligent DP slave under "Configured Stations"; drag it, with the DP master selected, to the network window and fill out the properties window that then appears (as in the Hardware Configuration tool).

With VIEW → HIGHLIGHT → MASTER SYSTEM, you graphically emphasize the assignment of the nodes of a DP master system. First select the master or a slave of this master system. With VIEW → REARRANGE, the DP slaves are optically assigned to their DP master.

PROFINET IO

You require the following in order to configure a PROFINET IO system:

▷ An Industrial Ethernet subnet (if not already available, drag the Industrial Ethernet subnet from the network object catalog to the network window),

▷ An IO controller in a station (if not already available, drag the station from the network object catalog to the network window, open the station and select an IO controller with the Hardware Configuration tool, either integrated in the CPU or as an autonomous module),

▷ The connection from the IO controller to the Industrial Ethernet subnet (either select the subnet in the Hardware Configuration tool or click on the network connection to the IO controller in the Network Configuration, "hold" and drag to the Industrial Ethernet network).

In the network window, select the IO controller to which the IO device is to be assigned. Find the IO device in the network object catalog under "PROFINET IO" and the relevant sub-catalog, drag it to the network window and fill out the properties window that appears.

You parameterize the IO device by selecting it and then selecting EDIT → OPEN OBJECT. The Hardware Configuration is started. Now you can set the user data addresses or select the I/O modules (see Chapter 2.3 "Configuring Stations").

With VIEW → HIGHLIGHT → PROFINET IO SYSTEM, you graphically emphasize the assignment of the nodes of a PROFINET IO system. First select the IO controller or an IO device. With VIEW → REARRANGE, the IO devices are optically assigned to their IO controller.

2.4.3 Configuring Connections

Connections describe the communications relationships between two devices. Connections must be configured if

▷ you want to establish S7 communications between two SIMATIC S7 devices ("Communication via configured connections") or

▷ the communications partner is not a device from the SIMATIC S7 family.

Note: you do not require a configured connection for direct online connection of a programming device to the MPI network for programming or debugging. If you want to reach other nodes arranged in other connected subnets with the programming device, you must configure the connection of the programming device: in the Network Object Catalog, select the object *PG/PC* under *Stations* by double-clicking, open *PG/PC* in the network window by double-clicking, and select the interface and assign it to a subnet.

Connection table

The communications connections are configured in the connection table. Requirement: you have created a project with all stations that are to exchange data with each other, and you have assigned the modules with communications capability to a subnet.

The object *Connections* in the *CPU* container represents the connection table. A double-click on *Connections* starts the Network Configuration in the same way as a double-click on a subnet in the project container.

Table 2.1 Connection Table Example

Local ID	Partner ID	Partner	Type	Active Connection Buildup	Send Operating Mode Messages
1	1	Station 416 / CPU416(5)	S7 connection	Yes	No
2	2	Station 416 / CPU416(5)	S7 connection	Yes	No
3		Station 315 / CPU315(7)	S7 connection	Yes	No
4	1	Station 417 / CPU414(4)	S7 connection	Yes	No

To configure the connections, select e.g. an S7-400 CPU in the Network Configuration. In the lower section of the network window, you get the connection table (Table 2.1; if it is not visible, place the mouse pointer on the lower edge of the window until it changes shape and then drag the window edge up). You enter a new communication connection with INSERT → NEW CONNECTION or by double-clicking on an empty line.

You create a connection for each "active" CPU. Please note that you cannot create a connection table for an S7-300 CPU; S7-300 CPUs can only be "passive" partners in an S7 connection.

In the "Insert New Connection" window, you select in the graphics the communications partner or enter it in the "Station" and "Module" dialog boxes (Figure 2.6); the station and the module must already exist. You also determine the connection type in this window.

If you want to set more connection properties, activate the check box "Before Inserting: Show Properties Dialog".

The connection table contains all data of the configured connections. To be able to display this clearly, use VIEW → OPTIMIZE COLUMN WIDTH and VIEW → DISPLAY COLUMNS.

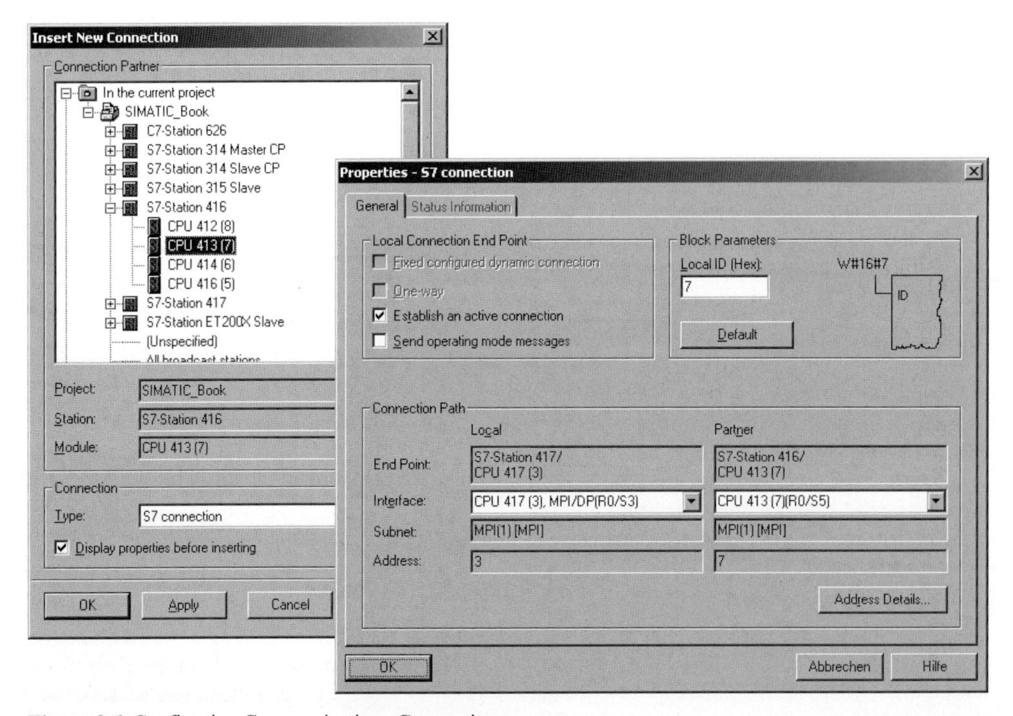

Figure 2.6 Configuring Communications Connections

Connection ID

The number of possible connections is CPU-specific. STEP 7 defines a connection ID for every connection and for every partner. You require this specification when you use communications blocks in your program.

You can modify the **local ID** (the connection ID of the currently opened module). This is necessary if you have already programmed communications blocks and you want to use the local ID specified there for the connection.

You enter the new local ID as a hexadecimal number. It must be within the following value ranges, depending on the connection type, and must not already be assigned:

- ▷ Value range for S7 connections: 0001_{hex} to $0FFF_{hex}$
- ▷ Value range for S7 connections with loadable S7 communication (S7-300): 0001_{hex} to $008F_{hex}$
- ▷ Value range for PtP connections: 1000_{hex} to 1400_{hex}

You change the **partner ID** by going to the connection table of the partner CPU and changing (what is then) the local ID: select the connection line and then EDIT → OBJECT PROPERTIES. If STEP 7 does not enter a partner ID, it is a one-way connection (see below).

Partner

This column displays the connection partner. If you want to reserve a connection resource without naming a partner device, enter "unspecified" in the dialog box under Station.

In a **one-way connection**, communication can only be initiated from one partner; example: S7 communications between an S7-400 and S7-300 CPU. Even without S7 communications functions in the S7-300 CPU, data can be exchanged by an S7-400 CPU with SFB 14 GET and SFB 15 PUT. In the S7-300, no user program runs for this communication but the data exchange is handled by the operating system.

A one-way connection is configured in the connection table of the "active" CPU. Only then does STEP 7 assign a "Local ID". You also load this connection only in the local station.

With a **two-way connection**, both partners can assume communication actively; e.g. two S7-400 CPUs with the communications functions SFB 8 SEND and SFB 9 BRCV.

You configure a two-way connection only once for one of the two partners. STEP 7 then assigns a "Local ID" and a "Partner ID" and generates the connection data for both stations. You must load each partner with its own connection table.

Connection type

The STEP 7 Basic Package provides you with the following connection types in the Network Configuration:

PtP connection, approved for the subnet PTP (3964(R) and RK 512 procedures) with S7 communications. A PtP (point-to-point) connection is a serial connection between two partners. These can be two SIMATIC S7 devices with the relevant interfaces or CPs, or a SIMATIC S7 device and a non-Siemens device, e.g. a printer or a barcode reader.

S7 connection, approved for the subnets MPI, PROFIBUS and Industrial Ethernet with S7 communications. An S7 connection is the connection between SIMATIC S7 devices and can include programming devices and human machine interface devices. Data are exchanged via the S7 connection, or programming and control functions are executed.

Fault-tolerant S7 connection, approved for the subnets PROFIBUS and Industrial Ethernet with S7 communications. A fault-tolerant S7 connection is made between fault-tolerant SIMATIC S7 devices and it can also be established to an appropriately equipped PC.

The software component "SIMATIC NCM", which is part of STEP 7, is available for **parameterizing CPs**. You can select the following types of connection: FMS connection, FDL connection, ISO transport connection, ISO-on-TCP connection, TCP connection, UDP connection and e-mail connection.

Active connection buildup

Prior to the actual data transfer, the connection must be built up (initialized). If the connection partners have this capability, you specify here which device is to establish the connection. You do this with the check box "Active Connection Buildup" in the properties window of the connection (select the connection and then EDIT → OBJECT PROPERTIES).

Sending operating state messages

Connection partners with a configured two-way connection can exchange operating state messages. If the local node is to send its operating state messages, activate the relevant check box in the properties window of the connection. In the user program of the partner CPU, these messages can be received with SFB 23 USTATUS.

Connection path

As the connection path, the properties window of the connection displays the end points of the connection and the subnets over which the connection runs. If there are several subnets for selection, STEP 7 selects them in the order Industrial Ethernet before Industrial Ethernet/TCP-IP before MPI before PROFIBUS.

The local and partner stations with the CPU over which the connection runs are displayed as the end points of the connection.

The modules with communications capability are listed under "Interface", with specification of the interface module, the rack number and the slot. If both peers can be accessed over several connection paths, you can set the preferred path here. STEP 7 automatically adapts the remaining settings. If both CPUs are located in the same rack (e.g. S7-400 CPUs in multiprocessor mode), the display box shows "PLC internal".

"Subnet" and "Address" then identify the subnet used and the node address set.

Connections between projects

For data exchange between two S7 modules belonging to different SIMATIC projects, you enter "unspecified" for connection partner in the connection table (in the local station in both projects).

Please ensure that the connection data agree in both projects (STEP 7 does not check this). After saving and compiling, you load the connection data into the local station in each project.

If a project is to subsequently be part of a multiproject, and if the connection partner is also in a project of the multiproject, select "In unknown project" as the connection partner, and enter an unambiguous connection name (reference) in the properties window.

Connection to non-S7 stations

Within a project, you can also specify stations other than S7 stations as connection partners:

- ▷ Other stations (non-Siemens devices and also S7 stations in another project)
- ▷ Programming devices/PCs
- ▷ SIMATIC S5 stations

A requirement for configuring the connection is that the non-S7 station exists as an object in the project container and you have connected the non-S7 station to the relevant subnet in the station properties (e.g. select the station in the Network Configuration, select EDIT → OBJECT PROPERTIES and connect the station with the desired subnet on the "Interfaces" tab).

2.4.4 Network Transitions

If the programming device is connected to a subnet, it can reach all other nodes on this subnet. For example, from one connection point, you can program and debug all S7 stations connected to an MPI network. If another subnet such as a PROFIBUS subnet is connected to an S7 station, the programming device can also reach the stations on the other subnet. The requirement for this is that the station with the subnet transition has routing capability, that is it will channel the transferred message frames.

When the network configuration is compiled, routing tables containing all the necessary information are automatically generated for the stations with subnet transitions. All accessible communications partners must be configured in a plant network within an S7 project and must be supplied with the "knowledge" of which stations can be reached via which subnets and subnet transitions.

If you want to reach all nodes in a subnet with a programming device from one connection point, you must configure the connection point. You enter a "placeholder", a PG/PC station from the Network Object Catalog in the network configuration at the relevant subnet. You configure a PG/PC station on every subnet to which you want to connect a programming device.

During operation, you connect the programming device to the subnet and select PLC → ASSIGN PG/PC. This adapts the interfaces of the programming device to the configured settings for the subnet. Before disconnecting the programming device again from the subnet, select PLC → UNDO PG/PC ASSIGNMENT.

If you go online with a programming device that does not contain the right project, you require the S7 subnet ID for network access. The S7 subnet ID comprises two numbers: the project number and the subnet number. You can obtain the subnet ID in the network configuration by selecting the subnet and then EDIT → OBJECT PROPERTIES on the "General" tab.

2.4.5 Loading the Connection Data

To activate the connections, you must load the connection table into the PLC following saving and compiling (all connection tables into all "active" CPUs).

Requirement: You are in the network window and the connection table is visible. The programming device is a node of the subnet over which the connection data are to be loaded into the modules with communications capability. All subnet nodes have been assigned unique node addresses. The modules to which connection data are to be transferred are in the STOP mode.

With PLC → DOWNLOAD TO CURRENT PROJECT → ... you transfer the connection and configuration data to the accessible modules. Depending on which object is selected and which menu command is selected, you can choose between the following

→ SELECTED STATIONS

→ SELECTED AND PARTNER STATIONS

→ STATIONS ON THE SUBNET

→ SELECTED CONNECTIONS

→ CONNECTIONS AND GATEWAYS

To delete all the connections of a programmable module, load an empty connection table into the respective module.

The compiled connection data are also a component part of the *System data* in the *Blocks* container. Transfer of the system data and the subsequent startup of the CPUs effectively also transfers the connection data to the modules with communications capability.

For online operation via MPI, a programming device requires no additional hardware. If you connect a PC to a network or if you connect a programming device to an Ethernet or PROFIBUS network, you require the relevant interface module. You parameterize the module with the application "Set PG/PC Interface" in the Windows Control Panel.

2.4.6 Adjusting Projects in the Multiproject

When opening a multiproject with the network configuration, a window is displayed with the projects present in the multiproject. This window is also displayed if you open a project linked in a multiproject and select VIEW → MULTIPROJECT. The window displays the projects present in the multiproject and the cross-project subnets which have already been combined. To permit further editing, you can now double click a project (Figure 2.7).

Projects usually contain communications connections between the individual stations. If projects are combined in a multiproject or if an existing project is incorporated into the multiproject, these connections can be combined and adjusted.

If you select VIEW → CROSS-PROJECT NETWORK VIEW in an open project which belongs to a multiproject, all stations of the multiproject as well as the current connections are shown in the overview. You cannot carry out any changes to the projects in the cross-project network view. You can leave the multiproject view by selecting VIEW → CROSS-PROJECT NETWORK VIEW again.

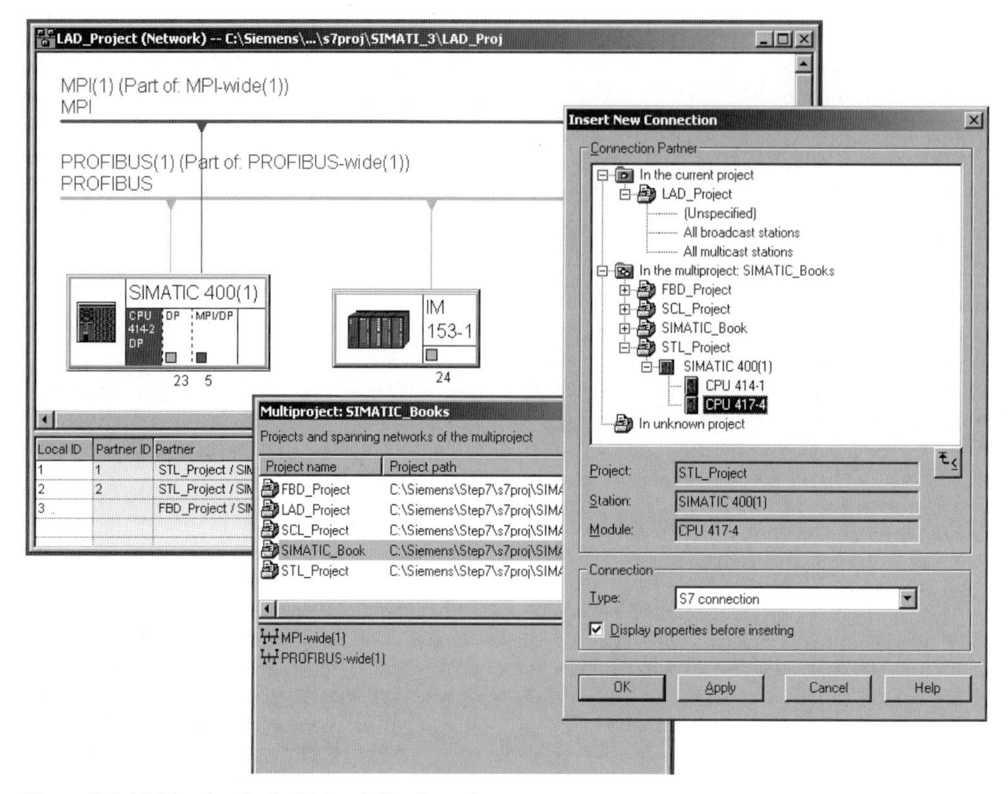

Figure 2.7 Multiproject in the Network Configuration

Combining subnets

The MPI, PROFIBUS and Industrial Ethernet subnets are initially combined. A prerequisite is that the subnets to be combined have the same subnet ID. With the network selected, you can set these in the network configuration with EDIT → OBJECT PROPERTIES.

With the multiproject open, you can use FILE → MULTIPROJECT → ADJUST PROJECTS to call a wizard in the SIMATIC Manager which supports you during the adjustment. In the network configuration, you access the dialog window with EDIT → MERGE/UNMERGE SUBNETWORKS → ...

Select the type of subnet, and click the "Run" button to obtain the subnets of the selected type present in the multiproject. You can now select individual subnets of the project and combine them in a multi-subnet. You can also use this dialog to separate subnets again from the multi-subnet.

Several multi-subnets of the same type can be created in one multiproject. The properties of the multi-subnet are determined by the first added subnet or the subnet selected using the "Select" button. Use "OK" or "Apply" to acknowledge the settings. Subnets which are part of a multi-subnet can be recognized by a different symbol in the SIMATIC Manager.

Combining connections

The connections configured in individual projects and leading to a partner in a different project can be combined in a multiproject. If you select the partner "In unknown project" in the window "Insert new connection" during configuration of the connection, you can subsequently enter a connection name (reference) in the window "Properties – S7 connection". Connections in different projects but with the same connection name can be automatically combined.

In the SIMATIC Manager, this is carried out by the wizard for project adjustment if you click "Merge connections" and "Execute". Connections are then combined which have an identical connection name (reference).

In the network configuration, you can also combine connections with "unspecified" partners. Select EDIT → MERGE CONNECTIONS to obtain a dialog box with all configured connections. Select one connection each in the windows "Connections without connection partners" and "Possible connection partners", and click "Assign". The assigned connections are listed in the bottom window "Assigned connections". Use "Merge" to then combine the connections. The connections are assigned the properties of the local module of the currently opened project. You are provided with the possibility for changing the connection properties when combining.

Configuring multi-project connections

Multi-project connections can be configured following the combination of subnets. The procedure is the same as with project-internal connections, expanded by specification of the project in the connection partner.

You can check the absence of network configuration inconsistencies in the multi-project with NETWORK → CHECK CROSS-PROJECT CONSISTENCY.

2.5 Creating the S7 Program

2.5.1 Introduction

The user program is created under the object *S7 Program*. You can assign this object in the project hierarchy of a CPU or you can create it independently of a CPU. It contains the object *Symbols* and the containers *Source Files* and *Blocks* (Figure 2.8).

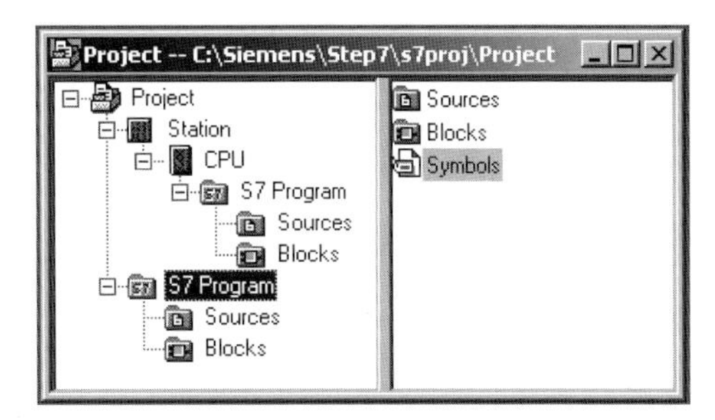

Figure 2.8 Objects for Program Generation

With **incremental** program creation, you enter the program direct block-by-block. Entries are checked immediately for syntax. At the same time, the block is compiled as it is saved and then stored in the container *Blocks*. With incremental programming, you can also edit blocks online in the CPU, even during operation. Incremental program creation is possible with all basic languages.

In the case of **source-oriented** program creation, you write one or more program sources and store these in the container *Source Files*. Program sources are ASCII text files that contain the program statements for one or more blocks, possibly even for the entire program. You compile these sources and you get the compiled blocks in the container *Blocks*. The compiled blocks consist of MC7 code and run on an S7 CPU. You apply source-oriented program creation for STL and SCL; although you cannot program source-oriented with LAD or FBD, you can save your programs created using LAD or FBD as source.

The signal states or the values of addresses are processed in the program. An address is, for example, the input I1.0 (*absolute addressing*). With the help of the **Symbol Table** under the object *Symbols*, you can assign a symbol (an alphanumeric name, e.g. "Switch motor on") to an address and then access it with this name (*symbolic addressing*). In the properties of the offline object container *Blocks*, you specify whether in the event of a change in the Symbol Table the absolute address or the symbol is to be definitive for the already compiled blocks when next saved (*address priority*).

2.5.2 Symbol Table

In the control program, you work with addresses; these are inputs, outputs, timers, blocks. You can assign absolute addresses (e.g. I1.0) or symbolic addresses (e.g. Start signal). Symbolic addressing uses names instead of the absolute address. You can make your program easier to read by using meaningful names.

In symbolic addressing, a distinction is made between *local* symbols and *global* symbols. A local symbol is known only in the block in which it has been defined. You can use the same local symbols in different blocks for different purposes. A global symbol is known throughout the entire program and has the same meaning in all blocks. You define global symbols in the symbol table (object *Symbols* in the container *S7 Program*).

A global symbol starts with an alpha character and can be up to 24 characters long. A global symbol can also contain spaces, special characters and national characters such as the Umlaut. Exceptions to this are the characters 00_{hex}, FF_{hex} and the inverted commas ("). You must enclose symbols with special characters in inverted commas when programming. In the compiled block, the program editor displays all global symbols in inverted commas. The symbol comment can be up to 80 characters long.

In the symbol table you can assign names to the following addresses and objects:

- ▷ Inputs I, outputs Q, peripheral inputs PI and peripheral outputs PQ
- ▷ Memory bits M, timer functions T and counter functions C
- ▷ Code blocks OBs, FBs, FCs, SFCs, SFBs and data blocks DBs
- ▷ User-defined data types UDTs
- ▷ Variable table VAT

Data addresses in the data blocks are included among the local addresses; the associated symbols are defined in the declaration section of the data block in the case of global data blocks and in the declaration section of the function block in the case of instance data blocks.

When creating an S7 program, the SIMATIC Manager also creates an empty symbol table *Symbols*. You open this and can then define the global symbols and assign them to absolute addresses (Figure 2.9). There can be only one single symbol table in an S7 program.

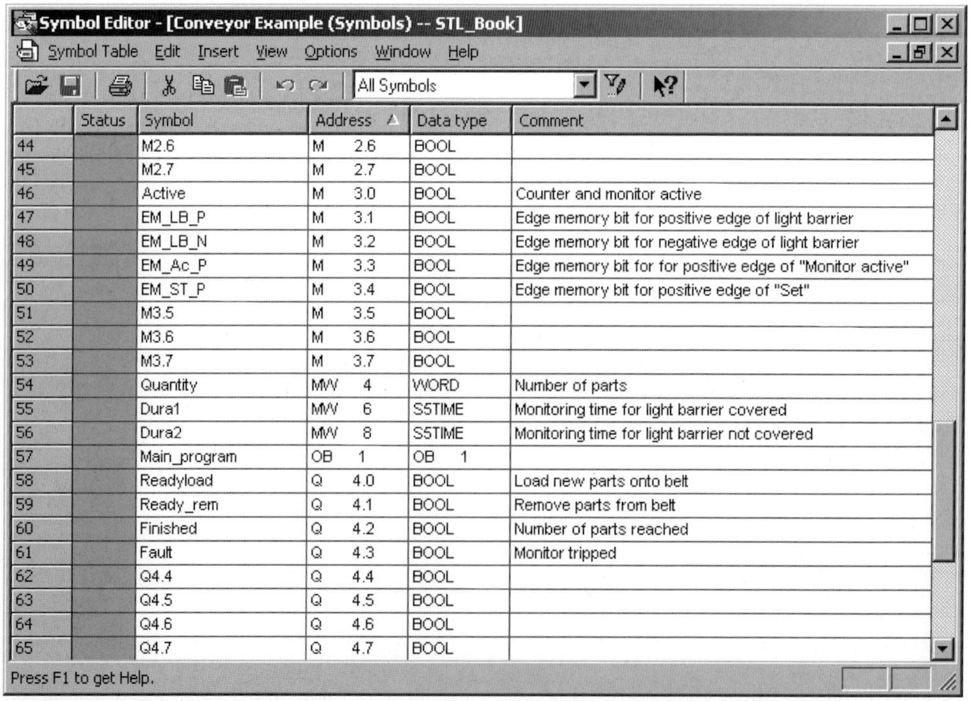

Symbol Editor - [Conveyor Example (Symbols) -- STL_Book]

Symbol Table Edit Insert View Options Window Help

All Symbols

	Status	Symbol	Address	Data type	Comment
44		M2.6	M 2.6	BOOL	
45		M2.7	M 2.7	BOOL	
46		Active	M 3.0	BOOL	Counter and monitor active
47		EM_LB_P	M 3.1	BOOL	Edge memory bit for positive edge of light barrier
48		EM_LB_N	M 3.2	BOOL	Edge memory bit for negative edge of light barrier
49		EM_Ac_P	M 3.3	BOOL	Edge memory bit for for positive edge of "Monitor active"
50		EM_ST_P	M 3.4	BOOL	Edge memory bit for positive edge of "Set"
51		M3.5	M 3.5	BOOL	
52		M3.6	M 3.6	BOOL	
53		M3.7	M 3.7	BOOL	
54		Quantity	MW 4	WORD	Number of parts
55		Dura1	MW 6	S5TIME	Monitoring time for light barrier covered
56		Dura2	MW 8	S5TIME	Monitoring time for light barrier not covered
57		Main_program	OB 1	OB 1	
58		Readyload	Q 4.0	BOOL	Load new parts onto belt
59		Ready_rem	Q 4.1	BOOL	Remove parts from belt
60		Finished	Q 4.2	BOOL	Number of parts reached
61		Fault	Q 4.3	BOOL	Monitor tripped
62		Q4.4	Q 4.4	BOOL	
63		Q4.5	Q 4.5	BOOL	
64		Q4.6	Q 4.6	BOOL	
65		Q4.7	Q 4.7	BOOL	

Press F1 to get Help.

Figure 2.9 Symbol Table Example

The data type is part of the definition of a symbol. It defines specific properties of the data behind the symbol, essentially the representation of the data contents. For example, the data type BOOL identifies a binary variable and the data type INT designates a digital variable whose contents represent a 16-bit integer. Please refer to Chapter 3.7 "Variables and Constants"; for an overview of the data types used in STEP 7; Chapter 24 "Data Types" contains a detailed description.

With incremental programming, you create the symbol table before entering the program; you can also add or correct individual symbols during program input. In the case of source-oriented programming, the complete symbol table must be available when the program source is compiled.

Importing, exporting

Symbol tables can be imported and exported. "Exported" means a file is created with the contents of your symbol table. You can select here either the entire symbol table, a subset limited by filters or only selected lines. For the data format you can choose between pure ASCII text (extension *.asc), sequential assignment list (*.seq), System Data Format (*.sdf for Microsoft Access) and Data Interchange Format (*.dif for Microsoft Excel). You can edit the exported file with a suitable editor. You can also import a symbol table available in one of the formats named above.

Special object properties

With EDIT → SPECIAL OBJECT PROPERTIES → ..., you set attributes for each symbol in the symbol table. These attributes or properties are used in the following:

- ▷ Process monitoring with S7-PDIAG
- ▷ Human machine interface functions for monitoring with WinCC
- ▷ Configuring messages
- ▷ Configuring communications with the NCM software
- ▷ Control at contact with inputs and bit memories in the program editor

VIEW → COLUMNS R, O, M, C, CC makes the settings visible. With OPTIONS → CUSTOMIZE, you can specify whether or not the special object properties are to be copied and you can define behavior when importing symbols.

2.5.3 STL-Program Editor

For creating the user program, the STEP 7 Basic Package contains a program editor for the LAD, FBD and STL programming languages. In the STL programming language, you can enter the program incrementally (directly) or generate a source program and compile it later. Figure 2.10 shows the possible actions associated with STL program creation.

If you use symbolic addressing for global addresses, the symbols must already be assigned to an absolute address in the case of incremental programming; however, you can enter new symbols or change symbols during program input. In the case of source-oriented programming, the complete symbol table must be available at the time of compiling.

STL blocks can be "decompiled", i.e. a readable block can be created again from the MC7 code without an offline database (you can read any block from a CPU using a programming device without the associated project). In addition, an STL program source can be recreated from any compiled block.

Starting the STL program editor

You reach the program editor when you open a block in the SIMATIC Manager, e.g. by double-clicking on the automatically generated symbol of the organization block OB 1, or via the Windows taskbar with START → SIMATIC → STEP 7 → LAD, STL, FBD - PROGRAM S7 BLOCKS.

You can customize the properties of the program editor with OPTIONS → CUSTOMIZE. On the "Editor" tab, select the properties with which a new block is to be generated and displayed, such as creation language, pre-selection for comments and symbols.

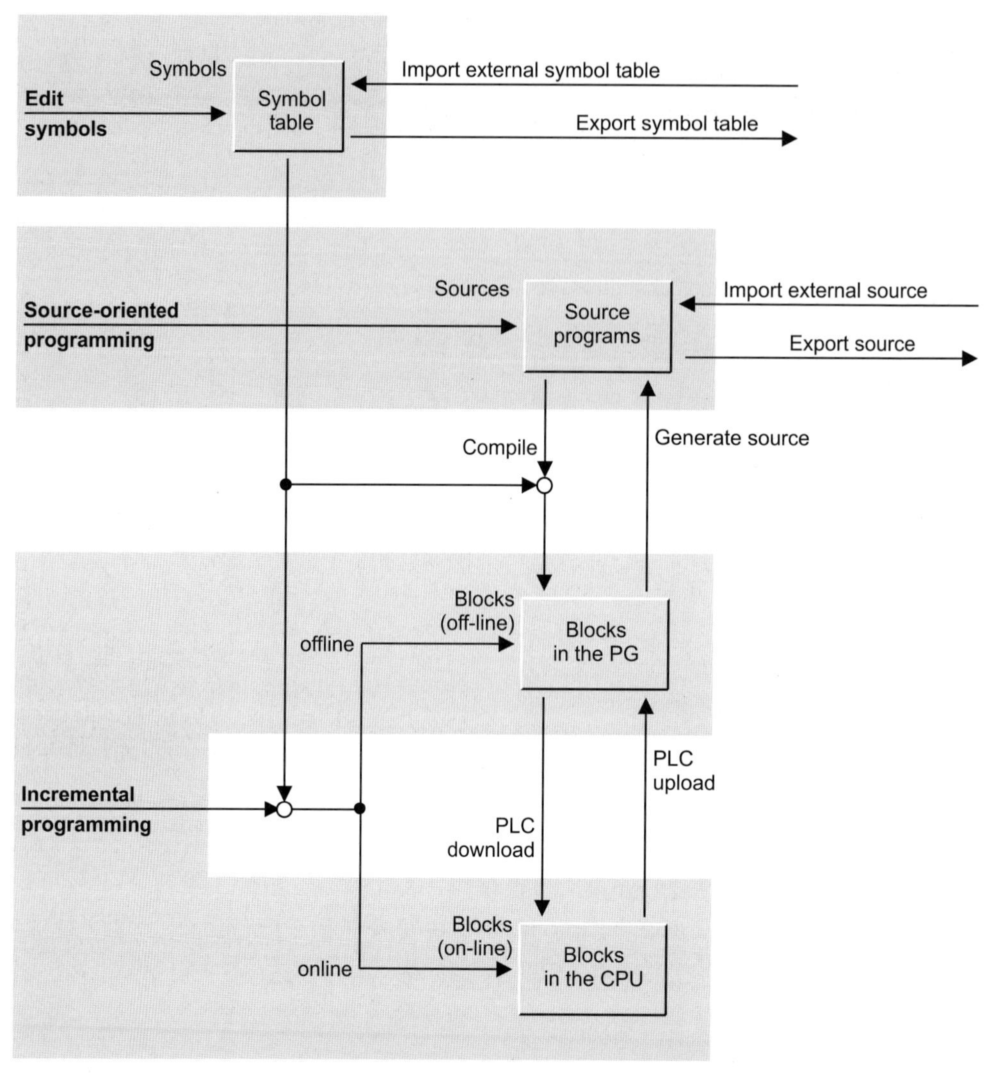

Figure 2.10 Writing Programs with the STL Editor

Program editor window

Further windows can be displayed in the window of the program editor: the block window, the *Details* and *Overviews* windows, and the window with the AS tabs (Figure 2.11).

The *block window* is automatically displayed when a block is opened, and contains the block interface at the top, i.e. the block parameters as well as the static and dynamic local data. You can program the block in the bottom program area. The block windows and the contents are described in Chapter 3.2 "Blocks".

The *Overviews* window displays the program elements and the call structure. If it is not visible, display it on the screen using VIEW → OVERVIEWS.

The *Details* window can be switched on and off using VIEW → DETAILS. It contains the following tabs:

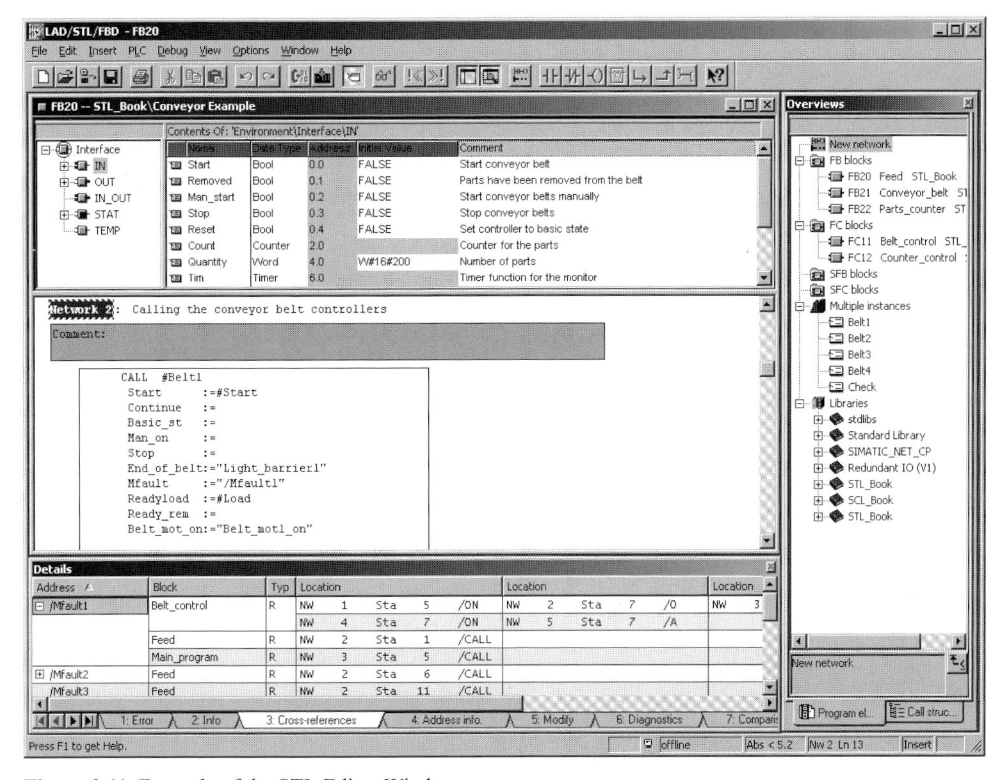

Figure 2.11 Example of the STL Editor Window

▷ 1: Error
Contains the errors found in the block by the program editor, e.g. following compilation. With OPTIONS → CUSTOMIZE in the tab "Sources" you can set whether warnings are also to be displayed here.

▷ 2: Info
Contains information on the currently marked addresses.

▷ 3: Cross-references
Contains the cross-references of the operations present in the current network (see Chapter 2.5.7 "Reference Data").

▷ 4: Address info
Contains the symbol information of the addresses present in the current network (see Chapter 2.5.2 "Symbol Table"). You can edit existing symbols here, add new ones, and view the status of the address.

▷ 5: Modify
Contains an empty table of variables into which you can enter the addresses to be controlled (see Chapter 2.7.3 "Monitoring and Modifying Variables").

▷ 6: Diagnostics
Contains a list with the existing monitoring functions for process diagnostics using the S7-PDIAG options package.

▷ 7: Comparison
Contains the results of a previous block comparison (see Chapter "Comparing blocks").

You can "dock" the *Overviews* and *Details* windows on the edge of the editor window by double clicking the respective title bar and then releasing it again.

The *PLC register contents* window shows the contents of the CPU registers (accumulators, address registers and DB registers).

Programming modes

When you open a compiled block in the *Blocks* container (e.g. with a double-click), it is opened for incremental programming. For source-oriented programming, you must open a program source file in the *Source files* container.

You can also mix: enter some blocks direct and program some blocks via a source file. It is also possible to call blocks that have been written with another programming language such as FBD or SCL. The user program is generated block by block and every block finally contains executable MC7 code regardless of the programming language used to write it.

Source-oriented programming generation with symbolic addressing is recommended. Editing is simpler, there are fewer syntax errors and another text editor can be used. Via the symbol table, you can specify different absolute addresses each time the program is compiled so that in this way you can create reusable "standard programs" independently of a hardware configuration.

Source-oriented program creation is the only possible method of providing your program with block protection (KNOW_HOW_PROTECT.

Incremental programming is optimal for "fast checking" of a program change direct in the CPU. If the change checks out, update it in the program source and compile again. This way, you always have the current version of the program available as an ASCII text file. Incremental programming is also well suited for testing the program with a few online statements that you then no longer require.

Source-oriented programming

Source-oriented programming is used to edit an STL source file in the *Source Files* object container. An STL source file is a pure ASCII text file. It may contain the source program for one or more code or data blocks as well as the definitions of the user data types.

In the SIMATIC Manager, select the source program container *Source Files* and create a new source file with INSERT → S7 SOFTWARE → STL SOURCE. You can open and edit this file.

You can make the settings for editing in the source file in the program editor using OPTIONS → CUSTOMIZE in the tab "Source Text". In order to display the program text more clearly, you can e.g. indent text blocks, display line numbers, and select a different font or color for the text for e.g. operations, addresses or keywords.

You can make the creation of new blocks much simpler for yourself by using INSERT → BLOCK TEMPLATE → ... (in the editor). The editor uses the templates from the directory ...\Step7\S7ska that are contained in the text files S7kaf*nn*x.txt. You may adapt these templates to suit your requirements. With INSERT → OBJECT → BLOCK, the program editor inserts a previously compiled block following the cursor as ASCII source into the source file.

You also have the editor option of generating a new STL source file from one or more compiled blocks with FILE → GENERATE SOURCE.

If you generated a source file with another text editor, you can use the SIMATIC Manager's INSERT → EXTERNAL SOURCE menu item to place that file in the *Source Files* container. You can copy the selected source file to a directory of your choice with EDIT → EXPORT SOURCE.

In source-file-oriented programming, you must observe certain rules and use keywords intended for the compiler. Chapter 3.4.5 "Source-oriented programming of an STL code block" and Chapter 3.6.2 "Source-Oriented Data Block Programming" shows the structure of an STL source file.

Compiling the STL source file

You can save the program source at any point during editing, even if the program is not yet complete. Only when the source file has been compiled does the program editor generate executable blocks that it stores in the *Blocks* container. If you have used global symbols in the STL source file, the completed symbol table must be available at the time of compiling.

On the "Sources Files" tab under OPTIONS → CUSTOMIZE, you can set the properties of the compiler, such as, should existing blocks be overwritten or should blocks be generated only when the entire program source is free from errors? On the "Generate Block" tab, you can set automatic updating of the reference data when compiling a block.

With FILE → CHECK CONSISTENCY, you can check the program source for correct syntax without having to compile the blocks.

When the program source is open, you start compiling with FILE → COMPILE. All error-free blocks located in the program source are compiled. Any block containing errors is not compiled. If warnings occur, the block is compiled anyway; however, execution in the CPU may not be error-free.

Called blocks must already exist as compiled blocks or they must exist in the program source before calling (see Chapter 3.4.5 "Source-oriented programming of an STL code block" for more details of the sequential order of blocks).

Updating or generating STL source files

On the "Sources" tab under OPTIONS → CUSTOMIZE, you can select the option "Generate source automatically" so that when you save an (incrementally created) block, the program source file is updated or created, if it does not already exist. You can derive the name of a new source file from the absolute address or the symbolic address. The symbolic addresses are imported if the address priority is set to "Symbol has priority" in the properties of the block container *Blocks* (see Chapter 2.5.6 "Address Priority"). With the setting "Absolute value has priority", you can select either absolute or symbolic addressing in the source.

With the "Execute" button, you select, in the subsequent dialog box, the blocks from which you want to generate a program source file.

Incremental programming

With incremental programming, you edit the blocks both in the offline and online *Blocks* container. The editor checks your entries in incremental mode as soon as you have terminated a program line. When the block is closed it is immediately compiled, so that only error-free blocks can be saved.

On the "Create Block" tab under OPTIONS → CUSTOMIZE, you set automatic updating of the reference data when saving a block.

The blocks can be edited both offline in the programming device's database and online in the CPU, generally referred to as the "programmable controller", or "PLC". For this purpose, the SIMATIC Manager provides an offline and an online window; the one is distinguished from the other by the labeling in the title bar.

In the offline window, you edit the blocks right in the PG database. If you are in the editor, you can store a modified block in the offline database with FILE → SAVE and transfer it to the CPU with PLC → DOWNLOAD. If you want to save the opened block under another number or in a different project, or if you want to transfer it to a library or to another CPU, use the menu command FILE → SAVE AS.

With the menu command FILE → STORE READ-ONLY... in the program editor, you can save a write-protected copy of the currently opened (and saved) block in a different block container.

To edit a block in the CPU, open that block in the online window. This transfers the block from the CPU to the programming device so that it can be edited. You can write the edited block back to the CPU with PLC → DOWNLOAD. If the CPU is in RUN mode, the CPU will process the edited block in the next program scan cycle. If you want to save a block that you edited online in the offline database as well, you can do so with FILE → SAVE.

Chapter 2.6.4 "Loading the User Program into the CPU" and Chapter 2.6.5 "Block Handling" contain further information on online programming. Chapter 3.4.2 "Programming STL Code Blocks Incrementally" and Chapter 3.6.1 "Programming Data Blocks Incrementally" show you how to enter an STL block.

Comparing blocks

You can use the block comparison function to determine the differences between two blocks. The blocks can be present in different projects, in different target systems (CPUs) or in one project and one target system.

In the program editor, you can compare the open block with the same block in the CPU or in the project using OPTIONS → COMPARE ON-/OFFLINE PARTNERS. The result is displayed in the details area of the editor window in tab "7:Comparison".

In the SIMATIC Manager, mark the object *Blocks* or only the blocks to be compared, and select OPTIONS → COMPARE BLOCKS. The comparison is carried out either between the online and offline data (ONLINE/offline) or between two projects (Path1/Path2). When comparing the complete program – which can also contain tables of variables and user-defined data types (UDTs) – you can incorporate the system data. Use "Execute code comparison" to additionally compare the program code of the blocks, also of blocks with different programming languages.

The comparison comprises all data of a block, also its time tag for program code and interface. If you wish to know whether the program code is identical – independent of the block properties – compare the checksum of the block. To do this, select the "Details" button in the result window of the block comparison.

2.5.4 SCL Program Editor

The S7-SCL optional software provides you with its own program editor for programming in SCL. At installation, this is integrated into the SIMATIC Manager. You use it in exactly the same way as the program editor for the standard languages. With SCL you use source-file-oriented programming (Figure 2.12).

You create an SCL source file that you then compile. You can also call already compiled blocks, located in the *Blocks* container here (integrate them into your program, so to speak). These blocks can also be written with another programming language such as STL.

If you use symbolic addressing in the program for global addresses, the complete symbol table must be available when you compile the program.

You cannot generate an SCL source file from a compiled block; for example, if you have deleted the program source file by mistake. (Note: an already compiled SCL block will execute in the CPU even if the program source file is not available.)

Starting the SCL program editor

You start the SCL program editor in the SIMATIC Manager by opening a compiled SCL block or an SCL source file, or via the Windows taskbar with START → SIMATIC → STEP 7 → S7-SCL - PROGRAM S7 BLOCKS.

If the program editor does not find the relevant program source file when starting via a compiled block, e.g. because it has been deleted or shifted, the block is opened with the STL program editor. However, as soon as you write the block back, even without changes, it is "unusable" for the SCL program editor.

You can set the properties of the SCL program editor to suit your own requirements with OPTIONS → CUSTOMIZE. On the "Editor" tab, select the properties with which a new block is to be created and displayed, such as, display with line numbers.

Creating the SCL source file

You create a new SCL source file in the SIMATIC Manager when you select the *Source files* container and then INSERT → S7 SOFTWARE → SCL SOURCE FILE. A double-click on the source file opens it.

Make the settings for editing in the source file in the program editor with OPTIONS → CUSTOMIZE in the tabs "Editor" and "Format". To display the program text more clearly, you can e.g indent text blocks, display line numbers, and select a different font or color for the text for e.g. keywords, comments or global symbols.

With INSERT → BLOCK CALL you can insert the call for an existing block (user or system block) into the program source at the cursor position. INSERT → BLOCK TEMPLATE → ... makes creation of new blocks easier and with INSERT → CONTROL STRUCTURE → ..., you can insert off-the-shelf program structures into the program source file at the cursor position.

Bookmark makes navigation easier in the source program, e.g. in order to rapidly jump between different parts of the source. Set the bookmark in the line you wish to mark, and select EDIT → BOOKMARK ON/OFF. If several bookmarks are present, you can swap between the marked lines with EDIT → GO TO → NEXT

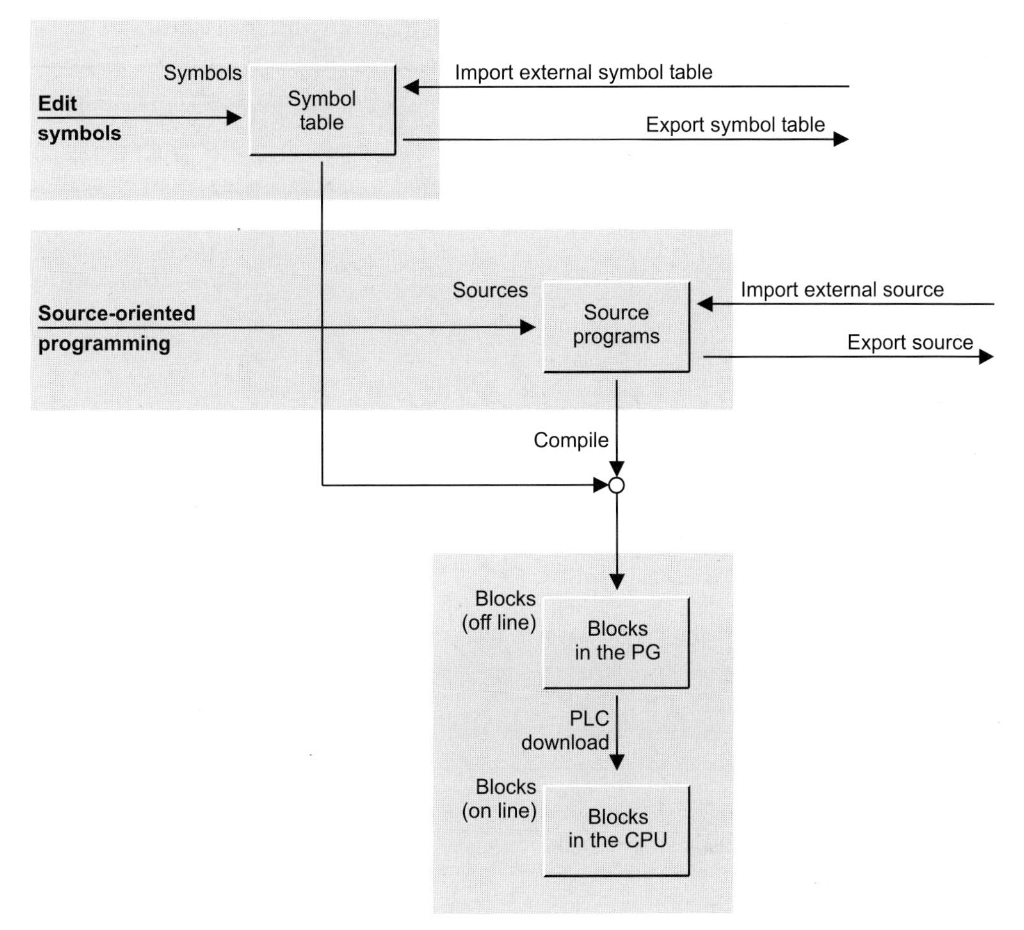

Figure 2.12 Program Creation with the SCL Program Editor

BOOKMARK and ... → PREVIOUS BOOKMARK. You can delete a bookmark if the cursor is located in a marked line and you select EDIT → BOOKMARK ON/OFF. Use EDIT → DELETE ALL BOOKMARKS to delete all of the bookmarks.

A separate menu bar which is made visible using VIEW → BOOKMARK BAR makes handling easier. Please note that the bookmarks are only available during the current input; they are not saved together with the source program.

If you have created an SCL source file with another text editor, you can fetch it to the *Source files* container with INSERT → EXTERNAL SOURCE under the SIMATIC Manager. With EDIT → EXPORT SOURCE, you can copy the selected source file to a folder (directory) of your choice.

In the case of source-file-oriented programming, you must observe certain rules and use keywords intended for the compiler. Chapter 3.5.2 "Programming SCL Code Blocks" and Chapter 3.6.2 "Source-Oriented Data Block Programming" shows you the structure of an SCL source file.

Compiling the SCL source file

You can save the program source file at any time during editing, even if the program is not yet complete. Only after the source file has been compiled does the program editor generate blocks that it then stores in the *Blocks* con-

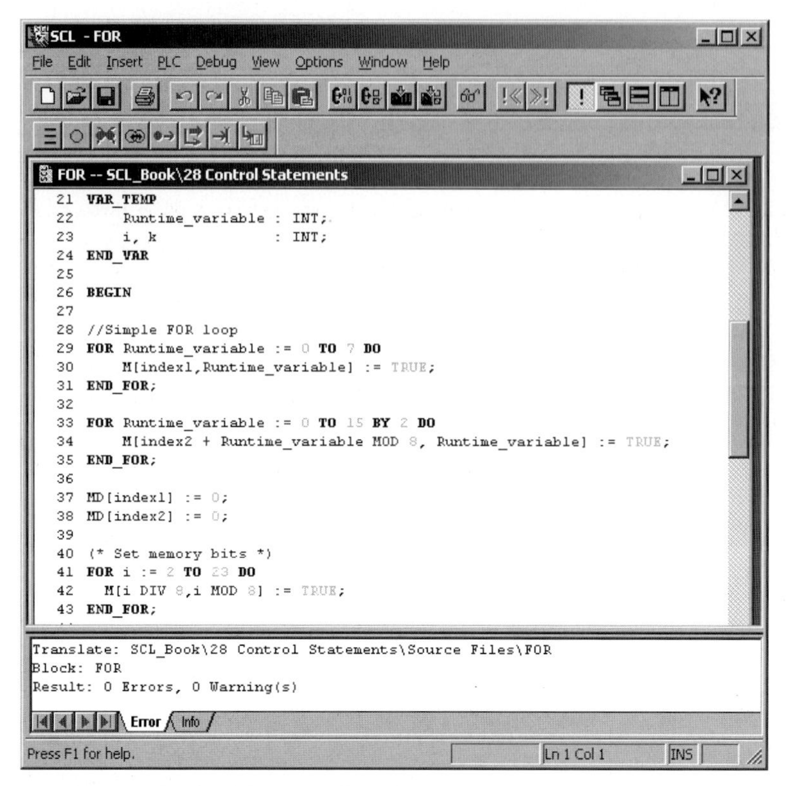

Figure 2.13 Example of the SCL Editor Window

tainer. If you have used global symbols in the SCL source file, the completed symbol table must be available at the time of compiling.

You can make the following setting, among others, on the "Compiler" tab under OPTIONS → CUSTOMIZE:

▷ Create object code
If this option is selected, blocks are generated following error-free compiling; otherwise, you can check the program source file for correct syntax, without generating blocks.

▷ Optimize object code
The generated blocks are optimized with regard to memory requirement and runtime.

▷ Monitor array limits
This causes the compiler to generate additional program code that allows checking of, e.g., array limits during runtime.

▷ Create debug info
If you still have to debug the compiled program with the program status, select this option (however, this increases the memory requirements and the program execution time).

▷ Set OK flag
You must set this option if you use the OK variable or the EN/ENO mechanism in the program.

If the compiler settings are only to apply to one program source or one compilation control file, you can also write the settings in the source program (see Chapter 3.5.2 "Programming SCL Code Blocks").

With the program source file open, you start compiling with FILE → COMPILE. All error-free blocks located in the program source are compiled. Any block containing errors is not compiled. If warnings occur, the block is compiled anyway; however, execution in the CPU may not be error-free. If you want to compile the selected blocks of the source file, select FILE → COMPILE SELECTED BLOCKS.

Called blocks must already exist as compiled blocks or they must exist in the program source before calling (see Chapter 3.5.2 "Programming SCL Code Blocks" for more details of the sequential order of blocks). The SCL compiler automatically creates any missing instance data blocks in the case of function block calls. The DB number is taken from the Symbol Table or the smallest free number is selected.

The standard blocks called in the first call level, such as an IEC function, are copied from the standard library to the *Blocks* container at the time of compiling.

PLC → DOWNLOAD loads into the connected CPU all blocks that have been generated or automatically copied from a standard library to the *Blocks* container the last time the program was compiled.

Compilation control file

SCL provides the facility of compiling several program source files in a specific order in one run. You create a compilation control file with the SIMATIC Manager by selecting INSERT → S7 SOFTWARE → SCL COMPILE CONTROL FILE with the *Source files* container selected.

Open the compilation control file and specify the names of the program source files in the order desired for compiling. You start the compile procedure with FILE → COMPILE.

2.5.5 Rewiring

The *Rewiring* function allows you to replace addresses in individually compiled blocks or in the entire user program. For example, you can replace input bits I0.0 to I 0.7 with input bits I 16.0 to I 16.7. Permissible addresses are inputs, outputs, memory bits, timers and counters as well as functions FCs and function blocks FBs.

In the SIMATIC Manager, you select the objects in which you wish to carry out the rewiring; select a single block, a group of blocks by holding Ctrl and clicking with the mouse, or the entire *Blocks* user program. OPTIONS → REWIRE takes you to a table in which you can enter the old addresses to be replaced and the new addresses. When you confirm with "OK", the SIMATIC Manager then exchanges the addresses.

When "rewiring" blocks, first change the numbers of the blocks and then carry out the rewiring which changes the calls. If you "rewire" a function block, its instance data block is then automatically assigned to the rewired function block; the data block number is not changed.

A subsequently displayed info file shows you in which block changes were made, and how many.

The reference data are no longer up-to-date following rewiring, and must be regenerated.

Please note that the "rewiring" only takes place in the compiled blocks; program source, if present, is not modified.

Further possible methods of rewiring are:

▷ With compiled blocks, you can also use the *Address priority* function.

▷ In the case of source-oriented programming and symbolic addressing, you change the symbol table prior to compiling, and after compiling you receive a "rewired" program.

2.5.6 Address Priority

In the properties window of the offline object container *Blocks* on the "Address Priority" tab, you can set whether the absolute address or the symbol is to have priority for already saved blocks when they are displayed and saved again following a change to the symbol table or to the declaration or assignment of global data blocks.

The default is "Absolute value has priority" (the same behavior as in the previous STEP 7 versions). This default means that when a change is made in the symbol table, the absolute address is retained in the program and the symbol changes accordingly. If "Symbol has priority" is set, the absolute address changes and the symbol is retained.

Example: The symbol table contains the following:

```
I 1.0  "Limit_switch_up"
I 1.1  "Limit_switch_down"
```

In the program of an already compiled block, input I1.0 is scanned:

```
A    I 1.0 "Limit_switch_up"
```

If the assignments for inputs I1.0 and I1.1 are now changed in the symbol table to:

```
I 1.0  "Limit_switch_down"
I 1.1  "Limit_switch_up"
```

and the already compiled block is read out, then the program contains

```
A    I 1.1"Limit_switch_up"
```

if "Symbol has priority" is set,

and if "Absolute value has priority" is set, the program contains

```
A    I 1.0 "Limit_switch_down"
```

If, as a result of a change in the symbol table, there is no longer any assignment between an absolute address and a symbol, the statement will contain the absolute address if "Absolute value has priority" is set (even with symbolic display because the symbol would, of course, be missing); if "Symbol has priority" is set, the statement is rejected as errored (because the mandatory absolute address is missing).

If "Symbol has priority" is set, incrementally programmed blocks with symbolic addressing will retain their symbols in the event of a change to the symbol table. In this way, an already programmed block can be "rewired" by changing the address assignment.

Please note that this "rewiring" is not carried out automatically because the already compiled blocks contain the executable MC7 code of the statements with absolute addresses. The change is only made in the relevant blocks - following the relevant message - after they have been opened and saved again.

In order to implement the modification in the complete block folder, select EDIT → CHECK BLOCK CONSISTENCY for the marked object *Blocks*.

2.5.7 Reference Data

As a supplement to the program itself, the SIMATIC Manager shows you the reference data, which you can use as the basis for corrections or tests. These reference data include the following:

- ▷ Cross references
- ▷ Assignment (I, Q, M and T, C)
- ▷ Program structure
- ▷ Unused symbols
- ▷ Addresses without symbols

To generate reference data, select the *Blocks* object in a project and the menu command OPTIONS → REFERENCE DATA → DISPLAY. The representation of the reference data can be changed specifically for each work window with VIEW → FILTER; you can save the settings for later with WINDOW → SAVE ARRANGEMENT ON EXIT. You can display and view several lists at the same time (Figure 2.14).

With OPTIONS → CUSTOMIZE in the program editor, specify on the "Block" tab whether or not the reference data are to be updated when compiling a program source file or when saving an incrementally written block.

Please note that the reference data are only available when the data of a project are managed offline; the offline reference data are displayed even if the function is called in a block opened online.

Cross references

The cross-reference list shows the use of the addresses and blocks in the user program. It includes the absolute address, the symbol (if any), the block in which the address was used, how it was used (read or write) and positions at which the address was used. Click on a column header to sort the table by column contents.

EDIT → GO TO → LOCATION with the position marked, or double clicking the position of use starts the program editor and displays the address in the programmed environment.

The cross-reference list shows the addresses you selected with VIEW → FILTER (for instance bit memory). STEP 7 then uses the filter saved as "Standard" every time it opens the cross-reference list.

Advantage: the cross references show you whether the referenced addresses were also scanned or reset. They also show you in which blocks addresses are used (possibly more than once).

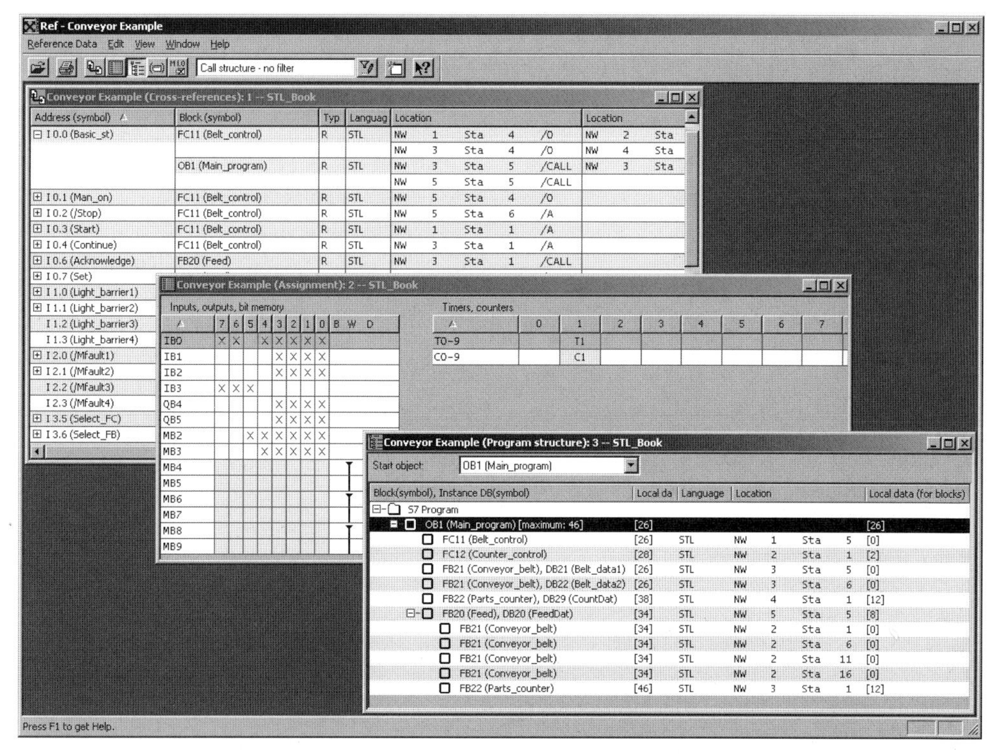

Figure 2.14 Examples22 of Reference Data (Cross-references, Assignment, Program Structure)

Assignments

The I/Q/M reference list shows which bits in address areas I, Q and M are assigned in the program. One byte, broken down into bits, appears on each line. Also shown is byte-, word- or doubleword-oriented access by a blue bar. The T/C reference list shows the timers and counters used in the program. Ten timers or counters are displayed on a line.

Advantage: the list shows you whether certain address areas were (improperly) assigned or where there are still addresses available.

Program structure

The program structure shows the call hierarchy of the blocks in a user program. You can select the start object for the call hierarchy from a selection list. With VIEW → FILTER you have a choice between two different views in the program structure:

The *Call structure* shows all nesting levels of the block calls. You control the display of nesting levels with the "+" and "–" boxes. The requirements for temporary local data are shown for a block or for the entire path up to the associated block. With the block marked, change using EDIT → GO TO → LOCATION to call the block, or open the block with EDIT → GO TO → BLOCK.

The display as *Dependency structure* shows two call levels. The blocks are displayed (indented) in which the block positioned on the left is called.

Advantage: which blocks were used? Were all programmed blocks called? What are the blocks' temporary local data requirements? Is the specified local data requirement per priority class (per organization block) sufficient?

Unused symbols

This list shows all addresses which have symbol table allocations but were not used in the program. The list shows the symbol, the address, the data type, and the comment from the symbol table.

Advantage: were the addresses in the list inadvertently forgotten when the program was being written? Are they spares? Or are they perhaps superfluous, and not really needed?

Addresses without symbol

This list shows all the addresses used in the program to which no symbols were allocated. The list shows these addresses and how often they were used.

Advantage: were addresses used inadvertently (by accident, or because of a typing error)?

2.5.8 Language Settings

STEP 7 offers several possibilities for working with different languages:

- ▷ The language of the operating system (character set)
- ▷ The STEP 7 language
- ▷ The language for comments and displays

The settings for the various languages are independent of one another.

Language setting of the operating system

Use the Windows' control panel to select the character set for use with the Windows operating system. The character sets tested with the multi-language version (MUI version) and the limitations when operating with STEP 7 can be found in the current Readme file or the STEP 7 help under "Creating and editing the project".

Project language

The project language is the language set when creating the project in the Windows' control panel. In the SIMATIC Manager, select EDIT → OBJECT PROPERTIES to show the language in which the selected project or library was created. The display "not yet set" means that language-independent use of the project or library is possible, e.g. in multi-projects. These are always language-independent. Only ASCII characters ($2A_{hex}$ to $7F_{hex}$) may be used in language-independent projects or libraries. You can find further information in the STEP 7 help under "Creating and editing the project".

STEP 7 language

The session language of the SIMATIC Manager, which e.g. defines the texts of the menus and error messages, is referred to as the STEP 7 language. You can set this language in the SIMATIC Manager using OPTIONS → CUSTOMIZE on the "Language" tab. "National Language" lists those which have been installed with STEP 7 and are available for selection. Also set the programming mnemonics on this tab, i.e. the language used by STEP 7 for addresses and operations, e.g. "A I" (AND Input) for English or "U E" (UND Eingang) for German.

Multilingual comments and display texts

Comments and display texts can be implemented in several languages. You have entered the texts in the original language, such as English, and you want to generate a German version of your program. To do so, export the desired texts or text types. The export file is a *.csy or *.xls. file that you can edit with Microsoft Excel. You can enter the translation for each text. You import the finished translation table back into your project. Now you can switch between the languages. You can do this with several languages.

Use OPTIONS → LANGUAGE FOR DISPLAY UNITS in the SIMATIC Manager to select the

Table 2.2
Text Types of the Translated Texts (Selection)

Text type	Meaning
BlockTitle	Block title
BlockComment	Block comment
NetworkTitle	Network title
NetworkComment	Network comment
LineComment	Line comment
InterfaceComment	Comment in ▷ the declaration table of code blocks ▷ data blocks ▷ user data types UDT
SymbolComment	Symbol comment

languages available in your project and to set the standard language for the display units.

Exporting and importing texts

Select the object in the SIMATIC Manager containing the comments you want to translate, e.g. the symbol table, the block container, several blocks or a single block. Select OPTIONS → MANAGE MULTILINGUAL TEXTS → EXPORT. In the dialog window that then appears, enter the storage location of the export file and the target language. Select the text types that you want to translate (Table 2.2).

A separate file is generated for every text type, e.g. the file SymbolComment.csv for the comments from the symbol table. Existing export files can be expanded. A log file provides information on the types of exported text and any errors.

Open the export file(s) with the FILE → OPEN dialog box in Microsoft Excel (not by double-clicking). The exported texts are displayed in the first column and you can translate the texts in the second column.

You can fetch the translated texts back to the project with OPTIONS → MANAGE MULTILINGUAL TEXTS → IMPORT. A log file provides information about the imported texts and any errors that may have occurred.

Please note that the name of the import file must not be changed since there is a direct relation between this and the text types contained in the file.

Selecting and deleting a language

You can change to all imported languages in the SIMATIC Manager with OPTIONS → MANAGE MULTILINGUAL TEXTS → CHANGE LANGUAGE. The language change is executed for the objects (blocks, symbol table) for which the relevant texts have been imported. This information is contained in the log file. Further settings, e.g. the importing of multilingual comments when copying a block, are made with OPTIONS → MANAGE MULTILINGUAL TEXTS → SETTINGS FOR COMMENT MANAGEMENT. You can delete the imported language again with OPTIONS → MANAGE MULTILINGUAL TEXTS → DELETE LANGUAGE.

2.6 Online Mode

You create the hardware configuration and the user program on the programming device, generally referred to as the “engineering system” (ES). The S7 program is stored offline on the hard disk here, also in compiled form.

To transfer the program to the CPU, you must connect the programming device to the CPU. You establish an “online” connection. You can use this connection to determine the operating state of the CPU and the assigned modules, i.e., you can carry out diagnostics functions.

2.6.1 Connecting a PLC

The connection between the programming device’s MPI interface and the CPU’s MPI interface is the mechanical requirement for an online connection. The connection is unique when a CPU is the only programmable module connected. If there are several CPUs in the MPI subnet, each CPU must be assigned a unique node number (MPI address). You set the MPI address when you initialize the CPU. Before linking all the CPUs to one network, connect the programming device to only one CPU at a time and transfer the *System Data* object from the offline container *Blocks* or direct with the Hardware Configuration editor using the menu command PLC → DOWNLOAD. This assigns a CPU its own special MPI address (“naming”) along with the other properties.

The MPI address of a CPU in the MPI network can be changed at any time by transferring a new parameter data record containing the new MPI address to the CPU. Note carefully: the new MPI address takes effect immediately. While the programming device adjusts immediately to the new address, you must adapt other applications, such as global data communications, to the new MPI address.

The MPI parameters are retained in the CPU even after a memory reset. The CPU can thus be addressed even after a memory reset.

A programming device can always be operated online on a CPU, even with a module-independent program and even though no project has been set up.

If no project has been set up, you establish the connection to the CPU with PLC → DISPLAY ACCESSIBLE NODES. This screens a project window with the structure "*Accessible Nodes*" – "Module (MPI=n)" – "Online User Program *(Blocks)*". When you select the *Module* object, you may utilize the online functions, such as changing the operational status and checking the module status. Selecting the *Blocks* object displays the blocks in the CPU's user memory. You can then edit (modify, delete, insert) individual blocks.

You can fetch back the system data from a connected CPU for the purpose of, say, continuing to work on the basis of the existing configuration, without having the relevant project in the programming device data management system. Create a new project in the SIMATIC Manager, select the project and then PLC → UPLOAD STATION TO PG. After specifying the desired CPU in the dialog box that then appears, the online system data are loaded onto the hard disk.

If there is a **CPU-independent program** in the project window, create the associated online project window. If several CPUs are connected to the MPI and accessible, select EDIT → OBJECT PROPERTIES with the online S7 program selected and set the number of the mounting rack and the CPU's slot on the "Addresses Module" tab.

If you select the *S7 Program* in the online window all the online functions to the connected CPU are available to you. *Blocks* shows the blocks located in the CPU's user memory. If the blocks in the offline program agree with the blocks in the online program you can edit the blocks in the user memory with the information from the data management system of the programming device (symbolic address, comments).

When you switch with VIEW → ONLINE a **CPU-assigned program** into online mode, you can carry out program modifications just as you would in a CPU-independent program. In addition, it is now possible for you to configure the SIMATIC station, that is, to set CPU parameters and address and parameterize modules.

2.6.2 Protection of the user program

With appropriately equipped CPUs, access to the user program can be protected with a password. Everyone in possession of the password has unrestricted access to the user program. For those who do not know the password, you can define 3 protection levels. You set the protection levels with the "Protection" tab of the Hardware Configuration tool when parameterizing the CPU.

The access privilege using the password applies until the SIMATIC Manager has been exited, or if the password protection is canceled again with PLC → ACCESS RIGHTS → CANCEL.

Protection level 1: mode selector

This protection level is set as default (without password). With CPUs with a keylock switch, the user program is protected in level 1 by the mode selector switch on the front of the CPU. In the RUN-P and STOP positions, you have unrestricted access to the user program; in the RUN position, only read access via the programming device is possible. In this position, you can also remove the keylock switch so that the mode can no longer be changed via the switch.

You can bypass protection via the keylock switch RUN position by selecting the option "Removable with password", e.g. if the CPU, and with it the keylock switch, are not easily accessible or are located at a distance.

If the mode selector is designed as a toggle switch, protection level 1 results in no limitation in access to the user program.

In protection level 1, the write protection (protection level 2) can be switched on and off again by the program using system function SFC 109 PROTECT (see Chapter 20.3.8 "Changing the Program Protection").

Protection level 2: write protection

At this protection level, the user program can only be read, regardless of the position of the keylock switch.

Protection level 3: read/write protection

No access to the user program, regardless of the keylock switch position. Exception: read diagnostics buffer and monitor variables in tables are possible in every protection level.

Password protection

If you select protection level 2 or 3 or protection level 1 with "Removable with password", you will be prompted to define a password. The password can be up to 8 characters long.

If you try to access a user program that is protected with a password, you will be prompted to enter the password. Before accessing a protected CPU, you can also enter the password via PLC → ACCESS RIGHTS → SETUP. First, select the relevant CPU or the S7 program.

In the "Enter Password" dialog box, you can select the option "Use password as default for all protected modules" to get access to all modules protected with the same password.

Password access authorization remains in force until the last S7 application has been terminated.

Everyone in possession of the password has unrestricted access to the user program in the CPU regardless of the protection level set and regardless of the keylock position.

2.6.3 CPU Information

In online mode, the CPU information listed below is available to you. The menu commands are screened when you have selected a module (in online mode and without a project) or S7 program (in the online project window).

▷ PLC → DIAGNOSTICS/SETTING

→ HARDWARE DIAGNOSTICS
(see Chapter 2.7.1 "Diagnosing the Hardware")

→ MODULE INFORMATION
General information (such as version), diagnostics buffer, memory (current map of work memory and load memory, compression), cycle time (length of the last, longest, and shortest program cycle), timing system (properties of the CPU clock, clock synchronization, run-time meter), performance data (available data handling blocks and system blocks, sizes of the address areas), communication (data transfer rate and communication links), stacks in STOP state (B stack, I stack, and L stack)

→ OPERATING MODE
Display of the current operating mode (for instance RUN or STOP), modification of the operating mode

→ CLEAR/RESET
Resetting of the CPU in STOP mode

→ SET TIME OF DAY
Setting of the internal CPU clock and, in expanded dialog, difference in time from a particular zone

▷ PLC → CPU MESSAGES
Reporting of asynchronous system errors and of user-defined messages generated in the program with SFC 52 WR_USMSG, SFC 18 ALARM_S, SFC 17 ALARM_SQ, SFC 108 ALARM_D and SFC 107 ALARM_DQ.

▷ PLC → DISPLAY FORCE VALUES, PLC → MONITOR/MODIFY VARIABLES,
(see Chapters 2.7.3 "Monitoring and Modifying Variables" and 2.7.4 "Forcing Variables")

2.6.4 Loading the User Program into the CPU

When you transfer your user program (compiled blocks and configuration data) to the CPU, it is loaded into the CPU's load memory. Physically, load memory can be a memory integrated in the CPU, a memory card or a micro memory card (see Chapter 1.1.6 "CPU Memory Areas").

With a micro memory card or a flash EPROM memory card, you can write to it in the programming device and use it as data medium. You plug the memory card into the CPU in the off-circuit state; on power up following memory reset, the relevant data of the memory card are transferred to the work memory of the CPU. With appropriately equipped CPUs, you can also overwrite a flash EPROM memory card or also write a micro memory card if it is plugged into the CPU, but only with the entire program.

In the case of a RAM load memory (integrated in the CPU, as a memory card or as a micro memory card) you transfer a complete user program by switching the CPU to the STOP state, performing memory reset and transferring the user program. The configuration data are also transferred.

If you only want to change the configuration data (CPU properties, the configured connections, GD communications, module parameters, and so on), you need only load the *System Data* object into the CPU (select the object and transfer it with menu command PLC → DOWNLOAD. The parameters for the CPU go into effect immediately; the CPU transfers the parameters for the remaining modules to those modules during startup.

Please note that the entire configuration is loaded onto the PLC with the *System data* object. If you use PLC → DOWNLOAD... in an application, e.g. in global data communications, only the data edited by the application are transferred.

Note: select PLC → SAVE TO MEMORY CARD to load the compressed archive file (see Chapter 2.2.2 "Managing, Rearranging and Archiving"). The project in the archive file cannot be edited direct either with the programming device or from the CPU.

2.6.5 Block Handling

Transferring blocks

In the case of a RAM load memory, you can also modify, delete or reload individual blocks in addition to transferring the entire program online.

You transfer individual blocks to the CPU by selecting them in the offline window and selecting PLC → DOWNLOAD. With the offline and online windows opened at the same time, you can also drag the blocks with the mouse from one window and drop them in the other.

Special care is needed when transferring individual blocks during operation. If blocks that are not available in the CPU memory are called within a block, you must first load the "lower-level" blocks. This also applies for data blocks whose addresses are used in the loaded block. You load the "highest-level" block last. Then, provided it is called, it will be executed immediately in the next program cycle.

The SIMATIC Manager also allows you to transfer individual blocks or the entire program from the offline container Blocks to the CPU in SCL. Transfer back from the CPU to the hard disk makes little sense since compiled blocks can no longer be edited by the SCL editor. You can only edit the SCL program source file and, form it, generate the compiled blocks.

Modifying or deleting blocks online

You can edit blocks incrementally in the online user program (on the CPU), in exactly the same way as in the offline user program. Using a programming device online on the CPU, you can read, modify or delete blocks in the load memory.

If the RAM section of the load memory is sufficiently large to accommodate the complete user program and also the modified blocks, you can edit blocks without limitation.

If the user program is on a flash EPROM memory card, you can edit blocks as long as the RAM section of the load memory is large enough to accommodate the modified blocks. The modified blocks in the RAM are valid during runtime, those in the FEPROM are marked as being invalid. Please note that following an overall reset or unbuffered switching-on the original blocks are loaded from the FEPROM into the work memory.

If you use a micro memory card as e.g. with the compact CPUs, all blocks in the load memory are non-volatile. You can modify individual blocks online, and these blocks retain their modifications even following an overall reset or unbuffered switching-on. Deleted blocks are then no longer present.

In incremental programming mode you can modify blocks in offline data management on the programming device and in online data management on the CPU independent of one another. However, if the online and offline data management diverge, it may be the case that the editor can no longer display the additional information of the offline data management; they may then be lost (symbolic names, jump labels, comments, user-defined data types).

Blocks that have been modified online are best stored offline on the hard disk to avoid data inconsistency (e.g. a "time stamp conflict" when the interface of the called block is later than the program in the calling block).

The following still applies even if you work with the program editor online: with FILE → SAVE you store the current block in the offline user program in the programming device data management; with PLC → LOAD you write the block back to the user program in the CPU.

Compressing

When you load a new or modified block into the CPU, the CPU places the block in load memory and transfers the relevant data to work memory. If there is already a block with the same number, this "old block" is declared invalid (following a prompt for confirmation) and the new block "added on at the end" in memory. Even a deleted block is "only" declared invalid, not actually removed from memory.

This results in gaps in user memory which increasingly reduce the amount of memory still available. These gaps can be filled only by the *Compress* function. When you compress in RUN mode, the blocks currently being executed are not relocated; only in STOP mode can you truly achieve compression without gaps.

The current memory allocation can be displayed in percent with the menu command PLC → DIAGNOSTICS/SETTING → MODULE INFORMATION, on the "Memory" tab. The dialog box which then appears also has a button for preventive compression.

You can initiate event-driven compressing per program with the call SFC 25 COMPRESS.

Data blocks offline/online

When programming, you assign a default value and an initial value to the data addresses in a data block (see also Chapter 3.6 "Programming Data Blocks"). If a data block is loaded into the CPU, the initial values are transferred to load memory and the actual values are subsequently transferred to work memory. Every value change made to a data address per program corresponds to a change to the actual value in the work memory (Figure 2.15).

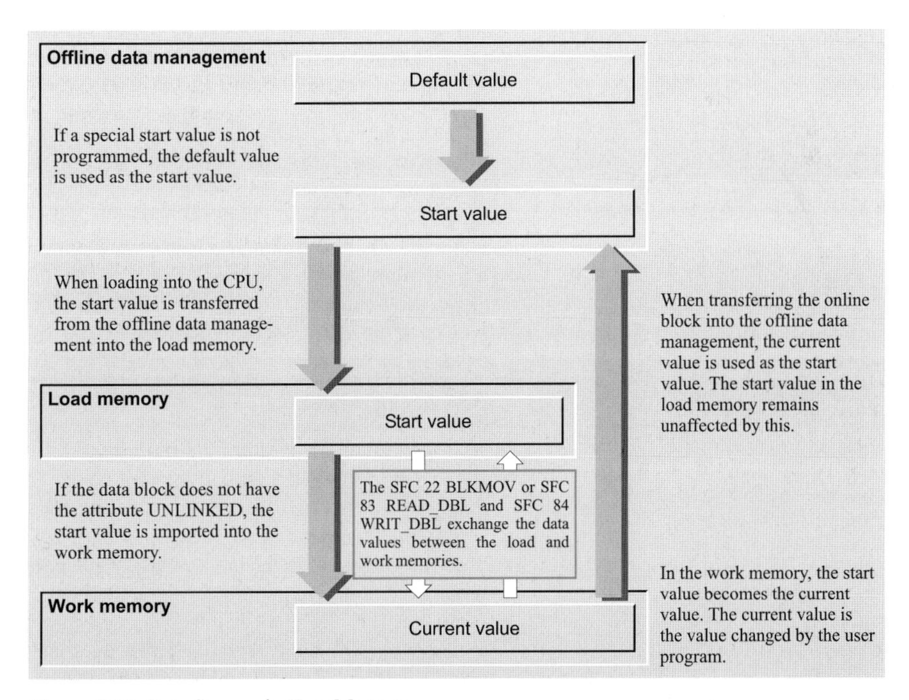

Figure 2.15 Data Storage in User Memory

You can download the current values generated in the work memory from a programmed (loaded) data block into the offline data management by opening the data block online and importing into the offline data management with FILE → SAVE. The variable names and the data types saved in the offline data management are then retained. If you upload the online data block into the offline data management using the SIMATIC Manager with PLC → UPLOAD TO PG or by dragging the data block from the online window into the offline window, the description of the addresses, e.g. variable name and data type, are lost.

If you transfer a data block from the CPU back into the offline data management, the current values present in the work memory are imported into the offline data management as initial values. The initial values in the load memory are not changed by this. Following an overall reset or an unbuffered switch-on, and with use of a flash EPROM memory card or a micro memory card, the (old) initial values present in the load memory are imported into the work memory as (new) current values.

If you wish to import the current values into the load memory when using a RAM load memory or a micro memory card, load the data block from the CPU into the programming device and then back again into the CPU. CPUs with micro memory card provide the system function SFC 84 WRIT_DBL with which you can directly write current values into the load memory. With appropriately designed CPUs, you can transmit the complete work memory contents into the ROM section of the load memory with PLC → COPY RAM TO ROM.

A data block generated with the property *unlinked* is not transferred to work memory; it remains in load memory. A data block with this property can be read with SFC 20 BLKMOV or – with appropriately designed CPUs – with SFC 83 READ_DBL.

In incremental programming mode, you can create data blocks directly in the work memory of the CPU. It is recommendable to also save these data blocks offline immediately following creation.

With the system functions SFC 22 CREAT_DB, SFC 85 CREA_DB and SFC 82 CREA_DBL you can generate data blocks during runtime where the description of the addresses, e.g. variable name and data type, is missing. When reading with the programming device, a BYTE field is therefore displayed with a name and index assigned by the program editor. If you transmit such a data block to the offline data management, this declaration is also imported. If the data block has the property *Unlinked*, the initial values from the load memory are imported into the offline data management as new initial values, otherwise the current values from the work memory.

When transferring to the offline data management, the checksum of the (offline) program is changed.

2.7 Testing the Program

After establishing a connection to a CPU and loading the user program, you can test (debug) the program as a whole or in part, such as individual blocks. You initialize the variables with signals and values, e.g. with the help of simulator modules and evaluate the information returned by your program. If the CPU goes to the STOP state as a result of an error, you can get support in finding the cause of the error from the CPU information among other things.

Extensive programs are debugged in sections. If, for example, you only want to debug one block, load this block into the CPU and call it in OB 1. If OB 1 is organized is in such a way that the program can be debugged section by section "from beginning to end", you can select the blocks or program sections for debugging by using jump functions to skip those calls or program sections that are not to be debugged.

With the S7-PLCSIM optional software, you can simulate a CPU on the programming device and so debug your program without additional hardware.

2.7.1 Diagnosing the Hardware

In the event of a fault, you can fetch the diagnostics information of the faulty modules with the help of the function "Diagnose Hardware".

You connect the programming device to the MPI bus and start the SIMATIC Manager.

If the project associated with the plant configuration is available in the programming device database, open the online project window with VIEW → ONLINE. Otherwise, select PLC → DISPLAY ACCESSIBLE NODES and select the CPU.

Now you can get a quick overview of the faulty modules with PLC → DIAGNOSTICS/SETTING → HARDWARE DIAGNOSTICS (default in the SIMATIC Manager with OPTIONS → CUSTOMIZE in the tab "View"). If the fast view is deselected, you obtain detailed diagnostics information on all modules.

If you are in the Hardware Configuration, switch the online view on with VIEW → ONLINE. You can now display the existing diagnostics information of the selected module with PLC → MODULE INFORMATION.

2.7.2 Determining the Cause of a STOP

If the CPU goes to STOP because of an error, the first measure to take in order to determine the reason for the STOP is to output the diagnostics buffer. The CPU enters all messages in the diagnostic buffer, including the reason for a STOP and the errors which led to it.

To output the diagnostic buffer, switch the PG to online, select an S7 program, and choose the *Diagnostics Buffer* tab with the menu command PLC → DIAGNOSTICS/SETTING → MODULE INFORMATION. The last event (the one with the number 1) is the cause of the STOP, for instance "STOP because programming error OB not loaded". The error which led to the STOP is described in the preceding message, for example "FC not loaded". By clicking on the message number, you can screen an additional comment in the next lower display field. If the message relates to a programming error in a block, you can open and edit that block with the "Open Block" button.

If the cause of the STOP is, for example, a programming error, you can ascertain the surrounding circumstances with the "Stacks" tab. When you open "Stacks", you will see the B stack (block stack), which shows you the call path of all non-terminated blocks up to the block containing the interrupt point. Use the "I stack" button to screen the interrupt stack, which shows you the contents of the CPU registers (accumulators, address register, data block register, status word) at the interrupt point at the instant the error occurred. The L stack (local data stack) shows the block's temporary local data, which you select in the B stack by clicking with the mouse.

2.7.3 Monitoring and Modifying Variables

One excellent resource for debugging user programs is the monitoring and modifying of variables with VAT variable tables. Signal states or values of variables of elementary data types can be displayed. If you have access to the user program, you can also modify variables, i.e. change the signal state or assign new values.

Please note that you can only control data addresses if the write protection for the data block is switched off, i.e. the block property *DB is write-protected in the AS* is not activated.

Addresses in data blocks with the block property *Unlinked* cannot be monitored since these data blocks are present in the load memory of the Micro Memory Card. A single updating operation is carried out when the data block is opened online.

Caution: you must ensure that no dangerous states can result from modifying variables!

Creating a variable table

For monitoring and modifying variables, you must create a VAT variable table containing the variables and the associated data formats. You can generate up to 255 variable tables (VAT 1 to VAT 255) and assign them names. The maximum size of a variable table is 1024 lines with up to 255 characters (Figure 2.16).

You can generate a VAT offline by selecting the user program *Blocks* and then INSERT → S7 BLOCK → VARIABLE TABLE, and you can generate an unnamed VAT online by selecting *S7 Program* and selecting PLC → MONITOR/MODIFY VARIABLES.

You can specify the variables with either absolute or symbolic addresses and choose the data type (display format) with which a variable is to

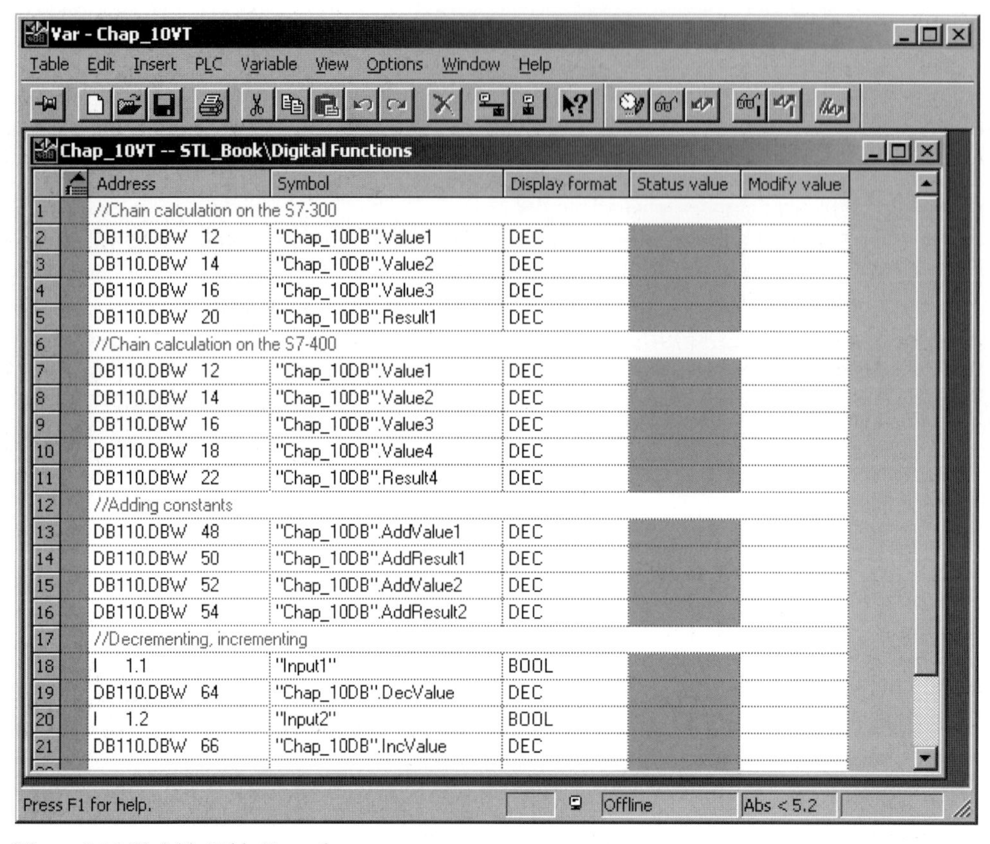

	Address	Symbol	Display format	Status value	Modify value
1	//Chain calculation on the S7-300				
2	DB110.DBW 12	"Chap_10DB".Value1	DEC		
3	DB110.DBW 14	"Chap_10DB".Value2	DEC		
4	DB110.DBW 16	"Chap_10DB".Value3	DEC		
5	DB110.DBW 20	"Chap_10DB".Result1	DEC		
6	//Chain calculation on the S7-400				
7	DB110.DBW 12	"Chap_10DB".Value1	DEC		
8	DB110.DBW 14	"Chap_10DB".Value2	DEC		
9	DB110.DBW 16	"Chap_10DB".Value3	DEC		
10	DB110.DBW 18	"Chap_10DB".Value4	DEC		
11	DB110.DBW 22	"Chap_10DB".Result4	DEC		
12	//Adding constants				
13	DB110.DBW 48	"Chap_10DB".AddValue1	DEC		
14	DB110.DBW 50	"Chap_10DB".AddResult1	DEC		
15	DB110.DBW 52	"Chap_10DB".AddValue2	DEC		
16	DB110.DBW 54	"Chap_10DB".AddResult2	DEC		
17	//Decrementing, incrementing				
18	I 1.1	"Input1"	BOOL		
19	DB110.DBW 64	"Chap_10DB".DecValue	DEC		
20	I 1.2	"Input2"	BOOL		
21	DB110.DBW 66	"Chap_10DB".IncValue	DEC		

Figure 2.16 Variable Table Example

be displayed and modified (select the lines to be changed then select VIEW → SELECT DISPLAY FORMAT or click with the right mouse button on "Display Format").

Use comment lines to give specific sections of the table a header. You may also stipulate which columns are to be displayed. You can change variables or display formats or add or delete lines at any time. You save the variable table in the *Blocks* object container with TABLE → SAVE.

Establishing an online connection

To operate a variable table that has been created offline, switch it online with PLC → CONNECT TO → ... You must switch each individual VAT online and you can clear down the connection again with PLC → DISCONNECT.

Trigger conditions

In the variable table, select VARIABLE → TRIGGER to set the trigger point and the trigger conditions separately for monitoring and modifying. The trigger point is the point at which the CPU reads values from the system memory or writes values to the system memory. You specify whether reading and writing is to take place once or periodically.

If monitoring and modifying have the same trigger conditions, monitoring is carried out before modifying. If you select the trigger point "Start of cycle" for modifying, the variables are modified after updating of the process input image and before calling OB 1. If you select the trigger point "End of cycle" for monitoring, the status values are displayed after termination of OB 1 and before output of the process output image.

Monitoring variables

Select the Monitor function with the menu command VARIABLE → MONITOR. The variables in the VAT are updated in accordance with the specified trigger conditions. Permanent monitoring allows you to follow changes in the values on the screen. The values are displayed in the data format which you set in the Display Format column. The ESC key terminates a permanent monitor function.

VARIABLE → UPDATE MONITOR VALUES updates the monitor values once only and immediately without regard to the specified trigger conditions.

Modifying variables

Use VARIABLE → MODIFY to transfer the specified values to the CPU dependent on the trigger conditions. Enter values only in the lines containing the variables you want to modify. You can expand the commentary for a value with "//" or with VARIABLE → MODIFY VALUE AS COMMENT; these values are not taken into account for modification. You must define the values in the data format which you set in the Display Format column. Only the values visible on starting the modify function are modified. The ESC key terminates a permanent modify function.

VARIABLE → ACTIVATE MODIFY VALUES transfers the modify values only once and immediately, without regard to the specified trigger conditions.

2.7.4 Forcing Variables

With appropriately equipped CPUs, you can specify fixed values for certain variables. The user program can no longer change these values ("forcing"). Forcing is permissible in any CPU operating state and is executed immediately.

Caution: you must ensure that no dangerous states can result from forcing variables!

The starting point for forcing is a variable table (VAT). Create a VAT, enter the addresses to be forced and establish a connection to the CPU. You can open a window containing the force values by selecting VARIABLE → DISPLAY FORCE VALUES.

If there are already force values active in the CPU, these are indicated in the force window in bold type. You can now transfer some or all addresses from the variable table to the force window or enter new addresses. You save the contents of a force window in a VAT with TABLE → SAVE AS.

The following address areas can be provided with a force value:

- ▷ Inputs I (process image) [S7-300 and S7-400]
- ▷ Outputs Q (process image) [S7-300 and S7-400]
- ▷ Peripheral inputs PI [S7-400]
- ▷ Peripheral outputs PQ [S7-400]
- ▷ Memory bits M [S7-400]

You start the force job with VARIABLE → FORCE. The CPU accepts the force values and permits no more changes to the forced addresses.

While the force function is active, the following applies:

- ▷ All read accesses to a forced address via the user program (e.g. load) and via the system program (e.g. updating of the process image) always yield the force value.
- ▷ On the S7-400, all write accesses to a forced address via the user program (e.g. transfer) and via the system program (e.g. via SFCs) remain without effect. On the S7-300, the user program can overwrite the force values.

Forcing on the S7-300 corresponds to cyclic modifying: after the process input image has been updated, the CPU overwrites the inputs with the force value; before the process output image is output, the CPU overwrites the outputs with the force value.

Note: forcing is not terminated by closing the force window or the variable table, or by breaking the connection to the CPU! You can only delete a force job with VARIABLE → STOP FORCING.

Forcing is also deleted by memory reset or by a power failure if the CPU is not battery-backed. When forcing is terminated, the addresses retain the force values until overwritten by either the user program or the system program.

Forcing is effective only on I/O assigned to a CPU. If, following restart, forced peripheral inputs and outputs are no longer assigned (e.g. as a result of reparameterizing), the relevant peripheral inputs and outputs are no longer forced.

Error handling

If the access width when reading is greater than the force width (e.g. forced byte in a word), the unforced component of the address value is read as usual. If a synchronization error occurs here (access or area length error) the "error substitute value" specified by the user program or by the CPU is read or the CPU goes to STOP.

If, when writing, the access width is greater than the force width (e.g. forced byte in a word), the unforced component of the address value is written to as usual. An errored write access leaves the forced component of the address unchanged, i.e. the write protection is not revoked by the synchronization error.

Loading forced peripheral inputs yields the force value. If the access width agrees with the force width, input modules that have failed or have not (yet) been plugged in can be "replaced" by a force value.

The input I in the process image belonging to a forced peripheral input PI is not forced; it is not preassigned and can still be overwritten. When updating the process image, the input receives the force value of the peripheral input.

When forcing peripheral outputs PQ, the associated output Q in the process image is not updated and not forced (forcing is only effective "externally" to the module outputs). The outputs Q are retained and can be overwritten; reading the outputs yields the written value (not the force value). If an output module is forced and if this module fails or is removed, it will receive the force value again immediately on reconnection.

The output modules output signal state "0" or the substitute value with the OD signal (disable output modules at STOP, HOLD or RESTART) – even if the peripheral outputs are forced (exception: analog modules without OD evaluation continue to output the force value). If the OD signal is deactivated, the force value becomes effective again.

If, in STOP mode, the function *Enable PQ* is activated, the force values also become effective in STOP mode (due to deactivation of the OD signal). When *Enable PQ* is terminated, the modules are set back to the "safe" state (signal state "0" or substitute value); the force value becomes effective again at the transition to RUN.

2.7.5 Enabling Peripheral Outputs

In STOP mode, the output modules are normally disabled by the OD signal; with the Enable peripheral outputs function, you can deactivate the OD signal so that you can modify the output modules even at CPU STOP. Modifying is carried out via a variable table. Only the peripheral outputs assigned to a CPU are modified. Possible application: wiring test of the output at STOP and without user program.

Caution: you must ensure that no dangerous states can result from enabling the peripheral outputs!

Create a variable table and enter the peripheral outputs (PQ) and the modify values. Switch the variable table online with PLC → CONNECT TO → ... online and stop the CPU if necessary, e.g. with PLC → OPERATING MODE and "STOP".

You deactivate the OD signal with VARIABLE → ENABLE PERIPHERAL OUTPUTS; the module outputs now have signal state "0" or the substitute value or force value. You modify the peripheral outputs with VARIABLE → ACTIVATE MODIFY VALUES. You can change the modify value and modify again.

You can switch the function off again by selecting VARIABLE → ENABLE PERIPHERAL OUTPUTS again, or by pressing the ESC key. The OD signal is then active again, the module outputs are set to "0" and the substitute value or the force value is reset

If STOP is exited while "enable peripheral outputs" is still active, all peripheral inputs are deleted, the OD signal is activated at the transition to RESTART and deactivated again at the transition to RUN.

2.7.6 Test and process mode

Recording of the program status information requires additional execution time in the program cycle. For this reason, you can choose two operating modes for debugging purposes: test mode and process mode. In *test mode*, all debugging functions can be used without restriction. You would select this, for example, to debug blocks without connection to the system, because this can significantly increase the cycle execution time. In *process mode*, care is taken to keep the increase in the cycle time to a minimum and this results in debugging restrictions, e.g. the status display is aborted at the return position in the case of program loops. Tests with breakpoints and single-step program execution cannot be carried out in process mode.

For the S7-300 CPUs, test mode is factory-set. You can set test or process mode for these CPUs with the Hardware Configuration in the "Protection" tab. The configuration must then be recompiled and downloaded to the CPU.

For the S7-400 CPUs, process mode is factory-set. You can change the mode online using the program editor. The set mode is displayed using DEBUG → OPERATION..., and it is possible to change this online.

2.7.7 STL Program Status

With the *Program status* function, the program editor provides an additional test method for the user program. The editor shows you line by line the register assignments you have selected on the "STL" tab under OPTIONS → CUSTOMIZE ("Standard status" means accumulator 1 here or timer value or counter value). The amount of status information displayed depends on the CPU and the data type of the variables. You can improve the display through smaller networks or by reducing the block window.

The block whose program you want to debug is in the CPU's user memory and is called and edited there. Open this block, for example by double-clicking on it in the SIMATIC Manager's online window. The editor is started and shows the program in the block.

Select the program section you want to debug. Activate the Program Status function with DEBUG → MONITOR. Now you can see the address statuses, the result of the logic operation and the register assignments (Figure 2.17). You can deactivate the Program Status function again by selecting DEBUG → MONITOR again.

You set the trigger conditions with DEBUG → CALL ENVIRONMENT (see further under "Breakpoints, single-step mode"). You require this setting if the block to be debugged is called more than once in your program. You can initiate status recording either by specifying the call path (determined from the reference data or manually) or by making it dependent on the opened data block when calling the block to be tested. If the call environment is not set, observe the block when it is called for the first time.

Controlling addresses

You can control addresses in the program status. If the address is of data type BOOL, set the cursor to the statement line and select DEBUG → MODIFY ADDRESS TO 0 or DEBUG → MODIFY ADDRESS TO 1. With a different data type, select DEBUG → MODIFY ADDRESS and enter the control value for the marked address in the displayed dialog box.

Operator inputs on the contact

In the program status, you can directly control binary inputs and bit memories in the user program using a button. The following prerequisites are necessary for this function:

- ▷ You assign the attribute CC (Control at Contact, see "Special object properties" in Chapter 2.5.2 "Symbol Table") to the inputs and bit memories in the symbol table.
- ▷ You have enabled control at contact in the program editor with OPTIONS → CUSTOMIZE on the tab "General".
- ▷ You are online in the program status with DEBUG → MONITOR and additionally select DEBUG → CONTROL AT CONTACT.

The symbols and addresses of the binary inputs and bit memories are displayed as buttons which you can click with the mouse. Addresses programmed as NO contacts or addresses with

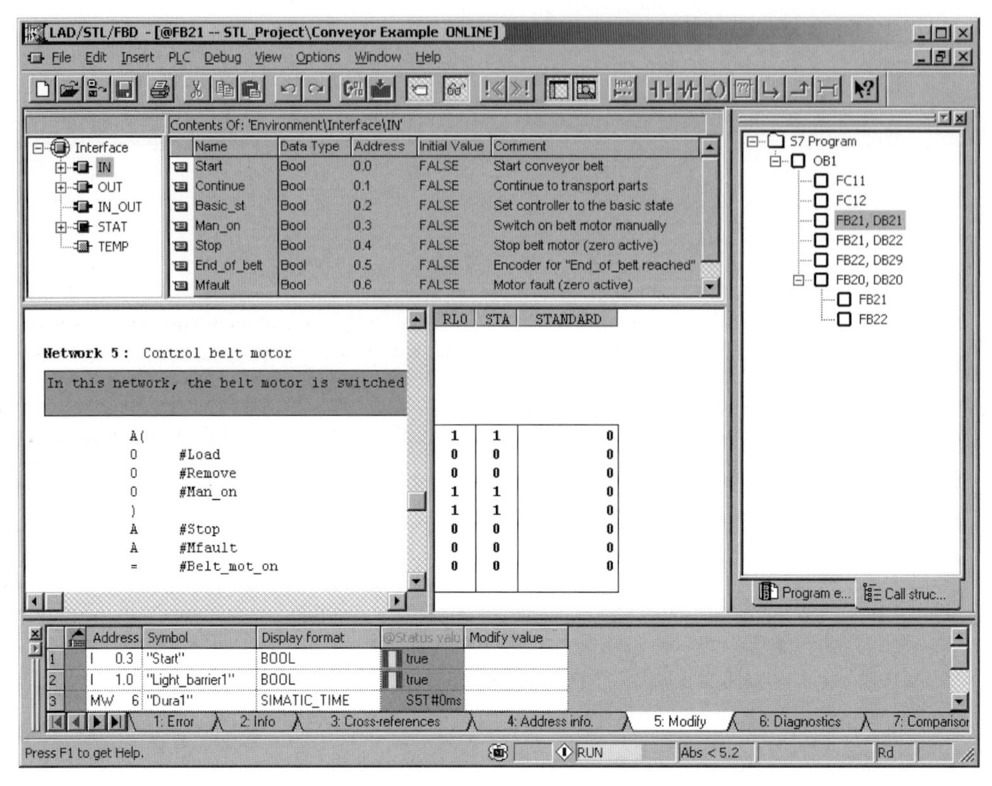

Figure 2.17 STL Program Status

scanning for signal status "1" then deliver the address status "1"; addresses programmed as NC contacts or addresses with scanning for signal status "0" deliver the status "0". Using the Ctrl key and the mouse, you can select several addresses and access them simultaneously when operating on the contact. You can deselect addresses in the same manner.

Breakpoints, single-step mode

If the block was written in the STL language, some CPUs allow you to debug the program statement by statement in single-step mode. The CPU is in HOLD mode; for reasons of safety, the peripheral outputs are disabled. Using breakpoints, you can stop the program at any location and debug it step by step.

The test mode must be set to test in single-step mode (see Chapter 2.7.6 "Test and process mode").

To set the breakpoint, position the cursor in the relevant statement line and select DEBUG → SET BREAKPOINT. To debug, select DEBUG → BREAKPOINTS ACTIVE; this causes the breakpoints to be transferred to the CPU and activated. If the CPU is not already operating, it will now restart and go to the HOLD mode when it encounters a breakpoint. Then, in a window reserved for this purpose, the current register contents are displayed at the statement.

You can now cause the program to be executed line by line by selecting DEBUG → EXECUTE NEXT STATEMENT. Program execution stops at each statement and displays the register contents. At a block call, you can continue execution in the called block by selecting DEBUG → EXECUTE CALL.

With DEBUG → RESUME, the program is executed at normal speed until the next breakpoint is reached.

Blocks containing breakpoints cannot be changed or reloaded online. All breakpoints must first be deleted. You must also delete all breakpoints in order to terminate debugging with breakpoints. With DEBUG → RESUME, the CPU switches back to the RUN state.

2.7.8 Monitoring and Controlling Data Addresses

If the variables to be tested are present in data blocks, you can also monitor and control them directly: mark the data block in the SIMATIC Manager and select EDIT → OPEN OBJECT. With STEP 7 V5.2 or higher, you are asked in the standard setting whether you wish to open the data block with the program editor or with the application "Parameter assignment for data blocks".

In the program editor, switch the data view on with VIEW → DATA VIEW and select DEBUG → MONITOR. You can now monitor the current values in the work memory, and also set (control) them if necessary. With PLC → DOWNLOAD you can write the modified current values back into the work memory, and with FILE → SAVE can import the modified values into the offline data management (first switch off DEBUG → MONITOR).

With "Parameter assignment for data blocks" you can directly monitor and control the current values in the CPU's user memory. You can also monitor and set the current values here with DEBUG → MONITOR. With PLC → DOWNLOAD PARAMETER DATA you have the opportunity for only writing the current values into the work memory and not the complete data block. With DATA BLOCK → SAVE you can import the data block into the offline data management.

The advantage of the application "Parameterization of data blocks" is the facility for displaying and parameterizing the data blocks in the parameterization view. Prerequisite: the system attribute *S7-techparam* (technological functions) is set and a parameterization desktop is available, e.g. from an option package. Figure 2.18 shows a comparison between parameterization view and data view using an example of the instance data block for the controller function block FB 58 TCONT_CP from the

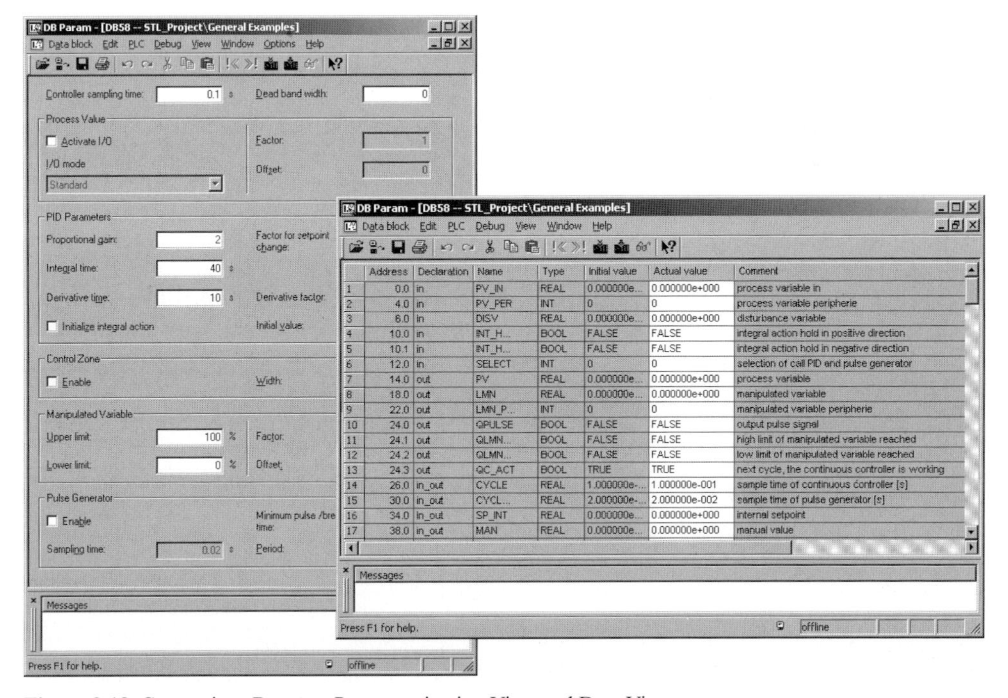

	Address	Declaration	Name	Type	Initial value	Actual value	Comment
1	0.0	in	PV_IN	REAL	0.000000e...	0.000000e+000	process variable in
2	4.0	in	PV_PER	INT	0	0	process variable peripherie
3	6.0	in	DISV	REAL	0.000000e...	0.000000e+000	disturbance variable
4	10.0	in	INT_H...	BOOL	FALSE	FALSE	integral action hold in positive direction
5	10.1	in	INT_H...	BOOL	FALSE	FALSE	integral action hold in negative direction
6	12.0	in	SELECT	INT	0	0	selection of call PID and pulse generator
7	14.0	out	PV	REAL	0.000000e...	0.000000e+000	process variable
8	18.0	out	LMN	REAL	0.000000e...	0.000000e+000	manipulated variable
9	22.0	out	LMN_P...	INT	0	0	manipulated variable peripherie
10	24.0	out	QPULSE	BOOL	FALSE	FALSE	output pulse signal
11	24.1	out	QLMN...	BOOL	FALSE	FALSE	high limit of manipulated variable reached
12	24.2	out	QLMN...	BOOL	FALSE	FALSE	low limit of manipulated variable reached
13	24.3	out	QC_ACT	BOOL	TRUE	TRUE	next cycle, the continuous controller is working
14	26.0	in_out	CYCLE	REAL	1.000000e-...	1.000000e-001	sample time of continuous controller [s]
15	30.0	in_out	CYCL...	REAL	2.000000e-...	2.000000e-002	sample time of pulse generator [s]
16	34.0	in_out	SP_INT	REAL	0.000000e...	0.000000e+000	internal setpoint
17	38.0	in_out	MAN	REAL	0.000000e...	0.000000e+000	manual value

Figure 2.18 Comparison Between Parameterization View and Data View

standard library *PID Control Blocks*. Its parameterization desktop is provided with STEP 7.

2.7.9 Debugging SCL Programs

If you want to debug an SCL program, you must compile it with the option "Create debug info". You can set this option on the "Compiler" tab under OPTIONS → CUSTOMIZE in the SCL Editor. Following compiling with "Create object code", transfer the program to the CPU with PLC → DOWNLOAD.

The SCL debugger is an integral component of the SCL program editor.

SCL program status

With this test function you can test a group of statements, the "monitoring range", during running operation. The monitoring range has a variable length that depends on the statements used. The values of the variables in this range are updated and displayed cyclically.

If the monitor area is in a program section that is executed in every program cycle, the variable values from concatenous cycles cannot usually be acquired. Values that have changed during the current pass are represented in black type, and values that have not changed are represented in light gray.

To debug the SCL program, switch the CPU to RUN or RUN-P and open the program source file containing the program section to be debugged. Select the operating mode DEBUG → OPERATION → TEST OPERATION.

Position the cursor at the start of the area to be debugged. Activate debugging with DEBUG → MONITOR. The names and values of the variables in monitor range are displayed by line in the right-hand section of the window that then appears.

You can interrupt the debug run by selecting DEBUG → MONITOR again; DEBUG → FINISH DEBUGGING terminates the debug run.

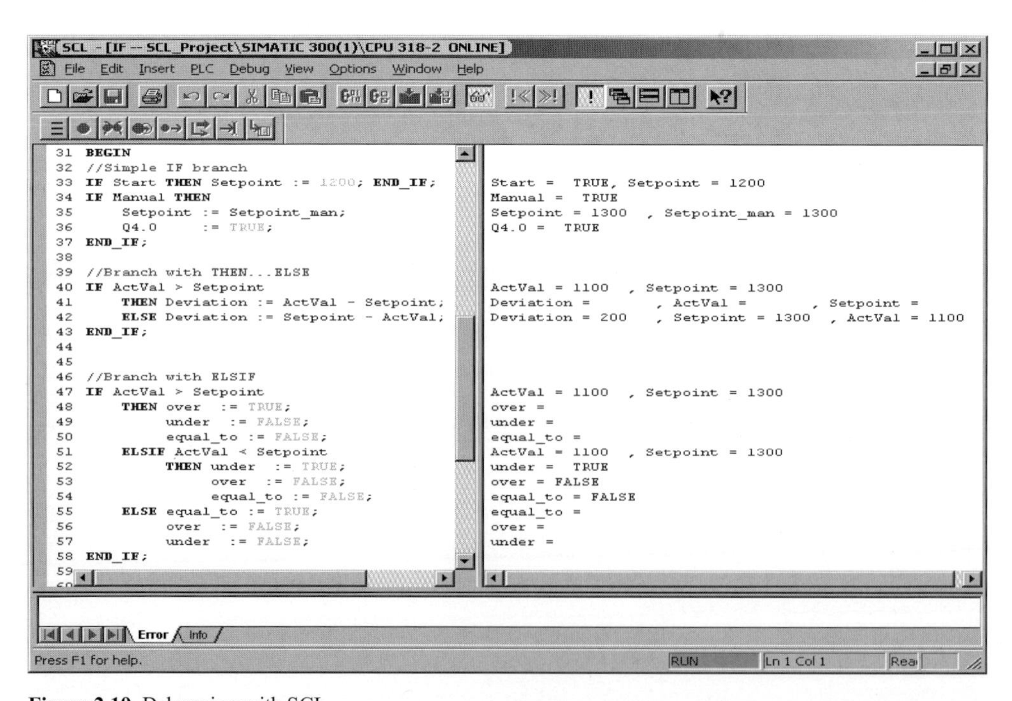

Figure 2.19 Debugging with SCL

Breakpoints, single-step mode

When debugging in single-step mode, you can execute the SCL program line by line and monitor the variable values. After setting breakpoints, you can execute the program first up to a breakpoint and then with step-by-step monitoring from there.

The following requirements must be met for debugging in single-step mode: the block to be debugged must not be protected, it must be opened online and the opened block must not have been modified in the editor. Single-step mode only functions on CPUs that support it. The operating mode "Test mode" must be set, and debugging with the Program Status function must be deactivated. The CPU carries out holding at a breakpoint and step-by-step execution only in HOLD mode.

To debug, open the program source file and define the breakpoints by positioning the cursor at the desired statement and selecting DEBUG → SET BREAKPOINT. Please ensure that no dangerous states can result and select DEBUG → BREAKPOINTS ACTIVE. If the CPU is not already operating, it will now go to RUN mode and then to HOLD mode at the next breakpoint (Figure 2.19).

The CPU goes to RUN mode with DEBUG → NEXT STATEMENT and stops again at the statement immediately following. The variable values of the executed statement line are displayed in the right-hand section of the editor window. Use DEBUG → EXECUTE CALL if the CPU stops at a block call and you wish to continue the single-step execution in the called SCL block. Display of the symbol names can be toggled on and off with VIEW → SYMBOLIC REPRESENTATION.

With DEBUG → RESUME, the CPU goes to RUN mode and stops at the next breakpoint. With DEBUG → TO CURSOR, the CPU goes to RUN mode and stops at the program section selected by the cursor.

You can manage the breakpoints in the program with DEBUG → EDIT BREAKPOINTS. You interrupt the debug process with a repeated DEBUG → BREAKPOINTS ACTIVE; DEBUG → FINISH DEBUGGING terminates the debug process.

Note: the menu commands DEBUG → NEXT STATEMENT and DEBUG → TO CURSOR set and activate a breakpoint. Please ensure that the CPU-specific number of breakpoints is not exceeded when you activate this function.

3 SIMATIC S7 Program

This chapter shows you the structure of the user program for the SIMATIC S7-300/400 CPUs starting from the different priority classes (program execution types) via the component parts of a user program (blocks) right up to the variables and data types. The focus of this chapter is the description of block programming with STL and SCL. The data types are dealt with in detail in Chapter 24 “Data Types”.

You define the structure of the user program right back at the design phase when you adapt the technological and functional conditions; it is decisive for program creation, program test and startup. To achieve effective programming, it is therefore necessary to devote special attention to the program structure.

3.1 Program Processing

The overall program for a CPU consists of the operating system and the user program.

The operating system is the totality of all instructions and declarations which control the system resources and the processes using these resources, and includes such things as data backup in the event of a power failure, the activation of priority classes, and so on. The operating system is a component of the CPU to which you, as user, have no write access. However, you can reload the operating system from a memory card, for instance in the event of a program update.

The user program is the totality of all instructions and declarations (in this case program elements) for signal processing, through which a plant (process) is affected in accordance with the defined control task.

3.1.1 Program Processing Methods

The user program may be composed of program sections which the CPU processes in dependence on certain events. Such an event might be the start of the automation system, an interrupt, or detection of a program error (Figure 3.1). The programs allocated to the events are divided into *priority classes*, which determine the program processing order (mutual interruptibility) when several events occur.

The lowest-priority program is the main program, which is processed cyclically by the CPU. All other events can interrupt the main program at any location, the CPU then executes the associated interrupt service routine or error handling routine and returns to the main program.

A specific organization block (OB) is allocated to each event. The organization blocks represent the priority classes in the user program. When an event occurs, the CPU invokes the assigned organization block. An organization block is a part of a user program which you yourself may write.

Before the CPU begins processing the main program, it executes a startup routine. This routine can be triggered by switching on the mains power, by actuating the mode switch on the CPU’s front panel, or via the programming device. Program processing following execution of the startup routine always starts at the beginning of the main program in the case of a cold or warm restart; in S7-400 systems, it is also possible to resume the program scan at the point at which it was interrupted (hot restart).

The main program is in organization block OB 1, which the CPU always processes. The start of the user program is identical to the first network in OB 1. After OB 1 has been processed (end of program), the CPU returns to the operating system and, after calling for the execution

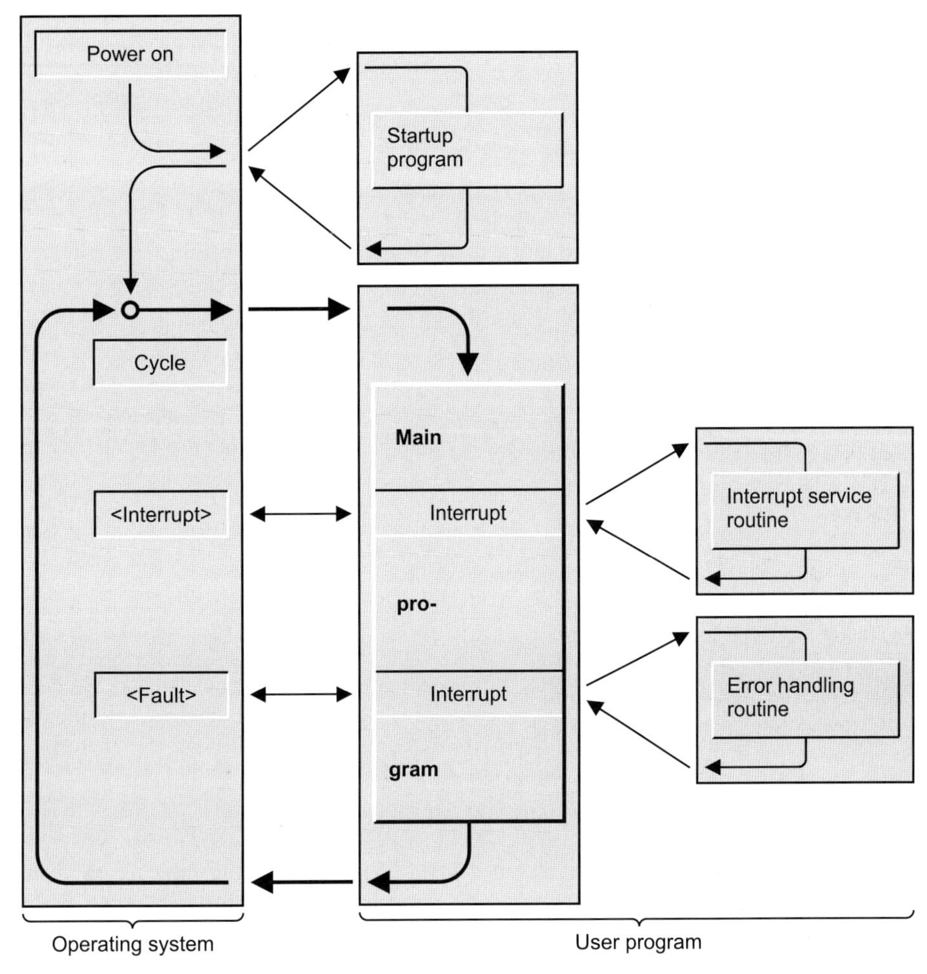

Figure 3.1 Methods of Processing the User Program

of various operating system functions, such as the updating of the process images, it once again calls OB 1.

Events which can intervene in the program are interrupts and errors. Interrupts can come from the process (process interrupts) or from the CPU (watchdog interrupts, time-of-day interrupts, etc.). As far as errors are concerned, a distinction is made between synchronous and asynchronous errors.

An asynchronous error is an error which is independent of the program scan, for example failure of the power to an expansion unit or an interrupt that was generated because a module was being replaced.

A synchronous error is an error caused by program processing, such as accessing a non-existent address or a data type conversion error. The type and number of recorded events and the associated organization blocks are CPU-specific; not every CPU can handle all possible STEP 7 events.

3.1.2 Priority Classes

Table 3.1 lists the available SIMATIC S7 organization blocks, each with its priority. In some priority classes, you can change the assigned priority when you parameterize the CPU. The Table shows the lowest and highest possible

Table 3.1 SIMATIC S7 Organization Blocks

Organization block	Called under the following circumstances	Priority	
		Default	Modifiable
Free cycle OB 1	Cyclically via the operating system	1	No
TOD interrupts OB 10 to OB 17	At a specific time of day or at regular intervals (e.g. monthly)	2	0, 2 to 24
Time-delay interrupts OB 20 to OB 23	After a programmable time, controlled by the user program	3 to 6	0, 2 to 24
Watchdog interrupts OB 30 to OB 38	Regularly at programmable intervals (e.g. every 100 ms)	7 to 15	0, 2 to 24
Process interrupts OB 40 to OB 47	On interrupt signals from I/O modules	16 to 23	0, 2 to 24
DPV1 interrupts OB 55 to OB 57	With status, update and manufacturer specific interrupts from PROFIBUS DPV1 slaves	2	0, 2 to 24
Multiprocessor interrupt OB 60	Event-driven via the user program in multiprocessor mode	25	No
Synchronous cycle interrupts OB 61 to OB 64	With an synchronous cycle interrupt of the DP master (synchronous with DP cycle)	25	0, 2 to 26
Technology synchronous interrupt OB 65	Synchronous following update of the technology data blocks of a CPU 317T	25	No
Redundancy errors OB 70 OB 72 OB 73	 In the case of loss of redundancy resulting from I/O errors In the case of CPU redundancy error In the case of communications redundancy error	 25 28 25	 2 to 26 2 to 28 2 to 26
Asynchronous error interrupts OB 80 OB 81 to OB 84, 86, 87 OB 85 OB 88	In the case of errors not involved in program execution (e.g. time error, SE error, diagnostics interrupt, insert/remove module interrupt, rack/station failure, processing abort)	 26 [2)] 25 [2)] 25 [2)] 28	 No 2 to 26 24 to 26 No
Background processing OB 90	Minimum cycle time duration not yet reached	29 [1)]	No
Startup routine OB 100, OB 101, OB 102	At programmable controller startup	27	No
Synchronous errors OB 121, OB 122	In the case of errors connected with program execution (e.g. I/O access error)	Priority of the OBs causing the errors	

[1)] see text [2)] at startup: 28

priority classes; each CPU has a different low/high range; a specific CPU occupies a section of this overview.

Organization block OB 90 (background processing) executes alternately with organization block OB 1, and can, like OB 1, be interrupted by all other interrupts and errors.

The startup routine may be in organization block OB 100 (warm restart), OB 101 (hot restart) or OB 102 (cold restart), and has priority 27. Asynchronous errors occurring in the startup routine have priority class 28. Diagnostic interrupts are regarded as asynchronous errors.

You determine which of the available priority classes you want to use when you parameterize the CPU. Unused priority classes (organization blocks) must be assigned priority 0.

The relevant organization blocks must be programmed for all priority classes used; otherwise the CPU will invoke OB 85 ("Program Processing Error") or go to STOP.

For each priority class selected, temporary local data (L stack) must be available in sufficient volumes (see Chapter 18.1.5 "Temporary Local Data" for more details).

3.1.3 Specifications for Program Processing

The CPU's operating system normally uses default parameters. You can change these defaults when you parameterize the CPU (in the Hardware Configuration) to customize the system to suit your particular requirements. You can change the parameters at any time.

Every CPU has its own specific number of parameter settings. The following list provides an overview of all STEP 7 parameters and their most important settings.

▷ General
Name of CPU, plant identifier, location identifier, settings for the MPI interface (if it is not an interface combined with DP), comment

▷ Startup
Specifies the type of startup (cold restart, warm restart, hot restart); monitoring of Ready signals or module parameterization; maximum amount of time which may elapse before a hot restart

▷ Cycle/Clock Memory
Enable/disable cyclic updating of the process image; specification of the cycle monitoring time and minimum cycle time; amount of cycle time, in percent, for communication; number of the clock memory byte; size of the process images

▷ Retentive Memory
Number of retentive memory bytes, timers and counters; specification of retentive areas for data blocks

▷ Memory
Max. number of temporary local data in the priority classes (organization blocks); max size of the L stack and number of communications jobs

▷ Interrupts
Specification of the priority for process interrupts, time-delay interrupts, asynchronous errors and DPV1 interrupts; assignment of partial process images with process interrupts and time-delay interrupts

▷ Time-of-Day Interrupts
Specification of the priority and assignment of partial process images; specification of the start time and periodicity

▷ Cyclic Interrupts
Specification of the priority, the time cycle and the phase offset; assignment of partial process images

▷ Synchronous cycle interrupts
Specification of the priority; assignment of the DP master system and the partial process images

▷ Diagnostics/Clock
Settings for system diagnostics; type and interval for time synchronization, correction factor

▷ Protection
Specification of the protection level; defining a password; setting of process and test modes

▷ Multicomputing
Specification of the CPU number

▷ Integrated Functions
Activation and parameterization of the integrated functions

▷ Communication
Definition of connection resources

▷ Web
Activation of Web server, language selection

On startup, the CPU puts the user parameters into effect in place of the defaults, and they remain in force until changed.

Program length, memory requirements

The memory requirements of a compiled block are listed in the block properties. If you mark the block in the SIMATIC Manager and select the tab "General - Part 2" with EDIT → OBJECT PROPERTIES, you will be provided with the load and work memory requirements for this block.

The length of the user program is listed in the properties of the offline container *Blocks* (mark *Blocks* and EDIT → OBJECT PROPERTIES). On the tab "Blocks" you can find the data "Size in work memory" and "Size in load memory".

Note that the configuration data (system data blocks) are missing for the size data for the load memory. With the container open, the SIMATIC Manager shows you blocks in the detail view (display as table) and, with the object *System data* marked, the memory requirements in the status line (at bottom right in window).

With the programming device switched online, the SIMATIC Manager shows you the current assignment of the CPU memory under PLC → DIAGNOSTIC/SETTING → MODULE INFORMATION, tab "Memory".

Checksum

The program editor generates a checksum for all blocks of the user program, and stores it in the object properties of the container *Blocks*. Identical programs have the same checksum, each change in the program also changes the checksum. A checksum is also generated using the system data. You can view the checksums in the SIMATIC Manager with the marked container *Blocks* and EDIT → OBJECT PROPERTIES.

The checksum of the user program is generated from the program code and the default and initial values of the data blocks. Writing of data addresses in the work memory (current values) does not change the checksum. The checksum is only changed when the data blocks are loaded back into the offline data management, during which the current values become the initial values. This also applies to data blocks generated using system functions.

If a data block generated using system functions is written or deleted, the checksum is not changed. The checksum is adapted if a programmed (loaded) data block is deleted, or if the initial values in the load memory are changed by the system function SFC 84 WRIT_DBL.

3.2 Blocks

You can subdivide your program into as many sections as you want to in order to make it easier to read and understand. The STEP 7 programming languages support this by providing the necessary functions. Each program section should be self-contained, and should have a technological or functional basis. These program sections are referred to as "Blocks". A block is a section of a user program which is defined by its function, structure or intended purpose.

3.2.1 Block Types

STEP 7 provides different types of blocks for different tasks:

▷ User blocks
Blocks containing user program and user data

▷ System blocks
Blocks containing system program and system data

▷ Standard blocks
Turnkey, off-the-shelf blocks, such as drivers for FMs and CPs

User blocks

In extensive and complex programs, "structuring" (subdividing) of the program into blocks is recommended, and in part necessary. You may choose among different types of blocks, depending on your application:

Organization blocks (OBs)

These blocks serve as the interface between operating system and user program. The CPU's operating system calls the organization blocks when specific events occur, for example in the event of a hardware or time-of-day interrupt. The main program is in organization block OB 1. The other organization blocks have permanently assigned numbers based on the events they are called to handle.

Function blocks (FBs)

These blocks are parts of the program whose calls can be programmed via block parameters. They have a variable memory which is located in a data block. This data block is permanently allocated to the function block, or, to be more precise to the function block *call*. It is even possible to assign a different data block (with the same data structure but containing different values) to each function block call. Such a permanently assigned data block is called an instance data block, and the combination of function block call and instance data block is referred to as a call instance, or "instance" for short. Function blocks can also save their variables in the instance data block of the calling function block; this is referred to as a "local instance".

Functions (FCs)

Functions are used to program frequently recurring or complex automation functions. They can be parameterized, and return a value (called the function value) to the calling block. The function value is optional, in addition to the function value, functions may also have other output parameters. Functions do not store information, and have no assigned data block.

Data blocks (DBs)

These blocks contain your program's data. By programming the data blocks, you determine in which form the data will be saved (in which block, in what order, and in what data type). There are two ways of using data blocks: as global data blocks and as instance data blocks. A global data block is, so to speak, a "free" data block in the user program, and is not allocated to a code block. An instance data block, however, is assigned to a function block, and stores part of that function block's local data.

The number of blocks per block type and the length of the blocks is CPU-dependent. The number of organization blocks, and their block numbers, are fixed; they are assigned by the CPU's operating system. Within the specified range, you can assign the block numbers of the other block types yourself. You also have the option of assigning every block a name (a symbol) via the symbol table, then referencing each block by the name assigned to it.

System blocks

System blocks are components of the operating system. They can contain programs (system functions (SFCs) or system function blocks (SFBs) or data (system data blocks (SDBs)). System blocks make a number of important system functions accessible to you, such as manipulating the internal CPU clock, or various communications functions.

You can call SFCs and SFBs, but you cannot modify them, nor can you program them yourself. The blocks themselves do not reserve space in user memory; only the block calls and the instance data blocks of the SFBs are in user memory.

SDBs contain information on such things as the configuration of the automation system or the parameterization of the modules. STEP 7 itself generates and manages these blocks. You, however, determine their contents, for instance when you configure the stations. As a rule, SDBs are located in load memory and only read from your program using special system blocks, e.g. when parameterizing modules.

Double-click on the object *System blocks* in the container *Blocks* to display a list of current system data blocks which have been generated by the Hardware Configuration (in the offline container) or are present on the CPU (in the online container). Table 3.2 shows an overview of the number scheme for system data blocks.

Standard blocks

In addition to the functions and function blocks you create yourself, off-the-shelf blocks (called "standard blocks") are also available. They can either be obtained on a storage medium or are contained in libraries delivered as part of the STEP 7 package (for example IEC functions, or functions for the S5/S7 converter).

Chapter 33 "Block Libraries" contains an overview of the system and standard blocks supplied in the *Standard Library*.

Table 3.2 Number ranges for system data blocks

SDB No.	Meaning, content
0	CPU parameters which control the response of the operating system and the CPU-internal default settings; overwrite the SDB 2 on transmission to the CPU
1	Module addressing for central I/O (reference configuration), e.g. assignment of logical and geographical addresses, address area for modules, etc.
2	CPU parameters (default settings in the CPU's operating system; become effective following overall reset if a configuration has not yet been transmitted)
3 to 7	Various CPU and module parameters, e.g. for saving the consistency of the transmitted configuration data
20 to 89	Module addressing for distributed I/O (reference configuration), e.g. assignment of logical and geographical addresses, address area for modules, etc.
90 to 99	Configuration data for fault-tolerant and fail-safe systems
100 to 149	Parameters for central and distributed modules assigned to the CPU
150 to 152	Parameters for interface modules
153 to 189	Parameters for distributed modules
200 to 998	Parameters for configuration of communication (e.g. global data communication, symbol-based messages, configuration of connections)
999	Configuration data for routing of connections
1000 and larger	Parameters for distributed I/O, parameters for CP and FM modules, parameters for fault-tolerant/fail-safe and TD/OP systems

3.2.2 Block Structure

Essentially, code blocks consist of three parts (Figure 3.2):

▷ The block header,
which contains the block properties, such as the block name

▷ The declaration section
in which the block-local variables are declared, that is, defined

▷ The program section
which contains the program and program commentary

A data block is similarly structured:

▷ The block header contains the block properties

▷ The declaration part contains the specification of the local block variables, here: data operands with data type

▷ The initialization section, in which initial values can be specified for individual data addresses

In incremental programming (direct program input without source text file), the declaration section and the initialization section are combined. You define the data addresses and their data types in the "declaration view", and you can initialize each data address individually in the "data view".

3.2.3 Block Properties

The block properties, or attributes, are contained in the block header. You can view and modify the block attributes with the menu command EDIT → OBJECT PROPERTIES in the SIMATIC Manager with the block marked or FILE → PROPERTIES from the Editor (Figure 3.3).

Tab "General - Part 1"

This tab contains under *Name* the absolute address of the block with block type and number as well as a symbolic name and symbol comment from the symbol table.

With function blocks, it is indicated, in addition to the name, whether the block has multi-instance capability. If the property *"Multiple Instance Capability"* is activated, which is normally the case, you can call the block as a local

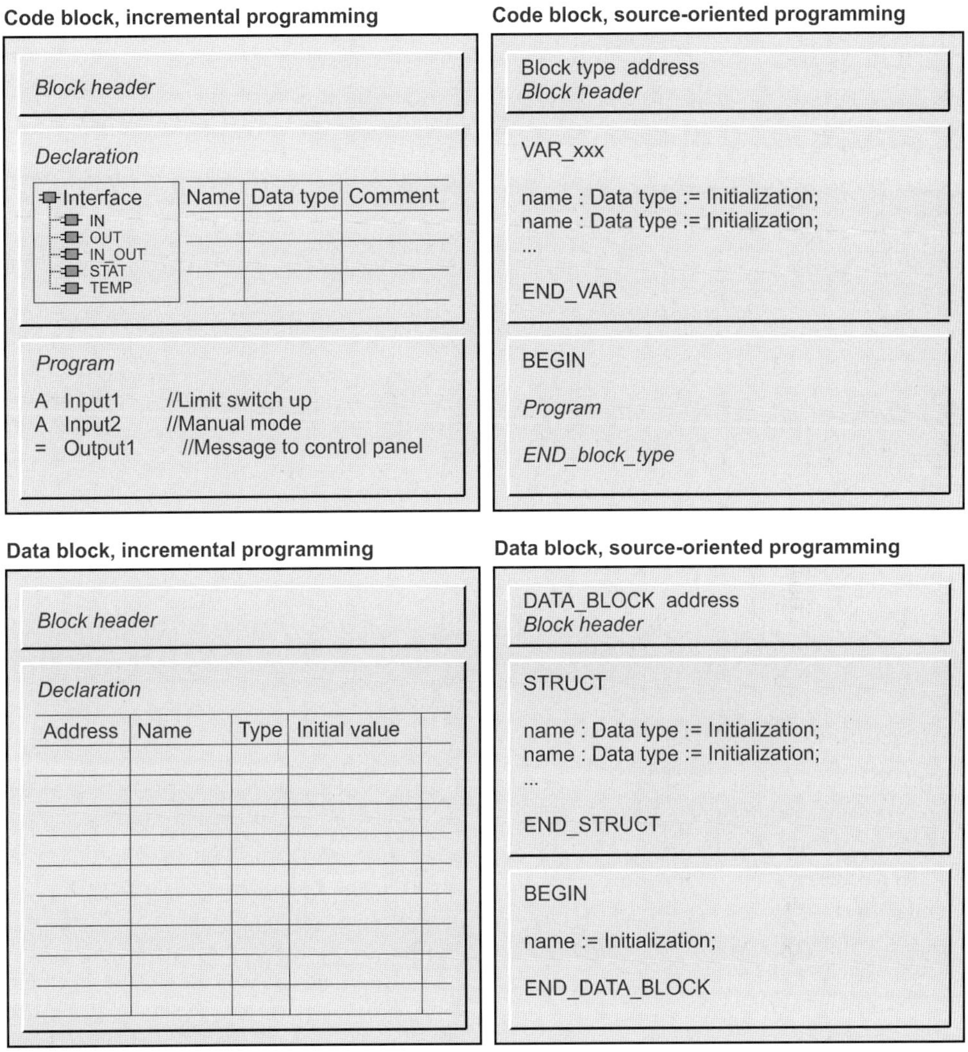

Code block, incremental programming

Block header

Declaration

Interface
- IN
- OUT
- IN_OUT
- STAT
- TEMP

Name	Data type	Comment

Program

```
A  Input1     //Limit switch up
A  Input2     //Manual mode
=  Output1    //Message to control panel
```

Code block, source-oriented programming

Block type address
Block header

```
VAR_xxx

name : Data type := Initialization;
name : Data type := Initialization;
...

END_VAR
```

BEGIN

Program

END_block_type

Data block, incremental programming

Block header

Declaration

Address	Name	Type	Initial value	

Data block, source-oriented programming

DATA_BLOCK address
Block header

```
STRUCT

name : Data type := Initialization;
name : Data type := Initialization;
...

END_STRUCT

BEGIN

name := Initialization;

END_DATA_BLOCK
```

Figure 3.2 Structure of a Block

instance and you can also call further function blocks with multi-instance capability within it as local instances. You can deselect the property *Multiple Instance Capability* when creating the function block; with a source-oriented program input, the keyword for deselection is CODE_VERSION1. The advantage of a function block without multi-instance capability is the unlimited use of instance data in the case of indirect addressing (only significant in STL programming).

The tab also displays the programming language of the block which you set when creating the block, as well as the storage locations of the block and the project.

The program editor stores the creation and modification dates of the block in two time stamps: these are the block parameters and the static local data for the program code and for the interface. Note that the modification date of the interface must be the same or smaller (older) than the modification date of the pro-

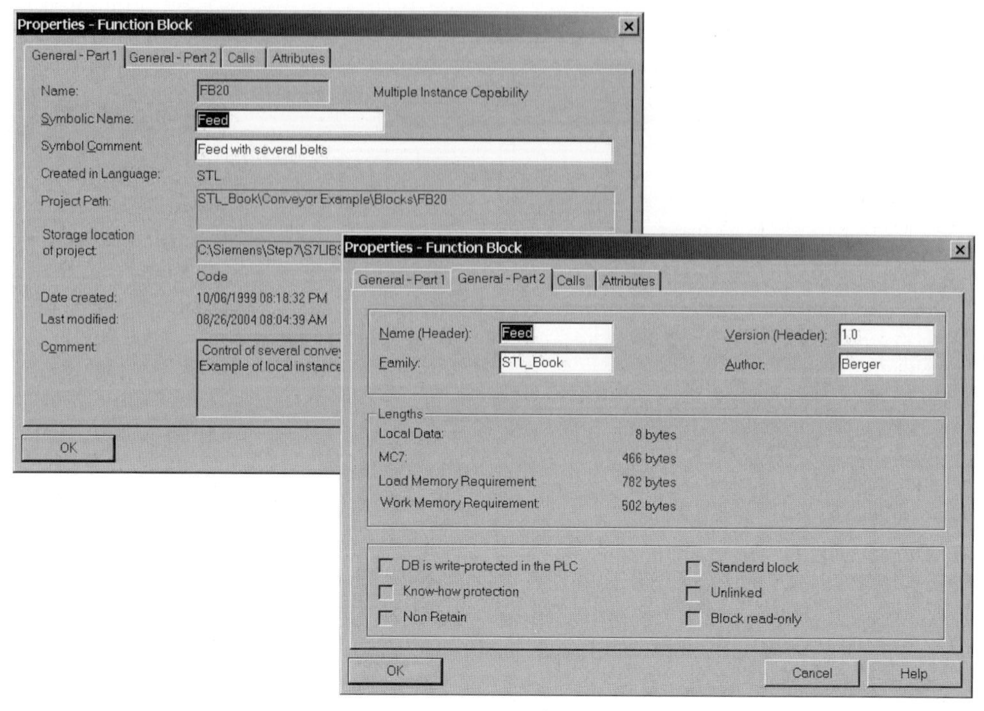

Figure 3.3 Block Properties

gram code in the calling block. If this is not the case, the program editor signals a "time stamp conflict" during output of the calling block.

The displayed comment consists of the block title and the block comment which you have entered when programming the block.

Tab "General - Part 2"

The *Name (Header)* displayed on this tab is the block name; this is not the same as the symbolic address. Different blocks can have the same name. With *Family*, you assign a common feature to a group of blocks. The block name and family are displayed when you insert blocks and when you select a block in the dialog window of the program elements catalog. *Author* is the creator of the block. Name, Family and Author can consist of up to 8 characters. Alphanumeric characters and the underscore are permissible. *Version* is entered twice, with two digits from 0 to 15.

The length data indicate the memory allocation for the block in bytes:

- ▷ Local data: Allocation in the local data stack (temporary local data)
- ▷ MC7: size of the block (code only)
- ▷ Load memory requirement
- ▷ Work memory requirement

A block occupies more space in the load memory since the data not relevant to the execution are additionally stored here.

The *Know-how protection* attribute is used for block protection. If a block is KNOW HOW-protected, the program in that block can not be viewed, printed out or modified. The Editor shows only the block header and the declaration table with the block parameters. You assign this property with the source-oriented input of the block with the keyword KNOW_HOW_PROTECT. When you do this to a block, no one can view the compiled version of that block, not even you (make sure you keep the source file in a safe place!).

DB is write-protected in the PLC is a property only for data blocks. It means that you can only read from this data block by means of a program. Overwriting of the data is prevented and an error message is generated. The write protection applies to the data relevant to execution (actual values) in the work memory; the data in the load memory (initial values) can be overwritten even if the data block is provided with write protection. The write protection must not be confused with the block protection: A data block with block protection can be read and written by the program; however, its data can no longer be viewed using a programming or monitoring device. The property *DB is write-protected in the PLC* is switched off as standard and can be changed at any time with the program editor. The keyword in source-oriented program entry for switching on write protection is READ_ONLY.

The block header of any *standard block* which comes from Siemens contains the "*Standard block*" attribute.

Data blocks can be assigned the property *Unlinked*. Data blocks of this type are only in the load memory, they are execution-relevant. Since their data are not in the work memory, direct access is not possible. Data in the load memory can be read using system functions and – if the load memory is a micro memory card – also written. Data blocks with the property *Unlinked* are suitable for accommodating data which are only seldom accessed, e.g. recipes or archives. This property is switched off in the default setting, but can be changed at any time using the program editor. The keyword for source-oriented program input for switching on this property is UNLINKED.

The property *Non Retain* means "non-retentive" and is assigned to data blocks for appropriately designed CPUs. If *Non Retain* is switched on, the data block accepts the initial values from the load memory into the work memory when the power is switched off/on and with a RUN-STOP transition (response as with a cold restart). If *Non Retain* is switched off, i.e. the corresponding data block is retentive, it retains its current values when the power is switched off/on and with a RUN-STOP transition (response as with a warm start/restart). This property is switched off in the default setting, but can be changed at any time using the program editor. The keyword for source-oriented program input for switching on this property is NON_RETAIN.

Blocks which have been saved in the program editor, e.g. for reference purposes, using the menu command FILE → STORE READ-ONLY are assigned the block property *Block read-only*. These can be all code blocks, data blocks and user-defined data types. This property can only be set using the program editor, a keyword for source-oriented programming is not available for this.

Tab "Calls"

This tab shows a list of all blocks called in this block with the time stamps for the code and the interface. With instance data blocks, the basic function block and the local instances (function blocks) called in this instance are present here, in each case with the time stamps for code and interface.

Tab "Attributes"

Blocks may have system attributes. System attributes control and coordinate functions between applications, for example in the SIMATIC PCS7 control system.

3.2.4 Block Interface

The declarations section contains the interface of the block to the rest of the program. It consists of the block parameters (input, output and in/out parameters) and also – in the case of function blocks – the static local data. The temporary local data, which really do not belong to the block interface, are also handled at this point. The block interface is defined in the interface window when programming the block and is initialized with variables when the block is called (see Chapter 19 "Block Parameters").

The Program Editor checks that the block parameter initialization in the called block agrees with the interface of the called block. The Editor also uses the time stamp for this: the interface of the called block must be older than the code in the calling block, that is, the last interface modification must have been made prior to integration of the block. The Program Editor updates the interface time stamp when the number of parameters changes or when a data type or a default value changes.

Time stamp conflict

A time stamp conflict occurs when the interface of the called block has a later time stamp than the code of the calling block. You will notice a time stamp conflict if you open an already compiled block again. The Program Editor then indicates the incorrect block call in red. A time stamp conflict can be caused, for example, if you modify the interfaces of blocks that are already called in other blocks, or if you combine blocks from different programs into a new program, or if you re-compile a section of the overall program with a source file.

However, the interface conflict generally described as a "time stamp conflict" can also have other causes. It also occurs if a called or referenced block is younger than the calling block. Examples of the occurrence of time stamp conflicts include the following:

- ▷ The interface of a called block is younger than the code of the calling block.
- ▷ The interface initialization does not agree with the block interface.
- ▷ A function block is younger than its instance data block (the instance data block is generated from the interface description of the function block and should therefore be younger than or the same age as the function block).
- ▷ A local instance is younger than the calling instance (affects function blocks).
- ▷ A user data type UDT is younger than the block whose variables are declared with the UDT; this can be any block including a data block or another UDT.

Correcting invalid block calls

There are several possibilities for using the program editor to find and correct invalid block calls. How you can check the block consistency in a complete program is described in the next section ("Checking block consistency").

You can check block calls which have become invalid with the block open (with the cursor at the invalid block call) using the menu command EDIT → BLOCK CALL → UPDATE. Block calls can become invalid following inserting, deleting or shifting block parameters, as well as when changing names and types.

With EDIT → BLOCK CALL → CHANGE TO MULTI-INSTANCE CALL and EDIT → BLOCK CALL → CHANGE TO FB/DB CALL you transfer calls of function blocks into local instance calls or into calls with data blocks. Following modification of the block calls, you must regenerate the associated instance data blocks.

A further possibility is provided by the menu command FILE → CHECK AND UPDATE ACCESSES. The invalid block calls are updated in an opened block, or made available for modification.

Checking block consistency

The Program Editor only indicates a time stamp conflict when you open a block containing a time stamp conflict. If you want to check an entire program, you can use the function "Check block consistency" in the SIMATIC Manager. This purges a majority of interface conflicts and directs you to the program locations that require editing.

To carry out a consistency check, select the *Blocks* container in the SIMATIC Manager and then select EDIT → CHECK BLOCK CONSISTENCY. If a call tree is not displayed in the window, because e.g. the program was compiled using an earlier version of STEP 7, select PROGRAM → COMPILE in this window.

Please note that the instance data blocks and the data blocks generated from the UDT are again assigned the initial values in the compiled program following checking of the block consistency.

You can see the progress and result of the consistency check in the output window (VIEW → ERRORS AND WARNINGS). The consistency check cannot be used on programs in libraries.

The dependencies in the case of called or referenced blocks are displayed in the form of a tree diagram (Figure 3.4). You can choose between two representations:

The view *Call tree: References* displays the dependencies in a similar way to the program structure: on the left are the calling blocks, fur-

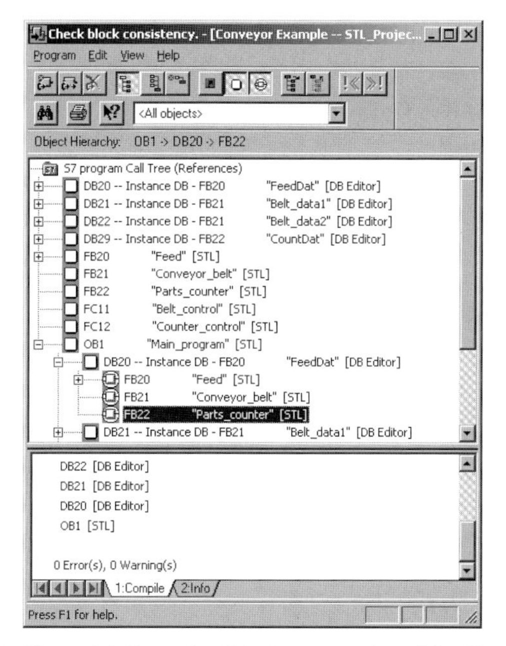

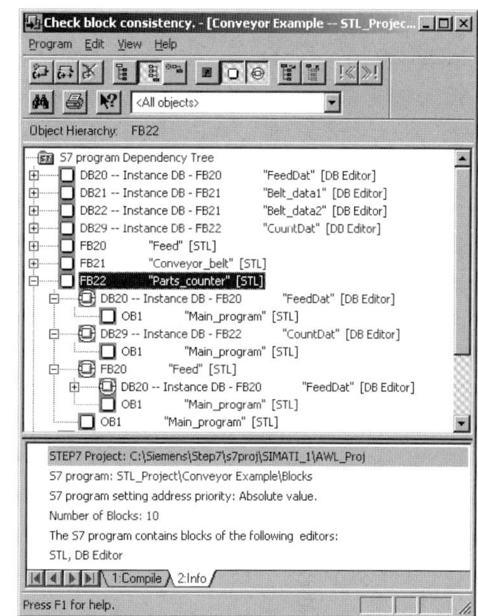

Figure 3.4 Example of the Representation of the Check Block Consistency Dependencies

ther to the right are the blocks called in the blocks on the left. Example: instance DB 20/FB20 is called in OB1 and local instances FB 21 and FB 22 are called in FB 20.

The view *Dependency tree* displays the dependencies starting from all called or referenced blocks. They are located in the left-hand column and to the right of this are listed the calling blocks. Example: FB 22 stores its data in instance DB 20/FB 20 that is called in OB 1. It also has its own DB 29 and it is called as a local instance in FB 20.

The determined information is displayed in compact form by symbols. An exclamation mark, for example, indicates that the object caused a time stamp conflict. A white cross on a red background indicates that the associated block has to be recompiled.

If you select a block in the tree diagram or in the output window, you can edit it with EDIT → OPEN OBJECT, e.g. correct an incorrect call.

The consistency check is not supported by the SCL option package in Version 5.1 or earlier (the blocks are not compiled by PROGRAM → COMPILE or PROGRAM → COMPILE ALL). Blocks programmed with SCL must therefore be compiled "manually" in the event of an error.

3.3 Addressing Variables

When addressing variables, you may choose between absolute addressing and symbolic addressing.

- ▷ Absolute addressing uses numerical addresses beginning with zero for each address area.
- ▷ The symbolic addressing uses alphanumerical names that you specify in the symbol table for the global operands or in the declaration part for local block operands.

An extension of absolute addressing is indirect addressing, in which the addresses of the memory locations are not computed until runtime.

3.3.1 Absolute Addressing of Variables

Variables of elementary data type can be referenced by absolute addresses.

The absolute address of an input or output is computed from the module start address which you set or had set in the configuration table and the type of signal connection on the module. A distinction is made between binary signals and analog signals.

Binary signals

A binary signal contains one bit of information. Examples of binary signals are the input signals from limit switches, momentary-contact switches and the like which lead to digital input modules, and output signals which control lamps, contactors, and the like via digital output modules.

Analog signals

An analog signal contains 16 bits of information. An analog signal corresponds to a "channel", which is mapped in the controller as a word (2 bytes) (see below). Analog input signals (such as voltages from resistance thermometers) are carried to analog input modules, digitized, and made available to the controller as 16 information bits. Conversely, 16 bits of information can control an indicator via an analog output module, where the information is converted into an analog value (such as a current).

The information width of a signal also corresponds to the information width of the variable in which the signal is stored and processed. The information width and the interpretation of the information (for instance the positional weight), taken together, produce the *data type* of the variable. Binary signals are stored in variables of data type BOOL, analog signals in variables of data type INT.

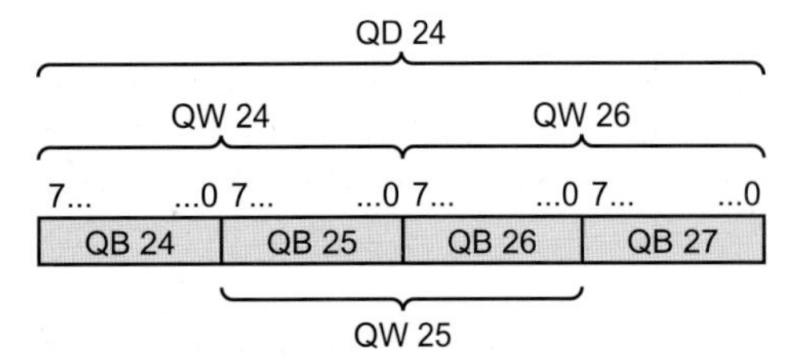

Figure 3.5
Bit and Byte Contents in Words and Doublewords

The only determining factor for the addressing of variables is the information width. In STEP 7, there are four widths which can be accessed with absolute addressing:

- ▷ 1 bit Data type BOOL
- ▷ 8 bits Data type BYTE or another data type with 8 bits
- ▷ 16 bits Data type WORD or another data type with 16 bits
- ▷ 32 bits Data type DWORD or another data type with 32 bits

Variables of data type BOOL are referenced via an address identifier, a byte number, and - separated by a decimal point - a bit number. Numbering of the bytes begins at zero for each address area. The upper limit is CPU-specific. The bits are numbered from 0 to 7.
Examples:

I 1.0 Input bit no. 0 in byte no. 1
Q 16.4 Output bit no. 4 in byte no. 16

Variables of data type BYTE have as absolute address the address identifier and the number of the byte containing the variable. The address identifier is supplemented by a B.
Examples:

IB 2 Input byte no. 2
QB 18 Output byte no. 18

Variables of data type WORD consist of two bytes (a word). They have as absolute address the address identifier and the number of the low-order byte of the word containing the variable. The address identifier is supplemented by a W.
Examples:

IW 4 Input word no. 4, contains bytes 4 and 5
QW 20 Output word no. 20, contains bytes 20 and 21

Variables of data type DWORD consist of four bytes (a doubleword). They have as absolute address the address identifier and the number of the low-order byte of the word containing the variable. The address identifier is supplemented by a D.

Examples:

ID 8 Input doubleword no. 8, contains bytes 8, 9, 10 and 11

QD 24 Output doubleword no. 24, contains bytes no. 24, 25, 26 and 27

Addresses for the data area include the data block. Examples:

DB 10.DBX 2.0
Data bit 2.0 in data block DB 10

DB 11.DBB 14
Data byte 14 in data block DB 11

DB 20.DBW 20
Data word 20 in data block DB 20

DB 22.DBD 10
Data doubleword 10 in data block DB 22

Additional information on addressing the data area can be found in Chapter 18.2.2 "Accessing Data Addresses".

3.3.2 Indirect Addressing

Indirect addressing allows you to wait until runtime to compute an address in the data area. STL and SCL use different methods for indirect addressing. In STL, a distinction is made between memory-indirect addressing and register-indirect addressing:

▷ Memory-indirect addressing,
IW [MD 200]
The address is in the memory doubleword

▷ Register-indirect area-internal addressing,
IW [AR1, P#2.0]
The address is in address register AR1, and is incremented by the offset P#2.0 when the statement is executed

▷ Register-indirect area-crossing addressing,
W [AR1, P#0.0]
The address area and the address itself are in address register AR 1

Doublewords from the address areas for data (DBD and DID), bit memory (MD) and temporary local data (LD) are available for saving addresses when using memory-indirect addressing. You can implement register-indirect addressing with two address registers (AR 1 and AR 2).

Indirect addressing is described in detail in the Chapter 25 "Indirect Addressing".

For SCL, the address areas consist of a field whose elements are then accessed indirectly and individually. For example, MW[*index*] addresses a memory word whose address is located in the variable *index*. The variable *index* can be modified at runtime. See Chapter 27.2.3 "Indirect Addressing in SCL" for more detailed information.

3.3.3 Symbolic Addressing of Variables

Symbolic addressing uses a name (called a symbol) in place of an absolute address. You yourself choose this name. Such a name must begin with a letter and may comprise up to 24 characters. A keyword is not permissible as a symbol in STL; to use a keyword as a symbol in SCL, insert the hash character (#) before the name.

A distinction is made between upper and lower case for input. For output, the editor uses the case and notation used at declaration of the symbol.

The name, or symbol, must be allocated to an absolute address. A distinction is made between global symbols and symbols that are local to a block.

Global symbols

You may assign names in the symbol table to the following objects:

▷ Data blocks and code blocks

▷ Inputs, outputs, peripheral inputs and peripheral outputs

▷ Memory bits, timers and counters

▷ User data types

▷ Variable tables

A global symbol may also include spaces, special characters and country-specific characters such as the umlaut. Exceptions to this rule are the characters 00_{hex} and FF_{hex}. When using symbols containing special characters, you must put the symbols in quotation marks in the program. In compiled blocks, the STL Editor always shows global symbols in quotation marks.

You can use global symbols throughout the program; each such symbol must be unique within a program.

Editing, importing and exporting of global symbols is described in Chapter 2.5.2 "Symbol Table".

Block-local symbols

The names for the local data are specified in the declaration section of the relevant block. These names may contain only letters, digits and the underline character.

Local symbols are valid only within a block. The same symbol (the same variable name) may be used in a different context in another block. The Editor shows local symbols with a leading "#". When the Editor cannot distinguish a local symbol from an address, you must precede the symbol with a "#" character during input.

Local symbols are available only in the programming device database (in the offline container *Blocks).* If this information is missing on decompilation, the Editor inserts a substitute symbol.

Using symbol names

If you use symbolic names while programming with the incremental Editor, they must have already been allocated to absolute addresses. You also have the option of entering new symbolic names in the symbol table while programming with the incremental Editor and can use these symbols for further programming.

If you are using a source text file to input your program, the complete assignment of symbolic names to absolute addresses need not be available until compilation starts.

In the case of arrays, the individual components are accessed via the array name and a subscript, for example MSERIES[1] for the first component. In STL, the index is a constant INT value, and in SCL it can also be an INT variable or an INT expression.

In structures, each subidentifier is separated from the preceding subidentifier by a decimal point, for instance FRAME.HEADER.CNUM. Components of user data types are addressed exactly like structures. For further details see Chapter 24 "Data Types".

Data addresses

Symbolic addressing of data uses complete addressing including the data block. Example: the data block with the symbolic address MVALUES contains the variables MVALUE1, MVALUE2 and MTIME.

These variables can be addressed as follows:

```
"MVALUES".MVALUE1
"MVALUES".MVALUE2
"MVALUES".MTIME
```

Please refer to Chapters 18.2.2 "Accessing Data Addresses" (STL) and 27.2.2 "Symbolic Addressing" (SCL) for further information on assigning data addresses.

3.4 Programming Code Blocks with STL

3.4.1 Structure of an STL Statement

An STL program consists of a sequence of individual statements. A statement is the smallest autonomous unit of a user program. It represents a work specification for the CPU. Figure 3.6 shows the structure of an STL statement.

An STL statement consists of the following

- ▷ A label (optional) comprising up to 4 characters and ending with a colon (see Chapter 16 "Jump Functions")
- ▷ An operation that describes what the CPU should do (e.g. load, scan and link according to AND operation, compare, etc.)
- ▷ An address providing the information needed to execute the operation (for instance an absolute address such as IW 12, the symbolic address of a variable such as ANALOGVALUE_1 or of a constant such as W#16#F001, and so on). Some operations require no address specification.
- ▷ A comment (optional), which must begin with two slashes and may extend up to the end of the line).

When inputting to a source file, you must terminate each statement (before beginning a comment, if any) with a semicolon. An STL line may contain no more than 200 characters, a comment no more than 160 characters.

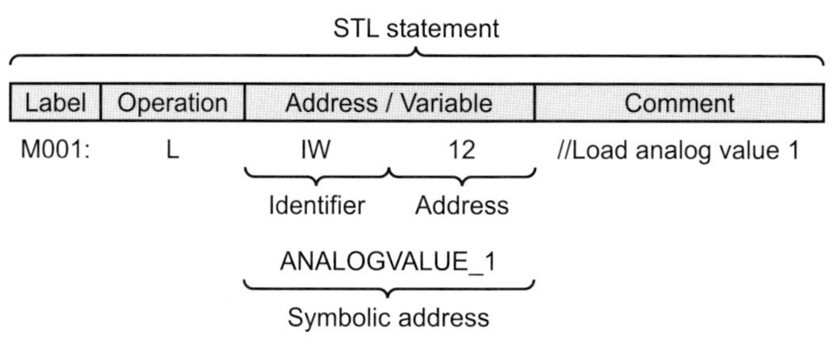

Figure 3.6 Structure of an STL statement

3.4.2 Programming STL Code Blocks Incrementally

Chapter 2.5 "Creating the S7 Program" gives an introduction to program creation and use of the program editor.

Open block

You begin block programming by opening a block. You open an existing block either by double-clicking on the block in the SIMATIC Manager's project window or by selecting FILE → OPEN in the Editor.

If you open a compiled block in the container *Blocks* e.g. by means of a double click, it is opened for incremental programming. This is the case both for offline and online programming.

If the block does not yet exist, you can generate it in the following ways:

- In the SIMATIC Manager by selecting the *Blocks* object in the left half of the project window and generating a new block with INSERT → S7 BLOCK → ... You see the Properties window of the block. Select the block number under *Name* and "STL" language on the "General - Part1" tab. You can also enter the remaining block properties later.
- In the Program Editor: with FILE → NEW you obtain a dialog box in which you can enter the desired block under *Object name*. After closing the dialog box you can program the contents of the block. The Program Editor uses the language set on the "Block" tab with OPTIONS → CUSTOMIZE.

You can enter the information for the block header when you generate the block or you can enter the block attributes later in the Editor by opening the block and selecting the menu command FILE → PROPERTIES.

Block window

The Program Editor shows the variables declaration table (block parameters and local data) of an opened code block, as well as the program window (code and comments). In addition, the program elements can be displayed in the overview window.

Variable declaration table

The variable declaration table is in the window above the program window. If it is not visible, position the mouse pointer to the upper line of demarcation for the program window, click on the left mouse button when the mouse pointer changes its form, and pull down. The overview of the types of variable is then displayed on the left, and to the right of this is the variable declaration table, which is where you define the block-local variables (see Table 3.3).

To declare a variable, select its type in the left area, and fill in the right area of the table. Not every type of variable can be programmed in every code block. The corresponding table remains empty if you do not use a type of variable.

The declaration for a variable consists of the name, the data type, a default value, if any, and a variable comment (optional). Not all variables can be assigned a default value (for instance, it is not possible for temporary local data). The default values for functions and function blocks

Table 3.3 Variable Types in the Declaration Section

Variable Type	Declaration	Possible in Block Type		
Input parameters	IN	-	FC	FB
Output parameters	OUT	-	FC	FB
In-out parameters	IN_OUT	-	FC	FB
Static local data	STAT	-	-	FB
Temporary local data	TEMP	OB	FC	FB
Return value	RETURN	-	FC	-

are described in detail in Chapter 19 "Block Parameters".

The order of the declarations in code blocks is fixed (as shown in the table above), while the order within a variable type is arbitrary. You can save room in memory by bundling binary variables into blocks of 8 or 16 and BYTE variables into pairs. The Editor stores a (new) BOOL or BYTE variable at a byte boundary and a variable of another data type at a word boundary (beginning at a byte with an even address).

Program window

In the program window, you will see - depending on the Editor's default settings - the fields for the block title and the block comment and, if it is the first network, the fields for the network title, the network comment, and the field for the program entry. In the program section of a code block, you control the display of comments and symbols with the menu commands VIEW → DISPLAY WITH. You can change the size of the display with VIEW → ZOOM IN, VIEW → ZOOM OUT and VIEW → ZOOM FACTOR.

3.4.3 Overview Window

The overview window contains the program element catalog and the call structure.

If the overview window is not visible, fetch it onto the screen with VIEW → OVERVIEWS or with INSERT → PROGRAM ELEMENTS (Figure 3.7).

The other views are present in a separate window which you can dock on the edge of the editor window, and release again (double click in each case on the title bar of the overview window).

Program elements catalog

The program element catalog supports you in programming in the languages LAD and FBD by offering the available graphic elements. In the STL view, it shows the blocks that are already in the offline *Blocks* container, as well as the multiple instances that are already programmed and the libraries that are available. By right-clicking on a block or a block type, you can choose whether the blocks are to be sorted by type and number or by block family.

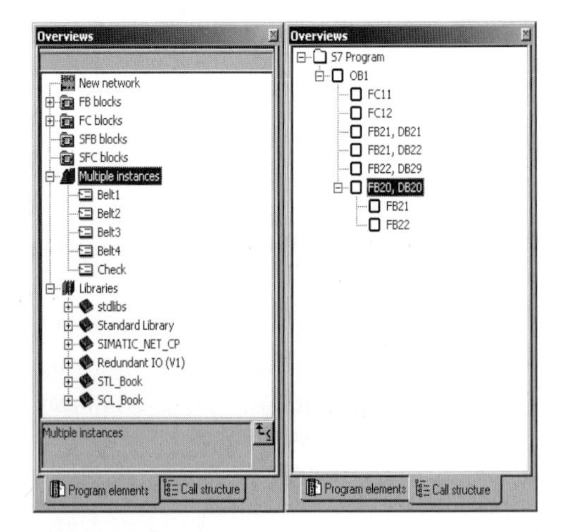

Figure 3.7
Program elements catalog and call structure

Call structure

The call structure indicates the block hierarchy in the current user program. You see the call environment of the currently open block and the use blocks.

3.4.4 Programming Networks

You can subdivide an STL program into networks. The Editor numbers the networks automatically, beginning with 1. Each block can accommodate up to 999 networks. You may give each network a network title and a network comment. During editing, you can select each network and each program line directly with the menu command EDIT → GO TO → NETWORK/ROW.

To enter the program code, click once below the window for the network comment, or, if you have set "Display with Comments", click once below the shaded area for network comments. You will see a framed empty window. You can begin entering your program anywhere within this window. Refer to Chapter 3.4.1 "Structure of an STL Statement" to review the structure of an STL statement. Separate the OP code (operator) and the address (operand) from one another by one or more spaces or tabs. Following the address, you can enter two slashes and a statement comment. Terminate a statement by pressing RETURN. You can also enter a line comment by beginning a line with two slashes.

You program a new network with INSERT → NETWORK. The Editor then inserts an empty network behind the currently selected network.

If you want to use symbolic names when entering the program incrementally, these names must already have absolute addresses assigned to them. You can call up a selection of the symbols entered in the Symbol Table with INSERT → SYMBOL and then transfer the desired symbol per mouse-click.

While entering the program with the incremental editor, you can also add symbolic names to the Symbol Table or correct them. You can call up the complete Symbol Table with OPTIONS → SYMBOL TABLE, and you can call up one line with the currently marked symbol from the table with EDIT → SYMBOLS. After editing the symbol, you continue entering the program with the new or modified symbol.

You need not terminate a block with a special statement, simply stop making entries. However, you can program a last (empty) network with the title "Block End", providing an easily seen visual end of the block (an advantage, particularly in the case of exceptionally long blocks).

When the program editor opens a compiled block, it "decompiles" it back into STL. To do this, the Editor uses the program sections in the programming device database which are not relevant to the program's execution in order, for example, to represent symbols, comments and jump labels. If information needed from the offline programming device database is missing when the program editor decompiles the program, it uses substitute symbols.

You can create new blocks or open and edit existing ones in the program editor without having to return to the SIMATIC Manager.

Network templates

Just like you can save blocks in a library in order to reuse them in other programs, you can also save network templates in order e.g. to repeatedly copy them into other blocks.

To save the network templates, create a library which contains at least one S7 program and the container *Sources*.

You program the networks you wish to use as templates completely "normally" in any block. You then replace the addresses you wish to modify by the dummy characters %00 to %99. You can also design the network title and the network comments variably in this manner.

The lines with a dummy character are displayed in red since this type of block cannot be saved. This is of no importance since this block can be rejected (closed without saving) following saving of the network template(s).

Following input of the dummy characters, mark the network by clicking the network number at the top left in front of the network title. You can also combine several networks in a template; hold the Ctrl key pressed while you click further network numbers.

Now select EDIT → CREATE NETWORK TEMPLATE. In the displayed dialog field, assign meaningful comments to the network and all dummy characters. In the next dialog box, assign the network template a name and define the storage location (container *Sources* in a library).

When using the network templates, open the corresponding library in the program element catalog and then select the desired network template (double click or drag into editor window). Replace the dummy values by valid entries in the automatically displayed dialog window. The network template is appended to the marked network.

3.4.5 Source-oriented programming of an STL code block

Chapter 2.5 "Creating the S7 Program" gives an introduction to program creation and the use of the program editor.

You begin source-file-oriented programming by generating an empty STL program source file in the SIMATIC Manager (see Chapter 2.5.3 "STL-Program Editor" under "Source-File-Oriented Programming"). You start the editor by opening the program source file and you can start entering the program immediately, for example, with the keyword for a function block or you use a block template with INSERT → BLOCK TEMPLATE.

Table 3.4 shows you which keywords you need for the block programming and the sequence in which you use the keywords.

Block header

You program the properties of a block in the block header after the block type and before the variable declaration. All information in the block header is optional; you can omit individual specifications or all of them. Please refer to Chapter 3.2.3 "Block Properties" for a description and the assignment of block properties.

With the keyword "TITLE=" immediately after the line for the block type, you can enter a block title of up to 64 characters. You can then add a block comment in the form of one or more comment lines beginning with a double slash. The block comment can be up to 18 KB long.

Variable declaration

The declaration section contains the definition of the block-local variables, that is, of the variables which you use only in that block. You cannot program every variable type in every block (see Table 3.4). If you do not use a variable type, omit the declaration, including keywords.

The declaration for a variable consists of the name, the data type, a default value, if any, and a variable comment (optional). Example:

```
Quantity: INT := +500; //Units per batch
```

Not all variables can be assigned a default value (for instance temporary local data cannot). The defaults for functions and function blocks are described in detail in Chapter 19 "Block Parameters".

The order of the declarations for code blocks is fixed (as shown in the table). The order within a variable type is arbitrary, and also determines, in conjunction with the data type, the amount of room required in memory; Chapter 24 "Data Types", shows you how you can optimize memory requirements by skillfully planning the order.

Program section

The program section of a code block begins with the keyword BEGIN and ends with END_xxx, with block type ORGANIZATION_BLOCK, FUNCTION_BLOCK or FUNCTION taking the place of xxx. The keyword END_xxx replaces Block End BE.

In both keywords and program code, the Editor accepts upper and lower case. Details on statement syntax can be found in Chapter 3.4.1 "Structure of an STL Statement". The OP code (operator) must be separated from the address (operand) by one or more spaces or tabs. To improve the readability of the source text, you can leave one or more spaces and/or tabs between words. You can set the font and color for different types of text in the program editor with OPTIONS → CUSTOMIZE on the tab "Source text".

You must terminate each statement with a semicolon. After the semicolon you can write a statement comment, but it must begin with two slashes; it may extend up to the end of the line.

Table 3.4 Keywords for programming of STL code blocks

Block type	Organization block	Function block	Function
Block type	ORGANIZATION_BLOCK	FUNCTION_BLOCK	FUNCTION : *Function value*
Header	TITLE = *block title* *//Block comment* KNOW_HOW_PROTECT NAME : *Block name* FAMILY : *Block family* AUTHOR : *Originator* VERSION : *Version*	TITLE = *block title* *//Block comment* CODE_VERSION1 KNOW_HOW_PROTECT NAME : *Block name* FAMILY : *Block family* AUTHOR : *Originator* VERSION : *Version*	TITLE = *block title* *//Block comment* KNOW_HOW_PROTECT NAME : *Block name* FAMILY : *Block family* AUTHOR : *Originator* VERSION : *Version*
Declaration		VAR_INPUT *Input parameters* END_VAR	VAR_INPUT *Input parameters* END_VAR
		VAR_OUTPUT *Output parameters* END_VAR	VAR_OUTPUT *Output parameters* END_VAR
		VAR_IN_OUT *In/out parameters* END_VAR	VAR_IN_OUT *In/out parameters* END_VAR
		VAR *Static local data* END_VAR	
	VAR_TEMP *Temporary local data* END_VAR	VAR_TEMP *Temporary local data* END_VAR	VAR_TEMP *Temporary local data* END_VAR
Program	BEGIN NETWORK TITLE = *Network title* *//Network comment* ... STL statements *//Line comment* NETWORK ... etc.	BEGIN NETWORK TITLE = *Network title* *//Network comment* ... STL statements *//Line comment* NETWORK ... etc.	BEGIN NETWORK TITLE = *Network title* *//Network comment* ... STL statements *//Line comment* NETWORK ... etc.
Block end	END_ORGANIZATION_ BLOCK	END_FUNCTION_ BLOCK	END_FUNCTION

You may also program several statements on one line, separating each from its predecessor by a semicolon.

You begin a line comment with two slashes at the beginning of the line. A line comment may comprise no more than 160 characters; it may contain no tabs and no non-printable characters.

For better readability and logic, you can divide the program in a block into networks. In the graphic languages, a subdivision into networks is necessary; in STL, it is not. Networks have no functional purpose; they are simply used in STL to divide the program into more logically related sections and to improve its readability, and to make it easier and more efficient to write comments. In very extensive programs, it is an advantage to be able to directly address the networks in the compiled block, thus reaching a particular program location quickly (with EDIT → GO TO → NETWORK/ROW you can specify the network number or the line number relative to the beginning of the network).

Networks begin with the keyword NETWORK; with the keyword in the next line, "TITLE =", you can give each network a heading of up to 64 characters. The line comments immediately after the network title form the network comment; it may be up to 18 KB long. STL numbers the networks automatically beginning with 1; there is a maximum of 999 networks per block. There are 64 KB available per block for block and network comments.

Please note when calling a block that the transferred block parameters are listed in the same order in which they have been declared in the called block.

Order of blocks in source-file-oriented programming

To call a block, the editor requires the information in the block header, the block parameters to be initialized and which declaration type and data type the block parameter has in each case. This means that you must first program the called functions and function blocks, or that you start programming with the "lowest-level" blocks (position them accordingly at the start of the source text file).

However, it is also sufficient if you program only the block header with the parameter declaration (only as an "interface description" as it were). You can then provide this interface description with a program at a later time. (Please ensure, however, that you do not modify the interface of an already called block! Otherwise, the editor will report a time stamp conflict when outputting the block call.)

The following order is recommended for the blocks in a source file:

- ▷ User-defined data types UDTs
- ▷ Global data blocks
- ▷ Functions and function blocks beginning with the blocks of the "lowest" call level
- ▷ Instance data blocks (can also be located directly after the assigned function block)
- ▷ Organization blocks

For comprehensive user programs, you will probably want to divide the entire program source into "convenient" individual files; for example, "program standards" that you use throughout the program, individual technologically or functionally distinguishable sub-programs, or a "main program" that contains e.g. organization blocks.

When creating individual source files, you must keep sight of the order of compilation – for the block call reasons listed above.

Example of a function block with instance data block

Figure 3.8 shows an example for a function block with static local data, followed by the programmed instance data block associated with that function block.

3.5 Programming Code Blocks with SCL

3.5.1 Structure of an SCL Statement

The SCL program consists of a sequence of individual STL statements. A statement is the smallest independent unit of the user program. It represents a procedural specification for the CPU. Figure 3.9 shows several examples of SCL statements.

An SCL statement comprises

- ▷ A jump label (optional) comprising up to 24 characters and terminated with a colon; jump labels must be declared.
- ▷ An instruction that describes what the CPU is to do (e.g. value assignments, control statements, etc.).
- ▷ A comment (optional) beginning with two slashes and going to the end of the line (only printable characters and no tabs).

You must terminate every statement with a semicolon (before any comment). An SCL statement can contain up to 126 characters.

```
FUNCTION_BLOCK W_Memory_STL
TITLE = Intermediate memory for 4 values
//Example of a function block with static local data in STL
AUTHOR  : Berger
FAMILY  : STL_Book
NAME    : Memory
VERSION : 01.00
VAR_INPUT
  Import  : BOOL := FALSE;           //Import with positive edge
  Input value : REAL := 0.0;         //in data format REAL (fraction)
END_VAR
VAR_OUTPUT
  Output value : REAL := 0.0;        //in data format REAL (fraction)
END_VAR
VAR
  Value1 : REAL := 0.0;              //First saved REAL value
  Value2 : REAL := 0.0;              //Second value
  Value3 : REAL := 0.0;              //Third value
  Value4 : REAL := 0.0;              //Fourth value
  Edge trigger flag : BOOL := FALSE;//Edge trigger flag for importing
END_VAR
BEGIN
NETWORK
TITLE = Program for importing and output
//Importing and output take place with a positive edge at import
      U    Import;                   //If import changes to "1",
      FP   Edge trigger flag;        //the RLO = "1" following FP
      SPBN end;                      //Jump if no positive edge is present
//Transfer of values starting with the last value
      L    Value4;
      T    Output value;             //Output of last value
      L    Value3;
      T    Value4;
      L    Value2;
      T    Value3;
      L    Value1;
      T    Value2;
      L    Input value;              //Importing of input value
      T    Value1;
End: BE;
END_FUNCTION_BLOCK

DATA_BLOCK Memory1_STL
TITLE = Instance data block for "W-Memory_STL"
//Example of an instance data block
AUTHOR  : Berger
FAMILY  : STL_Book
NAME    : W_SP_DB1
VERSION : 01.00
W_Memory_STL                         //Instance for the FB "W_Memory_STL"
BEGIN
  Value1 := 1.0;                     //Individual preallocation
  Value2 := 1.0;                     //of selected values
END_DATA_BLOCK
```

Figure 3.8
Example of programming of an STL function block and the associated instance data block

Value Assignments

```
Power     := Voltage * Current;
TooLarge  := Volt_Act > Volt_Set;
Switch_on := Manual_on OR Auto_on;
```

Control Statements

```
IF Input_value > Maximum
   THEN Delimiter := Maximum;
   ELSIF Input_value < Minimum
      THEN Delimiter := Minimum;
   ELSE Delimiter := Input_value;
END_IF;
FOR i := 1 TO 32 DO
   Measure_value[i] := 0;
END_FOR;
```

Function Calls

```
Result := Delimiter(
      Input_value:= Actual_value,
      Minimum    := Lower_limit,
      Maximum    := Upper_limit);
```

Figure 3.9 SCL Statement Examples

3.5.2 Programming SCL Code Blocks

An introduction to program creation and the operation of the program editor are described in Chapter 2.5 "Creating the S7 Program".

You begin programming by generating an empty SCL program source file in the SIMATIC Manager (see Chapter 2.5.4 "SCL Program Editor" under "Creating the SCL source file"). You start the editor by opening the program source file and you can start entering the program immediately, for example, with the keyword for a function block or you insert a block template with INSERT → BLOCK TEMPLATE.

Table 3.5 shows which keywords you require for block programming and the order in which you use the keywords.

Block header

You program the properties of a block in the block header after the block type and before the variable declaration. All information in the block header is optional; you can omit individual specifications or all of them. Please refer to Chapter 3.2.3 "Block Properties" for a description and the assignment of block properties.

With the keyword "TITLE=" immediately after the line for the block type, you can enter a block title of up to 64 characters. You can then add a block comment in the form of one or more comment lines beginning with a double slash. The block comment can be up to 18 KB long.

Variable declaration

The declaration section contains the definition of the block-local variables, that is, of the variables that you use only in that block. You cannot program every variable type in every block (see table). If you do not use a variable type, omit the declaration, including keywords.

The declaration for a variable consists of the name, the data type, a default value, if any, and a variable comment (optional). Example:

```
Quantity : INT := +500;//Units per batch
```

At declaration, SCL allows variables of the same data type to be combined in one line:

```
Value1, Value2, Value3, Value4 : INT;
```

Not all variables can be assigned a default value (for instance temporary local data cannot). The defaults for functions and function blocks are described in detail in Chapter 19 "Block Parameters".

The order of the individual declarations and the order within a variable type is arbitrary. It determines, in conjunction with the data type, the amount of memory space required; Chapter 24 "Data Types" shows you how you can optimize memory requirements by skillfully planning the order.

In SCL, you can declare constants, i.e. you assign a symbol to a fixed value. If you use jump labels in the block, you must declare them.

Program section

The program section of an SCL code block begins (optionally) with the keyword BEGIN and ends with END_xxx, where xxx stands for block type ORGANIZATION_BLOCK, FUNCTION_BLOCK or FUNCTION.

In both keywords and program code, the Editor accepts upper and lower case. Details on statement syntax can be found in Chapter 3.5.1

Table 3.5 Keywords for Programming SCL Code Blocks

Block	Organization Block	Function Block	Function
Block type	ORGANIZATION_BLOCK	FUNCTION_BLOCK PROGRAM [3]	FUNCTION : *Function value*
Header	TITLE = '*Block title*' *//Block comment* KNOW_HOW_PROTECT NAME : *Block name* FAMILY : *Block family* AUTHOR : *Originator* VERSION : '*Version*'	TITLE = '*Block title*' *//Block comment* KNOW_HOW_PROTECT NAME : *Block name* FAMILY : *Block family* AUTHOR : *Originator* VERSION : '*Version*'	TITLE = '*Block title*' *//Block comment* KNOW_HOW_PROTECT NAME : *Block name* FAMILY : *Block family* AUTHOR : *Originator* VERSION : '*Version*'
Declaration		VAR_INPUT *Input parameters* END_VAR	VAR_INPUT *Input parameters* END_VAR
		VAR_OUTPUT *Output parameters* END_VAR	VAR_OUTPUT *Output parameters* END_VAR
		VAR_IN_OUT *In-out parameters* END_VAR	VAR_IN_OUT *In-out parameters* END_VAR
		VAR *Static local data* END_VAR	VAR [1] *Temporary local data* END_VAR
	VAR_TEMP *Temporary local data* END_VAR	VAR_TEMP *Temporary local data* END_VAR	VAR_TEMP *Temporary local data* END_VAR
	CONST *Constants* END_CONST	CONST *Constants* END_CONST	CONST *Constants* END_CONST
	LABEL *Jump labels* END_LABEL	LABEL *Jump labels* END_LABEL	LABEL *Jump labels* END_LABEL
Program	BEGIN [2] ... SCL statements *//Line comment* *(* Block comment ...* *... Block comment *)* ... etc.	BEGIN [2] ... SCL statements *//Line comment* *(* Block comment ...* *... Block comment *)* ... etc.	BEGIN [2] ... SCL statements *//Line comment* *(* Block comment ...* *... Block comment *)* ... etc.
Block end	END_ORGANIZATION_ BLOCK	END_FUNCTION_ BLOCK END_PROGRAM [3]	END_FUNCTION

[1] The local data agreed under VAR in an SCL function FC are handled like temporary local data (VAR_TEMP).
[2] Not required for SCL
[3] Alternative to FUNCTION_BLOCK or END_FUNCTION_BLOCK

"Structure of an SCL Statement". The OP code must be separated from the address (operand) by one or more spaces or tabs. To improve the readability of the source text, you can leave one or more spaces and/or tabs between words. You can set the font and color for different types of text in the program editor with OPTIONS → CUSTOMIZE on the tab "Format".

You must terminate each statement with a semicolon. After the semicolon you can write a statement comment, but it must begin with two slashes; it may extend up to the end of the line. You may also program several statements on one line, separating each from its predecessor by a semicolon.

An SCL block must contain at least one statement (one semicolon). SCL does not have networks like STL.

You begin a line comment with two slashes at the beginning of the line. A line comment may comprise no more than 160 characters; it may contain no tabs and no non-printable characters.

SCL has a block comment that can extend over several lines. It begins with an open bracket and star and ends with a star and closed bracket. The block comment may also be placed within an SCL statement; however, it must not interrupt either a symbolic name or a constant (exception: character string).

Compiler properties

You can set the compiler properties globally in the SCL program editor with OPTIONS → CUSTOMIZE on the tabs "Generate blocks" and "Compiler". The properties apply to all compilations unless other properties have been defined in the source programs or in the compilation control file. The compiler settings defined in the source program apply starting at their position in the source program or compilation control file until they are overwritten by other settings, or until the end of the source program or compilation control file.

Table 3.6 shows the keywords available for the compiler settings. You write the keywords – enclosed in curly brackets – in a separate line outside a block. If several keywords are present in one line, they must be separated by a semicolon. Upper-case and lower-case letters are not differentiated.

Table 3.6 Keywords for the compiler properties

Keyword	Value	Property
[SCL_]ResetOptions	(none)	Import settings from menu dialog
[SCL_]OverwriteBlocks	'y[es]' or 'n[o]'	Overwrite blocks
[SCL_]GenerateReferenceData	'y[es]' or 'n[o]'	Generate reference data
[SCL_]S7ServerActive	'y[es]' or 'n[o]'	Consider system attribute "S7_server"
[SCL_]CreateObjectCode	'y[es]' or 'n[o]'	Create object code
[SCL_]OptimizeObjectCode	'y[es]' or 'n[o]'	Optimize object code
[SCL_]MonitorArrayLimits	'y[es]' or 'n[o]'	Monitor array limits
[SCL_]CreateDebugInfo	'y[es]' or 'n[o]'	Create debug information
[SCL_]SetOKFlag	'y[es]' or 'n[o]'	Set OK flag
[SCL_]SetMaximumStringLength	'1' .. '254'	Set maximum STRING length

Examples:

```
//Create debug info
//Set OK flag
{CreateDebugInfo:='yes'; SetOKFlag:='yes'}

FUNCTION_BLOCK example
...
END_FUNCTION_BLOCK

//Reset settings
{ResetOptions}
```

Order of blocks in source-file-oriented programming

To call a block, the editor requires the information in the block header, the block parameters to be initialized and which declaration type and data type the block parameter has in each case. This means that you must first program the called functions and function blocks, or that you start programming with the "lowest-level" blocks (position them accordingly at the start of the source text file).

However, it is also sufficient if you program only the block header with the parameter declaration (only as an "interface description" as it were). You can then provide this interface description with a program at a later time. (Please ensure, however, that you do not modify the interface of an already called block! Otherwise, the editor will report a time stamp conflict when outputting the block call.)

The following order is recommended for creation of a source file:

- ▷ User-defined data types UDTs
- ▷ Global data blocks
- ▷ Functions and function blocks beginning with the blocks of the "lowest" call level
- ▷ Instance data blocks (can also be located immediately after the assigned function block)
- ▷ Organization blocks

If you subdivide extensive user programs into individual program source files, you must keep sight of the order of compilation – for the block call reasons listed above.

Example of an SCL Function Block with Instance Data Block

Figure 3.10 shows an example of a function block with static local data. Subsequently, the associated instance data block is programmed.

3.6 Programming Data Blocks

Chapter 2.5 "Creating the S7 Program" gives an introduction to program creation and the use of the program editor.

Data blocks are programmed in the same way in STL and SCL. You use the STL program editor for incremental programming; both the STL program editor and the SCL program editor are available to you for source-file-oriented programming.

3.6.1 Programming Data Blocks Incrementally

Creating data blocks

You begin block programming by opening a block, either with a double-click on the block in the project window of the SIMATIC Manager or by selecting FILE → OPEN in the editor. If the block does not yet exist, create it as follows:

- ▷ In the SIMATIC Manager: select the object *Blocks* in the left-hand portion of the project window and create a new data block with INSERT → S7 BLOCK → DATA BLOCK. You see the properties window of the block. Specify the number and the type of the data block on the "General – Part 1" tab (see below). "Instance DB" or "DB of type" can only be selected if function blocks FB or system function blocks SFB or user-defined data types UDT are present in the block container. You can also enter the remaining block properties later.
- ▷ In the STL program editor: with FILE → NEW, you get a dialog box in which you can enter the desired block under "Object name". In the subsequently displayed dialog window "New data block" you will be requested to define the type of data block (see below). After closing the dialog box, you can program the block contents.

```
FUNCTION_BLOCK W_Memory_SCL
TITLE = 'Intermediate memory for 4 values'
//Example of a function block with static local data in SCL

AUTHOR  : Berger
FAMILY  : SCL_Book
NAME    : Memory
VERSION : '01.00'

VAR_INPUT
  Import  : BOOL := FALSE;          //Import with positive edge
  Input value : REAL := 0.0;        //in data format REAL (fraction)
END_VAR

VAR_OUTPUT
  Output value : REAL := 0.0;       //in data format REAL (fraction)
END_VAR

VAR
  Value1 : REAL := 0.0;             //First saved REAL value
  Value2 : REAL := 0.0;             //Second value
  Value3 : REAL := 0.0;             //Third value
  Value4 : REAL := 0.0;             //Fourth value
  Edge trigger flag : BOOL := FALSE;//Edge trigger flag for importing
END_VAR

BEGIN
//Importing and output take place with a positive edge at import
IF import = 1 AND edge memory bit = 0
THEN output value := Value4;
     //Transfer of values starting with the last value
     Value4 := Value3;
     Value3 := Value2;
     Value2 := Value1;
     Value1 := Input value;
     Edge memory bit := Import;     //Update edge memory bit
ELSE edge memory bit := Import;     //even if there is no edge
END_IF;
END_FUNCTION_BLOCK

DATA_BLOCK Memory1_SCL
TITLE = 'Instance data block for "W-Memory_SCL" '
//Example of an instance data block

AUTHOR  : Berger
FAMILY  : SCL_Book
NAME    : W_SP_DB1
VERSION : '01.00'

W_Memory_SCL                        //Instance for the FB "W_Memory_SCL"

BEGIN
  Value1 := 1.0;                    //Individual preallocation
  Value2 := 1.0;                    //of selected values
END_DATA_BLOCK
```

Figure 3.10
Example of programming of an SCL function block and the associated instance data block

You can fill out the header of a block as you create it or you can add the block properties at a later point. You program later additions to the block header in the editor by selecting FILE → PROPERTIES while the block is open.

Types of data blocks

When creating a new data block you will be requested to define the type of data block. When creating with the SIMATIC Manager, you set the type in the selection box of the properties window. When creating with the program editor, you set the type with the program editor in the window "New data block" by clicking one of the offered options.

Three different types of data block are differentiated depending on the creation and the application:

▷ "Data block" or "Shared DB"
Creation as a global data block; you declare the data addresses when programming the data block in this case.

▷ "Data block referencing a user-defined data type" or "DB of type"
Creation as a data block of user-defined data type; the data structure is used which you declared when programming the corresponding user-defined data type UDT.

▷ "Data block referencing a function block" or "Instance DB"
Creation as an instance data block; here, the data structure that you have declared when programming the relevant function block is transferred.

When creating a data block on the basis of a user-defined data type, you simultaneously define the UDT used as basis; i.e. the UDT must already be present in the block container. The same applies to creation of a data block with assigned function block.

Block window and views

When opening a data block whose structure is based on a user-defined data type or a (system) function block, you will be asked in the default setting whether you wish to open the data block using the program editor or the application "Parameter assignment for data blocks". The parameterization view displays the data values in technological groups and permits convenient parameterization (see Chapter 2.7.8 "Monitoring and Controlling Data Addresses"). The data views are described below.

The program editor provides two views for programming (creating) data blocks:

▷ The declaration view serves to define the data structure with global data blocks and the default values.

▷ In the data view you handle the online values.

A table is displayed in each view, and contains the data addresses with their absolute addresses and sequence, the names and data types, the initial values and comments (Figure 3.11). There is an additional column with the current value in the data view.

If you open a data block from the offline data management, you are presented with the offline window with which you can edit data in the programming device. If you open a data block which is present in the CPU's user memory, the editor presents the online window with which you can handle the data values on the CPU.

Offline window

You use the declaration view for global data blocks at the input. You declare the data operands in this view: You set the order of the data operands, you assign a name and data type for each data operand, and you may also add operand comments. Each data operand is preallocated with a default value. Depending on the data type, this can be zero, the minimum value, or a space. In the initial value column, you have the opportunity to change the default value.

The data operations and the default values are already fixed for data blocks derived from a user-defined data type or from a function block. They are obtained from the declaration of the user-defined data type or from the declaration of the function block.

The data view additionally shows the current value column. As standard, the default values

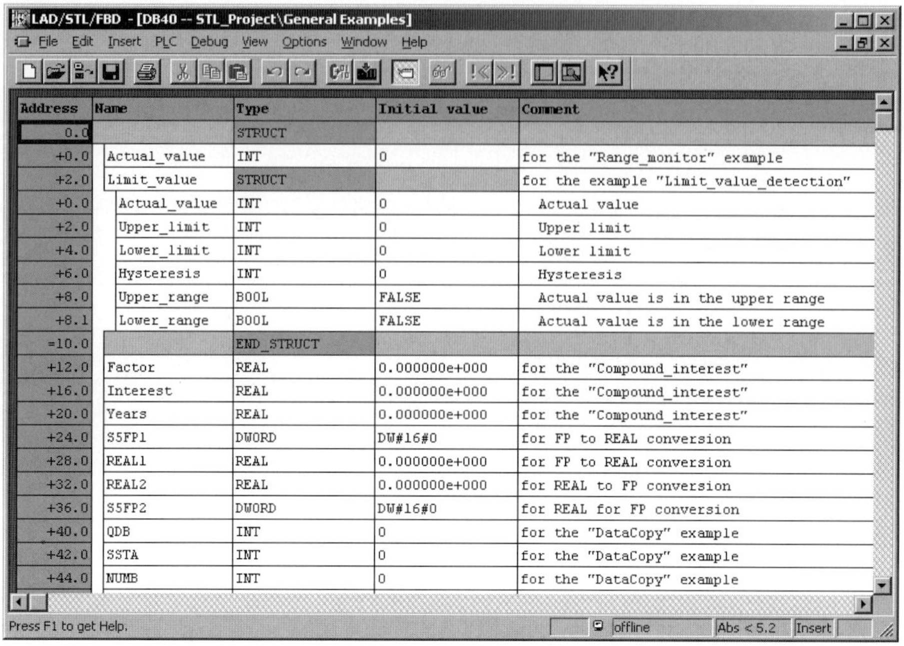

LAD/STL/FBD - [DB40 -- STL_Project\General Examples]

File Edit Insert PLC Debug View Options Window Help

Address	Name	Type	Initial value	Comment
0.0		STRUCT		
+0.0	Actual_value	INT	0	for the "Range_monitor" example
+2.0	Limit_value	STRUCT		for the example "Limit_value_detection"
+0.0	Actual_value	INT	0	Actual value
+2.0	Upper_limit	INT	0	Upper limit
+4.0	Lower_limit	INT	0	Lower limit
+6.0	Hysteresis	INT	0	Hysteresis
+8.0	Upper_range	BOOL	FALSE	Actual value is in the upper range
+8.1	Lower_range	BOOL	FALSE	Actual value is in the lower range
=10.0		END_STRUCT		
+12.0	Factor	REAL	0.000000e+000	for the "Compound_interest"
+16.0	Interest	REAL	0.000000e+000	for the "Compound_interest"
+20.0	Years	REAL	0.000000e+000	for the "Compound_interest"
+24.0	S5FP1	DWORD	DW#16#0	for FP to REAL conversion
+28.0	REAL1	REAL	0.000000e+000	for FP to REAL conversion
+32.0	REAL2	REAL	0.000000e+000	for REAL to FP conversion
+36.0	S5FP2	DWORD	DW#16#0	for REAL for FP conversion
+40.0	QDB	INT	0	for the "DataCopy" example
+42.0	SSTA	INT	0	for the "DataCopy" example
+44.0	NUMB	INT	0	for the "DataCopy" example

Press F1 to get Help. offline Abs < 5.2 Insert

Figure 3.11 Example of an opened data block (declaration view)

from the initial value column are entered into this column. In the data view you can define a different initial value for the load memory, and thus a current value for the work memory (Figure 3.12).

The possibility existing for each data block for individual default data is particularly applicable for the data blocks derived from a user-defined data type or from a function block. For example, if you produce several instance data blocks of a function block, all data blocks have the default setting made in the function block. You can then individually assign other default values to certain data addresses in the data view for each instance.

Online window

You usually use the online window to view the current data values in the CPU's user memory. However, you can also generate data blocks with it.

In the declaration view, the initial value column shows the initial value from the offline data management or the initial value from the load memory if the offline project belonging to the CPU program is not present. In the data view, the current value from the work memory is displayed in the current value column. You can leave the editor with EDIT → INITIALIZE DATA BLOCK, and all current values are replaced by the initial values again.

When writing back with PLC → DOWNLOAD, you write the value in the current value column into the work memory. You therefore have the possibility for manipulating the values of data addresses with the programming device during program execution. The value in the initial value column is rejected.

When writing back with FILE → SAVE, you write the value in the initial value column as the default value and the value in the current value column as the initial value into the offline data management.

Note that the complete information with respect to data addresses, e.g. the name, is only present in the offline data management. It is recommendable to also write the data blocks generated in the user memory of the CPU into the offline data management so that data consistency is guaranteed (Chapter 2.6.5 "Block Handling" under "Data blocks offline/online").

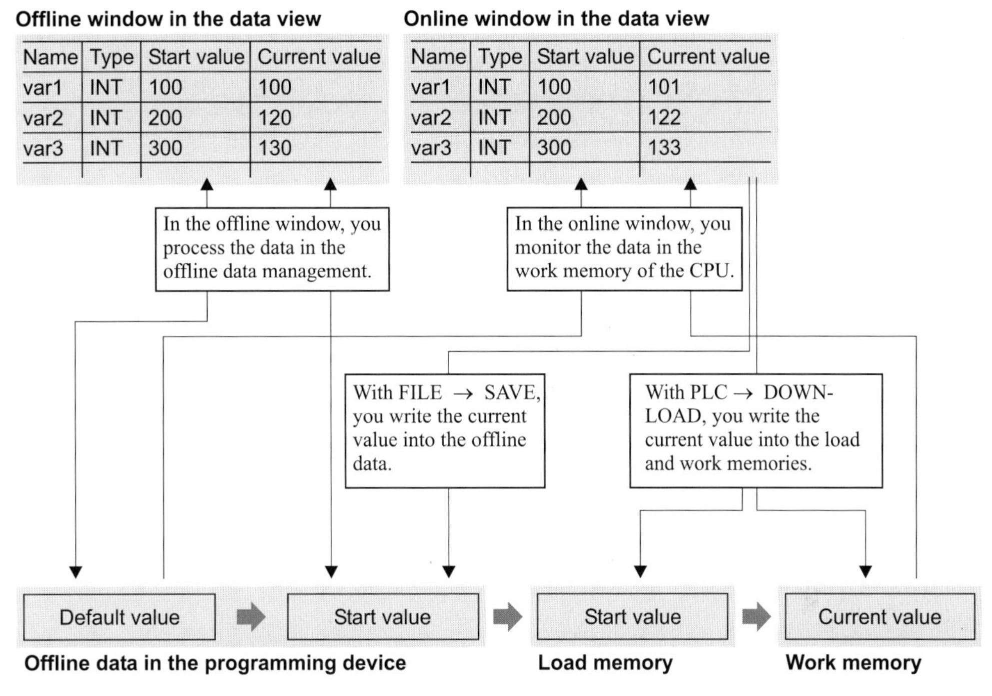

Fig. 3.12 Data storage in incremental programming

3.6.2 Source-Oriented Data Block Programming

When you create a source file for a data block, you must adhere to the structure or order shown in Table 3.7 for programming the block. This applies both for STL program source files and for SCL program source files.

Block header

You program the properties of a block in the block header after the block type and before the variable declaration. All information in the block header is optional; you can omit individual specifications or all of them. Please refer to Chapter 3.2.3 "Block Properties" for a description and the assignment of block properties.

With the keyword "TITLE =" immediately after the line for the block type, you can enter a block title of up to 64 characters. One or more comment lines beginning with two slashes can then be attached as block comments. The block comment can be a maximum of 18 KB long.

Declaration in the data block

The declaration section contains the definition of the block-local variables, that is, of the variables which you use only in that block. You can declare a data block as a global data block with "individual" variables, as a global data block with UDT and as an instance data block.

The declaration for a variable in a global data block consists of the name, the data type, a default value, if any, and a variable comment (optional).

Example:

```
Quantity : INT := +500;//Units per batch
```

All variables can be assigned a default value. The order of variables is arbitrary; it also determines the required memory space in conjunction with the data type. Chapter 24 "Data Types" shows the memory space occupied by the variables. In Chapter 26.2 "Data Storage of Variables" you can learn how the variables are stored in data blocks. You can optimize the memory requirement by skilful selection of the order.

Table 3.7 Keywords for programming of data blocks

Block type	Global data block	Global data block from UDT	Instance data block
Block type	DATA_BLOCK	DATA_BLOCK	DATA_BLOCK
Header	TITLE = *block title* //*Block comment* KNOW_HOW_PROTECT NAME : *Block name* FAMILY : *Block family* AUTHOR : *Originator* VERSION : *Version* UNLINKED READ_ONLY NON_RETAIN	TITLE = *block title* //*Block comment* KNOW_HOW_PROTECT NAME : *Block name* FAMILY : *Block family* AUTHOR : *Originator* VERSION : *Version* UNLINKED READ_ONLY NON_RETAIN	TITLE = *block title* //*Block comment* KNOW_HOW_PROTECT NAME : *Block name* FAMILY : *Block family* AUTHOR : *Originator* VERSION : *Version* UNLINKED READ_ONLY NON_RETAIN
Declaration	STRUCT *name : type := default setting;* END_STRUCT	*UDTname*	*FBname*
Initialization	BEGIN *name := Default;* ...etc.	BEGIN *KOMPname := Default;* ...etc.	BEGIN *KOMPname := Default;* ...etc.
Block end	END_DATA_ BLOCK	END_DATA_BLOCK	END_DATA_BLOCK

If you do not take the opportunity of assigning a default value, the editor will write zero or the smallest value into the variable or fill it with blanks – depending on the data type.

The declaration section of a data block derived from a UDT consists only of the UDT. You can use the absolute address (e.g. UDT 51) or the symbolic address (e.g. "Frame header").

The declaration section of an instance data block comprises only the specification of the assigned function block with either absolute or symbolic addressing.

Initialization in the data block

The initialization part begins with BEGIN and ends with END_DATA_BLOCK. Even if you do not preallocate anything in the initialization part, you must enter these keywords.

If you do not specify a value for a data operand in the initialization part, the editor takes the value from the declaration part. If you use user-defined data types in the declaration that are with preallocated default values, you can write over the default values in the initialization part. The same applies to instance data blocks that have the assigned function block (with its default value) as a data structure. Here, you can then set the start values individually for this instance (for calling the function block with this data block).

The preallocation of the data operands in the declaration part results in the default value in the offline data storage. This value is also accepted as start value. By preallocating values in the initialization part, you overwrite the start value (Figure 3.13).

When transferring to the CPU, the programming device writes the start value into the load memory. The CPU subsequently copies the start value from the load memory into the work memory, where it becomes the current value. The user program works with the current values of the data addresses in the work memory (Chapter 2.6.5 "Block Handling" under "Data blocks offline/online").

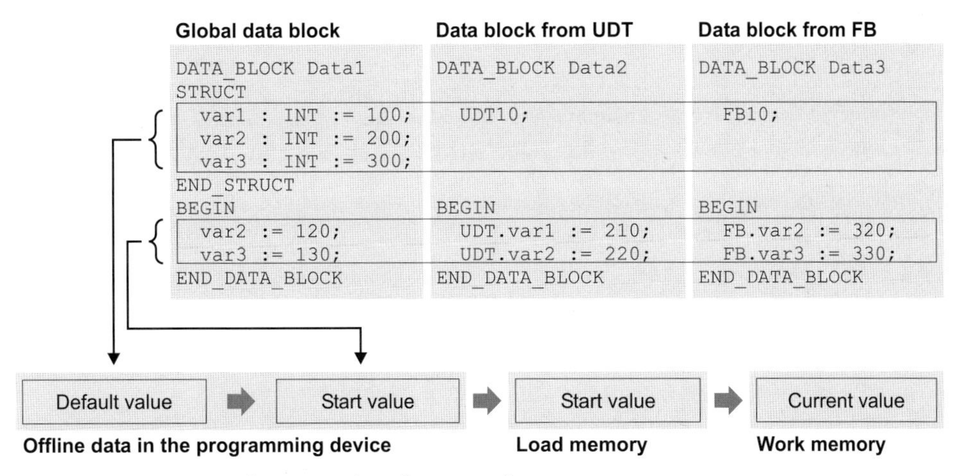

Figure 3.13 Data storage in source-oriented programming

3.7 Variables and Constants

3.7.1 General Remarks Concerning Variables

A variable is a value with a specific format (Figure 3.14). Simple variables consist of an address (such as input 5.2) and a data type (such as BOOL for a binary value). The address, in turn, comprises an address identifier (such as I for input) and an absolute storage location (such as 5.2 for byte 5, bit 2). You can also reference an address or a variable symbolically by assigning the address a name (a symbol) in the symbol table.

A bit of data type BOOL is referred to as a *binary address (*or *binary address).* Addresses comprising one, two or four bytes or variables with the relevant data types are called *digital addresses*.

Variables you declare within a block are called local (block) variables. These include block parameters, static and temporary local data, and also data operands in global data blocks. If these variables have an elementary data type, they can also be addressed as operands (e.g. static local data as DI operands, temporary local data as L operands, and data in global data blocks as DB operands).

Local variables, however, can also be of complex data type (such as structures or arrays). Variables with these data types require more than 32 bits, so that they can no longer, for example, be loaded into the accumulator. And for the same reason, they cannot be addressed with "normal" STL statements. There are special functions for handling these variables, such as the IEC functions, which are provided as a standard library with STEP 7 (you can generate variables of complex data type in block parameters of the same data type).

If variables of complex data type contain components of elementary data type, these components can be treated as though they were separate variables (for example, you can load a component of an array consisting of 30 INT values into the accumulator and further process it).

Constants are used to preset variables to a fixed value. The constant is given a specific prefix depending on the data type.

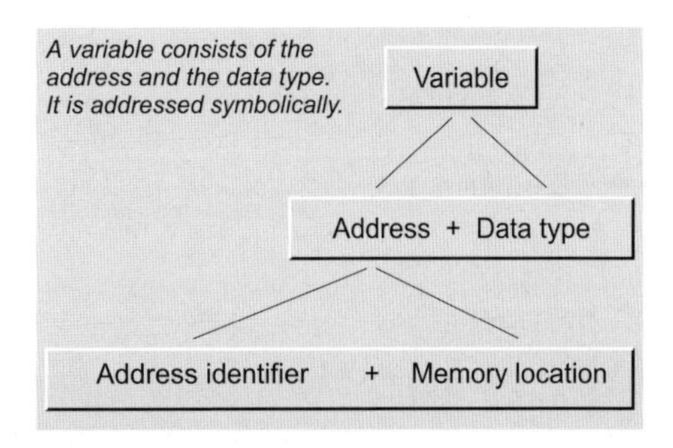

Figure 3.14 Structure of a tag

3.7.2 General Remarks Regarding Data Types

Data types define the properties of data, essentially the representation of the contents of a tag and the permissible ranges. STEP 7 provides predefined data types that you can also compile as self-defined data types. The data types are globally available and can be used in any block.

This section provides an overview of all data types and a brief introduction, particularly to the elementary data types. This knowledge will allow you to program a programmable controller.

Table 3.8 shows a rough overview of data types in STEP 7.

More depth of detail, such as the structure and format of variables of complex data type, are presented in Chapter 24 "Data Types", and information on data types in conjunction with block parameters is provided in Chapter 19 "Block Parameters".

Programming of user-defined data types is described in Chapter 24 "Data Types".

3.7.3 Elementary Data Types

Variables assigned this data type can be edited direct in STL since they represent either a bit or at the most an accumulator width (32 bits). The same applies for SCL in the case of value assignments.

Variables of elementary data type can be preassigned fixed values (constants) at the declaration stage. In this case, the STL (Table 3.9) and SCL (Table 3.10) notations differ from each other. For many data types, there is more than one constant notation and they can be used with equal validity (e.g. TIME# or T#).

Constant notation in STL

STL does not restrict operations (operators) to specific data types (with the exception of the differentiation between binary address and digital address). Comparison functions such as comparing the accumulator contents independently of the data type of the variables the accumulator contains.

Constant notation in SCL

In SCL you can only execute operations with variables of the permissible data types. Constants in SCL do not receive their data type until applied in conjunction with the operation.

Table 3.8 Splitting of data types

Elementary data types	Complex data types	User-defined data types	Parameter data types
BOOL, BYTE, CHAR, WORD, INT, DATE, DWORD, DINT, REAL, S5TIME, TIME, TOD	DT, STRING, ARRAY, STRUCT	UDT, global data blocks, instances	TIMER, COUNTER, BLOCK_DB, BLOCK_SDB, BLOCK_FC, BLOCK_FB, POINTER, ANY
Data types that have at most a double word (32 bits)	Data types that can be larger than a double word (DT, STRING) or that are made up of several components	Structures or data areas, that can be given a name	Block parameters
Can be mapped to absolutely and symbolically addressed operands	Can only be mapped to symbolically addressed tags		Can only be mapped to block parameters (only symbolic addressing)
Permitted in all operand ranges	Permissible in data blocks (as global data and instance data), as temporary local data, and as block parameters		Permitted in combination with block parameters

Table 3.9 Overview of elementary data types with STL notation for constants

Data type (width)	Description	Examples of STL notation for constants	
		Minimum value	Maximum value
BOOL (1 bit)	Bit	FALSE	TRUE
BYTE (8 bits)	8-bit hexadecimal number	B#16#00, 16#00	B#16#FF, 16#FF
CHAR (8 bits)	one character (ASCII)	printable character, e.g. 'A'	printable character, e.g. 'A'
WORD (16 bits)	16-bit hexadecimal number	W#16#0000, 16#0000	W#16#FFFF, 16#FFFF
	16-bit binary number	2#0000_0000_0000_0000	2#1111_1111_1111_1111
	Count value, 3 decades BCD	C#000	C#999
	2 × 8-bit decimal numbers without sign	B#(0,0)	B#(255,255)
DWORD (32 bits)	32-bit hexadecimal number	DW#16#0000_0000, 16#0000_0000	DW#16#FFFF_FFFF, 16#FFFF_FFFF
	32-bit binary number	2#0000_0000_..._0000_0000	2#1111_1111_..._1111_1111
	4 × 8-bit decimal numbers without sign	B#(0,0,0,0)	B#(255,255,255,255)
INT (16 bits)	Fixed-point number	–32 768	+32 767
DINT (32 bits)	Fixed-point number	L#–2 147 483 648 [1)]	L#+2 147 483 647 [1)]
REAL (32 bits)	Floating point number	Exponential representation: +1.234567E+02 [2)]	
		Decimal representation: 123.4567 [2)]	
S5TIME (16 bits)	Time value in SIMATIC format	S5T#0ms, S5TIME#0ms	S5T#2h46m30s, S5TIME#2h46m30s
TIME (32 bits)	Time value in IEC format	T#–24d20h31m23s648ms, TIME#–24d20h31m23s648ms	T#24d20h31m23s647ms, TIME#24d20h31m23s647ms
		T#–24.855134d, TIME#–24.855134d	T#24.855134d, TIME#24.855134d
DATE (16 bits)	Date	D#1990-01-01, DATE#1990-01-01	D#2168-12-31, DATE#2168-12-31
TIME_OF_DAY (32 bits)	Time of day	TOD#00:00:00.000, TIME_OF_DAY#00:00:00.000	TOD#23:59:59.999, TIME_OF_DAY#23:59:59.999

1) "L#" can be omitted if the number falls outside the INT numerical range

2) For range of values, see Chapter 24.1.3 “Number Representations”

Table 3.10 Overview of elementary data types with SCL notation for constants

Data type (width)	Description	Examples of SCL notation for constants
BOOL (1 bits)	Bit	FALSE, TRUE, BOOL#FALSE, BOOL#TRUE
	Binary number	2#0, 2#1, BOOL#0, BOOL#1
BYTE (8 bits)	8-bit decimal number	0, B#127, BYTE#255
	8-bit hexadecimal number	16#0, B#16#7F, BYTE#16#FF
	8-bit octal number	8#0, B#8#177, BYTE#8#377
	8-bit binary number	2#0, B#2#0111_1111, BYTE#2#1111_1111
CHAR (8 bits)	one printable character (ASCII)	' ', CHAR#' ', CHAR#20 'z', CHAR#'z', CHAR#122
WORD (16 bits)	16-bit decimal number	0, W#32767, WORD#65535
	16-bit hexadecimal number	16#0, W#16#7FFF, WORD#16#FFFF
	16-bit octal number	8#0, W#8#7_7777, WORD#8#17_7777
	16-bit binary number	2#0, W#2#0111_1111_... , WORD#2#1111_1111_...
DWORD (32 bits)	32-bit decimal number	0, DW#2147483647, DWORD#4294967295
	32-bit hexadecimal number	16#0, DW#16#7FFF_FFFF, DWORD#16#FFFF_FFFF
	32-bit octal number	8#0, DW#8#177_7777_7777, DWORD#8#377_7777_7777
	32-bit binary number	2#0, DW#2#0111_1111_... , DWORD#2#1111_1111_...
INT (16 bits)	16-bit decimal number	–32_768, 0, INT#+32_767
	16-bit hexadecimal number	INT#16#0, INT#16#7FFF, INT#16#FFFF
	16-bit octal number	INT#8#0, INT#8#7_7777, INT#8#17_7777
	16-bit binary number	INT#2#0, INT#2#0111_1111_... , INT#2#1111_1111_...
DINT (32 bits)	32-bit decimal number	–2_147_483_648, 0, DINT#+2_147_483_647
	32-bit hexadecimal number	DINT#16#0, DINT#16#7FFF_FFFF, DINT#16#FFFF_FFFF
	32-bit octal number	DINT#8#0, DINT#8#177_7777_7777, DINT#8#377_7777_7777
	32-bit binary number	DINT#2#0, DINT#2#0111_1111_... , DINT#2#1111_1111_...
REAL (32 bits)	Floating point number	Exponential representation: +1.234567E+02 [1)]
		Decimal representation: –123.4567 [1)]
		Integer: +1234567 [1)]
S5TIME (16 bits)	Time value for SIMATIC times	T#0ms, TIME#2h46m30s T#0.0s, TIME#24.855134d
TIME (32 bits)	Time value in IEC format	T#–24d20h31m23s648ms, T#0ms, TIME#24d20h31m23s647ms T#–24.855134d, T#0.0ms, TIME#24.855134d
DATE (16 bits)	Date	D#1990-01-01, D#2168-12-31 DATE#1990-01-01, DATE#2168-12-31
TIME_OF_DAY (32 bits)	Time of day	TOD#00:00:00.000, TOD#23:59:59.999 TIME_OF_DAY#00:00:00.000, TIME_OF_DAY#23:59:59.999

[1)] For range of values, see Chapter 24.1.3 "Number Representations"

Example: in SCL the constant 12345 has the data type class ANY_NUM so depending on the application, it is INT, DINT or REAL. With "type-defined" constant notation, you assign a specific data type direct to a constant, e.g. with DINT#12345 the data type DINT.

3.7.4 Complex Data Types

You can use complex data types (Table 3.11) in conjunction with variables in data blocks or in the L stack or in conjunction with variables which are block parameters.

Variables of complex data types can only be applied to block parameters as complete variables; individual sections cannot be processed with "normal" statements. However, with "direct variable access" and indirect addressing, STL provides a method of manipulating the variables if you know the internal structure.

In addition, there are IEC functions that can process DT and STRING variables (e.g. merging two character strings into one). The IEC functions are a component part of STEP 7; you can find them in the *Standard Library* in the *IEC Function Blocks* program. The IEC functions can be used in every programming language.

The length of a DT variable is fixed; you determine the length of STRING, ARRAY and STRUCT variables yourself when you define these variables.

A string can comprise up to 254 characters and reserves two bytes more in memory than the number of characters in the string.

An array can have as many as 65 536 elements per dimension (from –32 768 to 32 767).

3.7.5 Parameter Types

The parameter types are data types for block parameters (Table 3.12). The length specifications in the Table refer to the memory requirement for block parameters in the case of function blocks. Also use TIMER and COUNTER in the symbol table as data types for timers and counters.

Chapter 19 "Block Parameters" shows you how you can use the parameter types for declaring and assigning block parameters.

Table 3.11 Overview of Complex Data Types

Data Type	Description		Example
DATE_AND_TIME	Date and time	64 bits	DT#1990-01-01-00:00:00.000 DATE_AND_TIME#2168-12-31:23:59:59.999
STRING	String	Variable	Collection of ASCII characters, for instance "String 1"
ARRAY	Array	Variable	Collection of components with the same data type, as many as 6 dimensions possible
STRUCT	Structure	Variable	Collection of components of arbitrary data type, up to 8 nesting levels possible

Table 3.12 Overview of Parameter Types

Parameter Type	Description		Examples of Actual Addresses
TIMER	Timer	16 bits	T 15 or symbol
COUNTER	Counter	16 bits	Z 16 or symbol
BLOCK_FC	Function	16 bits	FC 17 or symbol
BLOCK_FB	Function block	16 bits	FB 18 or symbol
BLOCK_DB	Data block	16 bits	DB 19 or symbol
BLOCK_SDB	System data block	16 bits	(hitherto unused)
POINTER	DB pointer	48 bits	As pointer: P#M10.0 or P#DB20.DBX22.2 As address: MW 20 or I 1.0 or symbol
ANY	ANY pointer	80 bits	As area: P#DB10.DBX0.0 WORD 20 or any (complete) variable

Basic Functions

This part of the book describes the functions of the STL programming language that represent a certain "basic functionality". These functions allow you to program a PLC as you would contactor or relay controls.

The **binary logic operations** are used to simulate series and parallel circuits in a circuit diagram or to implement the AND and OR functions in electronic switching systems. Nesting functions make it possible to implement even complex binary logic operations.

The **memory functions** retain the result of a logic operation (RLO) so that it can, for example, be checked and further processed at another point in the program.

The **transfer functions** are the prerequisite for the handling of digital values. These functions are also required, for instance, to inform a timer of the time value.

The **timers** are to programmable controllers what timing relays are to contactor controls and timers are to electronic switching systems. The timers integrated in the CPU allow you to program such values as wait and monitoring times.

The **counters** are up and down counters that can count in the range from 0 to 999.

This part of the book describes the functions using the address areas for inputs, outputs, and memory bits. Inputs and outputs are the link to the process or plant. Memory bits correspond to auxiliary contactors that store binary states. The subsequent parts of the book deal with the remaining address areas which can be used in binary logic operations. Most importantly, these include the data bits in the global data blocks and the temporary and static local data bits.

Chapter 5 "Memory Functions" contains a programming example for the binary logic operations and memory functions; Chapter 8 "Counter Functions" provides an example of the use of timers and counters. In each case, the example is in an FC function without block parameters.

4 Binary Logic Operations

This chapter discusses the AND, OR and Exclusive OR functions as well as combinations of these functions for the STL programming language. AND, OR and Exclusive OR are used to check the signal states of binary locations and link them with one another.

A binary location can be checked (scanned) for signal state "1" or signal state "0". By negating the result of the logic operation and using nesting expressions, you can also program complex binary logic operations without saving the intermediate result.

The examples in this chapter can be found in function block FB 104 or in the source file Chap_4 in the STL_Book library under the "Basic Functions" program which you can download from the publisher's web site (see page 8).

4.1 Processing a Binary Logic Operation

Figure 4.1 shows, in broad outline, how a binary logic instruction is processed. An input module selects a sensor on the basis of the specified address, for instance the sensor at input I 1.2. The CPU checks the signal state (status) of that sensor, and links the result of the check (check result) with the result of the logic operation (RLO) saved from the preceding logic instruction. The result of this logic operation is saved and stored as the new RLO. The CPU then processes the next statement in the program, for instance storing the result of the logic operation in a specific memory location. The first check to follow the storing of the now "old" RLO is a new logic operation, in which the RLO is set to the check result.

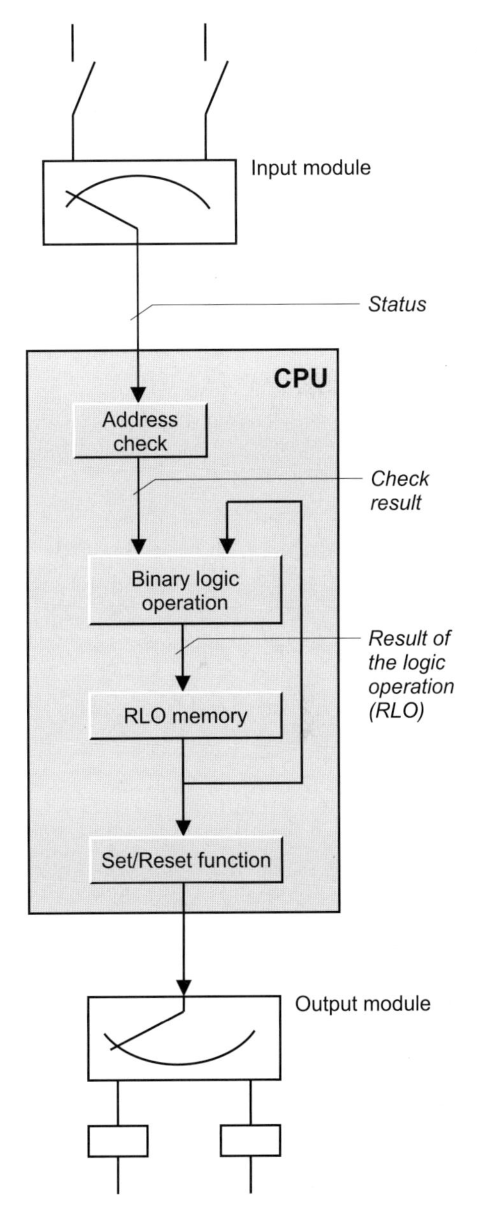

Figure 4.1
How a PLC Works, Using as Example a Binary Logic Operation

Status

The status of a bit is the same as its signal state, and can be "0" or "1". In SIMATIC S7, the signal state is "1" when voltage is present at the input (for instance 230 V AC or 24 V DC, depending on the module); if no voltage is present, the signal state of the input is "0".

A check statement queries the status of the bit. At the same time, it contains the rule of logic according to which the checked signal state is to be linked with the result of the logic operation stored in the processor. For example, the statement

```
A I 17.1
```

checks input I 17.1 for signal state "1" and links the checked signal state according to AND; the statement

```
ON M 20.5
```

checks memory bit M 20.5 for signal state "0" and links the checked signal state according to OR.

Check result

Strictly speaking, the CPU does not link the signal state of the bit checked, but rather first forms a check result. In checks for signal state "1", the check result is identical to the signal state of the bit checked. In checks for signal state "0", the check result is the negated signal state of the bit checked.

Result of the logic operation

The result of the logic operation (RLO) is the signal state in the CPU, which the CPU then uses for subsequent binary signal processing. The result of the logic operation is formed and modified by check statements. An RLO of "1" means that the condition of the binary logic operation was fulfilled; "0" means that the condition was not fulfilled. Bits are set or reset according to the result of the logic operation.

Logic step

Similar to a sequence step in a sequence control, it is possible to define a logic step in a logic control. A result of a logic operation is generated and evaluated in an operation step (further processed). An operation step consists of scan operations and conditional operations. The first scan operation following a conditional operation is the first check. The operation step is highlighted in the program section shown below:

```
...  ...
=    Q 4.0      Conditional operation
A    I 2.0      First check
A    I 2.1      Scan operation
...  ...
A    I 1.7      Scan operation
=    Q 5.1      Conditional operation
...  ...
=    Q 4.3      Conditional operation
O    I 2.6      First check
O    I 2.5      Scan operation
...  ...
```

First check

The first check statement following a conditional statement is called the first check. It has a special meaning because the CPU directly accepts the check result of this statement as the result of the logic operation. The "old" RLO is thus lost. The first check always represents the start of a logic operation. The rule of logic governing a first check (AND, OR, Exclusive OR) plays no role in this.

Check statements

The result of the logic operation is formed with check statements. These statements check the signal state of a bit for "1" or "0" and link it according to AND, OR, or Exclusive OR. The CPU then saves the result of this logic operation as new RLO.

Figure 4.2 shows how checks for signal states "1" and "0" are programmed. The check for "1" takes the status of the bit checked as the check result for the next link. The check for "0" forms the check result from the negated status.

Conditional statements

Conditional statements are statements whose execution depends on the result of the logic operation. They include statements for assign-

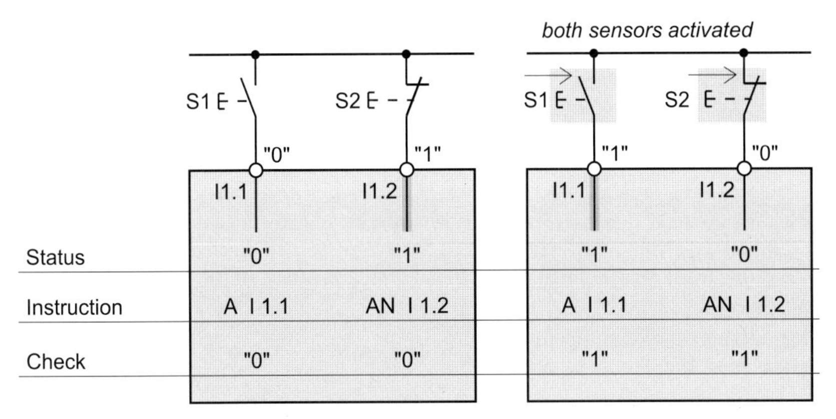

Figure 4.2 Checking for Signal States "1" and "0"

ing, setting and resetting binary locations, for starting timers and counters, and so on.

The conditional statements (with very few exceptions) are executed when the RLO is "1", and not executed when the RLO is "0". They do not affect the RLO (with very few exceptions), and the RLO thus remains the same over several contiguous statements.

Intelligible programming

The rule of logic governing a first check is irrelevant, as the result of the check is taken directly as the result of the logic operation. For the purpose of intelligible programming, the rule of logic for a first check should be identical to the desired function.

As an example, the sequence of statements

```
...
=    Q 15.3
O    I 18.5      First AND function
A    I 21.7
=    Q 15.4

A    I 18.4      Second AND function
A    I 21.6
=    Q 15.5
...
```

represents two AND functions, whereby the programming of the second AND function (in which both checks are programmed with AND) is to be preferred.

In the case of single scan statements, for example

```
...
=    Q 10.0
A    I 20.1      Assignment from
=    Q 10.1      I 20.1 to Q 10.1
...
```

the AND is preferable.

4.2 Elementary Binary Logic Operations

STL uses the binary functions AND, OR and Exclusive OR. These functions are linked to the check for signal state "1" or with the check for signal state "0".

A *Bit address*
Check for "1" and
combine according to AND

AN *Bit address*
Check for "0" and
combine according to AND

O *Bit address*
Check for "1" and
combine according to OR

ON *Bit address*
Check for "0" and
combine according to OR

X *Bit address*
Check for "1" and
combine according to Exclusive OR

XN *Bit address*
Check for "0" and
combine according to Exclusive OR

The checks for "1" set the result of the check to "1" when the signal state of the bit is "1". The checks for "0" produce a check result of "1" when the signal state of the bit is "0". This corresponds to an input which leads to the relevant function when negated.

The CPU then combines the result of the check with the current RLO as per the specified function and forms the new RLO. When a binary logic operation immediately follows a memory function, the result of the check is entered in the RLO buffer without a logic operation being performed.

The number of binary functions and the scope of a binary function are theoretically arbitrary; in practice, however, the restriction is given by the length of a block or the size of the CPU's work memory.

4.2.1 AND Function

The AND function links two binary states with one another and returns an RLO of "1" when both states (both results of the check) are "1". When you program the AND function several times in succession, all check results must be "1" in order for the common result of the logic operation to be "1". In all other cases, the AND function returns an RLO of "0".

Figure 4.3 shows an example for AND. In network 1, the AND function has three inputs; these may be arbitrary bit addresses. All of these bits are checked for signal state "1", so that the signal state of the bits will be directly linked according to AND. If all bits checked are "1", the Assign statement sets the bit *Output1* to "1". In all other cases, the AND condition is not fulfilled and bit *Output1* is set to "0".

Network 2 shows an AND function with a negated input. The input is negated by checking it for "0". The check result of a bit checked for "0" is "1" if that bit is "0", that is, the AND condition in the example is fulfilled when bit *Input4* is "1" and bit *Input5* is "0".

4.2.2 OR Function

The OR function combines two binary states with one another and returns an RLO of "1" when one of these states (one of the check results) is "1". When you program the OR function several times in succession, only one check result need be "1" for the common result of the logic operation to be "1". If all check results are "0", the OR function returns an RLO of "0".

Figure 4.3 shows an example of an OR function. In network 3, the OR function has three inputs; these may be arbitrary bit addresses. All bits are checked for "1", so that the signal state of the bits is linked directly according to OR. If at least one of the bits checked is "1", the subsequent Assign statement sets bit *Output3* to "1". If all the bits checked are "0", the OR condition is not fulfilled and *Output3* is reset to "0".

Network 4 shows an OR function with negated input. The input is negated by a check for "0". The check result for a bit that was checked for "0" is "1" when that bit is "0", that is, the OR condition in the example is fulfilled when bit *Input4* is "1" or bit *Input5* is "0".

4.2.3 Exclusive OR Function

The Exclusive OR function links two binary states with one another and returns an RLO of "1" when the two states (the two check results) are not the same. The function returns an RLO of "0" when the two states (the two check results) are the same.

Figure 4.3 shows an example of an Exclusive OR function. In network 5, two inputs (arbitrary bit addresses) lead to the Exclusive OR function. Both inputs are checked for "1". If the signal state of only one of these bits is "1", the Exclusive OR condition is fulfilled and the Assign statement sets bit *Output5* to "1". If both bits are "1" or "0", *Output5* is reset to "0".

Network 6 shows an Exclusive OR function with a negated input. The input is negated by a check for signal state "0". The check result for a bit checked for "0" is "1" when that bit is "0", that is, the Exclusive OR condition in the example is fulfilled when both input bits have the same signal state.

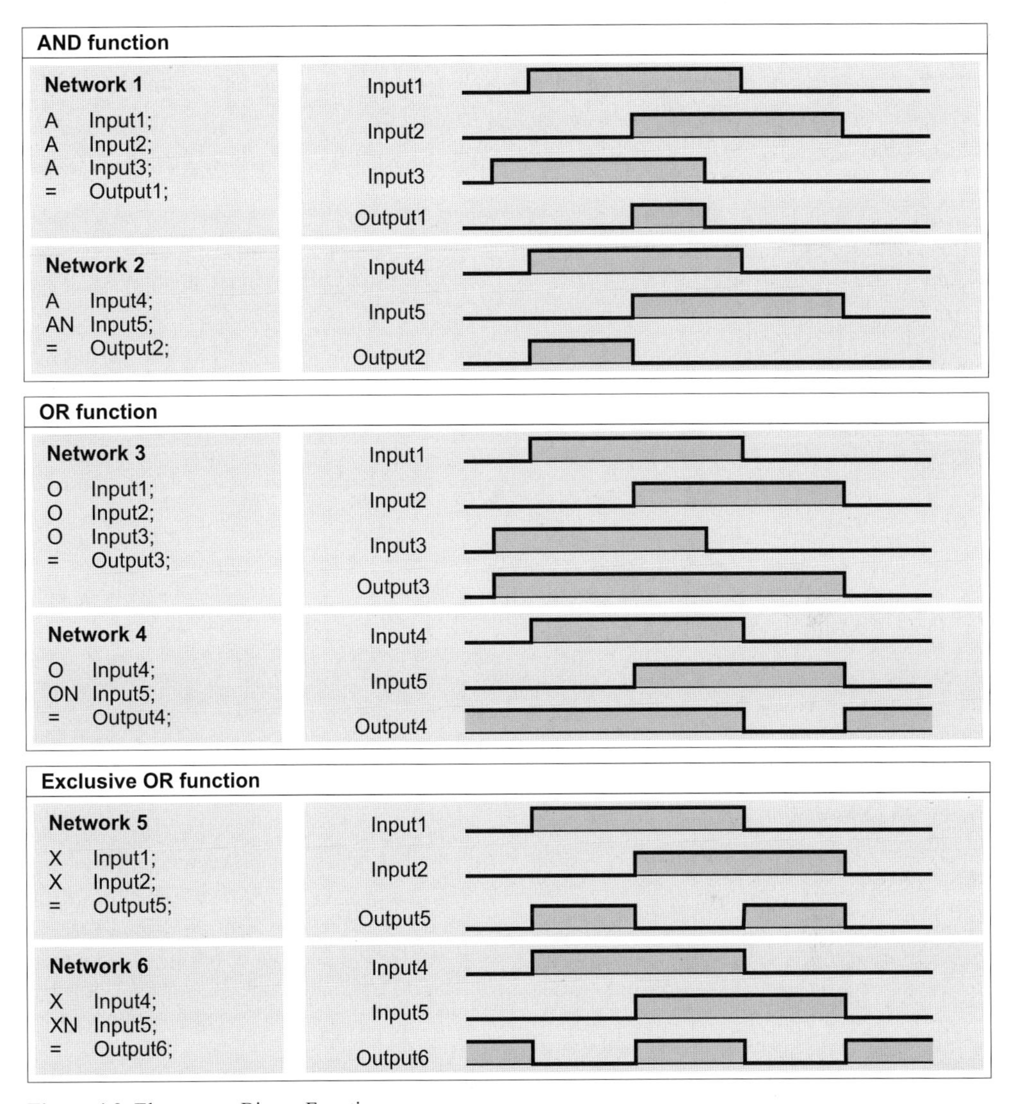

Figure 4.3 Elementary Binary Functions

You can also program the Exclusive OR function several times in succession, in which case the common RLO is "1" when an uneven number of the bits checked return a check result of "1".

Allowing for the Sensor Type

The binary functions AND, OR and Exclusive OR are described in preceding sections of this chapter as though normally open contacts were connected to the input modules (normally open contacts are sensors which return signal state "1" when activated). When implementing a control function, however, it is not always possible to use a normally open contact. In many cases, for example in the case of closed circuits, the use of normally closed contacts is absolutely essential (a normally closed contact is a sensor which returns signal state "0" when activated).

If the sensor connected to an input is a normally open contact, the input carries signal state "1" when the sensor is activated. If the sensor is a

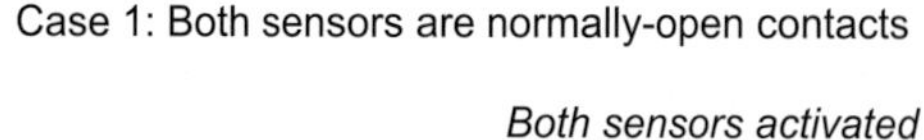
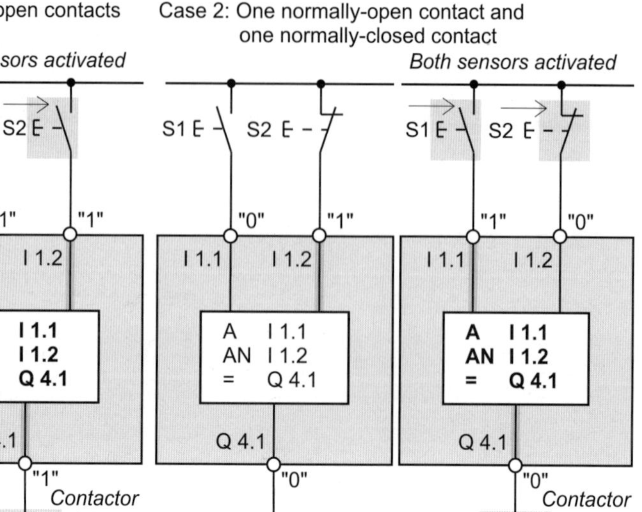

Figure 4.4 Allowing for the Sensor Type

normally closed contact, the input carries signal state "1" when the sensor is inactive. The CPU has no way of knowing whether an input is associated with a normally open or a normally closed contact; it can only differentiate between signal state "1" and signal state "0".

When developing the program, it is therefore necessary to take the sensor type into account. Before writing the program, you have to know whether the sensor is a normally closed contact or a normally open contact. Because the program is in part determined by the function of the sensor ("Sensor activated", "Sensor not activated"), it follows that you must check the input for signal state "1" or signal state "0", depending on the type of sensor used. In this way, you can also directly check inputs which are to execute various activities when "0" ("active when zero" inputs) and use the check result in subsequent links.

Figure 4.4 shows how to program in dependence on the sensor type. In the first case, two normally open contacts are connected to the programmable controller, in the second case, one normally open contact and one normally closed contact. In both cases, a contactor connected to an output is to pick up when both sensors are activated.

When a normally open contact is activated, the signal state of the input is "1"; in order to fulfill the AND condition with check result "1", the input is checked for "1". When a normally closed contact is activated, the signal state of the input is "0". In order to fulfill the AND condition with a check result of "1" in this case, the input must be checked for signal state "0".

4.3 Negating the Result of the Logic Operation

NOT negates the result of the logic operation. You can use NOT at any location in the program, even within a logic operation. You can use NOT, for example, to negate the AND condition for an output (NAND function, Figure 4.5 network 7). Network 8 shows the negation of an OR function, which is called a NOR function.

You will find additional examples for NOT in Chapter 4.4.6 "Negating Nesting Expressions".

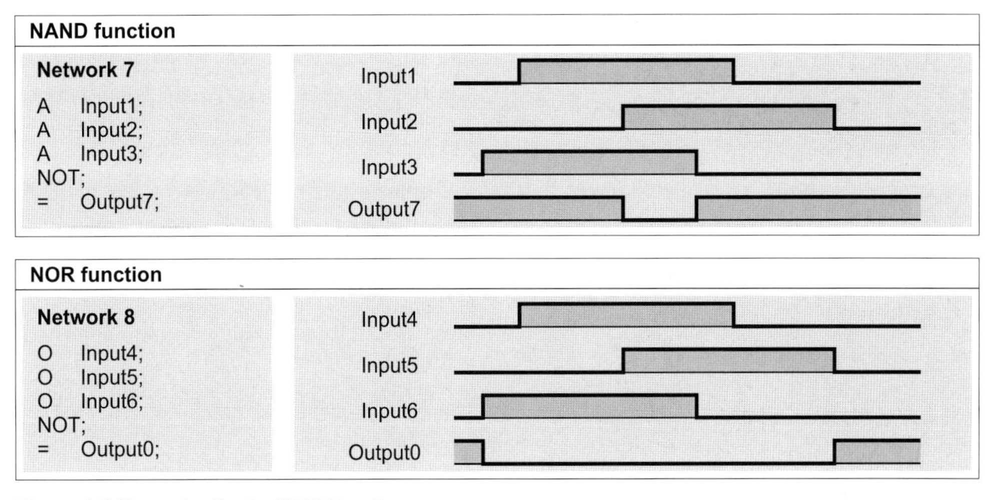

Figure 4.5 Examples for the NOT Function

4.4 Compound Binary Logic Operations

Binary logic operations can be combined, for instance AND and OR functions can be programmed in any order. When such functions are programmed in arbitrary order, the CPU's handling of them is very difficult to duplicate. It is better, for instance, to illustrate the problem solution in the form of a function block diagram, then program it in STL.

When compound binary logic operations are programmed, STL treats OR and Exclusive OR the same (they have the same priority). AND is executed "before" OR or Exclusive OR, and has a higher priority.

In order for the functions to be processed in the required order, it is sometimes necessary for the CPU to temporarily store the function value (the RLO that has been computed up to a certain point in the program). The nesting expressions are provided for this purpose. As in the case of the notation used for Boolean algebra, the nesting expressions cause one function to be executed "before" another. The nesting expressions also include OR.

The STL programming language provides the following binary nesting expressions:

O ORing of AND functions

A(Open bracket with AND function

O(Open bracket with OR function

X(Open bracket with Exclusive OR function

AN(Open bracket with negation and AND function

ON(Open bracket with negation and OR function

XN(Open bracket with negation and Exclusive OR function

) Close bracket

The rule of logic for the open bracket statement indicates how the result of the nesting expression is to be linked with the current RLO when the close bracket operation is encountered. Prior to this logic operation, the result of the nesting expression is negated if a negation character is specified.

4.4.1 Processing Nesting Expressions

In the STL programming language, the binary nesting expressions are used to define the order in which binary logic operations are processed. At runtime, the setting of brackets has the effect as the CPU processing the nesting expressions "first", that is, before executing the instructions outside the brackets.

When it encounters an open bracket statement, the CPU stores the current RLO internally, then processes the nesting expression, when it

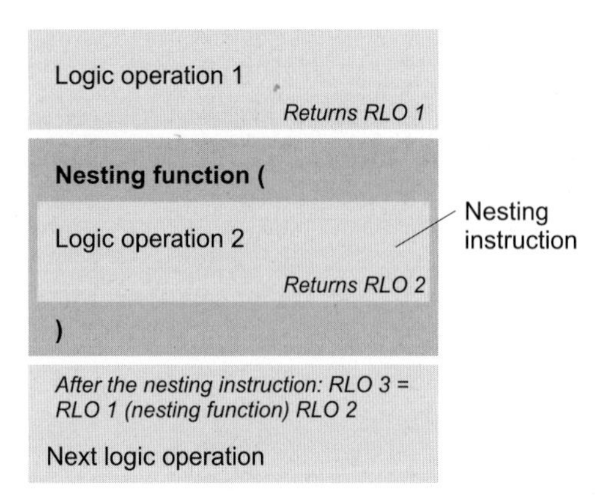

Figure 4.6 Processing Nesting Expressions

encounters the close bracket statement, it links the RLO from the nesting expression with the RLO it stored prior to processing the nesting expression as per the function given in the open bracket statement (Figure 4.6).

A check statement following a open bracket statement is always a first check because the CPU always regenerates the RLO within a nesting expression. A check statement following a close bracket statement is never a first check because, when a nesting expression is the first instruction in a logic operation, the CPU treats the RLO from the nesting statement like the result of a first check.

Nesting expressions can be nested, that is to say, you can program a nesting expression in a nesting expression (Figure 4.7). The nesting depth is seven, that is, you may begin a nesting expression seven times without first terminating one. Processing within the brackets is much as described above.

Saving intermediate results with the aid of the nesting stack

Internally, the CPU sets up a nesting stack in order to process nesting functions. In this stack it stores:

▷ The result of the logic operation (RLO) preceding the bracket,

▷ The binary result (BR) preceding the bracket,

▷ The status bit (OR) (indicating whether an OR condition was already fulfilled) and

▷ The nesting function (with which function the nesting expression is to be linked).

The CPU sets the binary result BR following the *close bracket* statement the signal state it had prior to the nesting expression.

Within a nesting expression, you can not only program binary logic operations but all statements in the STL programming language. Care must be taken, however, that nesting expression be terminated with the *"close bracket"* statement. It is thus possible, for example, to program several logic steps or memory and comparison functions within a nesting expression.

4.4.2 Combining AND Functions According to OR

These logic operations, which are a combination of OR and AND functions, can be written in Boolean algebra without brackets. It is the rule that the AND functions are processed "first". The results of the AND functions, together with additional OR checks, if any, are then linked with OR.

Example:

```
A     Input0;
A     Input1;
O     ;
A     Input2;
A     Input3;
=     Output8;
```

In the example, a single O (for OR) is between the first and the second AND function. This operation makes "AND before OR" possible, and is always necessary when an AND function is placed "before" an OR function. The single "O" precedes the AND function, and is no longer necessary after the AND function.

In the example, *Output8* is set when {*Input0* and *Input1*} or {*Input2* and *Input3*} are "1".

4.4.3 Combining OR and Exclusive OR Functions According to AND

These compound logic operations comprising AND and OR functions must be written with brackets in Boolean algebra to indicate that the OR functions are to be processed "before" the AND functions.

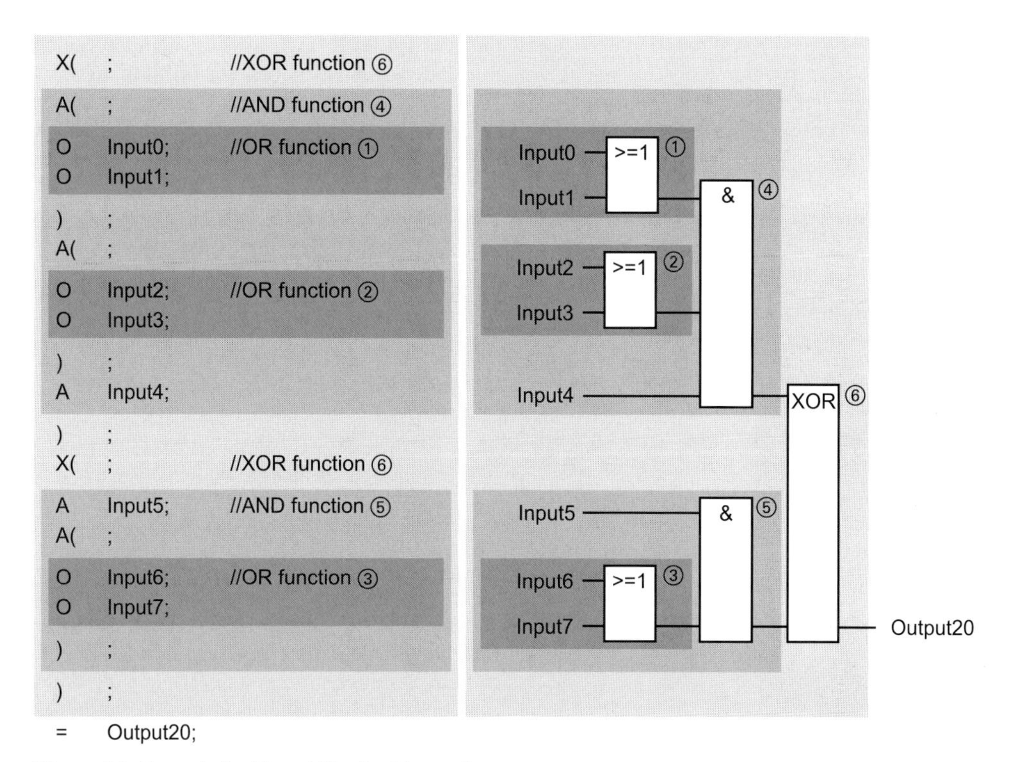

Figure 4.7 Example for Nested Nesting Expressions

Example:

```
A(      ;
O       Input0;
O       Input1;
)       ;
A(      ;
O       Input2;
O       Input3;
)       ;
=       Output10;
```

The *open bracket* statement is "combined" with an AND function. The OR function is within the nesting expression. The *close bracket* statement, in this case, links the result of the OR function (generally the result of the logic operation computed in the brackets) with additional checks, if any, according to AND.

In the example, *Output10* is set when {*Input0* or *Input1*} and {*Input2* or *Input3*} are "1".

The ANDing of Exclusive OR functions is programmed exactly the same way. The OR functions in the example could be replaced by Exclusive OR, since the two functions have the same priority.

4.4.4 Combining AND Functions According to Exclusive OR

An AND function before an Exclusive OR function is written in brackets. With the aid of the brackets, the CPU saves the result of the AND function and can then combine it, possibly with additional checks, according to the rules governing Exclusive OR.

Example:

```
X(      ;
A       Input0;
A       Input1;
)       ;
X(      ;
A       Input2;
A       Input3;
)       ;
=       Output12;
```

In the example, the first AND function need not be in brackets, as an AND function has a higher priority than an Exclusive OR. The brackets, however, make the program more readable.

Output12 in the example is set either when {*Input0* and *Input1*} or {*Input2* and *Input3*} are "1".

4.4.5 Combining OR Functions and Exclusive OR Functions

An OR function before an Exclusive OR function is written in brackets. With the aid of the brackets, the CPU saves the result of the OR function and can link it, possibly with additional checks, according to the rules governing Exclusive OR.

Example:

```
X(      ;
O       Input0;
O       Input1;
)       ;
X(      ;
O       Input2;
O       Input3;
)       ;
=       Output14;
```

In the example, *Output14* is set when one, and only one, of the two OR conditions is fulfilled.

An OR before an Exclusive OR is programmed in exactly the same way. An OR in the example can be replaced by an Exclusive OR and vice versa, since the two functions have the same priority.

4.4.6 Negating Nesting Expressions

Just as you can check a bit for signal state "0" (negate the status, as it were), you can also negate a nesting expression. This means that the CPU will post-process the result of the nesting expression in negated form. Negation is specified by an additional N in the open bracket statement.

Example:

```
AN(     ;
O       Input0;
O       Input1;
)       ;
AN(     ;
X       Input2;
X       Input3;
)       ;
=       Output16;
```

In the example, *Output16* is set if neither the OR condition nor the Exclusive OR condition is fulfilled.

A second way to negate nesting expressions is through the use of the NOT (negation) statement. A NOT written before the close bracket statement negates the result of the nesting statement prior to further processing.

Example:

```
A(      ;
O       Input0;
O       Input1;
NOT     ;
)       ;
A(      ;
X       Input2;
X       Input3;
NOT     ;
)       ;
=       Output17;
```

This logic operation has the same function as the preceding logic operation. Negation of the nesting expression is attained here within the brackets using NOT.

5 Memory Functions

This chapter describes the memory functions for the STL programming language; these include Assign for dynamic bit control and Set and Reset for static control. The memory functions also include edge evaluations.

The memory functions are used in conjunction with binary logic operations in order to affect the signal states of bits with the help of the RLO generated in the CPU.

You can use the memory functions to control all bit addresses: the process input/output image, the memory bits, the global data and the static and temporary local data.

The examples in this chapter can be found in function block FB 105 or in the source file Chap_5 in the STL_Book library under the "Basic Functions" program which you can download from the publisher's web site (see page 8).

5.1 Assign

= *Bit*
Assigns the result of the logic operation

The Assign statement "=" assigns the RLO in the processor directly to the bit specified in the statement. If the result of the logic operation is "1", the bit is set; if the result of the logic operation is "0", the bit is reset (Figure 5.1 network 1). If you want the bit to be set when the RLO is "0", you can negate the RLO prior to Assign with the NOT statement (network 2).

You will find additional examples for Assign in Chapter 4 "Binary Logic Operations".

Simultaneous execution of multiple Assigns

You can also assign the result of the logic operation to several different bits by programming successive Assign statements specifying the relevant bits (Figure 5.1 network 3).

All specified bits react the same, as the instructions used for bit control do not affect the RLO. The CPU does not generate a new result of logic operation (RLO) until it encounters the next check statement.

If you want to use the signal state of an output in another logic operation, simply check that output with the appropriate check statement (network 4).

5.2 Set and Reset

S *Bit*
Sets the bit when the result of the logic operation is "1"

R *Bit*
Resets the bit when the result of the logic operation is "1"

The Set S and Reset R instructions are executed only when the result of the logic operation is "1". The Set instruction then sets the specified bit to "1", and the Reset instruction sets it to "0". RLO "0" has no effect on the Set or Reset instruction; when the RLO is "0", the bit specified in a Set or Reset instruction retains it current signal state (Figure 5.1 networks 5 and 6).

Simultaneous execution of multiple memory functions

You can control multiple Set and Reset instructions, in any combination and together with Assigns, with the same RLO. Simply write successive statements specifying the relevant bits (Figure 5.1 network 7). As long as Set, Reset

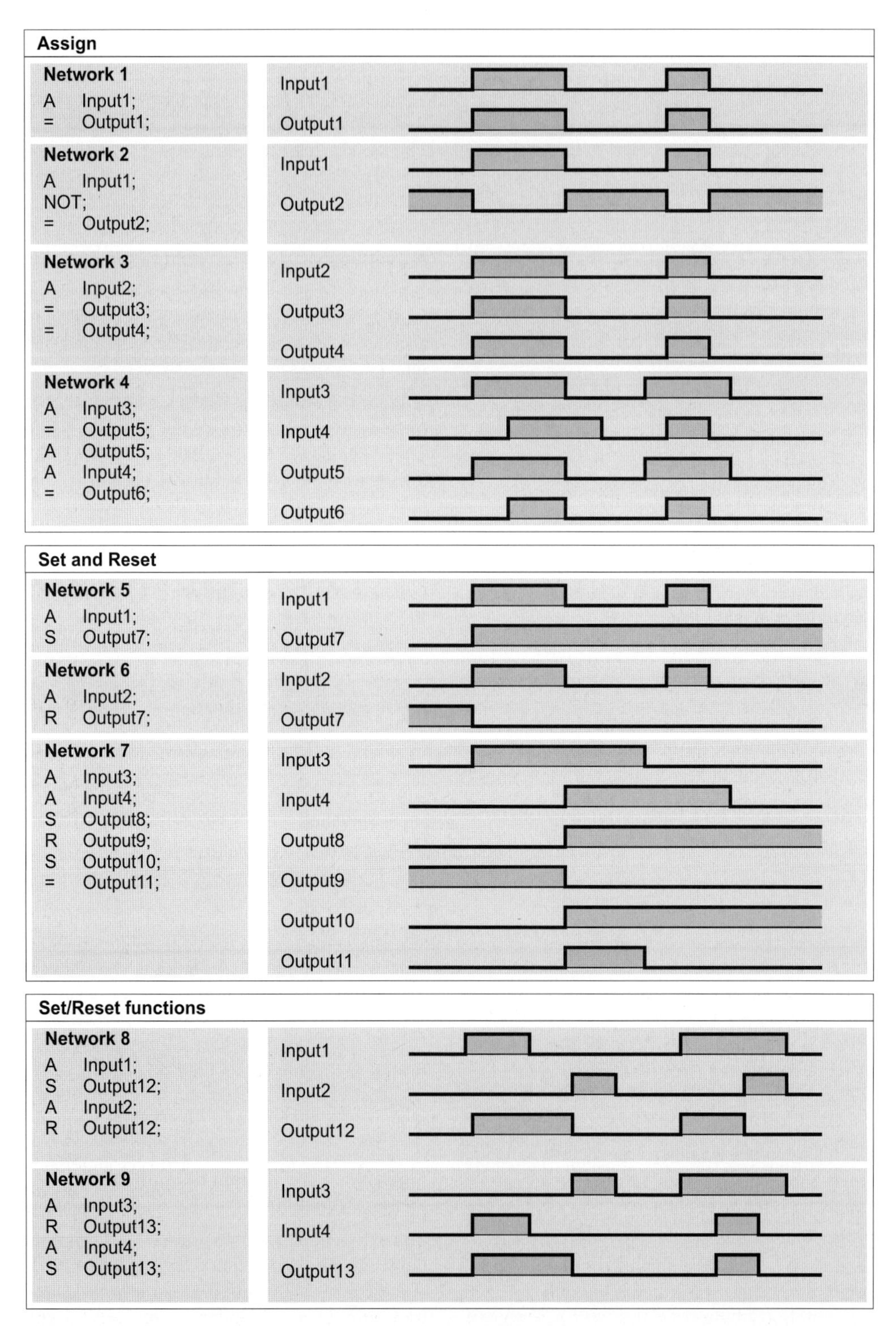

Figure 5.1 Assign, Set and Reset

and Assign statements are being processed, the RLO does not change. The CPU does not generate a new RLO until it encounters the next check statement.

Here, too, you can use NOT to negate the RLO within the sequence of memory statements.

To ensure the clarity and readability of the program, you should use the Set and Reset statements for a specific bit in pairs, and only once for a given bit.

5.3 RS Flipflop Function

The RS flipflop function consists of one Set and one Reset statement; there is no special identifier in STL. The RS flipflop function is implemented by programming successive Set and Reset statements specifying the same bit. Important to the RS flipflop function's functionality is the order in which you program the Set and Reset statements.

Note that the bits used in memory functions are normally reset on startup (warm restart). In special cases, the signal states of the bits specified in memory functions are retained; this depends on the type of startup (for instance a hot restart), the specified bit (for example a bit in the static local data area), and settings in the CPU (such as retentivity).

5.3.1 Memory Functions with Reset Priority

Reset priority means that the specified bit is reset when the Set and Reset instructions produce a signal state of "1" "simultaneously". The Reset instruction then takes priority over the Set instruction (Figure 5.1 network 8).

Because the statements are processed sequentially, the CPU initially sets the bit because it executes the Set instruction first, then resets it when it executes the Reset instruction. The output then remains reset for the remainder of the program scan cycle.

If the bit is an output, this brief setting only takes place in the process image, and the (external) output on the associated output module is not affected. The CPU does not transfer the process-image output table to the output modules until the end of the program scan cycle.

Reset priority is the "standard" form of this function, since the reset state is, as a rule, the safer or less hazardous state.

5.3.2 Memory Function with Set Priority

Set priority means that the specified bit is set when the Set and Reset instructions produce a signal state of "1" simultaneously. The Set instruction then takes priority over the Reset instruction (Figure 5.1 network 9).

When it processes the statement sequence, the CPU first sets the specified bit to "0"; then, when it processes the Set instruction, it sets the bit to "1". The output remains set for the remainder of the program scan cycle.

If the bit is an output, this brief setting only takes place in the process image and the "external" output on the associated output module remains unaffected.

Set priority is the exception rather than the rule for this function. It is used, for example, to implement a fault signal latch when, despite an acknowledgment at the Reset input, the current fault signal at the Set input is to continue to set the bit specified in the memory function.

5.3.3 Memory Function in a Binary Logic Operation

In the STL programming language, you can use the memory functions very freely. It is possible to save the RLO at any location in the program, and then reuse it later.

The example in Figure 5.2 does not use nesting statements to control the sequence of a binary logic operation, but rather to temporarily save the result of a logic operation.

Within the nested expression, an RS flipflop is used here. Its signal state is to be combined further. For this purpose, it is necessary for the flipflop to be scanned at the end of the nested expression in order to obtain the signal state of the flipflop before the close bracket statement. If this statement is missing, the signal state of the logic operation prior to the reset input would, in this case, be combined further.

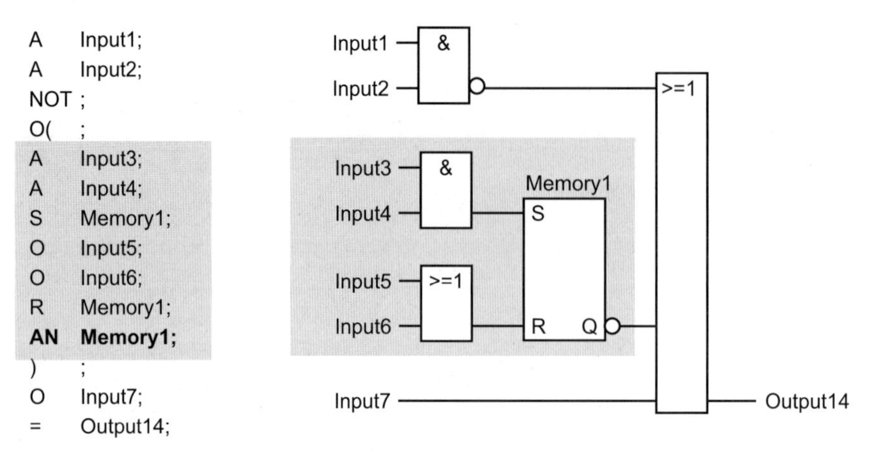

Figure 5.2 Nesting Statement as Intermediate Bit Buffer

You can program any STL statements you wish between the brackets; however, make sure that you have the RLO you want before writing the close bracket statement.

Intermediate binary results

Almost any bits can be used for the temporary storing of binary results:

▷ Temporary local data bits are the best suited when you need the intermediate result only within the block itself. All code blocks have temporary local data areas.

▷ Static local data bits are available only in function blocks, and save their signal states until set again.

▷ Memory bits are available globally in a CPU-specific quantity; for clarity of programming, avoid multiple use of memory bits (the same bits for different tasks).

▷ Data bits in global data blocks are also available throughout the whole program, but before they can be used, the relevant data block must first be opened (even if this is only implied through the use of complete addressing).

Note: You can replace the "scratchpad memory" used in STEP 5 with temporary local data, which are available in every block.

5.4 Edge Evaluation

FP *Bit*
Positive (rising) edge

FN *Bit*
Negative (falling) edge

Edge evaluation is the detection of a change in a signal state, a signal edge. A positive (rising) edge is present when the signal goes from "0" to "1". The opposite is referred to as a negative (falling) edge.

In a relay logic diagram, the equivalent of edge evaluation is a pulse contact element. If this pulse contact element emits a pulse when the relay is switched on, this corresponds to a rising edge. Emission of a pulse when the relay is switched off corresponds to a falling edge.

The bit specified in the edge evaluation is referred to as "edge memory bit" (it need not necessarily be a memory bit). However, it must be a bit whose signal state is once again available in the next program scan cycle and which is otherwise not used in the program. Suitable bits are memory bits, data bits in global data blocks, and static local data bits in function blocks.

This edge memory bit stores the "old" RLO, which is the one the CPU used for the last edge evaluation. In each edge evaluation, the CPU compares the current RLO with the signal state of the edge memory bit. An edge is present when they have different signal states. In this case, the CPU updates the signal state of the edge memory bit by assigning the current RLO to it, and sets the RLO to "1" on either a positive or negative edge, depending on the instruction, following edge evaluation. If the CPU does not detect an edge, it sets the RLO to "0".

Signal state "1" following an edge evaluation thus means "edge detected". The signal state remains at "1" only briefly, as a rule for only one scan cycle. Because the CPU does not detect an edge on the next edge evaluation (when the edge evaluation's "input RLO" does not change), it sets the RLO back to "0" following the edge evaluation.

You can process the RLO directly after an edge evaluation, for example with a Set operation, or you can store it in a bit (a "pulse memory bit"). Use a pulse memory bit when the RLO from the edge evaluation is to processed at another location in the program; it is effectively the intermediate buffer for a detected edge. Bits suitable as pulse memory bits are memory bits, data bits in global data blocks, and temporary and static local data bits.

You can also further process the RLO after an edge evaluation directly with the following check statements.

Note the response of the edge evaluation when you switch on the CPU. If an edge is to be detected, the RLO prior to edge evaluation and the signal state of the edge memory bit must be identical. Under certain circumstances, the edge memory bit must be reset on startup (depending on the desired response and on the bit used).

The following examples illustrate how edge evaluation works. In simplified form, an input represents the RLO prior to edge evaluation and a memory bit (the "pulse memory bit") the RLO following edge evaluation. Of course, the edge evaluation may also be preceded and followed by a binary logic operation.

5.4.1 Positive Edge

The CPU detects a positive (rising) edge when the result of the logic operation changes from "0" to "1" prior to the edge evaluation. The procedures involved are shown in Figure 5.3 above; the consecutive number stands for successive scan cycles:

① The first time around, the signal state of both the input and the edge memory bit is "0". The pulse memory bit remains reset.

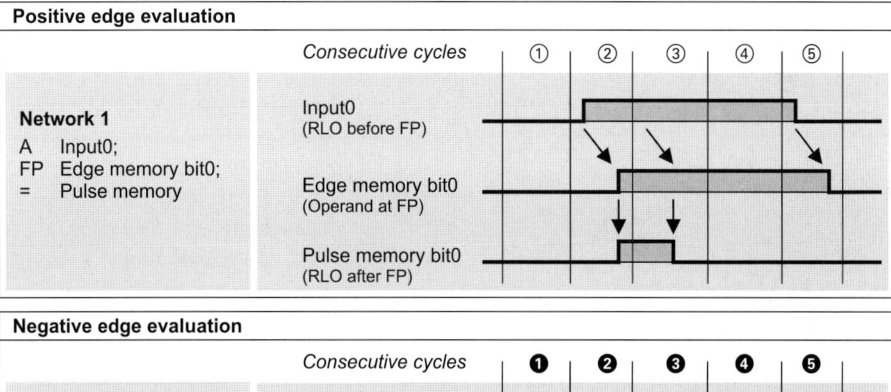

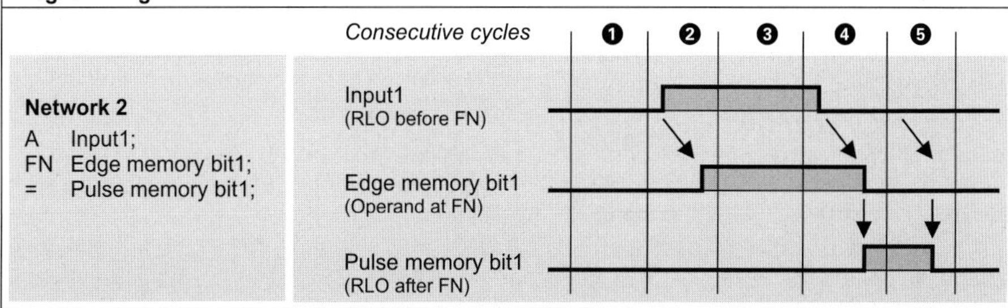

Figure 5.3 Edge Evaluations

② The second time around, the signal state of the input should have changed from “0” to “1”. The CPU detects the change by comparing the current RLO with the status of the edge memory bit. If the RLO is “1” and the edge memory bit “0”, the edge memory bit is set to “1”. The current RLO is also set to “1”.

③ The next time around, the CPU finds that the there is no different between the state of the input and that of the edge memory bit, and therefore sets the current RLO to “0”.

④ As long as there is no difference between the two states, the RLO remains at “0” and the edge memory bit remains set.

⑤ When the input once again has a signal state of “0”, the CPU corrects the edge memory bit. The RLO remains at “0”. The initial state is then reestablished.

5.4.2 Negative Edge

The CPU detects a negative (falling) edge when the RLO changes from “1” to “0” prior to the edge evaluation. The procedures involved are shown in Figure 5.3 below; the consecutive number stands for successive scan cycles:

❶ The first time around, the signal state of both the input and the edge memory bit is “0”. The pulse memory bit remains reset.

❷ The second time around, the signal state of the input should have changed from “0” to “1”. The CPU detects change by comparing the current RLO with the status of the edge memory bit. If the RLO is “1” and the edge memory bit “0”, the edge memory bit is set to “1”. Following edge evaluation, the RLO remains “0”.

❸ As long as there is no difference between the two states, the RLO remains at “0” and the edge memory bit remains set.

❹ When the input once again has a signal state of “0”, the CPU corrects the status of the edge memory bit and sets the RLO to “1” following edge evaluation.

❺ In the next scan cycle, the signal state of the input and that of the edge memory bit are the same. The CPU therefore sets the RLO back to “0”, thus reestablishing the original state.

5.4.3 Testing a Pulse Memory Bit

The signal states of the pulse memory bits are very difficult to monitor with the programming devices' test functions because they remain at “1” for only one scan cycle.

It is for this reason that an output is also unsuitable as pulse memory bit, as the signal amplifiers on the output module or the actuators are not capable of duplicating the signal changes all that quickly.

With a “flying restart circuit”, however, you can record the extremely brief signal states of the pulse memory bits in an RS flipflop. The pulse memory bit sets the RS flipflop, thus storing the “Edge Detected” signal. After you have evaluated this signal, you can reset the flipflop.

```
O       PMembit0;
O       PMembit1;
S       Flipflop2;
A       Input2;
R       Flipflop2;
```

After you have evaluated the saved edge, you can reset the flipflop again.

5.4.4 Edge Evaluation in a Binary Logic Operation

Edge evaluation in a binary logic operation can be used to serve a practical purpose only when you use the signal state following edge evaluation (the “pulse*”)* to control a memory, timer or counter function. Binary checks may lie between the edge evaluation and control of the relevant function.

```
O       Input3;
O       Input4;
FP      EMembit2
A       Input5;
S       Output15;
A       Input6;
FN      EMembit3;
R       Output15;
```

In the example, *Output15* is set at the instant at which the OR condition is fulfilled (when the bit in the OR statement goes from “0” to “1”*)* and *Input5* is “1”. *Output15* is reset on a falling edge at *Input6*.

An edge evaluation is effectively a first check, as the RLO generated by the edge evaluation can be post-processed. This also means that the logic operation up to the instant of the edge evaluation is regarded as "completed" (a fulfilled OR condition is not stored). Edge evaluation does not affect the processing of nesting instructions.

5.4.5 Binary Scaler

A binary scaler has one input and one output. If the signal at the binary scaler's input changes its state, for example from "0" to "1", the output changes its state as well (Figure 5.4). This (new) signal state is retained until the next, in our example positive, signal state change. Only then does the signal state of the output change again. This means that half the input frequency appears at the output of the binary scaler.

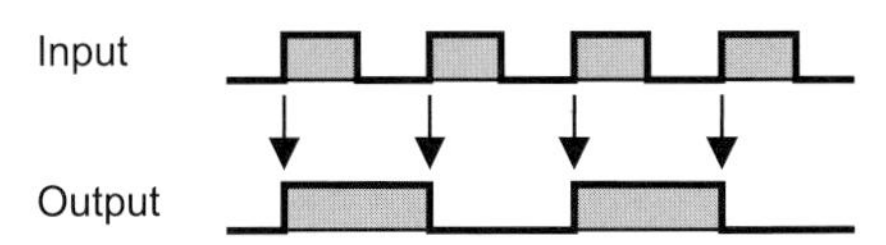

Figure 5.4 Pulse Diagram of a Binary Scaler

There are different methods of solving this task, two of which are presented below.

In the first solution, a pulse memory bit is used to set the output if it was reset and to reset it if it was set. The important thing to remember when programming this solution is that the pulse memory bit has to be reset once it has set the output (otherwise the output will be immediately reset again).

```
A       Input_1;
FP      EMembit_1;
=       PMembit_1;
A       PMembit_1;
AN      Output_1;
S       Output_1;
R       PMembit_1;
A       PMembit_1;
A       Output_1;
R       Output_1;
```

The second solution uses a conditional jump JCN to evaluate the edge. When the CPU does not detect an edge, the RLO is "0" and the program scan is resumed at the jump label.

In the case of a positive edge, the CPU does not execute the jump and executes the next two statements. If the output is reset, it is set; if it is set, it is reset. Although an Assign controls the output, the latter functions as a latch, as this program section is executed only when there is a positive edge.

```
        A       Input_2;
        FP      FMembit_2;
        JCN     M1;
        AN      Output_2;
        =       Output_2;
M1:     ...     ;
```

5.5 Example of a Conveyor Belt Control System

A functionally extremely simple conveyor belt control system is used here as an example to show how binary logic operations and memory functions work in conjunction with inputs, outputs, and memory bits.

Functional description

Parts are to be transported on a conveyor system, one crate or pallet per belt. The essential functions are as follows:

- ▷ When the belt is empty, the controller requests more parts with the "readyload" signal (ready to load)
- ▷ The "Start" signal starts the belt, and the parts are transported
- ▷ At the end of the belt, an "end-of-belt" sensor (a light barrier, for example) detects the parts, at which point the belt motor switches off and triggers the "ready_rem" signal (ready to remove)
- ▷ At the "Continue" signal, the parts are transported further until the "end-belt" (end-of-belt) sensor no longer detects them.

The function block diagram for the conveyor belt control system is shown in Figure 5.6. The example is programmed with inputs, outputs and memory bits. it can be loaded anywhere in any block. In the example, a function without a function value was chosen as block.

In the Chapter 19 "Block Parameters" the same example is programmed in a function block with block parameters; the function block can be called more than once (for more than one belt).

Signals and symbols

A number of additional signals supplement the functionality of the conveyor belt control system:

▷ Basic_st
Sets the controller to the basic state

▷ Man_on
Switches on the belt without regard to any conditions

▷ /Stop
Stops the belt as long as the "0" signal is present (an NC contact as sensor, "zero active")

▷ Light_barrier1
The parts have reached the end of the belt

▷ /Mfault
Fault signal from the belt motor (e.g. motor protection switch); designed as "zero active" signal so that other malfunctions, such as a wire break, will also generate a fault signal

We want to use symbolic addressing, that is, the addresses are assigned names which we then use when writing the program. Prior to incremental program input or to compilation, we generate a symbol table (Table 5.1), that contains the inputs, outputs, memory bits and blocks.

Program

The example is located in a function without block parameters. You can call this function, for instance, in organization block OB 1 as follows:

```
CALL Belt_control;
```

The example is in the form of source text with symbolic addressing. The global symbols can also be used without quotation marks when they contain no special characters. If a symbol contains a special character (such as a space), it must be enclosed in quotation marks. The STL editor displays all global symbols in the compiled block in quotation marks.

The program is subdivided into networks to improve clarity and readability. The last network, with the title BLOCK END, is not absolutely necessary, but serves as a visual end of the block, a very useful feature for extremely long blocks.

Table 5.1 Symbol Table for the Example "Conveyor Belt Control System"

Symbol	Address	Data Type	Comment
Belt_control	FC 11	FC 11	Belt control system
Basic_st	I 0.0	BOOL	Set controllers to the basic state
Man_on	I 0.1	BOOL	Switch on conveyor belt motor
/Stop	I 0.2	BOOL	Stop conveyor belt motor (zero-active)
Start	I 0.3	BOOL	Start conveyor belt
Continue	I 0.4	BOOL	Acknowledgment that parts were removed
Light barrier1	I 1.0	BOOL	(Light barrier) sensor signal "End of belt" for belt 1
/Mfault1	I 2.0	BOOL	Motor protection switch belt 1, zero-active
Readyload	Q 4.0	BOOL	Load new parts onto belt (ready to load)
Ready_rem	Q 4.1	BOOL	Remove parts from belt (ready to remove)
Beltmot1_on	Q 5.0	BOOL	Switch on belt motor for belt 1
Load	M 2.0	BOOL	Load parts command
Remove	M 2.1	BOOL	Remove parts command
EM_Rem_N	M 2.2	BOOL	Edge memory bit for negative edge of "remove"
EM_Rem_P	M 2.3	BOOL	Edge memory bit for positive edge of "remove"
EM_Loa_N	M 2.4	BOOL	Edge memory bit for negative edge of "load"
EM_Loa_P	M 2.5	BOOL	Edge memory bit for positive edge of "load"

```
FUNCTION Belt_control: VOID
TITLE = Control of a conveyor belt
//Example of binary logic operations and memory functions, without block parameters
NAME    : Belt1
AUTHOR  : Berger
FAMILY  : STL_Book
VERSION : 01.00
BEGIN
NETWORK
TITLE = Load parts
//This network generates the command "Load" that initiates transport of parts
//to the end of the belt.
      A     Start;                    //Start conveyor belt
      S     Load;
      O     Light_barrier1;           //Parts have reached end of belt
      O     Basic_st;
      ON    "/Mfault1";               //Motor protection switch (zero active)
      R     Load;
NETWORK
TITLE = Parts ready for removal
//When parts have reached end of belt, they are ready for removal.
      A     Load;                     //When end of belt has been reached,
      FN    EM_Loa_N;                 //"Load" is reset.
      S     Ready_rem;                //Parts are then "ready for removal"
      A     Remove;
      FP    EM_Rem_P;                 //The parts are removed
      O     Basic_st;
      ON    "/Mfault1";
      R     Ready_rem;
NETWORK
TITLE = Remove parts
//The "Remove" command initiates removal of the parts from the belt.
      A     Continue;                 //Switch belt back on
      S     Remove;
      ON    Ligh_barrier1;            //Parts leave the belt
      O     Basic_st;
      ON    "/Mfault1";               //Motor protection switch (zero active)
      R     Remove;
NETWORK
TITLE = Belt ready for loading
//The belt is ready for loading when the parts have left the belt.
      A     Remove;
      FN    EM_Rem_N:                 //Parts have left the belt
      O     Basic_st;
      S     Readyload;                //Belt is empty
      A     Load;
      FP    EM_Loa_P;                 //Belt is started
      ON    "/Mfault1";
      R     Readyload;
NETWORK
TITLE = Control belt motor
//The belt motor is switched on and off in this network.
      A(;
      O     Load;                     //Load parts onto belt
      O     Remove;                   //Remove parts from belt
      O     Man_on;                   //Start with "Man_on" (non-retentive)
      );
      A     "/Stop";                  //Stop and motor fault prevent
      A     "/Mfault1";               //belt motor from running
      =     Belt_motor1;
NETWORK
TITLE = Block end
      BE;
END_FUNCTION
```

Figure 5.5 Program of the Example of a Conveyor Belt Control System

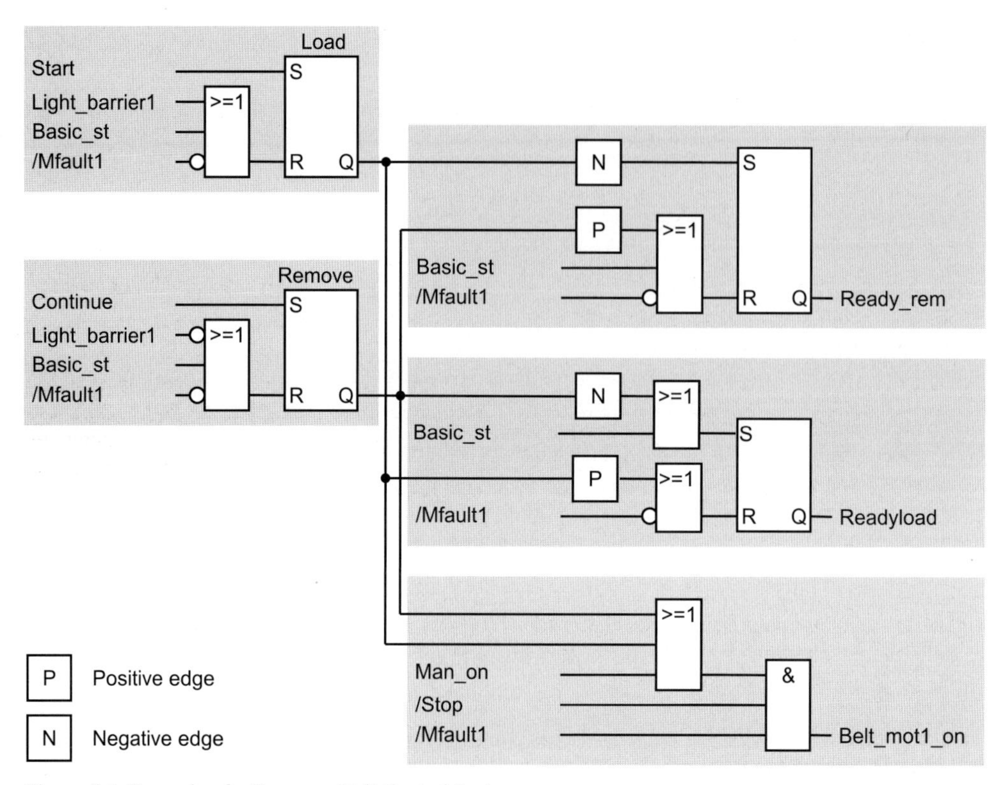

Figure 5.6 Example of a Conveyor Belt Control System

6 Move Functions

This chapter describes functions for the STL programming language which interchange data with the accumulators (registers). These include

- ▷ Load functions
  The load functions are used to fill the accumulators for subsequent digital post-processing, for instance compare, compute, and so on.

- ▷ Transfer functions
  The transfer functions transfer the digital results from accumulator 1 to memory areas in the CPU, for instance bit memory.

- ▷ Accumulator functions
  These functions transfer information from one accumulator to another or replace information in accumulator 1.

You also need load functions to specify initial values for timers and counters or to process current times and counts.

System functions SFC 20 BLKMOV, SFC 81 UBLKMOV, SFC 21 FILL, SFC 83 READ_DBL and SFC 84 WRIT_DBL are available for copying larger volumes of data in memory or to preset data areas.

You need the load and transfer functions to address modules via the user data area; when you address modules via the system data area, you must use system functions to transfer data records. You can also use these system functions to parameterize the modules.

The examples in this chapter can be found in function block FB 106 or in the source file Chap_6 in the STL_Book library under the "Basic Functions" program which you can download from the publisher's web site (see page 8).

6.1 General Remarks on Loading and Transferring Data

The load and transfer functions enable the exchange of information between different areas of memory. This information exchange does not take place directly, but instead is "routed through" accumulator 1. An accumulator is a special register in the processor, and serves as "intermediate buffer".

When information is exchanged, the direction in which the information flows is indicated by the instruction used to transfer that information. The information flow from a memory area to accumulator 1 is called *loading,* the reverse direction of flow is called *transferring* (the contents of the accumulator are "transferred" to the memory area).

Loading and transferring are the prerequisites for the use of the *digital functions,* which manipulate a digital value (convert or shift, for example) or combine two digital values (for instance compare or add). In order to combine two digital values, two intermediate buffers are needed, namely accumulator 1 and accumulator 2. All CPUs are equipped with these two special registers. The S7-400 CPUs have two additional intermediate buffers, accumulator 3 and accumulator 4, which are used primarily in conjunction with arithmetic functions. The group of functions called the *accumulator functions* is used to copy the contents of one accumulator to another.

These associations are illustrated graphically in Figure 6.1. The load function transfers information from system memory, work memory and the I/O to accumulator 1, shifting the "old" (that is to say, current) contents of accumulator 1 over to accumulator 2. The digital functions manipulate the contents of accumulator 1 or combine the contents of accumulators 1 and 2 and write the result back into accumulator 1.

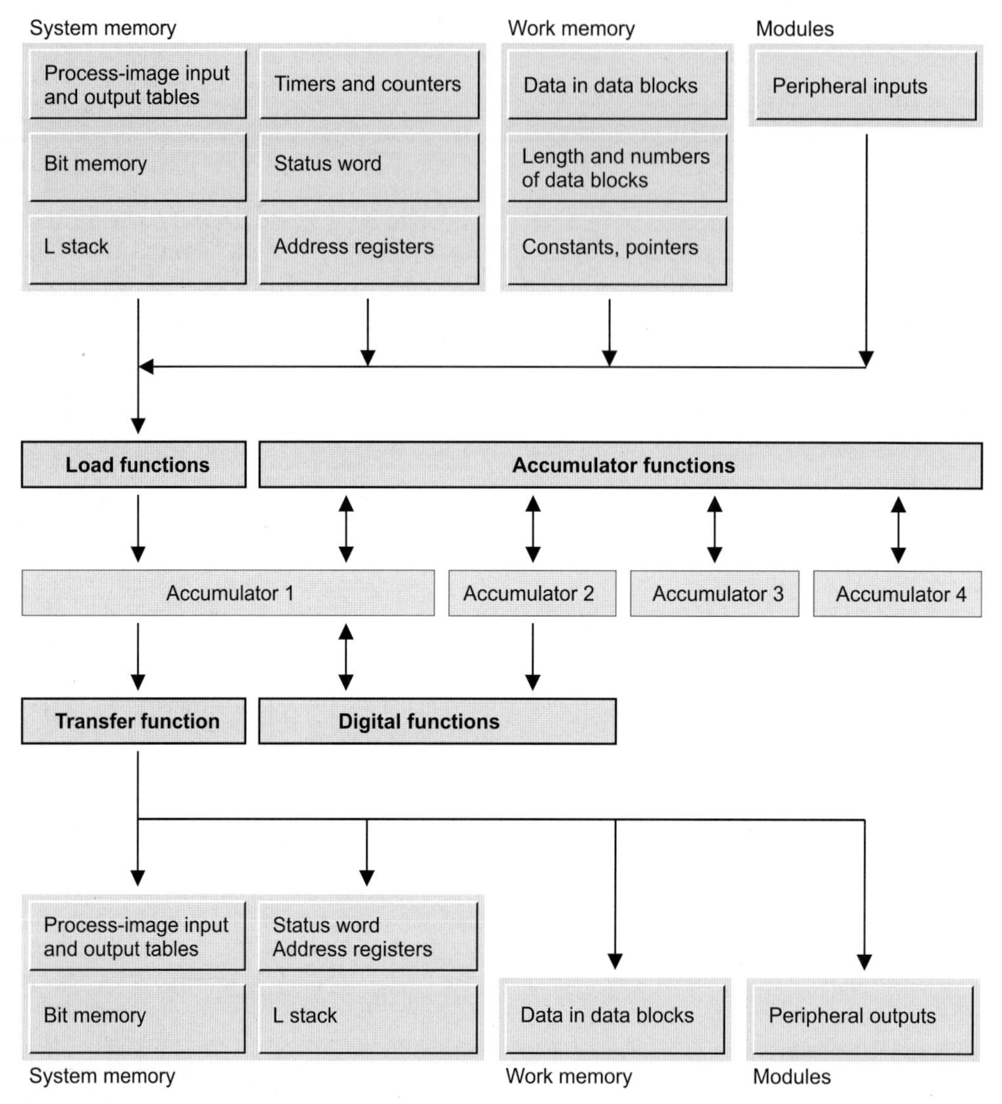

Figure 6.1 Memory Areas for Loading and Transferring

The accumulator functions can access the contents of all accumulators. The source for transferring information to system memory, work memory or the I/O is always and only accumulator 1.

Each accumulator comprises 32 bits, while all memory areas are byte-oriented. Information can be exchanged between the memory areas and accumulator 1 by byte, word, or doubleword.

In this chapter, the load and transfer functions are discussed in conjunction with the address areas for inputs, outputs, memory bits, I/O, and the loading of constants.

The load and transfer functions can also be combined with the following address areas:

▷ Timers and counters
(Chapter 7 “Timer Functions” and Chapter 8 “Counter Functions”)

- ▷ Status word (Chapter 15 "Status Bits")
- ▷ Temporary local data (L stack, Chapter 18.1.5 "Temporary Local Data")
- ▷ Data addresses, lengths and numbers of data blocks (Chapter 18.2 "Block Functions for Data Blocks")
- ▷ Address registers, pointers (Chapter 25 "Indirect Addressing")
- ▷ Variable address (Chapter 26 "Direct Variable Access")

6.2 Load Functions

6.2.1 General Representation of a Load Function

The load function consists of the operation code L (for load) and a constant, a variable, or an address with address identifier whose contents the function loads into accumulator 1.

L	+1200	Constant (immediate addressing)
L	IW 16	Digital memory location (direct addressing)
L	Actualvalue	Variable (symbolic addressing)

The CPU executes the load function without regard to the result of the logic operation or to the status bits. The load function affects neither the RLO nor the status bits.

Effect on accumulator 2

The load function also changes the contents of accumulator 2. While the value of the address, constant or variable specified in the load statement is loaded into accumulator 1, the current contents of accumulator 1 are transferred to accumulator 2. The load function transfers the entire contents of accumulator 1 to accumulator 2. The original contents of accumulator 2 are lost.

The load function does not affect the contents of accumulators 3 and 4 on the S7-400 CPUs and with the CPU 318.

Loading in general

The digital address specified in the load function may be that of a byte, a word, or a doubleword (Figure 6.2).

Loading a byte

When a byte is loaded, its contents are written right-justified into accumulator 1. The remaining bytes in the accumulator are padded with "0".

Loading a word

When a word is loaded, its contents are written right-justified into accumulator 1. The high-value byte of the word (n+1) is right-justified in the accumulator, the low-value byte of the word (byte n) is at its immediate left. The remaining bytes in the accumulator are padded with "0".

Loading a doubleword

When a doubleword is loaded, its contents are written into accumulator 1. The lowest-value byte (byte n) is at the far left in the accumulator, the highest-value byte (byte n+3) at the far right.

6.2.2 Loading the Contents of Memory Locations

Loading inputs

L	IB n	Loads an input byte
L	IW n	Loads an input word
L	ID n	Loads an input doubleword

With the S7-300 CPUs and, from 10/98 also with the S7-400 CPUs, loading inputs is also permissible if the relevant input modules are not available.

Loading outputs

L	QB n	Loads an output byte
L	QW n	Loads an output word
L	QD n	Loads an output doubleword

With the S7-300 CPUs and, from 10/98 also with the S7-400 CPUs, loading outputs is also permissible if the relevant output modules are not available.

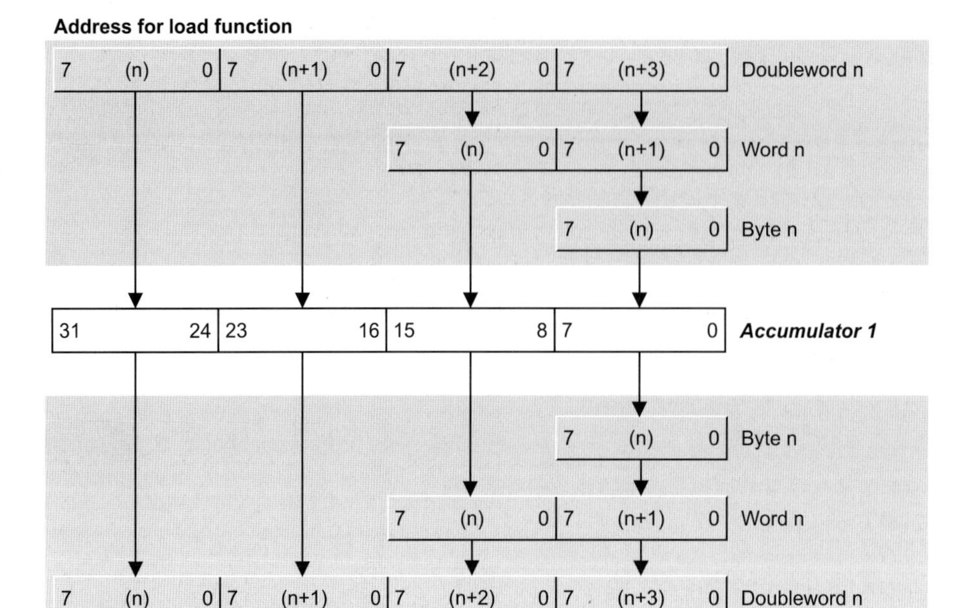

Figure 6.2 Loading and Transferring Bytes, Words and Doublewords

Loading from the I/O

L	PIB n	Loads a peripheral input byte
L	PIW n	Loads a peripheral input word
L	PID n	Loads a peripheral input doubleword

When loading from the I/O area, the input modules are referenced as peripheral inputs (PIs). Only the existing modules may be referenced.

Note that direct loading of an I/O module may produce a different value than the loading of inputs on a module with that same address, for whereas the signal states of the inputs are the same as they were at the start of the program scan cycle (when the CPU updated the process image), direct loading of the I/O modules loads the current value.

Loading bit memory

L	MB n	Loads a memory byte
L	MW n	Loads a memory word
L	MD n	Loads a memory doubleword

Loading from the bit memory is always allowed, as this entire area in the CPU. Note, however, that different CPUs have different-sized bit memory areas.

6.2.3 Loading Constants

Loading constants of elementary data type

You can load a constant, or fixed value, directly into the accumulator. For better readability, you can represent these constants in different formats. In Chapter 3 "SIMATIC S7 Program", you will find an overview of the permissible formats. All constants which can be loaded into the accumulator are of elementary data type.

Examples:

L	B#16#F1	Loads a 2-digit hexadecimal number
L	–1000	Loads an INT number
L	5.0	Loads a REAL number
L	S5T#2s	Loads an S5 timer
L	C#250	Loads a BCD number (count value)
L	TOD#8:30:00	Loads a time of day

Chapter 24 “Data Types” describes the bit assignments of the constants (structure of the data type).

Loading pointers

Pointers are a special form of constant, and are used for calculating memory locations. You can load the following pointers into the accumulator:

L	P#1.0	Loads an area-internal counter
L	P#M2.1	Loads an area-crossing counter
L	P#name	Loads the address of a local variable

You cannot load a DB pointer or an ANY pointer into the accumulator, as these pointers exceed 32 bits.

You will find further information on this topic in Chapters 25 “Indirect Addressing” and 26 “Direct Variable Access”.

6.3 Transfer Functions

6.3.1 General Representation of a Transfer Function

The transfer function comprises the operation code T (for transfer) and the digital address to which the contents of accumulator are to be transferred.

T	MW 120	Transfer accumulator contents to the specified memory location (absolute addressing)
T	Setpoint	Transfer accumulator contents to a variable (symbolic addressing)

The CPU executes the transfer statement without regard to the result of the logic operation or to the status bits. The function affects neither the RLO nor the status bits.

The transfer function transfers the contents of accumulator 1 by byte, word or doubleword to the specified address. The contents of accumulator 1 remain unchanged, making multiple transfers possible.

The transfer function may be used for accumulator 1 only. If you want to transfer a value from another accumulator, you must transfer that value to accumulator 1 using the accumulator functions, then transfer it to the desired address in memory.

Transferring in general

The digital address specified in the transfer statement may be that of a byte, a word, or a doubleword (Figure 6.2).

Transferring a byte

Transferring a byte shifts the rightmost byte in accumulator 1 to the byte specified in the transfer statement.

Transferring a word

Transferring a word shifts the contents of the rightmost word in accumulator 1 to the word specified in the transfer statement. The right byte of the word (low-order byte) is transferred to the byte with the higher address (n+1), the left word (high-order byte) to the byte with the lower address (n).

Transferring a doubleword

Transferring a doubleword shifts the contents of accumulator 1 to the doubleword specified in the transfer statement. The leftmost byte in the accumulator is transferred to the byte with the lowest address (n), the rightmost byte in the accumulator to the byte with the highest address (n+3).

6.3.2 Transferring to Various Memory Areas

Transferring to inputs

T	IB n	Transfer to input byte
T	IW n	Transfer to input word
T	ID n	Transfer to input doubleword

With the S7-300 CPUs and, from 10/98 also with the S7-400 CPUs, transferring to inputs is also permissible if the relevant input modules are not available.

Transfers to inputs affect only the bits in the process image, just like the setting and resetting of inputs. A possible application is in specifying values for debugging or startup: if you modify the inputs at the start of the program with the signal states you desire, the program then works with these new values and not with the values from the input modules.

Transferring to outputs

T QB n Transfer to output byte

T QW n Transfer to output word

T QD n Transfer to output doubleword

With the S7-300 CPUs and, from 10/98 also with the S7-400 CPUs, transferring to outputs is also permissible if the relevant output modules are not available.

Transferring to the I/O area

T PQB n Transfer to peripheral byte

T PQW n Transfer to a peripheral word

T PQD n Transfer to a peripheral doubleword

When transferring the contents of accumulator 1 to the I/O area, the output modules are referenced as peripheral outputs (PQs). Only addresses on output modules may be specified.

Transfers to I/O modules which have a process-image output table updates this table so that there is no difference between outputs and peripheral outputs with the same address.

Transferring to bit memory

T MB n Transfer to a memory byte

T MW n Transfer to a memory word

T MD n Transfer to a memory doubleword

Transfers to the memory areas are always permissible, since this entire area is located on the CPU. Note, however, that different CPUs have different-sized bit memory areas.

6.4 Accumulator Functions

The accumulator functions transfer values from one accumulator to another, or replace bytes in accumulator 1. Accumulator functions are executed without regard to the result of the logic operation or to the status bits. These functions affect neither the RLO nor the status bits.

6.4.1 Direct Transfers Between Accumulators

PUSH Shift accumulator contents "forward"

POP Shift accumulator contents "back"

ENT Shift accumulator contents "forward" (without accumulator 1)

LEAVE Shift accumulator contents "back" (without accumulator 1)

TAK Exchange contents of accumulators 1 and 2

CPUs with 2 rechargeable batteries (S7-300 and ET 200) only need the operations PUSH, POP, and TAK; with CPUs and 4 rechargeable batteries (S7-400), all operations are available (Figure 6.3).

PUSH

PUSH pushes the contents of accumulators 1 to 3 into the next higher accumulator (1 to 2, 2 to 3, and 3 to 4). The contents of accumulator 1 remain unchanged.

You can use PUSH to enter the same value into more than one accumulator.

POP

POP transfers the contents of accumulators 4 to 2 into the next lower accumulator (4 to 3, 3 to 2, 2 to 1). The contents of accumulator 4 remain unchanged.

POP puts the values in accumulators 2 to 4 into accumulator 1, from whence they can be transferred to memory.

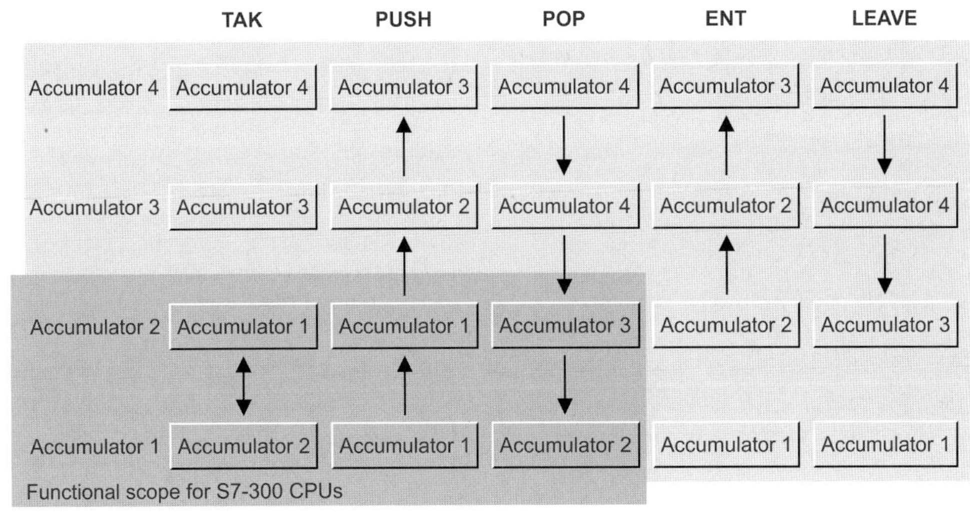

Figure 6.3 Direct Transfers between Accumulators on S7-300 and S7-400 CPUs

TAK

TAK exchanges the contents of accumulators 1 and 2. The contents of accumulators 3 and 4 remain unchanged.

ENT

ENT shifts the contents of accumulators 2 and 3 to the next higher accumulator. The contents of accumulators 1 and 2 remain unchanged.

If ENT is immediately followed by a load statement, the load shifts the contents of accumulators 1 to 3 "forward" (in a manner similar to that of PUSH); the new value is then in accumulator 1.

LEAVE

LEAVE shifts the contents of accumulators 3 and 4 into the next lower accumulator. The contents of accumulators 4 and 1 remain unchanged.

The arithmetic functions include LEAVE functionality. Using LEAVE, you can also simulate the same functionality in other digital logic operations (in word logic, for example).

When programmed after a digital logic operation, LEAVE places the contents of accumulators 3 and 4 into accumulators 2 and 3; the result of the digital logic operation remains unchanged in accumulator 1.

6.4.2 Exchange Bytes in Accumulator 1

CAW Exchange bytes in accumulator 1, low-order word

CAD Exchange bytes in entire accumulator 1

The CAW operation swaps the two low-order bytes in accumulator 1 (Figure 6.4). The high-order bytes remain unaffected.

The CAD operation swaps all bytes in accumulator 1. The highest-order byte then becomes the lowest following CAD; the two bytes in the middle change places.

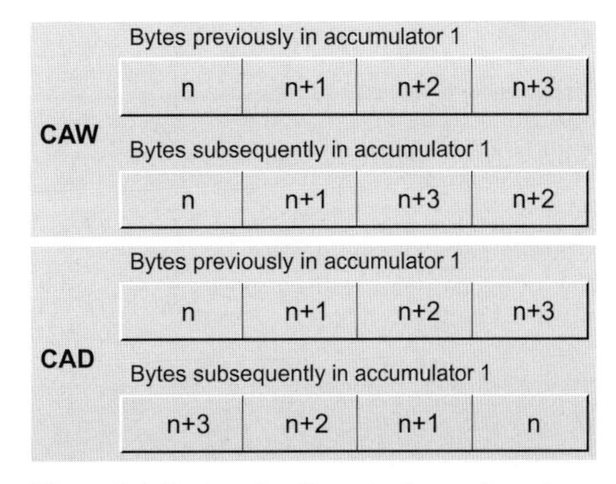

Figure 6.4 Exchanging Bytes in Accumulator 1

6.5 System Functions for Data Transfer

The following system functions are available for data transfer

- ▷ SFC 20 BLKMOV
  Copy memory area
- ▷ SFC 21 FILL
  Initializing a memory area
- ▷ SFC 81 UBLKMOV
  Uninterruptible copying of variables
- ▷ SFC 83 READ_DBL
  Read from load memory
- ▷ SFC 84 WRIT_DBL
  Write into load memory

ANY parameter with the SFC 20, 21 and 81

Each of these system functions has two parameters of data type ANY (Table 6.1). Theoretically, each of these parameters may specify an arbitrary address, variable, or absolute memory location.

If you specify a variable of complex data type, it must be a "complete" variable; components of a variable (such as individual array or structure components) are not permitted. For absolute addressing, use an ANY pointer; these pointers are discussed in detail in Chapter 25.1 "Pointers". With an ANY pointer of type BOOL (e.g. with a field), the length must be dividable by 8.

You can also copy individual variables of the data type STRING. However, the STL program editor and the SCL program editor behave differently in this case (see Chapter 6.5.4 "Copying STRING Variables").

If you use temporary local data as actual parameter in a block parameter of type ANY, the editor assumes that this actual parameter has the structure of an ANY pointer. In this way, you can generate an ANY pointer in the temporary local data that can be modified at runtime, that is, you can set up a variable area. The "Store Frame" example in Chapter 26 "Direct Variable Access", shows how to use this "variable ANY pointer".

ANY parameter with SFCs 83 and 84

The system functions SFC 83 READ_DBL and SFC 84 WRIT_DBL transmit data between data blocks present in the load and work memories. Complete data blocks or parts of data blocks are permissible as actual parameters for the block parameters SRCBLK and DSTBLK. With symbolic addressing, only "complete" variables are accepted which are present in a data block; single field or structure components are not permissible. Enter a memory area with absolute addressing as follows: P#Datablock.Dataoperand Type Number. With an ANY pointer of type BOOL (e.g. with a field), the length must be dividable by 8.

6.5.1 Copying Memory Area

System function SFC 20 BLKMOV copies, in the direction of ascending addresses (incrementally), the contents of a source area (SRCBLK

Table 6.1 Parameters for SFC 20, 21 and 81

SFC	Parameter	Declaration	Data Type	Contents, Description
20	SRCBLK	INPUT	ANY	Source area from which data are to be copied
	RET_VAL	RETURN	INT	Error information
	DSTBLK	OUTPUT	ANY	Destination to which data are to be copied
21	BVAL	INPUT	ANY	Source area to be copied
	RET_VAL	RETURN	INT	Error information
	BLK	OUTPUT	ANY	Destination to which the source area is to be copied (including multiple copies)
81	SRCBLK	INPUT	ANY	Source area from which data are to be copied
	RET_VAL	RETURN	INT	Error information
	DSTBLK	OUTPUT	ANY	Destination to which data are to be copied

parameter) to a target area (DSTBLK parameter).

The following actual parameters may be specified:

- ▷ Any variable from the address areas for inputs (I), outputs (Q), memory bits (M) or data blocks (variables from global data blocks and instance data blocks)
- ▷ Variables from the temporary local data (special case for data type ANY)
- ▷ Absolute-addressed data areas by specifying an ANY pointer

You cannot use SFC 20 to copy timers or counters, or to copy information from or to the modules (I/O area) or system data blocks (SDBs).

In the case of inputs and outputs, the specified area is copied regardless of whether or not the addresses specified are on the input or output modules. If the CPU does not possess an SFC 83 READ_DBL, you may also specify a variable or the address of an area in a data block in load memory.

Source and destination may not overlap. If source area and target area are of different lengths, the shorter of the two determines the transfer.

Example: The variable *Frame* (a structured variable as user data type, for example) in data block "Rec_mailb" is to be copied to variable *Frame1* (which is of the same data type as Frame) in data block "Buffer". The function value is to be entered in the variable *Copyerror* in data block "Evaluation".

```
CALL BLKMOV (
    SRCBLK  := Rec_mailb.Frame,
    RET_VAL := Evaluation.Copyerror,
    DSTBLK  := Buffer.Frame1);
```

6.5.2 Uninterruptible Copying of Variables

System function SFC 81 UBLKMOV copies the contents of a source area (SRCBLK parameter) to a target area (DSTBLK parameter) in the direction of rising addresses (incrementing). The copy operation cannot be interrupted, so under certain circumstances, the response times to interrupts may increase. Up to 512 bytes are copied.

The following actual parameters can be applied to the parameters:

- ▷ Any variables from the address areas inputs I, outputs Q, memory bits M, data blocks (variables from global data blocks and from instance data blocks)
- ▷ Variables and temporary local data (special case for data type ANY)
- ▷ Absolute-addressed memory areas with specification of an ANY pointer

You cannot copy the following with SFC 81: timer and counter functions, information from and to the modules (I/O address area), system blocks SDBs and data blocks in load memory (data blocks programmed with the keyword UNLINKED).

In the case of inputs and outputs, the area specified is copied regardless of the actual assignment to input modules or output modules.

Source and target areas must not overlap. If the source and target areas are of differing lengths, the transfer is carried out only to the length of the smaller area.

Example: From the data block "Buffer", the first component of the array *Data* is to be copied to the *Frame* variable in the data block "Send_mailb". The function value is to be stored in the *Copyerror* variable in the data block "Evaluation".

```
CALL UBLKMOV (
    SRCBLK  := Buffer.Data[1],
    RET_VAL := Evaluation.Copyerror,
    DSTBLK  := Send_mailb.Frame);
```

6.5.3 Initializing a Memory Area

System function SFC 21 FILL copies a specified value (source area) to a memory area (target area) until the target area is completely filled. The transfer is in the direction of ascending addresses (incremental). The parameters may be assigned the following actual values:

- ▷ Any variables from the address areas for inputs (I), outputs (Q), memory bits (M), or data blocks (variables from global data blocks and instance data blocks)
- ▷ Absolute-addressed data areas by specifying an ANY pointer

▷ Variables in the temporary local data of data type ANY (special case)

You cannot use SFC 21 to copy timers or counters, or to copy information from or to the modules (I/O area) or system data blocks (SDB).

In the case of inputs and outputs, the specified area is copied without regard to the whether or not the addresses are those on input or output modules.

Source and destination may not overlap. The target area is always completely filled, even when the source area is longer than the target area or when the length of the target area is not an integer multiple of the length of the source area.

Example: Data block DB 13 consists of 128 data bytes, all of which are to be set to the value of memory byte MB 80.

```
CALL SFC 21 (
    BVAL    := MB 80,
    RET_VAL := MW 32,
    BLK     := P#DB13.DBX0.0 BYTE 128);
```

6.5.4 Copying STRING Variables

You can copy individual STRING variables with the system functions SFC 20 BLKMOV and SFC 81 UBLKMOV. The STL program editor and the SCL program editor behave differently here.

The STL program editor handles the STRING variable like a BYTE array in this case, so that the SFC transfers the individual bytes 1:1 (including the two first bytes with the length specifications). If, for example, you transfer a BYTE array to a STRING variable, you must assign the correct length of the length bytes in the STRING variable to the BYTE array.

The SCL program editor writes the data type STRING to the ANY pointer. The SFC then transfers only the relevant "character positions" of the STRING variable. If the STRING variable is the destination, the actual length is corrected, if necessary. In this way, you can, for example, easily transfer a STRING variable to an ARRAY of CHAR and vice versa.

Both the STL and the SCL program editors copy a STRING variable correctly to another STRING variable.

6.5.5 Reading from Load Memory

The system function SFC 83 READ_DBL reads data from a data block present in the load memory, and writes them into a data block present in the work memory. The contents of the read data block are not changed. The block parameters are described in Table 6.2.

The system function SFC 83 READ_DBL operates asynchronously: you trigger the read

Table 6.2 Parameters for SFC 83 and 84

SFC	Parameter	Declaration	Data Type	Contents, Description
83	REQ	INPUT	BOOL	Trigger for reading with signal status "1"
	SRCBLK	INPUT	ANY	Data area in load memory which is read
	RET_VAL	RETURN	INT	Error information
	BUSY	OUTPUT	BOOL	With signal status "1": reading not yet finished
	DSTBLK	OUTPUT	ANY	Data area in work memory which is written
84	REQ	INPUT	BOOL	Trade for writing with signal status "1"
	SRCBLK	INPUT	ANY	Data area in work memory which is read
	RET_VAL	RETURN	INT	Error information
	BUSY	OUTPUT	BOOL	With signal status "1": writing not yet finished
	DSTBLK	OUTPUT	ANY	Data area in load memory which is written

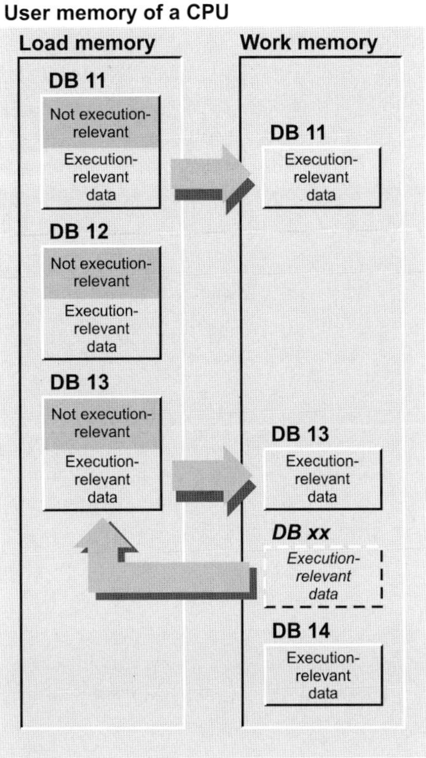

User memory of a CPU

Load memory | **Work memory**

DB 11: Not execution-relevant / Execution-relevant data → **DB 11**: Execution-relevant data

A "normal" data block created using the programming device is present twice in the CPU's user memory: it is completely present in the load memory, and the "execution-relevant" data, i.e. the data with which the program works, are present in the work memory.

DB 12: Not execution-relevant / Execution-relevant data

A data block created using the programming device with the property *Unlinked* is only present in the CPU's load memory. This data block does not occupy any space in the work memory.

DB 13: Not execution-relevant / Execution-relevant data → **DB 13**: Execution-relevant data; ***DB xx***: *Execution-relevant data*

A data block created using the SFC 82 CREA_DBL is present in the load memory and – if *Unlinked* is not activated – also in the work memory. A data block present in the work memory is the template according to which the new block is created.

DB 14: Execution-relevant data

A data block created using the SFC 22 CREAT_DB or SFC 85 CREA_DB is only present in the CPU's work memory.

Figure 6.5 Data Blocks in User Memory

operation by a signal status "1" on the parameter REQ. You may only access the read and written data areas again when the BUSY parameter has the signal status "0" again. Please also note the system resources of the CPU when using asynchronous system functions.

A data block is normally present twice in the user memory of a CPU: once in the load memory and – the sequence-relevant part – in the work memory. If a data block has the property *Unlinked*, it is only present in the load memory (Figure 6.5). The SFC 83 READ_DBL only reads the values from the load memory. The initial values of the data addresses are present here, and may differ from the current values in the work memory (see also Chapter 2.6.5 "Block Handling" under "Data blocks offline/online").

For the parameters SRCBLK and DSTBLK, it is possible to specify complete data blocks, e.g. DB 100 or "Recipe 1", variables from data blocks, or an absolute address data area as ANY pointers, e.g. P#DB100.DBX16.0 BYTE 64.

If the source area is smaller than the target area, the former is written completely into the latter. The remaining bytes of the target area are not changed. If the source area is larger than the target area, the latter is written completely; the remaining bytes of the source area are ignored.

6.5.6 Writing into the Load Memory

The system function SFC 84 WRIT_DBL reads data from a data block present in the work memory, and writes them into a data block present in the load memory. The contents of the read data block are not changed. The block parameters are described in Table 6.2.

The system function SFC 84 WRIT_DBL operates asynchronously: you trigger the write operation with signal status "1" for the parameter REQ. You can only access the read and written data areas again if the parameter BUSY has the signal status "0" again. Please also consider the system resources of the CPU when using asynchronous system functions.

A data block is normally present twice in the user memory of a CPU: once in the load memory and – the sequence-relevant part – in the work memory. If a data block has the property *Unlinked*, it is only present in the load memory (Figure 6.5). The SFC 84 WRIT_DBL only reads the values from the work memory. The initial values of the data addresses are present here, and may differ from the current values in the load memory (see also Chapter 2.6.5 "Block Handling" under "Data blocks offline/online").

For the parameters SRCBLK and DSTBLK, it is possible to specify complete data blocks, e.g. DB 200 or "Archive 1", variables from data blocks, or an absolute addressed data area as ANY pointers, e.g. P#DB200.DBX0.0 WORD 4.

If the source area is smaller than the target area, the former is written completely into the latter. The remaining bytes of the target area are not changed. If the source area is larger than the target area, the latter is written completely; the remaining bytes of the source area are ignored.

If the data block has been generated in the load memory by a system function during runtime, the checksum of the user program is not changed by writing with the SFC 84 WRIT_DBL. With (manual) creation of the data block, writing with the SFC 84 WRIT_DBL changes the user program checksum.

Please note that – for physical reasons – the load memory can usually only permit a limited number of write operations. Too frequent writing, e.g. cyclic, limits the service life of the load memory.

7 Timer Functions

Timer functions are used to implement timing sequences, such as, for example, waiting and monitoring times, measuring a period of time, or the generation of pulses.

The statements for the STL programming language are described in this chapter; in the SCL programming language, the timer functions are included among the standard functions (see Chapter 30.1 "Timer Functions").

The following timer types are available:

- ▷ Pulse timer
- ▷ Extended pulse timer
- ▷ On-delay timer
- ▷ Retentive on-delay timer
- ▷ Off-delay timer

When you start a timer, you specify the dynamic response and the duration, the latter being the length of time the timer is to run; you can also reset or enable ("retrigger") timers. Binary logic operations are used to check timers ("timer running"). Load functions are used to transfer the current time value, in binary or BCD, to accumulator 1.

The examples in this chapter and the IEC timer calls can be found in function block FB 107 or in the source file Chap_7 in the STL_Book library under the "Basic Functions" program, which you can download from the publisher's web site (see page 8).

7.1 Programming a Timer

7.1.1 Starting a Timer

A timer is started (begins running) when the result of the logic operation (RLO) changes prior to the start instruction. In the case of an off-delay timer, the RLO must change from "1" to "0"; in all other cases, the timer begins running when the RLO goes from "0" to "1".

You can start any timer as one of five possible types (Figure 7.1). However, it is not a good idea to use a given timer as more than one type.

You start the timer as	at AWL with	at SCL with	Start signal
Pulse timer	SP	S_PULSE	t
Extended pulse timer	SE	S_PEXT	t
On-delay timer	SD	S_ODT	t
Retentive on-delay timer	SS	S_ODTS	t
Off-delay timer	SF	S_OFFDT	t

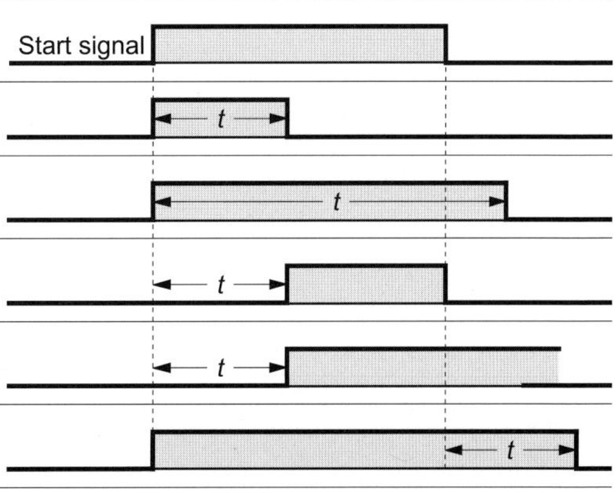

Figure 7.1 Start Instructions for Timers

7.1.2 Specifying the Time

When a timer is started, it takes the value in accumulator 1 as its running time, or duration. How and when the value gets into the accumulator is of no consequence. To make your program more readable, the best way would be to load the running time directly into the accumulator before starting the timer, either as a constant (direct specification of the value) or as a variable (for example a memory word containing the value).

Note: accumulator 1 must contain a valid time value even if the time is not started when the start instruction is processed.

Specifying the duration as a constant

```
L  S5TIME#10s;   //Duration 10 s
L  S5T#1m10ms;   //Duration 1 min + 10 ms
```

In the basic languages STL, LAD and FBD, the duration, or running time, is specified in hours, minutes, seconds and milliseconds. The defined number range extends from S5TIME#10ms to S5TIME#2h46m30s (which corresponds to 9990 s). You may use either S5TIME# or S5T# to identify a constant.

Specifying the duration as a variable

```
L  S5T#10m;    //Duration 10 min
T  MW 20;      //Save duration
... ;
L  MW 20;      //Load duration
```

Structure of the duration

Internally, the duration is composed of the time value and the time base. The duration is equal to time value × time base. The duration is the period of time during which the timer is active ("timer running"). The time value represents the number of timing periods the timer is to run. The time base specifies the timing period the CPU operating system is to use to decrement the timer (Figure 7.2).

You can also set up the duration directly in the word. The smaller the time base, the more accurately the actual duration processed. For example, if you want to implement a duration of 1 s, you can do so in one of three different ways:

Duration = 2001_{hex} Time base 1 s

Duration = 1010_{hex} Time base 100 ms

Duration = 0100_{hex} Time base 10 ms

The last of the three is the preferred method in this case.

When a timer is started, the CPU uses the programmed time value as the timer's running time. The operating system updates timers at fixed intervals and independently of the user program scan, that is, it decrements an active timer's time value as per the timing period indicated by the time base.

When the value reaches zero, the timer is regarded as expired. The CPU then sets the timer status (signal state "0" or "1", depending on the type of timer involved) and drops all further activities until the timer is started again.

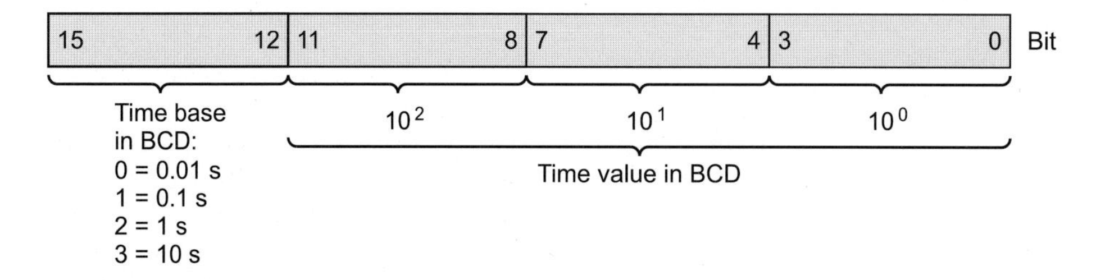

Figure 7.2 Bit Assignments in the Duration

When a time value of zero is specified, the timer remains active until the CPU processes it and discovers that the timer has expired.

Timers are updated asynchronously to the program scan. It is therefore possible that the status of the timer at the start of the cycle may differ from its status at the end of the cycle. If you use timer instructions at only one location in the program and in the suggested order (see below), there will be no errors due to asynchronous timer updating.

7.1.3 Resetting a Timer

R T n Resets a timer

A timer is reset when the RLO is "1" when the reset statement is encountered. As long as the RLO is "1", timer checks for "1" will return a check result of "0" and timer checks for "0" will return a check result of "1". Resetting a timer sets the time value and the time base to zero.

Note: resetting a timer function does not reset the internal edge memory bit for starting. To start again, the start instruction must be processed with RLO "0" before the timer function can be started with a signal edge.

7.1.4 Enabling a Timer

FR T n Enables a timer

The enable instruction is used to "retrigger" an active timer, that is, to restart it.

A timer is enabled if the enable instruction is processed with a positive (rising) edge. Then the internal edge memory bit is reset for starting the timer. If the RLO is then "1" when the start instruction is next processed, the timer is started even if there is no signal edge at the start instruction.

An enable instruction is not required to start or reset a timer, that is to say, it is not necessary to normal timer operation.

7.1.5 Checking a Timer

Checking the timer status

A	T n	Check for signal state "1" and combine according to AND
O	T n	Check for signal state "1" and combine according to OR
X	T n	Check for signal state "1" and combine according to Exclusive OR
AN	T n	Check for signal state "0" and combine according to AND
ON	T n	Check for signal state "0" and combine according to OR
XN	T n	Check for signal state "0" and combine according to Exclusive OR

You can check a timer as you would an input, for instance, and further process the result. Depending on the type of timer, a check for signal state "1" produces different variations in the timing sequence (see the description of the dynamic response in the coming chapters).

As it does in the case of inputs, a check for signal state "0" returns precisely the reverse result as does the check for signal state "1".

Checking the time value

L	T n	Loads a binary time value
LC	T n	Loads a BCD time value

Load functions L T and LC T check the specified time value and make it available in accumulator 1 in binary (L) or in binary-coded decimal (LC). The value loaded into the accumulator is the value current at the instant of the check (in the case of an active timer, the time value, in this case the value loaded into the accumulator, is counted down in the direction of zero).

Loading a time value (direct load)

The value specified in the timer instruction is in binary, and can be loaded into accumulator 1 in this form. The time base is lost in this case, and in its place in accumulator 1 is the value "0".

The value in accumulator 1 therefore corresponds to a positive number in INT format, and can be further processed with, for example, compare functions. Please note that it is the *time value* that is in the accumulator, not the *duration*.

Example:

```
L   T 15;     //Load current time value
T   MW 34;    //and save
```

Loading a time value (coded load)

You can also use a "coded load" instruction to load a binary value into accumulator 1. In this case, both the time value and the time base are available in binary-coded decimal (BCD). The contents of the accumulator are the same as when a time value is specified (see above), that is, the left-hand word (high-order word) in the accumulator contains zero.

Example:

```
LC  T 16;     //Load current time value
              in BCD
T   MW 122;   //and save
```

7.1.6 Sequence of Timer Instructions

When you program a timer, you do not need to use all of the statements that are available for timers, but only those statements applicable to the timer you want to implement. Normally, this would include starting the timer with the specified duration, and binary checking of the timer.

In order for a timer to perform as described in the preceding sections, a certain order must be observed when programming timer operations.

Table 7.1 shows the optimum order for all timer operations. Simply omit the statements that are not needed, such as enabling the timer.

Table 7.1 Sequence of Timer Operations

Timer Operation	Examples:
Enable timer	A I 16.5 FR T 5
Start timer	A I 17.5 L S5T#1s SI T 5
Reset timer	A I 18.0 R T 5
Digital timer check	L T 5 T MW 20 LC T 5 T MW 22
Binary timer check	A T 5 = Q 2.0

If a timer is started and reset "simultaneously" in the statement sequence shown, the timer will start, but the subsequent reset statement will immediately reset it. When the timer is then checked, the fact that it was started will therefore go unnoticed.

7.1.7 Clock Generator Example

The example shows a clock generator with a different pulse-pause ratio, implemented with a single timer.

Start_input starts the clock generator. If the time is not running or if it has run out, it is started as an extended pulse. At every start, the binary scaler *Output* changes its signal state and with this also determines the duration with which the time is to be started.

```
      AN    Start_input;
      R     Timer;
      R     Output;
      SPB   M1;
      A     Timer;
      SPB   M2;
      AN    Output;
      =     Output;
      L     Pulse_duration;
      SPB   M2;
      L     Pause_duration;
M2:   AN    Timer;
      SE    Timer;
M1:   ;     //remaining program
```

7.2 Pulse Timers

The complete STL sequence of statements for starting a timer as a pulse is as follows:

```
A       Enable_input;
FR      Timer;
A       Start_input;
L       Duration;
SI      Timer;
A       Reset_input;
R       Timer;
L       Timer;
T       Binary_time_value;
LD      Timer;
T       BCD_time_value;
A       Timer;
=       Timer_status;
```

For SCL, calling a timer as a pulse is programmed as follows:

```
BCD_time_value:= S_PULSE (
      T_NO := Timer,
      S := Start_input,
      TV := Duration,
      R := Reset_input,
      Q := Timer_status,
      BI := Binary_time_value);
```

Starting a pulse timer

The diagram in Figure 7.3 describes the dynamic response of a timer that was started as a pulse timer, and its behavior when it is reset. The description is valid if you adhere to the sequence of operations shown opposite for STL (starting before resetting before checking). Enabling is not required for the "normal" sequence, and in SCL it is also not available.

① The timer is started when the signal state at its Start input changes from "0" to "1" (positive edge). It runs for the programmed duration as long as the signal state at the Start input remains at "1". Checks for signal state "1" (timer status) return a check result of "1" as long as the timer is running.

The time is counted down from the initial value with the set time base.

② The timer stops if the signal state at its Start input goes to "0" before the time has elapsed. Checking the timer for signal state "1" (timer status) returns a check result of "0". The time value shows the time remaining, which also shows at what point the timing sequence was prematurely interrupted.

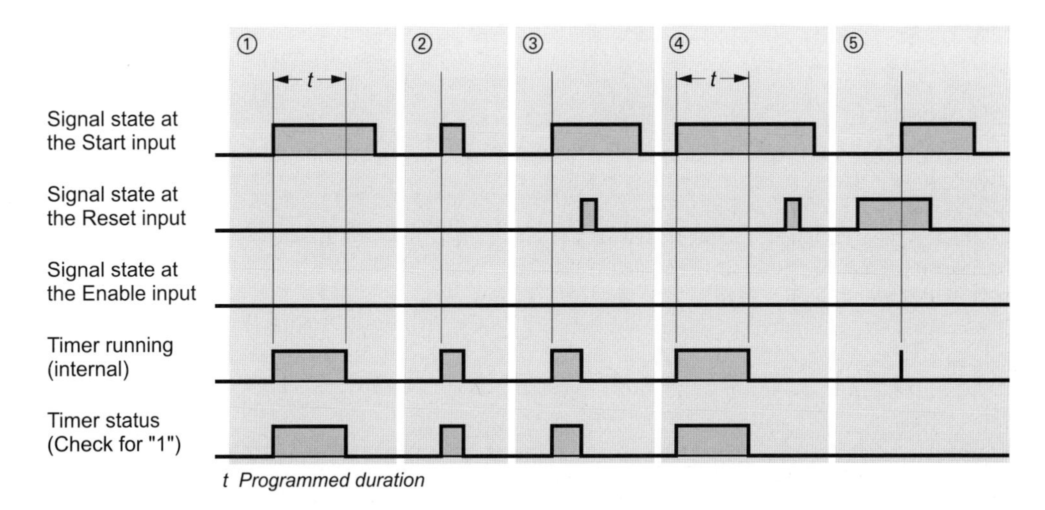

Figure 7.3 Response when Starting and Resetting a Pulse Timer

Resetting a pulse timer

Resetting of a pulse timer has a static effect and takes precedence over the starting of a timer (Figure 7.3).

③ Signal state "1" at the Reset input of an active timer resets that timer. A check for signal state "1" (timer status) then returns a check result of "0". The time value and the time base are also set to zero. A change in the signal state from "1" to "0" at the Reset input while signal state "1" is present at the Start input has no effect on the timer.

④ If a timer is not running, signal state "1" at its Reset input has no effect.

⑤ If the signal state at the Start input changes from "0" to "1" (positive edge) while the Reset signal is present, the timer is started but the subsequent reset instruction resets it immediately (indicated by a line in the diagram). If checking of the timer status is programmed after resetting, the brief starting of the timer will have no effect on the check.

Enabling a pulse timer

The timer is "retriggered", i.e. prompted to restart, with a positive edge at the enable input. Enabling is only possible in the STL programming language.

Figure 7.4 shows enabling of a timer started as a pulse timer.

❶ If the signal state goes from "0" to "1" (positive edge) at the enable input, the timer, if active, will be restarted when the start instruction is processed, provided the start input still has signal state "1". The programmed duration is taken as the current time value for the restart. A change in the signal state from "1" to "0" at the enable input has no effect.

❷ If the signal state at the enable input changes from "0" to "1" (positive edge) and if the start input still has signal state "1", the timer is also started with the programmed duration as a pulse timer.

❸ If the signal state at the Start input is "0", a positive signal edge at the enable input will have no effect.

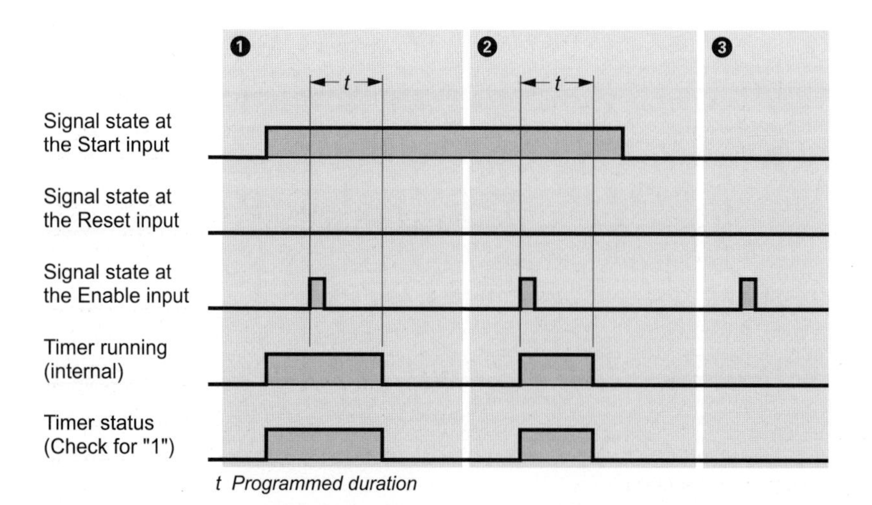

Figure 7.4 Enabling a Pulse Timer

7.3 Extended Pulse Timers

The complete STL statement list for starting a timer as an extended pulse is as follows:

```
A       Enable_input;
FR      Timer;
A       Start_input;
L       Duration;
SE      Timer;
A       Reset_input;
R       Timer;
L       Timer;
T       Binary_time_value;
LD      Timer;
T       BCD_time_value;
A       Timer;
=       Timer_status;
```

For SCL, calling a timer as an extended pulse is programmed as follows:

```
BCD_time_value:= S_PEXT (
      T_NO := Timer,
      S := Start_input,
      TV := Duration,
      R := Reset_input,
      Q := Timer_status,
      BI := Binary_time_value);
```

Starting an extended pulse timer

The diagram in Figure 7.5 describes a timer's performance after it is started as an extended pulse timer and when it is reset. The description is valid if you adhere to the sequence of operations shown opposite for STL (starting before resetting before checking). Enabling is not required for the "normal" sequence, and in SCL it is also not available.

①② When the signal state at the timer's Start input changes from "0" to "1" (positive edge), the timer is started. It runs for the programmed duration, even if the signal state at the start input changes back to "0". The checks for signal state "1" (timer status) return a check result of "1" as long as the timer is running.

The time is counted down from the initial value with the set time base.

③ If the signal state at the Start input goes from "0" to "1" (positive edge) while the timer is running, the timer is restarted with the programmed time value (that is, the timer is "retriggered"). It can be restarted as often as required without the time having to expire first.

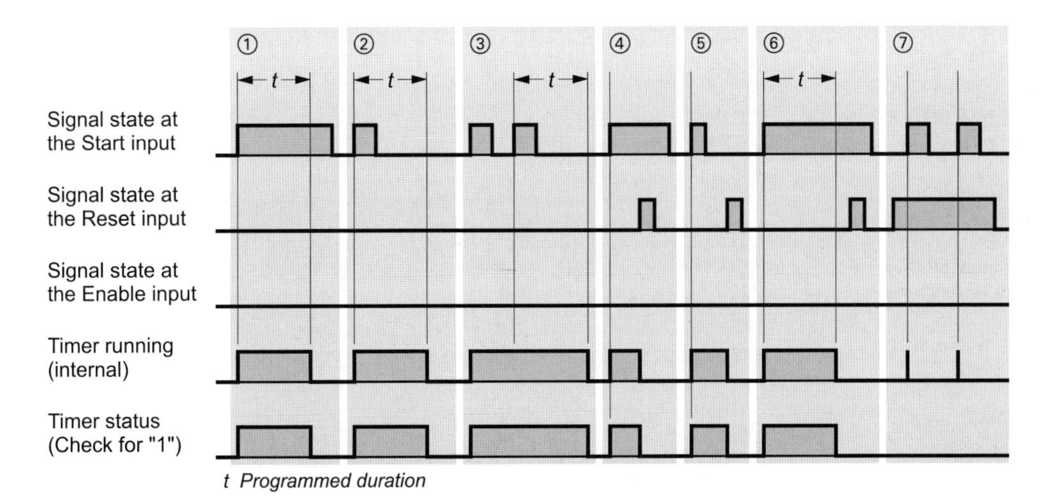

Figure 7.5 Response of an Extended Pulse Timer

Resetting an extended pulse timer

Resetting of an extended pulse timer has a static effect and takes precedence over the starting of a timer (Figure 7.5).

④⑤ Signal state "1" at an active timer's Reset input resets the timer. A check for signal state "1" (timer status) returns a check result of "0" if a timer is reset. The time value and the time base are also set to zero.

⑥ If the timer is not running, processing of the reset input with signal state "1" has no effect.

⑦ If the signal state at the Start input goes from "0" to "1" (positive edge) while the Reset signal is present, the timer is started but the subsequent reset resets it immediately (indicated by a line in the diagram). If checking of the timer status is programmed after resetting, this brief starting of the timer will not affect a check.

Enabling an extended pulse timer

The timer is "retriggered", i.e. prompted to restart, with a positive edge at the enable input. Enabling is only possible in the STL programming language.

Figure 7.6 shows enabling of a timer started as an extended pulse timer.

❶ When the signal state at an active timer's enable input goes from "0" to "1" (positive edge), the timer is restarted when the start instruction is processed, provided the start input still has signal state "1". The programmed duration is taken as the current time value for the restart. A change in the signal state from "1" to "0" at the enable input has no effect.

❷ If the signal state at the enable input of an inactive timer goes from "0" to "1" (positive edge) and if the start input still has signal state "1", the timer is also started with the programmed duration as an extended pulse timer.

❸❹ If the signal state at the Start input is "0", a positive signal edge at the enable input will have no effect.

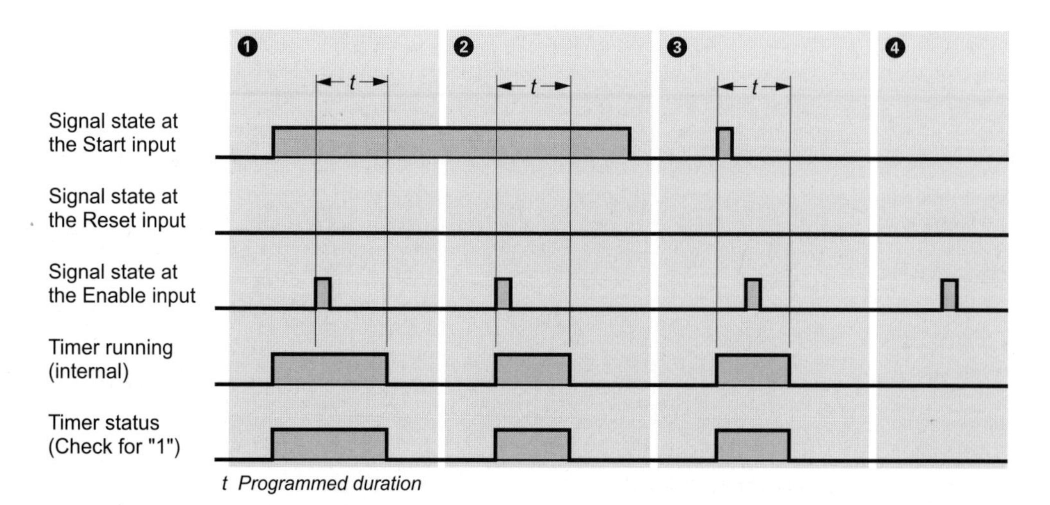

Figure 7.6 Enabling an Extended Pulse Timer

7.4 On-Delay Timers

The complete STL statement list for starting a timer as an on-delay is as follows:

```
A       Enable_input;
FR      Timer;
A       Start_input;
L       Duration;
SD      Timer;
A       Reset_input;
R       Timer;
L       Timer;
T       Binary_time_value;
LD      Timer;
T       BCD_time_value;
A       Timer;
=       Timer_status;
```

For SCL, calling a timer as an on-delay is programmed as follows:

```
BCD_time_value:= S_ODT (
      T_NO := Timer,
      S := Start_input,
      TV := Duration,
      R := Reset_input,
      Q := Timer_status,
      BI := Binary_time_value);
```

Starting an on-delay timer

The diagram in Figure 7.7 describes a timer's dynamic performance after it is started as an on-delay timer and when it is restarted. The description is valid if you adhere to the sequence of operations shown opposite for STL (starting before resetting before checking). Enabling is not required for the "normal" sequence, and in SCL it is also not available.

① When the signal state at the timer's Start input changes from "0" to "1" (positive edge), the timer is started. The timer runs with the programmed duration as the time value. Checks for signal state "1" (timer status) return a check result of "1" if the time expires without incident and the signal state at the Start input is still "1" (on delay).

The time is counted down from the initial value with the set time base.

② If the signal state at the Start input of a running timer goes from "1" to "0", the timer stops. In such cases, a check for signal state "1" (timer status) always returns a check result of "0". The time value shows the time remaining, that is, the period by which the timer was prematurely interrupted.

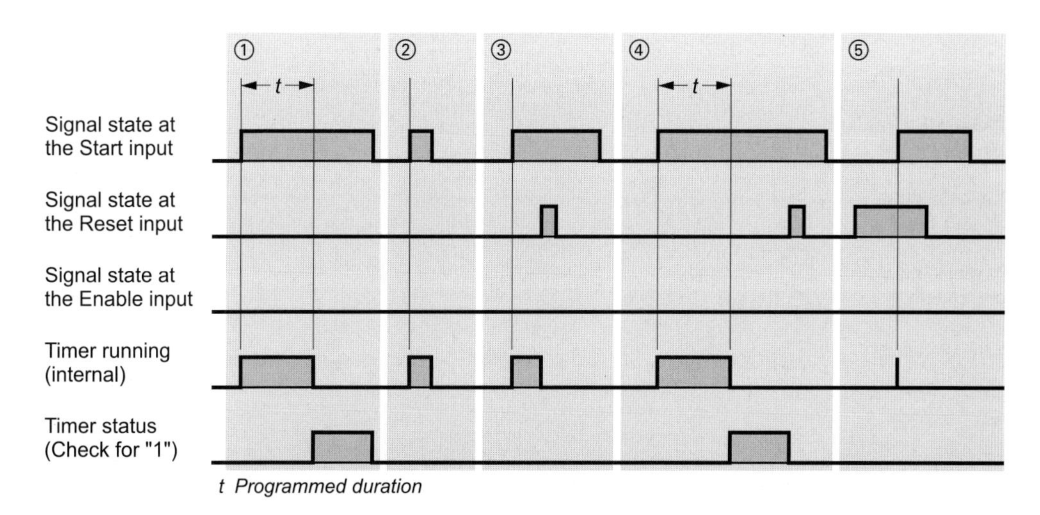

Figure 7.7 Response of an On-Delay Timer

Resetting an on-delay timer

Resetting of an on-delay timer has a static effect and takes precedence over the starting of a timer (Figure 7.7).

③④ Signal state "1" at the Reset input resets the timer whether the time is running or not. A check for signal state "1" (timer status) returns a check result of "0", even when the timer is not running and signal state "1" is still present at the Start input. Time value and time base are also set to zero.

A change in the signal state at the Reset input from "1" to "0" while signal state "1" is still present at the Start input has no effect on the timer.

⑤ If the signal state at the Start input changes from "0" to "1" (positive edge) while the Reset signal is still present, the timer is started but the subsequent reset resets it again immediately (indicated by a line in the diagram). If checking of the timer status is programmed after resetting, this brief starting of the timer will not affect a check.

Enabling an on-delay timer

The timer is "retriggered", i.e. prompted to restart, with a positive edge at the enable input (in STL only). Figure 7.8 shows enabling of a timer as an on-delay timer.

❶ If the signal state at a running timer's enable input changes from "0" to "1" (positive edge), the time is restarted when the start operation is processed, provided the start input still has signal state "1". The programmed duration is taken as the current time value for the restart. A change in the signal state from "1" to "0" at the enable input has no effect.

❷ If the signal state at the enable input goes from "0" to "1" (positive edge) after the time has elapsed without incident, processing of the start operation does not affect the timer.

❸❹ If there is a positive signal edge at the enable input and the timer is reset, the timer restarts if the start input still has signal state "1". The timer restarts with the programmed duration as the current time value.

If the signal state at the Start input is "0", a positive edge at the enable input has no effect.

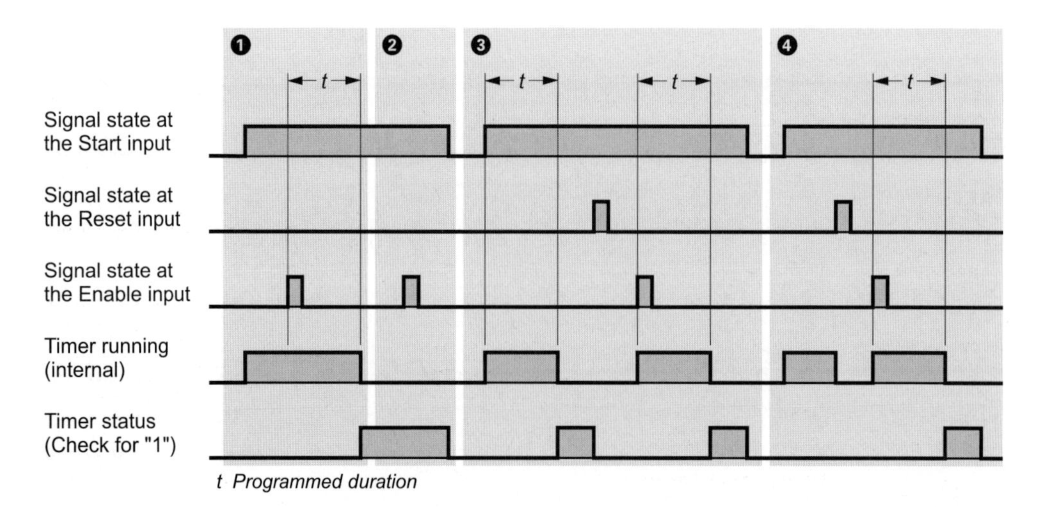

Figure 7.8 Enabling an On-Delay Timer

7.5 Retentive On-Delay Timers

The complete STL statement list for starting a timer as a retentive on-delay is as follows:

```
A       Enable_input;
FR      Timer;
A       Start_input;
L       Duration;
SS      Timer;
A       Reset_input;
R       Timer;
L       Timer;
T       Binary_time_value;
LD      Timer;
T       BCD_time_value;
A       Timer;
=       Timer_status;
```

For SCL, calling a timer as a retentive on-delay is programmed as follows:

```
BCD_time_value:= S_ODTS (
      T_NO := Timer,
      S := Start_input,
      TV := Duration,
      R := Reset_input,
      Q := Timer_status,
      BI := Binary_time_value);
```

Starting a retentive on-delay timer

The diagram in Figure 7.9 describes a timer's dynamic performance after it is started as a retentive on-delay timer, and after it is restarted. The description is valid if you adhere to the sequence of operations shown opposite for STL (starting before resetting before checking). Enabling is not required for the “normal” sequence, and in SCL it is also not available.

①② When the signal state at the timer's Start input changes from “0” to “1” (positive edge), the timer is started. It runs with the programmed duration even if the signal state at the start input changes back to “0”. When the time has elapsed, checks for signal state “1” (timer status) return a check result of “1”, regardless of the signal state at the Start input. The check result is not “0” until the timer has been reset, regardless of what the signal state at the Start input is. The time is counted down from the initial value with the set time base

③ If the signal state at the Start input goes from “0” to “1” (positive edge) while the timer is running, the timer is restarted with the programmed time value (that is, the timer is “retriggered”). The timer may be restarted as often as required without the time having to elapse first.

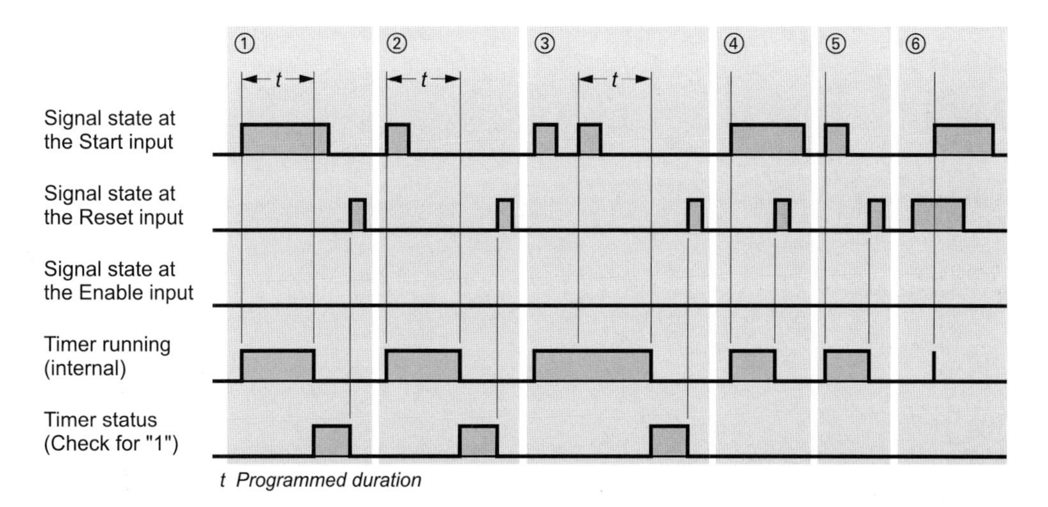

Figure 7.9 Response of a Retentive On-Delay Timer

Resetting a retentive on-delay timer

Resetting of a retentive on-delay timer has a static effect and takes precedence over the starting of a timer (Figure 7.9).

④⑤ Signal state "1" at the Reset input resets the timer without regard to the signal state at the Start input. Checks for signal state "1" (timer status) then return a check result of "0". Time value and time base are set to zero.

⑥ If the signal state at the Start input changes from "0" to "1" (positive edge) while the Reset signal is present, the timer is started, but the subsequent reset resets it again immediately (indicated by a line in the diagram). If checking of the timer status is programmed after resetting, the brief start will have no effect on the check.

Enabling a retentive on-delay timer

The timer is "retriggered", i.e. prompted to restart, with a positive edge at the enable input. Enabling is only possible in the STL programming language.

Figure 7.10 shows enabling of a timer as a retentive on-delay timer.

❶ If the signal state at a running timer's enable input changes from "0" to "1" (positive edge), the timer is restarted when the start instruction is processed, provided the start input still has signal state "1". The timer restarts with the programmed duration as the time value. A change in the signal state at the enable input from "1" to "0" has no effect.

❷ When the signal state at the enable input goes from "0" to "1" (positive edge) after the timer has expired without incident, the start instruction has no effect on the timer.

❸ When the signal state at the Start input is "0", a positive signal edge at the enable input has no effect.

❹❺ If there is a positive edge at the enable input when the timer is reset and the start input has signal state "1", the timer is restarted. The timer restarts with the programmed duration as the current time value.

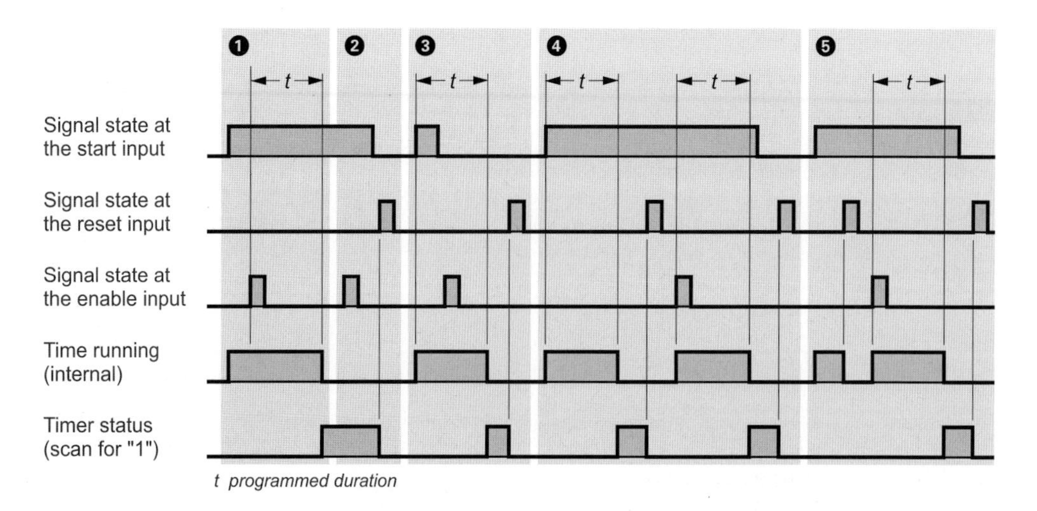

Figure 7.10 Enabling a Retentive On-Delay Timer

7.6 Off-Delay Timers

The complete STL statement list for starting a timer as an off-delay is as follows:

```
A       Enable_input;
FR      Timer;
A       Start_input;
L       Duration;
SF      Timer;
A       Reset_input;
R       Timer;
L       Timer;
T       Binary_time_value;
LD      Timer;
T       BCD_time_value;
A       Timer;
=       Timer_status;
```

For SCL, calling a timer as a retentive off-delay is programmed as follows:

```
BCD_time_value:= S_OFFDT (
      T_NO := Timer,
      S := Start_input,
      TV := Duration,
      R := Reset_input,
      Q := Timer_status,
      BI := Binary_time_value);
```

Starting an off-delay timer

The diagram in Figure 7.11 describes the dynamic performance of a timer after it is started as off-delay timer, and when it is restarted. The description is valid if you adhere to the sequence of operations shown opposite for STL (starting before resetting before checking). Enabling is not required for the "normal" sequence, and in SCL it is also not available.

①③ When the signal state at the timer's Start input goes from "0" to "1" (negative edge), the timer is started. It runs with the programmed duration as the time value. Checks for signal state "1" (timer status) return a check result of "1" when the signal state at the Start input is "1" or when the timer is running (off delay).

The time is counted down from the initial value with the set time base.

② When the signal state at the Start input changes from "0" to "1" (positive edge) while the timer is running, the timer is reset. The timer is not restarted until there is a negative edge at the Start input.

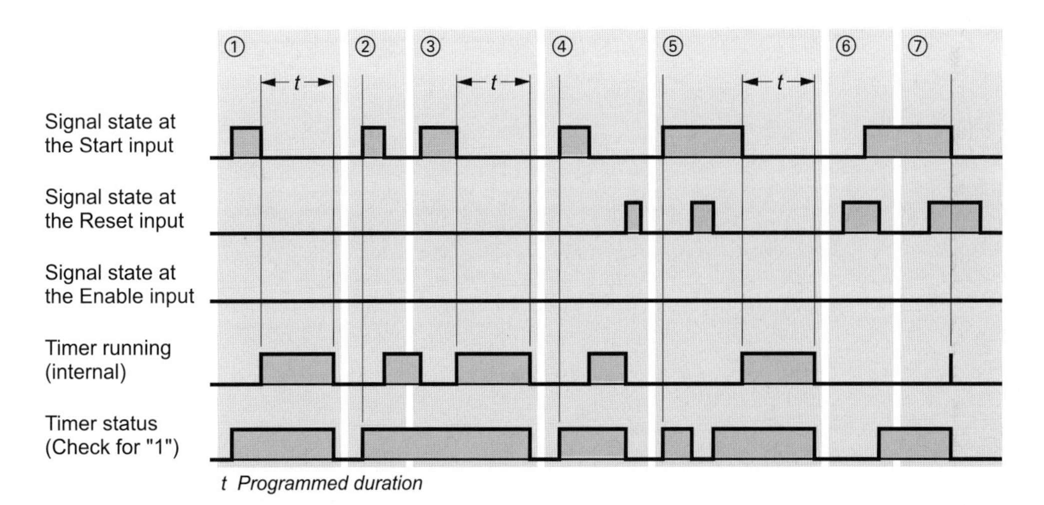

Figure 7.11 Response of an Off-Delay Timer

Resetting an off-delay timer

Resetting of an off-delay timer has a static effect and takes precedence over the starting of a timer (Figure 7.11).

④ Signal state "1" at the Reset input of a running timer resets the timer. Checks for signal state "1" (timer status) then return a check result of "0". Time value and time base are also set to zero.

⑤⑥ Signal state "1" at the Start input and at the Reset input resets the timer's binary output (a check for signal state "1" (timer status) then returns a check result of "0"). If the signal state at the Reset input then changes back to "0", the timer's output goes back to "1".

⑦ If the signal state at the Start input changes from "1" to "0" (negative edge) while the Reset signal is present, the timer is started but the subsequent reset resets it again immediately (indicated by a line in the diagram). A check for signal state "1" (timer status) then immediately returns a check result of "0".

Enabling an off-delay timer

The timer is "retriggered", i.e. prompted to restart, with a positive edge at the enable input. Enabling is only possible in the STL programming language.

Figure 7.12 shows enabling of a timer as an off-delay timer.

❶ If the signal state at the enable input of an inactive timer changes from "0" to "1" (positive edge), the timer is not affected by the execution of the start operation. A change in the signal state at the enable input from "1" to "0" also has no effect.

❷ If the signal state at the enable input of a running (active) timer changes from "0" to "1" (positive edge), the timer is restarted when the start operation is processed. The timer uses the programmed duration as the current time value for the restart.

❸ There is no effect from a change in the signal state at the enable input from "0" to "1" (positive edge) or a change in the signal state from "1" to "0" (negative edge) at the enable input when the time is not running.

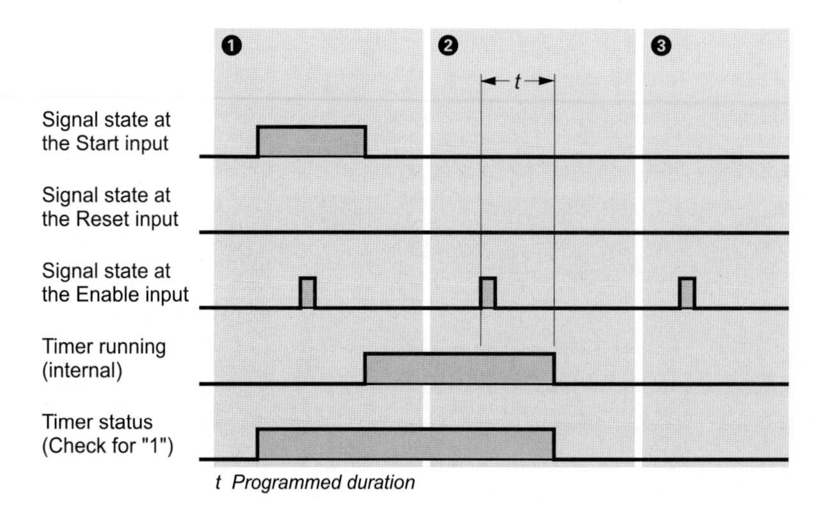

Figure 7.12 Enabling an Off-Delay Timer

7.7 IEC Timer Functions

The IEC timer functions are integrated into the operating system of the CPU as system function blocks SFBs.

In appropriately equipped CPUs, the following functions are available:

▷ SFB 3 TP
Pulse generation

▷ SFB 4 TON
On delay

▷ SFB 5 TOF
Off delay

Figure 7.13 shows the dynamic response of these timers.

You call these SFBs with an instance data block or you use these SFBs as local instances in a function block.

You can find the interface description for offline programming in the *Standard Library* under the *System Function Blocks* program.

Call examples can be found in the STL_Book library under the "Basic Functions" program in function block FB 107, or in the source file Chap_7, or in the SCL_Book library under the "30 SCL Functions" program. You can download the library from the publisher's web site.

Table 7.2 Parameters for the IEC Timer Functions

Name	Declaration	Data type	Description
IN	INPUT	BOOL	Start input
PT	INPUT	TIME	Pulse length or delay duration
Q	OUTPUT	BOOL	Timer status
ET	OUTPUT	TIME	Elapsed time

7.7.1 Pulse Generation SFB 3 TP

The IEC timer SFB 3 TP has the parameters shown in Table 7.2.

If the RLO at the start input of the timer changes from "0" to "1", the timer starts. It runs with the programmed duration regardless of any further changes in the RLO at the start input. Output Q returns signal state "1" as long as the time runs.

Output ET returns the time duration set at output Q. This duration begins at T#0s and ends at the set time PT. If PT has elapsed, ET remains at the elapsed value until input IN changes back to "0". If input IN has signal state "0" before PT has elapsed, output ET changes to T#0s immediately after PT has elapsed.

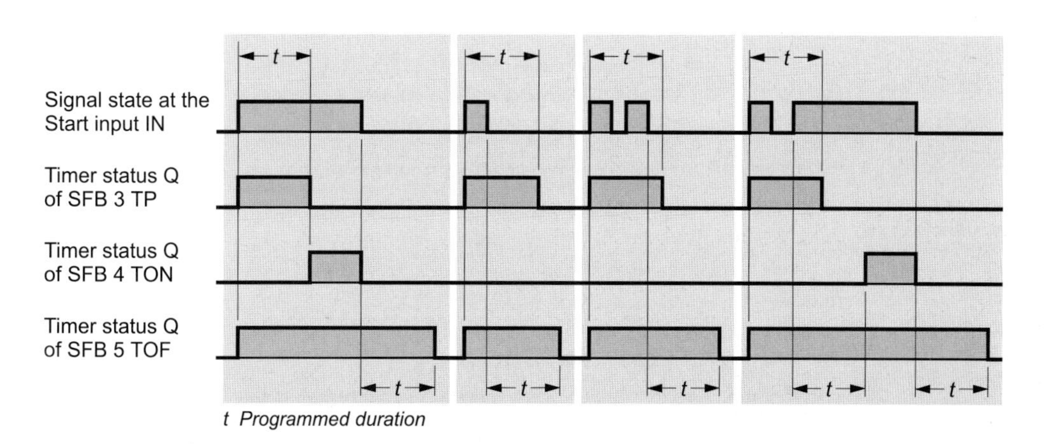

Figure 7.13 Dynamic Response of the IEC Timer Functions

If you want to reinitialize the timer, start with the time duration PT = T#0s.

SFB 3 TP runs in the operating states RESTART and RUN. It is reset (initialized) at cold restart.

7.7.2 On Delay SFB 4 TON

The IEC timer SFB 4 TON has the parameters shown in Table 7.2.

If the RLO at the start input of the timer changes from "0" to "1", the timer starts. It runs with the programmed duration. Output Q returns signal state "1" if the time has elapsed. If the RLO at the start input changes from "1" to "0" before the time has elapsed, the running time is reset. It starts again with the next positive edge.

Output ET returns the time duration run by the timer. This duration begins at T#0s and ends at the set time PT. If PT has elapsed, ET remains at the elapsed value until input IN changes back to "0". If input IN has signal state "0" before PT has elapsed, output ET changes to T#0s immediately.

If you want to reinitialize the timer, start with the time duration PT = T#0s.

SFB 4 TON runs in the operating states RESTART and RUN. It is reset at cold restart.

7.7.3 Off Delay SFB 5 TOF

The IEC timer SFB 5 TOF has the parameters shown in Table 7.2.

If the RLO at the start input of the timer changes from "0" to "1", output Q has signal state "1". If the RLO at the start input changes back to "0", the timer starts. Output Q remains at signal state "1" while the timer runs. When the time has elapsed, output Q is reset. If the RLO at the start input changes again to "1" before the time has elapsed, the timer is reset and output Q remains "1".

Output ET returns the time duration run by the timer. This duration begins at T#0s and ends at the set time PT. If PT has elapsed, ET remains at the elapsed value until input IN changes back to "1". If input IN has signal state "1" before PT has elapsed, output ET changes to T#0s immediately after PT has elapsed.

If you want to reinitialize the timer, start with the time duration PT = T#0s.

SFB 5 TOF runs in the operating states RESTART and RUN. It is reset at cold restart.

8 Counter Functions

Counter functions allow you to have counting tasks carried out directly by the central processor. The counters can count up and down, and the counting range extends over three decades (from 000 to 999).

This chapter describes the statements for the STL programming language. In the SCL programming language, the counter functions are included among the SCL standard functions (see Chapter 30.2 "Counter Functions").

The counting rate of these counters depends on your program's scan time! In order to count, the CPU must detect a signal state change in the input pulse, that is to say, an input pulse (or space (interpulse period)) must be present for at least one program scan cycle. The longer the program scan cycle, the lower the counting rate.

Note: the integrated functions of the compact CPUs with S7-300 (CPU 3xxC) also contain counter functions that can count via a special counter input at up to 10, 30 or 60 kHz depending on the CPU.

The counters described in this chapter are stored in the system memory of the CPU. You can set counters to an initial value, reset them, count up, and count down. You can find out whether the count is zero or not zero by checking a counter. The current count can be loaded into accumulator 1 in binary or in binary-coded decimal (BCD).

The examples in this chapter and the calls for IEC counters can be found in the STL_Book library under the "Basic Functions" program in function block FB 108 or source file Chap_8. You can download the library from the publisher's web site (see page 8).

8.1 Setting and Resetting Counters

Setting a counter

```
S   C n        Sets a counter
```

A counter is set when the RLO goes from "0" to "1" prior to the Set operation S. A positive edge is always required to set a counter.

"Set counter" means to load the counter with an initial value. The initial value with which it is to be loaded is in accumulator 1 (see below). The range extends from 0 to 999.

Specifying the count

The "set counter" statement takes the value in accumulator 1 as count value. How and when that value got into accumulator 1 is of no concern.

To make your program more readable, you should load the count value into the accumulator immediately before the Set statement, either in the form of a constant (direct specification of a count value) or a variable (such as a memory word containing the count value).

Note: accumulator 1 must contain a valid counter value even if the counter is not set when the set statement is processed.

Specifying a count in the form of a constant

```
L   C#100;          //Count value 100

L   W#16#0100;      //Count value 100
```

A count comprises three decades, and may be in the range from 000 to 999. Only positive BCD values are permitted; the counters cannot process negative values. You may use C# or

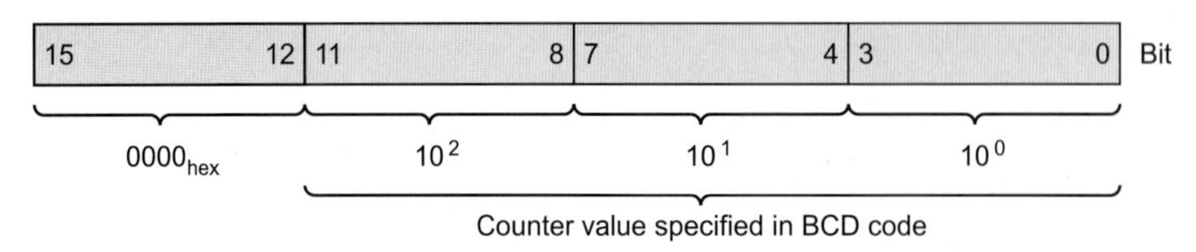

Figure 8.1 Bit Assignments of the Counter Value

W#16# (in conjunction with decimal digits only) to identify a constant.

Specifying a count in the form of a variable

```
L  C#200;     //Count value 200
T  MW 56;     //Save count value
.. ;
L  MW 56;     //Load count value
```

The Set operation expects there to be a count in accumulator 1 consisting of three right-justified decades. The meanings of the bits in the count (data type C#) are described in detail in Chapter 24 "Data Types".

Resetting a counter

R   C n   Resets a counter

A counter is reset when the RLO is "1" when the Reset statement is encountered. As long as RLO "1" is present, counter checks for "1" return a check result of "0" and counter checks for "0" return a counter result of "1". Resetting a counter sets the count value to "0".

Note: Resetting a counter function does not reset the internal edge memory bit for setting, up counting and down counting. For setting or counting again, the relevant statement must first be processed with RLO "0" before the counter function can be set again or counting can begin again. You can also use enabling of the counter function for this purpose.

8.2 Counting

Counting up

CU   C n   Count up

A counter is counted up (incremented) when the RLO changes from "0" to "1" prior to the CU (count up) statement. Up counting always requires a positive signal edge.

Each positive edge preceding the CU operation increases the count value by one unit until the upper limit value of 999 is reached. A positive edge at the CU input then has no further effect.

There is no carry.

Counting down

CD   C n   Count down

A counter is counted down (decremented) when the RLO changes from "0" to "1" prior to the CD (count down) statement. Down counting always requires a positive signal edge.

Each positive edge preceding the CD statement decreases the count value by one unit until the lower limit value of 0 has been reached. A positive edge at the CD input then has no further effect.

The count value does not go into the negative range.

8.3 Checking a Counter

Binary counter check

A	C n	Check for signal state "1" and combine according to AND
O	C n	Check for signal state "1" and combine according to OR
X	C n	Check for signal state "1" and combine according to Exclusive OR
AN	C n	Check for signal state "0" and combine according to AND
ON	C n	Check for signal state "0" and combine according to OR
XN	C n	Check for signal state "0" and combine according to Exclusive OR

You can check a logical counter combination as you would an input, for instance, and further combine the result of the check. Checks for signal state "1" return a check result of "1" when the count is greater than zero, and a check result of "0" when the count is zero.

Direct loading of a count value

L C n Direct loading of a count value

The load function L C transfers the count specified in the counter function into accumulator 1 in the form of a binary number. This value is the value current at the instant of the check. The value now in accumulator 1 corresponds to a positive number in INT format, and can be further processed, for example with arithmetic functions.

Example:

```
L  C 99;     //Load current count
T  MW 76;  //and save
```

Coded loading of a count value

LD C n Coded loading of a count value

The load function LC C transfers the count specified in the counter function to accumulator 1 in the form of a binary-coded decimal number. This value is the value current at the instant of the check. The count is subsequently available in the accumulator as a right-justified BCD number. It has the same structure as the specified count.

Example:

```
LD  C 99;     //Load current count value
T   MW 50;  //and save
```

8.4 Enabling a Counter

FR C n Enable counter

When you enable a counter, you can set the counter and use it for counting without a positive signal edge having to precede the relevant operation. However, this is possible only when the relevant operation is processed while the RLO is "1".

The enable is active when the RLO goes from "0" to "1" before the enable instruction is encountered. A positive signal edge is always required to enable a counter.

A counter need not be enabled in order for it to be set, reset, or used for counting (that is to say, for normal operation of a counter).

Note: Enable affects setting, counting up and counting down simultaneously! A positive edge at the time of the enable instruction causes all subsequent instructions (S, CU and CD) which have signal state "1" to be executed.

The counter function example below is designed to show the functional principle of the enable instruction on the remaining inputs (the diagram is shown in Figure 8.2):

```
A      "Enable";
FR     "Counter";
A      "Count up";
CU     "Counter";
A      "Count down";
CD     "Counter";
A      "Set";
L      C#020;
S      "Counter";
A      "Reset";
R      "Counter";
A      "Counter";
=      "Counter status";
```

① The positive edge at the set input sets the counter to the initial value of 20.

② A positive edge at the CU input increments the counter by one unit.

③ Because the signal state at the set input is "1", an enable instruction increments the count by one unit.

④ The positive edge at the reset input decrements the count by one unit.

⑤ The enable instruction causes the statements for up counting and down counting to be executed, as signal state "1" is present at both inputs.

⑥ The positive edge at the set input sets the counter to the initial value of 20.

⑦ Signal state "1" at the reset input resets the counter. A check for signal state "1" returns a check result of "0".

⑧ Because signal state "1" is still present at the set input, the enable instruction again sets the counter to 20. A check for signal state "1" returns a check result of "1".

8.5 Sequence of Counter Instructions

When programming a counter, you do not need all the statements available for it. You need program only those statements required for the timer in question. For example, all that is needed for a down counter is the setting of the initial value, down counting, and a binary check for "0".

For a counter to perform as described in the last several sections, you must observe a specific sequence when programming the counter instructions. Table 8.1 shows the optimum sequence for all counter instructions. Simply omit the unneeded statements when you write your program, for example, enabling the counter function.

If a Reset is to have a "static" effect on the CU, CD and S statements and be independent of the result of the logic operation (RLO), you have to write the Reset statement for the counter in question after these statements and before the check statement for the counter.

If the counter is then set and reset "simultaneously", it will still be assigned a value, but is then immediately reset by the Reset statement. The subsequent check therefore does not recognize the fact that the counter had been briefly set.

If the setting of a counter is to have a "static" effect on the counter statements and be independent of the RLO, the Set instruction for that counter must be programmed after the counting instructions. If a counter is set and reset "simul-

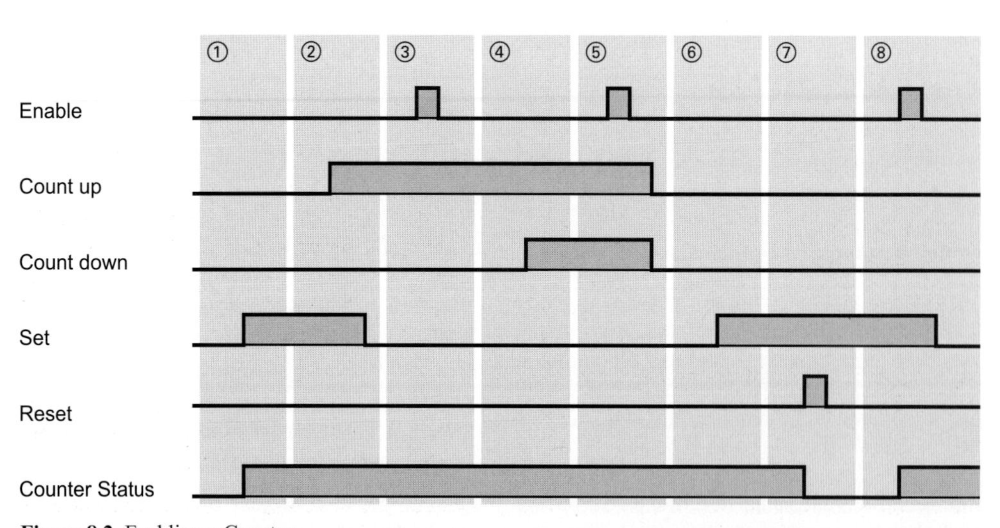

Figure 8.2 Enabling a Counter

Table 8.1 Sequence of Counter Instructions

Counter Instruction	Examples
Enable counter	A I 22.0 FR C 17
Count up	A I 22.1 CU C 17
Count down	A I 22.2 CD C 17
Set counter	A I 22.3 L C#500 S C 17
Reset counter	A I 22.4 R C 17
Digital check	L C 17 T MW 30 LC C 17 T MW 32
Binary check	A C 17 = Q 13.0

taneously", the counting instructions still affect the count, but the count is subsequently set to the programmed value, which it retains for the remainder of the program scan.

The sequence of statements for up and down counting is not relevant.

8.6 IEC Counter Functions

The IEC counter functions are integrated into the operating system of the CPU as system function blocks SFBs. In appropriately equipped CPUs, the following functions are available:

- SFB 0 CTU
  Up counter
- SFB 1 CTD
  Down counter
- SFB 2 CTUD
  Up-down counter

You call these SFBs with an instance data block or you use these SFBs as local instances in a function block.

You can find the interface description for offline programming in the *Standard Library* under the *System Function Blocks* program.

Call examples can be found in the STL_Book library under the "Basic Functions" program in function block FB 108, or in the source file Chap_8, and in the SCL_Book library under the "30 SCL Functions" program. You can download both libraries from the publisher's web site (see page 8).

8.6.1 Up Counter SFB 0 CTU

The IEC counter function SFB 0 CTU has the parameters shown in Table 8.2.

If the signal state at the up counter input CU changes from "0" to "1" (positive edge), the current counter value is incremented by 1 and displayed at output CV. When called for the first time (with signal state "0" at reset input R), the counter value corresponds to the preset value at input PV.

If the current counter value reaches the upper limit of 32,767, it is no longer incremented. CU then remains without effect.

The counter value is reset to zero if reset input R has signal state "1". A positive edge at CU remains without effect while input R has signal state "1".

Output Q has signal state "1" if the value at CV is greater than or equal to the value at PV.

SFB 0 CTU runs in the operating states RESTART and RUN. It is reset at cold restart.

8.6.2 Down Counter SFB 1 CTD

The IEC counter SFB 1 CTD has the parameters shown in Table 8.2.

If the signal state at the down counter input CD changes from "0" to "1" (positive edge), the current counter value is decremented by 1 and displayed at output CV. When called for the first time (with signal state "0" at the LOAD input), the counter value corresponds to the preset value at input PV.

If the current counter value reaches the lower limit of –32,768, it is no longer decremented. CD then remains without effect.

The counter value is reset to the preset value PV if the LOAD input has signal state "1". A posi-

Table 8.2 Parameters for the IEC Counter Functions

Name	Present in SFB			Declaration	Data Type	Description
CU	0	-	2	INPUT	BOOL	Up count input
CD	-	1	2	INPUT	BOOL	Down count input
R	0	-	2	INPUT	BOOL	Reset input
LOAD	-	1	2	INPUT	BOOL	Load input
PV	0	1	2	INPUT	INT	Preset value
Q	0	1	-	OUTPUT	BOOL	Counter status
QU	-	-	2	OUTPUT	BOOL	Count counter status up
QD	-	-	2	OUTPUT	BOOL	Count counter status down
CV	0	1	2	OUTPUT	INT	Current counter value

tive edge at CD remains without effect while the LOAD input has signal state “1”.

Output Q has signal state “1” if the value at CV is less than or equal to zero.

SFB 1 CTD runs in the operating states RESTART and RUN. It is reset at cold restart.

8.6.3 Up-Down Counter SFB 2 CTUD

The IEC counter SFB 2 CTUD has the parameters shown in Table 8.2.

If the signal state at the up count input CU changes from “0” to “1” (positive edge), the counter value is incremented by 1 and displayed at output CV. If the signal state at the down count input CD changes from “0” to “1” (positive edge), the counter value is decremented by 1 and displayed at output CV. If both counter inputs have a positive edge, the current counter value does not change.

If the current counter value reaches the upper limit of 32,767, it is no longer incremented if there is a positive edge at up count input CU. CU then remains without effect.

If the current counter value reaches the lower limit of –32,768, it is no longer decremented if there is a positive edge at down counter input CD. CD then remains without effect.

The counter value is reset to the preset value PV if the LOAD input has signal state “1”. Positive signal edges at the counter inputs remain without effect while the LOAD input has signal state “1”.

The counter value is reset to zero if reset input R has signal state “1”. Positive signal edges at the counter inputs and signal state “1” at the LOAD input remain without effect while input R has signal state “1”.

Output QU has signal state “1” if the value at CV is greater than or equal to zero.

Output QD has signal state “1” if the value at CV is less than or equal to zero.

SFB 2 CTUD runs in the operating states RESTART and RUN. It is reset at cold restart.

8.7 Parts Counter Example

The example illustrates the handling of timers and counters. It is programmed with inputs, outputs and memory bits so that it can be programmed at any point in any block. A function without block parameters has been used in the example.

Functional description

Parts are transported on a conveyor belt. A light barrier detects and counts the parts. After a set number, the counter sends the signal “Finished”. The counter is equipped with a monitoring circuit. If the signal state of the light barrier does not change within a specified time, the monitoring circuit emits a signal.

The *Set* input gives the counter its initial value (the number of parts to be counted). A positive edge at the light barrier decrements the counter by one unit. When the count reaches zero, the counter sends the "Finished" signal. Prerequisite is that the parts are lying individually (at intervals from one another) on the conveyor belt (Figure 8.3).

The *Set* input also sets the "Active" signal. The controller monitors a signal state change of the light barrier only in the active state. The *Active* signal is reset when counting is finished and the last item counted has exited the light barrier.

In the active state, a positive edge of the light barrier starts the timer with the time value *Dura1* as retentive pulse timer. If the timer's Start input is processed with "0" in the next scan cycle, it nevertheless continues to run. A new positive signal edge "retriggers", that is, restarts, the timer. The next positive edge for restarting the timer is generated when the light barrier signals a negative edge. The timer is then started with time value *Dura2*. If the light barrier is broken for a length of time exceeding *Dura1*, or free for a period of time which exceeds *Dura2*, the time elapses and the timer signals *Fault*. The first *Active* signal starts the timer with the time value *Dura2*.

The *Set* signal activates the counter and the monitoring circuit. The light barrier uses positive and negative signal edges to control the counter, the *Active* state, selection of the time value, and the starting (retriggering) of the watchdog timer.

Evaluation of the light barrier's positive and negative edge is required often, and temporary local data are suitable here as "scratchpad memory". Temporary local data are block-local variables, and are declared in the block (not in the symbol table). In the example, the edge evaluation's pulse memory bits are stored in temporary local

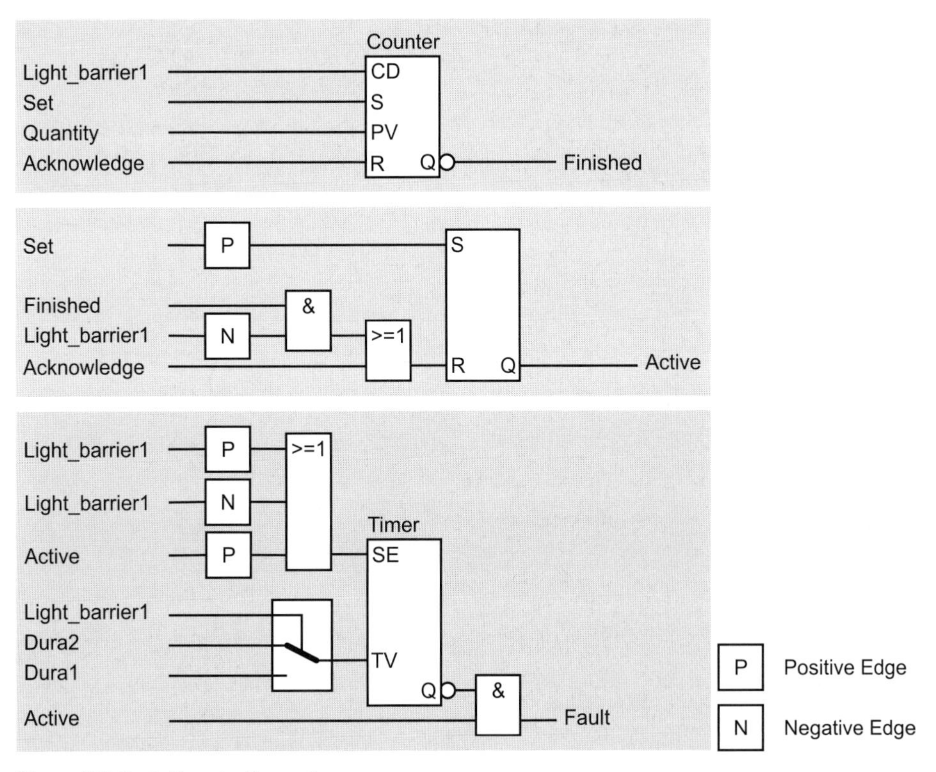

Figure 8.3 Parts Counter Example

(see page 195 for program)

data. (The edge memory bits also require their signal states in the next scan cycle, and must therefore not be temporary local data.)

The program is located in a function without block parameters. You can call this function, in OB 1 for example, as follows:

```
CALL "Counter_control";
```

The program is available as source text with symbolic addressing. The global symbols can also be used without quotation marks as long as they contain no special characters. If a special character (such as an umlaut or a space) is located in a symbol, then it must be enclosed in quotation marks. The editor shows all global symbols with quotation marks in the compiled block.

The program is subdivided into networks for better readability. The last network, which has the network title BLOCK END, is not absolutely necessary. However, it is a visual sign of the end of the block, which is very useful, particularly in the case of extremely long blocks.

You will find this example in the STL_Book library in the "Conveyor Example" program which you can download from the publisher's web site (see page 8). The *Symbols* object contains the symbol table, the *Source Files* container source the program "Conveyor", and the *Blocks* container the compiled program in the function FC 12.

```
FUNCTION "Counter_control" : VOID
TITLE = Parts counter with monitoring circuit
//Example of timer and counter functions
NAME    : Count
AUTHOR  : Berger
FAMILY  : STL_Book
Version : 01.00
VAR_TEMP
  PM_LB_P : BOOL;            //Pulse positive edge light barrier
  PM_LB_N : BOOL;            //Pulse negative edge light barrier
END_VAR
BEGIN
NETWORK
TITLE = Counter_control
     A    Light_barrier1;    //When light barrier is tripped,
     CD   Count;             //decrement counter by 1
     A    Set;
     L    Quantity;          //Preset count with "Quantity"
     S    Count;
     A    Acknowledge;
     R    Count;
     AN   Count;             //When count reaches zero,
     =    Finished;          //output "Finished" signal
NETWORK
TITLE = Activate monitor
     A    Light_barrier1;
     FP   EM_LB_P;           //Generate pulse memory bit
     =    PM_LB_P;           //on positive edge of light barrier
     A    Light_barrier1;
     FN   EM_LB_N;           //Generate pulse memory bit
     =    PM_LB_N;           //on negative edge of light barrier
     A    Set;
     FP   EM_ST_P;
     S    Active;            //Activate monitoring circuit
     A    Finished;
     A    PM_LB_N;
     O    Acknowledge;
     R    Active;            //Deactivate monitoring circuit
NETWORK
TITLE = Monitoring circuit
     L    Dura1;             //If light barrier is "1"
     A    Light_barrier1;    //jump JC to D1 is executed and the
     JC   D1;                //accumulator contains "Dura1" otherwise
     L    Dura2;             //the accumulator contains "Dura2"
D1:  A    Active;
     FP   EM_Ac_P;           //If there is a positive edge at "active"
     O    PM_LB_P;           //or a positive edge at the light barrier,
     O    PM_LB_N;           //or a negative edge at the light barrier,
     SE   Monitor;           //the timer is started or retriggered
     AN   Monitor;
     A    Active;            //If time elapses while "active",
     =    Fault;             //"Fault" is signaled
NETWORK
TITLE = Block End
     BE;
END_FUNCTION
```

Digital Functions

The digital functions process digital values predominantly of data type INT, DINT and REAL, and thus extend the functionality of the PLC. The digital functions for the STL programming language are described at this point. In the SCL programming language, comparisons, word logic operations and arithmetic functions are implemented with operators (Chapter 27.4 "Expressions"); the remaining digital functions are included in SCL among the standard functions (Chapters 30.3 "Math Functions", 30.4 "Shifting and Rotating" and 30.5 "Conversion Functions").

The **comparison functions** form a binary result from the comparison of two values. They take into account the data types INT, DINT and REAL.

You use the **arithmetic functions** to make calculations in your program. All basic arithmetic operations in data types INT, DINT and REAL are available.

The **math functions** extend the calculation options beyond the basic arithmetic operations to include such additions as trigonometric functions.

Before and after performing calculations, you match the digital values to the required data type using the **conversion functions**.

The **shift functions** make it possible to align the contents of the accumulator by shifting them to the right or left. It is always possible to scan the last bit shifted.

Word logic is used to mask digital values by targeting individual bits and setting them to "1" or "0".

Digital logic operations work mainly with values in data blocks. These can be global data blocks or instance data blocks if static local data are used. Chapter 18.2 "Block Functions for Data Blocks" deals with the use of data blocks and the addressing options for data.

With the exception of the accumulators, the temporary local data are extremely well suited for storing temporary results.

9 Comparison Functions

The comparison functions compare two digital values, one of which is in accumulator 1, the other in accumulator 2. As the result of the comparison, the comparison function sets the result of the logic operation (RLO) and status bits CC0 and CC1. The result can be post-processed with binary logic operations, memory functions, or jump statements. Table 9.1 provides an overview of the available comparison functions.

In Chapter 15 “Status Bits” you will learn how the comparison functions set status bits CC0 and CC1.

The comparison functions for the STL programming language are described in this chapter. In the SCL programming language, the comparison functions are formulated with comparison expressions (Chapter 27.4.2 “Comparison Expressions”).

The examples in this chapter can be found in the download files (download address: see pages 8-9) in the STL_Book library under the “Digital Functions” program in function block FB 109 or source file Chap_9.

Table 9.1
General Representation of a Comparison Function

	Comparison According To Data Type		
Comparison for	INT	DINT	REAL
equal to	==I	==D	==R
not equal to	<>I	<>D	<>R
greater than	>I	>D	>R
greater than or equal to	>=I	>=D	>=R
less than	<I	<D	<R
less than or equal to	<=I	<=D	<=R

9.1 General Representation of a Comparison Function

You program a comparison function according to the following general scheme:

```
Load                  Operand1;
Load                  Operand2;
Comparison function;
Assign                Result;
```

To begin with, the first of the addresses to be compared is loaded into accumulator 1. When the second address is loaded, the contents of accumulator 1 are shifted into accumulator 2 (see Chapter 6.2 “Load Functions”). Now the contents of accumulators 1 and 2 can be compared with each other using the comparison function.

The comparison function returns a binary result (data type BOOL) that can be assigned to a binary address or combined with other binary checks.

A comparison function does not modify the contents of the accumulators. It is always executed without regard to any conditions.

Table 9.2 shows an example for the various data types. The comparison instruction carries out the comparison according to the specified characteristic, without regard to the contents of the accumulators.

In the case of data type INT, the CPU compares only the right-hand (low-order) words of the accumulators; the contents of the left-hand (high-order) words are not taken into account.

In comparisons involving data type REAL, a check is made to make sure that the accumulators contains valid REAL numbers. If they do not, the CPU sets the RLO to “0” and status bits CC0, CC1, OV and OS to “1”.

Table 9.2 Examples of Comparison Functions

Comparison According to INT	Memory bit M99.0 is reset if the value in memory word MW 92 is equal to 120, otherwise it is not.	`L MW 92;`
		`L 120;`
		`==I ;`
		`R M 99.0;`

Comparison According to DINT	The variable "CompResult" in data block "Global_DB" is set if variable "CompVal1" is less than "CompVal2"; otherwise it is reset.	`L "Global_DB".CompVal1;`
		`L "Global_DB".CompVal2;`
		`<D ;`
		`= "Global_DB".CompResult;`

Comparison According to REAL	If variable #Actval is greater than or equal to variable #Calibra, #Recali is set, otherwise not.	`L #Actval;`
		`L #CALIBRA;`
		`>=R ;`
		`S #Recali;`

9.2 Description of the Comparison Functions

Comparison for equal to

The "comparison for equal to" instruction interprets the contents of the accumulators in accordance with the data type specified in the instruction and checks to see if the two values are equal. The RLO is "1" following the operation in the following cases:

- ▷ Data type INT
  If the contents of the low-order word of accumulator 2 is equal to the contents of the low-order word of accumulator 1.
- ▷ Data type DINT
  If the contents of accumulator 2 are equal to the contents of accumulator 1.
- ▷ Data type REAL
  If the contents of accumulator 2 are equal to the contents of accumulator 1, on condition that both accumulators contains valid REAL numbers.

If two REAL numbers are equal but invalid, the "equal to" condition is not fulfilled (RLO = "0").

Comparison for not equal to

The "comparison for not equal to" instruction interprets the contents of the accumulators in accordance with the data type specified in the comparison instruction and checks to see whether the two values differ. The RLO is "1" following the comparison operation in the following cases

- ▷ Data type INT
  If the contents of the low-order word of accumulator 2 are not equal to the contents of the low-order word of accumulator 1.
- ▷ Data type DINT
  If the contents of accumulator 2 are not equal to the contents of accumulator 1.
- ▷ Data type REAL
  If the contents of accumulator 2 are not equal to the contents of accumulator 1, on condition that both accumulators contain valid REAL numbers.

If two REAL numbers are not equal but one or both of them is invalid, the "not equal to" condition is not fulfilled (RLO = "0").

Comparison for greater than

The "comparison for greater than" instruction interprets the contents of the accumulators in accordance with the data type specified in the comparison instruction and checks to see whether the value in accumulator 2 is greater than the value in accumulator 1. Following the operation, the RLO is "1" in the following instances:

- ▷ Data type INT
  If the contents of the low-order word of accumulator 2 are greater than the contents of the low-order word of accumulator 1.

- ▷ Data type DINT
  If the contents of accumulator 2 are greater than the contents of accumulator 1.
- ▷ Data type REAL
  If the contents of accumulator 2 are greater than the contents of accumulator 1, on condition that both accumulators contain valid REAL numbers.

Comparison for greater than or equal to

The "comparison for greater than or equal to" instruction interprets the contents of the accumulators in accordance with the data type specified in the comparison statement and checks to see whether the value in accumulator 2 is greater than or equal to the value in accumulator 1. Following the comparison, the RLO is "1" in the following instances:

- ▷ Data type INT
  If the contents of the low-order word of accumulator 2 is greater than the contents of the low-order word of accumulator 1 or if the bit patterns of the two words are equal.
- ▷ Data type DINT
  If the contents of accumulator 2 are greater than the contents of accumulator 1 or if the bit patterns in the two accumulators are equal.
- ▷ Data type REAL
  If the contents of accumulator 2 are greater than the contents of accumulator 1 or if the contents of the two accumulators are equal on condition that both accumulators contain valid REAL numbers.

Comparison for less than

The "comparison for less than" instruction interprets the contents of the accumulators in accordance with the data type specified in the comparison operation and checks whether the value in accumulator 2 is less than the value in accumulator 1. Following the comparison, the RLO is "1" in the following instances:

- ▷ Data type INT
  If the contents of the low-order word of accumulator 2 are less than the contents of the low-order word of accumulator 1.
- ▷ Data type DINT
  If the contents of accumulator 2 are less than the contents of accumulator 1.
- ▷ Data type REAL
  If the contents of accumulator 2 are less than the contents of accumulator 1 on condition that both accumulators contain valid REAL numbers.

Comparison for less than or equal to

The "comparison for less than or equal to" operation interprets the contents of the accumulators in accordance with the data type specified in the comparison instruction and checks whether the value in accumulator 2 is less than or equal to the value in accumulator 1. Following the comparison operation, the RLO is "1" in the following instances:

- ▷ Data type INT
  If the contents of the low-order word of accumulator 2 are less than the contents of the low-order word of accumulator 1 or if the bit patterns of the two words are equal.
- ▷ Data type DINT
  If the contents of accumulator 2 are less than the contents of accumulator 1 or if the bit patterns of the values in the two accumulators are equal.
- ▷ Data type REAL
  If the contents of accumulator 2 are less than the contents of accumulator 1 or if the contents of the two accumulators are equal on condition that both contain valid REAL numbers.

9.3 Comparison Function in a Logic Operation

The comparison function returns a binary RLO and can therefore be used in conjunction with other binary functions. The comparison function sets status bit FC, that is to say, in binary logic operations, a comparison function is always a first check.

Comparison at the start of a logic operation

At the start of a logic operation, a comparison function is always a first check. The RLO returned by the comparison function can be directly combined using binary checks.

```
L       MW 120;
L       512;
>I      ;
A       Input1;
=       Output1;
```

In the example, *Output1* is set if the comparison condition is fulfilled and *Input1* has a signal state of "1".

Comparison within a logic operation

A comparison function within a binary logic operation must be enclosed, as the comparison function begins a new logic step (first check).

```
O       Input2;
O(      ;
L       MW 122;
L       200;
<=I     ;
)       ;
O       Input3;
=       Output2;
```

In the example, *Output2* is set if *Input2* or *Input3* has a signal state of "1" or if the compare condition is fulfilled.

Multiple comparisons

Because a comparison function does not alter the contents of the accumulators, multiple successive comparisons are possible in STL.

```
L       MW 124;
L       1200;
>I      ;
JC      GREA;
==I     ;
JC      EQUA;
```

In the example, two comparison functions are applied to the same accumulator contents. The first comparison generates RLO = "1" if MW 124 is greater than 1200, so that the jump to GREA is executed. Without reloading the accumulators, the second comparison function compares for equal to and generates a new RLO.

The comparison function sets the status bits based on the relationship between the values compared, that is, independently of the condition on which the comparison is based. You can make use of this fact by checking the status bits with the relevant jump functions. The example above can also be programmed as follows:

```
L       MW 124;
L       1200;
>I      ;
JP      GREA;
JZ      EQUA;
```

In this example, the comparison is evaluated on the basis of status bits CC0 and CC1. The comparison itself, in this case "greater than", does not affect the setting of the status bits; a different comparison, for example for "less than", could also have been used. JP scans to see whether the first comparison value is greater than the second, JZ to see whether they are equal.

10 Arithmetic Functions

The arithmetic functions combine two digital values in accumulators 1 and 2 in accordance with one of the basic arithmetic operations. The result is placed in accumulator 1. Status bits CC0, CC1, OV and OS provide information about the result and the progress of the calculation (see Chapter 15 "Status Bits"). Table 10.1 provides an overview of the available arithmetic functions.

In addition to applying the basic arithmetic operations to values in accumulator 2, you can also add constants directly to the contents of accumulator 1 or modify the contents of accumulator 1 by a fixed amount.

The statements for the STL programming language are described in this chapter. In the SCL programming language, the arithmetic functions are formulated using arithmetic expressions (Chapter 27.4.1 "Arithmetic Expressions").

The examples in this chapter can be found in the download files (download address: see pages 8-9) in the STL_Book library under the "Digital Functions" program in function block FB 110 or source file Chap_10.

Table 10.1
Overview of the Arithmetic Functions

Arithmetic Functions	With Data Type		
	INT	DINT	REAL
Addition	+I	+D	+R
Subtraction	–I	–D	–R
Multiplication	*I	*D	*R
Division with quotient as result	/I	/D	/R
Division with remainder as result	-	MOD	-

10.1 General Representation of an Arithmetic Function

You program an arithmetic function according to the following general scheme:

```
Load                    Operand1;
Load                    Operand2;
Arithmetic function     ;
Transfer                Result;
```

To begin with, the first of the addresses to be combined is loaded into accumulator 1. When the second address is loaded, the contents of accumulator 1 are shifted into accumulator 2 (see Chapter 6.2 "Load Functions"). Now the contents of accumulators 1 and 2 can be combined with each other using the arithmetic function. The result is stored in accumulator 1.

An arithmetic function executes the calculation according to the characteristics specified regardless of the contents of the accumulators and regardless of conditions. Table 10.2 shows an example for each of the different data types.

For data type INT, an arithmetic function uses only the low-order accumulator words; the high-order words are ignored. In the case of data type REAL, the accumulators are checked to make sure that both contain valid REAL numbers.

On the S7-300 CPUs, execution of an arithmetic function does not alter the contents of accumulator 2; on S7-400 CPUs, the contents of accumulator 2 are overwritten by the contents of accumulator 3. The contents of accumulator 4 are then "shifted over" to accumulator 3 (Figure 10.1).

Table 10.2 Examples for Arithmetic Functions

INT	The value in memory word MW 100 is divided by 250; the integer result is stored in memory word MW 102.	`L MW 100;`
		`L 250;`
		`/I ;`
		`T MW 102;`

DINT	The values in variables "ArithVal1" and "ArithVal2" are added and the result stored in variable "ArithResult". All variables are in data block "Global_DB".	`L "Global_DB".ArithVal1;`
		`L "Global_DB".ArithVal2;`
		`+D ;`
		`T "Global_DB".ArithResult;`

REAL	Variables #Actval and #Factor are multiplied; the product is transferred to variable #Display.	`L #Actval;`
		`L #FACTOR;`
		`*R ;`
		`T #DISPLAY;`

10.2 Calculating with Data Type INT

INT addition

The +I function interprets the values in the low-order words of accumulators 1 and 2 as numbers of data type INT. It adds the two numbers and stores the sum in accumulator 1.

After the calculation has been performed, status bits CC0 and CC1 indicate whether the sum is negative, zero, or positive. Status bits OV and OS flag any range violations.

The high-order word of accumulator 1 remains unchanged.

INT subtraction

The –I function interprets the values in the low-order words of accumulators 1 and 2 as numbers of data type INT. It subtracts the value in accumulator 1 from the value in accumulator 2 and stores the difference in accumulator 1.

After the calculation has been performed, status bits CC0 and CC1 indicate whether the differ-

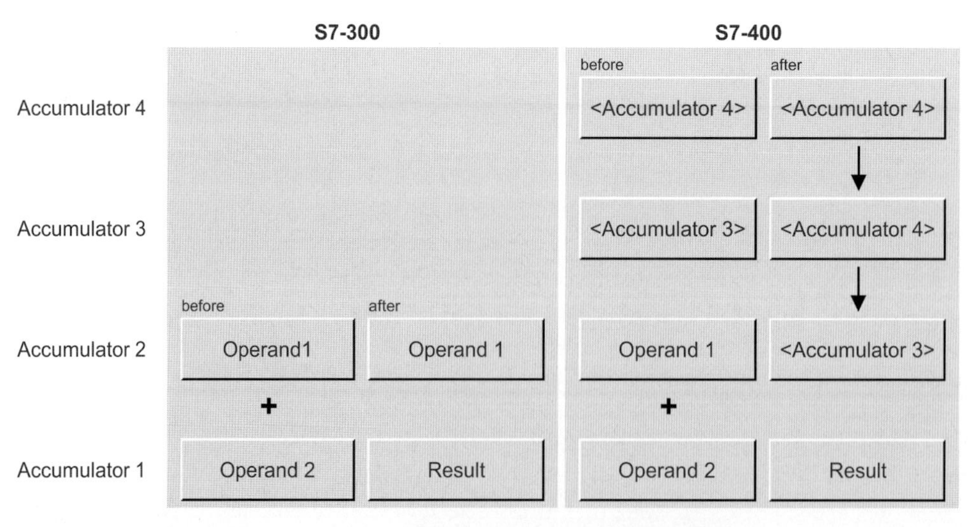

Figure 10.1 Contents of the Accumulators When Executing Arithmetic Functions

ence is negative, zero, or positive. Status bits OV and OS flag any range violations.

The high-order word of accumulator 1 remains unchanged.

INT multiplication

The *I function interprets the values in the low-order words of accumulators 1 and 2 as numbers of data type INT. It multiplies the two numbers and stores the product as a number of data type DINT in accumulator 1.

After the calculation has been performed, status bits CC0 and CC1 indicate whether the product is negative, zero, or positive. Status bits OV and OS flag any INT range violations.

Following execution of the *I function, the product is available as a DINT number in accumulator 1.

INT division

The /I function interprets the values in the low-order words of accumulators 1 and 2 as numbers of data type INT. It divides the value in accumulator 2 (dividend) by the value in accumulator 1 (divisor) and returns two results: the quotient and the remainder, both of which are of data type INT (Figure 10.2).

After the function has executed, the low-order word of accumulator 1 contains the quotient. The quotient is the integer result of the division operation. The quotient is zero when the dividend is zero and the divisor is not zero or when the dividend is smaller than the divisor. The quotient is negative if the divisor is negative.

After the /I function has executed, the high-order word contains the remainder of the division (not the places after the decimal point!). If the dividend is negative, the remainder is also negative.

After the calculation has been performed, status bits CC0 and CC1 indicate whether the quotient is negative, zero, or positive. Status bits OV and OS flag any range violations.

Division by zero returns a quotient of zero and a remainder of zero, and sets status bits CC0, CC1, OV and OS to “1”.

10.3 Calculating with Data Type DINT

DINT addition

The +D function interprets the values in accumulators 1 and 2 as numbers of data type DINT. It adds the two numbers and stores the sum in accumulator 1.

Following execution of the function, status bits CC0 and CC1 indicate whether the sum is negative, zero, or positive. Status bits OV and OS flag any range violations.

DINT subtraction

The –D function interprets the values in accumulators 1 and 2 as numbers of data type DINT. It subtracts the value in accumulator 1 from the value in accumulator 2 and stores the difference in accumulator 1.

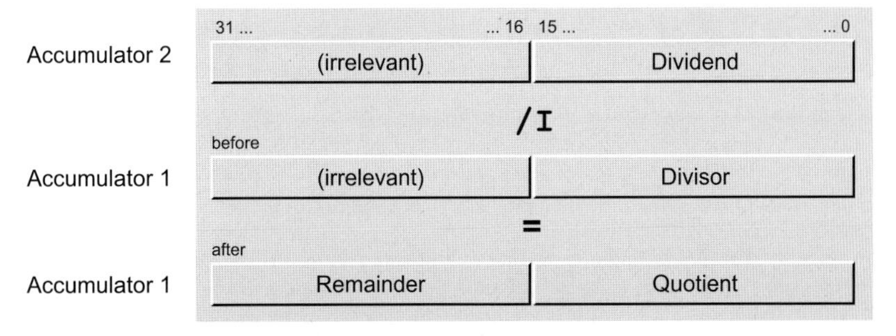

Figure 10.2 Results Returned by the Arithmetic Function /I

Following execution, status bits CC0 and CC1 indicate whether the difference is negative, zero, or positive. Status bits OV and OS flag any range violations.

DINT multiplication

The *D function interprets the values in accumulators 1 and 2 as numbers of data type DINT. It multiplies the two numbers and stores the product in accumulator 1.

Following execution of the function, status bits CC0 and CC1 indicate whether the product is negative, zero, or positive. Status bits OV and OS flag any range violations.

DINT division with quotient as result

The /D function interprets the values in accumulators 1 and 2 as numbers of data type DINT. It divides the value in accumulator 2 (dividend) by the value in accumulator 1 (divisor) and stores the quotient in accumulator 1.

The quotient is the integer result of the division. It is zero if the dividend is zero and the divisor is not zero or when the dividend is smaller than the divisor. The quotient is negative if the divisor is negative.

Following execution of the function, status bits CC0 and CC1 indicate whether the quotient is negative, zero, or positive. Status bits OV and OS flag any range violations.

Division by zero returns a quotient of zero and sets status bits CC0, CC1, OV and OS to "1".

DINT division with remainder as result

The MOD function interprets the values in accumulators 1 and 2 as numbers of data type DINT. It divides the value in accumulator 2 (dividend) by the value in accumulator 1 (divisor) and stores the remainder of the division in accumulator 1.

The remainder is what is left over from the division; it does not correspond to the decimal places. If the dividend is negative, the remainder is also negative.

Following execution of the function, status bits CC0 and CC1 indicate whether the remainder is negative, zero, or positive. Status bits OV and OS flag any range violations.

Division by zero returns a remainder of zero and sets status bits CC0, CC1, OV and OS to "1".

10.4 Calculating with Data Type REAL

REAL addition

The +R function interprets the values in accumulators 1 and 2 as numbers of data type REAL. It adds the two numbers and stores the sum in accumulator 1.

Following execution of the function, status bits CC0 and CC1 indicate whether the sum is negative, zero, or positive. Status bits OV and OS flag any range violations.

In the case of an impermissible calculation (one of the input values is an invalid REAL number or you tried to add $+\infty$ and $-\infty$), +R returns an invalid value in accumulator 1 and sets status bits CC0, CC1, OV and OS to "1".

REAL subtraction

The –R function interprets the values in accumulators 1 and 2 as numbers of data type REAL. It subtracts the number in accumulator 1 from the number in accumulator 2 and stores the difference in accumulator 1.

Following execution of the function, status bits CC0 and CC1 indicate whether the difference is negative, zero, or positive. Status bits OV and OS flag any range violations.

In the case of an impermissible calculation (one of the input values is an invalid REAL number or you attempted to subtract $+\infty$ and $+\infty$), –R returns an invalid value in accumulator 1 and sets status bits CC0, CC1, OV and OS to "1".

REAL multiplication

The *R function interprets the values in accumulators 1 and 2 as numbers of data type REAL. It multiplies the two numbers and stores the product in accumulator 1.

Following execution of the statement, status bits CC0 and CC1 indicate whether the product is negative, zero, or positive. Status bits OV and OS flag any range violations.

In the case of an impermissible calculation (one of the input values is an invalid REAL number or you attempted to multiply ∞ and 0), *R returns an invalid value in accumulator 1 and sets status bits CC0, CC1, OV and OS to "1".

REAL division

The /R function interprets the values in accumulators 1 and 2 as numbers of data type REAL. It divides the number in accumulator 2 (dividend) by the number in accumulator 1 (divisor) and stores the quotient in accumulator 1.

Following execution of the function, status bits CC0 and CC1 indicate whether the quotient is negative, zero, or positive. Status bits OV and OS flag any range violations.

In the case of an impermissible calculation (one of the input values is an invalid REAL number or you attempted to divide ∞ by ∞ or 0 by 0), /R returns an invalid value in accumulator 1 and sets status bits CC0, CC1, OV and OS to "1".

10.5 Successive Arithmetic Functions

You can program one arithmetic function immediately behind another, in which case the result of the first function is post-processed by the second, the accumulators serving as temporary storage.

Note: Please note that S7-300 CPUs, S7-400 CPUs and CPU 318 handle successive arithmetic functions differently (the S7-300 CPUs have only 2 accumulators while the S7-400 CPUs and the CPU 318 have 4).

Chain calculation with S7-300

A chain calculation is performed by following an arithmetic function with the loading and subsequent combining of the next value.

Example: Result1 := Value1 + Value2 – Value3

```
L      Value1;
L      Value2;
+I     ;              //Value1 + Value2
L      Value3;
-I     ;              //Sum - Value3
T      Result1;
```

On CPUs with two accumulators, the first value loaded remains unchanged in accumulator 2 during execution of the arithmetic function, and can be reused without having to be reloaded.

Example: Result2 := Value5 + 2 × Value6

```
L      Value6;
L      Value5;
+R     ;              //Value5 + Value6
+R     ;              //Sum + Value6
T      Result2;
```

Example: Result3 := Value7 x $(\text{Value8})^2$

```
L      Value8;
L      Value7;
*D     ;              //Value7 * Value8
*D     ;              //Product * Value8
T      Result3;
```

Chain calculation with S7-400

A chain calculation is performed by following an arithmetic function with the loading and subsequent combining of the next value. On CPUs with four accumulators, the value in accumulator 3 "shifts over" to accumulator 3 following execution of the arithmetic function. Beforehand, you can store an intermediate result in accumulator 3 (in the case of a dot-before-line calculation, for instance) with an ENT instruction (see Chapter 6.4 "Accumulator Functions").

Example:
Result4 := (Value1 + Value2) x (Value3 – Value4)

```
L      Value1;
L      Value2;
+I     ;
L      Value3;
ENT    ;
L      Value4;
-I     ;
*I     ;
T      Result4;
```

First, the sum of *Value1* and *Value2* is computed. While *Value3* is being loaded into accumulator 1, this sum is moved over to accumulator 2. From there, the ENT instruction copies it into accumulator 3. After *Value4* is loaded, the contents of *Value3* are in accumulator 2. When the two values are subtracted, the sum is "fetched back" into accumulator 2 from accumulator 3. The sum and the difference can now be multiplied.

10.6 Adding Constants to Accumulator 1

+	B#16#bb	Adds a byte constant
+	±w	Adds a word constant
+	L#±d	Adds a doubleword constant

You program the addition of a constant according to the following general scheme:

```
Load                 Operand;
Addition of constants;
Transfer             Result;
```

The constant addition is preferred for calculating addresses because, in contrast to an arithmetic function, it affects neither the contents of the remaining accumulators nor the status bits.

The "Add Constant" instruction adds the constant specified in the instruction to the contents of accumulator 1. You may specify this constant as a hexadecimal byte constant or as a decimal word or decimal doubleword constant. If you want to add a word constant using DINT, precede the constant with #L. If a decimal constant exceeds the permissible INT range, the calculation automatically becomes DINT.

You may write a decimal number with a minus sign, thus making it possible to subtract constants. Before a byte constant is added, it is expanded into a signed INT number.

As does a calculation with data type INT, the addition of a byte constant or word constant affects only the low-order word of accumulator 1; there is no carry to the high-order word.

If the INT value range is exceeded, bit 15 (which is the sign bit) is overwritten. The addition of a doubleword constant affects all 32 bits of accumulator 1, corresponding to a DINT calculation.

Execution of these statements is independent of any conditions.

Examples for adding constants:

```
L       AddValue1;
+       B#16#21;
T       AddResult1;
```

The value of variable *AddValue1* is increased by 33 and transferred to variable *AddResult1*.

```
L       AddValue2;
+       -33;
T       AddResult2;
```

The value of variable *AddValue2* is decreased by 33 and stored in variable *AddResult2*.

```
L       AddValue3;
+       L#-1;
T       AddResult3;
```

The value of variable *AddValue3* is decreased by 1 and stored in variable *AddResult3*. The subtraction is as for a DINT calculation.

10.7 Decrementing and Incrementing

DEC n	Decrement
INC n	Increment

You program decrementing and incrementing according to the following general scheme:

```
Load            Operand;
Decrementing    Decrement;
Transfer        Result;
```

```
Load            Operand;
Incrementing    Increment;
Transfer        Result;
```

The Decrement and Increment statements alter the value in accumulator 1. That value is decreased (decremented) or increased (incremented) by the number of units specified in the statement parameter. The parameter may assume a value between 0 and 255.

Only the value in the low-order byte of the accumulator is altered. There is no carry to the high-order byte. The calculation is carried out in "modulo 256", that is, when incrementation produces a value which exceeds 255, the "count" begins again from the beginning, or if decrementation produces a value which falls below 0, the "count" begins again at 255.

The Decrement and Increment instructions are executed without regard to the RLO. They are always executed when encountered, and affect neither the RLO nor the status bits.

Examples:

```
L      IncValue;
INC    5;
T      IncValue;
```

The value of variable *IncValue* is incremented by 5.

```
L      DecValue;
DEC    7;
T      DecValue;
```

The value of variable *DecValue* is decremented by 7.

11 Math Functions

"Math functions" include the following:

- ▷ Sine, cosine, tangent
- ▷ Arc sine, arc cosine, arc tangent
- ▷ Squaring, square-root extraction
- ▷ Exponential function to base e, natural logarithm

All math functions process numbers in data format REAL. Depending on the result, a math function sets status bits CC0, CC1, OV and OS as described in Chapter 15 "Status Bits".

The statements for the STL programming language are described in this chapter. In the SCL programming language, the math functions are included among the SCL standard functions (Chapter 30.3 "Math Functions").

The examples in this chapter are presented in the STL_Book library under the "Digital Functions" program in function block FB 111 or in the source file Chap_11. You can download the library from the publisher's web site (see page 8).

11.1 Processing a Math Function

A mathematical function takes the value present in accumulator 1 as an input value for the function to be performed and stores the result in accumulator 1. You program a mathematical function according to the following general scheme:

```
Load                   Operand;
Mathematical function  ;
Transfer               Result;
```

A math function alters only the contents of accumulator 1; the contents of all other accumulators remain unchanged. A math function is executed without regard to any conditions.

Table 11.1 shows three examples of math functions. A math function computes in accordance with the rules governing REAL numbers, even when absolute addressing is used and no data types are declared.

If accumulator 1 contains an invalid REAL number at the time the function is executed, the math function returns an invalid REAL number and sets status bits CC0, CC1, OV and OS to "1".

Table 11.1 Examples of Math Functions

Sine	The value in memory doubleword MD 110 contains an angle in radian measure. The sine of this angle is generated and stored in memory doubleword MD 104.	`L MD 110;`
		`SIN ;`
		`T MD 104;`

Square root	The square root of the value in variable "MathValue1" is generated and stored in the variable "MathRoot".	`L "Global_DB".MathValue1;`
		`SQRT ;`
		`T "Global_DB".MathRoot;`

Exponent	Variable #Result contains the power of e and #Exponent.	`L #Exponent;`
		`EXP ;`
		`T #Result;`

11.2 Trigonometric Functions

The trigonometric functions

- ▷ SIN Sine,
- ▷ COS Cosine and
- ▷ TAN Tangent

assume an angle in radian measure in form of a REAL number in accumulator 1.

Two units are normally used for the size of an angle: degrees from 0° to 360° and radian measure from 0 to 2π (where π = +3.141593e+00). Both can be converted proportionally. For example, the radian measure for a 90° angle is $\pi/2$ or +1.570796e+00.

With values greater than 2π (+6.283185e+00), 2π or a multiple thereof is subtracted until the input value for the trigonometric function is less than 2π.

Example:
Computing the idle power
Ps = U × I × sin(φ)

```
L      PHI;
SIN    ;
L      Current;
*R     ;
L      Voltage;
*R     ;
T      I_Power;
```

Please note that the angle must be specified in radian measure. If an angle is available in degrees, you must multiply it by the factor

$\pi/180$ = +1.745329e–02

before you can process it with a trigonometric function.

Table 11.2 Value Ranges for Arc Functions

Function	Permissible Value Range	Value Returned
ASIN	–1 to +1	$-\pi/2$ to $+\pi/2$
ACOS	–1 to +1	0 to π
ATAN	Entire range	$-\pi/2$ to $+\pi/2$

11.3 Arc Functions

The arc functions (inverse trigonometric functions)

- ▷ ASIN Arc sine,
- ▷ ACOS Arc cosine and
- ▷ ATAN Arc tangent

are the inverse functions of the corresponding trigonometric functions. They assume a REAL number in a specific range in accumulator 1, and return an angle in radian measure (Table 11.2).

If the permissible range is exceeded, the arc function returns an invalid REAL number and sets status bits CC0, CC1, OV and OS to “1”.

Example: In a right-angled triangle, one of the short sides of the triangle and the hypotenuse form an aspect ratio of 0.343. How big is the angle between them in degrees?

Arcsin (0.343) returns the angle in radian measure; multiplication with factor $360/2\pi$ (= 57.2958) gives you the angle in degrees (approx. 20°).

```
L      0.343;
ASIN   ;
L      57.2958;
*R     ;
T      Angle_Degree;
```

11.4 Other Math Functions

The following math functions are also available

- ▷ SQR Squaring,
- ▷ SQRT Square-root extraction,
- ▷ EXP Exponential function to base e and
- ▷ LN Compute natural logarithm (logarithm to base e).

Squaring

The SQR function squares the value in accumulator 1.

Example:
Computing the volume of a cylinder $V = r^2 \pi h$

```
L     Radius;
SQR   ;
L     Height;
*R    ;
L     3.141592;
*R    ;
T     Volume;
```

Square-root extraction

The SQRT function extracts the square root of the value in accumulator 1. If the value in accumulator 1 is less than zero, SQRT sets status bits CC0, CC1, OV and OS to “1” and returns an invalid REAL number. If accumulator 1 contains –0 (minus zero), –0 is returned.

Example: $c = \sqrt{a^2 + b^2}$

```
L     #a;
SQR   ;
L     #b;
SQR   ;
+R    ;
SQRT  ;
T     #c;
```

(If *a* or *b* is declared as a local variable, it must be preceded by # if the compiler is to recognize it as a local variable; if *a* or *b* is a global variable, it must be enclosed in quotation marks.)

Exponentiation to base e

The EXP function computes the power from base e (= 2.718282e+00) and the value in accumulator 1 (e^{Accu1}).

Example: Any power can be computed with the formula

$$a^b = e^{b \ln a}$$

```
L     Value_a;
LN    ;
L     Value_b;
*R    ;
EXP   ;
T     Power;
```

Computing the natural logarithm

The LN function computes the natural logarithm to base e (= 2.718282e+00) from the number in accumulator 1. If accumulator 1 contains a value less than or equal to zero, LN sets status bits CC0, CC1, OV and OS to “1” and returns an invalid REAL number.

The natural logarithm is the inverse of the exponential function: If $y = e^x$ then $x = \ln(y)$.

Example: Computing a logarithm to base 10 and to any other base.

The basic formula is

$$\log_b a = \frac{\log_n a}{\log_n b}$$

where b or n is any base. If you make n = e, you can compute a logarithm to any base using the natural logarithm:

$$\log_b a = \frac{\ln a}{\ln b}$$

In the special case for base 10, the formula is:

$$\lg a = \frac{\ln a}{\ln 10} = 0.4342945 \cdot \ln a$$

12 Conversion Functions

The conversion functions convert the data type of the value in accumulator 1. Figure 12.1 provides an overview of the data type conversions described in this chapter.

The statements for the STL programming language are described in this chapter. In the SCL programming language, the conversion functions are included among the SCL standard functions (Chapter 30.5 "Conversion Functions").

You will find details on the bits of the data formats in Chapter 24 "Data Types", and information on how the conversion functions set the status bits in Chapter 15 "Status Bits".

The examples in this chapter are presented in the STL_Book library under the "Digital Functions" program in function block FB 112 or in the source file Chap_12. You can download the library from the publisher's web site (see page 8).

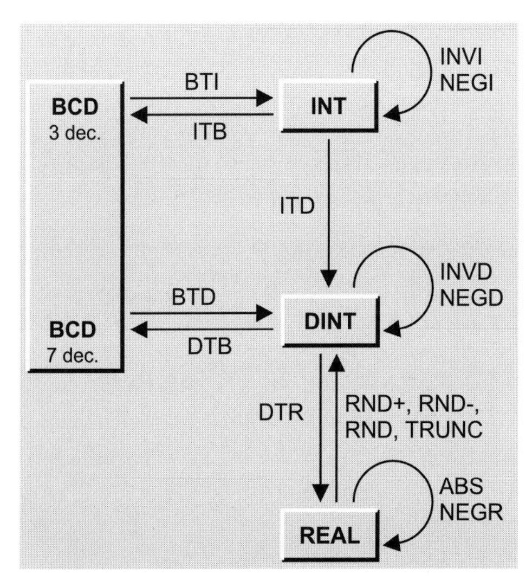

Figure 12.1 Overview of conversion functions

12.1 Processing a Conversion Function

The conversion functions affect only accumulator 1. Some functions affect only the low-order word (bits 0 to 15), others the entire accumulator. The conversion functions do not change the contents of any other accumulator.

You program a conversion function according to the following general scheme:

```
Load                Operand;
Conversion function;
Transfer            Result;
```

The Table 12.1shows an example for each of the various data types. A conversion function is carried out according to the defined characteristic even if no data types have been declared when using absolutely addressed operands. A conversion function is carried out independent of conditions.

Successive conversion functions

You can subject the contents of accumulator 1 to several successive conversions and so carry out conversions in stages without having to temporarily store the converted values.

Example:

```
L    BCD_Number;
BTI  ;           //BCD to INT
ITD  ;           //INT to DINT
DTR  ;           //DINT to REAL
T    REAL_Number;
```

This example converts a BCD number with 3 decades to a REAL number.

Table 12.1 Examples for Conversion Functions

Converting INT Numbers	The value in memory doubleword MW 120 is interpreted as an INT number and stored in memory word MW 122 as a BCD number.	`L MW 120;`
		`ITB ;`
		`T MW 122;`

Converting DINT Numbers	The value in variable "ConvertDINT" is interpreted as a DINT number and stored as a REAL number in the variable "ConvertREAL".	`L "Global_DB".ConvertDINT;`
		`DTR ;`
		`T "Global_DB".ConvertREAL;`

Converting REAL Numbers	The absolute value is generated from the variable #Display.	`L #Display;`
		`ABS ;`
		`T #Display;`

12.2 Converting INT and DINT Numbers

The following functions are provided for converting INT and DINT numbers:

▷ ITD Converts INT to DINT

▷ ITB Converts INT to BCD

▷ DTB Converts DINT to BCD

▷ DTR Converts DINT to REAL

Converting INT to DINT

The ITD statement interprets the value in the low-order word of accumulator 1 (bits 0 to 15) as a number of data type INT and transfers the signal state of bit 15 (the sign) to the high-order word, that is, bits 16 to 31.

The conversion of INT to DINT sets no status bits.

Converting INT to BCD

The ITB statement interprets the value in the low-order word of accumulator 1 (bits 0 to 15) as a number of data type INT and converts it to a 3-decade BCD number. The three decades are right-justified in accumulator 1 and represent the value of the decimal number. The sign is in bits 12 to 15. If all bits are "0", the sign is positive; if all bits are "1", it is negative. The contents of the high-order word (bits 16 to 31) remain unchanged.

If the INT number is too large to be converted to BCD (> 999), the ITB statement sets status bits OV and OS. The conversion is not performed in this case.

Converting DINT to BCD

The DTB statement interprets the value in accumulator 1 as a number of data type DINT and converts it to a 7-decade BCD number. The seven decades are right-justified in accumulator 1 and represent the value of the decimal number. The sign is in bits 28 to 31. If all bits are "0", the sign is positive; if all bits are "1", it is negative.

If the DINT number is too large to be converted to BCD (> 9 999 999), status bits OV and OS are set and the conversion is not carried out.

Converting DINT to REAL

The DTR statement interprets the contents of accumulator 1 as a number in DINT format and converts it to a number in REAL format.

Since a number in DINT format has a higher accuracy than a number in REAL format, rounding may take place during conversion, but only to the next whole number (as per the RND statement).

DTR sets no status bits.

12.3 Converting BCD Numbers

The following functions are available for converting BCD numbers:

▷ BTI Converts BCD to INT

▷ BTD Converts BCD to DINT

Converting BCD to INT

The BTI statement interprets the value in the low-order word of accumulator 1 (bits 0 to 15) as a 3-decade BCD number. The three decades are right-justified in the accumulator and represent the value of the decimal number. The sign is in bits 12 to 15. If all bits are "0", the sign is positive; if all bits are "1", it is negative. During conversion, only the signal state of bit 15 is taken into account. The contents of the high-order word of accumulator 1 (bits 16 to 31) remain unchanged.

If there is a pseudo tetrad in the BCD number (numerical value 10 to 15 or A to F in hexadecimal), the CPU reports a parameter assignment error and calls organization block 121 (synchronous errors). If OB 121 has not been programmed, the CPU goes to STOP.

The BTI statement sets no status bits.

Converting BCD to DINT

The BTD statement interprets the value in accumulator 1 as a 7-decade BCD number. The seven decades are right-justified in the accumulator and represent the value of the decimal number. Bits 28 to 31 contain the sign. If these bits are all "0", the sign is positive; if they are all "1", the sign is negative. During conversion, only the signal state of bit 31 is taken into account.

If the BCD number contains a pseudo tetrad (numerical value 10 to 15 or A to F in hexadecimal), the CPU reports a parameter assignment error and calls organization block OB 121 (synchronous errors). If OB 121 is not available, the CPU goes to STOP.

The BTD statement sets no status bits.

12.4 Converting REAL Numbers

There are several statements for converting a number in REAL format to DINT format (conversion of a fractional value to an integer value). They differ from one another in the way they perform rounding.

Table 12.2 shows the different effect of the REAL conversion functions according to DINT. The range between –1 and +1 has been selected as example.

- ▷ RND+ With rounding to the next higher integer
- ▷ RND– With rounding to the next lower integer
- ▷ RND With rounding to the next integer
- ▷ TRUNC No rounding

Rounding to the next higher integer number

The RND+ statement interprets the contents of accumulator 1 as a number in REAL format and converts it to a number in DINT format.

The RND+ statement returns an integer greater than or equal to the number to be converted.

If the value in accumulator 1 exceeds or falls short of the permissible range for a DINT number or if it is not a REAL number, RND+ sets status bits OV and OS. The conversion is not carried out.

Rounding to the next lower integer

The RND– statement interprets the contents of accumulator 1 as a number in RAL format and converts it to a number in DINT format.

The RND– statement returns an integer less than or equal to the number to be converted.

If the value in accumulator 1 exceeds or falls short of the permissible range for a DINT number or if it is not a REAL number, RND– sets status bits OV and OS. The conversion is not carried out.

Rounding to the next integer

The RND statement interprets the contents of accumulator 1 as a number in REAL format and converts it to a number in DINT format. The RND statement returns the next higher or next lower integer, whichever is closest to the true result. If the result lies exactly between an even and an odd number, the even number takes priority.

If the value in accumulator 1 exceeds or falls short of the permissible range for a DINT number or if it is not a REAL number, RND sets sta-

Table 12.2 Rounding Modes for the Conversion of REAL Numbers

Input Value		Result			
REAL	DW#16#	RND	RND+	RND-	TRUNC
1.0000001	3F80 0001	1	2	1	1
1.00000000	3F80 0000	1	1	1	1
0.99999995	3F7F FFFF	1	1	0	0
0.50000005	3F00 0001	1	1	0	0
0.50000000	3F00 0000	0	1	0	0
0.49999996	3EFF FFFF	0	1	0	0
5.877476E–39	0080 0000	0	1	0	0
0.0	0000 0000	0	0	0	0
–5.877476E–39	8080 0000	0	0	–1	0
–0.49999996	BEFF FFFF	0	0	–1	0
–0.50000000	BF00 0000	0	0	–1	0
–0.50000005	BF00 0001	–1	0	–1	0
–0.99999995	BF7F FFFF	–1	0	–1	0
–1.00000000	BF80 0000	–1	–1	–1	–1
–1.0000001	BF80 0001	–1	–1	–2	–1

tus bits OV and OS. The conversion is not carried out.

No rounding

The TRUNC statement interprets the contents of accumulator 1 as a number in REAL format and converts it to a number in DINT format. The TRUNC statement returns the integer component of the number to be converted; the fractional component is "truncated".

If the value in accumulator 1 exceeds or falls short of the permissible range for a DINT number or if it is not a REAL number, TRUNC sets status bits OV and OS. The conversion is not carried out.

12.5 Other Conversion Functions

Other available conversion functions are:

- ▷ INVI One's complement INT
- ▷ INVD One's complement DINT
- ▷ NEGI Negation of an INT number (two's complement)
- ▷ NEGD Negation of a DINT number (two's complement)
- ▷ NEGR Negation of a REAL number
- ▷ ABS Generation of an absolute REAL number

One's complement INT

The INVI statement negates the value in the low-value word of accumulator 1 (bits 0 to 15) bit for bit. It replaces the zeroes with ones and vice versa. The contents of the high-order word (bits 16 to 31) remain unchanged.

The INVI statement sets no status bits.

One's complement DINT

The INVD statement negates the value in accumulator 1 bit for bit. It replaces the zeroes with ones and vice versa.

The INVD statement sets no status bits.

Negation INT

The NEGI statement interprets the value in the low-order word of accumulator 1 (bits 0 to 15) as an INT number and changes the sign by generating the two's complement. This operation is identical to multiplication with –1. The high-order word of accumulator 1 (bits 16 to 31) remains unchanged.

The NEGI statement sets status bits CC0, CC1, OV and OS.

Negation DINT

The NEGD statement interprets the value in accumulator 1 as DINT number and changes the sign by generating the two's complement. This operation is identical to multiplication with –1.

The NEGD statement sets status bits CC0, CC1, OV and OS.

Negation REAL

The NEGR statement interprets the value in accumulator 1 as a REAL number and multiplies this number by –1 (it changes the sign of the mantissa, even if the number in the accumulator is not a valid REAL number).

The NEGR statement sets no status bits.

Absolute-value generation REAL

The ABS statement interprets the value in accumulator 1 as a REAL number and generates the absolute value of this number (it sets the sign of the mantissa to “0”, even if the number in the accumulator is not a valid REAL number).

The ABS statement sets no status bits.

13 Shift Functions

The shift functions shift the contents of accumulator 1 bit by bit to the left or right. Table 13.1 provides an overview of the available shift functions.

The statements for the STL programming language are described in this chapter. In the SCL programming language, the shift functions are included among the SCL standard functions (Chapter 30.4 "Shifting and Rotating").

The examples in this chapter can be found in the download files (download address: see pages 8-9) in the STL_Book library under the "Digital Functions" program in function block FB 113 or source file Chap_13.

13.1 Processing a Shift Function

The contents of accumulator 1 are shifted bit by bit to the left or right; depending on the function, the accumulator contains either a word or a doubleword. The bits that are shifted out are either lost (Shift operations) or are added at the other side of the word or doubleword (Rotate operations). The shift functions have no effect on the other accumulators.

The shift functions are executed without regard to any conditions. They affect only the contents of accumulator 1. The result of the logic operation (RLO) is not affected.

You can program a shift function in two ways:

- ▷ With the number of positions in accumulator 2
- ▷ With the number of positions as a parameter

The general schemes are as follows:

```
Load            Shift number;
Load            Operand;
Shift function  ;
Transfer        Result;
```

```
Load            Operand;
Shift function  Shift number;
Transfer        Result;
```

The shift functions set status bit CC0 to "0" and status bit CC1 to the signal state of the last bit shifted (Figure 13.1). The status bits are evaluated with binary check or jump instructions as described in Chapter 15 "Status Bits", and Chapter 16 "Jump Functions".

Table 13.1 Overview of the Shift Functions

Shift Functions	Word		Doubleword	
	with no. of positions as a parameter	with no. of positions in accum2	with no. of positions as a parameter	with no. of positions in accum2
Shift left	SLW n	SLW	SLD n	SLD
Shift right	SRW n	SRW	SRD n	SRD
Shift with sign	SSI n	SSI	SSD n	SSD
Rotate left	-	-	RLD n	RLD
Rotate right	-	-	RRD n	RRD
Rotate left through CC1	-	-	RLDA [1)]	-
Rotate right through CC1	-	-	RRDA [1)]	-

[1)] Without parameter, as only one bit is shifted

Table 13.2 shows several examples of shift functions. A word shift changes only the low-order word of accumulator 1; the contents of the high-order word are not affected. Rotation through status bit CC1 shifts the contents of the accumulator by one bit position.

Successive shift functions

Shift functions can be applied as often as required to the contents of the accumulator.

Example:

```
L    Value1;
SSD  4;
SLD  2;
T    Result1;
```

In the example, a value is shifted, with sign, to the right by (effectively) 2 positions, whereby the two right bit positions are reset to "0".

13.2 Shifting

Shift left word

SLW n Shift left word by n bits

SLW Shift left word by the number of bits in accumulator 2

Shift function SLW shifts bits 0 to 15 of accumulator 1 bit by bit to the left. The bit positions freed by the shift operation are padded with zeroes. The high-order word of accumulator 1 remains unchanged; there is no carry to bit 16.

The number of positions may be specified as a parameter in the SLW statement or loaded as a positive number in INT format right-justified into accumulator 2. If the number of positions is = 0, the statement is not executed (no operation, or NOP); if it is greater than 15, the low-order word of accumulator 1 contains zero following execution of the SLW statement.

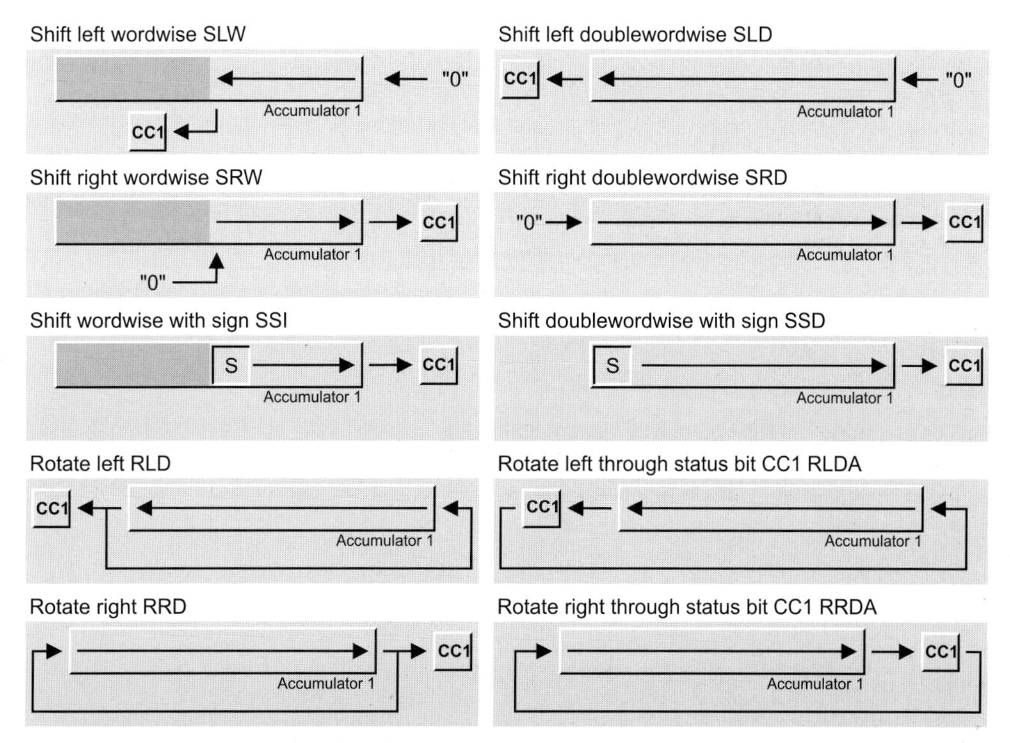

Figure 13.1 How Shift Functions Work

Table 13.2 Examples for Shift Functions

Shifting word variables	The value in memory word MW 130 is shifted 4 positions to the left and stored in memory word MW 132. Here, the number of positions appears as a parameter in the shift operation.	`L MW 130;`
		`SLW 4;`
		`T MW 132;`

Shifting doubleword variables	The value in variable "ShiftOn" is shifted right by "ShiftPos" positions and stored in "ShiftOff". Here, the number of positions is in accumulator 2.	`L "Global_DB".ShiftPos;`
		`L "Global_DB".ShiftOn;`
		`SRD ;`
		`T "Global_DB".ShiftOff;`

Shifting with sign	The variable #Actval is shifted, with sign, 2 positions to the right and transferred to the variable #Display.	`L #Actval;`
		`SSI 2;`
		`T #Display;`

If the contents of accumulator 1 (low-order word) are interpreted as a number in INT format, a shift to the left would be equivalent to multiplication with a power to base 2. The exponent is then the number of positions.

Shift left doubleword

SLD n Shift left doubleword by n bits

SLD Shift left doubleword by the number of bits in accumulator 2

Shift function SLD shifts the entire accumulator bit by bit to the left. The bit positions freed by the shift operation are padded with zeroes.

The number of positions may be specified as a parameter in the SLD statement or as a positive number in INT format right-justified in accumulator 2. If the number of positions is = 0, the statement is not executed (no operation, or NOP); if it exceeds 31, accumulator 1 contains zero following execution of the SLD statement.

If the contents of accumulator 2 are interpreted as a number in DINT format, it would correspond to multiplication with a power to base 2, with the exponent as the number of positions.

Shift right word

SRW n Shift right word by n bits

SRW Shift right word by the number of bits in accumulator 2

Shift function SRW shifts bits 0 to 15 of accumulator 1 bit by bit to the right. The bit positions freed by the shift operation are padded with zeroes. Bits 16 to 31 are not affected.

The number of positions may be specified as a parameter in the SRW operation or as a positive number in INT format right-justified in accumulator 2. If the number of positions is = 0, the statement is not executed (no operation, or NOP); if it exceeds 15, the low-order word of accumulator 1 contains zero following execution of the SRW statement.

If the contents of accumulator 1 (low-order word) are interpreted as a number in INT format, a shift to the right is equivalent to division by a power to base 2, with the exponent as the number of positions. Because the freed bits are padded with zeroes, this applies only to positive numbers. The result of such a division corresponds to the rounded integer component.

Shift right doubleword

SRD n Shift right doubleword by n bits

SRD Shift right doubleword by the number of bits in accumulator 2

Shift function SRD shifts the entire accumulator bit by bit to the right. The bit positions freed by the shift operation are padded with zeroes.

The number of positions may be specified as a parameter in the SRD statement or as a positive number in INT format right-justified in accumulator 2. If the number of positions is = 0, the statement is not executed (no operation, or NOP); if it exceeds 31, accumulator 1 contains zero following execution of the SRD statement.

If the contents of accumulator 1 are interpreted as a number in DINT format, a shift to the right is equivalent to division with a power to base 2. The exponent is the number of positions. Because the freed bits are padded with zeroes, this applies only to a positive number. The result of such a division corresponds to the rounded integer component.

Shift word with sign

SSI n	Shift word with sign by n bits
SSI	Shift word with sign by the number of bits in accumulator 2

Shift function SSI shifts bits 0 to 15 of accumulator 1 bit by bit to the right. The bit positions freed by the shift operation are padded with the signal state of bit 15 (which is the sign of an INT number), that is to say, with "0" in the case of a positive number and "1" in the case of a negative number.

Bits 16 to 31 are not affected.

The number of positions may be specified as a parameter in the SSI statement or as a positive number in INT format right-justified in accumulator 2. If the number of positions is = 0, the operation is not executed (no operation, or NOP); if it exceeds 15, the sign is in all bit positions of the low-order word of accumulator 1 following execution of the statement.

If the contents of accumulator 1 (low-order word) are interpreted as a number in INT format, a shift to the right is equivalent to division by a power to base 2, with the exponent as the number of positions. The result of such a division corresponds to the rounded integer component.

Shift doubleword with sign

SSD n	Shift doubleword with sign by n bits
SSD	Shift doubleword with sign by the number of bits in accumulator 2

Shift function SSD shifts the entire accumulator 1 bit by bit to the right. The bit positions freed by the shift operation are padded with the signal state of bit 31 (which is the sign of a DINT number), that is to say, with "0" in the case of a positive number and with "1" in the case of a negative number.

The number of positions may be specified as a parameter in the SSD statement or as a positive number in INT format right-justified in accumulator 2. If the number of positions is = 0, the statement is not executed (no operation, or NOP); if it exceeds 31, the sign is in all bit positions of accumulator 1 following execution of the statement.

If the contents of accumulator 1 are interpreted as a number in DINT format, a shift to the right is equivalent to division by the power to base 2, with the exponent as the number of positions. The result of such a division corresponds to the rounded integer component.

13.3 Rotating

Rotate left

RLD n	Rotate left by n bits
RLD	Rotate left by the number of bits in accumulator 2

Shift function RLD shifts the entire accumulator 1 bit by bit to the left. The bit positions freed by the shift operation are padded with the contents of the bit positions that were shifted out.

The number of positions may be specified as a parameter in the RLD statement or as a positive number in INT format right-justified in accumulator 2. If the number of positions is = 0, the statement is not executed (no operation, or NOP); if it exceeds 32, the contents of accumulator 1 remain unchanged and status bit CC1 assumes the signal state of the last bit shifted (bit 0). If the number of positions is 33, the accumulator is shifted by one bit position, if 34 by two bit positions, and so on (the shift is executed modulo 32).

Rotate right

RRD n	Rotate right by n bits

RRD Rotate right by the number of bits in accumulator 2

Shift function RRD shifts the entire accumulator 1 bit by bit to the right. The bit positions freed by the shift operation are padded with the values of the bit positions that were shifted out.

The number of positions may be specified as a parameter in the RRD statement or as a positive number in INT format right-justified in accumulator 2. If the number of positions is = 0, the statement is not executed (no operation, or NOP); if it exceeds 32, the contents of accumulator remain unchanged and status bit CC1 assumes the signal state of the last bit shifted (bit 31). If the number of positions is 33, the contents of the accumulator are shifted by one position, if 34 by two positions, and so on (the shift is executed modulo 32).

Rotate left through CC1

RLDA Rotate left through status bit 1 by one position

The RLDA function shifts the entire contents of accumulator 1 one bit to the left. The bit position freed by the shift (bit 0) assumes the signal state of status bit CC1. Status bit CC1 assumes the signal state of the bit that was shifted out (bit 31); status bit CC0 is set to “0”.

Rotate right through CC1

RRDA Rotate right through status bit CC1 by one position

The RRDA function shifts the entire contents of accumulator 1 one bit to the right. The bit position freed by the shift (bit 31) assumes the signal state of status bit CC1. Status bit CC1 assumes the signal state of the bit that was shifted out (bit 0); status bit CC0 is set to “0”.

14 Word Logic

Word logic operations combine the value in accumulator 1 bit by bit with a constant or with the contents of accumulator 2 and store the result in accumulator 1. The logic operation can be performed on a word or a doubleword.

The following word logic operations are available:

▷ AND

▷ OR

▷ Exclusive OR

The operations for the STL programming language are described in this chapter. In the SCL programming language, the word logic operations are formulated using logic expressions (Chapter 27.4.3 "Logical Expressions").

Chapter 15 "Status Bits" provides information on the status bits set by these instructions.

The examples in this chapter can be found in the download files (download address: see pages 8-9) in the STL_Book library under the "Digital Functions" program in function block FB 114 or source file Chap_14.

14.1 Processing a Word Logic Operation

You perform a word logic operation according to one of the two general schemes below:

```
Load                Operand1;
Load                Operand2;
Word logic operation;
Transfer            Result;
```

```
Load                Operand;
Word logic operationConstant;
Transfer            Result;
```

Word logic operations execute without regard to any conditions. They do not affect the RLO.

Table 14.1
Generating the Result of Word Logic Operations

Contents of accumulator 2 or bit in the constant	0	0	1	1
Contents of accumulator 1	0	1	0	1
Result of AW, AD	0	0	0	1
Result of OW, OD	0	1	1	1
Result of XOW, XOD	0	1	1	0

Generating the result of a word logic operation

A word logic operation generates the result bit by bit, exactly as described in Chapter 4 "Binary Logic Operations" (Table 14.1).

The operation combines bit 0 of accumulator 1 with bit 0 of accumulator 2 or the constant specified in the instruction; the result is stored in bit 0 of the accumulator. The same logic is used on bit 1, bit 2, and, up to and including bit 15 (word instructions) or 31 (doubleword instructions). The contents of accumulator 2 remain unchanged.

Word logic with the contents of accumulator 2

The actual word logic operation is preceded by two load operations, one for each of the two values to be combined. When the word logic operation has executed, the result is in accumulator 1.

Example:

```
L     MW 142;    //Address 1
L     MW 144;    //Address 2
AW    ;          //Logic operation
T     MW 146;    //Result
```

Word logic with a constant

The address to be combined is loaded into accumulator 1 and then combined with the value specified as a constant in the instruction. Fol-

Table 14.2 Examples for Word Logic Operations

AND logic	The four high-order bits of memory word MW 138 are set to "0"; the result is stored in memory word MW 140.	`L MW 138;`
		`AW W#16#0FFF;`
		`T MW 140;`

OR logic	Variables "WLogicVal1" and "WLogicVal2" are combined bit for bit according to OR and the result stored in "WLogicReslt".	`L "Global_DB".WLogicVal1;`
		`L "Global_DB".WLogicVal2;`
		`OD ;`
		`T "Global_DB".WLogicReslt;`

Exclusive OR	The value generated by combining variables #Input and #Mask with Exclusive OR is in variable #Buffer.	`L #Input;`
		`L #Mask`
		`XOW ;`
		`T #Buffer;`

lowing execution, the result of the word logic operation is in accumulator 1.

Example:

```
L     MW 148;
AW    W#16#807F;
T     MW 150;

L     MD 152;
OD    DW#16#8000_F000;
T     MD 156;
```

In the example above, the logic operation is performed on a word; in the example below, it is performed on the entire accumulator, that is, on a doubleword.

Performing word logic operations on words

The logic operations for words affect only the low-order word (bits 0 to 15) of the two accumulators. The high-order word (bits 16 to 31) remains unchanged (Figure 14.1).

Successive word logic operations

Following execution of a word logic operation, you can proceed immediately to the next (load the addresses and execute the word logic operation or execute the word logic operation using a constant) without having to store the intermediate result (in the local data area, for instance). The accumulators serve here as temporary stores. Examples:

```
L     Value1;
L     Value2;
AW    ;
L     Value3;
OW    ;
T     Result1;
```

The result of the AW instruction is in accumulator 1, and is moved to accumulator 2 when Value3 is loaded. The two values can now be combined with OW.

```
L     Value4;
L     Value5;
XOW   ;
AW    W#16#FFF0;
T     Result2;
```

The result of the XOW instruction is in accumulator 1. Bits 0 to 3 of accumulator 1 are set to "0" with the AW statement.

Table 14.2 shows one example for each of the different word logic operations.

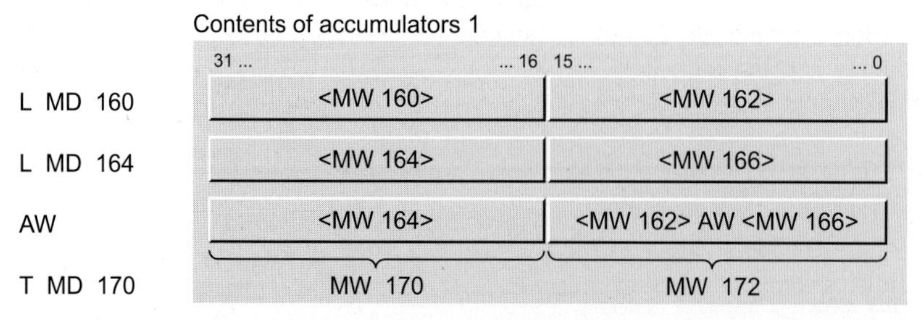

Figure 14.1 Performing Word Logic Operations on Words

14.2 Description of the Word Logic Operations

Digital AND operation

AW	AND operation (word) with accum2 and accum1
AW W#16#	AND operation (word) with constant and accum1
AD	AND operation (doubleword) with accum2 and accum1
AD DW#16#	AND operation (doubleword) with constant and accum1

The digital AND operation combines the bits of the value in accumulator 1 with the corresponding bits of the value in accumulator 2 or the constant according to AND. A bit in the result word is "1" only when the corresponding bits in both of the values being ANDed (combined according to logic AND) are also "1".

Since those bits in accumulator 2 or the constant which are "0" also set the corresponding bits in the result to "0", regardless of their signal state in accumulator 1, one also says of these bits that they are "masked". This so-called "masking" is the main purpose of the digital AND operation.

Digital OR operation

OW	OR operation (word) with accum2 and accum1
OW W#16#	OR operation (word) with constant and accum1
OD	OR operation (doubleword) with accum2 and accum1
OD DW#16#	OR operation (doubleword) with constant and accum1

The digital OR operation combines the bits of the value in accumulator 1 with the corresponding bits of the value in accumulator 2 according to OR. A bit in the result word is "0" only when the corresponding bits in both of the values being ORed (combine according to logic OR) are also "0".

Since those bits in accumulator 2 or the constant which are "1" also set the corresponding bits in the result to "1", regardless of their signal state in accumulator 1, one also says of these bits that they are "masked". This so-called "masking" is the main purpose of the digital OR operation.

Digital Exclusive OR operation

XOW	Exclusive OR operation (word) with accum2 and accum1
XOW W#16#	Exclusive OR operation (word) with constant and accum1
XOD	Exclusive OR operation (doubleword) with accum2 and accum1
XOD DW#16#	Exclusive OR operation (doubleword) with constant and accum1

The digital Exclusive OR operation combines the bits of the value in accumulator 1 with the corresponding bits of the value in accumulator 2 according to Exclusive OR. A bit in the result word is "1" only when precisely one of the corresponding bits being combined is "1". If a bit in accumulator 2 or the constant has signal state "1", the result at this point contains the reverse of the previous signal state of the bit in accumulator 1.

In the result, only those bits are "1" that had different signal states in the two accumulators or in accumulator 1 and the constant prior to execution of the Exclusive OR instruction. Ascertaining these bits or "negating" the signal states of individual bits is the primary purpose of the digital Exclusive OR operation.

Program Flow Control

STEP 7 provides you with a variety of possibilities for controlling the flow of the program. You can exit linear program execution within a block or you can structure the program with parameterizable block calls. You can influence program execution depending on values calculated at runtime, or depending on process parameters, or according to your plant status.

The **status bits** provide information on the result of an arithmetic or math function and on errors (for example, number range violation in a calculation). You can incorporate the signal state of the status bits direct into your program using binary logic combinations.

You can use the **jump functions** to branch within your program either unconditionally or dependent on the status bits, the RLO or the binary result. With STL, you can execute the jumps with calculated jump width (jump distributor) or you can easily implement program loops (loop jump).

A further method of influencing program execution is provided by the **Master Control Relay** (MCR). Originally developed for relay contactor controls, STL offers a software version of this program control method.

STL provides the **block functions** as a means for you to structure your program. You can use functions and function blocks again and again by defining **block parameters**.

For details of how to program blocks in STL (with the keywords for source-file-oriented programming), see Chapter 3.4 "Programming Code Blocks with STL". Chapters 18 "Block Functions" and 19 "Block Parameters" continue this topic. The corresponding references for the SCL programming language are Chapter 3.5 "Programming Code Blocks with SCL" and Chapter 29 "SCL Blocks".

Chapter 26 "Direct Variable Access" contains further information on the block parameters, such as how they are stored in memory and how they can be used in conjunction with complex data types.

15 Status Bits

The status bits are binary flags (indicator bits). The CPU uses them for controlling the binary logic operations and sets them in digital processing. You can check these status bits (for example, as a result check in calculations) or you can influence specific bits. The status bits are combined into a word, the status word.

The examples in this chapter can be found in the download files (download address: see pages 8-9) in the STL_Book library under the "Program Flow Control" program in function block FB 115 or source file Chap_15.

15.1 Description of the Status Bits

Table 15.1 shows the status bits available with STL. The first column shows the bit number in the status word. The CPU uses the binary flags for controlling the binary functions; the digital flags indicate primarily results of arithmetic and math functions.

Table 15.1 Status Bits

Bit	Binary Flags	
0	/FC	First check
1	RLO	Result of logic operation
2	STA	Status
3	OR	OR status bit
8	BR	Binary result
	Digital Flags	
4	OS	Stored overflow
5	OV	Overflow
6	CC0	CC0 (condition code) status bit
7	CC1	CC1 (condition code) status bit

First check

The /FC status bit steers the binary logic operation within a logic control system. A binary logic step always starts with /FC = "0" and a binary check instruction, the first check, as shown in the description of binary logic operations. The first check sets /FC = "1". A binary logic step ends with a binary value assignment or with a conditional jump or a block change. These set /FC = "0". The next binary check is then the start of a new binary logic combination.

Result of logic operation (RLO)

The RLO status bit is the intermediate buffer in binary logic operations. In first check, the CPU transfers the check result to the RLO, combines the check result with the stored RLO at each subsequent check, and stores the result, in turn, in the RLO (as described in Chapter 4 "Binary Logic Operations"). You can also set, reset or negate the RLO direct or store it in the BR. Memory, timer and counter functions are controlled using the RLO and certain jump functions are executed.

Status

The STA status bit corresponds to the signal state of the addressed binary address or of the checked condition in the case of binary logic operations (A, AN, O, ON, X, XN).

In the case of memory functions (S, R, =), the value of STA is the same value as the written value or (if no write operation takes place, for example, in the case of RLO = "0" or MCR active), STA corresponds to the value of the addressed (and unmodified) binary address.

With edge evaluations FP or FN, the value of the RLO prior to the edge evaluation is stored in STA. All other binary statements set STA = "1"; also the binary flag-dependent jumps JC, JCN, JBI, JNBI (Exception: CLR sets STA = "0").

The STA status bit has no effect on the processing of STL statements. It is displayed in the programming device test functions (such as program status) so that you can use it to trace binary logic sequences or for troubleshooting.

Table 15.2 Example of Influencing the Status Bits

STL Statements	Binary Flags: /FC	RLO	STA	OR	Remark	
...						
= M 10.0	0	x	x	0		
A I 4.0	1	1	1	0	I 4.0 has "1"	
AN I 4.1	1	1	0	1	I 4.1 has "0"	*The shaded area is a binary logic step*
O	1	1	1	0		
O I 4.2	1	1	0	0	I 4.2 has "0"	
ON I 4.3	1	1	1	0	I 4.3 has "1"	
= Q 8.0	0	1	1	0	Q 8.0 to "1"	
R Q 8.1	0	1	0	0	Q 8.1 to "0"	
S Q 8.2	0	1	1	0	Q 8.2 to "1"	
A I 5.0	1	x	x			
...						

STL Statements	Digital Flags: CC0	CC1	OV	OS	Remark	
...						
T MW 10	x	x	x	x		
L +12	x	x	x	x		
L +15	x	x	x	x		
-I	1	x	0	0	Result negative	
L +20000	1	0	0	0		
*I	0	0	1	1	Overflow	OV and OS to "1"
L +20	0	1	1	1		
+I	0	1	0	1	OV becomes "0"	OS remains "1"
T MW 30	0	1	0	1		
L MW 40	1	1	0	1		
...						

OR status bit

The OR status bit stores the result of a fulfilled AND operation ("1") and indicates to a subsequent OR operation that the result is already fixed (in conjunction with the O statement in an AND before OR operation). All other binary statements reset the OR status bit.

Table 15.2 (under "Binary Flags") uses the example of a binary logic step to show how the binary flags are affected. The binary logic step starts with the first check following a memory function and ends with the last memory function prior to a check.

Overflow

The OV status bit indicates a number range overflow or the use of invalid REAL numbers.

The following functions influence the OV status bit: Arithmetic functions, math functions, some conversion functions, REAL comparison functions.

You can evaluate the OV status bit with check statements or with JO jump statement.

Stored overflow

The OS status bit stores an OV status bit setting: When the CPU sets the OV status bit, it also always sets the OS status bit. However, while the next properly executed operation resets OV, OS remains set. This provides you with the opportunity of evaluating, even at a later point in your program, a number range overflow or an operation with an invalid REAL number.

You can evaluate the OS status bit with check statements or with the JOS jump statement. JOS or a block change reset the OS status bit.

CC0 and CC1 status bits (condition code bits)

The CC0 and CC1 status bits provide information on the result of a comparison function, an arithmetic or math function, a word logic operation or on the shifted out bit in the case of a shift function.

You can evaluate all digital flags with jump functions and check statements (see below in this chapter). Table 15.2 shows an example of setting digital flags in the lower section "Digital Flags".

Binary result

The BR status bit helps in the implementation of the EN/ENO mechanism for block calls (in conjunction with graphical languages). Chapter 15.4 "Using the Binary Result" shows you how STEP 7 uses the binary result. You can also set or reset the BR status bit yourself and check it with binary checks or with jump statements.

Status word

The status word contains all status bits. You can load it into accumulator 1 or write it out of accumulator 1 with a value.

```
L STW;    //Load the status word
          //...
T STW;    //Transfer to the status word
```

See Chapter 6 "Move Functions" for a description of the load and transfer statements; Table 15.1 contains the assignment of the status word with the status bits.

You can use the status word to check the status bits or to set them according to your wishes. In this way, you can store a current status word or begin a program section with a specific assignment of status bits.

Please note that the S7-300-CPUs do not load status bits /FC, STA and OR into the accumulator; the accumulator contains "0" at these locations.

15.2 Setting the Status Bits and the Binary Flags

The digital functions affect the CC0, CC1, OV and OS status bits as shown in Table 15.3. There are special STL statements for influencing the RLO and BR status bits.

Status bits with INT and DINT calculation

The arithmetic functions with data formats INT and DINT set all digital flags (status bits). A result of zero sets CC0 and CC1 to "0". CC0 = "0" and CC1 = "1" indicates a positive result, CC0 = "1" and CC1 = "0" indicates a negative result. A number range overflow sets OV and OS (please note the other meaning of CC0 and CC1 in the case of overflow). Division by zero is indicated by "1" at all digital status bits.

Status bits with REAL calculation

The arithmetic functions with data format REAL and the math functions set all digital status bits. A result of zero sets CC0 and CC1 to "0". CC0 = "0" and CC1 = "1" indicates a positive result, CC0 = "1" and CC1 = "0" indicates a negative result. A number range overflow sets OV and OS (please note the other meaning of CC0 and CC1 in the case of overflow). An invalid REAL number is indicated with "1" at all digital status bits.

A REAL number is referred to as "denormalized" if it is represented with reduced accuracy. the exponent is then zero; the absolute value of a denormalized REAL number is less than $1.175\,494 \times 10^{-38}$ (see also the Chapter 24 "Data Types"). S7-300 CPUs treat denormalized REAL numbers as equal to zero.

Status bits with the conversion functions

Of the conversion functions, the two's complements affect all digital status bits. In addition, the following conversion functions set status bits OV and OS in the event of an error (number range overflow or invalid REAL number):

- ▷ ITB and DTB:
  Conversion of INT or DINT to BCD
- ▷ RND+, RND–, RND, TRUNC:
  Conversion of REAL to DINT

Table 15.3 Setting the Status Bits

INT Calculation

The result is:	CC0	CC1	OV	OS
< –32 768(+I, –I)	0	1	1	1
< –32 768(*I)	1	0	1	1
–32 768 to –1	1	0	0	-
0	0	0	0	-
+1 to +32 767	0	1	0	-
> +32 767(+I, –I)	1	0	1	1
> +32 767(*I)	0	1	1	1
32 768(/I)	0	1	1	1
(–) 65 536	0	0	1	1
Division by zero	1	1	1	1

DINT Calculation

The result is:	CC0	CC1	OV	OS
< –2 147 483 648 (+D, –D)	0	1	1	1
< -2 147 483 648(*D)	1	0	1	1
–2 147 483 648 to –1	1	0	0	-
0	0	0	0	-
+1 to +2 147 483 647	0	1	0	-
> +2 147 483 647 (+D, –D)	1	0	1	1
> +2 147 483 647(*D)	0	1	1	1
2 147 483 648(/D)	0	1	1	1
(–) 4 294 967 296	0	0	1	1
Division by zero (/D, MOD)	1	1	1	1

REAL Calculation

The result is:	CC0	CC1	OV	OS
+ normalized	0	1	0	-
± denormalized	0	0	1	1
± zero	0	0	0	-
– normalized	1	0	0	-
+ infinite (division by zero)	0	1	1	1
– infinite (division by zero)	1	0	1	1
± invalid REAL number	1	1	1	1

Comparison

The result is:	CC0	CC1	OV	OS
equal to	0	0	0	-
greater than	0	1	0	-
less than	1	0	0	-
invalid REAL number	1	1	1	1

Conversion NEG_I

The result is:	CC0	CC1	OV	OS
+1 to +32 767	0	1	0	-
0	0	0	0	-
–1 to –32 767	1	0	0	-
(–) 32 768	1	0	1	1

Conversion NEG_D

The result is:	CC0	CC1	OV	OS
+1 to +2 147 483 647	0	1	0	-
0	0	0	0	-
–1 to –2 147 483 647	1	0	0	-
(–) 2 147 483 648	1	0	1	1

Shift function

The shifted out bit is:	CC0	CC1	OV	OS
"0"	0	0	0	-
"1"	0	1	0	-
with number of positions 0	-	-	-	-

Word logic

The result is:	CC0	CC1	OV	OS
zero	0	0	0	-
not zero	0	1	0	-

Status bits with comparison functions

The comparison functions set the CC0 and CC1 status bits. The flags are set independently of the executed comparison function; it depends only on the relation between the two values involved in the comparison function. A REAL comparison checks for valid REAL numbers.

Status bits with word logic operations and shift functions

Word logic operations and shift functions set the CC0 and CC1 status bits. OV is reset.

Setting and resetting the RLO

SET sets the RLO to "1" and CLR sets it to "0". In parallel with this, the STA status bit is also set to "1" or to "0". Both statements are executed unconditionally.

SET and CLR also reset the OR and /FC status bits, that is, after SET or CLR a new logic operation starts with the next scan (check).

You can program an absolute set or reset of binary addresses with SET:

```
SET  ;
S     M 8.0;      //Memory bit is set
R     M 8.1;      //Memory bit is reset
CLR  ;
S     C 1;        //Reset edge memory bit
                  //for "Set counter"
```

Direct setting and resetting of the RLO is also useful in conjunction with timers and counters. To start a timer or counter, you require a change of the RLO from "0" to "1" (please note that you also require a positive edge for enabling). In program sections with predominantly digital logic operations, the RLO is generally not defined, for example, following the jump functions for evaluating the digital flags (status bits). Here you can use SET and CLR for defined setting or resetting of the RLO or for programming an RLO change.

See Chapter 4 "Binary Logic Operations" for details of how to negate the RLO with NOT.

Setting and resetting the BR

With SAVE you can save the RLO in the binary result. SAVE transfers the signal state from the RLO to the BR status bit. SAVE operates unconditionally and does not affect any other status bits.

```
SET  ;
SAVE ;     //Set BR to "1"
...
AN   OV;
SAVE ;     //Set BR to "0" on overflow
```

15.3 Evaluating the Status Bit

The status bits RLO and BR and all digital flags can be evaluated with binary checks and jump functions. It is also possible to further process

A	-	Check for fulfilled condition and logic AND	
O	-	Check for fulfilled condition and logic OR	
X	-	Check for fulfilled condition and logic exclusive OR	
AN	-	Check for unfulfilled condition and logic AND	
ON	-	Check for unfulfilled condition and logic OR	
XN	-	Check for unfulfilled condition and logic exclusive OR	
	>0	Result greater than zero	[(CC0=0) & (CC1=1)]
	>=0	Result greater than or equal to zero	[(CC0=0)]
	<0	Result less than zero	[(CC0=1) & (CC1=0)]
	<=0	Result less than or equal to zero	[(CC1=0)]
	<>0	Result not equal to zero	[(CC0=0) & (CC1=1) v (CC0=1) & (CC1=0)]
	==0	Result equal to zero	[(CC0=0) & (CC1=0)]
	UO	Result invalid (unordered)	[(CC0=1) & (CC1=1)]
	OV	Overflow	[OV=1]
	OS	Stored overflow	[OS=1]
	BR	Binary result	

all status bits after loading the status word into the accumulator.

Evaluation with binary checks

You can use all checks described in Chapter 4 "Binary Logic Operations" to check the digital flags and the binary result (see before). The principle of operation is the same as for checking an input, for example.

Evaluation with jump functions

You can evaluate the RLO and BR status bits, all combinations of CC0 and CC1 and the OV and OS status bits with the relevant jump functions (Table 15.4). Chapter 16 "Jump Functions" contains a detailed description.

Notes on evaluating a number range overflow

A calculation result outside the number range defined for the data type sets the OV status bit and the OS (stored overflow) status bit in parallel.

If the result of a subsequent function (in the case of a chain calculation, for example) is within the permissible number range, the OV flag is reset. The OS flag, however, remains set, so that a result overflow within a chain calculation can also be detected at the end of the calculation.

OS is not reset until the JOS jump function or a block change (call or block end).

You can evaluate an overflow with:

Binary checks

```
L     Value1;
L     Value2;
+I    ;
A     OV;          //Individual evaluation
=     Status1;
L     Value3;
+I    ;
A     OV;          //Individual evaluation
=     Status2;
L     Value4;
+I    ;
A     OS;          //Overall evaluation
=     Status_overall;
T     Result;
```

Jump functions

```
L     Value1;
L     Value2;
+I    ;
JO    ST1;         //Individual evaluation
L     Value3;
+I    ;
JO    ST2;         //Individual evaluation
L     Value4;
+I    ;
JOS   STOV;        //Overall evaluation
T     Result;
```

You can evaluate a number overflow either after every calculation operation (check the OV status bit) or after the overall calculation (check the OS status bit).

Table 15.4 Evaluating the Status Bits Using Jump Functions

RLO	BR	CC0	CC1	OV	OS	Executed jump functions
"1"	-	-	-	-	-	JC, JCB
"0"	-	-	-	-	-	JCN, JNB
-	"1"	-	-	-	-	JBI
-	"0"	-	-	-	-	JNBI
-	-	0	0	-	-	JZ, JMZ, JPZ
-	-	0	1	-	-	JN, JP, JPZ
-	-	1	0	-	-	JN, JM, JMZ
-	-	1	1	-	-	JUO
-	-	-	-	1	-	JO
-	-	-	-	-	1	JOS

15.4 Using the Binary Result

STEP 7 uses the binary result to represent the EN/ENO mechanism in the ladder diagram LAD and function block diagram FBD programming languages. You can ignore this if you program only in STL; you then have the binary result at your disposal as an additional RLO memory.

However, you can use BR as a group error flag, even with pure STL programming, in order to indicate errors in block processing (as used by the SFB and SFC system blocks and some standard blocks).

EN/ENO mechanism

In the LAD and FBD programming languages, all boxes have an enable input EN and an enable output ENO. If the enable input has "1", the function in the box is executed. If the box is executed properly, the enable output then also has signal state "1". If an error occurs during execution of the box, (for example, overflow in the case of an arithmetic function), ENO is set to "0". If EN has signal state "0", ENO is also set to "0".

You can use these characteristics of EN and ENO to link several boxes in a chain, with the

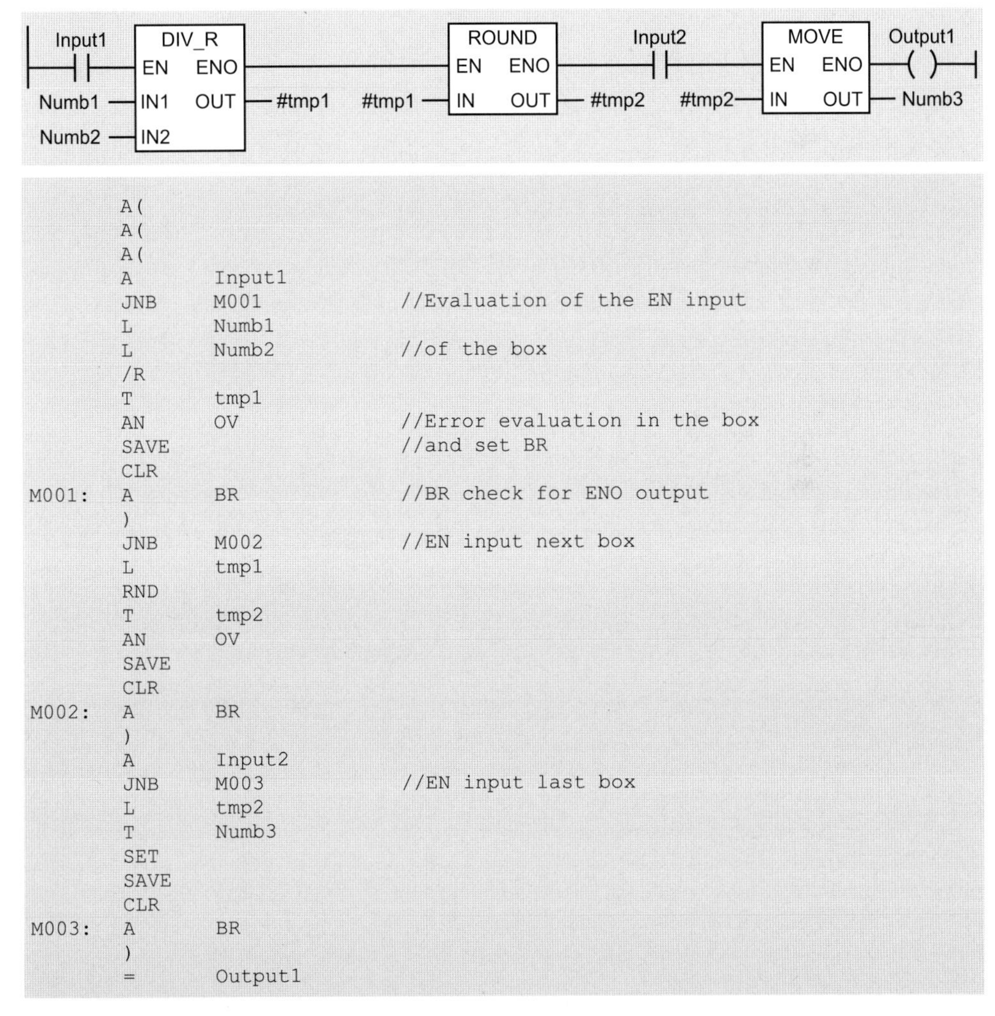


```
      A(
      A(
      A(
      A      Input1
      JNB    M001           //Evaluation of the EN input
      L      Numb1
      L      Numb2          //of the box
      /R
      T      tmp1
      AN     OV             //Error evaluation in the box
      SAVE                  //and set BR
      CLR
M001: A      BR             //BR check for ENO output
      )
      JNB    M002           //EN input next box
      L      tmp1
      RND
      T      tmp2
      AN     OV
      SAVE
      CLR
M002: A      BR
      )
      A      Input2
      JNB    M003           //EN input last box
      L      tmp2
      T      Numb3
      SET
      SAVE
      CLR
M003: A      BR
      )
      =      Output1
```

Figure 15.1 Example of the EN/ENO Mechanism

enable output ENO leading to the enable input EN of the next box (Figure 15.1). This makes it possible to "switch off" the entire chain (no box is processed if *input I* E 1.0 in the example has signal state "0") or the remainder of the chain is no longer processed if one box signals an error.

The EN input and the ENO output are not block parameters but statement results that the LAD/FBD Editor generates itself prior to and following all boxes (even in the case of functions and function blocks). Here, the LAD/FBD Editor uses the binary result to store the signal state at EN during block processing or to check the error message from the box.

You can find the statement sequence shown in Figure 15.1 in Network 8 of FB 115 in the "Program Flow Control" program (STL_Book library). If you view this network FB 115 on screen, you can switch to Ladder diagram representation with VIEW → LAD. The Editor then displays the LAD graphics.

If you write your own functions or function blocks and you want to use these in, for example, ladder or function block diagram representation, you must influence the binary result in such a way that BR will be set to "0" if an error is detected (see below).

Group error message in blocks

You can use the binary result as a group error message in blocks. If a block has been executed properly, set BR to "1". BR is set to "0" if a block signals an error.

Example: At the start of the block, BR is set to "1". If an error now occurs during execution of the block, for example, a result exceeds the defined range, so that further processing must be stopped, set the binary result to "0" with JNB and jump to the block end, for example (in the event of an error, the condition must supply signal state "0").

```
SET   ;
SAVE  ;           //BR = "1"
...
L     10_000;
L     Result;     //if result>10000
<=I   ;           //then BR = "0"
JNB   ERR;        //and jump to ERR
...
```

The "Clock entry" example in Chapter 26.4 "Brief Description of the Message Frame Example" also uses BR as a group error message.

16 Jump Functions

With jump functions you can interrupt linear execution of the program and continue it at another point in the block. This program branching can be executed either conditionally or unconditionally.

The jump distributor (case branching) and the loop jump are available as special forms of the jump functions.

Overview

JU	*label*	Unconditional jump
JC	*label*	Jump if RLO = "1"
JCN	*label*	Jump if RLO = "0"
JCB	*label*	Jump if RLO = "1" and save RLO
JNB	*label*	Jump if RLO = "0" and save RLO
JBI	*label*	Jump if BR = "1"
JBI	*label*	Jump if BR = "0"
JZ	*label*	Jump if result is zero
JN	*label*	Jump if result is not zero
JP	*label*	Jump if result is greater than zero (positive)
JPZ	*label*	Jump if result is greater than or equal to zero
JM	*label*	Jump if result is less than zero (negative)
JMZ	*label*	Jump if result is less than or equal to zero
JUO	*label*	Jump if result is invalid
JO	*label*	Jump if overflow
JOS	*label*	Jump if stored overflow
JL	*label*	Jump distributor
LOOP	*label*	Loop jump

This chapter describes the jump functions for the STL programming language. In the SCL programming language, there are various methods for branching within the program, e.g. with the IF statement (see Chapter 28 "Control Statements").

The examples in this chapter can be found in the download files (download address: see pages 8-9) in the STL_Book library under the "Program Flow Control" program in function block FB 116 or source file Chap_16.

16.1 Programming a Jump Function

A jump function consists of a jump operation that defines the checked condition, and a jump label that indicates the program location at which program execution is to be continued if the condition is met.

A jump label consists of up to 4 characters which can include alphanumeric characters and the underscore. A jump label must not start with a numeric character. A jump label followed by a colon indicates the statement (line) that is to be processed after the executed jump statement.

Figure 16.1 gives an example.The condition for the jump here is a comparison operation; it supplies an RLO. This RLO is the jump condition for the JC jump. If the comparison is fulfilled,

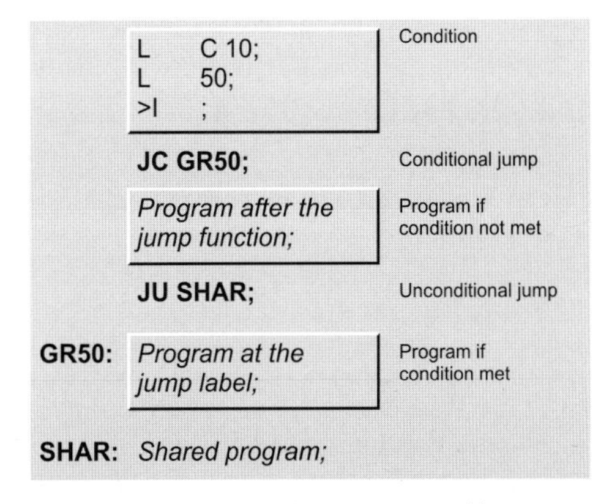

Figure 16.1 Example of Program Branching

the RLO also equals “1” and the jump to jump label GR50 is executed. Program execution is then continued here. An unfulfilled comparison supplies RLO = “0”, so that in this example the jump function is not executed. The program is continued at the next statement. A jump can be made forward (in the direction of program processing; in the direction of higher line numbers) as well as backward. The jump can only take place within a block; that is, the jump destination must be in the same block as the jump statement. Subdivision into networks has no effect on the jump function.

The jump destination must have a unique ID, that is, you can assign any given jump label only once in a block. The jump destination can be jumped to from several locations. If you use the Master Control Relay MCR, the jump destination must be in the same MCR zone or the same MCR area as the jump statement.

STL stores the jump label designations in the non-execution-relevant section of the block on the data medium of the programming device. Only the jump widths are stored in the work memory of the CPU (in the compiled block). For this reason, program modifications made to blocks online at the CPU must also always be updated on the programming device data medium in order to retain the original designations. If this update is not made or if blocks are transferred from the CPU to the programming device, the non-execution-relevant block sections will be overwritten or deleted. The Editor then generates its own jump label designations (M001, M002 etc.) on screen or in the printout.

16.2 Unconditional Jump

The JU jump function is always executed, that is, regardless of any conditions. JU interrupts linear execution of the program and continues it at the location indicated by the jump label.

The JU jump function does not affect the status bits. If there are check statements, for example, AI, OI, etc., both immediately prior to the jump function and at the jump destination, these are treated as a single logic operation.

16.3 Jump Functions with RLO and BR

A program branch can be made dependent on the signal states of the RLO and BR status bits (Table 16.1). In addition, it is also possible to store the RLO in the BR status bit at the same time that it is checked.

Setting the status bits

The jump functions conditional on the RLO set the STA and RLO status bits to “1” and OR and /FC to “0” whether the condition is fulfilled or not.

This results in the following consequences for the use of these jump functions: The RLO is always set to “1”. If the statements contain operations conditional on the RLO immediately following these jump functions, they will be executed if the jump does not take place. If there are check statements, such as AI, OI etc., immediately following these jump functions, these checks are treated as first checks, that is, a new logic operation starts.

The jump functions conditional on the binary result set the STA status bit to “1” and the OR and /FC status bits to “0” whether the condition is fulfilled or not. The RLO and BR status remain unchanged. This has the following consequences for use: These jump functions terminate a logic operation; a new logic operation starts following this jump function or at the jump destination. The RLO is retained and can be evaluated with a memory function following the jump function.

Table 16.1 Jump Functions with RLO and BR

RLO	BR	Executed Jumps	
“1”	-	JC	Jump if RLO = “1”
“1”	→ “1”	JCB	Jump if RLO = “1” and save RLO
“0”	-	JCN	Jump if RLO = “0”
“0”	→ “0”	JNB	Jump if RLO = “0” and save RLO
-	“1”	JBI	Jump if BR = “1”
-	“0”	JNBI	Jump if BR = “0”

Jump if RLO = "1"

The jump function JC is only executed if the RLO is "1" when the function is processed. If it is "0", the jump is not executed and program processing is continued with the next statement.

Jump if RLO = "0"

The jump function JCN is only executed if the RLO is "0" when the function is processed. If it is "1", the jump is not executed and program processing is continued with the next statement.

Jump if RLO = "1" and save the RLO

The jump function JCB is only executed if the RLO is "1" when the function is processed. Simultaneously, JCB sets the binary result to "1". If the RLO is "0", the jump is not executed and program processing is continued with the next statement. JCB sets the binary result in this case to "0" (the RLO is transferred in each case to the binary result).

Jump if RLO = "0" and save the RLO

The jump function JNB is only executed if the RLO is "0" when this function is processed. Simultaneously, JNB sets the binary result to "0". If the RLO is "1", the jump is not executed and program processing is continued with the next statement. JNB sets the binary result in this case to "1" (the RLO is transferred in each case to the binary result).

Jump if BR = "1"

The jump function JBI is only executed if the binary result is "1" when this function is processed. If the binary result is "0", the jump is not executed and program processing is continued with the next statement.

Jump if BR = "0"

The jump function JBIN is only executed if the binary result is "0" when this function is processed. If the binary result is "1", the jump is not executed and program processing is continued with the next statement.

16.4 Jump Functions with CC0 and CC1

A program branch can be made conditional on the CC0 and CC1 status bits (Table 16.2). This allows you to, for example, check to see if the result of a calculation is positive, zero or negative. See Chapter 15 "Status Bits" for details of when the CC0 and CC1 status bits are set.

Setting the status bits

The jump functions conditional on the CC0 and CC1 status do not change any status bits. When the jump is made, the RLO is "taken along" and can be combined further (no change to /FC).

The binary checks are another method of checking the status bits (see Chapter 15 "Status Bits").

Jump if result is zero

The jump function JZ is only executed if CC0 = "0" and CC1 = "0". This is the case if

- accumulator 1 contains zero after an arithmetic or math function,
- accumulator 2 contains the same as accumulator 1 in a comparison function,
- accumulator 1 contains zero after a digital logic operation and
- the value of the last shifted bit is "0" after a shift function.

Table 16.2 Jump Functions with CC0 and CC1

CC0	CC1	Executed Jumps	
0	0	JZ	Jump if zero
		JMZ	Jump if zero or less than zero
		JPZ	Jump if zero or greater than zero
1	0	JM	Jump if less than zero
		JMZ	Jump if zero or less than zero
		JN	Jump if not zero
0	1	JP	Jump if greater than zero
		JPZ	Jump if zero or greater than zero
		JN	Jump if not zero
1	1	JUO	Jump if invalid result

In all other cases, JZ continues program processing with the next statement.

Jump if result is not zero

The jump function JN is only executed if status bits CC0 and CC1 have different signal states. This is the case if

- ▷ accumulator 1 does not contain zero after an arithmetic or math function,
- ▷ the contents of accumulator 2 are not the same as the contents of accumulator 1 in a comparison function,
- ▷ accumulator 1 does not contain zero after a digital logic operation and
- ▷ the value of the last shifted bit is “1” after a shift function.

In all other cases, JN continues program processing with the next statement.

Jump if result is greater than zero

The jump function JP is only executed if CC0 = “0” and CC1 = “1”. This is the case if

- ▷ the contents of accumulator 1 are within the permissible positive number range following an arithmetic or math function (you check for a number range violation with JO or JOS),
- ▷ the contents of accumulator 2 are greater than the contents of accumulator 1 in a comparison function,
- ▷ accumulator 1 does not contain zero after a digital logic operation and
- ▷ the value of the last shifted bit is “1” after a shift function.

In all other cases, JP continues program processing with the next statement.

Jump if result is greater than or equal to zero

The jump function JPZ is only executed if CC0 = “0”. This is the case

- ▷ if the contents of accumulator 1 are within the permissible positive number range or are equal to zero following an arithmetic or math function (you check for a number range violation with JO or JOS),
- ▷ if the contents of accumulator 2 are greater than or equal to the contents of accumulator 1 in the case of a comparison function,
- ▷ after every digital logic operation and
- ▷ after every shift function.

In all other cases, JPZ continues program processing with the next statement.

Jump if result is less than zero

The jump function JM is only executed if CC0 = “1” and CC1 = “0”. This is the case if

- ▷ the contents of accumulator 1 are within the permissible negative number range following an arithmetic or math function (you check for a number range violation with JO or JOS) and
- ▷ the contents of accumulator 2 are less than the contents of accumulator 1 in a comparison operation.

In all other cases, JM continues program processing with the next statement.

Jump if result is less than or equal to zero

The jump function JMZ is only executed if CC1 = “0”. This is the case if

- ▷ the contents of accumulator 1 are within the permissible negative number range or are equal to zero following an arithmetic or math function (you check for a number range violation with JO or JOS), and
- ▷ the contents of accumulator 2 are less than or equal to the contents of accumulator 1 in the case of a comparison operation.

In all other cases, JMZ continues program processing with the next statement.

Jump if invalid result

The jump function JUO is only executed if CC0 = “1” and CC1 = “1”. This is the case if

- ▷ a division by zero is made in an arithmetic function and
- ▷ an invalid REAL number is specified as the input value or is produced as the result.

In all other cases, JUO continues program processing with the next statement.

16.5 Jump Functions with OV and OS

A program branch can be executed dependent on the OV and OS status bits. This is a check to see if the result of a calculation is still within the permissible number range. See Chapter 15 "Status Bits" for details of when the OV and OS status bits are set.

Jump if overflow

The jump function JO is only executed if the OV status bit has been set to "1". This is the case if the permissible number range has been exceeded following execution of an operation. The following functions can set the OV status bit:

▷ Arithmetic functions,

▷ Math functions,

▷ Two's complement,

▷ Comparison functions with REAL numbers and

▷ Conversion functions INT/DINT to BCD and REAL to DINT.

If OV = "0", JO continues program processing with the next statement.

In the case of a chain calculation with several calculations performed one after the other, the OV status bit must be evaluated after each calculation function since OV is reset again following the next calculation operation whose result is within the permissible number range. Check the OS status bit in order to evaluate a possible number range overflow at the end of the chain calculation.

Jump if stored overflow

The jump function JOS is only executed if the OS status bit has been set to "1". This is always the case if a number range overflow sets the OV status bit (see above). In contrast to OV, OS remains set if a result is then in the permissible number range.

The following functions reset OS:

▷ Block call and block end

▷ Jump if stored overflow JOS

If OS = "0", JOS continues program processing with the next statement.

16.6 Jump Distributor

The jump distributor JL allows specific (calculated) jumping to a program section in the block conditional on a number of positions.

JL works in conjunction with a list of JU jump functions. This list immediately follows JL and can contain up to 255 entries. There is a jump label at JL that points to the end of the list (to the first statement after the list).

You program a jump distributor according to the following general schematic:

```
L      Number_of_positions;
       JL   End;
       JU   M0;
       JU   M1;
       ...
       JU   Mx;
End:   ...
```

In the example, the variable *Number_of_positions* loads a number into accumulator 1. This is followed by the jump distributor JL with the jump label to the end of the list of JU statements.

The number of the jump to be executed is in the right-hand byte of accumulator 1. If accumulator 1 contains 0, the first jump statement is executed, and if it contains 1, the second is executed, and so on. If the number is greater than the length of the list, JL branches to the end of the list (to the statement following the last jump).

JL is not subject to conditions and does not change the status bits.

Only JU statements are permissible in the list without gaps. You can assign the jump label designations as you please within the general rules.

16.7 Loop Jump

The loop jump LOOP allows simplified programming of program loops.

LOOP interprets the right-hand word of accumulator 1 as a signless 16-bit number in the range 0 to 65535.

When processed, LOOP first decrements the contents of accumulator 1 by 1. If the value is then not zero, the jump is executed to the jump label specified.

If the value is equal to zero after decrementing, the jump is not executed and the next statement is processed.

The value in accumulator 1 thus corresponds to the number of program loops to be passed. You must store this number in a loop counter. You can use any digital address as the loop counter.

You program a loop jump according to the following general schematic:

```
      L       Number;
Next: T       Counter;
      ...
      ...
      ...
      L       Counter;
      LOOP    Next;
      ...
```

The variable *Number* contains the number of loop passes. The variable *Counter* contains the loop passes still to be executed.

At the first pass, *Counter* is preassigned with the number of loop passes. At the end of the program loop, the contents of *Counter* are loaded into the accumulator and decremented by the LOOP statement. If the accumulator does not contain zero following this, the jump to the specified jump label – here: Next – is executed and the variable *Counter* is updated.

The loop jump does not change the status bits.

17 Master Control Relay

With contact controls, a Master Control Relay activates or de-activates a section of the control that can consist of one or more rungs. A deactivated rung

▷ switches all non-retentive contactors off and

▷ retains the state of retentive contactors.

You can only change the state of the contactors again when the Master Control Relay (MCR) is active.

This chapter describes the statements required for implementing the Master Control Relay for the STL programming language. You can use these statements to emulate the properties of a Master Control Relay in the statement list. Master Control Relay examples can be found in the download files (download address: see pages 8-9) in the STL_Book library under the "Program Flow Control" program in function block FB 117 or source file Chap_17.

Please note that switching off with the "software" Master Control Relay is no substitute for an EMERGENCY OFF or safety facility! Treat Master Control Relay switching in exactly the same way as switching with a memory function!

STL provides the following statements for implementing the Master Control Relay (MCR):

▷ MCRA Activate MCR area

▷ MCR(Open MCR zone

▷)MCR Close MCR zone

▷ MCRD Deactivate MCR area

The statements MCRA and MCRD identify an area in your program in which MCR dependency is to take effect. Within this area, you use the statements MCR(and)MCR to define one or more zones in which MCR dependency can be switched on and off. You can also nest the MCR zones. The result of logic operation (RLO) immediately prior to an MCR zone switches MCR dependency on or off within this zone.

17.1 MCR Dependency

The MCR affects all operations that write a value back to memory. These MCR-dependent operations respond as follows when MCR dependency is switched on, regardless of any previous binary or digital logic operation:

▷ Assignment (=):
the address is reset to "0"

▷ Set (S) or reset (R):
the address remains unchanged

▷ Transfer (T):
Zero is transferred.

Some STL functions use transfer statements (invisible to the user) to write a value to an address register, for example. Since a transfer statement writes the value zero if MCR dependency is switched on, the corresponding function can no longer be guaranteed.

You must exclude the following program sections from MCR dependency otherwise the CPU will go to STOP or undefined runtime behavior can occur:

▷ Block calls with block parameters

▷ Accesses to block parameters that are parameter types (e.g. BLOCK_DB)

▷ Accesses to block parameters that are components or elements of complex data types or UDTs

If MCR dependency is switched off, the MCR-dependent operations respond in the "normal" way as described in the relevant chapters.

You switch on MCR dependency in a zone if the RLO is "0" immediately prior to opening the zone (analogous to switching off the Master Control Relay). If you open an MCR zone with RLO "1" (Master Control Relay switched on), processing within this MCR zone takes place without MCR dependency.

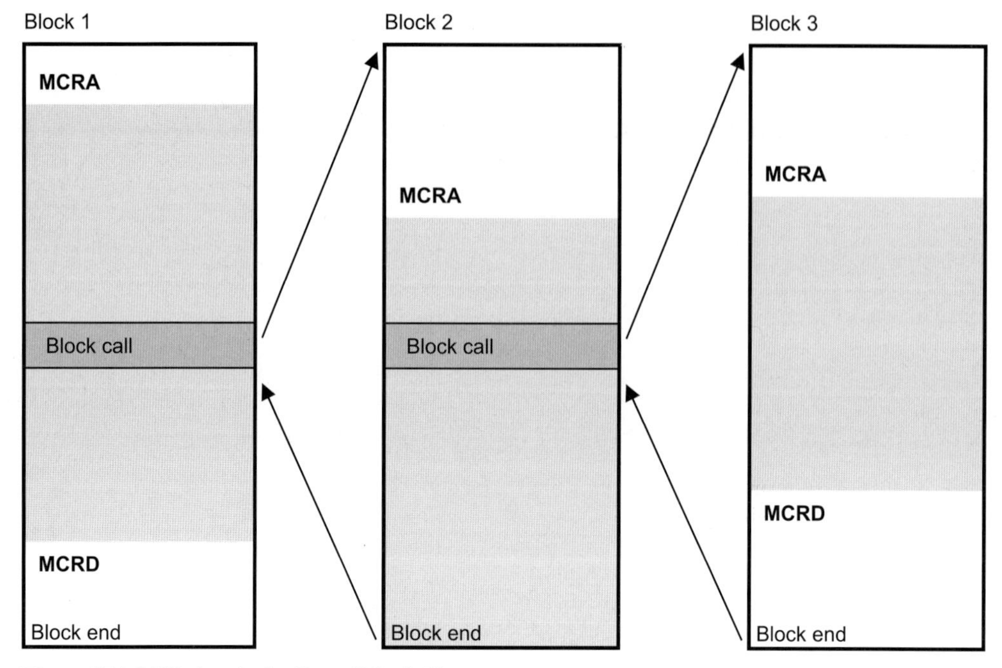

Figure 17.1 MCR Area in the Case of Block Change

Example:

```
MCRA  ;         //Activate MCR
A     Input0;
MCR(  ;         //Open MCR zone
A     Input1;
A     Input2;
=     Output0;
)MCR  ;         //Close MCR zone
MCRD  ;         //Deactivate MCR
```

In the example, *Input0* = “0” also sets the address *Output0* to “0”. If *Input0* has signal state “1”, you control the address *Output1* with *Input1* and *Input2*.

17.2 MCR Area

To be able to use the characteristics of the Master Control Relay, define an MCR area with the statements MCRA and MCRD. MCR dependency is active within an MCR area (but not yet switched on).

```
MCRA; //Activate MCR
...   //MCR area
MCRD; //Deactivate MCR
```

MCRA defines the start of an MCR area and MCRD its end. If you call a block within an MCR area, MCR dependency is de-activated in the called block (Figure 17.1). An MCR area only starts again with the MCRA statement. When a block is exited, MCR dependency is set as it was before the block was called, regardless of the MCR dependency with which the block was exited.

17.3 MCR Zone

You define an MCR zone with the statements MCR(and)MCR. Within this zone, you can switch MCR dependency on with RLO = “0” and off with RLO = “1”.

```
...             //Switch on MCR with "0"
A       Input3;
MCR(    ;       //Start of dependency
...
...             //MCR zone
...
)MCR    ;       //End of dependency
```

The statements MCR(and)MCR end a bit logic combination.

You can open another MCR zone within an MCR zone. The nesting depth for MCR zones has the value 8; that is, you can open up to eight zones before having to close one.

You control the MCR dependency of a switched on MCR zone with the RLO when opening the zone. However, if MCR dependency is switched on in a higher-level zone, you cannot switch MCR dependency off in a lower-level zone. The Master Control Relay of the first MCR zone controls the MCR dependency in all switched on zones (Figure 17.2).

A block call within an MCR zone does not change the nesting depth of an MCR zone. The program in the called block is still in the MCR zone that was open when the block was called (and is controlled form here). However, you must re-activate MCR dependency in a called block by opening the MCR area.

In Figure 17.3 the addresses *Input5* and *Input6* control the MCR dependencies. With *Input5*

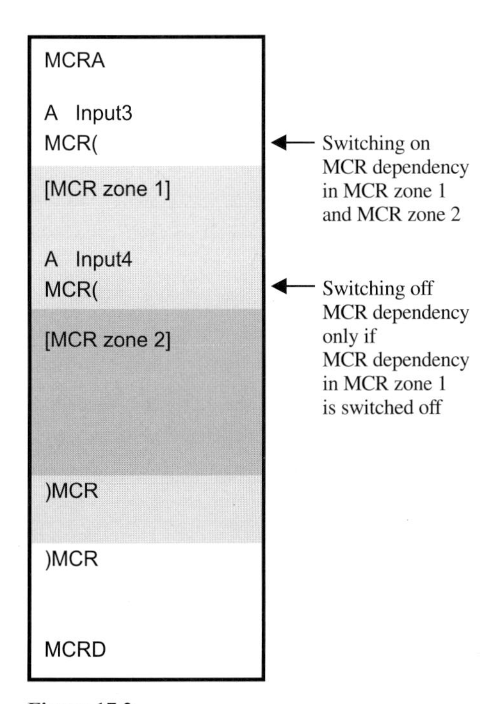

Figure 17.2
MCR Dependency in the Case of Nested MCR Zones

Table 17.1 MCR Dependency in the Case of Nested MCR Zones (Example)

Input5	Input6	Zone 1	Zone 2
"1"	"1"	No MCR dependency	
"1"	"0"	No MCR dependency	MCR dependency switched on
"0"	"1" or "0"	MCR dependency switched on	

you can switch MCR dependency on in both zones (with "0"), regardless of the signal state of *Input6*. If the MCR dependency of Zone 1 is switched of with *Input5* = "1", you can control the MCR dependency of zone 2 with *Input6* (Table 17.1).

17.4 Setting and Resetting I/O Bits

Despite MCR dependency being switched on, you can set or reset the bits of an I/O area with the system functions. A requirement for this is that the bits to be controlled are in the process-image output or a process-image output has been defined for the I/O area to be controlled.

The system function **SFC 79 SET** is available for setting the I/O bits, and **SFC 80 RSET** for resetting (Table 17.2). You call these system functions in an MCR zone. The system functions are only effective if MCR dependency is switched on; if MCR dependency is switched off, the calls of these SFCs remain without effect.

Setting and resetting the I/O bits also simultaneously updates the process-image output. The I/O are affected byte-by-byte. The bits not selected with the SFCs (in the first and in the last byte) retain the signal states as they are currently available in the process-image.

Example:

```
CALL SFC 79 (N       := 8,
             RET_VAL := MW 10,
             SA      := P#12.0);
CALL SFC 80 (N       := 16,
             RET_VAL := MW 12,
             SA      := P#13.5);
```

In the example, calling SFC 79 SET sets the I/O bits in accordance with outputs Q 12.0 to Q 12.7; calling SFC 80 RSET resets the I/O bits in accordance with outputs Q 13.5 to Q 15.5.

The parameter N determines the number of bits to be controlled and parameter SF determines the first bit (Data type POINTER). The SFC uses RET_VAL to return any detected error.

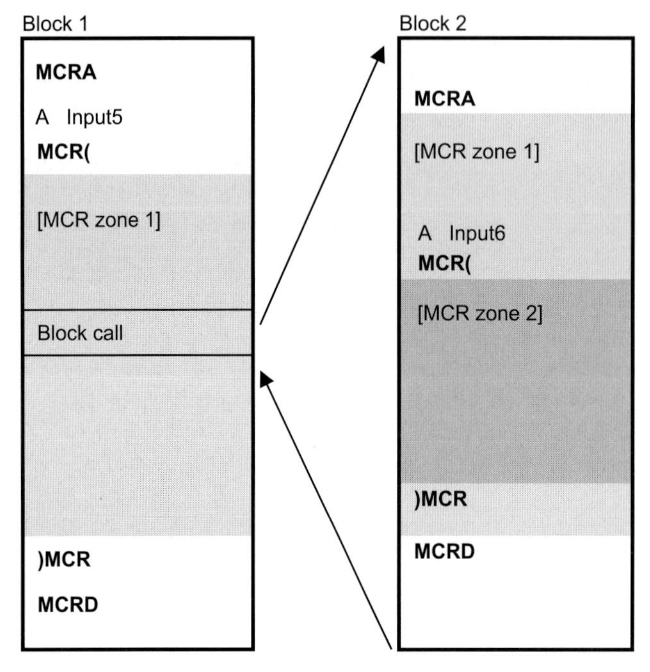

Figure 17.3 MCR Zones at Block Change

Table 17.2 Parameters of the SFCs for Controlling the I/O Bits

SFC	Parameter	Declaration	Data Type	Assignment, Description
79	N	INPUT	INT	Number of bits to be set
	RET_VAL	RETURN	INT	Error information
	SA	OUTPUT	POINTER	Pointer to the first bit to be set
80	N	INPUT	INT	Number of bits to be reset
	RET_VAL	RETURN	INT	Error information
	SA	OUTPUT	POINTER	Pointer to the first bit to be reset

18 Block Functions

In this chapter, you will learn how to call and terminate code blocks and how to work with addresses from data blocks in the STL programming language. The next chapter then deals with using block parameters. This chapter is the continuation of Chapter 3.4 "Programming Code Blocks with STL" and Chapter 3.6 "Programming Data Blocks".

Block calls in the SCL programming language are described in Chapter 29 "SCL Blocks".

Examples of the block functions can be found in the download files (download address: see pages 8-9) in the STL_Book library under the "Program Flow Control" program in function block FB 118 or source file Chap_18.

18.1 Block Functions for Code Blocks

Block functions for code blocks include instructions for calling and terminating blocks (Table 18.1). Code blocks are called and processed with CALL. You can pass data for processing to the called block and take over data from the called block. This data transfer is carried out via block parameters. CALL transfers the block parameter to the called block and also opens the instance data block in the case of function blocks. When code blocks have no block parameters, they can also be called with UC or CC. A block is terminated with a block end statement.

Table 18.1 Block Functions for Code Blocks

Calling a Function Block		
with data block and with block parameter	as local instance and with block parameter	without block parameter, unconditionally and conditionally

`CALL FB 10, DB 10 (`
`In1 := Number1,`
`In2 := Number2,`
`Out := Number3);`
	`CALL name (`
	`In1 := Number1,`
	`In2 := Number2,`
	`Out := Number3);`
		`UC FB 11;`
		`CC FB 11;`

Calling a function		
with function value and with block parameter	without function value and with block parameter	without block parameter, unconditionally and conditionally

`CALL FC 10 (`
`In1 := Number1,`
`In2 := Number2,`
`RET_VAL := Number3);`
	`CALL FC 10 (`
	`In1 := Number1,`
	`In2 := Number2,`
	`Out := Number3);`
		`UC FC 11;`
		`CC FC 11;`

Block end statements		
Conditional block end **BEC**	Unconditional block end **BEU**	Block end **BE**

18.1.1 Block Calls: General

If a code block is to be processed, it must be "called". Figure 18.1 gives an example of calling function FC 10 in organization block OB 1.

A block call consists of the call statement (here: CALL FC 10) and the parameter list. If the called block has no block parameters, there is no need for the parameter list. After the call statement has been executed, the CPU continues program processing in the called block (here: FC 10). The block is processed until a block end statement is encountered. Then the CPU returns to the calling block (here: OB 1) and continues processing this block after the call statement. If an organization block is terminated, the CPU continues in the operating system.

The information the CPU requires to make its return to the calling block is stored in the block stack (B stack). With each block call, a new stack element is created that includes the return address, the contents of the data register and the address of the local data stack of the calling block. If the CPU goes to the Stop state as a result of an error, you can use the programming device to see from the contents of the B stack which blocks were processed up to the triggering error.

The block parameters are the data interface to the called block. You are advised to avoid data transfer via internal registers (for example, accumulators, address registers, RLO) since the contents of these registers can be changed at a block change (as a result of "concealed" statements from the Editor).

18.1.2 CALL Call Statement

You call FBs, FCs, SFCs and SFCs with CALL. CALL is an unconditional call, that is, the specified block is always called and processed independently of any conditions. (You cannot call organization blocks; they are called by the operating system depending on events.)

Note that the CALL statement can change the data block registers DB and DI, the address registers AR1 and AR2, the status bits including RLO, and the contents of the accumulators.

Calling function blocks

You call a function block FB by specifying, after CALL, the function block and, separated with a comma, the instance data block associated with the call. You can address both blocks either absolutely or symbolically. The assignment of the absolute address to the symbol address is made in the symbol table, with an instance data block having the associated function block as data type.

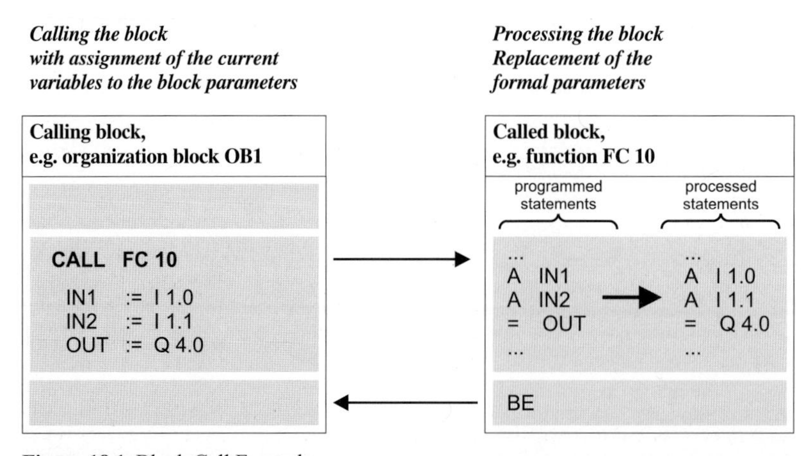

Figure 18.1 Block Call Example

The CALL operation is followed by the list with the block parameters. In source-oriented input, the list of the block parameters is placed between round brackets; the block parameters are each separated by a comma.

With function blocks, you need not initialize all block parameters when the block is called. The uninitialized block parameters retain their current value. If you do not specify any parameters, the brackets are also dispensed with in source-oriented input. However, block parameters saved as pointers should at least be initialized when called for the first time so that meaningful values are entered (see Chapter 19.3 "Actual Parameters").

If you have generated function blocks with the block attribute "multi-instance-capable", you can also call these as a local instance within other "multi-instance-capable" function blocks. Here, the called function block uses the instance data block of the calling function block to store its local data. You declare the local instance in the static local data of the calling function block and you can then call the function block in the program (without specifying an instance data block). The local instance is treated like a complex data type within the "higher-level" function block. You can find more information in Chapter 18.1.6 "Static Local Data".

Calling functions

You call a function FC by specifying the function, after CALL, either with absolute addressing or with symbolic addressing. This is followed by the parameter list, in brackets in the case of source-oriented input. You must initialize all existing parameters; the parameter sequence is defined by the declaration. Calling functions with function value takes exactly the same form as calling functions without function value. Only the first output parameter - corresponding to the function value - has the name RET_VAL.

Calling system blocks

The operating system of the CPU contains system functions SFCs and system function blocks SFBs, that you can use. The number and type of system blocks is CPU-specific. All system blocks are called with CALL.

You call a system function block in the same way as a function block you have written yourself; you set up the associated instance data block in the user memory with the data type of the SFB. You call a system function in the same way as a function you have written yourself.

System blocks are only available in the operating system of the CPU. When calling system blocks during offline programming, the Editor requires a description of the call interface in order to be able to initialize the parameters. The interface description is located in the *Standard Library* under *System Function Blocks*. From here, the Editor copies the relevant interface description into the offline container "Blocks" when you call a system block. The copied interface description then appears as a "normal" block object.

18.1.3 UC and CC Call Statements

You can call function blocks and functions with UC and CC. It is a requirement that the called function has no block parameters and the called function block has no instance data block – and therefore also no block parameters and no static local data. However, the Editor does not check this.

You can use the UC and CC operations if a block is too long or not clear enough for you, by simply "breaking down" the block into sections and then calling the sections in sequence. The UC and CC call operations do not distinguish between functions FCs and function blocks FBs. Both block types are handled in the same way.

The *UC call statement* is an unconditional statement, that is, UC always calls the block regardless of conditions.

The *CC call statement* is a conditional statement, that is, CC only calls the block if the result of logic operation (RLO) is "1". If the RLO is "0", CC does not call the block and sets the RLO to "1". The statement following CC is then processed.

Effect on the indicator bits (condition code bits): The OS status bit is reset at block change; the CC0, CC1 and OV status bits are not affected, the /FC status bit is reset; that is, a new logic operation begins with the first check statement in the new block or following a block call.

Binary nesting stack at a block change: You can also call a code block within a binary nesting expression. The current stack depth of the binary nesting stack does not change at a block change. The possible nesting stack depth in a block that can be called within a binary nest is therefore the difference between the maximum possible nesting depth and the current nesting depth at the block call.

Master Control Relay at a block change: MCR dependency is deactivated at a block call. The MCR is switched off in the called block, regardless of whether the MCR was switched on or off prior to the block call. When a block is exited, MCR dependency is set as it was prior to the block being called.

Accumulators and address registers at block change: The contents of accumulators and of the address registers are not changed at a block change with UC or CC.

Data blocks at a block change: Calling a block saves the data block register in the B stack; the block end statement restores its contents when the called block is exited. The global and the instance data block current prior to the block call are also opened following the block call. If no data block was opened prior to the block call (for example, no instance data block in OB 1), there will also be no data block open following the block call, regardless of the data blocks open in the called block.

Additional possibilities:

- ▷ Indirect addressing of FB and FC calls with UC and CC
- ▷ Calling via block parameters with UC
- ▷ Calling via block parameters with CC also in function blocks

18.1.4 Block End Functions

The BEC statement terminates program processing in a block conditional on the RLO, and the BEU and BE statements end a block unconditionally.

Conditional block end BEC

Execution of BEC depends on the RLO. If the RLO is "1" when BEC is processed, the statement is executed and the block currently being processed is terminated. A return jump is then made to the previously processed block containing the block call.

If the RLO is "0" when BEC is processed, the statement is not executed. The CPU sets the RLO to "1" and processes the statement following BEC. A subsequently programmed check statement is always a first check.

Unconditional block end BEU

When BEU is processed, the block currently being processed is exited. A return jump is then made to the previously processed block containing the block call.

In contrast to the BE statement, you can program BEU several times within one block. The program section following BEU is only processed if it is jumped to with a jump function.

Block end BE

When BE is processed, the block currently being processed is exited. A return jump is then made to the previously processed block containing the block call.

BE is always the last statement in a block.

Programming BE is a matter of choice. With incremental input, you terminate block programming by closing the block; with source-oriented input, the keyword replaces the block end, for example, END_FUNCTION_BLOCK instead of the BE statement.

18.1.5 Temporary Local Data

You use the temporary local data for intermediate storage of results occurring during block processing. Temporary local data are only available during block processing; after the block is terminated, its data are lost.

Temporary local data are addresses that are located in the local data stack (L stack) in the CPU system memory. The operating system of the CPU provides the temporary local data for every code block when that code block is called. The values in the L stack are semi random when the block is called. In order to make meaningful use of the local data, you must first write them before reading. After a block is terminated, the L stack is assigned to the next called block.

The volume of temporary local data required by a block is shown in the block header. In this way, the operating system learns how many bytes are available in the L stack when the block is called. You can also see from the entry in the block header how many local data bytes the block requires (in the Editor when the block is open with FILE → PROPERTIES or in the SIMATIC Manager when the block is selected with EDIT → OBJECT PROPERTIES, on "General - Part 2" tab in each case).

Declaration of temporary local data

You declare the temporary local data in the declaration section of the code block:

▷ with incremental programming under "temp" or

▷ with source-oriented programming between VAR_TEMP and END_VAR.

Figure 18.2 gives an example of declaring temporary local data. The variable *temp1* is located in the temporary local data and is of data type INT, the variable *temp2* is of data type REAL.

The temporary local data are stored in the L stack in the order of their declaration according to data type. Chapter 26.2 "Data Storage of Variables" contains more detailed information on data storage in the L stack.

Symbolic addressing of temporary local data

You address the temporary local data with their symbolic names. You assign the names in accordance with the rules for block-local symbols.

All operations that are also valid for the memory bits are permissible for the temporary local data. However, please note that a temporary local data bit is not suitable as an edge memory bit since it does not retain its signal state beyond processing of the block.

You can only access the temporary local data of a block in the block itself. (Exception: The temporary local data of the calling block can be accessed via block parameters.)

Size of the L stack

The size of the overall L stack is CPU-specific. The quantity of temporary local bytes available in a priority class, that is in the program of an organization block, is also fixed. On the S7-300 the quantity is fixed; for example, 510 bytes per priority class on the CPU 314. On the S7-400 you can adjust the quantity of local data bytes to your requirements when parameterizing the CPU. This quantity must be shared between the blocks called in the relevant organization block

Incremental programming

Address	Declaration	Name	Type	Initial value
0.0	IN	In	INT	0
	OUT			
	IN_OUT			
2.0	STAT	Total	INT	0
0.0 2.0	TEMP TEMP	temp1 temp2	INT REAL	

Source-oriented programming

```
VAR_INPUT
  In    : INT := 0;
END_VAR
VAR_OUTPUT ... END_VAR
VAR_IN_OUT ... END_VAR
VAR
  Total : INT := 0;
END_VAR
VAR_TEMP
  temp1 : INT;
  temp2 : REAL;
END_VAR
```

Figure 18.2 Example of Declaring Local Data in a Function Block

and the blocks called in turn from within these blocks.

Please note in this regard that the Editor also uses temporary local data, for example when transferring block parameters. You do not see these temporary local data at the programming interface.

Start information

The operating system of the CPU transfers start information in the temporary local data when an organization block is called. This start information is 20 bytes long in each organization block and has an almost identical structure in each block. Chapters 20 “Main Program”, 21 “Interrupt Handling”, 22 “Restart Characteristics” and 23 “Error Handling” describe the start information assignments for the individual organization blocks.

These 20 bytes of start information must always be available in every priority class used. If you program evaluation of synchronous errors (programming and access errors), you must provide an additional 20 bytes at least for the start information of these error organization blocks since these error OBs are processed in the same priority class.

You declare the start information when programming an organization block. This is mandatory. The standard library *Standard Library* contains templates for declaration in English under *Organization Blocks*. If you do not require the start information, it is sufficient to declare the first 20 bytes as, for example, a field (as shown in Figure 18.3).

Absolute addressing of temporary local data

Normally, you access the temporary local data via symbolic addressing, with absolute addressing being the exception. If you are familiar with data storage in the L stack, you can work out for yourself the addresses at which the static local are located. You can also see the addresses in the variable declaration table of the compiled block.

The address identifier for the temporary local data is L; a bit is addressed with L, a byte with LB, a word with LW and a doubleword with LD.

Example: For absolute addressing, you want to keep 16 bytes of temporary local data whose individual values you then want to access both as byte and as bit. Create this area as a field right at the start of the temporary local data so that the addressing starts at 0. In an organization block, you would locate this field declaration immediately following the declaration of the start information, so that in this case, addressing begins at 20.

Note: Absolute addressing of temporary local data is only possible in the basic languages STL, LAD and FBD. With SCL, you can only address temporary local data symbolically.

Chapter 26 “Direct Variable Access” describes how you learn the address of a variable in the temporary local data at runtime.

Data type ANY

A variable in the temporary local data can be declared – as an exception – with data type ANY.

With STL, you can thus generate an ANY pointer that can be changed at runtime. See Chapter 26.3.3 ““Variable” ANY Pointer” for more details.

Incremental programming

Address	Declaration	Name	Type
0.0	TEMP	SINFO	ARRAY [1..20]
*1.0	TEMP		BYTE
20.0	TEMP	LByte	ARRAY [1..16]
*1.0	TEMP		BYTE

Source-oriented programming

```
VAR_TEMP
 SINFO : ARRAY [1..20] OF BYTE;
 LByte : ARRAY [1..16] OF BYTE;
END_VAR
```

Figure 18.3 Example of Declaration of Temporary Local Data in an Organization Block

With SCL, you can assign the address of another (complex) variable to a temporary ANY variable at runtime. For more details, see Chapter 29.2.4 “Temporary Local Data”.

18.1.6 Static Local Data

Static local data are addresses that a function block stores in its instance data block.

The static local data are the “memory” of a function block. They retain their value until this is changed by the program, just like data addresses in global data blocks.

The volume of static local data is limited by the data type of the variables and by the CPU-specific length of a data block.

Declaration of static local data

You declare the static local data in the declaration section of the function block:

- ▷ with incremental programming under “stat” or
- ▷ with source-oriented programming between VAR and END_VAR.

Figure 18.2 in Chapter 18.1.5 “Temporary Local Data” gives an example of variable declaration in a function block. The block parameters are declared first, followed by the static local data and finally the temporary local data.

The static local are stored in the instance data block after the block parameters in order of declaration and according to data type. Chapter 26.2 “Data Storage of Variables” contains more detailed information on data storage in data blocks.

Symbolic addressing of static local data

You access the static local data with their symbolic names. You assign the names in accordance with the rules for block-local symbols.

You can access static local data with all operations that can also be used in conjunction with data addresses in global data blocks.

Example: The function block “Totalizer” adds an input value to a value stored in the static local data and then stores the total in the static local again. At the next call, the input value is added to this total again, and so on (Figure 18.4 top).

Total is a variable in the data block “TotalizerData”, the instance data block of the function block “Totalizer” (you can define the names of all blocks yourself in the symbol table within the permissible framework). The instance data block has the data structure of the function block; in the example, it contains two INT variables with the names *In* and *Total.*

Accessing static local data from outside the function block

The static local data are usually only processed in the function block itself. However, they are stored in a data block, you can access the static local data at any time in the same way as you access variables in a global data block with

“DataBlockName”.AddressName.

In our little example, the data block is called *“TotalizerData”* and the data address *Total.* An access could take the following form:

```
L    "TotalizerData".Total;
T    MW 20;
L    0;
T    "TotalizerData".Total;
```

Local instances

When you call a function block, you normally specify the instance data block provided for the call. The function block then stores its block parameters and its static local data in the instance data block.

From STEP 7 V2, you can generate “multi instances”, that is, you can call a function block in another function block. The static local data (and the block parameters) of the called function block are then a subset of the static local data of the calling block. A requirement for this is that both the calling and the called function block have block version 2, that is, they have “multi-instance capability”. In this way, you can “nest” up to eight function block calls.

Example (Figure 18.4 bottom): In the static local data of the function block “Evaluation”, you declare a variable *Memory* that corresponds to the function block “Totalizer” and has the same structure. Now you can call the function block “Totalizer” via the variable *Memory,* without,

FB "Totalizer"

Address	Declaration	Name	Type
+ 0.0	IN	In	INT
+ 2.0	STAT	Total	INT

```
        L       #In;
        L       #Total;
        +I      ;
        T       #Total;
```

DB "TotalizerData"

Address	Declaration	Name	Type
+ 0.0	IN	In	INT
+ 2.0	STAT	Total	INT

FB "Evaluation"

Address	Declaration	Name	Type
0.0	IN	Add	BOOL
0.1	IN	Delete	BOOL
2.0	STAT	EM_Add	BOOL
2.1	STAT	EM_Del	BOOL
4.0	STAT	Memory	Totalizer

```
        A       #Add;
        FP      #EM_Add;
        JCN     M1;
        CALL    #Memory
           (In := "Value2");

M1:     A       #Delete;
        FP      #EM_Del;
        JCN     End;
        L       #Memory.Total;
        T       "Result";
        L       0;
        T       #Memory.Total;
```

DB "EvaluationData"

Address	Declaration	Name	Type
0.0	IN	Add	BOOL
0.1	IN	Delete	BOOL
2.0	STAT	EM_Add	BOOL
2.1	STAT	EM_Del	BOOL
4.0	STAT:IN	Memory.In	INT
6.0	STAT	Memory.Total	INT

In the ***Data view****, the data block shows all individual variables so that the variables of a local instance appear with their full names.*

Simultaneously, you see the corresponding absolute addresses.

Figure 18.4 Example of Static Local Data and Local Instances

however, specifying a data block because the data for *Memory* are located 'block-local' in the static local data (*Memory* is the local instance of the FB "Totalizer").

You access the static local data of *Memory* in the program of the function block "Evaluation" in the same way as you access structure components by specifying the structure name (*Memory*) and the component name (*Total*).

The instance data block "EvaluationData" therefore contains the variables *Memory.In* and *Memory.Total*, that you can also access as global variables, for example as *"EvaluationData".Memory.Total*.

This example of the use of a local instance in function blocks FB 6, 7 and 8 in the "Program Flow Control" program can be found in the download files (download address: see pages 8-9). The example in Chapter 19.5.3 "Feed Example" contains further applications of local instances.

Absolute addressing of static local data

Normally you access static local data using symbolic addresses with absolute addressing being the exception. Within a function block, the instance data block is opened via the DI register. Addresses in this data block, static local data as well as block parameters, therefore have the address identifier DI. You address a bit with DIX, a byte with DIB, a word with DIW and a doubleword with DID.

If you are familiar with storing data in a data block, you can work out yourself the addresses at which the static local data are stored. You can also see the addresses in the variable declaration table of the compiled block. But a word of caution! *These addresses are relative to the start of the instance.* They are only valid if you call the function block with a data block. If you call the function block as a local instance, the local data of the local instance are located in the middle of the instance data block of the calling function block. You can see the absolute addresses in, for example, the compiled instance data block which contains all local instances. Select VIEW → DATA VIEW, if you want to read the addresses of individual local data addresses.

If we consider our example, we could access the variable *Total* in the function block "Totalizer" with DIW 2 if the FB "Totalizer" is called with a data block (cf. the address assignment in the DB "TotalizerData"), and with DIW 6, if the FB "Totalizer" is called as a local instance in the FB "Evaluate" (cf. the address assignment in the DB "EvaluateData").

However, if we program a function block without knowing whether it is called with a data block or as a local instance, that is, one that is to be "multi-instance-capable", how can we then assign absolute addresses to the static local data? Put briefly, we add to the address of the variable the offset of the local instance from address register AR2. See Chapter 25 "Indirect Addressing" and Chapter 26 "Direct Variable Access" for more detailed information.

Note: Absolute addressing of static local data is only possible in the basic languages STL, LAD and FBD. With SCL, you can only address static local data symbolically.

18.2 Block Functions for Data Blocks

You store your program data in the data blocks. In principle, you can also use the bit memory area for storing data; however, with the data blocks, you have significantly more possibilities with regard to data volume, data structuring and data types. This chapter shows you

- ▷ how to work with data addresses,
- ▷ how to call data blocks and
- ▷ how to create, delete and test data blocks at runtime.

You can use data blocks in two versions: as *global data blocks*, that are not assigned to any code block, and as *instance data blocks*, that are assigned to a function block. The data in the global data blocks are, in a manner of speaking, "free" data that every code block can make use of. You yourself determine their volume and structure direct through programming the global data block. An instance data block contains only the data with which the associated function block works; this function block then also determines the structure and storage location of the data in 'its' instance data block.

The number and length of data blocks are CPU-specific. The numbering of the data block begins at 1; there is no data block DB 0. You can use each data block either as a global data block or as an instance data block.

You must first create ("set up") the data blocks you use in your program, either by programming, such as code blocks, or at runtime using the system function SFC 22 CREAT_DB.

Data blocks must be stored in work memory so that they can be read and written to from the user program. You can also leave data blocks in load memory by using the block attribute "Unlinked" (keyword UNLINKED in source-oriented input).

Such data blocks do not occupy space in work memory, but do have an increased access time. This procedure is suitable for data blocks with parameterization data or recipe data that are required relatively rarely for controlling the plant or the process. The SFCs 20 BLKMOV and SFC 83 READ_DBL read data from the load memory, SFC 84 WRIT_DBL writes data to the load memory.

If you set the attribute *The data block is write-protected in the programmable controller* in the block properties (corresponds to the keyword READ_ONLY in source-oriented input) in the work memory, you can then only read from this DB.

18.2.1 Two Data Block Registers

The CPU provides two data block registers for processing data addresses. These registers contain the numbers of the current data blocks; these are the data blocks with whose addresses processing is currently taking place. Before accessing a data block address, you must open the data block containing the address. If you use

Table 18.2 Data Addresses

Data address	located in a data block opened via the	
	DB register	DI register
Data bit	DBX y.x	DIX y.x
Data byte	DBB y	DIB y
Data word	DBW y	DIW y
Data doubleword	DBD y	DID y

x = Bit address, y = Byte address

fully-addressed access to data addresses (with specification of the data block, see below), you need not be concerned with opening the data blocks and with the assignments of the data block register. The Editor generates the necessary instructions from your specifications.

The Editor uses the first data block register preferably for accessing global data blocks and the second data block register for accessing instance data blocks. For this reason, these registers are given the names "Global data block register" (DB register) and "Instance Data Block Register" (DI register). The handling of the registers by the CPU is absolutely identical. Each data block can be opened via one of the two registers (or also via both simultaneously).

When you load a data word, you must specify which of the two possible open data blocks contains the data word. If the data block has been opened via the DB register, the data word is called DBW; if the data word is in the data block opened via the DI register, it is called DIW. The other data addresses are named accordingly (Table 18.2).

18.2.2 Accessing Data Addresses

You can use the following methods for accessing data addresses:

▷ Symbolic addressing with full addressing,

▷ Absolute addressing with full addressing and

▷ Absolute addressing with part addressing.

See Chapter 25 "Indirect Addressing" for further addressing methods.

Symbolic access to the data addresses in global data blocks requires the minimum system knowledge. For absolute access or for using both data block registers, you must observe the notes described below.

Symbolic addressing of data addresses

I recommend you use symbolic addressing of data addresses as far as possible. Symbolic addressing

▷ makes it easier to read and understand the program (if meaningful terms are used as symbols),

▷ reduces write errors in programming (the Editor compares the terms used in the symbol table and in the program; "number switching errors" such as DBB 156 and DBB 165 that can occur when using absolute addresses, cannot occur here) and

▷ does not require programming knowledge at the machine code level (which data block has the CPU opened currently?).

Symbolic addressing uses fully-addressed access (data block together with data address), so that the data address always has a unique address.

You determine the symbolic address of a data address in two steps:

1) Assignment of the data block in the symbol table
Data blocks are global data that have unique addresses within a program. In the symbol table, you assign a symbol (e.g. Motor1) to the absolute address of the data block (e.g. DB 51).

2) Assignment of the data addresses in the data block
You define the names of the data addresses (and the data type) during programming of the data block. The name applies only in the associated block (it is "block-local"). You can also use the same name in another block for another variable.

Fully-addressed access to data addresses

In the case of fully-addressed access, you specify the data block together with the data address. This method of addressing can be symbolic or absolute:

```
L     MOTOR1.ACTVAL;
L     DB 51.DBW 20;
```

MOTOR1 is the symbolic address that you have assigned to a data block in the symbol table. ACTVAL is the data address you defined when programming the data block. The symbolic name MOTOR1.ACTVAL is just as unique a specification of the data address as the specification DB 51.DBW 20.

Fully-addressed data access is only possible in conjunction with the global data block register (DB register). With fully-addressed data addresses, the Editor executes two statements: First, the data block is opened via the DB register and this is then followed by access to the data addresses.

You can use fully-addressed access with all operations permissible for the data type of the addressed data address. These are the bit logic operations, the memory functions for binary addresses and the load and transfer functions for digital addresses. You can also specify fully-addressed data addresses at the block parameters, for example (strongly recommended, see Chapter 19 "Block Parameters").

Absolute addressing of data addresses

For absolute addressing of data addresses, you must know the addresses at which the Editor places the data addresses when setting up. You can find out the addresses by outputting them after programming and compiling the data

Table 18.3 Operations with Data Blocks

Statement		Meaning
A	-	Check for signal state "1" and combine according to logic AND of a
O	-	Check for signal state "1" and combine according to logic OR of a
X	-	Check for signal state "1" and combine according to logic exclusive OR of a
AN	-	Check for signal state "0" and combine according to logic AND of a
ON	-	Check for signal state "0" and combine according to logic OR of a
XN	-	Check for signal state "0" and combine according to logic exclusive OR of a
=	-	Assignment to a
S	-	Set a
R	-	Reset a
FP	-	Edge evaluation for positive edge with a
FN	-	Edge evaluation for negative edge with a
	DBX y.x	Data bit via the DB register
	DIX y.x	Data bit via the DI register
	DBz.DBX y.x	Fully-addressed data bit
L	-	Load a
T	-	Transfer a
	DBB y	Data byte via the DB register
	DBW y	Data word via the DB register
	DBD y	Data double word via the DB register
	DIB y	Data byte via the DI register
	DIW y	Data word via the DI register
	DID y	Data doubleword via the DI register
	DBz.DBB y	Fully-addressed data byte
	DBz.DBW y	Fully-addressed data word
	DBz.DBD y	Fully-addressed data doubleword

x = Bit address, y = Byte address, z = Number of the data block

block. You will then see from the address column the absolute address at which the relevant variable begins.

This procedure is suitable for all data blocks, both those you use as global data blocks as well as those you use as instance data blocks. In this way, you can also see where the Editor stores the block parameters and the static local data in the case of function blocks.

If you want to calculate the address, Chapter 26.2 "Data Storage of Variables" provides valuable information.

Data addresses are addressed bytewise like the bit memory, for example; they are also used in conjunction with the same operations (Table 18.3) and are executed in exactly the same way.

If you intend to assign exclusively absolute addresses to the addresses of a data block, reserve the required quantity of bytes via a field declaration.

18.2.3 Open Data Block

OPN	DB x	Open a data block via the DB register with absolute address
OPN	DI x	Open a data block via the DI register with absolute address
OPN	"name"	Open a data block via the DB register with symbolic address
OPN	#name	Open a data block via the DB register with a block parameter

Data blocks are opened regardless of any conditions. Opening does not affect the RLO and the contents of the accumulators; the nesting depth of the block calls does not change.

The opened data block must be in work memory.

Example: The value of data word DBW 10 from data block DB 12 is to be transferred to data word DBW 12 of data block DB 13 (Figure 18.5 left). The values in the data words DBW 14 from data blocks DB 12 and DB 13 should be added; the result should be saved in data word DBW 14 of data block DB 14.

You can program this example in two ways: with part addressing and with full addressing (Figure 18.5 on the right).

When a data block is opened it remains "valid" until another data block is opened. Under certain circumstances – not visible to you– this can be done by the Editor (see "Special Points in Data Addressing" below). For example, a block call with CALL in conjunction with parameter transfer can change the contents of the data block registers.

With a block change using UC or CC, the contents of the data block registers are retained. On returning to the calling block, the block end statement restores the old contents of the registers.

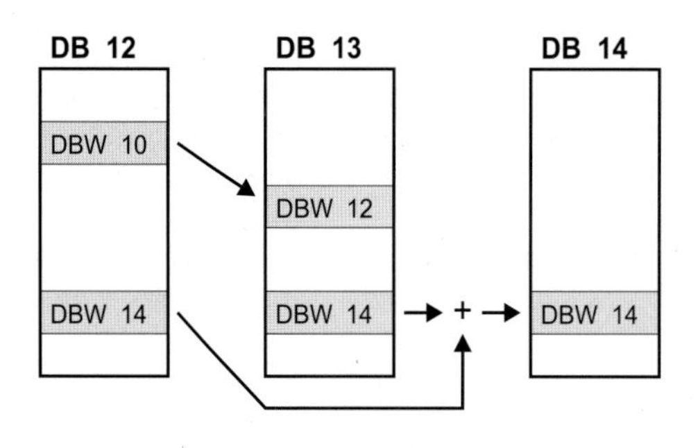

Programming with partial addressing	Programming with complete addressing
OPN DB 12; L DBW 10;	L DB 12.DBW 10;
OPN DB 13; T DBW 12;	T DB 13.DBW 12;
OPN DB 12; L DBW 14;	L DB 12.DBW 14;
OPN DB 13; L DBW 14;	L DB 13.DBW 14;
+I ;	+I ;
OPN DB 14; T DBW 14;	T DB 14.DBW 14;

Figure 18.5 Opening data blocks (example)

18.2.4 Exchanging the Data Block Registers

CDB Exchange data block registers

The statement CDB exchanges the contents of the data block registers. It is executed regardless of conditions and does not affect either the status bits or the other registers.

Example: With the statement CDB, you can take a "detour" via the DB register to open, via the DI register, a data block transferred as a block parameter (not possible direct).

```
CDB   ;
OPN   #Data2;
CDB  ;
```

With CDB, you transfer the contents of the DB register temporarily to the DI register. Then you open, via the block parameter *#Data2*, the data block transferred as an actual parameter; that is, you write its number into the DB register. After renewed exchange, the old value is again in the DB register and the DI register contains the number of the parameterized data block.

18.2.5 Data Block Length and Number

L DBLG Load the length of the data block opened via the DB register

L DBNO Load the number of the data block opened via the DB register

L DILG Load the length of the data block opened via the DI register

L DINO Load the number of the data block opened via the DI register

The statement L DBLG loads the length of the data block that was opened via the DB register into accumulator 1. The length is the same as the quantity of data bytes. The statement L DILG is the same for the DI register.

The statement L DBNO loads the number of the data block that was opened via the DB register into accumulator 1. L DINO shows you the number of the current data block that was opened via the DI register.

These statements transfer the previous contents of accumulator 1 into accumulator 2 in accordance with a "normal" load function. If no data block has been opened via the relevant register, zero is loaded both as the length and as the number.

It is not possible to write a number back to a data block register direct; you can only influence the data block registers via OPN DB or OPN DI and CDB (exchange data block registers).

18.2.6 Special Points in Data Addressing

Changing the assignments in the DB register

With the following functions, the Editor generates additional statements that can affect the contents of one of the two data block registers:

Full addressing of data addresses

Each time data addresses are addressed fully, the Program Editor first opens the data block with the statement OPN DB, and then accesses the data addresses. The DB register is overwritten each time here. This applies also when initializing block parameters with fully-addressed data addresses.

Access to block parameters

Access to the following block parameters changes the contents of the DB register: With functions, all block parameters of complex data type and with function blocks, in/out parameters of complex data type.

Block calls CALL FB and CALL SFB

Prior to the actual block call, CALL FB and CALL SFB store the number of the current instance data block in the DB register (by exchanging the data block registers) and open the instance data block for the called function block. In this way, the associated instance data block is always open in a called function block. Following the actual block call, CALL FB and CALL SFB exchange the data block registers again, so that the current instance data block is once again available in the calling function block. In this way, CALL FB and CALL SFB change the contents of the DB register.

DI register in function blocks

In function blocks the DI register is permanently assigned the number of the current instance data block. All accesses to block parameters or static local data are made via the DI register

```
VAR_TEMP
  ZW_DB : WORD;             //Intermediate buffer for global data block
  ZW_DI : WORD;             //Intermediate buffer for instance data block
END_VAR
//Save data block registers
  L    DBNO;                //Buffer global data block number
  T    ZW_DB;
  L    DINO;                //Buffer instance data block number
  T    ZW_DI;
//Working with part-addressed data addresses
//when using both data block registers
  OPN  DB 12;               //Open data block DB 12 via the DB register
  OPN  DI 13;               //Open data block DB 13 via the DI register
  L    DBW 16;              //##########
  T    DIW 28;              //# Be careful with symbolic addressing in this
  L    DID 30;              //# program section, e.g. when using block
  L    DBD 30;              //# parameters, block-local variables and
  +R   ;                    //# fully-addressed data addresses
  T    DID 26;              //##########
//Restore data block registers
  OPN  DB[ZW_DB];           //Open original global data block
  OPN  DI[ZW_DI]; //Open original instance data block
```

Figure 18.6 Example of the Direct Use of Both Data Block Registers

and, incidentally, also via the address register AR2 in the case of "multi-instance-capable" function blocks. Please note this permanent assignment if you change the contents of the DI register with CDB or OPN DI.

If, for example, you want to use both data block registers simultaneously for data exchange, you must first save the register contents and then restore them. The example shown in Figure 18.6 describes a relevant method.

Making changes to data block assignments at a later stage

On the "Blocks" tab in the properties window of the offline object container *Blocks* you can specify whether the absolute address or the symbol is to take precedence when a change is made to the data block assignment for the already saved code blocks when they are displayed and saved again.

The default is "Absolute value has priority" (the same characteristic as in the previous STEP 7 versions). This default means that when a change is made in the declaration, the absolute address is retained in the program and the symbol changes accordingly. With the setting "Symbol has priority", the absolute address changes and the symbol is retained.

Example: In the data block DB 1, data word DBW 10 is assigned the symbol *Actual_value*. In the program, you load this data word with, for example:

```
L "Data".Actual_value   DB1.DBW 10
```

if "Data" is the symbol for data block DB 1. If you now add an additional data word with the symbol *MaxCurrent* immediately in front of data word DBW 10, the program will then contain the following when the code block is next opened (and saved):

if "Absolute value has priority":

```
L "Data".MaxCurrent    DB1.DBW 10
```

if "Symbolic has priority":

```
L "Data".Actual_value  DB1.DBW 12
```

For accessing data addresses in global data blocks, the same thing applies as for accessing global addresses (e.g. inputs) for which a symbol is assigned in the symbol table. Chapter 2.5.6 "Address Priority" contains detailed information on this topic.

18.3 System Functions for Data Blocks

There are three system functions for handling data blocks. Their parameters are described in Table 18.4.

▷ SFC 22 CREAT_DB
Create data block in the work memory

▷ SFC 85 CREA_DB
Create data block in the work memory

▷ SFC 82 CREA_DBL
Create data block in the load memory

▷ SFC 23 DEL_DB
Delete data block

▷ SFC 24 TEST_DB
Test data block

18.3.1 Creating a Data Block in the Work Memory

System functions SFC 22 CREAT_DB and SFC 85 CREA_DB create a data block in the work memory. As the data block number, the system function takes the lowest free number in the number band given by the input parameters LOW_LIMIT and UP_LIMIT. The numbers specified at these parameters are included in the number band. If both values are the same, the data block is generated with this number. You cannot assign the number of a data block already present in the user program to another block, not even if the data block is only present in the load memory.

The output parameter DB_NUMBER supplies the number of the actually created data block. With the input parameter COUNT, you specify the length of the data block to be created. The length corresponds to the number of data bytes and must be an even number.

Creating the data block is not the same as calling it. The current data block is still valid. A data block created with the system function contains random data. For meaningful use, data must first be written to a data block created in this way before the data can be read.

The data blocks created with the SFCs 22 CREAT_DB and 85 CREA_DB are only present in the work memory. If a CPU differentiates between retentive and non-retentive work memory, the SFC 22 CREAT_DB generates a retentive data block and the SFC 85 CREA_DB a data block as specified by the parameter ATTRIB. "Retentive" data block means that its contents are retained following a cold restart/hot restart (see Chapter 22.2.4 "Retentivity").

The system function SFC 85 CREA_DB replaces the SFC 22 CREAT_DB.

A data block created with the SFC 22 CREAT_DB and SFC 85 CREA_DB does not change the checksum of the user program, not even if it is written or deleted again. If a data block created with an SFC is imported into the offline data management, this influences the checksum.

In the event of an error, no data block is created, the parameter DB_NUMBER is assigned zero, and an error number is signaled via RET_VAL.

18.3.2 Creating a Data Block in the Load Memory

The system function SFC 82 CREA_DBL creates a data block in the load memory, and also in the work memory if applicable. The system function assigns the lowest free number of the number band given by the input parameters LOW_LIMIT and UP_LIMIT to the data block. The numbers specified at these parameters are included in the number band. If both values are the same, the data block is generated with exactly this number. You cannot reassign a number of a data block already present in the user program, not even if the data block is only present in the work memory.

The output parameter DB_NUMBER supplies the number of the actually created data block. With the input parameter COUNT, you specify the length of the data block to be created. The length corresponds to the number of data bytes and must be an even number.

The created data block is preassigned the data area specified at the input parameter SRCBLK. Here you can specify a complete data block, e.g. DB 160 or "Archive 1", a variable from a data block, or an absolute addressed data area as the ANY pointer,

e.g. P#DB160.DBX16.0 BYTE 64

Table 18.4 SFCs for Handling Data Blocks

SFC	Name	Declaration	Data Type	Assignment, Description
22	LOW_LIMIT	INPUT	WORD	Lowest number of the data block to be created
	UP_LIMIT	INPUT	WORD	Highest number of the data block to be created
	COUNT	INPUT	WORD	Length of the data block in bytes (even number)
	RET_VAL	RETURN	INT	Error information
	DB_NUMBER	OUTPUT	WORD	Number of the created data block
85	LOW_LIMIT	INPUT	WORD	Lowest number of the data block to be created
	UP_LIMIT	INPUT	WORD	Highest number of the data block to be created
	COUNT	INPUT	WORD	Length of the data block in bytes (even number)
	ATTRIB	INPUT	BYTE	Block attributes: B#16#00 Retain B#16#04 Non_Retain
	RET_VAL	RETURN	INT	Error information
	DB_NUMBER	OUTPUT	WORD	Number of the created data block
82	REQ	INPUT	BOOL	Trigger for creating with signal status "1"
	LOW_LIMIT	INPUT	WORD	Lowest number of the data block to be created
	UP_LIMIT	INPUT	WORD	Highest number of the data block to be created
	COUNT	INPUT	WORD	Length of the data block in bytes (even number)
	ATTRIB	INPUT	BYTE	Properties of created data block Bit 0: UNLINKED "1" = the DB is only in the load memory. Bit 1: READ_ONLY "1" = the DB is write-protected. Bit 2: NON_RETAIN "1" = the DB is non-retentive. Bits 3 to 7: reserved
	SRCBLK	INPUT	ANY	Data area in work memory with which the created data block is initialized
	RET_VAL	RETURN	INT	Error information
	BUSY	OUTPUT	BOOL	With TRUE, creation has not yet been completed
	DB_NUM	OUTPUT	WORD	Number of the created data block
23	DB_NUMBER	INPUT	WORD	Number of the data block to be deleted
	RET_VAL	RETURN	INT	Error information
24	DB_NUMBER	INPUT	WORD	Number of the data block to be tested
	RET_VAL	RETURN	INT	Error information
	DB_LENGTH	OUTPUT	WORD	Length of the data block (in bytes)
	WRITE_PROT	OUTPUT	BOOL	TRUE = write-protected

The source must be a data area in the work memory.

If the source area is smaller than the target area, the former is written completely into the latter. The remaining bytes of the target area are filled with zeros. If the source area is larger than the target area, the latter is written completely; the remaining bytes of the source area are ignored.

You can assign the generated data block the following properties using the input parameter ATTRIB:

- Bit 0 = "1"
  The data block has the property *Unlinked.* Following transfer to the offline data management and loading back into the CPU, the data block is again only present in the load memory. If the bit has the signal status "0", the data block is created in both the load and work memories.

- Bit 1 = "1"
  The data block has the property *DB is write-protected in the AS.* You can only read the values of this data block.

- Bit 2 = "1"
  The data block has the property *Non_Retain.*

The remaining bits are not occupied at the moment. You can find further information on the block properties in Chapter 3.2.3 "Block Properties".

The system function SFC 82 CREA_DBL operates asynchronously: you trigger the create operation with signal status "1" at the input parameter REQ. You can only access the read and written data areas again if the parameter BUSY has the signal status "0" again.

Creating does not call the associated data block. The current data block is still valid.

A data block is not generated in the event of an error, the output parameters are undefined, and and an error message is output by the function value.

18.3.3 Deleting a Data Block

System function SFC 23 DEL_DB deletes the data block in RAM (work and load memory) whose number is specified at the input parameter DB_NUMBER. The data block must not be opened when this is done, otherwise the operating system of the CPU calls the organization block OB 121. If this is not present, the CPU changes to the STOP state.

If the load memory is a flash EPROM memory card, the deleted data block present on it is declared to be invalid, and is then practically no longer existent for the user program. Following a cold restart or an unbuffered power on/off, the data block is loaded from the load memory into the work memory and is then present again. If the load memory is a RAM memory card or a micro memory card, the data block is really deleted.

If the deleted data block has been created by a system function during runtime, the checksum of the user program is not changed by the deletion. The deletion of programmed (loaded) data blocks changes the checksum.

In the event of an error, the data block is not deleted and an error number is signaled in the function value.

18.3.4 Testing a Data Block

System function SFC 24 TEST_DB supplies information on a data block whose number you specify at the input parameter DB_NUMBER. The output parameter DB_LENGTH indicates the number of existing bytes, and the output parameter WRITE_PROT indicates whether the data block is write-protected.

If the tested data block is only present in the load memory, this is indicated as an error by RET_VAL; the parameters DB_LENGTH and WRITE_PROT are nevertheless correctly assigned.

If the specified data block is not present in the user memory of the CPU, RET_VAL = W#16#80B1 is returned.

18.4 Null Operations

Null operations have no effect when processed by the CPU. STL has NOP 0, NOP 1 and BLD statements as null operations.

18.4.1 NOP Statements

You can use the statements NOP 0 (bit pattern 16x "0") and NOP 1 (bit pattern 16x "1") to enter a statement that has no effect. Please note that null operations occupy memory space (2 bytes) and have an instruction execution time.

Example: There must always be a statement at a jump label. If you want to have a jump in your program but do not want anything further to be executed, use NOP 0.

```
        A         I 1.0
        JC        MXX1
        ...
MXX1:   NOP 0
        ...
```

You can enter an empty line for clarity more effectively by simply entering an (empty) line comment (this does not require user memory space and involves no loss of execution time since no code is entered).

18.4.2 Program Display Statements

The Editor uses the program display instruction BLD *nnn* to incorporate decompiling information into the program.

19 Block Parameters

This chapter describes how to use block parameters. You will learn

- ▷ how to declare block parameters,
- ▷ how to work with block parameters,
- ▷ how to initialize block parameters and
- ▷ how to "pass on" block parameters.

Block parameters represent the transfer interface between the calling and the called block. All functions and function blocks can be provided with block parameters.

19.1 Block Parameters in General

19.1.1 Defining the Block Parameters

You enable parameterization of the processing instruction (the block function) present in a block by means of block parameters. Example: You want to write a block as an adder that you want to repeatedly use in your program with different tags. You transfer the tags as block parameters – in our example, three input parameters and one output parameter (Figure 19.1). Since the adder does not have to save values internally, a function is suitable as the block type.

You define a block parameter as an *input parameter* if you only check or load its value in the block program. If you only write a block parameter (assign, set, reset, transfer), you use an *output parameter*. You must always use an *in/out parameter* if a block parameter is to be both checked and written. The Editor does not check the use of the block parameters.

19.1.2 Processing the Block Parameters

In the adder program, the names of the block parameters stand as place holders for the later current variables. You use the block parameters

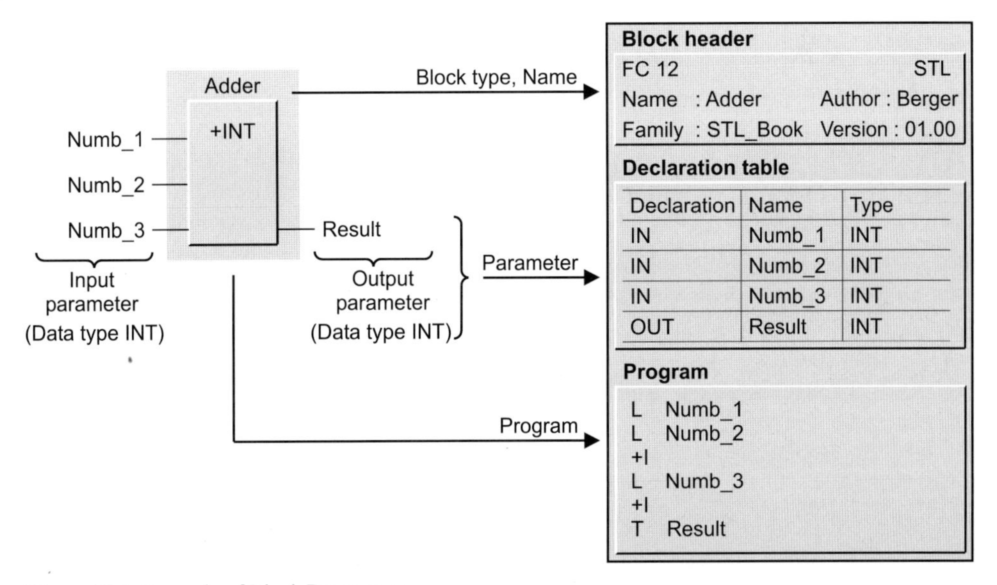

Figure 19.1 Example of Block Parameters

in the same way as symbolically addressed variables; in the program, they are called *formal parameters*.

You can call the "Adder" function several times in your program. With each call, you transfer other values to the adder at the block parameters (Figure 19.2). The values can be constants, addresses or variables; they are called *Actual parameters*.

At runtime, the CPU replaces the formal parameters with the actual parameters. The first call in the example adds the contents of memory words MW 30, MW 32 and MW 34 and stores the result in memory word MW 40. The same block with the actual parameters of the second call adds data words DBW 30, DBW 32 and DBW 34 of data block DB 10 and stores the result in data word DBW 40 of data block DB 10.

19.1.3 Declaration of the Block Parameters

You define the block parameters in the declaration section of the block when you program the block. With incremental input, you complete a list and with source-oriented input you define the block parameters in specific sections (Figure 19.3). The keyword is VAR_INPUT for input parameters, VAR_OUTPUT for output parameters and VAR_IN_OUT for in/out parameters.

The pre-assignment is optional and only makes sense with function blocks if the block parameter is stored as a value. This applies to all block parameters of elementary data type and to input and output parameters of complex data type. Specification of a parameter comment is optional and always possible.

The *Block parameter name* can be up to 24 characters in length. It must consist only of alphanumeric characters (without national characters such as the German Umlaut) and the underscore. A distinction is made between upper and lower case. The name must not be a keyword.

No distinction is made between upper and lower case when entering a block parameter name. At output, the editor uses the case established when the block parameter name was declared.

For the *Data type* of a block parameter all elementary, complex and user-defined data types are permissible. In addition, you can use the parameter types with block parameters.

STEP 7 stores the names of the block parameters in the non-execution-relevant section of the

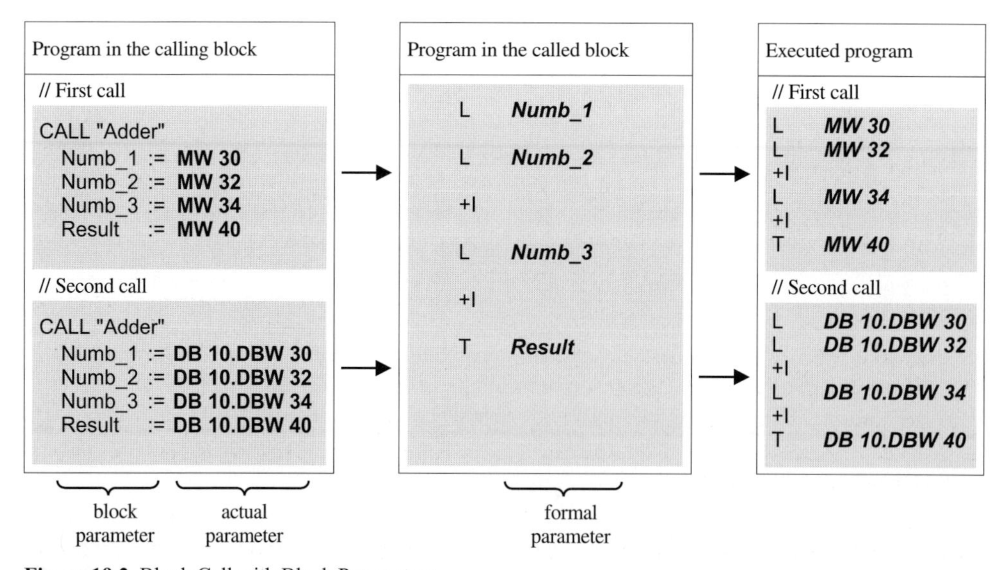

Figure 19.2 Block Call with Block Parameters

Incremental programming

Address	Declaration	Name	Type	Initial value	Comment
0.0	IN	Manual	BOOL	TRUE	Manual control
2.0	IN	Setpoint	INT	10000	Speed for Motor1
4.0	IN	Characteristic	ANY		Pointer to data area
14.0	OUT	Actual_value	INT	0	Speed of Motor1
16.0	OUT	Temperature	REAL	0.000000e+00	Temperature of Motor1
20.0	OUT	Message	WORD	W#16#0	Fault message
22.0	IN_OUT	EM	ARRAY[1..16]		Edge memory bit
*0.1	IN_OUT		BOOL		
28.0	IN_OUT	Interface	DWORD	DW#16#0	Interface to Motor2

Source-oriented programming

```
VAR_INPUT
  Manual          : BOOL   := TRUE;             //Manual control
  Setpoint        : INT    := 10_000;           //Speed for Motor1
  Characteristic  : ANY;                        //Pointer to data area
END_VAR
```

```
VAR_OUTPUT
  Actual_value    : INT    := 0;                //Speed of Motor1
  Temperature     : REAL   := 0.0;              //Temperature of Motor1
  Message         : WORD   := 16#0000;          //Fault message
END_VAR
```

```
VAR_IN_OUT
  EM              : ARRAY[1..8] OF BOOL;        //Edge memory bit
  Interface       : DWORD  := 16#0000_0000;     //Interface to Motor1
END_VAR
```

Figure 19.3 Examples for the Declaration of Block Parameters

blocks on the data medium of the programming device. The work memory of the CPU (in the compiled block) contains only the declaration types and the data types. For this reason, program changes made to blocks online at the CPU must always be updated on the data medium of the programming device, in order to retain the original names.

If the update is not made, and blocks are transferred from the CPU to the programming device, the non-execution-relevant block sections are overwritten or deleted. The Editor then generates replacement symbols for display or printout (IN*n* with input parameters, OUT*n* with output parameters and INOUT*n* with in/out parameters, with *n* beginning at 0).

19.1.4 Declaration of the Function Value

The function value in the case of functions is a specially treated output parameter. It has the name RET_VAL (or ret_val) and is defined as the first output parameter.

All elementary data types are permissible as the data type of the function value and, in addition, the data types DATE_AND_TIME, STRING, POINTER, ANY and user-defined data types UDT are also permissible. The data types ARRAY and STRUCT are not permissible.

The above-named example of the adder can also be programmed with the total as the function value.

Source-file-oriented programming

In source-file-oriented programming, you declare the function value by specifying the data type of the function value after the block type and separated from this by a colon.

```
FUNCTION FC 12 : INT
VAR_INPUT
  Numb_1 : INT;
  Numb_2 : INT;
  Numb_3 : INT;
END_VAR
BEGIN
  L    Numb_1;
  L    Numb_2;
  +I   ;
  L    Numb_3;
  +I   ;
  T    RET_VAL;
END_FUNCTION
```

In the example, the function value is of data type INT. With T RET_VAL, the function value is assigned the total from *Numb_1*, *Numb_2* and *Numb_3*.

Incremental programming

With incremental programming, the function value with the name RET_VAL is present under the declaration RETURN. Here you define the data type of the function value and enter a comment.

In the program, you treat the function value like an output parameter. In the example, you assign the total of *Numb_1*, *Numb_2* and *Numb_3* to the function value with the operation T RET_VAL.

19.1.5 Initializing Block Parameters

When calling a block, you initialize the block parameters with actual parameters. These can be constants, absolute addresses, fully-addressed data addresses or symbolically addressed variables. The actual parameter must be of the same data type as the block parameter (Chapter 19.3 "Actual Parameters").

From STEP 7 V5.1, you must specify the block parameters in the program source in exactly the order in which you defined them in the declaration of the block during programming.

You must initialize all block parameters of a function at every call. In the case of function blocks, initialization of individual or all block parameters is optional.

19.2 Formal Parameters

In this chapter, you will learn how to access the block parameters within a block. Table 19.1 shows that it is possible to access block parameters of elementary data types, components of a field or a structure, and timer and counter functions without restriction.

Access to complex data types and with parameter types POINTER and ANY is currently not supported by STL. However, you can initialize acquired blocks or system blocks that have such parameters with the relevant variables. Chapter 26 "Direct Variable Access" shows you how you can nevertheless use parameters with these data types in blocks you have written yourself.

Block parameters of data type BOOL

Block parameters of data type BOOL can be individual binary variables or binary components of fields and structures. You can check input parameters and in/out parameters with contacts or with binary box inputs, and you can influence output parameters and in/out parameters with memory functions.

With functions FCs, you must assign a value to a binary output parameter and to a function value in the block or you must set or reset it. You must not, for example, exit the block first.

Table 19.2 shows the permissible operations. When programming, you use the formal parameter in place of the block parameter xxxx.

After the CPU has used the actual parameter specified as the block parameter, it processes the statement as described in the Chapters 4 "Binary Logic Operations" and 5 "Memory Functions".

Block parameters of digital data type

Block parameters of digital data type occupy 8, 16 or 32 bits (all elementary data types except

Table 19.1 Access to Block Parameters (General)

Data Types	Permissible with			Access in the block possible with
	IN	I_O	OUT	
Elementary data types				
BOOL	x	x	x	Binary checks, memory operations
BYTE, WORD, DWORD, CHAR, INT, DINT, REAL, S5TIME, TIME, TOD, DATE	x	x	x	Load and transfer operations
Complex data types				
DT, STRING	x	x	x	Not possible direct in STL
ARRAY, STRUCT, UDT				
Individual binary components	x	x	x	Binary checks, memory operations
Individual digital components	x	x	x	Load and transfer operations
Complete variables	x	x	x	Not possible direct in STL
Parameter types				
TIMER	x	-	-	All operations for timer functions
COUNTER	x	-	-	All operations for counter functions
BLOCK_FC, BLOCK_FB	x	-	-	Calling with UC and CC[2)]
BLOCK_DB	x	-	-	Opening with OPN DB
BLOCK_SDB	x	-	-	Not possible[3)]
POINTER, ANY	x	x	x[1)]	Not possible direct in STL

1) Only with functions   2) CC not with functions   3) Only meaningful with system blocks

Table 19.2 Accessing Block Parameters of Data Type BOOL

A	-	AND logic operation with check for signal state "1"
AN	-	AND logic operation with check for signal state "0"
O	-	OR logic operation with check for signal state "1"
ON	-	OR logic operation with check for signal state "0"
X	-	Exclusive OR logic operation with check for signal state "1"
XN	-	Exclusive OR logic operation with check for signal state "0"
-	xxxx	of an input or in/out parameter of data type BOOL
-	xxxx	of an input parameter of data type TIMER
-	xxxx	of an input parameter of data type COUNTER
S	-	Set
R	-	Reset
=	-	Assignment
-	xxxx	of an output or in/out parameter of data type BOOL
FP	-	Edge evaluation positive
FN	-	Edge evaluation negative
-	xxxx	of an in/out parameter of data type BOOL

BOOL). They can be individual digital variables or digital components of fields and structures. You read input parameters and in/out parameters with the load function, and you write output parameters and input parameters with the transfer function.

With functions FCs you must transfer a value to a digital output parameter and to a function value. You must not, for example, exit the block first.

L	xxxx	Load an input or in/out parameter
T	xxxx	Transfer to an output or in/out parameter

When programming, you use the formal parameter in place of the block parameter xxxx.

After the CPU has used the actual parameter, it processes the statements as described in the Chapter 6 "Move Functions".

Block parameters of data type DT and STRING

Direct access to block parameters of data type DT and STRING is not currently possible. In function blocks, you can "pass on" parameters of data types DT and STRING to parameters of called blocks.

Chapter 26 "Direct Variable Access" shows you how to program access to parameters of a higher data type yourself.

Block parameters of data type ARRAY and STRUCT

Direct access to block parameters of data type ARRAY and STRUCT is possible on a component-wise basis, that is, you can access individual binary or digital components with the relevant operations (binary logic operations, memory functions, load and transfer functions).

Access to the complete variable (entire field or entire structure) is not currently possible and neither is access to individual components of combined or user-defined data type. In function blocks, you can 'pass on' parameters of data type ARRAY and STRUCT to parameters of called blocks".

Chapter 26 "Direct Variable Access" shows you how to program access to parameters of a higher data type yourself.

Block parameters of user-defined data type

You handle block parameters of user-defined data type in the same way as block parameters of data type STRUCT.

Direct access to block parameters of data type UDT is possible on a component-wise basis, that is, you can access individual binary or digital components with the relevant operations (binary logic operations, memory functions, load and transfer functions).

Access to the complete variable is not currently possible and neither is access to individual components of combined or user-defined data type. In function blocks, you can "pass on" parameters of data type UDT to parameters of called blocks.

Chapter 26 "Direct Variable Access" shows you how to program access to parameters of a higher data type yourself.

Block parameters of data type TIMER

In addition to the check statements listed in Table 19.2, you can program a block parameter of data type TIMER with the following statements:

SP	-	Start as pulse
SD	-	Start as ON delay
SE	-	Start as extended pulse
SS	-	Start as retentive ON delay
SF	-	Start as OFF delay
R	-	Reset
FR	-	Enable
-	xxxx	input parameter of data type TIMER

When programming, you use the formal parameter in place of the block parameter xxxx.

After using the formal parameter, the CPU processes this STL statement in exactly the same way as described in Chapter 7 "Timer Functions". When a timer is started, the time value can also be a block parameter of data type S5TIME.

Block parameters of data type COUNTER

In addition to the check statements listed in Table 19.2, you can program a block parameter of data type COUNTER with the following statements:

S - Set counter
CU - Count up
CD - Count down
R - Reset
FR - Enable

- xxxx of an input parameter of data type COUNTER

When programming, you use the formal parameter in place of the block parameter xxxx.

After using the formal parameter, the CPU processes this STL statement exactly as described in Chapter 8 "Counter Functions". When setting a counter, the count value can also be a block parameter of, for example, data type WORD.

Block parameters of data type BLOCK_xx

OPN - Open a data block (parameter type BLOCK_DB)

UC - Call a function (parameter type BLOCK_FC)

UC - Call a function block (parameter type BLOCK_FB)

CC - Conditional call of a function (parameter type BLOCK_FC)

CC - Conditional call of a function block (parameter type BLOCK_FB) (see text)

- xxxx via an input parameter

When programming, you use the formal parameter in place of the block parameter xxxx.

When opening a data block via a block parameter, the CPU always uses the global data block register (DB register).

The functions and function blocks transferred with block parameters must themselves not contain block parameters. A conditional block call via a block parameter is only possible if it is the block parameter of a function block.

As the instance data block in a function block call, you can also use a data block that you have transferred as a block parameter. Since the Editor has no means of checking the data type of the data block used at runtime, you must yourself ensure that the transferred data block is also suitable as an instance data block for the called function block.

Example: You can specify a block parameter of type BLOCK_DB with the name *#Data* as the instance data block in a function block call:

```
CALL FB 10, #Data
```

Block parameters of data type POINTER and ANY

Direct access to block parameters of data type POINTER and ANY is not possible.

Chapter 26 "Direct Variable Access" shows you how to program access to parameters of data types POINTER and ANY yourself.

19.3 Actual Parameters

When you call a block, you initialize its block parameters with constants, addresses or variables with which it is to operate. These are the actual parameters. If you call the block often in your program, you usually use different actual parameters each time it is called.

The actual parameter must agree in data type with the block parameter: You can only apply a binary actual parameter (for example, a memory bit) to a block parameter of data type BOOL; you can only initialize a block parameter of data type ARRAY with an identically dimensioned field variable. Table 19.3 gives an overview of which addresses you can use as actual parameters with which data type.

When calling functions, you must initialize all block parameters with actual parameters.

When calling function blocks, it is not necessary to initialize the block parameters. STEP 7 stores all block parameters of elementary data type, input and output parameters of complex data type and input parameters of data types TIMER, COUNTER and BLOCK_xx as a value or as a number. In/out parameters of complex data types and block parameters of data types POINTER and ANY are stored as pointers to the actual parameters. So that a meaningful value is entered here, you should initialize at

Table 19.3 Initialization with Actual Parameters

Data Type of the Block Parameter	Permissible Actual Parameters
Elementary data type	▷ Simple addresses, fully-addressed data addresses, constants ▷ Components of fields or structures of elementary data type ▷ Block parameter of the calling block ▷ Components of block parameters of the calling block of elementary data type
Complex data type	▷ Variables or block parameters of the calling block
TIMER, COUNTER and BLOCK_xx	▷ Timers, counters and blocks
POINTER	▷ Simple addresses, fully-addressed data addresses ▷ Range pointer or DB pointer
ANY	▷ Variables of any data type ▷ ANY pointer

least the last named block parameters – at least, at the first call.

You can also access the block parameters of the function block direct. Since they are located in a data block, you can handle the block parameters like data addresses.

Example: A function block with the instance data block "Lift_stat_1" controls a binary output parameter with the name *Up*. Following processing in the function block (after its call), you can check the parameter as follows, without having initialized the output parameter:

```
U     "Lift_stat_1".Up;
```

You program this check instead of initializing the parameter.

Initializing block parameters of elementary data types

The actual parameters listed in Table 19.4 are permissible as actual parameters of elementary data types.

You can assign either absolute or symbolic addresses to input, output and memory bit addresses. Input addresses should be placed only at input parameters and output addresses at output parameters (however, this is not mandatory). Memory bit addresses are suitable for all declaration types. You must apply peripheral inputs only to input parameters and peripheral outputs only to output parameters.

When you use part-addressed data addresses, you must ensure that when you access the block parameter (in the called block), the currently open data block is also the "correct" one. Since the Editor may in certain circumstances change the data block when the block is called, part addressing is not recommended for data addresses. Use only fully-addressed data addresses for this reason.

Temporary local data are usually symbolically addressed. They are located in the L stack of the calling block (and are declared in the calling block).

If the calling block is a function block, you can also use its static local data as actual parameters (see "Passing On Block Parameters" below). Static data are usually symbolically addressed. If you use absolute addressing via the DI register (DI addresses), you must ensure that when accessing the block parameter (in the called block) the data block currently opened via the DI register is also the "correct" one. Please note in this regard that when using the called block as a local instance, the absolute address of the block-local variable depends on the declaration of the local instance in the called block.

With a block parameter of data type BOOL, you can apply the constant TRUE (signal state "1") or FALSE (signal state "0"), and with block parameters of digital data type, you can apply all constants corresponding to the data type. Initialization with constants is only meaningful with input parameters.

You can also initialize a block parameter of elementary data type with components of fields

Table 19.4 Actual Parameters of Elementary Data Types

Addresses	Permissible with			Binary address or symbolic name	Digital address or symbolic name
	IN	I_O	OUT		
Inputs (process image)	x	x	x	I y.x	IB y, IW y, ID y
Outputs (process image)	x	x	x	Q y.x	QB y, QW y, QD y
Memory bits	x	x	x	M y.x	MB y, MW y, MD y
Peripheral inputs	x	-	-	-	PIB y, PIW y, PID y
Peripheral outputs	-	-	x	-	PQB y, PQW y, PQD y
Global data					
Part addressing	x	x	x	DBX y.x	DBB y, DBW y, DBD y
Full addressing	x	x	x	DB z.DBX y.x	DB z.DBB y, etc.
Temporary local data	x	x	x	L y.x	LB y, LW y, LD y
Static local data	x	x	x	DIX y.x	DIB y, DIW y, DID y
Constants	x	-	-	TRUE, FALSE	all digital constants
Components of ARRAY, STRUCT or UDT	x	x	x	Complete component name	Complete component name

x = bit number, y = byte address, z = data block number

and structures, provided such a component is of the same data type as the block parameter.

Initializing block parameters of complex data types

Every block parameter can be of the complex data type or of the user-defined data type. Variables of the same data type are permissible as actual addresses.

For initializing block parameters of data type DT or STRING, individual variables or components of fields or structures of the same data type are permissible. Initialization with constants is not possible in STL.

If you initialize a function block with a STRING variable, this variable must have the same maximum length as the STRING block parameter.

When creating the STRING variable in the temporary local data, pre-assignment is not possible, so that the STRING variable contains "random" values so to speak. If you use such a variable as an actual parameter for an IEC function, you must pre-assign "valid" values to this variable via the program (before writing to a STRING variable, the IEC function checks that the value to be written also "fits" this variable).

For initializing block parameters of data type ARRAY or STRUCT, variables with exactly the same structure as the block parameters are permissible.

Parameter assignment with complex data types is described in Chapter 26.4 "Brief Description of the Message Frame Example" in the examples "Composing the Message Frame" and "Read Time of Day".

Initializing block parameters of user-defined data type

With complex or extensive data structures, the use of user-defined data types (UDTs) is recommended. First, you define the UDT and then you use it, for example, to apply the variable in the data block or to declare the block parameter. Following this, you can use the variable when initializing the block parameter. It is also the case here, that the actual parameter (the variable) must be of the same data type (the same UDT) as the block parameter.

A complete data block with the same UDT type as the block parameter is not permissible as the actual parameter.

Parameter assignment with user-defined data types is shown in Chapter 26.4 "Brief Description of the Message Frame Example" in the example "Message Frame Data".

Initializing block parameters of type TIMER, COUNTER and BLOCK_xx

You initialize a block parameter of type TIMER with a timer function, and a block parameter of type COUNTER with a counter function. To block parameters of parameter types BLOCK_FC and BLOCK_FB, you can apply only blocks without their own parameters. These blocks are then called in the case of access with UC (and also CC in the case of function blocks). You initialize BLOCK_DBs with a data block that is opened in the called block via the DB register.

Block parameters of types TIMER, COUNTER and BLOCK_xx must only be input parameters.

Initializing block parameters of type POINTER

Pointers (constants) are permissible for block parameters of parameter type POINTER. These pointers are either range pointers (32-bit pointers) or DB pointers (48-bit pointers). The addresses are of elementary data type and can also be fully-addressed data addresses.

Output parameters of type POINTER are not permissible with function blocks.

Initializing block parameters of type ANY

Variables of all data types are permissible for block parameters of parameter type ANY. The programming within the called block determines which variables (addresses or data types) must be applied to the block parameters, or which variables are feasible. You can also specify a constant in the format of the ANY pointer "P#[Data_block.]Address Data_type Number" and so define an absolute-addressed area.

An exception is the initialization of an ANY parameter with a temporary local data item of data type ANY. In this case, rather than generating a pointer to the variable, the Editor assumes that a pointer of data type ANY already exists in the temporary local data. This gives you the ability to apply to an ANY parameter an ANY pointer that you can change at runtime. The "variable ANY pointer" can be particularly useful in conjunction with the system function SFC 20 BLKMOV (see the "Buffer Entry" example in 26.4 "Brief Description of the Message Frame Example").

Output parameters of type ANY are not permissible with function blocks.

19.4 "Passing On" Block Parameters

"Passing on" block parameters is a special form of access and of initializing block parameters. The block parameters of the calling block are "passed on" to the parameters of the called block. Here, the formal parameter of the calling block then becomes the actual parameter of the called block.

In general, it is also the case here that the actual parameter must be of the same type as the formal parameter (that is, the relevant block parameters must agree in their data types). In addition, you can apply an input parameter of the calling blocks only at an input parameter of the

Table 19.5 Permitted combinations for passing on block parameters

Calling → Called	FC calls FC			FB calls FC			FC calls FB			FB calls FB		
Declaration type	E	Z	P	E	Z	P	E	Z	P	E	Z	P
Input → Input	x	-	-	x	x	-	x	-	x	x	x	x
Output → Output	x	-	-	x	x	-	x	-	-	x	x	-
In/out → Input	x	-	-	x	-	-	x	-	-	x	-	-
In/out → Output	x	-	-	x	-	-	x	-	-	x	-	-
In/out → In/out	x	-	-	x	-	-	x	-	-	x	-	-

E = Elementary data types
Z = Complex data types, UDT
P = Parameter types TIMER, COUNTER, and BLOCK_xx

called block, and similarly, an output parameter at an output parameter. You can apply an in/out parameter of the calling block to all declaration types of the called block.

There are restrictions with regard to data types caused by the variations in block parameter storage between functions and function blocks. Block parameters of elementary data type can be passed on without restriction in accordance with the information in the previous paragraph. Complex data types at inputs and output parameters can only be passed on if the calling block is a function block. Block parameters of parameter types TIMER, COUNTER and BLOCK_xx can only be passed on from one input parameter to another if the calling block is a function block. These statements are represented in Table 19.5.

You can "pass on" the parameter types TIMER, COUNTER, and BLOCK_DB in functions using indirect addressing. The relevant parameter first receives the data type WORD or INT; you supply it with a constant or tag that has as its content the numerical value of the timer, counter, or block to be passed on. You can "pass on" this parameter to other blocks, because it is an elementary data type. In the "last" block, you transfer the content of the parameter to a temporary local data word using a load function and edit the time function, the counter, or the block memory-indirectly.

19.5 Examples

19.5.1 Conveyor Belt Example

The example shows the transfer of signal states via block parameters. For this purpose, we use the function of a conveyor belt control explained in Chapter 5 "Memory Functions". The conveyor belt control is to be located in a function block and all inputs and outputs are to be routed via block parameters, so that the function block can be used repeatedly (for several conveyor belts). Figure 19.4 shows the input and output parameters for the function block as well as the static local data used.

Distributing the parameters is quite simple in this case: All binary addresses that were inputs have become input parameters, all outputs have become output parameters and all memory bits have become static local data. You will also have noticed that the names have also been slightly changed because only alphanumeric characters and the underscore are permissible for block-local variables.

The function block "Conveyor_belt" is to control two conveyor belts. For this purpose, it will be called twice; the first time with the inputs and outputs of conveyor belt 1 and the second time with those of conveyor belt 2. For each call, the function block requires an instance data block where it stores the data for the conveyor belt in each case. The data block for conveyor belt 1 is to be called "Belt_data1" and the data block for conveyor belt 2 is to be called "Belt_data2".

The executed programming example can be found in the download files (download address: see pages 8-9) in the STL_Book library under the "Conveyor Example" program. The source program contains the programming of the function block with the input parameters, the output parameters and the static local data. This is followed by the programming of the instance data blocks; here, it is sufficient to specify the function block as the declaration section. You can use any data blocks as the instance block, for example, DB 21 for "Belt_data1" and DB 22 for "Belt_data2". In the symbol table, these data blocks have the data type of the function block.

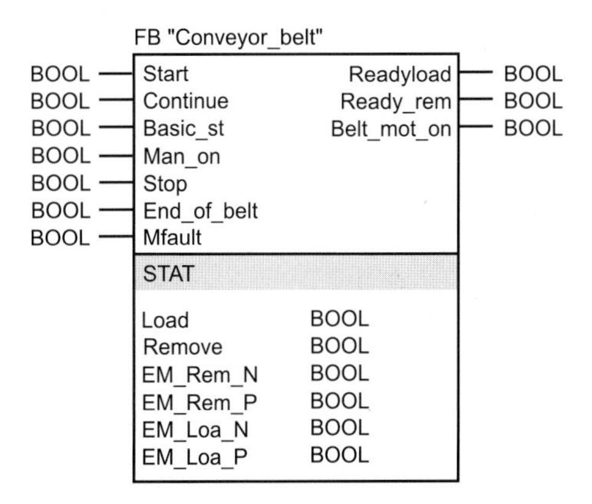

Figure 19.4
Function Block for the Conveyor Belt Example

At the end of the source program, you see another two complete calls of the function block, such as they might be found in OB1, for example. The inputs and output from the symbol table are used as the actual parameters.

In those cases where these global symbols contain special characters, you must place these symbols between quotation marks in the program.

19.5.2 Parts Counter Example

The example demonstrates the handling of block parameters of elementary data types. The "Parts Counter" example from Chapter 8 "Counter Functions" is the basis of the function. The same function is implemented here as a function block, with all global variables declared either as block parameters or as static local data (Figure 19.5).

Timer and counter functions are transferred via block parameters of parameter types TIMER and COUNTER. These block parameters must be input parameters. The initial values of the counter (Quantity) and the timer function (*Dura1* and *Dura2*) can also be transferred as block parameters; the data type of the block parameters corresponds here to the actual parameters.

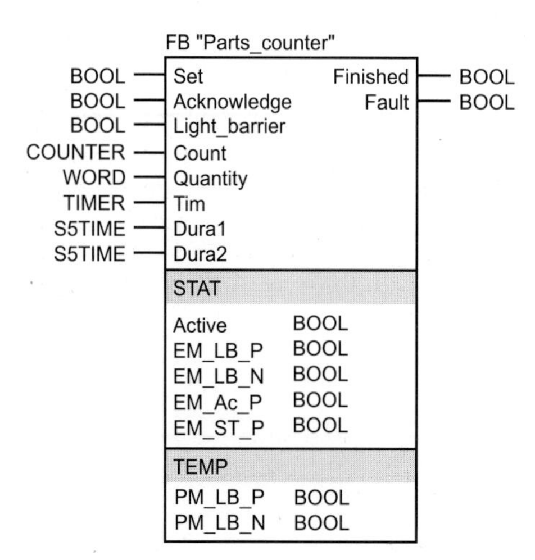

Figure 19.5
Function Block for the Parts Counter Example

The edge memory bits are stored in the static local data and the pulse memory bits are stored in the temporary local data.

The executed programming example can be found in the download files (download address: see pages 8-9) in the STL_Book library under the "Conveyor Example" program. The source program contains the function block "Parts_counter", the associated instance data block "CountDat" and the call of the function block with instance data block.

19.5.3 Feed Example

The same functions as described in the two previous examples can also be called as local instances. In our example, this means that we program a function block "Feed" that is to control four conveyor belts and count the conveyed parts. In this function block, the FB "Conveyor Belt" is called four times and the FB "Parts_counter" is called once. The call does not take place in each case with its own instance data block, but the called FBs are to store their data in the instance data block of the function block "Feed".

Figure 19.6 shows how the individual conveyor belt controls are connected together (the FB "Parts_counter" is not represented here). The start signal is connected to the *Start* input of the controller of belt 1, the *ready_rem* output is connected to the *Start* input of belt 2, etc. Finally, the *ready_rem* output of belt 4 is connected to the *Remove* output of "Feed". The same signal sequence leads in the reverse direction from *Removed* via *Continue* and *Readyload* to *Load*.

Belt_mot_on, *Light_barrier* and */Mfault* (motor fault) are individual signals of the conveyor belts; *Reset*, *Man_start* and *Stop* control all conveyor belts via *Basic_st*, *Man_on* and *Stop*.

The following program for the function block "Feed" is designed in the same way. The input and output parameters of the function block can be seen from the figure. In addition, the digital values for the parts counter *Quantity*, *Dura1* and *Dura2* are designed as input parameters here. We declare the data of the individual conveyor belt controls and the data of the parts counter in the static local data in exactly the same way as for a user-defined data type, i.e.

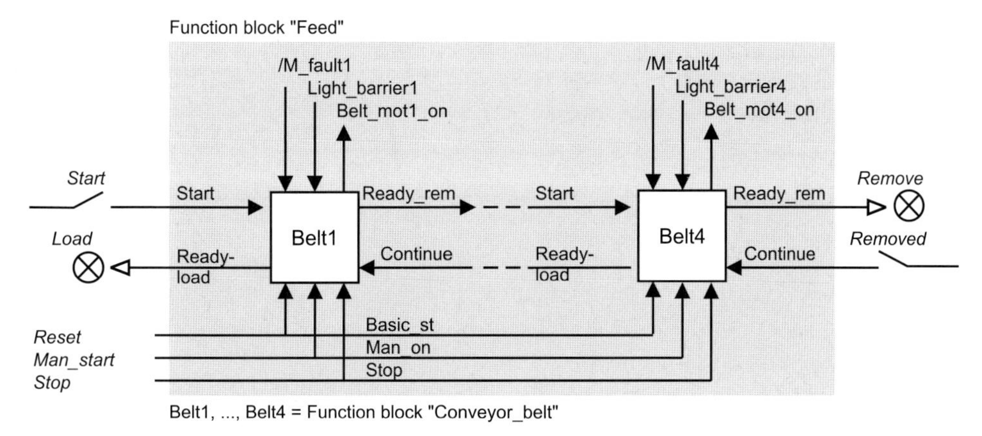

Figure 19.6 Feed Programming Example

with name and data type. The variable *Belt1* is to receive the data structure of the function block "Conveyor_belt", also the variable *Belt2*, etc.; the variable *Check* receives the data structure of the function block "Parts_counter".

The program in the function block starts with the initialization of the signals common to all conveyor belts. Here, we make use of the fact that the block parameters of the function blocks called as local instances are static local data in the current block and can be handled as such. The block parameter *Man_start* of the current function block controls the input parameter *Man_on* of all four conveyor belt controls with a simple assignment. We proceed in the same way with the signals *Stop* and *Reset*. And now the conveyor belt controls are initialized with the common signals. (You can, of course, also initialize these input parameters when the function block is called.)

The subsequent calls of the function blocks for conveyor belt control contain only the block parameters for the individual signals for each conveyor belt and the connection to the block parameters of "Feed". The individual signals are the light barriers, the commands for the belt motor and the motor faults. (We make use here of the fact that when a function block is called, not all block parameters have to be initialized.)

We program the connections between the individual belt controllers using assignments.

The FB "Parts_counter" is called as a local instance even if it has no closer connection with the signals of the conveyor belt controls. The instance data block of "Feed" takes the FB data.

The input parameters *Quantity*, *Dura1* and *Dura2* of "Feed" need to be set only once. This can be done with the default (as in the example) or in the restart program in OB 100 (through direct assignment, for example, if these three parameters are treated as global data).

The source program in the STL_Book library under the program "Example of conveyor technology" contains the "Feed" function block and the related "FeedData" instance data block. At the end, the "Feed" function block call with instance data block is shown for the main program.

```
FUNCTION_BLOCK "Feed"
TITLE = Control of several conveyor belts
//Example for local instances (declaration, calls)

NAME    : Feed
AUTHOR  : Berger
FAMILY  : STL_Book
VERSION : 01.00

VAR_INPUT
  Start           : BOOL     := FALSE;       //Start conveyor belts
  Removed         : BOOL     := FALSE;       //Goods removed from conveyor
  Start up        : BOOL     := FALSE;       //Start conveyor belts manually
  Hold            : BOOL     := FALSE;       //Stop conveyor belts
  Reset           : BOOL     := FALSE;       //Set control to initial state
  Counter         : COUNTER;                 //Counter for piece goods
  Quantity        : WORD     := W#16#0200;   //Quantity of goods
  Time            : TIMER;                   //Time function for monitoring
  Duration1       : S5TIME   := S5T#5s;      //Monitoring time for goods
  Duration2       : S5TIME   := S5T#10s;     //Monitoring time for gap
END_VAR

VAR_OUTPUT
  Load            : BOOL     := FALSE;       //Place new goods on belt
  Remove          : BOOL     := FALSE;       //Remove goods from belt
END_VAR

VAR
  Belt1 : "Conveyor belt";                   //Control for belt 1
  Belt2 : "Conveyor belt";                   //Control for belt 2
  Belt3 : "Conveyor belt";                   //Control for belt 3
  Belt4 : "Conveyor belt";                   //Control for belt 4
  Check : "Parts counter";                   //Control for counting and monitoring
END_VAR

BEGIN

NETWORK
TITLE = Supplying shared signals

     A    Startup;
     =    Belt1.Manual_on;
     =    Belt2.Manual_on;
     =    Belt3.Manual_on;
     =    Belt4.Manual_on;

     A    Stop;
     =    Belt1.Stop;
     =    Belt2.Stop;
     =    Belt3.Stop;
     =    Belt4.Stop;

     A    Reset;
     =    Belt1.Initial_state;
     =    Belt2.Initial_state;
     =    Belt3.Initial_state;
     =    Belt4.Initial_state;
```

(continued on next page)

```
NETWORK
TITLE = Call conveyor belt controls

  CALL Belt1 (
     Start             := Start,
     Belt end          := Light barrier1,
     Motor fault       := "/Motor fault1",
     Ready to accept   := Load,
     Belt motor_on     := Belt motor1_on);

  A  Belt2.ready to accept;
  =  Belt1.Continue;
  U  Belt1.ready for pickup;
  =  Belt2.Start;

  CALL Belt2 (
     Belt end          := Light barrier2,
     Motor fault       := "/Motor fault2",
     Belt motor_on     := Belt motor2_on);

  A  Belt3.ready to accept;
  =  Belt2.Continue;
  A  Belt2.ready for pickup;
  =  Belt3.Start;

  CALL Belt3 (
     Belt end          := Light barrier3,
     Motor fault       := "/Motor fault3",
     Belt motor_on     := Belt motor3_on);

  A  Belt4.ready to accept;
  =  Belt3.Continue;
  A  Belt3.ready for pickup;
  =  Belt4.Start;

  CALL Belt4 (
     Continue          := Removed,
     Belt end          := Light barrier4,
     Ready for pickup  := Remove
     Motor fault       := "/Motor fault4",
     Belt motor_on     := Belt motor4_on);

NETWORK
TITLE = Call for counting and monitoring

  CALL check (
     Set               := Set,
     Acknowledge       := Acknowledge,
     Light barrier     := Light barrier1,
     Counter           := #Counter,
     Quantity          := #Quantity,
     Time              := #Time,
     Duration1         := #Duration1,
     Duration2         := #Duration2,
     Finished          := Finished,
     Fault             := "Fault");

NETWORK
TITLE = Block end
  BE

END_FUNCTION_BLOCK
```

Program Processing

This section of the book discusses the various methods of program processing.

The **main program** executes cyclically. After each program pass, the CPU returns to the beginning of the program and executes it again. This is the "standard" method of processing PLC programs.

Numerous system functions support the utilization of system services, such as controlling the real-time clock or communication via bus systems. In contrast to the static settings made when parameterizing the CPU, system functions can be used dynamically at program run time.

The main program can be temporarily suspended to allow **interrupt servicing**. The various types of interrupts (time-of-day interrupts, time-delay interrupts, watchdog interrupts, process interrupts, DPV1 interrupts, multiprocessor interrupts, synchronous cycle interrupts) are divided into priority classes whose processing priority you may yourself, to a large degree, determine. Interrupt servicing allows you to react quickly to signals from the controlled process or implement periodic control procedures independently of the processing time of the main program.

Before starting the main program, the CPU initiates a **start-up program** in which you can make specifications regarding program processing, define default values for variables, or parameterize modules.

Error handling is also part of program processing. STEP 7 distinguishes between synchronous errors, which occur during processing of a statement, and asynchronous errors, which can be detected independently of program processing. In both cases you can adapt the error routine to suit your needs.

20 Main Program

The main program is the cyclically scanned user program; cyclic scanning is the "normal" way in which programs execute in programmable logic controllers. The large majority of control systems use only this form of program execution. If event-driven program scanning is used, it is in most cases only in addition to the main program.

The main program is invoked in organization block OB 1. It executes at the lowest priority level, and can be interrupted by all other types of program processing. The user program is executed in the RUN state which is set using the mode selector on the front of the CPU. The position is RUN in the case of a toggle switch as mode selector, and RUN and RUN-P in the case of a key-operated switch. In the RUN-P position, the CPU can be programmed using a connected programming device. In the RUN position, you can remove the key so that no one can change the operating mode without proper authorization; when the mode selector is at RUN, programs can only be read.

20.1 Program Organization

20.1.1 Program Structure

To analyze a complex automation task means to divide that task into smaller tasks or functions in accordance with the structure of the process to be controlled. You then define the individual tasks resulting from this dividing process by determining the functions and stipulating the interface signals to the process or to other tasks. This breakdown into individual tasks can be done in your program. In this way, the structure of your program corresponds to the division of the automation task.

A divided user program can be more easily configured, and can be programmed in sections (even by several people in the case of very large user programs). And finally, but not lacking in importance, dividing the program simplifies both debugging and service and maintenance.

The structuring of the user program depends on its size and its function. A distinction is made between three different "methods":

In a **linear program**, the entire main program is in organization block OB 1. Each current path is in a separate network. STEP 7 numbers the networks in sequence. When editing and debugging, you can reference every network directly by its number.

A **partitioned program** is basically a linear program which is divided into blocks. Reasons for dividing the program might be because it is too long for organization block OB 1 or because you want to make it more readable. The blocks are then called in sequence. You can also divide the program in another block the same way you would the program in organization block OB 1. This method allows you to call associated process-related functions for processing from within one and the same block. The advantage of this program structure is that, even though the program is linear, you can still debug and run it in sections (simply by omitting or adding block calls).

A **structured program** is used when the conceptual formulation is particularly extensive, when you want to reuse program functions, or when complex problems must be solved. Structuring means dividing the program into sections (blocks) which embody self-contained functions or serve a specific functional purpose and which exchange the fewest possible number of signals with other blocks. Assigning each program section a specific (process-related) function will produce easily readable blocks with simple interfaces to other blocks when programmed.

The STL and SCL programming languages support structured programming through functions with which you can create "blocks" (self-contained program sections). Chapter 3.2 "Blocks", discusses the different kinds of blocks and their uses. You will find a detailed description of the functions for calling and ending blocks in Chapter 18 "Block Functions". The blocks receive the signals and data to be processed via the call interface (the block parameters), and forward the results over this same interface. The options for passing parameters are described in detail in Chapter 19 "Block Parameters". Chapter 29 "SCL Blocks" contains a description of block handling with SCL.

20.1.2 Program Organization

Program organization determines whether and in what order the CPU will process the blocks which you have generated. To organize your program, you program block calls in the desired sequence in the supraordinate blocks. You should chose the order in which the blocks are called so that it mirrors the process-related or function-related division of the controlled plant.

Nesting depth

The maximum depth applies for a priority class (for the program in an organization block), and is CPU-dependent. On the CPU 314, for example, the nesting depth is eight, that is, beginning with one organization block (nesting depth 1), you can add seven more blocks in the "horizontal" direction (this is called "nesting"). If more blocks are called, the CPU goes to STOP with a "Block overflow" error. Do not forget to include system function block (SFB) calls and system function (SFC) calls when calculating the nesting depth.

A data block call, which is actually only the opening or selecting of a data area, has no effect on the nesting depth of blocks, nor is the nesting depth affected by calling several blocks in succession (linear block calls).

Practice-related program organization

In organization block OB 1, you should call the blocks in the main program in such a way as to roughly organize your program. A program can be organized on either a process-related or function-related basis.

The following points of discussion can give only a rough, very general view with the intention of giving the beginner some ideas on program structuring and on translating his control task into reality. Advanced programmers normally have sufficient experience to organize a program to suit the special control task at hand.

A **process-related program structure** closely follows the structure of the plant to be controlled. The individual program sections correspond to the individual parts of the plant or of the process to be controlled. Subordinate to this rough structure are the scanning of the limit switches and operator panels and the control of the actuators and display devices (in different parts of the plant). Bit memory or global data are used for signal interchange between different parts of the plant.

A **function-related program structure** is based on the control function to be executed. Initially, this method of program structuring does not take the controlled plant into account at all. The plant structure first becomes apparent in the subordinate blocks when the control function defined by the rough structure is divided further.

In practice, a hybrid of these two concepts is normally used. Figure 20.1 shows an example: A functional structure is mirrored in the operating mode program and in the data processing program which goes above and beyond the plant itself. Program sections Feeding Conveyor 1, Feeding Conveyor 2, Process and Discharging Conveyor are process-related.

The example also shows the use of different types of blocks. The main program is in OB 1; it is in this program that the blocks for the operating modes, the various pieces of plant equipment, and for data processing are called. These blocks are function blocks with an instance data block as data store. Feeding Conveyor 1 and Feeding Conveyor 2 are identically structured; FB 20, with DB 20 as instance data block for Feeding Conveyor 1 and with DB 21 as instance data block for Feeding Conveyor 2, is used for control

In the conveyor control program, function FC 20 processes the interlocks; it scans inputs

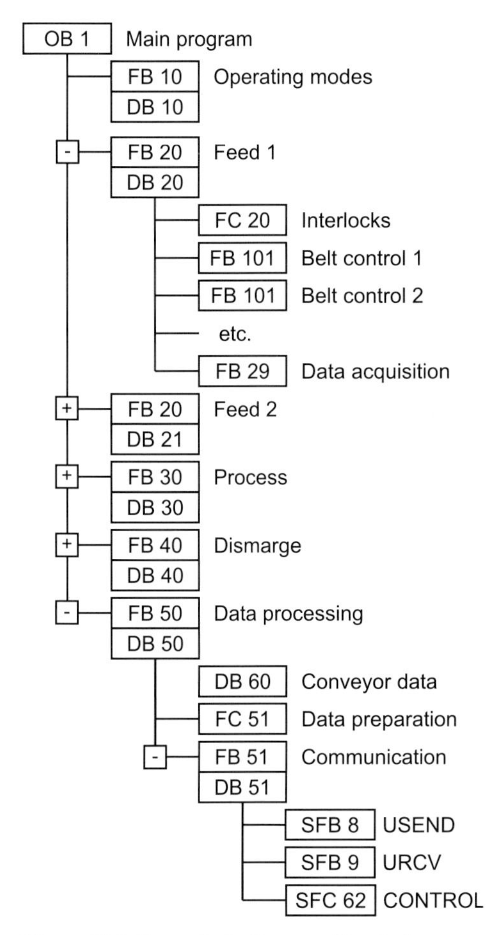

Figure 20.1 Example for Program Structuring

or memory bits and controls FB 20's local data. Function block FB 101 contains the control program for a conveyor belt, and is called once for each belt. The call is a local instance, so that its local data are in instance data block DB 20. The same applies for the data acquisition program in FB 29.

The data processing program in FB 50, which uses DB 50, processes the data acquired with FB 29 (and other blocks), which are located in global data block DB 60. Function FC 51 prepares these data for transfer. The transfer is controlled by FB 51 (with DB 51), in which system blocks SFB 8, SFB 9 and SFB 62 are called. Here, too, the SFBs save their instance data in "supraordinate" data block DB 51.

20.2 Scan Cycle Control

20.2.1 Process Image Updating

The process image is part of the CPU's internal system memory (Chapter 1.1.6 "CPU Memory Areas"). It begins at I/O address 0 and ends at an upper limit stipulated by the CPU. On appropriately equipped CPUs, you can define this limit yourself.

Normally, all digital modules lie in the process image address area, while all analog modules have addresses outside this area. If the CPU has free address allocation, you can use the configuration table to direct any module over the process image or address it outside the process image area.

The process image consists of the process-image input table (inputs I) and the process-image output table (outputs Q).

After CPU restart and prior to the first execution of OB 1, the operating system transfers the signal states of the process-image output table to the output modules and accepts the signal states of the input modules into the process-image input table. This is followed by execution of OB 1 where normally the inputs are combined with each other and the outputs are controlled. Following termination of OB 1, a new cycle begins with the updating of the process image (Figure 20.2).

If an error occurs during automatic updating of the process image, e.g. because a module is no longer accessible, organization block OB 85 "Program Execution Errors" is called. If OB 85 is not available, the CPU goes to STOP.

Subprocess images

With appropriately equipped CPUs, you can divide the process image into up to 30 partial process images. You make this division during parameterization of the signal modules by defining the partial process image via which the module is to be addressed when you assign addresses. You can separate the division according to process-image input table and process-image output table.

All modules that you do not assign to one of the partial process images 1 to 30 are stored in par-

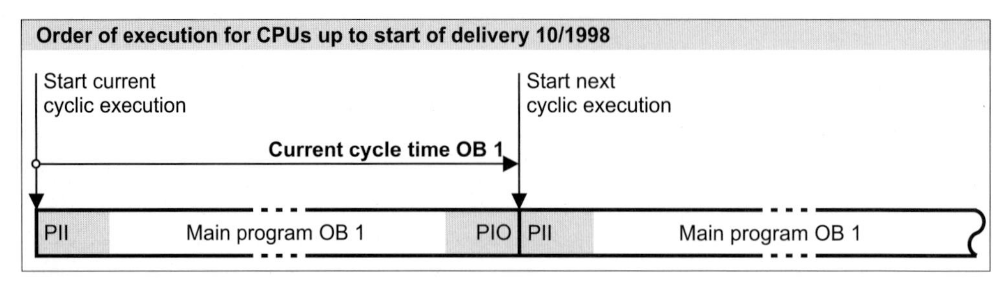

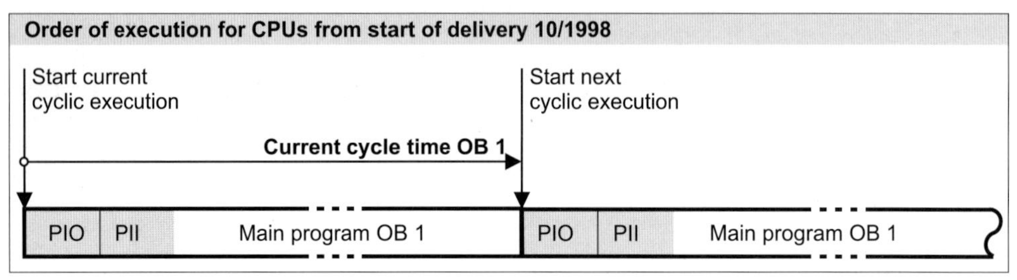

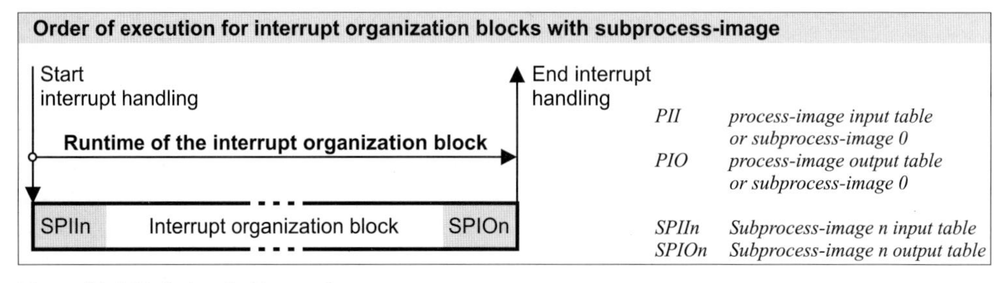

Figure 20.2 Updating the Process Image

tial process image 0, which is also called the OB1 process image (OB1-PI). This partial process image 0 is updated automatically by the operating system of the CPU as part of cyclic execution. You can also switch off this automatic updating for the S7-400 using CPU properties.

With appropriately equipped CPUs, you can also assign the partial process images to the interrupt organization blocks so that they are automatically updated when these OBs are called.

The system functions SFC 26 UPDAT_PI and SFC 27 UPDAT_PO are available for updating the partial process images by the user program. In the synchronous cycle interrupt organization blocks, you use the system functions SFC 126 SYNC_PI and SFC 127 SYNC_PO (Chapter 21.8.2 "Isochronous Updating of Process Image").

SFC 26 UPDAT_PI
SFC 27 UPDAT_PO

The system function SFC 26 UPDAT_PI updates a partial process image of the inputs, the system function SFC 27 UPDAT_PO a partial process image of the outputs. Table 20.1 shows the parameters of these SFCs. You can also update partial process image 0 with these SFCs.

You can carry out updating of individual partial process images by calling these SFCs at any time and at any location. For example, you can define a partial process image for a priority class (a program execution level) and you can then cause this partial process image to be

Table 20.1 Parameters for the SFCs for Process Image Updating

Parameter Name	SFC		Declaration	Data Type	Contents, Description
PART	26	27	INPUT	BYTE	Number of the partial process image (0 to 15)
RET_VAL	26	27	RETURN	INT	Error information
FLADDR	26	27	OUTPUT	WORD	On an access error: the address of the first byte to cause the error

updated at the start and at the end of the relevant organization block when this priority class is processed.

Updating of a process image can be interrupted by calling a higher priority class. If an error occurs during updating of a process image, e.g. because a module can no longer be accessed, this error is reported via the function value of the SFC.

20.2.2 Scan Cycle Monitoring Time

Program scanning in organization block OB 1 is monitored by the so-called "scan cycle monitor" or "scan cycle watchdog". The default value for the scan cycle monitoring time is 150 ms. You can change this value in the range from 1 ms to 6 s by parameterizing the CPU accordingly.

If the main program takes longer to scan than the specified scan cycle monitoring time, the CPU calls OB 80 ("Timeout"). If OB 80 has not been programmed, the CPU goes to STOP.

The scan cycle monitoring time includes the full scan time for OB 1. It also includes the scan times for higher priority classes which interrupt the main program (in the current cycle). Communication processes carried out by the operating system, such as GD communication or PG access to the CPU (block status!), also increase the runtime of the main program. The increase can be reduced in part by the way you parameterize the CPU ("Cyclic load from communication" on the "Cycle/Clock memory bits" tab).

Cycle statistics

If you have an online connection from a programming device to an operating CPU, select PLC → MODULE INFORMATION to call up a dialog box that contains several tabs. The "Cycle Time" tab shows the current cycle time as well as the shortest and longest cycle time. The parameterized minimum cycle duration and the scan cycle monitoring time are also displayed.

The cycle time for the last cycle and the minimum and maximum cycle time since the PLC was last started up can also be read in the temporary local data in the start information of OB 1.

SFC 43 RE_TRIGR
Restarting the scan cycle monitoring time

An SFC 43 RE_TRIGR system function call restarts the scan cycle monitoring time; the timer restarts with the new value set via CPU parameterization. SFC 43 has no parameters.

Operating system run times

The scan cycle time also includes the operating system run times. These are composed of the following:

- ▷ System control of cyclic scanning ("no-load cycle"), fixed value
- ▷ Updating of the process image; dependent on the number of bytes to be updated
- ▷ Updating of the timers; dependent on the number of timers to be updated
- ▷ Communications load

Communications functions for the CPU include the transfer of user program blocks or data exchange between CPU modules using system functions. The time the CPU is to use for these functions can be limited by parameterizing the CPU.

All values at operating system runtime are properties of the relevant CPU.

20.2.3 Minimum Scan Cycle Time, Background Scanning

With appropriately equipped CPUs, you may specify a minimum scan cycle time. If the main program (including interrupts) takes less time, the CPU waits until the specified minimum scan cycle time has elapsed before beginning the next cycle by recalling OB 1.

The default value for the minimum scan cycle time is 0 ms, that is to say, the function is disabled. You can set a minimum scan cycle time of from 1 ms to 6 s in "Cycle/Clock memory bits" tab when you parameterize the CPU.

Background scanning OB 90

In the interval between the actual end of the cycle and expiration of the minimum cycle time, the CPU executes organization OB 90 "Background scanning" (Figure 20.3). OB 90 is executed "in slices". When the operating system calls OB 1, execution of OB 90 is interrupted; it is then resumed at the point of interruption when OB 1 has terminated. OB 90 can be interrupted after each statement, any system block called in OB 90, however, is first scanned in its entirety.

The length of a "slice" depends on the current scan cycle time of OB 1. The closer OB 1's scan time is to the minimum scan cycle time, the less time remains for executing OB 90. The program scan time is not monitored in OB 90.

OB 90 is scanned only in RUN mode. It can be interrupted by interrupt and error events, just like OB 1. The start information in the temporary local data (Byte 1) also tells which events cause OB 90 to execute from the beginning:

- B#16#91
  After a CPU restart,
- B#16#92
  After a block processed in OB 90 was deleted or replaced,
- B#16#93
  After (re)loading of OB 90 in RUN mode,
- B#16#95
  After the program in OB 90 was scanned and a new background cycle begins.

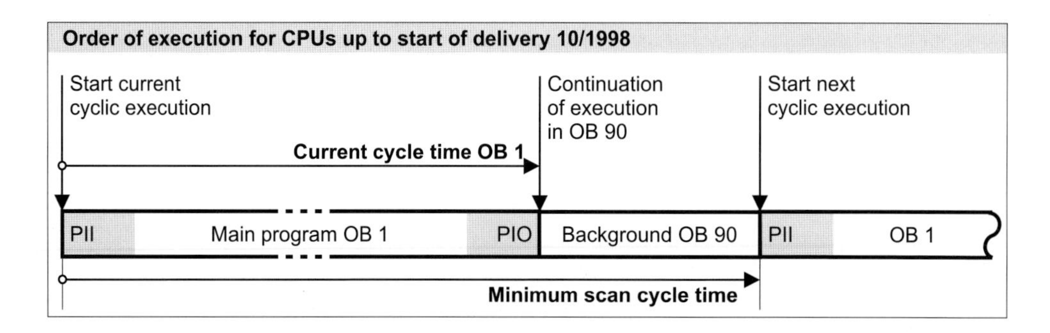

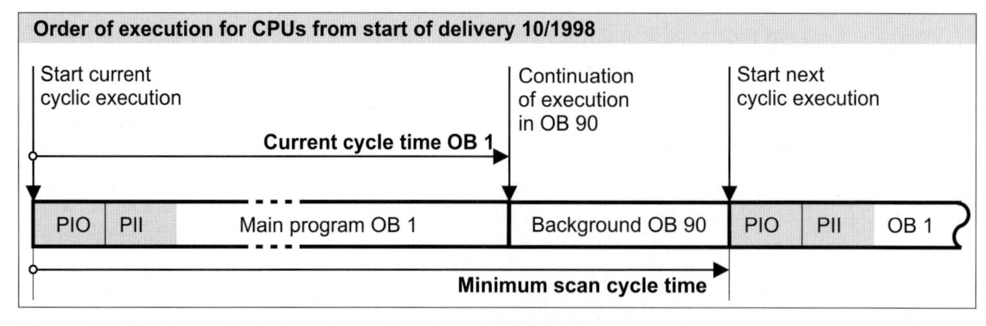

Figure 20.3 Minimum Cycle Duration and Background Scanning

20.2.4 Response Time

If the user program in OB 1 works with the signal states of the process images, this results in a response time which is dependent on the program execution time (scan cycle time). The response time lies between one and two scan cycles, as the following example explains.

When a limit switch is activated, for instance, it changes its signal state from "0" to "1". The programmable controller detects this change during the subsequent updating of the process image, and sets the inputs allocated to the limit switch to "1". The program evaluates this change by resetting an output, for example, in order to switch off the corresponding motor. The new signal state of the output that was reset is transferred at the end of the program scan; only then is the corresponding bit reset on the digital output module.

In a best-case situation, the process image is updated immediately following the change in the limit switch's signal (Figure 20.4). It would then take only one cycle for the relevant output to respond. In the worst-case situation, updating of the process image was just completed when the limit switch signal changed. It would then be necessary to wait approximately one cycle for the programmable controller to detect the signal change and set the input. After yet another cycle, the program can respond.

When so considered, the user program's execution time contains all procedures in one program cycle (including, for instance, the servicing of interrupts, the functions carried out by the operating system, such as updating timers, controlling the MPI interface and updating the process images).

The response time to a change in an input signal can thus be between one and two cycles. Added to the response time are the delays for the input modules, the switching times of contactors, and so on.

In some instances, you can reduce the response times by addressing the I/Os directly or calling program sections on an event-driven basis.

Uniform response times or equal time intervals in the process control can be achieved if a program section is always processed with the same time pattern, e.g. a watchdog interrupt program. Program execution that is isochronous with respect to the processing cycle of a PROFIBUS DP master system also results in calculable response times (Chapter 21.8 "Synchronous Cycle Interrupts").

20.2.5 Start Information

The CPU's operating system forwards start information to organization block OB 1, as it does to every organization block, in the first 20 bytes of temporary local data. You can generate the declaration for the start information yourself or you can use information from the *Standard Library* under *Organization Blocks*.

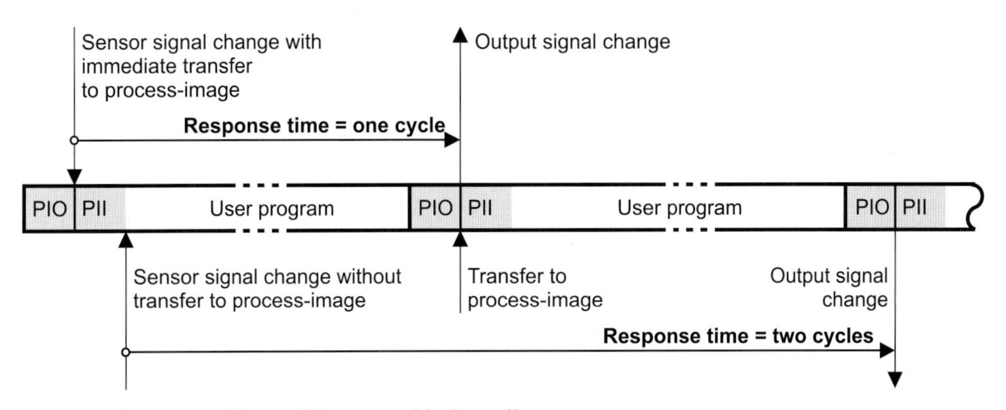

Figure 20.4 Response Times of Programmable Controllers

Table 20.2 Start Information for the OB 1

Byte	Variable Name	Data Type	Description	Contents
0	OB1_EV_CLASS	BYTE	Event class	B#16#11 = Call standard OB
1	OB1_SCAN_1	BYTE	Start information	B#16#01 = 1st cycle after warm restart B#16#02 = 1st cycle after hot restart B#16#03 = Every other cycle B#16#04 = 1st cycle after cold restart
2	OB1_PRIORITY	BYTE	Priority	B#16#01
3	OB1_OB_NUMBR	BYTE	OB Number	B#16#01
4	OB1_RESERVED_1	BYTE	Spared	-
5	OB1_RESERVED_2	BYTE	Spared	-
6..7	OB1_PREV_CYCLE	INT	Previous scan cycle time	in ms
8..9	OB1_MIN_CYCLE	INT	Minimum scan cycle time	in ms
10..11	OB1_MAX_CYCLE	INT	Maximum scan cycle time	in ms
12..19	OB1_DATE_TIME	DT	Event occurrence	Call time of the OB (cyclic)

Table 20.2 shows this start information for the OB1, the default symbolic designation, and the data types. You can change the designation at any time and choose names more acceptable to you. Even if you don't use the start information, you must reserve the first 20 bytes of temporary local data for this purpose (for instance in the form of a 20-byte array).

In SIMATIC S7, all event messages have a fixed structure which is specified by the event class. The start information for OB 1, for instance, reports event B#16#11 as a standard OB call. From the contents of the next byte you can tell whether the main program is in the first cycle after power-up and is therefore calling, for instance, initialization routines in the cyclic program.

The priority and OB number of the main program are fixed. With three INT values, the start information provides information on the cycle time of the last scan cycle and on the minimum and maximum cycle times since the last power-up. The last value, in DATE_AND_TIME format, indicates when the priority control program received the event for calling OB 1.

Note that direct reading of the start information for an organization block is possible only in that organization block because that information consists of temporary local data. If you require the start information in blocks which lie on deeper levels, call system function SFC RD_SINFO at the relevant location in the program.

SFC 6 RD_SINFO
Reading OB start information

System function SFC 6 RD_SINFO makes the start information on the current organization block (that is, the OB at the top of the call tree) and on the start-up OB last executed available to you even at a deeper call level (Table 20.3).

Output parameter TOP_SI contains the first 12 bytes of start information on the current OB, output parameter START_UP_SI the first 12 bytes of start information on the last start-up OB executed. There is no time stamp in either case.

SFC 6 RD_SINFO can not only be called at any location in the main program but in every priority class, even in an error organization block or in the start-up routine. If the SFC is called in an interrupt organization block, for example, TOP_SI contains the start information of the interrupt OB. In the case of a call at restart, TOP_SI and START_UP_SI have the same contents.

Table 20.3 Parameters for SFC 6 RD_SINFO

SFC	Parameter Name	Declaration	Data Type	Contents, Description
6	RET_VAL	RETURN	INT	Error information
	TOP_SI	OUTPUT	STRUCT	Start information for the current OB (with the same structure as START_UP_SI)
	START_UP_SI	OUTPUT	STRUCT	Start information for the last OB started:
	.EV_CLASS		BYTE	Event ID and event class
	.EV_NUM		BYTE	Event number
	.PRIORITY		BYTE	Execution priority (number of the execution level)
	.NUM		BYTE	OB number
	.TYP2_3		BYTE	ID of supplementary information 2_3
	.TYP1		BYTE	ID of supplementary information 1
	.ZI1		WORD	Supplementary information 1
	.ZI2_3		DWORD	Supplementary information 2_3

20.3 Program Functions

In addition to parameterizing the CPU with the Hardware Configuration, you can also select a number of program functions dynamically at runtime via the integrated system functions.

20.3.1 Time

Each SIMATIC CPU has a clock which you can set and scan using STEP 7 or system functions. The time is represented in the user program in the format DATE_AND_TIME, consisting of the date, time and day of week.

With newer CPUs with firmware version 3 or higher, the time status is added. You can then additionally set a difference from a time zone and a summer/winter time ID.

You can use the following system functions to set the CPU clock:

▷ SFC 0 SET_CLK
Set date and time

▷ SFC 1 READ_CLK
Read date and time

▷ SFC 48 SNC_RTCB
Synchronize Slave clocks

▷ SFC 100 SET_CLKS
Set date, time and clock status

You will find a list of system function parameters in Table 20.4.

Set and read time

Calling SFC 0 SET_CLK or SFC 100 SET_CLKS with MODE = B#16#01 or B#16#03 respectively sets the clock to the value specified at parameter PDT. SFC 0 SET_CLK sets the winter time for CPUs with summer/winter time ID. With the SFC 100 SET_CLKS you can use the parameter SUMMER to specify whether the time to be set is the winter time (with "0") or the summer time (with "1").

SFC 1 READ_CLK reads the current time and outputs it at parameter CDT. When setting and reading, the time has the format DATE_AND_TIME, i.e. it includes both the date and time.

Module time, local time

The time on the CPU is the module time. This is decisive for all timing processes controlled by the CPU, e.g. run-time meter, starting of time-of-day interrupts or entering of time stamps in the diagnostics buffer and in the OB start information. You set and read the module time when using the system functions for the CPU clock.

Appropriately designed CPUs additionally save a "time status". This contains a correction value which, added to the module time, results in the local time. The correction value is entered in intervals of 30 minutes, and can also be negative (parameter CORR of the SFC 100 SET_CLKS).

The local time can be used to visualize time zones.

Table 20.4 SFC Parameters for the CPU Clock

SFC	Parameter Name	Declaration	Data Type	Contents, Description
0	PDT	INPUT	DT	Date and time (new)
	RET_VAL	RETURN	INT	Error information
1	RET_VAL	RETURN	INT	Error information
	CDT	OUTPUT	DT	Date and time (current)
48	RET_VAL	RETURN	INT	Error information
100	MODE	INPUT	BYTE	Operating mode B#16#01: only set time B#16#02: only set time status B#16#03: set time and time status
	PDT	INPUT	DT	Defined time
	CORR	INPUT	INT	Difference from basic time in 0.5-hour cycle from –24 to +26
	SUMMER	INPUT	BOOL	Summer/winter time ID ("1" = summer time)
	ANN_1	INPUT	BOOL	Announcement of time switchover: with "1", a change is made from summer time to winter time or vice versa at the next time the hour changes
	RET_VAL	RETURN	INT	Error information

Time status

The time status is set when parameterizing the CPU with STEP 7 or with the SFC 100 SET_CLKS. You can read the time and the time status with the SFC 51 RDSYSST via the system status list (SZL_ID = W#16#0132 with INDEX W#16#0008). In the variable *status*, it includes:

- ▷ The correction value (bits 2 to 6) in interval of 30 minutes
- ▷ The sign of the correction value (bit 7)
- ▷ The summer/winter time ID (bit 14)
- ▷ The announcement hour (bit 15)

The *summer/winter time ID* indicates whether the local time calculated from the module time and the correction value is the summer time (with "1") or the winter time (with "0").

If the bit *Announcement hour* has the signal status "1", the conversion to summer or winter time is carried out when the hour changes for the next time.

Using the data in the time status, a local time can be generated from the module time in order to control the timing processes in the user program.

Loadable standard blocks help you with handling of the summer/winter time switching of the local time in the user program, especially the starting of time-of-day interrupts depending on the local time (see "Loadable time blocks")

Time synchronization

The clocks of all CPUs can be synchronized in automation networks with several SIMATIC stations which exchange data via subnets. You parameterize the clock of one CPU as the "master clock", and set the interval at which the synchronization is to be carried out. You parameterize the clocks to be synchronized as "slave clocks".

The synchronization can be carried out within an S7 station via the C bus (backplane bus) or between the stations via the MPI bus. It is carried out automatically at the parameterized interval following initial setting of the master clock. If you set a master clock with the SFC 0 SET_CLK or SFC 100 SET_CLKS, all other clocks in the subnet are automatically synchronized to this value.

By calling the SFC 48 SNC_RTCB in the master clock, you synchronize all slave clocks independent of the automatic interval.

If the master clock does not have a time status, the slave clocks are synchronized with the winter time. The correction factor is zero, the local time then corresponds to the module time.

If the master clock works with the time status, the complete time status is transmitted in addition to the time. The same local time (the same time zone) therefore also exists in the time network on all CPUs.

Set time using STEP 7

In the CPU parameterization, you set the synchronization type (master, slave, or none) and interval in the "Diagnostics/Clock" tab. The correction values set here are used to correct the accuracy of the clock.

You can set the time and the time status using STEP 7 if the programming device is connected online to a CPU. Select PLC → DIAGNOSTICS/SETTINGS → SET TIME OF DAY. In the extended dialog, you can set the local time as a difference from the module time, and define the summer/winter time. The box “Status” then shows the time status.

Loadable time blocks

The program *Miscellaneous Blocks* in the *Standard Library* contains loadable blocks for handling the summer/winter time switchover and the local time in the user program.

▷ FC 60 LOC_TIME
Determine local time

▷ FC 61 BT_LT
Convert module time into local time

▷ FC 62 LT_BT
Convert local time into module time

▷ FC 63 S_LTINT
Set time-of-day interrupt according to local time

▷ FB 60 SET_SW
Switch over summer/winter time

▷ FB 61 SET_SW_S
Switch over summer/winter time with time status

▷ UDT 60 WS_RULES
Rules for the summer/winter time switchover (e.g. time of switchover)

20.3.2 Read System Clock

A CPU's system clock starts running on power-up or on a warm restart. The system clock keeps running as long as the CPU is executing the restart routine or is in RUN mode. When the CPU goes to STOP or HOLD, the current system time is “frozen”.

If you initiate a hot restart on an S7-400 CPU, the system clock starts running again using the saved value as its starting time. Cold restart or warm restart reset the system time.

The system time has data format TIME, whereby it can assume only positive values:

TIME#0ms to
TIME#24d20h31m23s647ms.

In the event of an overflow, the clock starts again at 0. More recent CPUs update the system clock every millisecond, older S7-300 CPUs every 10 milliseconds.

SFC 64 TIME_TCK
Read system time

You can read the current system time with system function SFC 64 TIME_TCK. The RET_VAL parameter contains the system time in the TIME data format.

You can use the system clock, for example, to read out the current CPU runtime or, by computing the difference, to calculate the time between two SFC 64 calls. The difference between two values in TIME format is computed using DINT subtraction.

20.3.3 Run-Time Meter

A run-time meter in a CPU counts the hours. You can use the run-time meter for such tasks as determining the CPU runtime or ascertaining the runtime of devices connected to that CPU.

The reading of a run-time meter is retained also following a cold restart, failure of the backup voltage or an overall reset.

The range of values and the number of run-time meters per CPU depends on the CPU. The range of values is 16 bits ($2^{15} - 1$ hours) or

Table 20.5 Parameters of the SFCs for the Run-Time Meter

SFC	Parameter	Declaration	Data Type	Contents, Description
2	NR	INPUT	BYTE	Number of the run-time meter (B#16#00 to B#16#07)
	PV	INPUT	INT	New value for the run-time meter
	RET_VAL	RETURN	INT	Error information
3	NR	INPUT	BYTE	Number of the run-time meter (B#16#00 to B#16#07)
	S	INPUT	BOOL	Start (with “1”) or stop (with “0”) run-time meter
	RET_VAL	RETURN	INT	Error information
4	NR	INPUT	BYTE	Number of the run-time meter (B#16#00 to B#16#07)
	RET_VAL	RETURN	INT	Error information
	CQ	OUTPUT	BOOL	Run-time meter running (“1”) or stopped (“0”)
	CV	OUTPUT	INT	Current value of the run-time meter
101	NR	INPUT	BYTE	Number of the run-time meter (B#16#00 to B#16#0F)
	MODE	INPUT	BYTE	Job ID (see text)
	PV	INPUT	DINT	New value for the run-time meter
	RET_VAL	RETURN	INT	Error information
	CQ	OUTPUT	BOOL	Run-time meter running (“1”) or stopped (“0”)
	CV	OUTPUT	DINT	Current value of the run-time meter

32 bits ($2^{31} - 1$ hours). When the CPU is at STOP or HOLD, the run-time meter also stops running; when the CPU is restarted, the run-time meter begins again with the previous value.

When a run-time meter reaches the maximum duration limit value, it stops and reports an overflow. A run-time meter can be set to a new value or reset to zero only via an SFC call.

The following system functions are available to control a run-time meter:

▷ SFC 2 SET_RTM
Set run-time meter 16-Bit

▷ SFC 3 CTRL_RTM
Start or stop run-time meter 16-Bit

▷ SFC 4 READ_RTM
Read run-time meter 16-Bit

▷ SFC 101 RTM
Use 32-bit run-time meter

Table 20.5 shows the parameter for these system functions.

The NR parameter stands for the number of the run-time meter, and has the data type BYTE. It can be initialized using a constant or a variable (as can all input parameters of elementary data type). The PV parameter (data type INT) is used to set the run-time meter to an initial value. SFC 3's-S-parameter starts (with signal state “1”) or stops (with signal state “0”) the selected run-time meter. CQ indicates whether the run-time meter was running (signal state “1”) or stopped (signal state “0”) when scanned. The CV parameter records the hours in INT format.

By assigning the MODE parameter of the SFC 101 you can control a 32-bit run-time meter as follows:

B#16#00 Read current meter value

B#16#01 Start with the last meter value

B#16#02 Stop counter

B#16#04 Set to value specified at PV

B#16#05 Set to and start at value specified at PV

B#16#06 Set to and stop at value specified at PV

You can also use the SFCs for a 16-bit run-time meter to control a 32-bit run-time meter. The latter then responds like a meter with a 16-bit value range.

20.3.4 Compressing CPU Memory

Multiple deletion and reloading of blocks, which often occur during online block modification, can result in gaps in the CPU's work memory and in the RAM load memory which decrease the amount of usable space in memory. When you call the "Compress" function, you start a CPU program which fills these gaps by pushing the blocks together. You can initiate the "Compress" function via a programming device connected to the CPU or by calling system function **SFC 25 COMPRESS**. The parameters for SFC 25 are listed in Table 20.6.

The compression procedure is distributed over several program cycles. The SFC returns BUSY = "1" to indicate that it is still in progress, and DONE = "1" to indicate that it has completed the compression operation. The SFC cannot compress when an externally initiated compression is in progress, when the "Delete Block" function is active, or when PG functions are accessing the block to be shifted (for instance the Block Status function).

Note that blocks of a particular CPU-specific maximum length cannot be compressed, so that gaps would still remain in CPU memory. Only the Compress function initiated via the PG while the CPU is at STOP closes all gaps.

20.3.5 Waiting and Stopping

The system function **SFC 47 WAIT** halts the program scan for a specified period of time.

SFC 47 WAIT has input parameter WT of data type INT in which you can specify the waiting time in microseconds (ms).

The maximum waiting time is 32767 microseconds; the minimum waiting time corresponds to the execution time of the system function, which is CPU-specific.

SFC 47 can be interrupted by higher-priority events. On an S7-300, this increases the waiting time by the scan time of the higher-priority interrupt routine.

The system function **SFC 46 STP** terminates the program scan, and the CPU goes to STOP. SFC 46 STP has no parameters.

20.3.6 Multiprocessing Mode

The S7-400 enables multiprocessing. As many as four appropriately designed CPUs can be operated in one rack (universal rack UR) on the same P bus and K bus.

An S7-400 station is automatically in multiprocessor mode if you arrange more than one CPU in the central rack in the Hardware Configuration. The slots are arbitrary; the CPUs are distinguished by a number assigned automatically in ascending order when the CPUs are plugged in. You can also assign this number yourself on the "Multicomputing" tab.

The configuration data for all the CPUs must be loaded into the PLC, even when you make changes to only one CPU.

After assigning parameters to the CPUs, you must assign each module in the station to a CPU. This is done by parameterizing the module in the "Addresses" tab under "CPU assignment" (Figure 20.5). At the same time that you assign the module's address area, you also assign the module's interrupts to this CPU. With VIEW → FILTER → CPU No. x-MODULES, you can emphasize the modules assigned to a CPU in the configuration tables.

Table 20.6 Parameters for SFC 25 COMPRESS

SFC	Parameter	Declaration	Data Type	Contents, Description
25	RET_VAL	RETURN	INT	Error information
	BUSY	OUTPUT	BOOL	Compression still in progress (with "1")
	DONE	OUTPUT	BOOL	Compression completed (with "1")

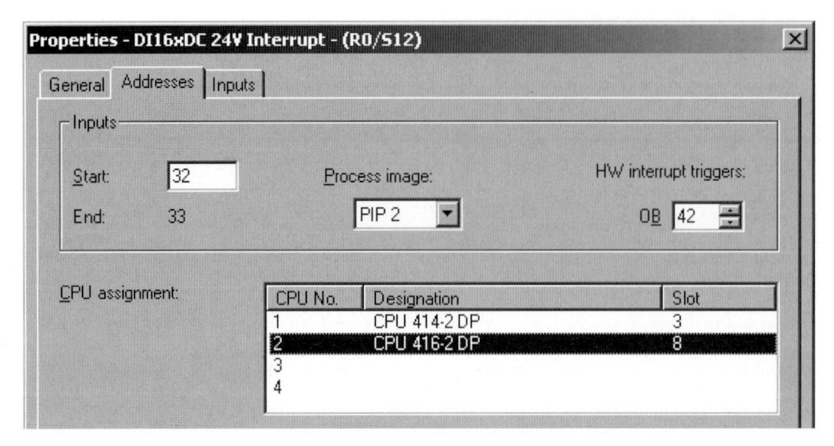

Figure 20.5 Module Assignments in Multiprocessor Mode

The CPUs in a multiprocessing network all have the same operating mode. This means

- ▷ They must all be parameterized with the same restart mode,
- ▷ They all go to RUN simultaneously,
- ▷ They all go to HOLD when you debug in single-step mode in one of the CPUs
- ▷ They all go to STOP as soon as one of the CPUs goes to STOP.

When one rack in the station fails, organization block OB 86 is called in each CPU.

The user programs in these CPUs execute independently of one another; they are not synchronized.

An SFC 35 MP_ALM call starts organization block OB 60 "Multiprocessor interrupt" in all CPUs simultaneously (see Chapter 21.7 "Multiprocessor Interrupt").

20.3.7 Determining OB Program Execution Time

The system function **SFC 78 OB_RT** determines the runtime of individual organization blocks over different periods. This enables you to determine the time loading of the user program (workload).

The operating system of a CPU designed for this purpose records the runtimes of the individual organization blocks and makes them available for reading by the SFC 78 OB_RT. The accuracy of the time recording depends on the CPU. The times are specified in microseconds. If a value is not available for the requested time, –1 (DW#16#FFFF FFFF) is returned.

Time measuring principle

A timer runs in the CPU's operating system with a relative time in microseconds from 0 to $2^{31}-1$. The timer is started on a transition from STOP to RUN, runs up to the upper limit, and then commences at zero again.

The operating system records the OB start event, the start and end of OB processing, and the interruptions caused by OBs of higher priority. The data of the last completed OB processing and of that being carried out when the SFC 78 is called are saved.

SFC call outside the OB to be measured

When using the SFC, a distinction is made between a call in the program of the requested OB and a call outside the requested OB. Example: the SFC 78 is called in OB 1 and has a value of 30 in parameter OB_NR. The last recorded times are then read for OB 30. Specification of the synchronous error OBs with numbers 121 and 122 is not permissible, since these are included in the priority class of the OBs causing the errors and thus to their program.

Figure 20.6 shows a number of examples of calling the SFC 78 outside the OB to be measured. The start values following a STOP/RUN transition are –1 (example ①).

Table 20.7 Parameters of the SFC 78

SFC	Parameter	Declaration	Data Type	Contents, Description
78	OB_NR	INPUT	INT	Number of the OB whose times are to be scanned
	RET_VAL	RETURN	INT	Error information or number of the OB with OB_NR=0
	PRIO	OUTPUT	INT	Priority class of scanned OB
	LAST_RT	OUTPUT	DINT	Runtime of last completed processing
	LAST_ET	OUTPUT	DINT	Time span between OB request and end of processing for the last completed processing
	CUR_T	OUTPUT	DINT	Time of OB request (relative time)
	CUR_RT	OUTPUT	DINT	Previous runtime of OB processing
	CUR_ET	OUTPUT	DINT	Time span between OB request and scanning by SFC 78 in the requested OB
	NEXT_ET	OUTPUT	DINT	Time span since next OB request and scanning by SFC 78

LAST_RT indicates the runtime in microseconds of the last completed OB processing (examples ② to ⑥). The "net" runtimes are output. The interruption times resulting from program execution levels of higher priority are not included in LAST_RT (④).

LAST_ET indicates the time span in microseconds between the start request and the end of processing for the last completed processing of the OB to be measured (examples ② to ⑥). In this case, the interruption times resulting from program execution levels of higher priority are included in LAST_ET (④).

CUR_T indicates the relative time in microseconds (counter value in operating system) of the start request of the OB. Following initialization, –1 is present in CUR_T (①). On completion of OB processing, CUR_T is set to zero. Since the SFC 78 is called outside the OB in these examples, it subsequently outputs zero in this parameter.

CUR_RT indicates the effective processing time of the OB in microseconds up to calling of the SFC 78. Following initialization, –1 is present in CUR_RT (①). On completion of OB processing, the value present in CUR_RT is imported into LAST_RT, and CUR_RT is set to zero. Since the SFC call is outside the OB in these examples, the value is always zero.

CUR_ET indicates the time span in microseconds from the OB start request up to calling of the SFC 78. On completion of OB processing, the value present in CUR_ET is imported into LAST_ET, and CUR_ET is set to zero. Since the SFC call is outside the OB in these examples, the value is always zero.

NEXT_RT indicates the time span in microseconds from the subsequent OB start request up to calling of the SFC if further unprocessed start requests are present. NEXT_RT is not determined in the currently supplied CPUs, and the value is always –1.

SFC call in program of OB to be measured

The SFC 78 can also be called in the program of the OB to be measured. The parameter OB_NR is then supplied with zero, and the parameter RET_VAL returns the current OB number if the processing is fault-free. The times are read for the OB in which the SFC 78 is called. When calling in one of the synchronous error OBs with the numbers 121 and 122, these are the data of the OB causing the error.

Figure 20.7 shows a number of examples of calling the SFC 78 in the program of the OB to be measured. This can be directly in the OB program or in one of the blocks called there. The start values following a STOP/RUN transition are –1.

LAST_RT indicates the runtime in microseconds of the last completed OB processing (ex-

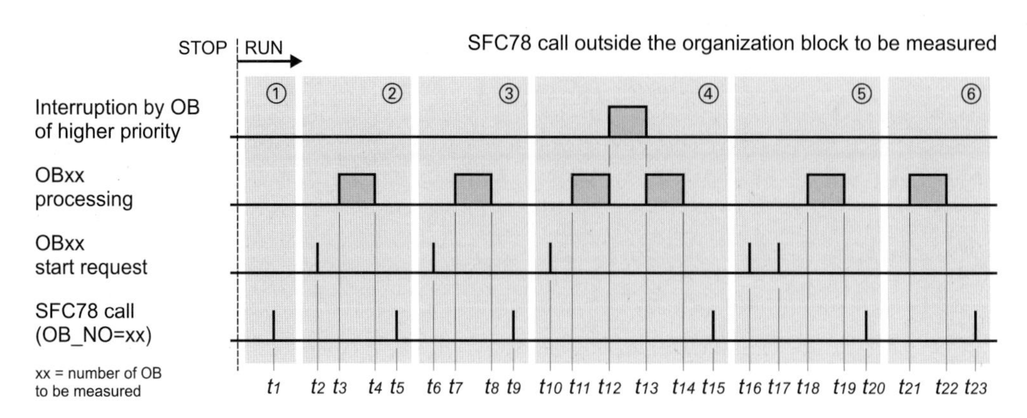

Values in the SFC78 parameters

Name	①	②	③	④	⑤	⑥
LAST_RT	–1	t4 - t3	t8 – t7	(t14 – t13) + (t12 – t11)	t19 – t18	t22 – t21
LAST_ET	–1	t4 - t2	t8 – t6	t14 – t10	t19 – t16	t22 – t17
CUR_T	–1	0	0	0	0	0
CUR_RT	–1	0	0	0	0	0
CUR_ET	–1	0	0	0	0	0
NEXT_ET	–1	–1	–1	–1	–1	–1

Figure 20.6 Call of SFC 78 Outside the Organization Block to be Measured

amples ⑧, ⑨ and ⑪). If the SFC is called more than once in the OB to be measured, –1 is output (example ⑩). The "net" runtimes are output; the interruption times resulting from program execution levels of higher priority are not included in LAST_RT (example ⑨).

LAST_ET indicates the time span in microseconds between the request and the end of processing for the last completed processing of the OB to be measured (examples ⑧ and ⑨). This also applies to the first call of the SFC 78 in the OB to be measured (⑨). If the SFC is called more than once in the OB to be measured, –1 is output (example ⑩). LAST_ET also contains the interruption times resulting from program levels of higher priority (⑨).

CUR_T indicates the relative time in microseconds (counter value in operating system) of the start request of the OB if – as in the following examples – the SFC 78 is called within the OB. On completion of OB processing, CUR_T is set to zero.

CUR_RT indicates the effective processing time of the OB in microseconds up to calling of the SFC 78. On completion of OB processing, the value present in CUR_RT is imported into LAST_RT, and CUR_RT is set to zero. The interruption times resulting from program levels of higher priority are not included in CUR_RT (⑨ and ⑩)

CUR_ET indicates the time span in microseconds from the OB start request up to calling of the SFC 78. On completion of OB processing, the value present in CUR_ET is imported into LAST_ET, and CUR_ET is set to zero. CUR_ET also includes the runtimes of the OBs of higher priority which interrupt the current OB to be measured.

NEXT_RT indicates the time span in microseconds from the subsequent OB start request up to calling of the SFC if further unprocessed start requests are present. NEXT_RT is not determined in the currently supplied CPUs, and the value is always –1.

20.3.8 Changing the Program Protection

The application program in a CPU can be protected against unauthorized access in three protection levels (see Chapter 2.6.2 "Protection of the user program"). You can use the system

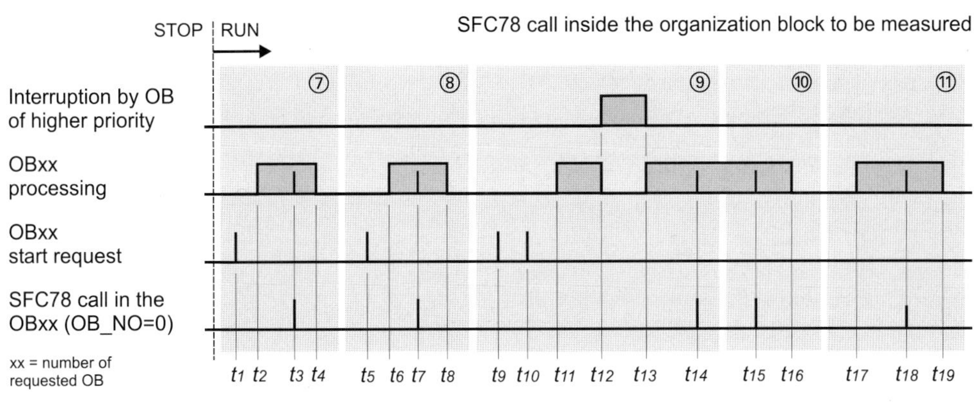

Values in the SFC78 parameters

Name	⑦	⑧	⑨	⑩	⑪
LAST_RT	–1	t4 – t2	t8 – t6	–1	(t16 – t13) + (t12 – t11)
LAST_ET	–1	t4 – t1	t8 – t5	–1	t16 – t9
CUR_T	t1	t5	t9	t9	t10
CUR_RT	t3 – t2	t7 – t6	(t14 – t13) + (t12 – t11)	(t15 – t13) + (t12 – t11)	t18 – t17
CUR_ET	t3 – t1	t7 – t5	t14 – t9	t15 – t9	t18 – t10
NEXT_ET	–1	–1	–1	–1	–1

Figure 20.7 Call of SFC 78 Inside the Organization Block to be Measured

function **SFC 109 PROTECT** to switch back and forth program-controlled between protection levels 1 and 2. The parameters of this system function are listed in Table 20.8.

Calling of the SFC 109 PROTECT is only effective if you have set protection level 1 in hardware configuration. It remains ineffective if protection level 2 or 3 is set or if a password has been entered in protection level 1 using the option "Can be canceled by password".

The protection level set using the SFC 109 PROTECT remains unchanged if

- ▷ the CPU goes to STOP as a result of a (program) error, by calling of SFC 46 STP, or by an operator input
- ▷ the CPU is buffered and the power supply returns
- ▷ a restart is carried out (S7-400)

In all other cases, the protection level is set to 1 when there is an operating mode transition. Even if you set the mode selector to the STOP position, protection level 1 is (re)set.

You can identify the current protection level online in the SIMATIC Manager by selecting the CPU selected and PLC → DIAGNOSTICS/SETTINGS → MODE. Using the program, you can identify the protection level with SFC 51 RDSYSST by means of the SSL parts list W#16#0232 with the index W#16#0004.

With a CPU with key-operated switch as mode selector, you can remove the switch in the RUN

Table 20.8 Parameters of the SFC 109 PROTECT

SFC	Parameter	Declaration	Data Type	Contents, Description
109	MODE	INPUT	WORD	Job ID: W#16#0000 Protection level 1 W#16#0001 Protection level 2
	RET_VAL	RETURN	INT	Error information

position, and only permit reading by the programming device during operation. This possibility does not exist if a toggle switch is used as mode selector. In this case, you can activate protection level 2 when switching on the CPU, i.e. write protection (only reading possible), by calling SFC 109 PROTECT in the startup organization blocks.

You can use SFC 109 PROTECT to change the protection level during ongoing operation without activating the mode selector. For example, you can use SFC 109 PROTECT and MODE = W#16#0000 to (re)set the protection level to 1 depending on the signal status of a binary variable, e.g. to download program sections. You can then activate write protection again with MODE = W#16#0001.

20.4 Communication via Distributed I/O

Distributed I/O is understood as modules connected over PROFIBUS DP or PROFINET IO.

With PROFIBUS DP, the DP master communicates with the DP slaves assigned to it over the PROFIBUS subnet. With PROFINET IO, it is the IO controller which exchanges data with the IO devices assigned to it over the Industrial Ethernet subnet.

Data exchange takes place "automatically", you need not be concerned with it. You configure and address the distributed I/O in a manner similar to the central modules using the Hardware Configuration. You address the inputs and outputs in the distributed I/O stations from the user program just like the inputs and outputs of the central modules.

20.4.1 Addressing PROFIBUS DP

Similarly to the way in which centrally arranged modules are assigned to a CPU and are controlled from the CPU, distributed modules with PROFIBUS DP (stations, DP slaves) are assigned to a DP master. The DP master with all its DP slaves is referred to as a DP master system. One S7 station can contain several DP master systems.

Currently available DP masters recognize two modes: DPV1 and S7-compatible. "S7-compatible" corresponds to the previous mode. In this case you can operate all DP standard slaves according to EN 50170, plus the "DP S7 slaves" from Siemens which were already able to send interrupts to the DP master. In DPV1 mode, DP slaves can be additionally used which exhibit the new properties according to IEC 61131, such as increased diagnostics and parameterization capabilities through the use of acyclically transmitted data records or the use of new types of interrupt. New system functions also exist for these new "DPV1 slaves" for transmission of data records, as well as new interrupt organization blocks.

The DP slaves occupy addresses in the I/O area of the CPU ("logical address area") just like the central modules. The DP master can be considered as being "transparent" for the addresses of the DP slaves; the CPU "sees" the addresses of the DP slaves so that their addresses do not overlap with those of the central modules, not even with those of DP slaves in other DP master systems assigned to the CPU.

Every DP slave has in addition to the node address a geographical address, a module starting address and at least a diagnostics address (Figure 20.8).

Node address

Every node on the PROFIBUS subnet has a unique address, the node address (station number) in that subnet that distinguishes it from the other nodes on the subnet. The station (the DP master or a DP slave) is accessed on PROFIBUS with this node address.

Please note that there must be a gap of at least 1 between the addresses of the active bus nodes (e.g. in the case of DP master and nodes in cross traffic). STEP 7 takes this into account when assigning node addresses automatically.

Geographical address

The geographical address identifies the module slot. With central modules, the geographical address contains the number of the rack and the

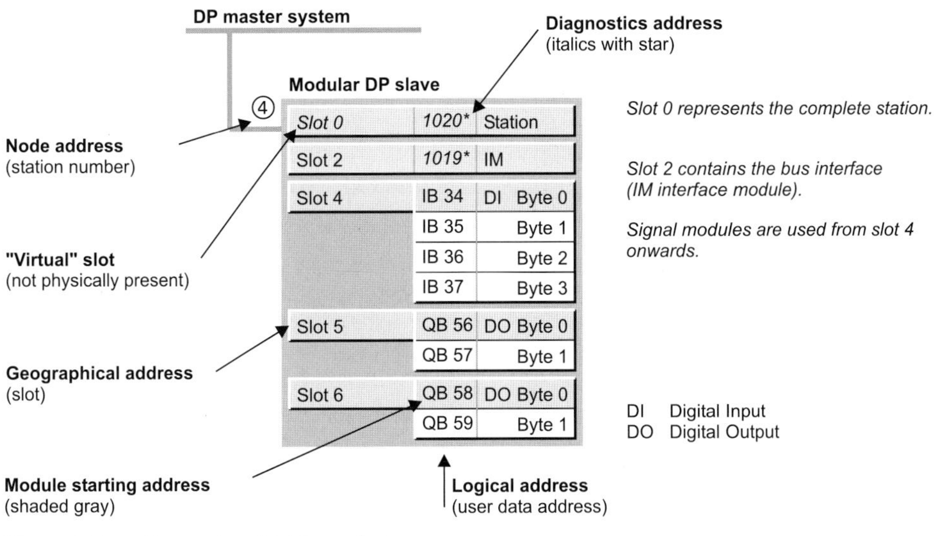

Figure 20.8 Addresses in a DP Master System

slot. Accordingly, with PROFIBUS DP, the geographical address contains the number of the DP master system, the station number and the slot number.

The slot number of a DP slave depends on its type. If it is integrated via a GSD file, the entries in the GSD file define the slot at which the I/O modules commence. With DP standard slaves, the slots for I/O modules commence at 1. The slot numbering of an S7 slave is based on the slots of an S7-300 station. Slots 1 (power supply) and 3 (expansion unit interface module) remain vacant. Slot 2 (CPU) corresponds to the interface module (header module) of the modular DP slave. The signal modules (SM) are positioned starting at slot 4. The "virtual" (not physically present) slot 0 exists in addition; it represents the complete station.

The case is similar with intelligent DP slaves. In this case, the transfer memory constitutes the interface to the DP master. Through configuration of the transfer memory, which you carry out with the Hardware Configuration, areas are formed which correspond to the modules or slots. The slots do not actually exist, and they are therefore referred to as "virtual" slots.

Virtual slot 0 represents the DP station, virtual slot 2 the bus interface, in this case the slave CPU as "header module" of the DP slave. From virtual slot 4 onwards, the user data areas are present in the transfer memory; they correspond to the signal modules. The virtual slots in the transfer memory are "seen" by both the master CPU and the slave CPU.

The definition of virtual slots permits direct assignment of diagnostics and interrupt events to the interface module or station (see "Diagnostics address" below). The system functions SFC 5 GADR_LGC and SFC 49 LGC_GADR and, also suitable for PROFINET IO, the system functions SFC 70 GEO_LOG and SFC 71 LOG_GEO are available for conversion of the geographical address into the logical address and vice versa.

Note that the DP slaves which are integrated into the Hardware Configuration via a GSD file according to EN 50170 equal to Revision 3 or higher (DPV1) can save the user data starting at slot 1.

Logical address, module starting address

With the logical address you access the user data of a station. Each byte of the user data is defined unambiguously by the logical address. The logical address corresponds to the absolute address; it can be assigned a symbol (a name) so that it is easier to read (symbolic addressing).

The smallest logical address of a module or station is the module starting address (see also Chapter 1.4 "Module Addresses").

Diagnostics address

The diagnostics address is used to identify modules and stations which can deliver diagnostics data and do not have a user data address themselves. The diagnostics address occupies one byte of peripheral inputs in the logical address volume. As standard, STEP 7 assigns the diagnostics address commencing with the highest address in the I/O area of the CPU. You can change the diagnostics address. The address overview in the Hardware Configuration identifies the diagnostics address by a asterisk.

Signal modules or user data areas in the transfer memory of intelligent DP slaves have logical addresses, also for scanning diagnostics data. The complete station delivers its diagnostics data via a diagnostics address which is assigned to the virtual slot 0. With modular and intelligent DP slaves, the bus interface can deliver its diagnostics data via the diagnostics address of slot 2.

Figure 20.9 shows an example of the diagnostics addresses in a DP master system. A compact DP slave has one diagnostics address for the complete station, a modular DP slave has one diagnostics address for the station and one for the interface module. Intelligent DP slaves have an additional diagnostics address for the DP interface.

The diagnostic addresses are assigned starting downwards from the highest I/O address. For example, the DP interface of the CPU 317-2PN/DP is assigned the diagnostic address 8191, the PN interface and the two ports the addresses 8190 to 8188 (not shown in Figure 20.9), and the virtual slots 0 and 2 in the transfer memory the addresses 8187 and 8186. It is similar with the master CPU: Here, the diagnostic addresses begin at 16383 for the DP interface and continue with 16382 for the MPI/DP interface (not shown), 16381 and 16380 for virtual slots 0 and 2 of the first intelligent DP slave, 16379 and 16378 for the second DP slave, etc. The diagnostic address for the DP slaves is assigned by the hardware configuration following the order of connections to the DP master system.

The diagnostics data are scanned by system blocks in the user program. The system function SFC 13 DPNRM_DG is available for this for conventional DP standard slaves. With DP S7 slaves, the SFC 59 RD_REC is used to read the data set DS1 with the diagnostics data. DPV1 slaves are able to deliver more comprehensive diagnostics data which can be read with the system function block SFB 52 RDREC. The modules are addressed using the logical module starting addresses of the user data or using the diagnostics addresses.

Transfer memory on intelligent DP slaves

In the case of compact and modular DP slaves, the addresses of the inputs and outputs together with the addresses for the central modules are located in the address volume of the master CPU. In the case of intelligent slaves, the master CPU has no direct access to the input/output modules of the DP slave. Every intelligent DP slave therefore has a transfer memory whose size depends on the CPU used. The transfer memory can be divided into several subsidiary areas of different length and data consistency. The individual areas then respond like modules whose lowest address is the module starting address. From the viewpoint of the master CPU, the intelligent DP slave then appears as a compact or modular DP slave, depending on the division (Figure 20.10).

During configuration of the slave, you can configure individual areas of the transfer memory as inputs or outputs with the "module starting address" and the area length. Exception: if the CP 342-5DP forms the DP interface for the intelligent slave, the division of its transfer memory is not configured until the slave is connected to the DP master system. The addresses of the transfer memory must not overlap with those of the central modules located in the intelligent DP slave. If the addresses are present in the process image, the areas can be handled by the user program like inputs and outputs, otherwise like peripheral inputs and peripheral outputs. If the slave CPU has partial process images, you can also assign a partial process image to each area.

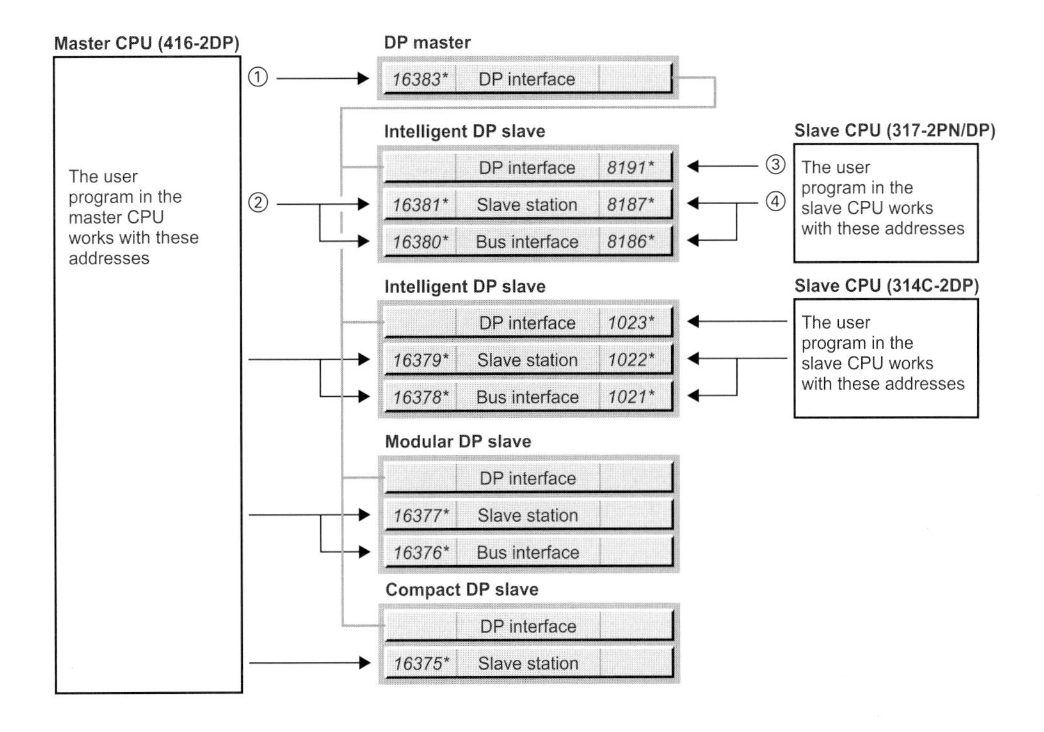

① Each bus interface has a diagnostic address which is set in the object properties of the interface in the "Addresses" tab (double-click on the "DP" line in the master CPU).

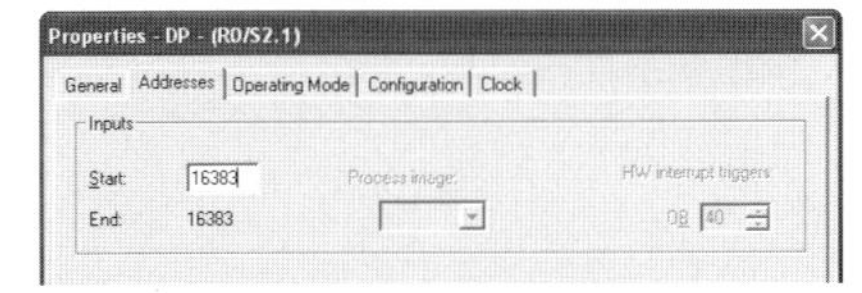

③ Each bus interface has a diagnostic address which is set in the object properties of the interface in the "Addresses" tab (double-click on the "DP" line in the slave CPU).

② The station and bus interface in the station have a diagnostic address which is set in the object properties of the station in the "General" tab (double-click in the hardware configuration of the DP master on the DP slave station).

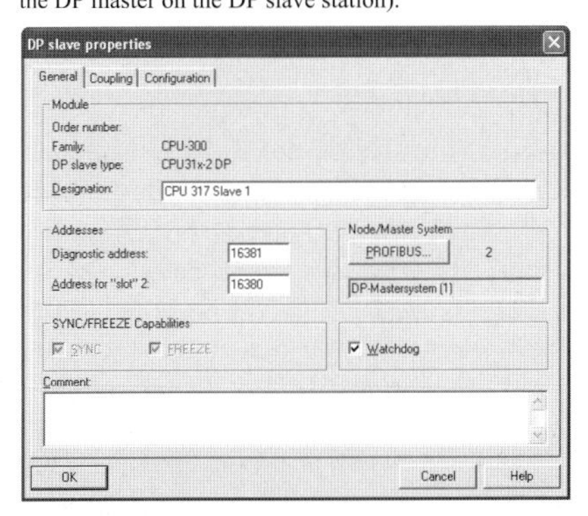

④ The station and bus interface in the station have a diagnostic address which is set in the object properties of the station in the "Mode" tab (double-click on the DP line in the slave CPU).

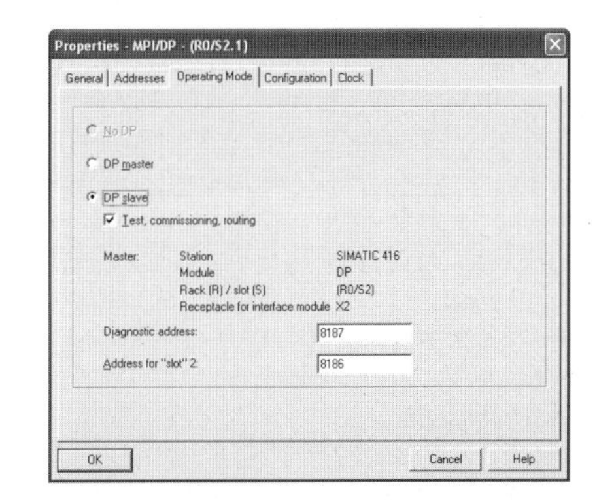

Figure 20.9 Diagnostic addresses in a DP master system

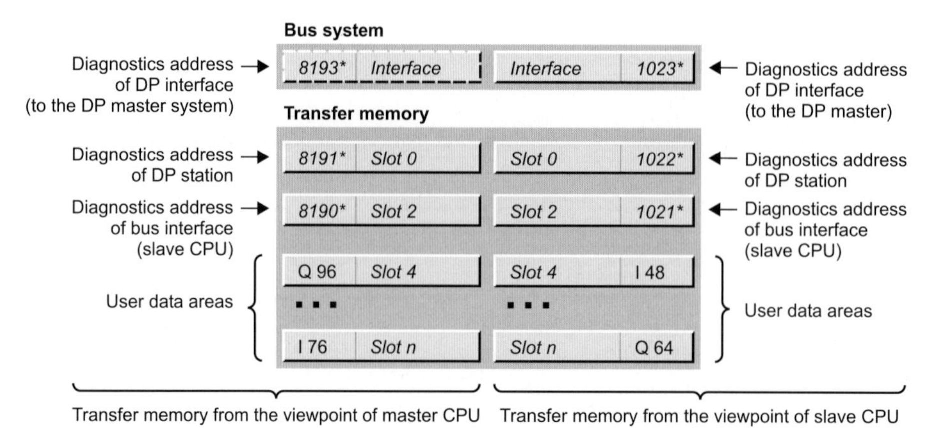

Figure 20.10 Transfer memory of an intelligent DP slave

When coupling to the DP master system, you supplement the configuration at the master end by the "module starting addresses" from the viewpoint of the master CPU and by the transmission direction. You assign inputs on the slave side to outputs on the master side and vice versa. If the addresses are present in the process image, the areas can be handled by the user program like inputs and outputs, otherwise like peripheral inputs and outputs. If the master CPU possesses process image partitions, you can also assign a process image partition to each area. From the viewpoint of the master CPU, the addresses of the transfer memory may not overlap with the addresses of other modules in the (central) S7 station, neither with the addresses of the centrally arranged modules, nor with the addresses in other DP master systems assigned to the master CPU.

You also define the diagnostics addresses from the viewpoint of the intelligent DP slave during configuration of the DP slave. You define the diagnostics addresses from the viewpoint of the master CPU when coupling of the intelligent DP slave to the DP master system.

20.4.2 Configuring PROFIBUS DP

General procedure

You configure the distributed I/O on PROFIBUS DP essentially in the same way as the centralized modules. Instead of arranging modules in a mounting rack, you assign DP stations (PROFIBUS nodes) to a DP master system. The following order is recommended for the necessary actions:

1) Create a new project or open an existing one with the SIMATIC Manager.

2) Create a PROFIBUS subnet in the project with the SIMATIC Manager and, if required, set the bus profile.

3) Use the SIMATIC Manager to create the master station in the project that is to accommodate the DP master, e.g. an S7-400 station.

   If your system contains intelligent DP slaves, you also create the relevant slave stations at this point, e.g. S7-300 stations.

   You start the Hardware Configuration by opening the master station.

4) With the Hardware Configuration, you place a DP master in the master station. This can be, for example, a CPU with integral DP interface. You assign the previously created PROFIBUS subnet to the DP interface and you then have a DP master system. You also define the DP mode in the “Operating mode” tab: DPV1 or S7-compatible. You can also configure the remaining modules later. Save and compile the station.

5) When you have created an S7 station for an intelligent DP slave, open it in the hardware configuration and "plug in" the module with the desired DP interface, e.g. an S7-300 CPU with integrated DP interface or an ET 200pro basic module IM 154-8/CPU. If necessary, set the DP interface to "DP slave", assign the previously created PROFIBUS subnet to the DP interface, and configure the transfer memory from the viewpoint of the DP slave. You can also configure the remaining modules later. Save and compile the station.

   Proceed in the same way for the remaining stations intended for intelligent DP slaves.

6) Open the master station with the DP master system and use the mouse to drag the PROFIBUS nodes (compact and modular DP slaves) from the hardware catalog to the DP master system. Assign node addresses and, if necessary, set the module starting address and the diagnostics address.

7) If you have created intelligent DP slaves, drag the relevant icon (in the hardware catalog under "PROFIBUS DP" and "Configured Stations") with the mouse to the DP master system.

   Open the icon and assign an already configured intelligent DP slave ("Couple"), assign a node address, and configure the transfer memory from the viewpoint of the DP master (or from the viewpoint of the central master CPU). Proceed in the same manner with each intelligent DP slave.

8) Save and compile all stations. The DP master system is now configured. You can now supplement the configuration with centralized modules or with further DP slaves.

You can also represent the DP master system configured in this way graphically with the Network Configuration tool. Open Network Configuration by, for example, double-clicking on a subnet. Select VIEW → DP SLAVES/IO DEVICES to display the slaves. You can also create a DP master system (or more precisely, assign the nodes to a PROFIBUS subnet) with the Network Configuration tool. You parameterize the stations after opening them with the Hardware Configuration. Here too, you must first set up an intelligent DP slave before you can integrate it into a DP master system.

Configuring the DP master

Prerequisite: You have created a project and an S7 station with the SIMATIC Manager. Open the S7 station and position the DP master module. This can be an S7 or ET 200 CPU with DP interface, a CP 342 communications processor (for S7-300), or an IM 467 interface module (for S7-400) (see Chapter 2.3.1 "Arranging Modules").

When placing the DP master module, you select in a window the PROFIBUS subnet to which the DP master system is to be assigned and the node address to be assigned to the DP master. You can also create a new PROFIBUS subnet in this window.

You also define the DP mode with which the DP master is to be operated in the "Operating mode" tab. This mode applies to the complete DP master system.

Together with the DP master, the hardware configuration shows a DP master system (broken black-and-white bar) in the station window. If there is no DP master system available (it may be that it is obscured behind an object or it is outside the visible area), create one by selecting the DP master in the configuration window and then selecting INSERT → MASTER SYSTEM. You can change the node address and the connection to the PROFIBUS subnet by selecting the module and then making your changes with the "Properties" button on the "General" tab under EDIT → OBJECT PROPERTIES.

CP 342-5DP as DP master

If a CP 342-5DP is the DP master, place it in the configuration table of the station, select it and then EDIT → OBJECT PROPERTIES. Set "DP Master" on the "Mode" tab.

The "Addresses" tab shows the user data address occupied by the CP in the address area of the CPU. From the viewpoint of the master CPU, the CP 342-5DP is an "analog module" with a module starting address and 16 bytes of user data.

Only DP standard slaves, or DP S7 slaves that behave like DP standard slaves, can be connected to a CP 342-5DP as DP master. You can find the suitable DP slaves in the hardware catalog under “PROFIBUS DP” and “CP 342-5 as DP Master”. Select the desired slave type and drag it to the DP master system.

The transfer memory as DP master has a maximum length of 240 bytes. It is transferred as one with the loadable blocks FC 1 DP_SEND and FC 2 DP_RECV (included in the *Standard Library* under the *Communication Blocks* program).

The data consistency covers the entire transfer memory.

You read the diagnostics data of the connected DP slaves with FC 3 DP_DIAG (e.g. station list, diagnostics data of a specific station). FC 4 DP_CTRL transfers control jobs to the CP 342-5DP (e.g. SYNC/FREEZE command, CLEAR command, set operating state of the CP 342-5DP).

If you select CPU or CP 342-5DP, VIEW → ADDRESS OVERVIEW shows you a list of the assigned addresses, inputs and/or outputs. You can also screen the existing address gaps.

Configuring compact DP slaves

The compact DP slaves can be found in the hardware catalog under "PROFIBUS DP" and the corresponding subcatalog, e.g. ET 200B. Click on the selected DP slave and use the mouse to drag it to the DP master system icon.

You will see the properties sheet of the station; here, you set the node address and any diagnostics address. Then the DP slave appears as an icon in the upper section of the station window and the lower section contains a configuration table for this station.

A double-click on the icon in the upper section of the station window opens a dialog box with one or more tabs in which you set the desired station properties. In the lower sub-window, you then see the input/output addresses. Double-clicking on an address line shows you a window where you can change the suggested addresses.

The lower sub-window shows optionally the configuration table of the selected DP slave or of the master system (toggle with the “arrow” button).

Configuring modular DP slaves

The modular DP slaves can be found in the hardware catalog under "PROFIBUS DP" and the corresponding subcatalog, e.g. ET 200M.

Click on the selected interface module (basic module) and drag it to the icon for the DP master system. This screens the properties sheet for the station; here, you set the node address and any diagnostics address. Then the DP slave appears as an icon in the upper section of the station window and the lower section contains configuration table for this station.

Now place the modules that you can find in the hardware catalog *under the selected interface module (!)* in the configuration table. Double-clicking on the line opens the properties sheet of the module and allows you to parameterize the module.

The bottom pane displays either the configuration table of the selected DP slave or of the master system (selectable using the “Arrow” button).

Combining modules

Digital electronic modules with two or four bit channels, e.g. with ET 200S or ET 200pro, initially occupy one address each with one byte in the configuration table. After all modules have been configured, you can use the "Pack addresses" button in the configuration table to remove the gaps between the bit channels of the selected modules; fewer addresses are thus occupied. The address areas for inputs, outputs, and motor starters are packed separately.

Please note the following special features of a “packed” module:

▷ Slot assignment is no longer possible; the CPU cannot determine a geographical address for this module.

▷ No module status information can be read for this module.

▷ Interrupts cannot be assigned to a “packed” address. A diagnostics address is therefore assigned to a module (italics with star in the configuration table), at which you can obtain interrupt information.

▷ “Pack addresses” and “Insert/remove module interrupts” are mutually exclusive.

Handling of options in general

With the handling of options you prepare an ET 200S or ET 200pro station for a future expansion. This means you can start operation with a small configuration and subsequently upgrade the station to a (previously) planned maximum configuration without modification of the hardware configuration. To do this, you require appropriately designed IM interface modules and PM power modules. The user program in an ET 200S can scan and control the slot assignment during runtime via a control and feedback interface.

Please note when calculating runtimes, as necessary e.g. for isochronous mode, that these must always be based on the configured maximum configuration.

Depending on the version of the interface module, you can use the handling of options with or without reserve modules. With the handling of options without reserve modules, you configure the envisaged maximum configuration of the ET 200 station, where you position the modules to be inserted later in the rear slots. Before the station commences with cyclic data exchange with I/O access operations following the initial startup, you use the user program to define the slots identified as "reserve" via the control interface, and then enable operation. Following replacement by the planned module, you cancel the identification as "reserve" using the user program.

With the option handling function with reserve modules, you can carry out the retrofitting at any slots – occupation of the slots without gaps must be ensured however. You configure the planned maximum configuration and identify the slots which are to be initially occupied by reserve modules in the hardware configuration. The reserve modules provide the signal status "0" as replacement value for digital inputs, and the value W#16#7FFF for analog channels. Following replacement by the planned module, you delete the "Reserve" identification in the user program via the control interface. The use of reserve modules provides the advantage that the wiring for the maximum configuration can be carried out completely right at the beginning.

Handling of options with ET 200S

Prerequisite: You have at least configured the ET 200S station with the power module. You activate the handling of options in both the IM interface module and the PM power module.

To activate the handling of options in the interface module, open the station's Properties window (select station and EDIT → OBJECT PROPERTIES). On the "Option handling" tab, switch on the handling of options using the "Option handling active" checkbox. With a correspondingly designed module, you can choose between "Without reserve module" and "With reserve module". When selecting "Without reserve module", terminate configuration using "OK".

If you have chosen "With reserve module" or if this selection is missing for the interface module, now select the slots in the "Parameters" window which are only to be equipped later with the envisaged modules. The reserve modules are not configured; you configure the station with the planned maximum configuration of electronic modules envisaged for later.

Open the Properties window of the power module (select module in the configuration table and EDIT → OBJECT PROPERTIES).

On the "Addresses" tab, check the "Option handling" checkbox. In the address area of the station, an additional eight bytes of digital inputs are then assigned for the feedback interface and an additional eight bytes of digital outputs for the control interface. The user program can query via the feedback interface whether the handling of options is active and whether the envisaged (configured) module is inserted in a slot and ready for operation.

If you activate the handling of options immediately following "insertion" of the power module and before configuration of the electronic modules, the addresses of these interfaces are set as standard to the beginning of the ET 200S address area.

Handling of options with ET 200pro

Prerequisite: You have at least configured the ET 200pro station with the power module. The power module must be designed for the handling of options (PM-O). You activate the handling of options in the IM interface module.

To activate the handling of options in the interface module, open the station's Properties window (select station and EDIT → OBJECT PROPERTIES). On the "Operating parameters" tab, check the "Option handling" checkbox.

Configuring a CPU with integral DP interface as an intelligent DP slave

With an appropriately equipped CPU, you can parameterize the station either as a DP master station or as a DP slave station. Before the station can be connected as a DP slave to a DP master system, it must be created. The procedure for doing this is exactly the same as that for a “normal” station; insert an S7 station into the project using the SIMATIC Manager and open the *Hardware* object. Drag a mounting rack to the window in the Hardware Configuration and place the desired modules. For configuring the DP slave, it is enough to place the CPU; you can add all other modules later.

When inserting the CPU, the properties sheet of the PROFIBUS interface is screened. Here, you must assign a subnet to the DP interface and you must assign an address. If the PROFIBUS subnet does not yet exist in the project, you can create a new one with the “New” button. This is the subnet to which the intelligent slave will later be connected.

EDIT→ OBJECT PROPERTIES with a selected DP interface or double-click on the interface to open the properties sheet of the interface. On the "Mode" tab, select the option "DP Slave". Now you can configure the transfer memory on the "Configuration" tab from the viewpoint of the DP slave. Select MS (master/slave configuration) as the mode and define the structure and addresses of the transfer memory from the viewpoint of the slave CPU. You can find information on the transfer memory in Chapter 20.4.1 “Addressing PROFIBUS DP” under “Transfer memory on intelligent DP slaves”.

If the intelligent DP slave is already coupled to a DP master system, you can also immediately enter the user data addresses from the viewpoint of the DP master (Figure 20.11).

The size and structure of the transfer memory depends on the CPU. In the case of the CPU 317-2DP, for example, you can divide the entire transfer memory into a total of 32 address areas that you can access separately. Such an address area can be up to 32 bytes in size. The entire transfer memory can encompass up to 244 input addresses and 244 output addresses.

The locally defined addresses are within the address volume of the slave CPU. These addresses must not overlap with addresses of the central or distributed modules in the DP slave station. The lowest address of an address area is the "module starting address".

The user program in the slave CPU gets diagnostics information from the DP master via the diagnostics addresses specified on this tab.

You terminate configuration of the intelligent DP slave with STATION → SAVE AND COMPILE. Connecting the intelligent DP slave into the DP master system is described below.

Configuring an ET200pro as an intelligent DP slave

If you want to create an ET 200pro as intelligent DP slave, first add a SIMATIC-300 Station in SIMATIC Manager under the project and open the *Hardware* object.

Now drag an interface module with CPU functionality *IM xxx CPU* under "PROFIBUS DP" and "ET 200pro" in the hardware catalog to the free window or select it by double-clicking on it. On the displayed properties sheet of the Ethernet interface, set "not networked".

EDIT → OBJECT PROPERTIES with the X1 interface selected or a double-click on the MPI/DP interface opens the properties window. On the “General” tab, select PROFIBUS as the interface, and select the station address and the PROFIBUS subnet in the properties window of the PROFIBUS interface. If the PROFIBUS subnet does not yet exist in the project, you can create one using the “New” button. This is the subnet to which the intelligent slave will be connected later.

Select the “DP slave” option on the “Mode” tab. The meanings of the addresses on this tab and of the address on the tab “Addresses” are described in Chapter 20.4.1 “Addressing PROFIBUS DP” under “Diagnostics address”.

Now you can configure the transfer memory on the "Configuration" tab from the viewpoint of the DP slave. Select MS (master/slave configu-

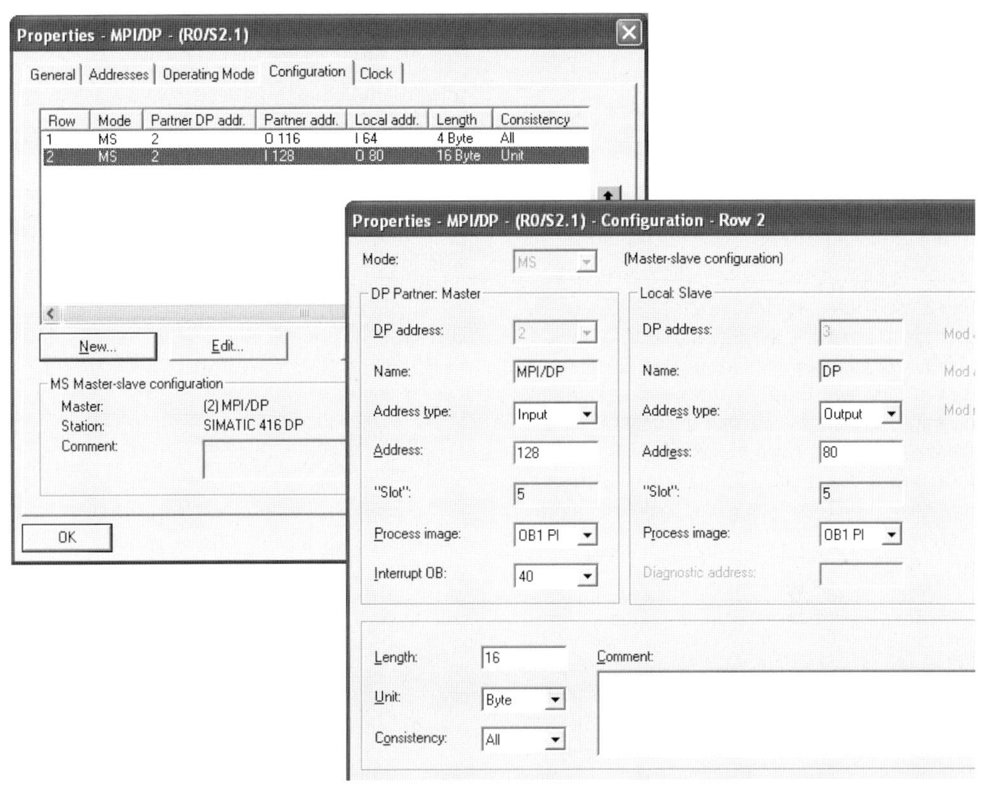

Figure 20.11 Configuring the transfer memory of an intelligent slave with integrated DP interface

ration) as the mode and define the structure and addresses of the transfer memory from the viewpoint of the ET 200pro CPU. If the intelligent DP slave is already coupled to a DP master system, you can also immediately enter the user data addresses from the viewpoint of the DP master (Figure 20.11). You can find information on the transfer memory in Chapter 20.4.1 “Addressing PROFIBUS DP” under “Transfer memory on an intelligent IO device”.

On the IM 154-8 CPU interface module, you can divide the entire transfer memory into up to 32 address areas that you can address separately. Such an address area can be up to 32 bytes in size. The entire transfer memory can encompass up to 244 input addresses and 244 output addresses.

The locally defined addresses are within the address volume of the ET 200pro CPU. These addresses must not overlap with addresses of the central or distributed modules in the ET 200pro station. The lowest address of an address area is the "module starting address".

The further configuration of the ET 200pro station is as with an S7-300 station with fixed slot addressing. You can only arrange modules that can be found in the hardware catalog under the interface module used.

STATION → SAVE AND COMPILE is used to finish the configuration of the ET 200pro station as an intelligent DP slave. The connection of the intelligent DP slave to the DP master system will be described further below.

Configuring an ET 200S as intelligent DP slave

If you want to create an ET 200S as an intelligent DP slave, first add a SIMATIC 300 station in SIMATIC Manager under the project and open the *Hardware* object.

Now drag an interface module with CPU functionality *IM xxx CPU* under "PROFIBUS DP" and "ET 200S" from the hardware catalog to the free window or select it by double-clicking on it.

You are given a configuration table as with a SIMATIC 300 station. Instead of the CPU, there is an IM 151 intelligent interface module of the ET 200S station.

Double-clicking on the IM 151 line opens the window for the IM properties; double-clicking on the DP interface opens the properties window of the interface. If you have not already done so, set the interface type "PROFIBUS", the node address, and the PROFIBUS subnet used on the "General" tab.

Set the address areas for the transfer memory from the viewpoint of the DP slave. If the intelligent DP slave is already coupled to a DP master system, you can also immediately enter the user data addresses from the viewpoint of the DP master (Figure 20.11). The maximum size of the user data area is 32 bytes of inputs and 32 bytes of outputs for the IM 151/CPU interface module. You can divide this area into eight subareas with different data consistencies. The slave program obtains diagnostic information from the DP master via the diagnostic address.

The further configuration of the ET 200S station is as with an S7-300 station with fixed slot addressing. You can only arrange modules that can be found in the hardware catalog under the interface module used.

You conclude configuring the intelligent DP slave with STATION → SAVE AND COMPILE. Integration of the intelligent DP slave into the DP master system is described further below.

Configuring an S7-300 station with CP 342-5DP as an intelligent DP slave

If you insert an S7-300 station in the SIMATIC Manager, open the *Hardware* object and configure a "normal" S7-300 station. Among other things, you arrange a CP 342-5DP communications processor in the configuration table.

When inserting the station, the properties sheet of the DP interface appears; the subnet to which the intelligent DP slave is later to be connected is to be assigned to the DP interface here and you must also assign the node address.

To open the properties window, select the CP 342-5DP and then EDIT → OBJECT PROPERTIES, or double-click on the CP 342-5DP. Select the option "DP Slave" on the "Mode" tab.

The "Addresses" tab shows the transfer memory from the viewpoint of the slave CPU. The maximum size of the transfer memory with the CP 342-5DP as DP slave is 240 bytes each for inputs and outputs, which you can divide into a maximum of 63 address areas following coupling to the master system.

STATION → SAVE AND COMPILE terminates configuration of the intelligent DP slave.

Connecting an intelligent DP slave to a DP master

You must have created a project and configured a DP master station and the intelligent DP slave (in each case at least with the DP interface). The DP master and the DP slave must be configured for the same PROFIBUS subnet.

Open the master station; a DP master station (black/white dashed rail) must be present, otherwise generate it with the DP interface selected with INSERT → MASTER SYSTEM.

In the hardware catalog under "PROFIBUS DP" and "Configured stations", you can find the objects which represent the intelligence slaves: "CPU31x" and "CPU41x" stand for S7-300 or S7-400 stations with an integrated DP slave, "ET 200pro/CPU" and "ET 200S/CPU" stand for stations configured as DP slave, and in the folder "S7-300 CP342-5 DP" you can find the proxies for S7-300 stations with CP 342-5 as DP slave interface. Select the desired slave type and drag it onto the DP master system.

CPU, ET 200pro or ET 200S as DP slave

"Dragging" to the DP master system or double-clicking on the DP slave opens the properties sheet. The "Coupling" tab lists the DP slaves already configured for this PROFIBUS subnet. Select the required DP slave and click on the "Couple" button. The result is shown in the same dialog box under the active coupling.

In the "General" tab, set the diagnostics address of the DP slave from the viewpoint of the master station.

Now, in the "Configuration" tab, set the addresses of the transfer memory from the viewpoint of the DP master. Output addresses at the master are input addresses at the slave and vice versa. You can find further information on the transfer memory in Chapter 20.4.1 "Addressing PROFIBUS DP" under "Transfer memory on intelligent DP slaves".

CP 342-5DP as DP slave

"Dragging" to the DP master system or double-clicking on the DP slave opens the properties sheet. The "Coupling" tab lists the DP slaves already configured for this PROFIBUS subnet. Select the required DP slave and click on the "Couple" button. The result is shown in the same dialog box under the active coupling.

The configuration table of the selected DP slave is shown in the lower part of the station window. You now configure the transfer memory: Drag the modules with the required properties from the selection under the used CP, or the universal module, into the configuration table.

EDIT → OBJECT PROPERTIES with the module selected in the bottom part of the window, or a double-click on the table line, opens a window in which you can set the start address. In the properties of the universal module, you can set whether an empty location, an input or output area, or both is to be displayed.

If a CP 342-5DP is the DP master, structuring of the transfer memory can be omitted because the CP 342-5DP transfers the entire transfer area in one piece.

When dividing the transfer memory, you arrange the address areas together without gaps starting from byte 0. You access the entire assigned transfer memory in the slave CPU with the loadable blocks FC 1 DP_SEND and FC 2 DP_RECV (included in the *Standard Library* under the *Communication Blocks* program).

The data consistency covers the entire transfer memory.

On the "General" tab, you set the diagnostics address of the DP slave from the viewpoint of the master station. The diagnostics data are read with FC 3 DP_DIAG (in the master station).

You can find further information on the transfer memory in Chapter 20.4.1 "Addressing PROFIBUS DP" under "Transfer memory on intelligent DP slaves".

Configuring the DP/DP coupler

The DP/DP coupler connects two PROFIBUS subnets. It is configured as a modular DP slave in both subnets.

Prerequisite: the two subnets, each with a DP master system, are configured. Open one of the stations with the DP master. In the hardware catalog under "PROFIBUS DP" and "Network components" you can find the *DP/DP Coupler, Release 2*, which you drag using the mouse to the DP master system.

The properties sheet of the PROFIBUS interface appears on which you set the node address. You set the diagnostic address and further parameters on the "General" and "Parameterize" tabs in the object properties of the DP/DP coupler.

When the DP/DP coupler is selected, the configuration table for the transfer memory is shown. Then "insert" the desired modules which are listed in the hardware catalog under the DP/DP coupler into the configuration table without gaps, starting with slot 1. You can set the universal module to the desired number of inputs and outputs. The user data addresses which you specify in the module properties are in the address space of the opened DP master CPU.

Configure the second part of the DP/DP coupler in the same manner. The structure of the transfer memory must agree with that of the first part. Note that inputs on one side are outputs on the other side, and vice versa. The addresses in both parts of the DP/DP coupler depend on the address assignments of the respective master CPU, and can differ.

Configuring the DP/AS-i link

You configure the DP/AS interface link like a modular DP slave. Under "PROFIBUS DP" and "DP/AS-i" in the hardware catalog, you can select the modules which you can "drag" to

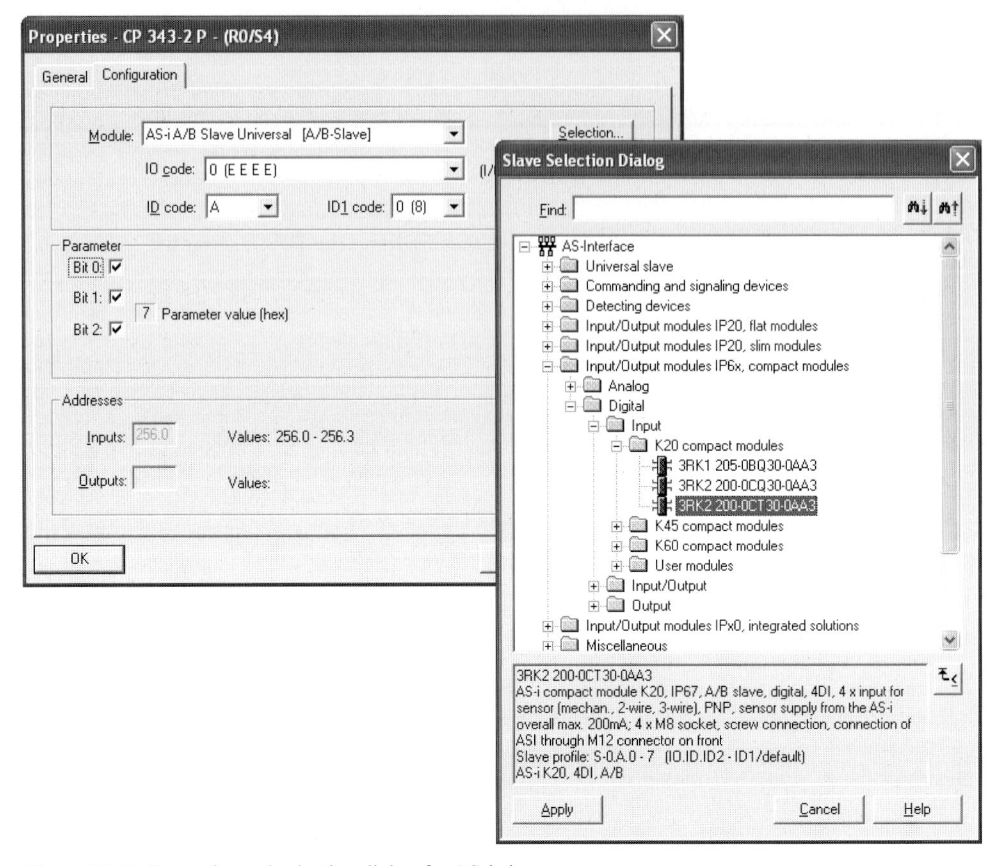

Figure 20.12 Properties and selection dialog for AS-i slaves

the DP master system. You subsequently set the properties of the DP part and configure the AS-i slaves – in the properties sheet or the configuration table depending on the link.

The AS-i master system with the AS-i slaves is not displayed as a subnet by the Hardware Configuration.

DP/AS-i Link 20E and DP/AS-i Link Advanced

"Drag" the DP/AS-i Link from the hardware catalog to the bar of the PROFIBUS DP master system. You can set the node address on the displayed window or in the properties of the DP slave under the PROFIBUS button.

A configuration table with the AS-i interfaces and AS-i slaves appears. If you have selected the double master for *DP/AS-i Link Advanced*, the configuration table is designed for both masters.

Now set the address area under which you want to access the AS-i slaves from the user program. EDIT→ OBJECT PROPERTIES with a selected AS-i master interface (in the configuration table), or double-click on the master line to open the properties dialog. On the "Digital Addresses" tab, you can set the start address and the inputs; this address also applies to the outputs. You can select different values for the reserved area lengths for inputs and outputs. This tab also contains the "Pack" and "Sort" buttons with which you can optimize the address assignments after you have configured the AS-i slaves.

Then "drag" the dummy value for an AS-i slave located under the link from the hardware catalog into the configuration table. Repeat the process for all envisaged AS-i slaves. EDIT → OBJECT PROPERTIES with the AS-i slave selected (in the configuration table) or double-clicking on the slaves line opens the properties dialog. Set the slave properties in the "Configuration" tab. Using the "Selection" button you are provided with all AS-i slaves known to the Hardware Configuration (Figure 20.12).

Configuring the DP/RS232C link

You configure the DP/RS232C link as with a modular DP slave. You can find the DP/RS232C link in the hardware catalog under "PROFIBUS DP", "Other field devices" and "Gateway" and can drag it to the DP master system.

The properties sheet of the PROFIBUS interface appears on which you set the node address. You set the diagnostic address and further parameters on the "General" and "Parameterize" tabs in the object properties of the DP/RS232C link.

With the DP/RS232C link selected, you are provided with the configuration table. Now "Connect" the desired modules listed in the hardware catalog under DP/RS232C link into the configuration table, without gaps and starting at slot 1. You can set the universal module to the desired number of inputs and outputs. The user data addresses that you specify in the module properties are in the address area of the opened DP master CPU.

20.4.3 Special Functions for PROFIBUS DP

GSD files

You can subsequently install DP slaves which are not included in the module catalog. To do this, you require the type file tailored to the slave (GSD file, General Station Description). As of GSD revision 3, DP slaves introduced with a GSD file support the DPV1 functionality.

To install, select OPTIONS → INSTALL GSD FILES in the hardware configuration and enter the directory of the GSD file or of a different STEP 7 project in the displayed window. STEP 7 imports the GSD file and shows the slave in the hardware catalog under "PROFIBUS DP" and "Other field devices".

STEP 7 saves the GSD files in the directory ...\Step7\S7DATA\GSD. The GSD files deleted when installing or importing at a later time are stored in the subdirectory ...\GSD\BKP*n*. From here, they can be restored with OPTIONS → INSTALL GSD FILE.

Configuring SYNC/FREEZE groups

The SYNC control command causes the DP slaves combined as a group to output their output states simultaneously (synchronously). The FREEZE control command causes the DP slaves combined as a group to "freeze" the current input signal states simultaneously (synchronously), in order to allow them to be then fetched cyclically by the DP master. The UNSYNC and UNFREEZE control commands revoke the effect of SYNC and FREEZE respectively.

It is a requirement that the DP master and the DP slaves have the relevant functionality. From the object properties of a slave, you can see which command it supports (select DP slave, EDIT → OBJECT PROPERTIES, "General" tab under "SYNC/FREEZE Capabilities").

Per DP master system, you can form up to 8 SYNC/FREEZE groups that are to execute either the SYNC command or the FREEZE command or both. You can assign any DP slave to a group; on the CP 342-5DP of a specific version, one DP slave can be represented in up to 8 groups.

When you call SFC 11 DPSYC_FR, you cause the user program to output a command to a group (see Chapter 20.4.7 "System Blocks for the Distributed I/O"). The DP master then sends the relevant command simultaneously to all DP slaves in the specified group.

You configure the SYNC/FREEZE groups following configuration of the DP master system (all DP slaves must be present in the DP master system). Select the DP master system (the black/white dashed rail) and choose EDIT → OBJECT PROPERTIES. In the window that appears, first specify the commands to be performed for the groups (Figure 20.13) in the "Group Properties" tab, and then assign the DP slaves to the individual groups in the "Group assignment" tab.

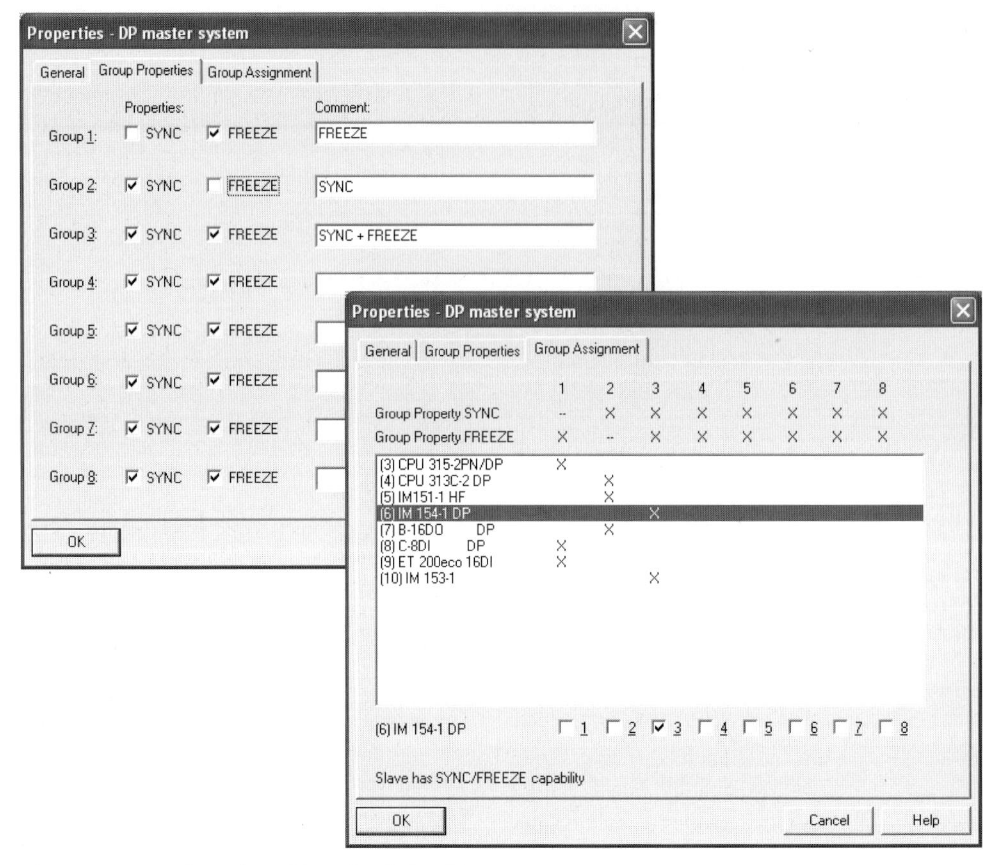

Figure 20.13 Configuration of SYNC and FREEZE groups

Here you select each of the DP slaves listed with its node number, one after the other, and select the group to which each should belong. If a DP slave cannot execute a specific command, e.g. FREEZE, it cannot be assigned to a group which contains this command. Finish configuring the SYNC/FREEZE groups by clicking OK.

Please note that when configuring bus cycles of the same length (equidistant), groups 7 and 8 acquire a special meaning.

Configuring PROFIBUS PA

To configure a PROFIBUS PA master system and to parameterize the PA field devices, use the Hardware Configuration for STEP 7 V5.1 SP3 and higher, or the SIMATIC PDM option software for an earlier version of STEP 7. Use the Hardware Configuration to establish the connection to the DP master system with the DP/PA Link: drag the IM 157 to the DP master system in the hardware catalog under "PROFIBUS DP" and "DP/PA Link". With the DP slave, a PA master system is simultaneously created in a separate PROFIBUS subnet (45.45 kbit/s); this is indicated by the broken black-and-white bar.

The DP/PA coupler exchanges the data between the bus systems unchanged and without interpreting them; it is therefore not parameterized. The PA field devices are addressed by the DP master. They can be integrated as a DP standard slave into the hardware configuration of STEP 7 by means of a GSD file. The PA field devices are subsequently found in the hardware catalog under "PROFIBUS DP" and "Other field devices".

Configuring direct data exchange (lateral communication)

In a DP master system, the DP master has exclusive control over the slaves assigned to it. With appropriately equipped stations, another node (master or intelligent slave, referred to as the receiver or subscriber) can monitor the PROFIBUS subnet to learn which input data a DP slave (sender or publisher) is sending to "its" master. This direct data exchange is also called "lateral communication". In principle all DP slaves from a specific revision level can function as senders in direct data exchange.

You configure the direct data exchange with the Hardware Configuration in the Properties window of the DP slave (receiver) if all stations are connected to the PROFIBUS subnet. Open the receiver station and, with a selected DP interface, select EDIT → OBJECT PROPERTIES. The "Configuration" tab contains the transfer memory between the DP slave and DP master. Click on "New" and set the DX mode (direct data exchange) in the configuration window that appears. In the same window, set the parameters for the DP partner (transmitter).

You can also use the direct data exchange between two DP master systems on the same PROFIBUS subnet. For example, the master in master system 1 can "monitor" the data of a slave in master system 2 in this manner.

Configuring constant bus cycle time and isochrone mode

Constant bus cycle time

Normally, the DP master controls the DP slaves assigned to it cyclically without a pause. With S7 Communication, such as when the programming device executes modify functions via the PROFIBUS subnet, this can result in variations in the time intervals. If, for example, the outputs are to be modified via DP slaves at a regular interval, you can set constant bus cycles with the appropriately equipped DP master. For this purpose, the DP master must be the only Class 1 master on the PROFIBUS subnet. Constant bus cycle time behavior is possible with the bus profiles "DP" and "User-Defined". The PROFIBUS must not be cross-project, and neither a fault-tolerant system nor a CiR (Configuration in Run) object may be connected.

If you configure SYNC/FREEZE groups in addition to the equidistant behavior, please note the following:

- ▷ For DP slaves in group 7, the DP master automatically initiates a SYNC/FREEZE command in every bus cycle. Initiation per user program is prevented.
- ▷ Group 8 is used for the equidistance signal and is disabled for DP slaves. You cannot configure equidistant behavior if you have already configured slaves for group 8.

Isochrone mode

We refer to isochrone mode if a program is executed synchronous to the PROFIBUS DP cycle. In association with equidistant bus cycles, the result is reproducible response times to the process I/O of equal duration which include the distributed signal acquisition, the signal transmission via PROFIBUS and the program execution including updating of the process image. The isochronous user program is present in organizations blocks OB 61 to OB 64. The system functions SFC 126 SYNC_PI and SFC 127 SYNC_PO are available for the isochronous updating of the process image (see Chapter 21.8 "Synchronous Cycle Interrupts").

Figure 20.14 shows the times involved in the isochronous mode. Ti is the time required for reading in the process values. It contains the execution time in the input modules or electronic modules and, in the case of modular DP slaves, the transfer time on the backplane bus. At the end of Ti, the input information for transfer using the global control command (GC) is available. Then the equidistance time begins. It is the time between two global control commands and encompasses the transfer to the subnet as well as the execution of the isochronous interrupt OB. Between completion of the execution of this OB to the next global control command there must be time for execution of the main program.

To is the time required to output the process values. It begins with the global control command and comprises the transfer time on the subnet as well as the processing time in the out-

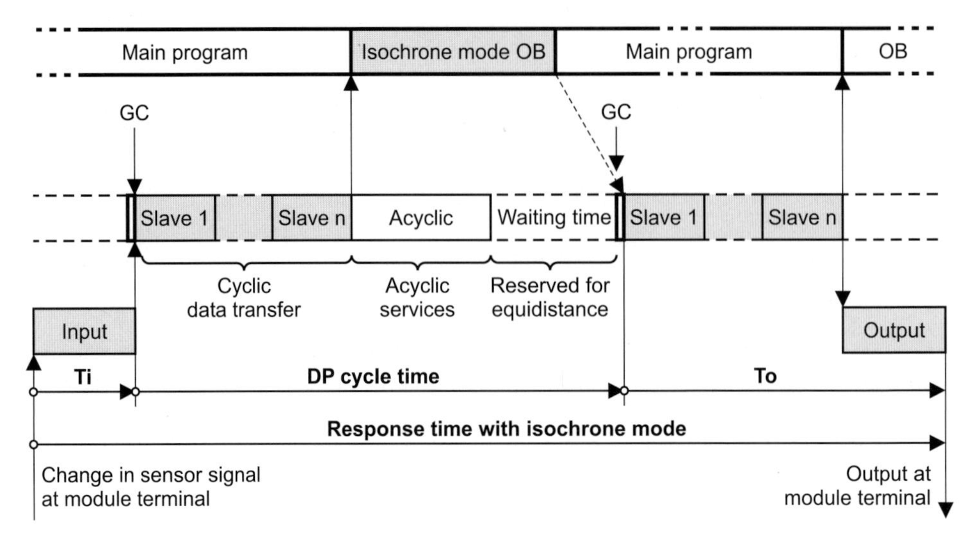

Figure 20.14 Response time with isochronous mode and constant bus cycle time

put modules or electronic modules. In the case of modular DP slaves, the transfer time on the backplane bus is also added. The response time in the case of isochronous mode is the total of the times Ti, equidistance time, and To.

Appropriately designed DP slaves permit shortening of the response time through so-called *overlapping isochrone mode*. The input and output signals are then updated overlapping (overlapping of Ti and To). In this case, you must deactivate the checkbox "Ti and To times same for all slaves" in the "Equidistance" tab, and enter the individual times for the modules involved. If modules operated in isochrone mode have both inputs and outputs, overlapping of Ti and To is not possible.

Configuration of isochrone mode

A prerequisite for configuration of isochronous mode is the constant bus cycle time and the corresponding functionality of the participating DP components.

Add a station and a CPU with integrated DP interface in the project, for example an S7-300 station with a CPU 317-2PN/DP. To insert a DP master system, set the type "PROFIBUS" on the "General" tab in the interface properties of the MPI/DP interface, and the option "DP Master" on the "Mode" tab. Click the "Properties..." button on the "General" tab, and connect the interface to a PROFIBUS subnet.

Click the "Properties..." button to activate the constant bus cycle times. Select the "Network Settings" tab in the properties window of the PROFIBUS interface. Note that the constant bus cycle times can only be set with the bus profiles "DP" and "User-defined". Click the "Options" button and the "Activate constant bus cycle times" checkbox in the options window which is then displayed.

For isochronous mode, specify additionally the times Ti and To on this tab. Either select the "Times Ti and To same for all slaves", or set the times individually in the slave properties.

Each *module* or each *submodule* participating in isochronous mode must have a user data address in a process image partition that is updated in isochronous mode by the system functions SFC 126 SNYC_PI and SFC 127 SNYC_PO. You make the assignment between the user data addresses and a process image partition in the module or submodule properties when setting the address.

The isochronous modules must be made known to the *DP interface module*. In the properties window of the DP slave, activate the option "Synchronize DP slave to equidistant DP cycle" on the "Isochronous mode" tab. Here you also select the modules or submodules involved in isochronous mode.

To *update Ti and To*, click in the properties of the DP master system on the "Properties" button on the "General" tab. In the displayed window, change to the "Network Settings" tab and click on the "Options" button. When the "Recalculate" button is activated, STEP 7 updates all times involved in isochronous mode. You can modify the suggested equidistance time but not below the displayed minimum time. The "Details" button shows the individual proportions of the equidistance time. Please note that the equidistance time increases the more programming devices are connected directly to the PROFIBUS subnet and the more intelligent DP slaves are in the DP master system.

EDIT → MASTER SYSTEM → ISOCHRONOUS MODE gives you an overview of all components involved in isochronous mode and the relevant parameters (Figure 20.15). If the checkbox "Times Ti and To identical for all slaves" under "Network settings" and "Options" has been deactivated, you can set the update times individually for each slave (prerequisite for overlapping isochronous mode). Select the DP slave in the "Isochronous mode" window. "Edit Parameters" provides you with a dialog window for entering the individual update times and the modules involved in isochronous mode.

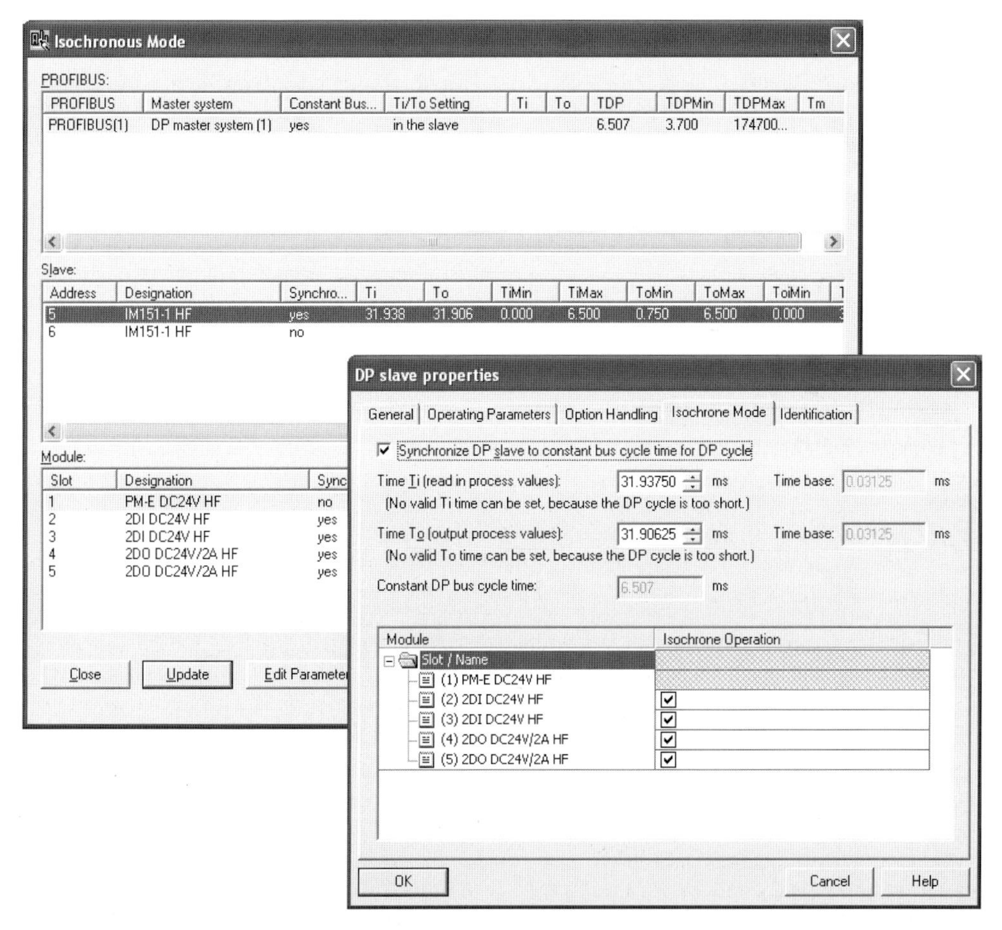

Figure 20.15 Isochronous mode: Overview and DP slave properties

20.4.4 Addressing PROFINET IO

Similar to the manner in which central modules are assigned to a CPU and are controlled by this, the distributed modules with PROFINET IO (stations, IO devices) are assigned to an IO controller. The IO controller together with all of "its" IO devices is referred to as a PROFINET IO system.

Like the central module, the I/O devices are assigned addresses in the I/O area of the CPU ("logical address area"). The IO controller is so to say "transparent" for the addresses of the IO devices; the CPU "sees" the addresses of the IO devices so that the addresses of the IO devices must not overlap with those of the central modules, not even with the addresses of further distributed modules.

Each station operated on the Industrial Ethernet has an IP address. This is assigned to the IO controller during configuration. The IP addresses for the IO devices are derived from the IP address of the IO controller. In addition, an IO device is assigned a device name, a device number (station number), a geographical address (slot) and at least one diagnostics address (Figure 20.16).

MAC address

The MAC assigned to the device is a unique global address. It comprises three bytes with the manufacturer's ID and three bytes with the device ID. The MAC address is usually printed on the device, and is assigned to it during configuration (unless already factory-assigned).

IP address

Each station on the Industrial Ethernet subnet which uses the TCP/IP protocol requires an IP address. The IP address must be unique on the subnet. For the nodes of a PROFINET IO system, it is assigned for the I/O controller once. Based on this, the hardware configuration assigns the IP addresses to the IO devices in ascending order.

The IP address consists of four bytes, each separated by a dot. Each byte is represented as a decimal number from 0 to 255.

The IP address consists of the subnet address and the station address. The contribution made by the network address to the IP address is determined by the subnet mask. This consists – like the IP address – of four bytes which normally have a value of 255 or 0. Those bytes with

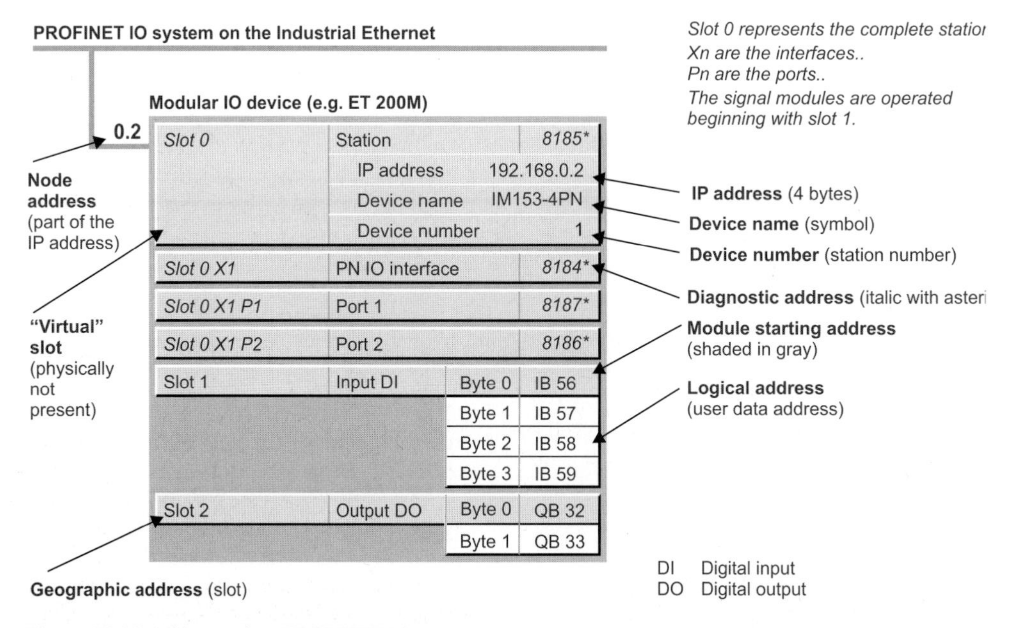

Figure 20.16 Addresses in a PROFINET IO system

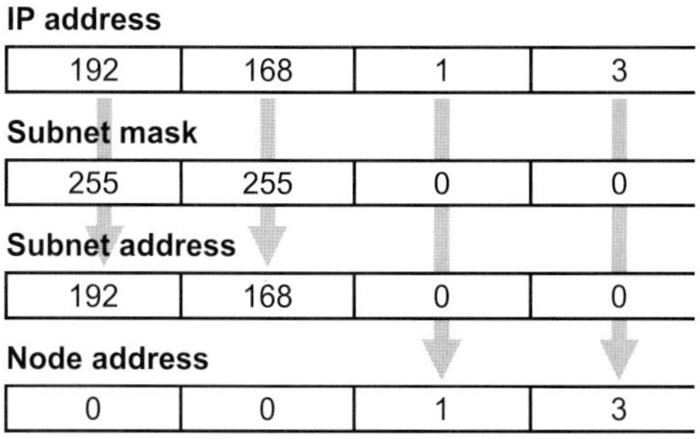

	Byte 1	Byte 2	Byte 3	Byte 4
IP address	192	168	1	3
Subnet mask	255	255	0	0
Subnet address	192	168	0	0
Node address	0	0	1	3

The subnet address is left-justified in the IP address, and is generated by ANDing the IP address with the subnet mask.

Figure 20.17
Example of the structure of the IP address

a value of 255 in the subnet mask determine the subnet address, those bytes with a value of 0 determine the node address (Figure 20.17).

Values other than 0 and 255 can also be assigned in a subnet mask, thereby dividing up the address area even further. For example, in a subnet 192.168.x.x, the subnet mask 255.255.128.0 divides the stations into the two address areas 192.168.0.0 to 192.168.127.254 and 192.168.128.0 to 192.168.255.254.

Please note that assignment of the IP addresses is carried out according to international, national and company rules. For example, the IP addresses 10.0.0.0 to 10.255.255.255, 172.16.0.0 to 172.31.255.255 and 192.168.0.0 to 192.168.255.255 are provided for private networks in accordance with RFC 1918. These addresses are not passed on in the Internet.

The value 255 in the node address is provided for broadcasting. With an address x.y.255.255, for example, all nodes in subnet x.y are addressed.

Device name, devices number

You assign a device name to the IO controller and each IO device during configuration. This name must not be longer than 127 characters, and may consist of letters, digits, hyphens and dots.

The name of the IO system can be appended to the device name, separated by a dot. To do so, check the "Use name in IO device/controller" checkbox in the properties of the PROFINET IO system.

As a supplement to the device name, the hardware configuration assigns a device number to each IO device which is independent of the IP address and which you can change. You can use this device number (station number) to address the IO device from the user program, e.g. as an actual parameter on a system block.

Geographical address

The geographical address identifies a module slot. With central modules, the geographical address contains the number of the rack and slot. Accordingly, the geographical address with PROFINET IO contains the number of the PROFINET IO system, the station number, the slot number, and possibly a subslot number.

The "virtual" (not physically present) slot 0 represents the IO device. The user data is stored beginning with slot 1. The system functions SFC 70 GEO_LOG and SFC 71 LOG_GEO are used to convert geographical addresses into logical addresses and vice versa.

Logical address, module starting address

You use the logical address to address the user data of a station. Each byte of the user data is uniquely defined by a logical address. The logical address corresponds to the absolute address; a symbol (a name) can be assigned to it to make it easier to read (symbolic addressing).

The smallest logical address of a module or station is the module starting address (see also Chapter 1.4 “Module Addresses”).

Diagnostics address

The diagnostics address is used to identify modules and stations which can deliver diagnostics data but do not have a user data address themselves. The diagnostics address occupies one byte of peripheral inputs in the logical address volume. As default, STEP 7 assigns the diagnostics address commencing with the highest address in the I/O area of the CPU. You can change the diagnostics address. The address overview in the Hardware Configuration identifies the diagnostics address by a star.

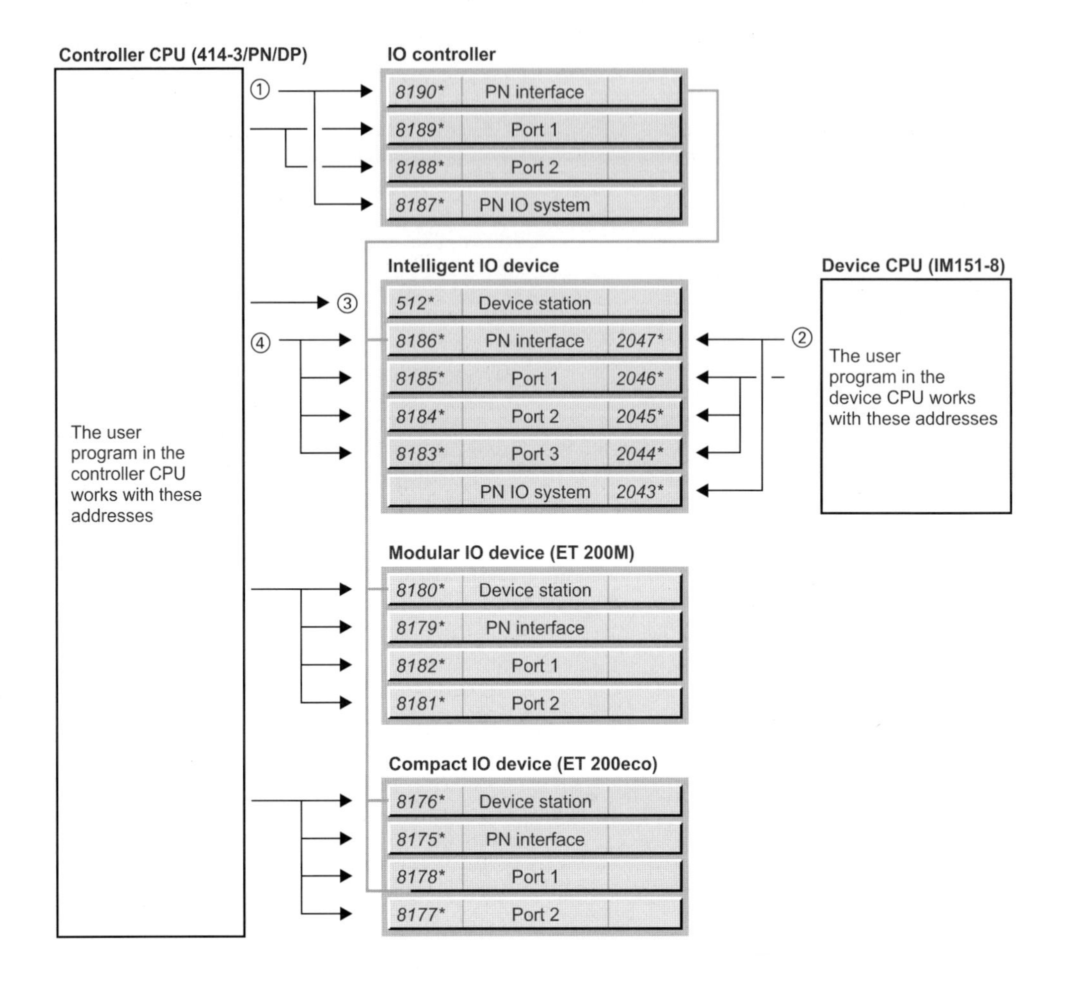

① Each bus interface has a diagnostic address which is set in the object properties of the interface in the "Addresses" tab (double-click on the "PN" line in the controller CPU).

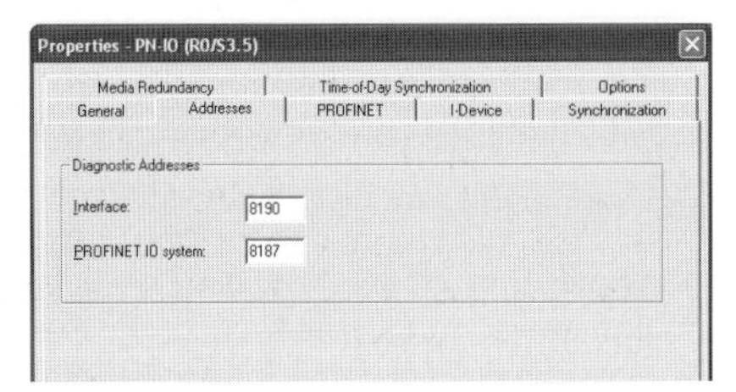

② Each bus interface has a diagnostic address which is set in the object properties of the interface in the "Addresses" tab (double-click on the "PN" line in the device CPU).

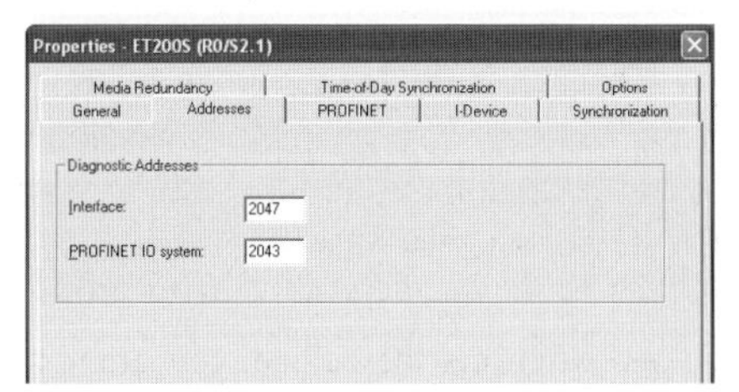

③ The diagnostic address of the station of an intelligent IO device corresponds to the user data address ("module starting address") of the first transfer area in the transfer memory.

④ The IO controller only assigns the diagnostic addresses of the PN interface and ports of the intelligent IO device if the "Parameter assignment for the PN interface and its ports on the higher-level IO controller" checkbox is selected.

Figure 20.18 Diagnostics addresses in a PROFINET IO system

In the example shown (Figure 20.18), the MPI/DP interface of the controller CPU 414-3 PN/DP is assigned the diagnostic address 8191 (not shown) and the PN interface is assigned address 8190. The diagnostic addresses for the ports and the PN IO system are assigned in subsequent order. The diagnostic addresses are assigned to the device CPU in a similar manner. From the viewpoint of the IO controller, the station of the device CPU is assigned the user data address of the first transfer area as the diagnostic address. It can then still address the station even if the parameters of the PN interface have been assigned by the device CPU (from the viewpoint of the controller CPU, no more diagnostic addresses are visible for the PN interface and the ports of the IO device).

Each compact and modular IO device is assigned the diagnostic addresses for the station, the PN interface, and the ports by the IO controller.

The diagnostics data are scanned in the user program by means of system blocks. For example, SFB 54 RALRM is used for a diagnostics interrupt, and reads the supplementary interrupt information. You can use the SFB 52 RDREC to scan the diagnostics record DS1.

Transfer memory on an intelligent IO device

In the case of compact and modular IO devices, the addresses of the inputs and outputs are located together with the addresses for the central modules in the address volume of the controller CPU. Intelligent IO devices have a transfer memory which can be divided into several transfer areas of different lengths. The individual areas then respond like modules whose lowest address is the module starting address. From the viewpoint of the controller CPU, the intelligent IO device then appears as a compact or modular IO device, depending on the division.

When configuring the device, you can configure the individual areas of the transfer memory as inputs or outputs with the “module starting address” and the area length. The addresses of the transfer memory must not overlap with the central modules in the device CPU. If the addresses are present in the process image, the areas can be handled by the user program like inputs and outputs, otherwise like peripheral inputs and outputs. If the device CPU possesses process image partitions, you can also assign a process image partition to each area.

When coupling to the PROFINET IO system, you supplement the configuration at the controller end by the "module starting addresses" from the viewpoint of the controller CPU and by the transmission direction. You assign inputs on the device side to outputs on the controller side and vice versa. If the addresses are present in the process image, the areas can be handled by the user program like inputs and outputs, otherwise like peripheral inputs and outputs. If the controller CPU possesses process image partitions, you can also assign a process image partition to each area. From the viewpoint of the controller CPU, the addresses of the transfer memory must not overlap with the addresses of other central or distributed modules in the controller station.

20.4.5 Configuring PROFINET IO

General procedure

You basically configure the distributed I/O on PROFINET IO like the central modules. Instead of arranging modules in a rack, you assign IO devices (nodes on the Industrial Ethernet subnet) in this case to a PROFINET IO system. The following sequence is recommended:

1) Use the SIMATIC Manager to create a new project or open an existing one.
2) Use the SIMATIC Manager to create an Industrial Ethernet subnet in the project.
3) Use the SIMATIC Manager to create the station in the project which is to accommodate the IO controller, e.g. an S7-300 station.
4) If the project contains intelligent IO devices, you can now also create the corresponding device stations, for example an ET 200pro station.
5) You start the hardware configuration by opening the controller station. Insert a CPU with integrated PN interface. You assign the previously created Ethernet subnetwork to the PN interface and create a PROFINET IO system. In the "General" tab, you can change the specified device name and IP address. You can also configure the remain-

ing modules later. Save and compile the station.

6) If you have created an S7 station for an intelligent IO device, open this in the hardware configuration and insert an appropriate CPU with integrated PN interface, if applicable (with an ET 200 station, the CPU is inserted by STEP 7). Activate the "I-device mode" of the PN interface, and configure the transfer memory from the viewpoint of the IO device. You can also configure the remaining modules later. Save and compile the station.

   Subsequently generate a GSD file from the intelligent IO device and install it.

   Proceed in the same manner with further stations envisaged for intelligent IO devices.

7) Now couple the IO devices to the PROFINET IO system. Open the controller station and drag the PROFINET nodes using the mouse from the hardware catalog to the PROFINET IO system, assign a device name and, if applicable, the device number (station number).

8) If you use IRT communication, create a new sync domain, import the involved PROFINET IO systems, and set the properties of the corresponding devices.

9) If you use IRT with the option "High performance", you must configure the network topology either in the Hardware Configuration directly on the port (interface connection) or centrally using the Topology Editor.

10) Saving and compiling stations. The PROFINET IO system is now configured. You can now supplement the configuration with centralized modules, with IO systems, or with further IO devices. If you change the assignment of an intelligent IO device, you must also generate a new GSD file, and in turn a station from this, which you can add to the PROFINET IO system.

You can also graphically display the configured PROFINET IO system using the network configuration. Open the network configuration, e.g. by double-clicking on a subnet. Select VIEW → DP-SLAVES/IO DEVICES to display the IO devices. You can also use the network configuration to create a PROFINET IO system (more exactly: assign the nodes to an Ethernet subnet). When the stations are open, set their parameters using the Hardware Configuration.

Before the configuration data is downloaded, the device name must be assigned to each IO device. In the STOP state, download the configuration data to the CPU which applies its own parameters such as the IP address. Use the Hardware Configuration to download the data of the currently opened station (PLC → DOWNLOAD TO MODULE), use the Network Configuration to send the data to several stations, e.g. using PLC → DOWNLOAD TO CURRENT PROJECT → STATIONS IN SUBNET.

During the startup, the CPU transfers the configuration information to the IO devices and monitors the parameterization. The IO devices are also assigned their IP address together with the parameters. Following successful parameterization, the useful data are subsequently exchanged cyclically between IO controller and IO devices in RUN.

Configuring IO controller

Prerequisite: You have created a project and an S7 station with the SIMATIC Manager. Using the hardware configuration, open the station and position a CPU with a PN interface (see Chapter 2.3.1 "Arranging Modules").

To set the properties, double-click in the configuration table on the line with the PN interface. You can change the device name of the IO controller on the "General" tab of the Properties window.

In order to assign an Ethernet subnet to the PN interface, click on the "Properties" button. Select the subnet in the displayed window, or create a new one using the "New" button. Adjust the IP address and possibly the subnet mask if necessary. Close the dialog box with "OK".

To create the PROFINET IO system, select the PN interface in the configuration table and then the command INSERT → PROFINET IO SYSTEM.

Select the black/white rail and then EDIT → OBJECT PROPERTIES. Assign a name and an IO system number (from 100 to 115) on the "General" tab of the Properties window. Here you can also choose whether the system name is to be a component of the device name for IO controllers and IO devices. You can access the S7 subnet ID using the "Properties..." button.

Configuring a compact IO device

The IO devices can be found in the hardware catalog under "PROFINET IO" in the corresponding subcatalog, e.g. "I/O". Select the desired IO device and drag it with the mouse to the PROFINET IO system.

Double-clicking on the IO device opens the Properties window. You can change the device name and number on the "General" tab. To change the IP address, click on the "Ethernet..." button.

The bottom part of the window shows the configuration table of either the selected IO device or the PROFINET IO system. You can switch over using the "Arrow" buttons.

In order to change a user data address, select the IO device and double-click on the address line in the configuration table. In the Properties window, you can then change the address on the "Addresses" tab.

Configuring a modular IO device

The IO devices can be found in the hardware catalog under "PROFINET IO" in the corresponding subcatalog, e.g. "I/O".

Click on the desired interface module (basic module) and drag it to the icon for the PROFINET IO system. By double-clicking on the IO device, you are provided with the properties sheet of the station; you can set the device name and number on the "General" tab. You can change the proposed IP address after clicking on the "Ethernet..." button.

The bottom part of the window shows the configuration table of either the selected IO device or the PROFINET IO system. You can switch over using the "Arrow" buttons.

Now place the modules that you can find in the hardware catalog *under the selected interface module (!)* into the configuration table. Double-clicking on the line opens the Properties window and allows the assignment of parameters to the module.

Special functions for ET 200S and ET 200pro

The *Combine modules* function optimizes the address assignments for modules with two or four bit channels. You can find a description of this function under “Combining modules” on Seite 300.

With the *Option handling* function, you prepare a future extension of an ET 200S or ET 200pro station. You can find a description of this function under “Handling of options in general” starting on Seite 300.

Configuring an intelligent IO device

You configure a station for an intelligent IO device like a controller station: In the SIMATIC Manager or in the hardware configuration, insert a new station into the project, position the CPU with the PN interface, and connect the PN interface to the Ethernet subnet. Examples of intelligent IO devices you can use are a CPU 400 with firmware release V6.0 or higher, a CPU 300, a CPU ET 200S, or a CPU ET 200pro with firmware release V3.2 or higher.

To set the operating mode, select the PN interface in the configuration table and then EDIT → OBJECT PROPERTIES. Select the "I device" tab in the Properties window, and check the "I device mode" checkbox there.

If the "Parameter assignment for the PN interface and its ports on the higher-level IO controller" checkbox is selected, the IO controller assigns the parameters for the PN interface. Otherwise the parameters are set by the IO device.

You must subsequently still configure the transfer memory. The transfer memory is the user data interface between the IO controller and the IO device. It can be divided into several transfer areas whose addresses correspond to individual modules. With correspondingly designed CPUs, you can also specify input modules of the IO device in the transfer area, which the IO controller can then access almost directly.

Following activation of the I device mode, you create a new transfer area in the transfer memory using the "New..." button. You define the type of transfer area in the transfer area's properties: "Application" if it is to be a freely-defined area and "I/O" if an I/O module is to be addressed.

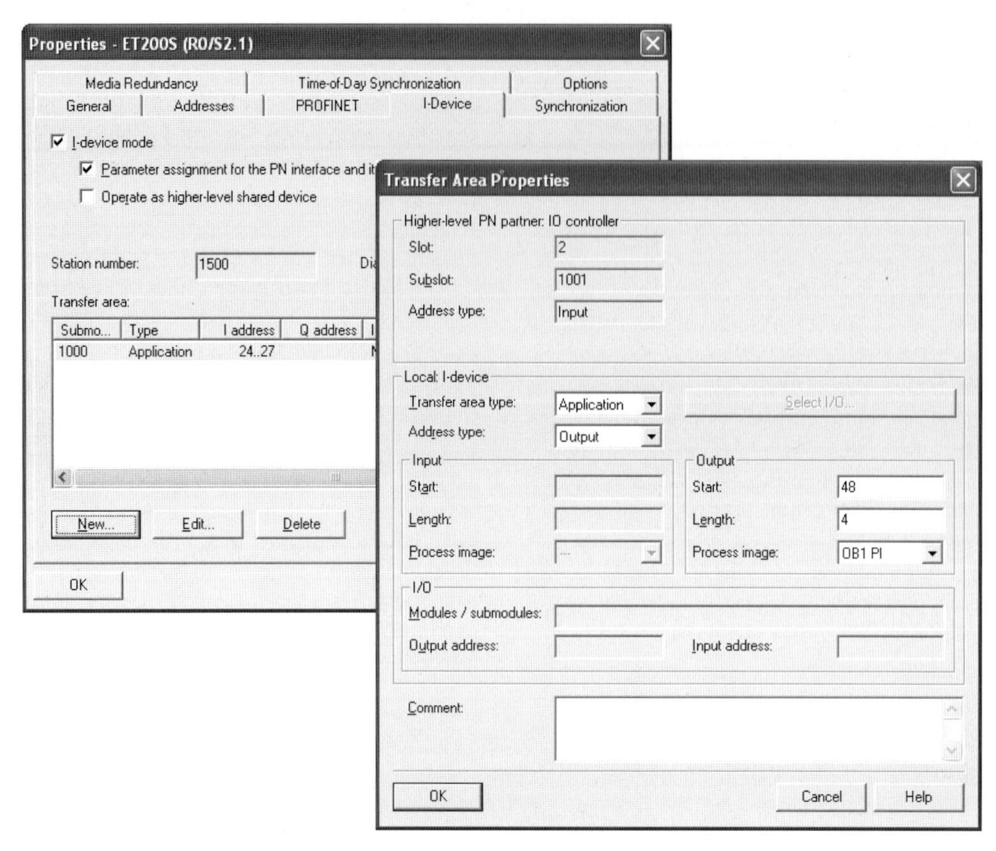

Figure 20.19 Configuring the transfer memory of an intelligent IO device

When selecting "Application", you define whether the area is to be an input or output from the viewpoint of the IO device, and define the start address, the length (in bytes), and also the process image partition if required (Figure 20.19).

When selecting "I/O", you click on the "Select I/O" button and then select the desired input module from the previously configured input modules in the dialog window. The I/O module appears automatically as the output transfer area whose start address (corresponds to the module starting address) you can change.

Once you have configured all transfer areas, save and compile the station, and generate a GSD file with OPTIONS → CREATE GSD FILE FOR I DEVICE ... from the IO device. In the "Designation for I device proxy" box in the dialog window, you can change the name of the PN interface as it is to be displayed later in the IO controller.

Click the "Create" button. Following creation of the GSD file, you can save it using "Export" and install it later. To install it immediately, click on the "Install" button and then on "Close". STEP 7 creates a folder "Preconfigured Stations" in the hardware catalog under "PROFINET IO", and inserts a folder with the symbol for the intelligent IO device just created.

The intelligent IO device is now coupled as with a compact IO device: Open the controller station and drag the symbol for the intelligent IO device to the PROFINET IO system using the mouse.

With the IO device selected, the transfer areas are displayed in the configuration table as a subslot for slot 2 with the user data address as "seen" by the controller CPU. To change an address, double-click on the address line and enter the new address on the "Addresses" tab in the Properties window.

PN/PN coupler configuration

The PN/PN coupler connects each of two Industrial Ethernet subnets to a PROFINET IO system. In each of the two PROFINET IO systems, one half of the PN/PN coupler appears as an IO device. Using Configuration of transfer memory, combine the two IO devices.

With a correspondingly designed PN/PN coupler, you can use other functions in addition to the cyclic I/O transmission, for example the shared device function (see section „Shared device“ on page 321).

Prerequisite: The two PROFINET IO systems are set up. You have opened one of the controller stations.

You can find the symbol of the PN/PN coupler in the hardware catalog under "PROFINET IO" and "Gateway" in the "PN/PN coupler" folder. Use the mouse to drag one half of the PN/PN coupler (e.g. X1) to the PROFINET IO system. Drag the second half to the second PROFINET IO system.

To combine the two halves, select the PN/PN coupler and then EDIT → OBJECT PROPERTIES. Configure the coupling partner on the "Coupling" tab in the Properties window. Specify the other half of the PN/PN coupler here (Figure 20.20).

You can set the device name and number for each half of the coupler on the "General" tab in the Properties window – just like with an IO device – and change the IP address after clicking on the "Ethernet..." button.

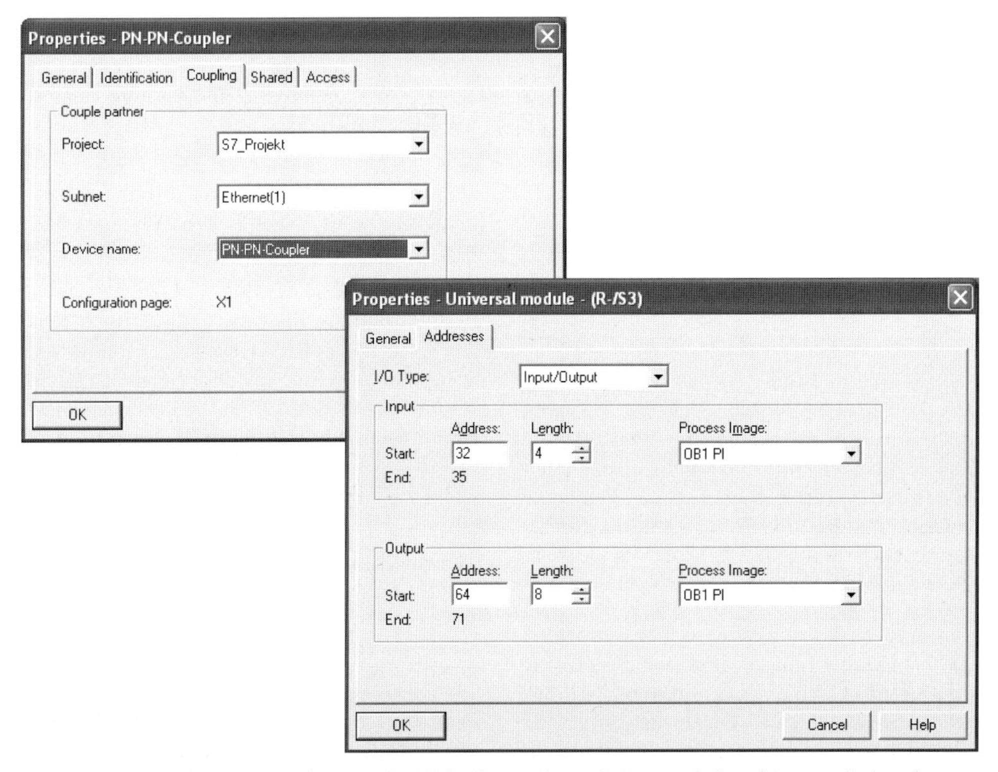

Figure 20.20 Configuring a PN/PN coupler V3.0: Connection and characteristics of the transfer interface

With the PN/PN coupler selected, the configuration table is displayed in the bottom part of the window. Double-clicking on the station line (slot 0) or on the interface line opens the Properties window in which you can set the parameters of the station or of the PN interface for the current part of the PN/PN coupler.

Both halves of the coupler have a transfer memory for data exchange which can be divided into several transfer areas. Each area corresponds to a signal module whose addresses are within the address volume of the respective IO controller. Each area has a start address ("module starting address") and a length in bytes.

If an input area is configured in one half of the PN/PN coupler, an output area of the same length must be created in the other half, and vice versa. If the other subnet is in the same project, STEP 7 creates the associated transfer area in the other half of the coupler.

You configure the transfer areas like signal modules: Input and output areas of various lengths are listed in the hardware catalog under the corresponding PN/PN coupler. Use the mouse to drag the symbol with the desired property to a slot in the configuration table, starting without gaps at slot 1. A double-click on the line of the transfer area opens the Properties window in which you can set the start address and the process image partition used on the "Addresses" tab.

If you are using the universal module, you must set the I/O type, the start address, the area length, and the process image partition in the properties (Figure 20.20).

On the "Connection" tab, you can also clear down the connection between the subnets again (enter "- - - - -" under "Subnet").

Configuring IE/PB Link

You can find the IE/PB Link PNIO in the hardware catalog under "PROFINET IO" and "Gateway" in the "IE/PB Link PN IO" folder. Use the mouse to drag the symbol for the IE/PB link to the PROFINET IO system. In the Properties window which is displayed, you set the PROFIBUS subnet and the station numbers in this subnet.

Selecting EDIT → OBJECT PROPERTIES with the IE/PB Link selected shows its properties window with the possibility for setting the device name and number on the Ethernet subnet. You can change the IP address using the “Ethernet...” button (Figure 20.18).

The IE/PB Link is quasi the DP master for the “subordinate” DP master system. Position the DP slaves from the hardware catalog in this master system, and assign them the desired properties (see Chapter 20.4.2 “Configuring PROFIBUS DP”).

You require a device number in order to address the DP slaves over PROFINET IO. You make the assignment between node number on the PROFIBUS and device number on the PROFINET in the “Device numbers” tab in the properties window of the IE/PB Link. As standard, STEP 7 uses the PROFIBUS address as the device number (indicated by a star on the number). Select a DP slave in the list, and click the “Change” button.

The IE/PB Link is able to pass on time frames and parameter records. You make the settings for these in the “Options” tab in the properties window.

Configuring IE/AS-i link

You can find the IE/AS-i Link PNIO in the hardware catalog under "PROFINET IO" and "Gateway" in the "IE/AS-i Link PN IO" folder. Use the mouse to drag the symbol for the IE/AS-i link to the PROFINET IO system.

EDIT → OBJECT PROPERTIES with the IE/AS-i link selected shows its properties window with the option for setting the device name and number on the Ethernet subnet. You can change the IP address using the “Ethernet…” button.

The IE/AS-i link is quasi the AS-i master. The AS-i master system with the AS-i slaves is not displayed as a subnet by the Hardware Configuration. With the IE/AS-i link selected, you are provided with the configuration table in which you “insert” the AS-i slaves which are arranged in the catalog under the link symbol.

The configuration of the AS-i master system is described in Chapter 20.4.2 “Configuring PROFIBUS DP” under “Configuring the DP/AS-i link”.

20.4.6 Special functions for PROFINET IO

GSD files

You can subsequently install IO devices which are not included in the module catalog. To do this, you require the type file tailored to the IO device (GSD file, Generic Station Description). Suitable for IO devices are GSD files with GSD version 5 or higher in XML format (GSDML, Generic Station Description Markup Language).

To install, select OPTIONS → INSTALL GSD FILES in the hardware configuration and enter the directory of the GSD file or of a different STEP 7 project in the displayed window. STEP 7 imports the GSD file and shows the IO device in the hardware catalog under "PROFINET IO" and "Other field devices".

STEP 7 saves the GSD files in the directory ...\Step7\S7DATA\GSD. The GSD files deleted during post-installation or importing are saved in the subdirectory ...\GSD\BKP*n*. They can be restored from there using OPTIONS → INSTALL GSD FILES.

Shared device

The "Shared device" function allows different IO controllers to access submodules (I/O modules and transfer areas) in one IO device. The associated IO device is used by the IO controllers together (shared device). Each submodule of the shared device is assigned to an IO controller.

The basic conditions for use of a shared device are:

- The IO controller and the IO device must be present in the same Ethernet subnet.
- When using isochronous real-time communication (IRT), a shared device can only be used with the IRT option "High performance".
- The shared device function can only be used with "even" send cycle times.
- A shared device cannot be operated in an isochronous manner with the constant PROFINET IO cycle.

The shared device function is available with a CPU 400 with firmware version 6.0 and higher and with a CPU 300 or CPU ET 200 with firmware version 3.2 and higher.

Prerequisite: A project has been created with two or more IO controllers and PROFINET IO systems on the same Ethernet subnet.

To create a (modular) shared device, open a controller station and drag the IO device from the hardware catalog to the PROFINET IO system with the mouse. Configure the modules by dragging from the hardware catalog to the slot in the configuration table. Position all modules for all IO controllers.

Following configuration, copy the IO device into the clipboard, for example using the "Copy" command from the shortcut menu. Save the controller station, and open another one.

To insert the saved IO device, click with the right mouse button on the PROFINET IO system and select the "Shared insert" command from the shortcut menu. Subsequently save the controller station. IO devices of identical design are now present in the two controller stations. Repeat inserting with the other IO controllers if applicable.

Double-click on the IO device in one of the controller stations. The shared device (of the other PN IO system) is entered in the Properties window in the "Coupled devices" table on the "Shared" tab. Here you can also delete the coupling again: Select the shared device in the "Coupled devices" table and click on the "Uncouple" button. This tab also permits coupling of the same type of IO devices which were not transferred with the "Shared insert" command: Select the IO device in the "Devices which can be coupled" table and click on the "Couple" button.

To assign the modules to an IO controller, open the "Access" tab. All modules are listed in a tree structure. A module has the value "Full" if it is assigned to the IO controller of the currently open PROFINET IO system. Otherwise it has the value "---". Open the shared devices in succession in each PROFINET IO system, and assign the modules to the associated IO controller by clicking in the "Value" column (Figure 20.21).

If the IO controller is in a different project, you must manually configure the shared device in

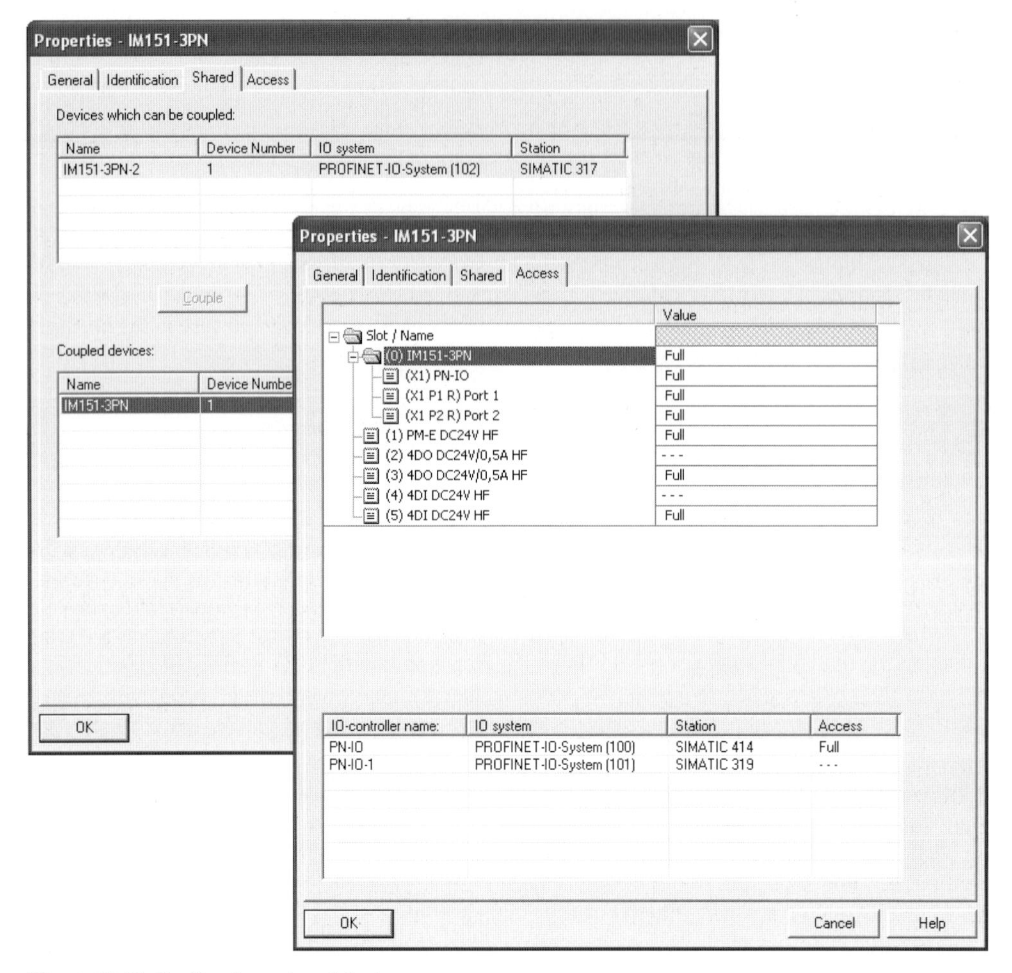

Figure 20.21 Configuring a shared device

the other project with exactly the same module assignment, but with the assignment referring to the current IO controller. Save the controller station following the assignment.

Real-time communication with PROFINET

PROFINET IO enables several types of data transmission:

▷ Non-time-critical data such as configuration and diagnostics information is transmitted acyclically with the standard TCP/IP communication.

▷ User data (input/output information) is exchanged cyclically within a defined time period – the updating time – between IO controller and IO device (real-time RT).

▷ Time-critical user data, e.g. for motion control applications, is transferred with hardware support isochronously (isochronous real time IRT).

A fixed communications channel is reserved on the Ethernet subnet for IRT communication. Within the updating time, the RT communication (the cyclic data exchange between IO controller and IO devices) and the non-real-time TCP/IP communication are carried out parallel to this. In this manner, all three types of communication can exist in parallel in the same subnet.

Send cycle time, updating time

Cyclic data exchange is carried out at a defined rate, the send cycle time. STEP 7 calculates the send cycle time from the configuration data for the PROFINET IO system. The send cycle time is the shortest possible updating time. This is the time period within which each IO device in the IO system exchanges its user data with the IO controller. The actual updating time for an IO device can be a multiple of the send cycle time. You can manually increase the updating time, e.g. to reduce the load on the bus.

As standard, the updating time is the same for all IO devices in the IO system. If necessary, one can shorten the updating time for individual IO devices if the time for other devices can be increased because the rate at which their user data are exchanged is non-critical.

You can configure the send cycle time (without IRT communication) centrally in the properties dialog of the PN interface on the "PROFINET" tab or in the properties dialog of the PROFINET IO system on the "Updating time" tab.

This tab also lists the IO devices with the update times. You can increase the time for an IO device by selecting it and clicking on the "Edit" button, or you can set the time in the Properties dialog box of the PN interface of the IO device on the "IO cycle" tab.

In addition to the updating time, you can also set the response monitoring time in the interface properties. This is the product of the updating time and the "Number of accepted updating cycles with missing IO data".

If there is at least one synchronized device in the IO system, the send frequency is determined by the sync master of the sync domain and can only be modified in the properties of the sync domain. Select the IO system in the Hardware Configuration or the subnet in the Network Configuration. Select EDIT → PROFINET IO → DOMAIN MANAGEMENT. The cycle time can then no longer be modified anywhere else.

Real-time

Real-time (RT) means that a system processes external events within a defined time. If it reacts predictably, it is referred to as deterministic. In the case of RT communication, the transmission takes place in a defined interval (updating time) at a defined point in time (Send cycle time). PROFINET IO permits the use of standard network components for RT communication.

If it is impossible to transmit all required data within the planned interval, e.g. because new network components have been added, some of the data is transferred to other send cycle times, and this can lead to an increase in the updating time for individual IO devices.

Isochronous real-time

Isochronous real-time (IRT) is hardware-supported real-time communication designed, for example, for motion control applications. IRT message frames are deterministically transmitted via planned communication paths in a specified order. IRT communication therefore requires network components that support this planned data transmission.

Isochronous real-time is available in the options "High flexibility" for simple configuration and plant expansion, and "High performance" for fast updating times.

To be able to configure IRT communication, set up a new sync domain (see next section) and determine a sync master to handle the synchronized distribution of the IRT message frames to the sync slaves. IRT with the "High performance" option requires a topology configuration (see section "Topology Editor") and thus a defined structure that takes account of the transmission properties of the cables and the switches used.

SYNC domain

A sync domain is a group of PROFINET IO nodes which exchange synchronized data with one another. A node, which can be an IO controller or IO device, takes on the role of the sync master, the others are the sync slaves.

A sync domain can include several IO systems, where one IO system is always completely assigned to a single sync domain. Several sync domains can exist on an Ethernet subnet.

When configuring an IO system, a special sync domain is created automatically: the *sync domain default*. All configured IO systems, IO controllers and IO devices are initially present in the sync domain *sync domain default*.

You create a new sync domain for IRT communication, and import the IO system (from the *sync domain default*) into the new sync domain. It is not necessary to synchronize all devices of an IO system, i.e. not all exchange data by means of IRT communication. When configuring, the non-synchronized nodes are initially also listed in the sync domain, but during runtime only the synchronized nodes are present in the sync domain.

Configuring a new sync domain

Prerequisite: You have configured the Ethernet subnet with one or more PROFINET IO systems. The nodes involved in IRT communication must also support this function.

To create a new sync domain, select the PROFINET IO system in the Hardware Configuration or the subnet in the Network Configuration, and select EDIT → PROFINET IO → DOMAIN MANAGEMENT. The "Domain Management" window that appears shows the sync domain *sync domain-default* and all IO systems on the same subnet that are located in the sync domain (Figure 20.22).

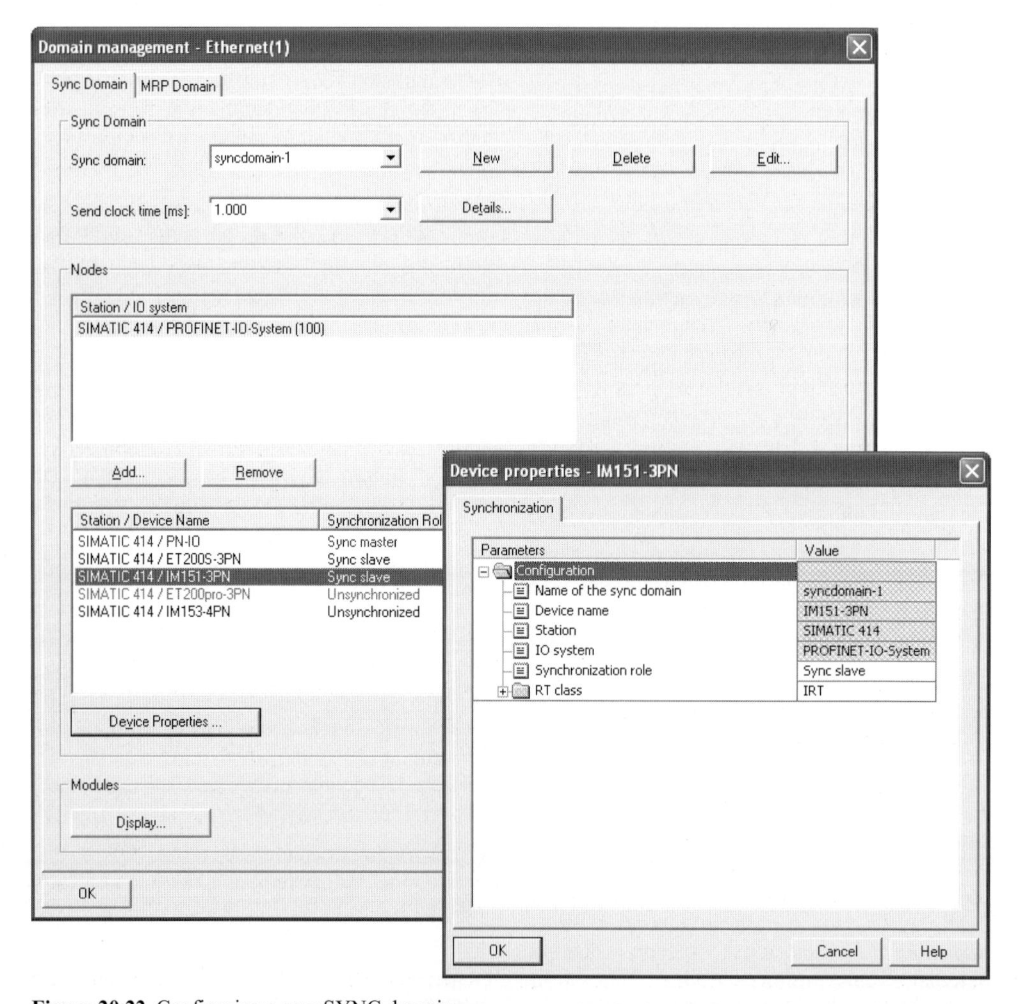

Figure 20.22 Configuring a new SYNC domain

Create a new sync domain using the "New" button, assign a name, and select an IO system using the "Add" button. Repeat the procedure if you wish to add further IO systems. IO systems added to the new sync domain are no longer part of the sync domain *sync domain default*.

The IO systems of the sync domain and the bus nodes of the selected IO system are displayed. Then select a device, select "Device properties", and set the type of synchronization (sync master or sync slave) and the RT class (RT, IRT "High flexibility" or IRT "High performance") in the parameters of the displayed properties window. Proceed in the same manner for the other bus nodes. Only one device can be the sync master, all other devices are sync slaves. In a sync domain, either devices of classes IRT "High flexibility" and RT or of classes IRT "High performance" and RT may be present.

In the "Send cycle time" box, select the send cycle time and use the "Details" button to select the portion of bus communication reserved for IRT. Save the settings using "OK".

Topology Editor

The topology editor allows you to configure the cabling of devices on the Industrial Ethernet subnet. The logic connections between the PROFINET devices are configured using the Hardware Configuration and Network Configuration tools. The topology editor is used to configure the physical connections with the length and cable type properties with which the signal propagation times are determined. Application of the topology editor is a prerequisite for use of IRT communication (isochronous real-time) with the "High performance" option.

The physical connections between devices on the Ethernet subnet are point-to-point connections. The connections on a PN interface are called ports. The Ethernet cable connects a port of one device to a port of the partner device.

To enable several nodes to communicate with each other, they are connected to a switch that has several connections (ports) and that distributes signals. There are also S7 devices featuring a PN interface that has two or more ports connected by an integral switch. With this interface you can cable communication devices in a linear bus topology without external switches.

The connection between two ports can be configured using the Hardware Configuration. Select the port in the configuration table, and then EDIT → OBJECT PROPERTIES. On the "Topology" tab in the display properties window, you can now determine the partner port and edit the cable properties.

Before you call up the Topology Editor, use the Hardware Configuration or the Network Configuration to configure the communication partners on the Ethernet subnet including the necessary switches. In the Hardware Configuration, open the controller station, select the PROFINET IO system, and choose EDIT → PROFINET IO → TOPOLOGY. In the Network Configuration, select an Ethernet subnet and then EDIT → PROFINET IO → TOPOLOGY.

The interconnection table in the *tabular view* shows the pairs of ports of all configured active and passive components. By setting filters, you can display all ports, only the interconnected ones, or only the non-interconnected ones (Figure 20.23).

In the object properties of the port you can set the connection to the partner port. You can cancel an interconnection by selecting the port followed by clicking SEPARATE PORT INTERCONNECTION with the right mouse button.

If there is a connection to the plant, you can use the "Online" button to check whether the devices configured offline are available and what their status is. The comparison is based on the device name, the IP address and the device ID. The data determined online is displayed in the "Status" and "Attenuation value" columns.

The topology window in the *graphic view* shows the devices, their ports and the interconnection. The offline view shows the configured devices, the online view – with an existent online connection to the system – shows those actually present in the system.

To improve editing, you can "shut" the display of the stations, suppress the miniature view and the catalog of the passive components ("Options" button and "Options" tab), and zoom the view by scrolling with the center mouse button.

To interconnect two ports, select a port and "drag" a connection to the partner port with the

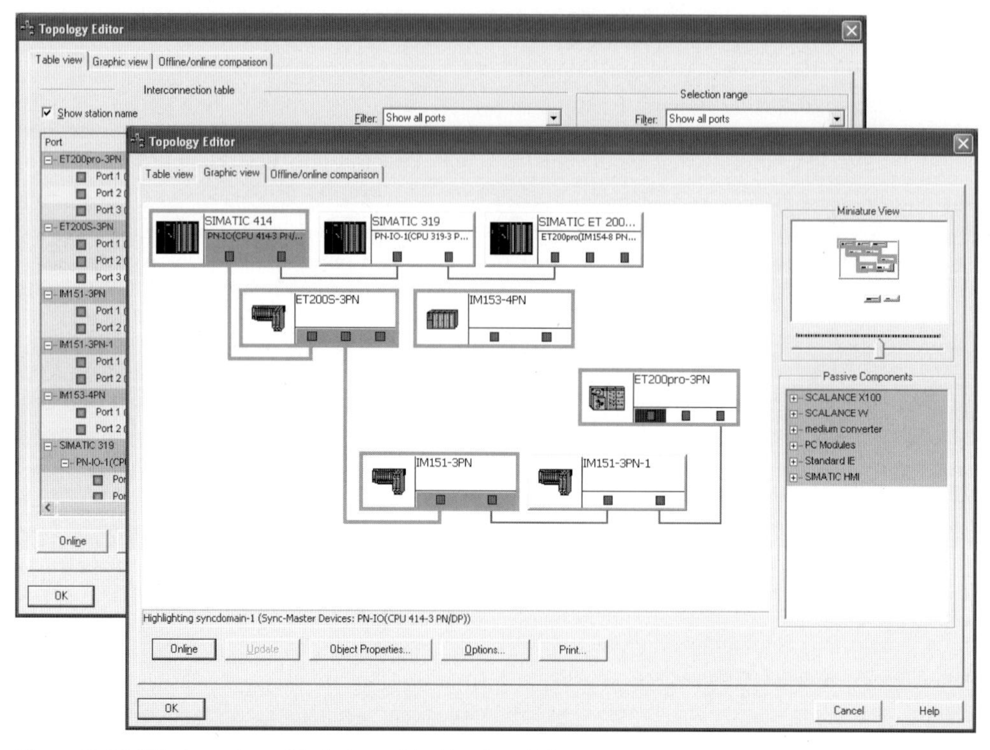

Figure 20.23 Tabular and graphical view of the Topology Editor

right mouse button pressed. You can cancel an interconnection by selecting it with the right mouse button and clicking SEPARATE PORT INTERCONNECTION.

The topology configured offline is displayed compared to the topology determined online on the *Offline/Online comparison* tab. All stations and modules with their ports and the respective partner port and the cable data are displayed. You can thus check the configuration with the connections and cables, and supplement any system components which are missing.

A selection can be made with the relevant filter settings. In the overviews the determined differences, e.g. the modules that the Topology Editor could not assign, are highlighted in color. You can now undertake manual assignment.

Isochronous mode

The "Isochronous mode" function permits synchronous reading, processing and output of I/O signals in a fixed (equidistant) cycle. A prerequisite for isochronous mode is isochronous realtime (IRT) with the "High performance" option.

The basis of the time pattern is the cycle time and the data cycle derived from this (the update time, Figure 20.24).

The data cycle is the interval at which the IRT transmission takes place on the subnet. The application cycle is the interval at which the isochronous mode OB is called. This is a multiple of the data cycle.

Ti is the time required for reading the I/O signals. It includes the times for preparation of the I/O signals in the input modules or electronic modules, and for processing in the IO device.

Ti is followed by the data cycle. This begins with transmission of the I/O signals over the subnet. Transmission takes place in both directions; the input signals are transmitted to the controller station, and the output signals (from the previous application cycle) are transmitted to the IO devices.

Mode 1: The execution time of the isochronous mode program is shorter than a data cycle.

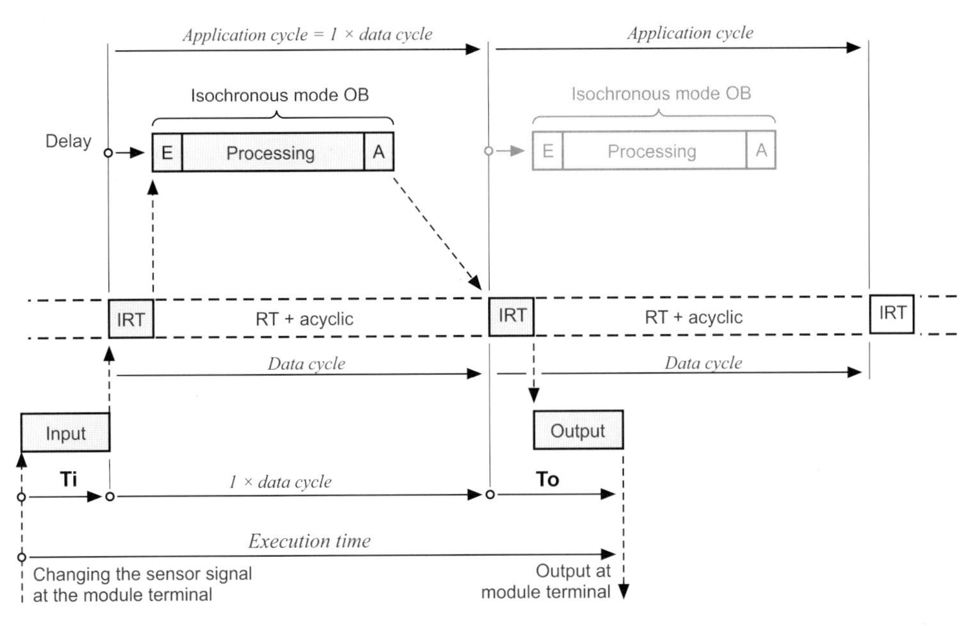

Figure 20.24 Isochronous mode in the PROFINET IO system (mode 1)

The isochronous mode organization block assigned to the PROFINET IO system is called following a delay time during which the IRT transmission takes place. System function SFC 126 SYNC_PI must be called in the organization block in order to read the input signals in isochronous mode, and system function SFC 127 SYNC_PO in order to write the output signals in isochronous mode. The processing time of the isochronous mode OB must be (significantly) shorter than the application cycle time, for the main program is further processed during the differential time.

To begins at the end of the data cycle. This is the time required for output of the I/O signals. It is made up of the transmission time on the subnet, the time for processing in the IO device, and the times for preparation of the I/O signals in the output modules or electronic modules.

With isochronous mode, a distinction is made between two types: The processing time of the isochronous mode program is (significantly) shorter than the time for one data cycle, or it is longer. In the first case, the isochronous mode OB can be called in every data cycle (shown in Figure 20.24); in the second case, the cycle in which the isochronous mode OB is called – the application cycle – is a multiple of the data cycle (shown with factor 2 in Figure 20.25).

If the isochronous mode OB is called in every data cycle – the "Application cycle factor" is then 1 – the system function SFC 126 SYNC_PI for isochronous updating of the input signals is called first in the isochronous mode program. This is followed by processing of the signals, and subsequent output with the system function SFC 127 SYNC_PO.

With this mode, the shortest response time between an input signal and the corresponding output signal is therefore the total of Ti, the data cycle time, and To. The longest response time occurs if the input signal changes shortly after the time for reading-in, and is the total of Ti, To, and twice the data cycle time.

With an application cycle which takes longer than the data cycle (Figure 20.25), you should select a different sequence for updating of the process image: Updating of the output signals first, then of the input signals, and then the processing. In this manner it is possible that the

Mode 2: The execution time of the isochronous mode program is longer than a data cycle

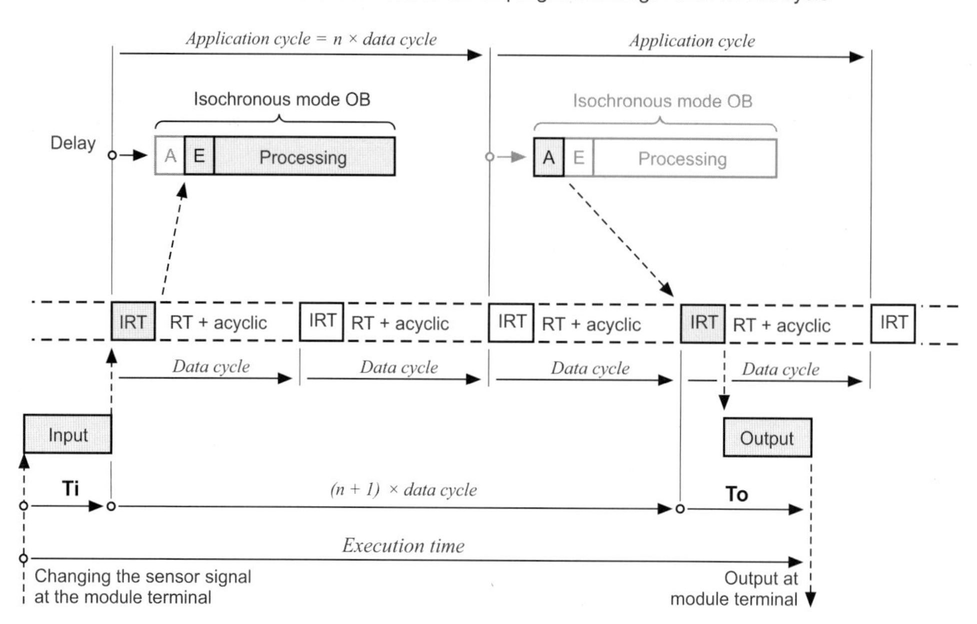

Figure 20.25 Isochronous mode in the PROFINET IO system (mode 2)

output signals are transmitted with the next possible data cycle (in the next application cycle) even if the data cycle time is short compared to the process image updating time.

With this mode, the shortest response time between an input signal and the corresponding output signal is therefore the total of Ti, the application cycle time, the data cycle time, and To. The longest response time occurs if the input signal changes shortly after the time for reading-in, and is the total of Ti, To, the data cycle time, and twice the application cycle time.

In order to configure isochronous mode, you create a PROFINET IO system with the controller station and the IO devices, import the stations into a SYNC domain with the IRT option "High performance", and configure networking between the stations using the Topology Editor. In the device stations, you assign a process image partition, for example the TPA1, to the modules in their properties on the "Addresses" tab.

You assign the isochronous mode organization block to the PROFINET IO system in the CPU properties: Open the controller station and double-click on the CPU to open the CPU properties window; select the "Synchronous cycle interrupts" tab and set the PROFINET IO system for the organizational block, for example the IO system number 100 for OB 61. Click on the "Details" button (Figure 20.26).

The duration of the application cycle is calculated from the data cycle, multiplied by a factor which you specify on this tab. It is therefore necessary to estimate the processing time of the isochronous mode program and to compare this with the data cycle time.

If applicable, you set the delay time on this tab with which the isochronous mode OB is to start, and assign the process image partition which you have set for the module addresses in the IO devices. The following methods are available for determining the times Ti and To:

- "Automatic" – STEP 7 determines the times and sets them the same for all IO devices
- "Fixed" – you enter the times which then apply to all IO devices
- "In IO device" – the times are then set individually in the respective IO device.

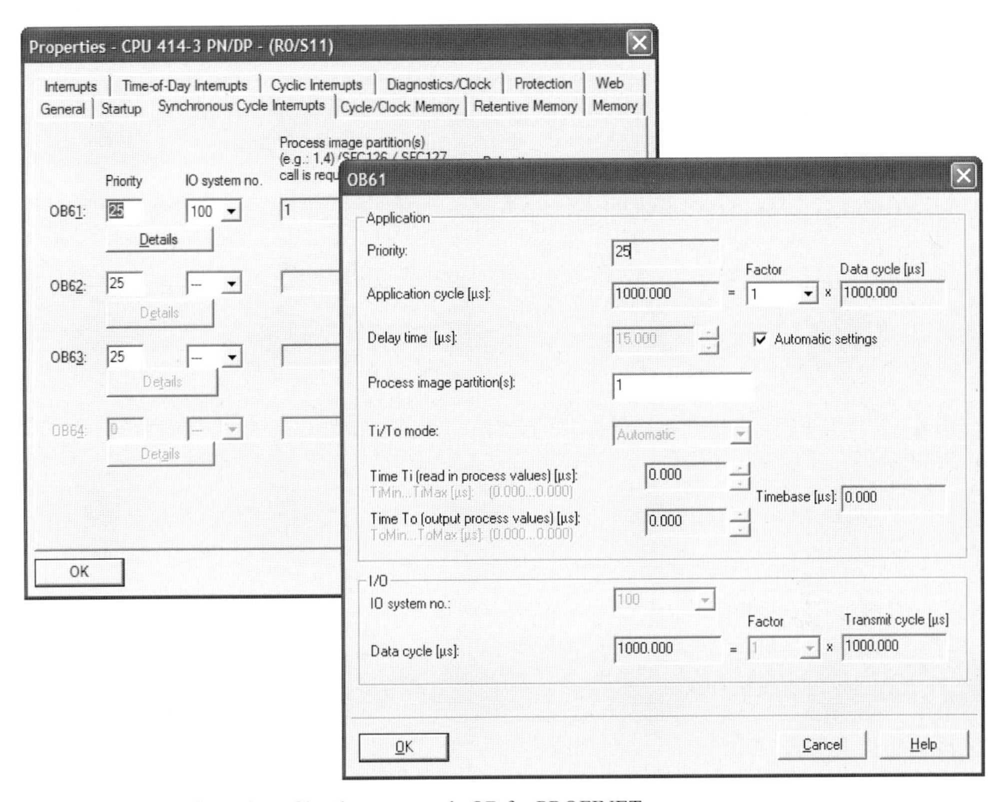

Figure 20.26 Configuration of isochronous mode OB for PROFINET

To assign the modules to isochronous mode, select the IO device, double-click on the PN interface in the configuration table, and select the "IO cycle" tab in the Properties dialog. In the section "Isochronous mode", assign the isochronous mode OB to the IO device, and click on the "Isochronous module/submodule..." button. You can activate or deactivate the individual modules of the IO device for isochronous mode in the displayed window. Proceed in the same manner for the other IO devices.

You are provided with an overview of the configuration if, with the controller station selected, you then select the EDIT → PROFINET IO → ISOCHRONOUS MODE command.

20.4.7 System Blocks for the Distributed I/O

Read and write I/O signals

The following blocks transmit I/O signals from and to stations of the distributed I/O:

- FB 20 GETIO
  Read all inputs of a station
- FB 21 SETIO
  Write all outputs of a station
- FB 22 GETIO_PA
  Read some inputs of a station
- FB 23 SETIO_PA
  Write some outputs of a station
- SFC 14 DPRD_DAT
  Read user data
- SFC 15 DPWR_DAT
  Write user data

You can find the block parameters in the Tables 20.9 (FBs) and 20.12 (SFCs).

The loadable function blocks FB 20 to FB 23 have interfaces compliant with PI (PROFIBUS International) and can be used in conjunction with DP standard slaves and IO devices.

Table 20.9 Parameters of the function blocks for transferring I/O signals

FB	Parameter	Declaration	Data type	Assignment, description
20	ID	INPUT	DWORD	Logical user data start address
	STATUS	OUTPUT	DWORD	Error information of SFC 14 DPRD_DAT [1)]
	LEN	OUTPUT	INT	Amount of bytes to be read
	INPUTS	IN_OUT	ANY	Destination area for the user data read (only data type BYTE permissible in the ANY pointer)
21	ID	INPUT	DWORD	Logical user data start address
	LEN	INPUT	INT	Irrelevant
	STATUS	OUTPUT	DWORD	Error information of SFC 15 DPWR_DAT [1)]
	OUTPUTS	IN_OUT	ANY	Source area for the user data to be written (only data type BYTE permissible in the ANY pointer)
22	ID	INPUT	DWORD	Logical user data start address
	OFFSET	INPUT	INT	Number of the first byte to be read (from number 0)
	LEN	INPUT	INT	Amount of bytes to be read
	STATUS	OUTPUT	DWORD	Error information of SFC 81 UBLKMOV [1)]
	ERROR	OUTPUT	BOOL	Error occurred at signal state "1"
	INPUTS	IN_OUT	ANY	Destination area for the user data read (only data type BYTE permissible in the ANY pointer)
23	ID	INPUT	DWORD	Logical user data start address
	OFFSET	INPUT	INT	Number of the first byte to be written (from number 0)
	LEN	INPUT	INT	Amount of bytes to be written
	STATUS	OUTPUT	DWORD	Error information of SFC 81 UBLKMOV [1)]
	ERROR	OUTPUT	BOOL	Error occurred at signal state "1"
	OUTPUTS	IN_OUT	ANY	Source area for the user data to be written (only data type BYTE permissible in the ANY pointer)

[1)]Contains error information of the SFC used in the form DW#16#40xx xx00

You can find the function blocks in the *Standard Library* supplied with STEP 7 in the program *Communication Blocks*.

You can find the system functions SFC 14 and SFC 15 in the *Standard Library* supplied with STEP 7 in the program *System Function Blocks*.

FB 20 GETIO
Read all inputs of a station

With use of the SFC 14 DPRD_DAT, the FB 20 GETIO consistently reads all input data or all data of an input area of stations with a modular design from a DP standard slave or an IO device. The right word of the parameter ID is the start address of the input area to be read.

The target area specified by the parameter INPUTS must be exactly as long as the configured length of the input area read, which is additionally output in the parameter LEN.

FB 21 SETIO
Write all outputs of a station

With use of the SFC 15 DPWR_DAT, the FB 21 SETIO consistently writes all output data to a DP standard slave or an IO device or all data of an output area of stations with a modular design. The right word of parameter ID is the start address of the output area to be written.

The source area specified by the parameter OUTPUTS must be exactly as long as the configured length of the output area to be written, and therefore values for the parameter LEN are irrelevant.

FB 22 GETIO_PA
Read some inputs of a station

With use of the SFC 81 UBLKMOV, the FB 22 GETIO_PA consistently reads some of the input data or some of the data of an input area of stations with a modular design from a DP standard slave or an IO device. The right word of the parameter ID is the start address of the input area. The parameter OFFSET is the number of the first byte to be read, and the parameter LEN is the number of bytes.

A prerequisite for use of the FB 22 GETIO_PA is that the input bytes to be read are addressed in the process input image. If possible, it is preferable to use a partial process image. Make sure that no borders to adjacent data of other stations are violated when using the parameters OFFSET and LEN.

If the destination area specified by the INPUTS parameter is smaller than the input area read, the function only transfers as many bytes as can be written to the destination area. If the destination area is larger, only the first LEN bytes of the area are written to. In both cases, no error is indicated on the ERROR parameter. ERROR only has signal state "1" if an error is reported when calling SFC 81 BLKMOV.

FB 23 SETIO_PA
Write some outputs of a station

With use of the SFC 81 UBLKMOV, the FB 23 SETIO_PA consistently writes some of the output data to a DP standard slave or an IO device or some of the data of an output area of stations with a modular design. The right word of the parameter ID is the start address of the output area. The parameter OFFSET is the number of the first byte to be written, and the parameter LEN is the number of bytes.

A prerequisite for use of the FB 23 SETIO_PA is that the output bytes to be written are addressed in the process output image. If possible, it is preferable to use a partial process image. Make sure that no borders to adjacent data of other stations are violated when using the parameters OFFSET and LEN.

If the source area specified by the parameter OUTPUTS is smaller than the input area to be written, only as many bytes are transmitted as are present in the source area. If the source area is larger, only the first LEN bytes are transferred. An error is not indicated on the parameter ERROR in either case. ERROR only has signal state "1" if an error is signaled when calling the SFC 81 BLKMOV.

SFC 14 DPRD_DAT
Read user data

SFC 14 DPRD_DAT reads consistent user data with a length of exactly 3 bytes or greater than 4 bytes (means 5 bytes or higher) from a DP slave or an IO device. You specify the length of the data consistency when you parameterize the station. Table 20.10 shows the parameters of the SFCs 14 and 15.

The LADDR parameter receives the module starting address of the user data (input area). The RECORD parameter describes the area in which the read data is saved. Variables of data type ARRAY and STRUCT or an ANY pointer of data type BYTE (e.g. P#DBzDBXy.x BYTE nnn) are permissible as actual parameters.

Table 20.10 Parameters of the SFCs for consistent transfer of user data

SFC	Parameter	Declaration	Data type	Assignment, description
14	LADDR	INPUT	WORD	Configured start address (from the I area)
	RET_VAL	RETURN	INT	Error information
	RECORD	OUTPUT	ANY	Destination area for the read user data
15	LADDR	INPUT	WORD	Configured start address (from the Q area)
	RECORD	INPUT	ANY	Source area for the user data to be written
	RET_VAL	RETURN	INT	Error information

Please note: If peripheral inputs (PI) are addressed whose addresses are in the process image image (I), the process image is not updated.

SFC 15 DPWR_DAT
Write user data

SC 15 DPWR_DAT writes consistent user data with a length of 3 bytes or greater than 4 bytes to a DP slave or an IO device. You specify the length of the data consistency when you parameterize the station.

The LADDR parameter receives the module starting address of the user data (output area).

The RECORD parameter describes the area from which the transferred data is read. Variables of data type ARRAY and STRUCT or an ANY pointer of data type BYTE (e.g. P#DBz-DBXy.x BYTE nnn) are permissible as actual parameters.

If peripheral outputs (PQ) are addressed whose addresses are in the process image output (Q), the process image is updated as with the transfer instruction (STL) or the MOVE box (LAD, FBD).

Activate/deactivate distributed station

The following system function activates or deactivates a station of the distributed I/O (DP slave or IO device):

▷ SFC 12 D_ACT_DP
Activate/deactivate distributed station

The parameters of this system function are shown in Table 20.11.

SFC 12 D_ACT_DP
Activate/deactivate distributed station

SFC 12 D_ACT_DP deactivates and activates stations of the distributed I/O and allows scanning of the deactivated or activated status. A distributed station can be a DP slave or an IO device.

SFC 12 D_ACT_DP is called in the cyclic program; a call in the start-up routine is not supported. The SFC works in asynchronous mode, i.e. processing of a job can extend over several program cycles. An activation or deactivation job is started by "1" in the REQ parameter. The REQ parameter must remain "1" for as long as the BUSY parameter has signal state "1". The job has been completed if BUSY = "0".

After deactivation, a configured (and existing) station is no longer addressed by the DP master or the IO controller. The output terminals of deactivated output modules carry zero or a substitute value. The process image input of deactivated input modules is set to "0".

A deactivated station can be removed from the bus without generating an error message; it is not signaled as faulty or missing. The calls of the asynchronous error organization blocks OB 85 (program execution error if the user data of the deactivated station is present in an automatically updated process image), and OB 86 (station failure) are omitted. They must no longer address the station from the program following deactivation, otherwise an I/O access error with calling of OB 122 will result in the case of direct access, or the station will be signaled as not being present when reading the data record with SFC 59 RD_REC or SFB 52 RDREC.

Table 20.11 Parameters of the SFC for activation and deactivation of I/O stations

SFC	Parameter	Declaration	Data type	Assignment, description
12	REQ	INPUT	BOOL	Request for activation/deactivation if REQ = "1"
	MODE	INPUT	BYTE	Function mode 0 Scan whether the station is activated or deactivated 1 Activate station 2 Deactivate station
	LADDR	INPUT	WORD	Any logic address of the station
	RET_VAL	RETURN	INT	Result of scan or error information
	BUSY	OUTPUT	BOOL	Job still running if BUSY = "1"

Use SFC 12 D_ACT_DP to reactivate a deactivated station. The station is configured and parameterized by the DP master or IO controller as with a return of station. The asynchronous error OBs 85 and 86 are not started when activating. If the BUSY parameter has signal state "0" following activation, the station can be addressed from the user program.

In the case of a cold or warm restart, the CPU's operating system automatically activates the deactivated stations. An S7-300 CPU does not start up until all stations have been activated. An S7-400 CPU starts up and reports I/O access errors until the stations have been activated. The station status is retained during a hot restart.

Triggering and interrupt with PROFIBUS DP

The following system blocks trigger an interrupt with PROFIBUS DP:

▷ SFB 75 SALRM
Trigger interrupt

▷ SFC 7 DP_PRAL
Initiate hardware interrupt

The parameters of the system blocks are listed in Table 20.12.

SFB 75 SALRM
Trigger interrupt

With SFB 75 SALRM, you initiate a diagnostics or process interrupt from the user program

Table 20.12 Parameters of the system blocks for triggering an interrupt with PROFIBUS DP

Block	Parameter	Declaration	Data type	Assignment, description
SFB 75	REQ	INPUT	BOOL	Request for triggering if REQ = "1"
	ID	INPUT	DWORD	Address of an address area in the transfer memory
	ATYPE	INPUT	INT	Interrupt type: 1 = diagnostics interrupt 2 = hardware interrupt
	ASPEC	INPUT	INT	Interrupt ID: 0 = no additional information 1 = slot faulty (UP) 2 = slot no longer faulty (DOWN) 3 = slot still faulty (DOWN)
	LEN	INPUT	INT	Length (in bytes) of additional interrupt information to be sent (max. 16)
	DONE	OUTPUT	BOOL	Interrupt has been transmitted if DONE = "1"
	BUSY	OUTPUT	BOOL	Interrupt transfer still running if BUSY = "1"
	ERROR	OUTPUT	BOOL	Error occurred if ERROR = "1"
	STATUS	OUTPUT	DWORD	Error information
	AINFO	IN_OUT	ANY	Source area for the additional interrupt information
SFC 7	REQ	INPUT	BOOL	Request for triggering if REQ = "1"
	IOID	INPUT	BYTE	B#16#54 = input ID B#16#55 = output ID
	LADDR	INPUT	WORD	Start address of an address area in the transfer memory
	AL_INFO	INPUT	DWORD	Interrupt ID (transfer to the start information of the interrupt OB)
	RET_VAL	RETURN	INT	Error information
	BUSY	OUTPUT	BOOL	Still no acknowledgment from DP master if BUSY = "1"

of an intelligent slave in the associated DP master. You determine the type of interrupt using the parameter ATYPE.

The interrupt request is initiated with REQ = "1"; the parameters DONE, BUSY, ERROR and STATUS indicate the job status. The job is complete (BUSY = "0"), when the interrupt OB in the DP master has been executed.

The transfer memory between the DP master and the intelligent DP slave can be divided into individual address areas that represent individual modules from the viewpoint of the master CPU. You can trigger an interrupt in the master for each of these address areas ("virtual" modules). You specify the address area with the parameter ID which you occupy with a user data address from the viewpoint of the slave CPU. Bit 15 contains the I/O code: "0" corresponds to an input address, "1" to an output address. The start information of the interrupt OB then contains the addresses of the interrupt-triggering "modules" from the viewpoint of the master CPU.

You can use parameter AINFO to transfer additional interrupt information you have defined which can be evaluated in the interrupt OB of the master CPU. The data at AINFO is an ANY pointer to a data area. The length of the transmitted information is determined by the parameter LEN and by the area length of the ANY pointer (the shorter length is decisive). The first 4 bytes are displayed in the start information of the interrupt OB in the master CPU in bytes 8 to 11 (the variable OBxx_POINT_ADDR for process interrupts, the data record DS 0 for diagnostics interrupts). You can read out the complete additional interrupt information in the master CPU with the SFB 54 RALRM.

SFC 7 DP_PRAL
Initiate a process interrupt

With SFC 7 DP_PRAL you trigger a hardware interrupt on the DP master from the user program of an intelligent slave.

At the parameter AL_INFO you transfer an interrupt ID defined by you that is transferred to the start information of the interrupt OB called in the DP master (variable OBxx_POINT_ADDR). The interrupt request is initiated with REQ = "1"; the parameters RET_VAL and BUSY indicate job status. The job is complete when the interrupt OB in the DP master has been executed.

The transfer memory between the DP master and the intelligent DP slave can be divided into individual address areas that represent individual "modules" from the viewpoint of the master CPU. The lowest address of an address area is taken as the module starting address. You can initiate a process interrupt in the master for each of these address areas ("virtual" modules).

You specify an address area at SFC 7 with the parameters IOID and LADDR from the viewpoint of the slave CPU (the I/O ID and the starting address of the slave side). The start information of the interrupt OB then contains the addresses of the "module" initiating the interrupt from the viewpoint of the master CPU.

System blocks for PROFIBUS DP

You can additionally use the following system functions with PROFIBUS DP:

▷ SFC 11 DPSYC_FR
Send SYNC/FREEZE commands

▷ SFC 13 DPNRM_DG
Read diagnostic data from a DP standard slave

▷ SFC 103 DP_TOPOL
Determine bus topology

The parameters of these system functions are shown in Table 20.13.

With DPV1 mode set and DP slaves which support the DPV1 functionality, you can use other system blocks to parameterize and read the diagnostic data (see Chapters 21.9.3 "Reading Additional Interrupt Information" and 22.5 "Parameterizing Modules").

SFC 11 DPSYC_FR
Send SYNC/FREEZE commands

Using SFC 11 DPSYC_FR you send the SYNC, UNSYNC, FREEZE, and UNFREEZE commands to a SYNC/FREEZE group which you have configured in the hardware configuration. The send procedure is triggered by REQ = "1" and is finished when BUSY signals "0".

In the parameter GROUP each group occupies one bit (from bit 0 = group 1 to bit 7 = group 8).

Table 20.13 Parameters of the SFCs for addressing the distributed I/O

SFC	Parameter	Declaration	Data type	Assignment, description
11	REQ	INPUT	BOOL	Request for sending if REQ = "1"
	LADDR	INPUT	WORD	Configured diagnostics address of the DP master
	GROUP	INPUT	BYTE	DP slave group (from the hardware configuration)
	MODE	INPUT	BYTE	Command (see text)
	RET_VAL	RETURN	INT	Error information
	BUSY	OUTPUT	BOOL	Job still running if BUSY = "1"
13	REQ	INPUT	BOOL	Request for reading if REQ = "1"
	LADDR	INPUT	WORD	Configured diagnostics address of the DP slave
	RET_VAL	RETURN	INT	Error information
	RECORD	OUTPUT	ANY	Destination area for the read diagnostic data
	BUSY	OUTPUT	BOOL	Read operation still running if BUSY = "1"
103	REQ	INPUT	BOOL	Triggering of topology determination if REQ = "1"
	R	INPUT	BOOL	Cancelation of topology determination if R = "1"
	DP_ID	INPUT	INT	ID of the DP master system whose topology is to be determined
	RET_VAL	RETURN	INT	Error information for the SFC
	BUSY	OUTPUT	BOOL	Determination still running if BUSY = "1"
	DPR	OUTPUT	BYTE	PROFIBUS address of the diagnostics repeater reporting the error
	DPRI	OUTPUT	BYTE	Measuring segment and error information of the diagnostics repeater reporting the error

The commands in the parameter MODE are also organized by bit:

▷ UNFREEZE, if bit 2 = “1”

▷ FREEZE, if bit 3 = “1”

▷ UNSYNC, if bit 4 = “1”

▷ SYNC, if bit 5 = “1”

A SYNC and an UNSYNC command or a FREEZE and an UNFREEZE command must not be triggered simultaneously in a call.

In this way, SYNC mode and FREEZE mode on the DP slaves are first switched off. The inputs of the DP slaves are scanned in sequence by the DP master and the outputs of the DP slaves are modified; the DP slaves pass the received output signals immediately to the output terminals.

If you want to “freeze” the input signals of several DP slaves at a specific time, you output the command FREEZE to the relevant group. The input signals then read in sequence by the DP master have the signal states they had when “frozen”. These input signals retain their value until you output another FREEZE command to cause the DP slaves to read in and hold the current input signals, or until you switch the DP slaves back to the “normal” mode with the UNFREEZE command.

If you wish to output the output signals of several DP slaves synchronously at a certain time, first output the SYNC command to the associated group. The addressed DP slaves then retain the current signals at the output terminals. You can then transfer the desired signal states to the DP slaves. Output the SYNC command again following completion of transfer; in this manner you request the DP slaves to connect the received output signals simultaneously to the output terminals. The DP slaves retain the signals at the output terminals until you connect the new output signals using a further SYNC command, or until you switch the DP slaves back to "normal" mode using the UNSYNC command.

SFC 13 DPNRM_DG
Read diagnostic data

SFC 13 DPNRM_DG reads diagnostic data from a DP slave. The read procedure is initiated with REQ = “1”, and is terminated when BUSY = “0” is returned. Function value RET_VAL then contains the number of bytes read. Depending on the slave, diagnostic data may comprise from 6 to 240 bytes. If there are more than 240 bytes, the first 240 bytes are transferred and the relevant overflow bit is then set in the data.

The parameter RECORD writes to the area in which the read data are stored. Variables of data type ARRAY and STRUCT or an ANY pointer of data type BYTE (e.g. P#DBzDBXy.x BYTE nnn) are permissible as actual parameters.

Please note that the SFC 13 DPMRM_DG is an asynchronous system function. It must be executed until the BUSY parameter has a signal state “0”. With more recent CPUs, the SFB 54 RALRM is available which provides the data synchronously, i.e. directly when it is called.

SFC 103 DP_TOPOL
Identify bus topology

With the assistance of a diagnostics repeater, the SFC 103 DP_TOPOL determines the bus topology of the DP master system whose ID you specify at parameter DP_ID. The determination is triggered by REQ = “1” and is terminated when BUSY signals “0”. You can abort determination of the topology with R = “1”.

If an error is signaled by a diagnostics repeater which prevents determination of the bus topology, it is indicated in parameters DPR and DPRI. If there are several diagnostics repeaters signaling errors, the error message of the first one is indicated, and the complete diagnostics information can be read with the SFC 13 DPNRM_DG.

In the case of the error information at parameter DPRI, a differentiation is made between temporary and permanent faults. It may not be possible to unambiguously identify temporary faults such as a loose contact which may disappear on their own. You must eliminate permanent faults before you call the SFC 103 DP_TOPOL again to determine the topology.

After SFC 103 DP_TOPOL is called, the determined data is available on the diagnostics repeater and can be read with the SFC 59 RD_REC or the SFB 52 RDREC. The data comprises the topology of the bus segment (stations and cable lengths), the contents of the segment diagnostic buffers (last ten events with fault information, location, and cause), and the statistics data (information on the quality of the bus system)

System block for PROFINET

The following system function block sets the IP configuration during runtime:

- ▷ SFB 104 IP_CONF
  Setting the IP configuration

The parameters of this block are shown in Table 20.14.

SFB 104 IP_CONF
Setting the IP configuration

SFB 104 IP_CONF overwrites the IP address, the subnet mask, the router address, and – if the station is an IO device – the device name during runtime.

The job is triggered with REQ = "1"; the DONE, BUSY, ERROR, and STATUS parameters indicate the job status. The job has been completed if BUSY = "0".

Changing the IP configuration during runtime must already be prepared in the hardware configuration: In the properties of the PN interface, click on the "Properties" button and check the "Obtain IP address in different manner" on the "Parameters" tab in the displayed window.

The ANY pointer in the CONF_DB parameter points to a data area which contains the new values for the IP configuration. It consists of a header containing the field type (= 0), the field ID (= 0), and the number of following subfields. The header is followed by the subfields.

Subfields are currently defined for the IP parameters and the device name (Figure 20.27).

Table 20.14 Parameters of the SFB for setting the IP configuration

SFB	Parameter	Declaration	Data type	Assignment, description
104	REQ	INPUT	BOOL	Request for setting if REQ = "1"
	LADDR	INPUT	WORD	Diagnostics address of the PROFINET interface
	CONF_DB	INPUT	ANY	Pointer to the configuration data
	DONE	OUTPUT	BOOL	Job completed without error if "1"
	BUSY	OUTPUT	BOOL	Job still running if BUSY = "1"
	ERROR	OUTPUT	BOOL	An error has occurred if "1"
	STATUS	OUTPUT	DWORD	Error information
	ERR_LOC	OUTPUT	DWORD	Error source

CONF_DB data area

The CONF_DB data area comprises a header and several subfields. Subfields are currently defined for the IP parameters and for the device name.

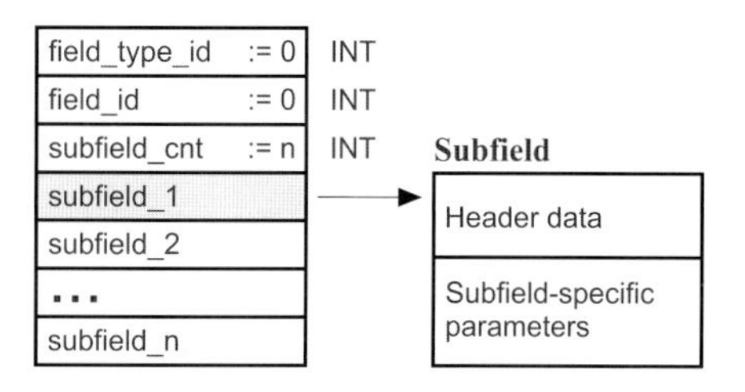

*) If the length of the device name field is defined to be shorter, the length must be adapted in the header.
If the device name is shorter than the field, B#16#:=00 must be assigned to the byte after the device name.
If B#16#:=00 is at the beginning of the device name, the name will be deleted.

Subfield for the IP parameters

id	:= 30	INT	Header
len	:= 18	INT	
mode	:= 1	INT	
ipaddr_3		BYTE	IP address
...			
ipaddr_0		BYTE	
snmask_3		BYTE	Subnet mask
...			
snmask_0		BYTE	
router_3		BYTE	Router address
...			
router_0		BYTE	

Subfield for the device name

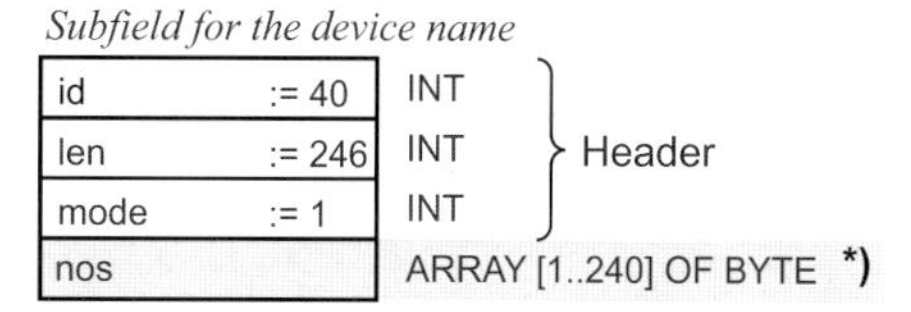

Figure 20.27 Data structure for the IP configuration

20.5 Global Data Communication

20.5.1 Fundamentals

Global data communication (GD communication) is a communications service integrated into the operating system of the CPU and used for exchanging small volumes of non-time-critical data via the MPI bus.

The transferable global data include

- ▷ Inputs and outputs (process images)
- ▷ Memory bits
- ▷ Data in data blocks
- ▷ Timer and counter values as data to be sent.

It is a requirement that the CPUs are networked together via the MPI interface or connected via the K bus as in the S7-400 mounting rack. All CPUs must exist in the same STEP 7 project in order to be able to configure GD communication.

The cyclic GD communication service does not require an operating system: there are system

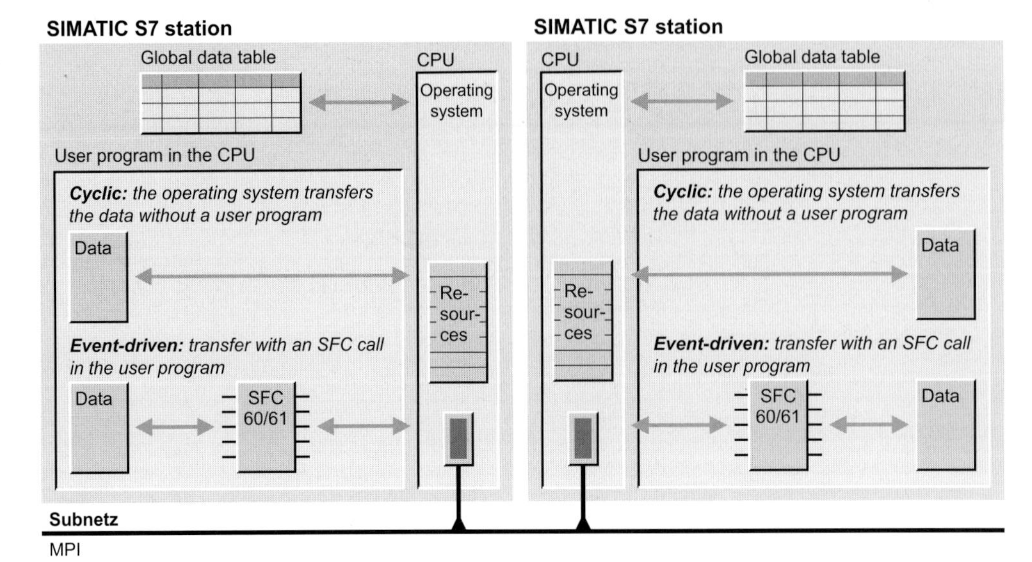

Figure 20.28 Global Data Communication

functions available for event-driven GD communication on the S7-400.

Please note that a receiver CPU does not acknowledge receipt of global data. The sender therefore does not receive any response to tell it if a receiver has received data and if so, which one. However, you can screen the communication status between two CPUs as well as the overall status of all GD circles of a CPU.

Sending and receiving global data is controlled with what are known as scan rates. These specify the number of (user program) cycles after which the CPU sends or receives the data. Sending and receiving takes place synchronously between the sender and the receiver at the cycle control point in each case, i.e. following cyclic program execution and before a new program cycle begins (like process image updating, for example).

Data is exchanged in the form of data packets (GD packets) between CPUs grouped into GD circles.

GD circle

The CPUs that exchange a shared GD packet form a GD circle. A GD circle can be any of the following

- The one-way connection of a CPU that sends a GD packet to several other CPUs that then receive that packet.
- The two-way connection between two CPUs where each of the two CPUs can send a GD packet to the other.
- The two-way connection between three CPUs where each of the three CPUs can send one GD packet to the other two CPUs (S7-400 CPUs only)

Up to 15 CPUs can exchange data with each other in one GD circle. One CPU can also belong to several GD circles. See Table 20.14 for the resources of each individual CPU here.

GD packet

A GD packet comprises the packet header and one or more global data elements (GD elements):

- Packet header (8 bytes)
- ID of 1st GD element (2 bytes)
- User data of 1st GD element (x bytes)
- ID of 2nd GD element (2 bytes)
- User data of 2nd GD element (x bytes)
- etc.

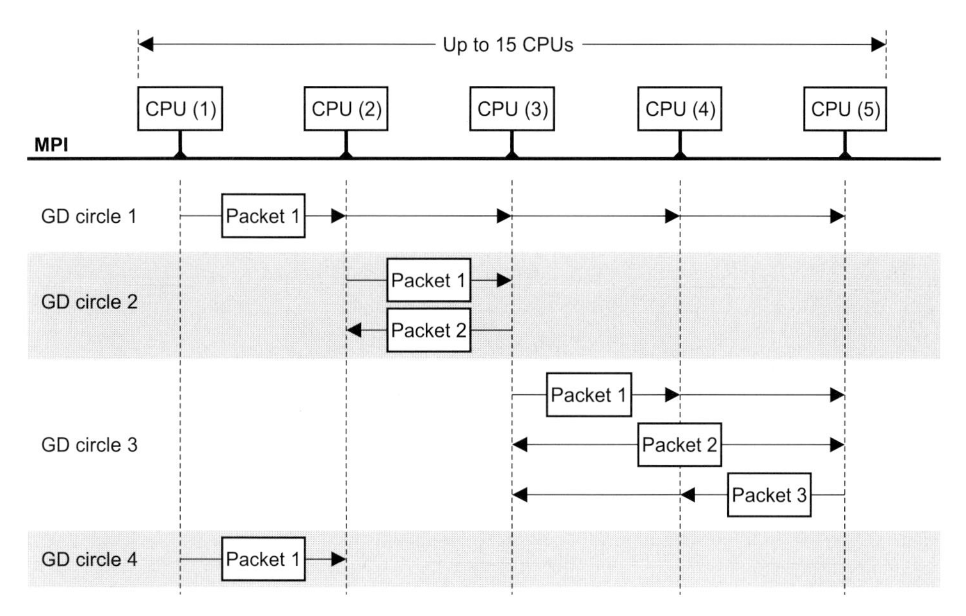

Figure 20.29 Example for GD circles

Each GD element consists of 2 bytes of description and the actual net data. 3 bytes are required in the GD packet to transfer a memory byte, 4 bytes are required for a memory word, and 6 bytes for a memory doubleword. A Boolean variable occupies 1 byte of net data; it therefore requires the same space as a byte-sized variable. Timer and counter values with 2 bytes each occupy 4 bytes each in the GD packet.

A GD element can also be an address area. MB 0:15, for example, represents the area from memory byte MB 0 to MB 15, and DB20.DBW14:8 represents the data area located in DB 20 that starts from data word DBW 14 and comprises 8 data words.

The maximum size of a GD package is 32 bytes for the S7-300 and 64 bytes for the S7-400. You can achieve the maximum number of net data bytes per packet by transmitting only one GD element which can contain max. 22 bytes of net data with S7-300 and max. 54 bytes with S7-400.

Data consistency

The data consistency covers one GD element. If a GD element overwrites a CPU-specific variable, the areas specified in Table 20.15 apply.

If a GD element is greater than the length of the data consistency, blocks with consistent data of the relevant length are formed, starting with the first byte.

20.5.2 Configuring GD Communication

Requirements

You must have created a project, there must be an MPI subnet available and you must have configured the S7 stations. The CPU, at least, must be available in the stations. Under the "Properties" button of the MPI interface, on the "General" tab of the properties window of the CPU (double-click on the CPU line in the Hardware Configuration or on the line with the MPI interface submodule), you can set the MPI address and select the MPI subnet with which the CPU is connected.

Global data table

You configure GD communication by filling out a table. With the icon for the MPI subnet selected in the SIMATIC Manager or in the Network Configuration, you can call up an empty table with OPTIONS → DEFINE GLOBAL DATA. Select a column and then EDIT → CPU.

Table 20.15 CPU resources for global data communication

GD resources Maximum number of:	CPU 312 CPU 313 CPU 314 CPU3xxC	CPU 315 CPU 317 CPU 319	CPU 412 CPU 414	CPU 416 CPU 417
GD circles per CPU	4	8	8	16
Receive GD packages per CPU	4	8	16	32
Receive GD packages per circle	1	1	2	2
Send GD packages per CPU	4	8	8	16
Send GD packages per circle	1	1	1	1
Max. size of a GD package	22 bytes	22 bytes	64 bytes	64 bytes
Max. data consistency	22 bytes	22 bytes	1 tag	1 tag

Select the station in the left half of the project selection window that then opens, and select the CPU in the right half. This CPU is accepted into the global data table with "OK".

Proceed in exactly the same way with the other CPUs participating in GD communication. A global data table can contain up to 15 CPU columns.

To configure data transfer between CPUs, select the first line under the send CPU and specify the address whose value is to be transferred (terminate with RETURN).

With EDIT → SENDER, you define this value as the value to be sent, indicated by a prefixed character ">" and shading. In the same line under Receiver CPU, you enter the address that is to accept the value (the property "Receiver" is set as default). You may use timer and counter functions only as senders; the receiver must be an address of word width for each timer or counter function.

A line can contain several receivers but only one sender (Table 20.16). After filling this in, you select GD TABLE → COMPILE.

After initial compilation (phase 1), the system data produced is sufficient for global data communication. If you also configure the GD status (the status of the GD connection) and the reduction ratios, you must then compile the GD table a second time.

Table 20.16 Example for a GD table with status and reduction ratios

GD identifier	Station 417 \ CPU417 (3)	Station 417 \ CPU414 (4)	Station 416\ CPU 416 (5)	Station 315 Slave\ CPU315 (7)	Station 314CP\ CPU314 (10)
GST	MD100	MD100	MD100	DB10.DBD200	DB10.DBD200
GDS 1.1	DB9.DBD0		MD92	DB10.DBD204	DB10.DBD204
SR 1.1	44	0	44	8	8
GD 1.1.1	>DB9.DBW10		MW90	DB10.DBW208	DB10.DBW208
GDS 2.1	MD96	MD96			
SR 2.1	44	23	0	0	0
GD 2.1.1	>Z10:10	DB3.DBW20:10			
GDS 3.1			MD96		
SR 3.1	0	0	44	8	8
GD 3.1.1			>MW98	DB10.DBW220	DB10.DBW210

GD ID

Following error-free compiling, STEP 7 completes the "GD ID" column. The GD ID shows you how the transferred data are structured into GD circles, GD packets and GD elements. For example, the GD ID "GD 2.1.3" corresponds to GD circle 2, GD packet 1, GD element 3. You can then find the resource assignment (number of GD circles) per CPU in the CPU column of the global data table.

GD status

Following compiling, you can enter the addresses for the communication status into the global data table with VIEW → GD STATUS. The overall status (GST) shows the status of all communications connections in the table. The status (GDS) shows the status of a communications connection (a transmitted GD packet). The status is shown in a doubleword in each case.

Scan rates

The GD communication service requires a significant portion of execution time in the CPU operating system and demands transmission time on the MPI bus. To keep this "communications load" to a minimum, it is possible to specify a "scan rate". A scan rate specifies the number of program cycles after which the data (or more precisely, a GD packet) are to be sent or received.

Since the data are not updated in every program cycle with a scan rate, you should avoid sending time-critical data via this form of communication.

After the first (error-free) compilation, you can use VIEW → SCAN RATES to define the scan rates (SRs) yourself for each GD packet and each CPU. The scan rate is set as standard in such a way that with an "empty" CPU (no user program) the GD packets are sent and received approximately every 10 ms. If a user program is then loaded, the time interval increases.

You can enter the scan rates in the area between 1 and 255. Please note, that as the scan rates decrease, the communications load on the CPU increases. To keep the communications load within tolerable limits, set the scan rate in the send CPU in such a way that the product of scan rate and cycle time on the S7-300 is greater than 60 ms and on the S7-400 greater than 10 ms. In the receive CPU, this product must be less than that in the send CPU to avoid the loss of any GD packets.

With reduction ratio 0, you switch off cyclic data exchange of the relevant GD packet with S7-400 CPUs if you only want to send or receive it event-driven with SFCs.

After configuring the GD status and the scan rates, you must compile the GD table for a second time. Then STEP 7 enters the compiled data in the *System data* object. GD communication becomes effective when you transfer the GD table to the connected CPUs with PLC → DOWNLOAD.

GD communication also becomes effective when the *System data* object, that contains all hardware settings and parameter settings, is transferred.

20.5.3 System Functions for GD Communication

In S7-400 systems, you can also control GD communication in your program. Additionally or alternatively to the cyclic transfer of global data, you can send or receive a GD packet with the following SFCs:

▷ SFC 60 GD_SND
Send GD packet

▷ SFC 61 GD_RCV
Receive GD packet

The parameters for these SFCs are listed in Table 20.17. The prerequisite for the use of these SFCs is a configured global data table. After compiling this table, STEP 7 shows you, in the "GD Identifier" column, the numbers of the GD circles and GD packets which you need for parameter assignments.

SFC 60 GD_SND enters the GD packet in the system memory of the CPU and initiates transfer; SFC 61 GD_RCV fetches the GD packet from the system memory. If a scan rate greater than 0 has been specified for the GD packet in the GD table, cyclic transfer also takes place.

If you want to ensure data consistency for the entire GD packet when transferring with SFCs 60 and 61, you must disable or delay higher-priority interrupts and asynchronous errors on

Table 20.17 Parameters of the SFCs for GD communication

Parameter	Available in SFC		Declaration	Data type	Assignment, description
CIRCLE_ID	60	61	INPUT	BYTE	Number of the GD circle
BLOCK_ID	60	61	INPUT	BYTE	Number of the GD package to be sent or received
RET_VAL	60	61	RETURN	INT	Error information

both the Send and Receive side during processing of SFC 60 or SFC 61.

The SFCs need not be called in pairs; "mixed" operation is also possible. For example, you can use SFC 60 GD_SND to have event-driven transmission of GD packets but then receive cyclically.

20.6 S7 Basic Communication

20.6.1 Station-Internal S7 Basic Communication

Fundamentals

With station-internal S7 basic communication, you can exchange data between programmable modules within a SIMATIC station. The communication functions required here are SFCs in the operating system of the CPU. These SFCs establish the communication connections themselves, if necessary. For this reason, these station-internal connections are not configured via the connection table ("Communication via non-configured connections").

Station-internal S7 basic communication can, for example, take place in parallel to the cyclical data exchange via PROFIBUS DP between the master CPU and the slave CPU. Event-triggered data is transferred (Figure 20.30).

Addressing the nodes, connections

Node identification is derived from the I/O address: at the LADDR parameter, you specify the module starting address and at the IOID parameter, you specify whether this address is in the input area or the output area.

These system functions establish the necessary communication connections dynamically and they clear the connections down again following completion of the job (programmable). If a connection buildup cannot be executed due to lack of resources either in the sending device or in the receiving device, "Temporary lack of resources" is signaled. Transfer must then be reinitiated. There can only be one connection between two communication partners in each direction.

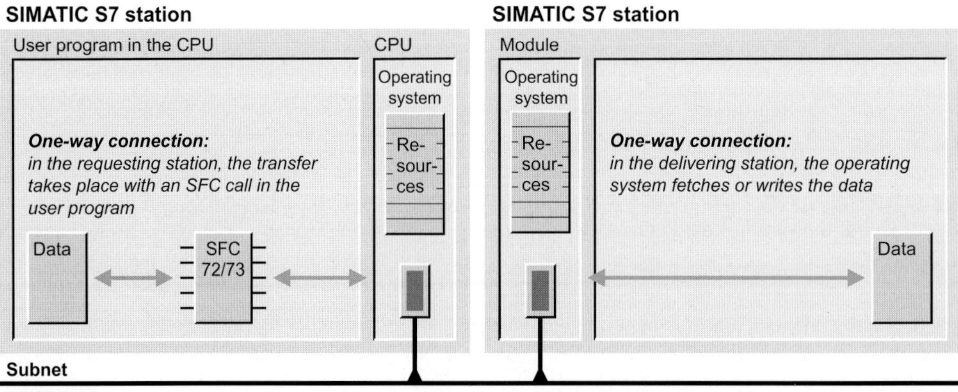

Figure 20.30 Station-internal S7 basic communication

You can use one system function for different communication connections by modifying the block parameters at runtime. An SFC cannot interrupt itself. A program section in which one of these SFCs is used can only be modified in the STOP mode; following this, a cold or warm restart is executed.

User data, data consistency

These SFCs transfer up to 76 bytes as user data. Regardless of the direction of transfer, the operating system of a CPU arranges the user data in blocks that are consistent within themselves. On the S7-300, these blocks have a length of 8 bytes, on the CPU 412/413, they have a length of 16 bytes and on the CPU 414/416, they have a length of 32 bytes. If two CPUs exchange data, the block size of the "passive" CPU is decisive for data consistency.

Configuring station-internal S7 basic communication

Station-internal S7 basic communication is a special case in that it requires no configuring since data transfer is handled via dynamic connections. You simply use an existing PROFIBUS subnet or you create one either in the SIMATIC Manager (select the *Project* object and then INSERT → SUBNET → PROFIBUS) or in the Network Configuration (see Chapter 2.4 "Configuring the Network").

Example: you have configured distributed I/O with a CPU 315-2DP as master. You use another CPU 315-2DP as an "intelligent" slave. You can now use station-internal S7 basic communication from both controllers to read and write data.

20.6.2 System Functions for Data Interchange within a Station

The following system functions handle data transfers between two CPUs in the same station:

▷ SFC 72 I_GET
Read data

▷ SFC 73 I_PUT
Write data

▷ SFC 74 I_ABORT
Disconnect

The parameters for these SFCs are listed in Table 20.18.

SFC 72 I_GET
Read data

A job is initiated with REQ = "1" and BUSY = "0" ("first call"). While the job is in progress, BUSY is set to "1". Changes to the REQ parameter no longer have any effect. When the job is completed, BUSY is reset to "0". If REQ is still "1", the job is immediately restarted.

Table 20.18 Parameters of the SFCs for station-internal S7 basic communication

Parameter	Available in SFC			Declaration	Data type	Assignment, description
REQ	72	73	74	INPUT	BOOL	Trigger job with REQ = "1"
CONT	72	73	-	INPUT	BOOL	CONT = "1": Connection remains after job is completed
IOID	72	73	74	INPUT	BYTE	B#16#54 = input range, B#16#55 = output range
LADDR	72	73	74	INPUT	WORD	Module starting address
VAR_ADDR	72	73	-	INPUT	ANY	Data area in partner device
SD	-	73	-	INPUT	ANY	Data area in the own CPU containing send data
RET_VAL	72	73	74	RETURN	INT	Error information
BUSY	72	73	74	OUTPUT	BOOL	Job running if BUSY = "1"
RD	72	-	-	OUTPUT	ANY	Data area in the own CPU that accepts receive data

When the read procedure has been initiated, the operating system in the partner CPU assembles and sends the requested data. An SFC call transfers the Receive data to the target area. RET_VAL then shows the number of bytes transferred.

If CONT is = "0", the communication link is broken. If CONT is = "1", the link is maintained. The data are also read when the communication partner is in STOP mode.

The RD and VAR_ADDR parameters describe the area from which the data to be transferred are to be read or to which the receive data are to be written. Actual parameters may be addresses, variables or data areas addressed with an ANY pointer. The Send and Receive data are not checked for identical data types.

SFC 73 I_PUT
Write data

A job is initiated with REQ = "1" and BUSY = "0" ("first call"). While the job is in progress, BUSY is set to "1". Changes to the REQ parameter no longer have any effect. When the job is completed, BUSY is reset to "0". If REQ is still "1", the job is immediately restarted.

Following triggering of the write operation, the operating system accepts all data from the source area into an internal buffer during the initial call and sends it to the partner device. The received data is written there by its operating system into the data area VAR_ADDR. BUSY is subsequently set to "0". The data is even written if the communication partner is at STOP.

The SD and VAR_ADDR parameters describe the area from which the data to be transferred are to be read or to which the receive data are to be written. Actual parameters may be addresses, variables or data areas addressed with an ANY pointer. The Send and Receive data are not checked for identical data types.

SFC 74 I_ABORT
Disconnect

REQ = "1" breaks a connection to the specified communication partner. With I_ABORT, you can break only those connections established in the same station with I_GET or I_PUT.

While the job is in progress, BUSY is set to "1". Changes to the REQ parameter no longer have any effect. When the job is completed, BUSY is reset to "0". If REQ is still "1", the job is immediately restarted.

20.6.3 Station-External S7 Basic Communication

Fundamentals

With station-external S7 basic communication, you can have event-driven data exchange between SIMATIC S7 stations. The stations must be connected to each other via an MPI subnet. The communications functions required for this are SFCs in the operating system of the CPU. These SFCs establish the communication connections themselves, if necessary. For this reason, these station-external connections are not configured via the connection table ("Communication via non-configured connections").

Station-external S7 basic communication can execute event-driven data transfer, for example, parallel to cyclic global data communication.

Addressing the nodes, connections

These functions address nodes that are on the same MPI subnet. The node identification is derived from the MPI address (DEST_ID parameter).

These system functions establish the necessary communication connections dynamically and they terminate the connections following completion of the job (programmable). If a connection cannot be established because resources are missing either in the sender or receiver, "Temporary shortage of resources" is signaled. Triggering of the transmission must then be repeated. There can only be one connection in each direction between two communication partners.

On a transition from RUN to STOP, all active connections (all SFCs except X_RECV) are cleared.

By modifying the block parameters at run time, you can utilize a system function for different communication links. An SFC may not interrupt itself. You may modify a program section in which one of these SFCs is used only in STOP

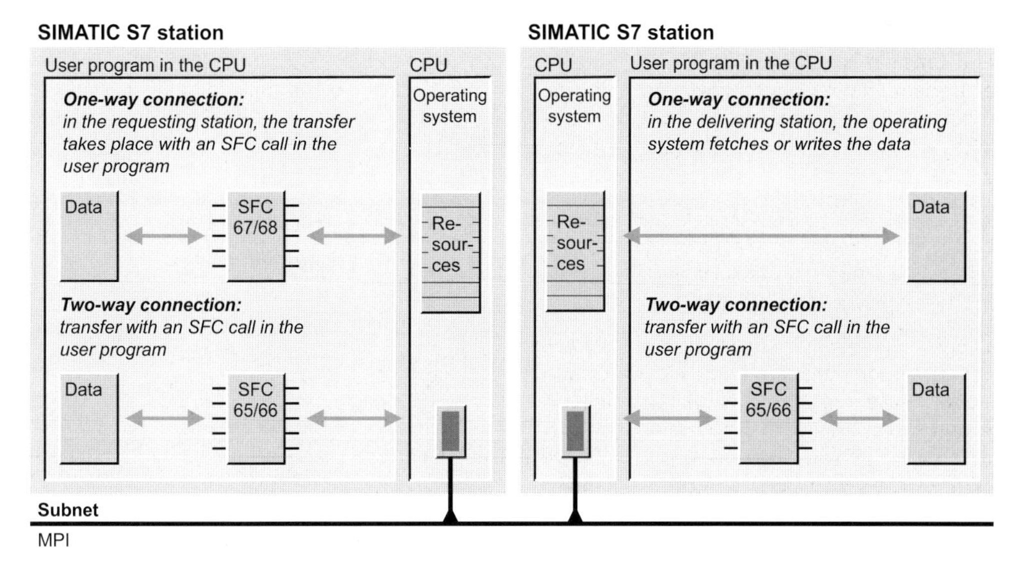

Figure 20.31 Station-external S7 basic communication

mode; a cold or warm restart must then be executed.

User data, data consistency

These SFCs transfer a maximum of 76 bytes of user data. A CPU's operating system combines the user data into blocks consistent within themselves, without regard to the direction of transfer. In S7-300 systems, these blocks have a length of 8 bytes, in systems with a CPU 412/413 a length of 16 bytes, and in systems with a CPU 414/416 a length of 32 bytes.

If two CPUs exchange data via X_GET or X_PUT, the block size of the "passive" CPU is decisive to data consistency of the transferred data.

In the case of a SEND/RECEIVE connection, all data are consistent.

Configuring station-external S7 basic communication

Station-external S7 basic communication is a special case in that it requires no configuring since data transfer is handled via dynamic connections. You simply use an existing PROFIBUS subnet or you create one.

Example: you have a divided S7-400 mounting rack with one CPU 416 in each section. In addition, you connect an S7-300 station with a CPU 314 via an MPI cable to one of the S7-400s. You configure all three CPUs in the Hardware Configuration, for example, as "networked" via an MPI subnet. You can now use station-external S7 basic communication from all three controllers to exchange data.

20.6.4 System Functions for Station-External S7 Basic Communication

The following system functions handle data transfers between partners in different stations:

▷ SFC 65 X_SEND
Send data

▷ SFC 66 X_RCV
Receive data

▷ SFC 67 X_GET
Read data

▷ SFC 68 X_PUT
Write data

▷ SFC 69 X_ABORT
Disconnect

The parameters for these SFCs are listed in Table 20.19.

Table 20.19 SFC Parameters for Station-External S7 Basic Communication

Parameter	Present in SFC					Declaration	Data Type	Contents, Description
REQ	65	-	67	68	69	INPUT	BOOL	Job initiation with REQ = "1"
CONT	65	-	67	68	-	INPUT	BOOL	CONT = "1": Connection is maintained when job is completed
DEST_ID	65	-	67	68	69	INPUT	WORD	Partner's node identification (MPI address)
REQ_ID	65	-	-	-	-	INPUT	DWORD	Job identification
VAR_ADDR	-	-	67	68	-	INPUT	ANY	Data area in partner CPU
SD	65	-	-	68	-	INPUT	ANY	Data area in own CPU which contains the Send data
EN_DT	-	66	-	-	-	INPUT	BOOL	If "1": Accept Receive data
RET_VAL	65	66	67	68	69	RETURN	INT	Error information
BUSY	65	-	67	68	69	OUTPUT	BOOL	Job in progress when BUSY = "1"
REQ_ID	-	66	-	-	-	OUTPUT	DWORD	Job identification
NDA	-	66	-	-	-	OUTPUT	BOOL	When "1": Data received
RD	-	66	67	-	-	OUTPUT	ANY	Data area in own CPU which will accept the Receive data

SFC 65 X_SEND
Send data

A job is initiated with REQ = "1" and BUSY = "0" ("first call"). While the job is in progress, BUSY is set to "1"; changes to the REQ parameter now no longer have any effect. When the job terminates, BUSY is set back to "0". If REQ is still "1", the job is immediately restarted.

On the first call, the operating system transfers all data from the source area to an internal buffer, then transfers the data to the partner CPU.

BUSY is "1" for the duration of the send procedure. When the partner has signaled that it has fetched the data, BUSY is set to "0" and the send job terminated.

If CONT is = "0", are available to other communication links. If CONT is = "1", the connection is maintained. The REQ_ID parameter makes it possible for you to assign an ID to the Send data which you can evaluate with SFC X_RCV.

The SD parameter describes the area from which the data to be sent are to be read. Actual parameters may be addresses, variables, or data areas addressed with an ANY pointer. Send and Receive data are not checked for matching data types.

SFC 66 X_RCV
Receive data

The Receive data are placed in an internal buffer. Multiple packets can be put in a queue in the chronological order of their arrival.

Use EN_DT = "0" to check whether or not data were received; if so, NDA is "1", RET_VAL shows the number of bytes of Receive data, and REQ_ID is the same as the corresponding parameter in SFC 65 X_SEND. When EN_DT is = "1", the SFC transfers the first (oldest) packet to the target area; NDA is then "1" and RET_VAL shows the number of bytes transferred. If EN_DT is "1" but there are no data in the internal queue, NDA is "0".

On a cold or warm restart, all data packets in the queue are rejected.

In the event of a broken connection or a restart, the oldest entry in the queue, if already "queried" with EN_DT = "0", is retained; otherwise, it is rejected like the other queue entries.

The RD parameter describes the area to which the Receive data are to be written. Actual parameters may be addresses, variables, or data areas addressed with an ANY pointer.

Send and Receive data are not checked for matching data types. When the Receive data are irrelevant, a "blank" ANY pointer (NIL pointer) as RD parameter in X_RCV is permissible.

SFC 67 X_GET
Read data

A job is initiated with REQ = "1" and BUSY = "0" ("first call"). While the job is in progress, BUSY is set to "1"; changes to the REQ parameter now no longer have any effect.

When the job terminates, BUSY is set back to "0". If REQ is still "1", the job is immediately restarted.

When the read procedure has been initiated, the operating system in the partner CPU assembles and sends the data required under VAR_ADDR. On an SFC call, the Receive data are entered in the target area specified at the RD parameter. RET_VAL then shows the number of bytes transferred.

If CONT is "0", the communication link is broken. If CONT is "1", the connection is maintained. The data are then read even when the communication partner is in STOP mode.

The RD and VAR_ADDR parameters describe the area from which the data to be sent are to be read or to which the Receive data are to be written. Actual parameters may be addresses, variables, or data areas addressed with an ANY pointer. Send and Receive data are not checked for matching data types.

SFC 68 X_PUT
Write data

A job is initiated with REQ = "1" and BUSY = "0" ("first call"). While the job is in progress, BUSY is set to "1"; changes to the REQ parameter now no longer have any effect.

When the job terminates, BUSY is set back to "0". If REQ is still "1", the job is immediately restarted.

When the write procedure has been initiated, the operating system transfers all data from the source area specified at the SD parameter to an internal buffer on the first call, then sends the data to the partner CPU. There, the partner CPU's operating system writes the Receive data to the data area specified at the VAR_ADDR parameter. BUSY is then set to "0".

The data are written even if the communication partner is in STOP mode.

The RD and VAR_ADDR parameters describe the area from which the data to be sent are to be read or to which the Receive data are to be written. Actual parameters may be addresses, variables, or data areas addressed with an ANY pointer. Send and Receive data are not checked for matching data types.

SFC 69 X_ABORT
Disconnect

REQ = "1" breaks an existing connection to the specified communication partner. The SFC X_ABORT can be used to break only those connections established in the CPU's own station with the SFCs X_SEND, X_GET or X_PUT.

20.7 S7 Communication

20.7.1 Fundamentals

With S7 communication, you transfer larger volumes of data between SIMATIC S7 stations. The stations are connected to each other via a subnet; this can be an MPI subnet, a PROFIBUS subnet or an Ethernet subnet. The communications connections are static; they are configured in the connection table ("Communication via configured connections").

The communications functions are S7-400 system function blocks SFBs integrated in the operating system of the CPUs. The associated instance data block is located in the user memory. If you want to use S7 communication, copy the interface description of the SFBs from the *Standard Library* under *System Function Blocks* to the *Blocks* container, generate an instance data block for each call and call the SFB with the associated instance data block. With incremental input, you can also select the SFB from the program element catalog and have the instance data block generated automatically.

With S7-300, the communications functions are standard function blocks FB which you can find

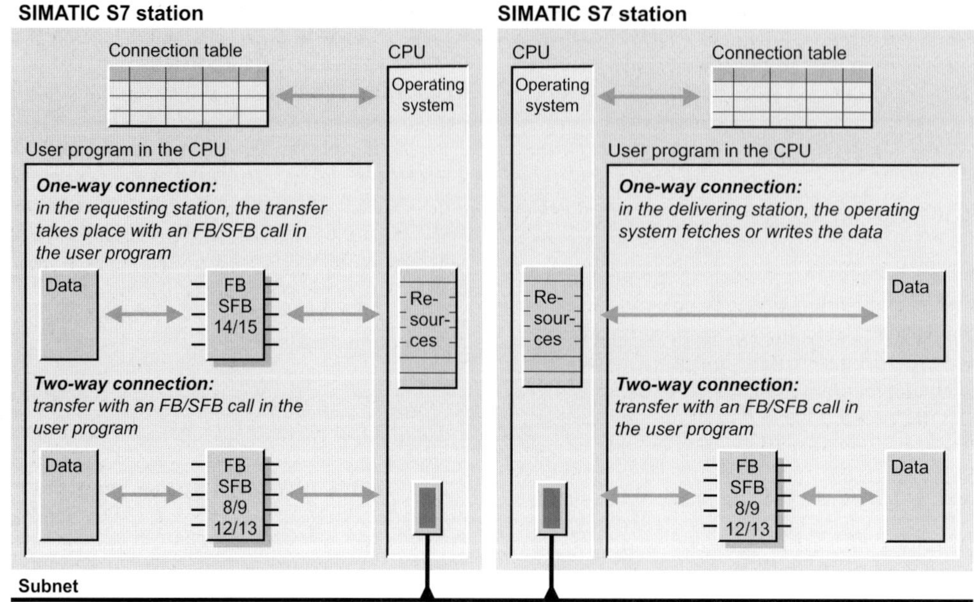

Figure 20.32 S7 Communication

under *Communication Blocks* in the *Standard Library*. Copy the function blocks you wish to use into the container *Blocks* and use them exactly like "normal" function blocks.

Configuring S7 communication

The prerequisite for S7 communication is a configured connection table in which the communication links are defined.

A communication link is specified by a connection ID for each communication partner. STEP 7 assigns the connection IDs when it compiles the connection table. Use the "local ID" to initialize the FB or SFB in the local or "own" module and the "remote ID" to initialize the FB or SFB in the partner module.

The same logical connection can be used for different Send/Receive requests. To distinguish between them, you must add a job ID to the connection ID in order to define the relationship between the Send block and Receive block.

Initialization

S7 communication must be initialized at restart so that the connection to the communication partner can be established. Initialization takes place in the CPU that receives the attribute "Active connection buildup = Yes" in the connection table. You call the communication SFBs used in cyclic operation in a restart OB and initialize the parameters (provided they are available) as follows:

▷ REQ = FALSE

▷ ID = local connection ID from the connection table (data type WORD W#16#xxxx)

▷ R_ID = request ID which you define for a "block pair" (data type DWORD DW#16#xxxx xxxx)

▷ PI_NAME = variable with the contents 'P_PROGRAM' in ASCII coding (e.g. ARRAY[1..9] OF CHAR).

The blocks must continue to be called in a program loop until the DONE parameter has signal state "1". The parameters ERROR and STATUS give information concerning the errors that have occurred and the job status.

You do not need to switch the data areas at start-up (concerns the ADDR_n, RD_n, and SD_n parameters). Exceptions for S7-400: For the SFB 8 USEND, SFB 9 URCV, SFB 14 GET, and SFB 15 PUT the communication buffers are created on the first call to ensure consistency; these define the maximum data amount per transfer for all subsequent calls.

Cyclic operation

In cyclic operation, you call the communication blocks absolutely and you control data transfer via the parameters REQ and EN_R. You must evaluate the results at the parameters NDR, DONE, ERROR and STATUS immediately following each processing of a communication block since they are only valid up to the next call.

With S7-300, the parameters with data type ANY (SD_1, RD_1, ADDR_1) must only be assigned with bit memory and data address areas.

20.7.2 Two-Way Data Exchange

For two-way data exchange, you require one SEND block and one RECEIVE block each at the ends of a connection. Both blocks carry the connection IDs that are located in the connection table in the same line. You can also use several "block pairs" which are then distinguished from each other by the job ID.

The following blocks are available for two-way data interchange:

- FB/SFB 8 USEND
  Uncoordinated sending of a data packet of CPU-specific length
- FB/SFB 9 URCV
  Uncoordinated receiving of a data packet of CPU-specific length
- FB/SFB 12 BSEND
  Sending of a data block of up to 32 or 64 Kbytes in length
- FB/SFB 13 BRCV
  Receiving of a data block of up to 32 or 64 Kbytes in length

FB/SFB 8 and FB/SFB 9 or FB/SFB 12 and FB/SFB 13 must always be used as a pair.

The parameters of these blocks are listed in Table 20.20.

FB 8 USEND and FB 9 URCV
FB 28 USEND_E and FB 29 URCV_E
SFB 8 USEND and SFB 9 URCV
Uncoordinated sending and receiving

The SD_n and RD_n parameters are used to specify the variable or the area you want to transfer. The send area SD_n must correspond to the respective receive area RD_n. Use the parameters without gaps, beginning with 1. Unneeded parameters are not assigned (as in an FB, an SFB does not have to have values for all parameters).

A positive edge at the REQ (request) parameter starts the data exchange, a positive edge at the R (reset) parameter aborts it. A "1" in the EN_R (enable receive) parameter signals that the partner is ready to receive data, "0" can be used to abort a current request.

When the parameter NDR has assumed the value "1" following the data transfer, call the block again, this time with EN_R = "0", to prevent the receive area from being overwritten by new data during the data evaluation.

Initialize the ID parameter with the connection ID, which STEP 7 enters in the connection table for both the local and the partner (the two IDs may differ). R_ID allows you to choose a specifiable but unique job ID which must be identical for the Send and Receive block. This allows several pairs of Send and Receive blocks to share a single logical connection (as each has a unique ID).

With S7-400, the system function blocks import the ID and R_ID values to their instance data block on the first call. The first call establishes the communication relationship (for this instance) until the next warm restart. With S7-300, you can change the assignment of the ID and R_ID parameters following each completed job.

With signal state "1" in the DONE or NDR parameter, the block signals that the job terminated without error. An error, if any, is flagged in the ERROR parameter. A value other than zero in the STATUS parameter indicates either a warning (ERROR = "0") or an error (ERROR = "1").

Table 20.20 FB/SFB parameters for sending and receiving data

Parameter	Available for FB/SFB						Declaration	Data type	Assignment, description
REQ	8	28	12	-	-	-	INPUT	BOOL	Start data exchange
EN_R	-	-	-	9	29	13	INPUT	BOOL	Ready to receive
R	-	-	12	-	-	-	INPUT	BOOL	Cancel data exchange
ID	8	28	12	9	29	13	INPUT	WORD	Connection ID
R_ID	8	28	12	9	29	13	INPUT	DWORD	Job ID
DONE	8	28	12	-	-	-	OUTPUT	BOOL	Job completely processed
NDR	-	-	-	9	29	13	OUTPUT	BOOL	New data accepted
ERROR	8	28	12	9	29	13	OUTPUT	BOOL	Fault occurred
STATUS	8	28	12	9	29	13	OUTPUT	WORD	Job status
SD_1	8	28	12	-	-	-	IN_OUT	ANY	First send area
SD_2 [1)]	8	28	-	-	-	-	IN_OUT	ANY	Second send area
SD_3 [1)]	8	28	-	-	-	-	IN_OUT	ANY	Third send area
SD_4 [1)]	8	28	-	-	-	-	IN_OUT	ANY	Fourth send area
RD_1	-	-	-	9	29	13	IN_OUT	ANY	First receive area
RD_2 [1)]	-	-	-	9	29	-	IN_OUT	ANY	Second receive area
RD_3 [1)]	-	-	-	9	29	-	IN_OUT	ANY	Third receive area
RD_4 [1)]	-	-	-	9	29	-	IN_OUT	ANY	Fourth receive area
LEN	-	-	12	-	-	13	IN_OUT	WORD	Data block length in bytes

[1)] Not for FB 8 or FB 9

FB 12 BSEND and FB 13 BRCV
SFB 12 BSEND and SFB 13 BRCV
Block-oriented sending and receiving

At the SD_n or RD_n parameter you specify a pointer to the first byte of the data area (when called for the first time, the length of this actual parameter determines the maximum size of the communication buffer, it is not evaluated with further calls); the number of bytes of the data to be currently sent or received is present at the LEN parameter.

Up to 64 Kbytes (32 Kbytes with S7-300 without integrated interface) may be transferred; the data are transferred in blocks (sometimes called frames), and the transfer itself is asynchronous to the user program scan. The LEN parameter is updated following each received block.

A positive edge at the REQ (request) parameter starts the data exchange, a positive edge at the R (reset) parameter aborts it. A "1" in the EN_R (enable receive) parameter signals that the partner is ready to receive data, "0" can be used to abort a current job.

If the NDR parameter has assumed the value "1" following the data transmission, call the block again, this time with EN_R = "0", to prevent the received area from being overwritten by new data during the data evaluation.

Supply the ID parameter with the connection ID which STEP 7 defines in the connection table both for the local device and for the partner device (the two IDs can be different). R_ID allows you to choose a specifiable but unique job ID which must be identical for the Send and Receive block. This allows several pairs of Send and Receive blocks to share a single logical connection (as each has a unique ID.

With S7-400, the system function blocks import the ID and R_ID values to their instance data block on the first call. The first call establishes the communication relationship (for this instance) until the next warm or cold restart. With S7-300, you can change the assignment of the ID and R_ID parameters following each finished job.

With signal state "1" in the DONE or NDR parameter, the block signals that the job terminated without error. An error, if any, is flagged in the ERROR parameter. A value other than zero in the STATUS parameter indicates either a warning (ERROR = "0") or an error (ERROR = "1").

20.7.3 One-Way Data Exchange

In one-way data exchange, the communication block call is located in only one CPU. In the partner CPU, the operating system handles the necessary communication functions.

The following blocks are available for one-way data interchange:

▷ FB 14 GET
FB 34 GET_E
SFB 14 GET
Read data from a partner CPU

▷ FB 15 PUT
FB 35 PUT_E
SFB 15 PUT
Write data to a partner CPU

Table 20.21 lists the parameters for these blocks.

The operating system in the partner CPU collects the data read with FB/SFB 14; the operating system in the partner CPU distributes the data written with FB/SFB 15. A Send or Receive (user) program in the partner CPU is not required. The partner CPU can provide the required communications services both in RUN and STOP. The size of the consistent data blocks transmitted depends on the (server) CPU used.

A positive edge at parameter REQ (request) starts the data interchange. Set the ID parameter to the connection ID entered by STEP 7 in the connection table.

With a "1" in the DONE or NDR parameter, the block signals that the job terminated without error. An error, if any, is flagged with a "1" in the ERROR parameter. A value other than zero in the STATUS parameter is indicative of either a warning (ERROR = "0") or an error (ERROR = "1"). You must evaluate the DONE, NDR, ERROR and STATUS parameters after *every* block call.

Table 20.21 FB/SFB parameters for reading and writing data

Parameter	Available for FB/SFB				Declaration	Data type	Assignment, description
REQ	14	34	15	35	INPUT	BOOL	Start data exchange
ID	14	34	15	35	INPUT	WORD	Connection ID
NDR	14	34	-	-	OUTPUT	BOOL	New data accepted
DONE	-	-	15	35	OUTPUT	BOOL	Job completely processed
ERROR	14	34	15	35	OUTPUT	BOOL	Fault occurred
STATUS	14	34	15	35	OUTPUT	WORD	Job status
ADDR_1	14	34	15	35	IN_OUT	ANY	First data area in partner device
ADDR_2 [1)]	14	34	15	35	IN_OUT	ANY	Second data area in partner device
ADDR_3 [1)]	14	34	15	35	IN_OUT	ANY	Third data area in partner device
ADDR_4 [1)]	14	34	15	35	IN_OUT	ANY	Fourth data area in partner device
RD_1	14	34	-	-	IN_OUT	ANY	First receive area
RD_2 [1)]	14	34	-	-	IN_OUT	ANY	Second receive area
RD_3 [1)]	14	34	-	-	IN_OUT	ANY	Third receive area
RD_4 [1)]	14	34	-	-	IN_OUT	ANY	Fourth receive area
SD_1	-	-	15	35	IN_OUT	ANY	First send area
SD_2 [1)]	-	-	15	35	IN_OUT	ANY	Second send area
SD_3 [1)]	-	-	15	35	IN_OUT	ANY	Third send area
SD_4 [1)]	-	-	15	35	IN_OUT	ANY	Fourth send area

[1)] Not for FB 14 or FB 15

At the ADDR_n parameter, you specify the tag or area in the partner device from which you wish to fetch data or to which you wish to send data. The areas at ADDR_n must agree with the corresponding areas at SD_n or RD_n. Use the parameters without gaps, beginning with 1. Unneeded parameters are not assigned (as in an FB, an SFB does not have to have values for all parameters).

20.7.4 Transferring Print Data

SFB 16 PRINT allows you to transfer a format description and data to a printer via a CP 441 communications processor. Table 20.22 lists the parameters for this SFB.

A positive edge at the REQ parameter starts the data exchange with the printer specified by the ID and PRN_NR parameters. The block signals an error-free transfer by setting DONE to "1". An error, if any, is flagged by a "1" in the ERROR parameter. A value other than zero in the STATUS parameter is indicative of either a warning (ERROR = "0") or an error (ERROR = "1"). You must evaluate the DONE, ERROR and STATUS parameters after *every* block call.

Enter the characters to be printed in STRING format in the FORMAT parameter. You can integrate as many as four format descriptions for variables in this string, defined in parameters SD_1 to SD_4. Use the parameters without gaps, beginning with 1; do not specify values for unneeded parameters. You can transfer up to 420 bytes (the sum of FORMAT and all variables) per print request.

20.7.5 Control Functions

The following SFBs are available for controlling the communication partner

- ▷ SFB 19 START
  Execute a cold or warm restart in the partner controller
- ▷ SFB 20 STOP
  Switch the partner controller to STOP
- ▷ SFB 21 RESUME
  Execute a hot restart in the partner controller

These SFBs are for one-way data exchange; no user program is required in the partner device for this purpose. The parameters for them are listed in Table 20.23.

A positive edge at the REQ parameter starts the data exchange. Enter as ID parameter the connection ID which STEP 7 entered in the connection table.

With a "1" in the DONE parameter, the block signals that the job terminated without error. An error, if any, is flagged by a "1" in the ERROR parameter. A value other than zero in the STATUS parameter is indicative of either a warning (ERROR = "0") or an error (ERROR = "0"). You must evaluate the DONE, ERROR and STATUS parameters after *every* block call.

Table 20.22 Parameters for SFB 16 PRINT

Parameter	Declaration	Data Type	Contents, Description
REQ	INPUT	BOOL	Start data exchange
ID	INPUT	WORD	Connection ID
DONE	OUTPUT	BOOL	Job terminated
ERROR	OUTPUT	BOOL	Error occurred
STATUS	OUTPUT	WORD	Job status
PRN_NR	IN_OUT	BYTE	Printer number
FORMAT	IN_OUT	STRING	Format description
SD_1	IN_OUT	ANY	First variable
SD_2	IN_OUT	ANY	Second variable
SD_3	IN_OUT	ANY	Third variable
SD_4	IN_OUT	ANY	Fourth variable

Table 20.23 SFB Parameters for Partner Controller

Parameter	Present in SFB			Declaration	Data Type	Contents, Description
REQ	19	20	21	INPUT	BOOL	Start data exchange
ID	19	20	21	INPUT	WORD	Connection ID
DONE	19	20	21	OUTPUT	BOOL	Job terminated
ERROR	19	20	21	OUTPUT	BOOL	Error occurred
STATUS	19	20	21	OUTPUT	WORD	Job status
PI_NAME	19	20	21	IN_OUT	ANY	Program name (P_PROGRAM)
ARG	19	-	21	IN_OUT	ANY	In the case of value "C" a cold start is initiated in the partner device – if permitted
IO_STATE	19	20	21	IN_OUT	BYTE	Irrelevant

Specify as PI_NAME an array variable with the contents "P_PROGRAM" (ARRAY [1..9] OF CHAR). If you leave the ARG parameter unassigned, a warm restart is triggered in the partner controller; if ARG is assigned "C", a cold restart is triggered in the partner controller if permissible. The IO_STATE parameter is currently irrelevant, and need not be assigned a value.

SFB 19 START executes a cold or warm restart of the partner CPU. Prerequisite is that the partner CPU is at STOP and that the mode selector is positioned to either RUN or RUN-P.

SFB 20 STOP sets the partner CPU to STOP. Prerequisite for error-free execution of this job request is that the partner CPU is not at STOP when the request is submitted.

SFB 21 RESUME executes a hot restart of the partner CPU. Prerequisite is that the partner CPU is at STOP, that the mode selector is set to either RUN or RUN-P, and that a hot restart is permissible at this time.

20.7.6 Monitoring Functions

The following system blocks are available for monitoring functions

▷ SFB 22 STATUS
Check partner status

▷ SFB 23 USTATUS
Receive partner status

▷ SFC 62 CONTROL
Check status of a communications instance

▷ FC 62 C_CNTRL
Scan status of a connection

▷ SFC 87 C_DIAG
Determine connection status

The parameters of these blocks are described in Tables 20.24, 20.25 and 20.26.

The following applies for these system blocks: an error is indicated with "1" at the ERROR parameter. If the STATUS parameter has a value not equal to zero, this indicates either a warning (ERROR = "0") or an error (ERROR = "1").

SFB 22 STATUS
Check the status of the partner device

SFB 22 STATUS fetches the status of the partner CPU and displays it in the PHYS (physical status), LOG (logical status) and LOCAL (operating status if the partner is an S7 CPU) parameters.

A positive edge at the REQ (request) parameter starts the query. Enter as ID parameter the connection ID which STEP 7 entered in the connection table.

With a "1" in the NDR parameter, the block signals that the job terminated without error. You must evaluate the NDR, ERROR and STATUS parameters after *every* block call.

SFB 23 USTATUS
Receive the status of the partner device

SFB 23 USTATUS receives the status of the partner, which it sends, unbidden, in the event of a change. The device status is displayed in the PHYS, LOG and LOCAL parameters.

Table 20.24 SFB Parameters for Querying Status

Parameter	Present in SFB		Declaration	Data Type	Contents, Description
REQ	22	-	INPUT	BOOL	Start data exchange
EN_R	-	23	INPUT	BOOL	Ready to receive
ID	22	23	INPUT	WORD	Connection ID
NDR	22	23	OUTPUT	BOOL	New data fetched
ERROR	22	23	OUTPUT	BOOL	Error occurred
STATUS	22	23	OUTPUT	WORD	Job status
PHYS	22	23	IN_OUT	ANY	Physical status
LOG	22	23	IN_OUT	ANY	Logical status
LOCAL	22	23	IN_OUT	ANY	Status of an S7 CPU as partner

Table 20.25 Parameters of the FC 62 C_CNTRL and SFC 62 CONTROL blocks

Parameter	Present in		Declaration	Data Type	Contents, Description
EN_R	FC	SFC	INPUT	BOOL	Ready to receive
I_DB	-	SFC	INPUT	BLOCK_DB	Instance data block
OFFSET	-	SFC	INPUT	WORD	Number of the local instance
ID	FC	-	INPUT	WORD	Connection ID
RET_VAL	-	SFC	RETURN	INT	Error information
RETVAL	FC	-	RETURN	INT	Error information
ERROR	FC	SFC	OUTPUT	BOOL	Error detected
STATUS	FC	SFC	OUTPUT	WORD	Status word
I_TYP	-	SFC	OUTPUT	BYTE	Block type identifier
I_STATE	-	SFC	OUTPUT	BYTE	Current status identifier
I_CONN	-	SFC	OUTPUT	BOOL	Connection status (“1” = connection exists)
I_STATUS	-	SFC	OUTPUT	WORD	STATUS parameter for communications instance
C_CONN	FC	-	OUTPUT	BOOL	Connection status (“1” = connection exists)
C_STATUS	FC	-	OUTPUT	WORD	Connection status

Table 20.26 Parameters of the SFC 87 C_DIAG

Parameter	Declaration	Data Type	Contents, Description
REQ	INPUT	BOOL	Trigger request with signal status “1”
MODE	INPUT	BYTE	Operating mode, see text
RET_VAL	RETURN	INT	Error information
BUSY	OUTPUT	BOOL	With “1”, the request is still busy
N_CON	OUTPUT	INT	Index of the last structure
CON_ARR	OUTPUT	ANY	Target area for the read connection data

A "1" in the EN_R (enable receive) parameter signals that the partner is ready to receive data. Initialize the ID parameter with the connection ID, which STEP 7 enters in the connection table.

With a "1" in the NDR parameter, the block signals that the request terminated without error. You must evaluate the NDR, ERROR and STATUS parameters after *every* block call.

SFC 62 CONTROL
Check the status of a communications instance

With S7-400, SFC 62 CONTROL determines the status of a communications instance and the associated connection in the local controller. Enter the SFB's instance data block in the I_DB parameter. If the SFB is called as local instance, specify the number of the local instance in the OFFSET parameter (zero when no local instance, 1 for the first local instance, 2 for the second, and so on).

A "1" in the EN_R (enable receive) parameter signals that the partner is ready to receive data specified at the I_DB parameter. You must evaluate the NDR, ERROR and STATUS parameters after *every* block call.

The parameters I_TYP, I_STATE, I_CONN and I_STATUS provide information concerning the status of the local Kommunikationsinstanz.

FC 62 C_CNTRL
Scan status of a connection

With S7-300, FC 62 C_CNTRL determines the status of a connection in the local device. At the ID parameter, enter the connection ID which STEP 7 defines in the connection table for the local device.

With signal status "1" at the EN_R parameter (enable to receive), the current connection status is displayed. The ERROR and STATUS parameters must be evaluated following each block call.

The C_CONN and C_STATUS parameters provide information on the current connection status.

SFC 87 C_DIAG
Determine connection status

The system function SFC 87 C_DIAG determines the current status of connections with a fixed configuration, i.e. all S7 connections and all fault-tolerant S7 connections. With each call, the SFC 87 C_DIAG reads the connection data from the operating system and enters them into the user memory for evaluation. The SFC subsequently acknowledges reading of the data in the operating system, so that a change in status since the last read request can be recorded. If you wish to monitor the connections permanently, call the SFC at regular intervals, e.g. every 10 seconds in a watchdog interrupt organization block.

The SFC 87 C_DIAG is an asynchronous system function. It triggers a request with signal status "1" at the REQ parameter. If the request can be executed immediately, the SFC returns the signal status "0" at the BUSY parameter when called for the first time, otherwise the request is still being processed if BUSY = "1".

Table 20.26 shows the parameters of the SFC 87 C_DIAG.

The SFC 87 C_DIAG can work in various operating modes which you set at the MODE parameter:

▷ MODE = B#16#00
The SFC acknowledges reading, without copying the connection data.

▷ MODE = B#16#01
The SFC copies the connection data and acknowledges reading.

▷ MODE = B#16#02
The SFC only copies the connection data if they have changed, and acknowledges reading even if a change has not taken place.

▷ MODE = B#16#03
The SFC copies the connection data without acknowledging.

The SFC 87 C_DIAG transmits the current connection data from the operating system to the target area specified at the CON_ARR parameter. The target area is an array of structures; each array components contains the data for one connection. The number of array elements (structures) must correspond to the number of

```
DATA_BLOCK con_data          //Data block with the connection data
...
STRUCT
...
con_req          : BOOL;     //Job start
con_busy         : BOOL;     //Job running
con_error        : INT;      //Error information of SFC
con_index        : INT;      //Number of field elements read from the SFC
con_status       : ARRAY [1..12] OF STRUCT
      CON_ID     : WORD;     //Connection ID
      STAT_CON   : BYTE;     //Connection status
      PROD_CON   : BYTE;     //Partial connection number of the productive
connection
      STBY_CON   : BYTE;     //Partial connection number of the standby connection
      DIS_PCON   : BOOL;     //High availability status change
      DIS_CON    : BOOL;     //Conn. status change (without high availability)
      RES0       : BYTE;     //Reserve
      RES1       : BYTE;     //Reserve
      END_STRUCT;
...
END_STRUCT
END_DATA_BLOCK
```

You can use your own terms for tag and component names.
The call might then appear as follows:

```
CALL C_DIAG (
      REQ       := con_data.con_req,
      MODE      := B#16#02,
      RET_VAL   := con_data.con_error,
      BUSY      := con_data.con_busy,
      N_CON     := con_data.con_index,
      CON_ARR   := con_data.con_status);
```

Figure 20.33 Programming example for the SFC 87 C_DIAG

possible connections. Figure 20.33 shows how a corresponding array variable can be structured with the connection data.

The array with the connection data is not arranged according to the connection IDs; the individual connections can be arbitrarily assigned to the array elements. Frame elements with invalid connections may even be present in between arrays with valid connections. The data of a connection are consistent with one another.

20.8 IE Communication

20.8.1 Fundamentals

By means of "Open communication over Industrial Ethernet" (abbreviated to IE communication), you can transmit data between two devices connected to the Ethernet subnet. Communication can be implemented using the TCP native protocol in accordance with RFC 793, the ISO-on-TCP protocol in accordance with RFC 1006 or the UDP protocol in accordance with RFC 768.

The communication functions are loadable function blocks (FB) in STEP 7 present in the *Standard Library* under *Communication Blocks*. Also included are user-defined data types (UDT) with the structure of the connection data and the address of the communication partner.

Configuring IE communication

The following is necessary before data can be transmitted with IE communication:

- ▷ With the TCP native and ISO-on-TCP protocols, a connection must be established to the communication partner ("connection-oriented protocols") or

▷ With the UDP protocol, a connection must be established to the communication layer of the CPU operating system ("connectionless protocol"). The partner is then addressed when calling the corresponding function block.

The connection is configured using a data area (not via the connection table). The data structures required are stored in the user-defined data type UDT 65 TCON_PAR, which the function blocks use to establish and cancel the connection. The data contains the connection ID which specifies a particular connection and the associated function block calls, as well as information on the protocol used.

Establishment of the connection to the partner or setting up of the communication access point is handled by the function block FB 65 TCON that you call in the main program of both partner devices. Data can be transferred in parallel in both directions over an established connection. Several connections can exist on one physical line. The function block FB 66 TDIS_CON cancels the connection again and thus releases the resources used (Figure 20.34).

With the function blocks FB 63 TSEND and FB 64 TRCV, you can transfer data with the protocols TCP native or ISO-on-TCP. Data transfer with the UDP protocol requires the function blocks FB 67 TUSEND and FB 68 TURCV. When calling these function blocks, specify the address of the partner device in a data area. The structure of this address is in the user-defined data type UDT 66 TADD_PAR.

Executing function blocks

The function blocks for IE communication are executed asynchronously, i.e. the processing of a job can extend over several program cycles. You call the communication blocks in the main program, and control data transmission using the parameters REQ and EN_R. The results on the parameters BUSY, NDR, DONE, ERROR and STATUS must be evaluated immediately following each processing of the communication block, since they are only valid until the next call.

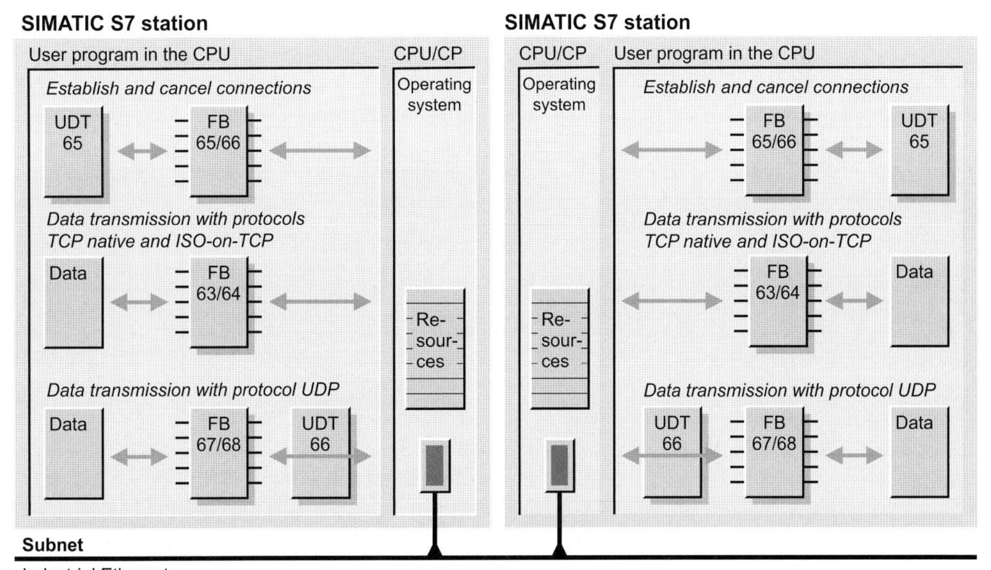

Figure 20.34 IE communication

20.8.2 Establishment and Cancellation of Connections

Before data can be transmitted using IE communication, it is necessary to establish a connection to the partner device (with TCP native and ISO-on-TCP) or to the communication layer of the operating system (with UDP). The following blocks are available for this purpose:

- ▷ FB 65 TCON
  Establish connection to the communication partner or the communication layer of the operating system
- ▷ FB 66 TDISCON
  Cancel connection
- ▷ UDT 65 TCON_PAR
  Structure for the connection data

The parameters of the function blocks can be found in Table 20.27, the data structure is shown in Table 20.28.

FB 65 TCON
Establish connection

The function block FB 65 TCON establishes the prerequisites for IE communication. The parameters required for this are located in a data area which has the structure of the user-defined data type UDT 65 TCON_PAR.

When using the TCP native and ISO on TCP protocols, a connection is established to the communication partner. Establishment of the connection is handled by the station for which "Active establishment of connection" is entered. The partner station must then be identified as "Passive". This identification is independent of the data transmission direction.

The connection is monitored and handled by the CPU's operating system. If a connection is canceled, the active partner attempts to reestablish the connection without the FB 65 TCON having to be executed again. The connection is canceled by the FB 66 TDISCON, in the operating state STOP of the CPU, or with POWER OFF/ON.

When using the UDP protocol, the FB 65 TCON initializes a local communication access point which represents the connection between user program and the communication layer of the operating system. A connection is not established to the partner.

By assigning the parameter ID, you identify the communication connection. The data must correspond to the variable *id* in the connection data. You specify the connection data using the pointer at parameter CONNECT.

In the initial state, the parameters REQ, BUSY, DONE and ERROR have the signal state "0". You start establishment of the connection with a rising edge at parameter REQ.

As long as the job is being executed, BUSY = "1". The job has been completed successfully if BUSY = "0", DONE = "1" and ERROR = "0". If the job is executed with an error, BUSY = "0", DONE = "0" and ERROR = "1". The error is then specified at parameter STATUS. BUSY, DONE and ERROR are set to "0" again if REQ is returned to "0".

Table 20.27 Parameters for FB 65 TCON and FB 66 TDISCON

Parameter	With FB		Declaration	Data type	Contents, Description
REQ	65	66	INPUT	BOOL	Start job (with rising edge)
ID	65	66	INPUT	WORD	Reference to communication connection
DONE	65	66	OUTPUT	BOOL	Job running ("0") or executed without error ("1")
BUSY	65	66	OUTPUT	BOOL	Job being processed ("1") or is finished ("0")
ERROR	65	66	OUTPUT	BOOL	Error occurred (with "1")
STATUS	65	66	OUTPUT	WORD	Job status, error information with ERROR = "1"
CONNECT	65	-	IN_OUT	ANY	Pointer to connection description

Table 20.28 Structure of Connection Description UDT 65 TCON_PAR

Byte	Parameter	Data type	Default value	Contents, Description
0 to 1	block-length	WORD	W#16#0040	Length of UDT 65 (64 bytes)
2 to 3	id	WORD	W#16#0000	Connection reference Range of values: W#16#0001 to W16#0FFF
4	connection_ type	BYTE	B#16#01	Connection type B#16#01: TCP/IP native (compatibility mode) B#16#11: TCP/IP native B#16#12: ISO on TCP B#16#13: UDP
5	active_est	BOOL	"0"	Type of establishment of connection: "1" active; "0" passive; with UDP: always "0"
6	local_ device_id	BYTE	B#16#02	Device ID: identification of communication device (see manual)
7	local_tsap_ id_len	BYTE	B#16#02	Length of parameter local_tsap_id
8	rem_sub_ net_id	BYTE	B#16#00	Currently not used
9	rem_staddr_ len	BYTE	B#16#00	Length of address of remote connection point; is not used with UDP
10	rem_tsap_ id_len	BYTE	B#16#00	Length of parameter rem_tsap_id; is not used with UDP
11	next_staddr_ len	BYTE	B#16#00	Length of parameter next_staddr; is not used with UDP
12 to 27	local_tsap_ id	ARRAY [1..16] OF BYTE	16(B#16#00)	Local port number or local TSAP
28 to 33	rem_subnet_ id	ARRAY [1..6] OF BYTE	6(B#16#00)	Currently not used
34 to 39	rem_staddr	ARRAY [1..6] OF BYTE	6(B#16#00)	IP address of remote connection end point; is not used with UDP
40 to 55	rem_tsap_id	ARRAY [1..16] OF BYTE	16(B#16#00)	Remote port number or remote; is not used with UDP
56 to 61	next_staddr	ARRAY [1..6] OF BYTE	6(B#16#00)	Rack and slot of local CP; is not used with UDP
62 to 63	spare	WORD	W#16#0000	Must be assigned with W#16#0000

FB 66 TDISCON Cancel connection

The function block FB 66 TDISCON terminates the prerequisites for IE communication. It cancels the connection to the communication partner or cancels the communication access point.

You identify the communication connection by assigning the parameter ID. The data must correspond to the variable *id* in the connection data.

In the initial state, the parameters REQ, BUSY, DONE and ERROR have the signal state "0". You start cancellation of the connection with a rising edge at parameter REQ.

As long as the job is being executed, BUSY = "1". The job has been completed successfully if BUSY = "0", DONE = "1" and ERROR = "0". If the job is executed with an error, BUSY = "0", DONE = "0" and ERROR = "1". The error is then specified at parameter STATUS. BUSY,

DONE and ERROR are set to "0" again if REQ is returned to "0".

UDT 65 TCON_PAR
Structure of connection data

The user data type UDT 65 TCON_PAR contains the structure of the connection data either for the communication connection to the partner device (TCP native and ISO-on-TCP protocols) or for the connection to the communication layer of the local operating system (UDP protocol).

You require a data block with this structure for each connection. You can use a separate global data block for each connection in which you define the UDT when creating the data block, or combine the data blocks in a common global data block.

The assignment of the variables depends on the protocol and the devices used (see online help of STEP 7). UDTs with different default settings are present in the library:

- ▷ UDT 651: for TCP active
- ▷ UDT 652: for TCP passive
- ▷ UDT 653: for ISO-on-TCP active
- ▷ UDT 654: for ISO-on-TCP passive
- ▷ UDT 655: for ISO-on-TCP active with CP
- ▷ UDT 656: for ISO-on-TCP passive with CP
- ▷ UDT 657: for open UDP local

20.8.3 Data Transmission with TCP Native or ISO-on-TCP

The following function blocks are available for the data transmission with the connection-oriented TCP native and ISO-on-TCP protocols:

- ▷ FB 63 TSEND
  Send data with logic connection
- ▷ FB 64 TRCV
  Receive data with logic connection

The parameters of these function blocks are listed in Table 20.29.

Prior to transmission of the data, a connection must be established to the partner station using FB 65 TCON (say Chapter 20.8.2 "Establishment and Cancellation of Connections"). With FB 63 TSEND and FB 64 TRCV, data can be exchanged in both directions simultaneously.

FB 63 TSEND
Send data with logic connection

Function block FB 63 TSEND sends data with the TCP native or ISO on TCP protocol over an existing communication connection.

You identify the communication connection by assigning the parameter ID. The data must correspond to the variable *id* in the connection data. You specify the send mailbox using the pointer at parameter DATA.

Table 20.29 Parameters for FB 63 TSEND and FB 64 TRCV

Parameter	With FB		Declaration	Data type	Contents, Description
REQ	63	-	INPUT	BOOL	Start sending data (with rising edge)
EN_R	-	64	INPUT	BOOL	FB ready to receive (with "1")
ID	63	64	INPUT	WORD	Reference to communication connection
LEN	63	64	INPUT	INT	Number of bytes to be sent or received
DONE	63	-	OUTPUT	BOOL	Job running ("0") or executed without error ("1")
NDR	-	64	OUTPUT	BOOL	Job running ("1") or completed ("1")
BUSY	63	-	OUTPUT	BOOL	Job being processed ("1") or is finished ("0")
ERROR	63	64	OUTPUT	BOOL	Error occurred (with "1")
STATUS	63	64	OUTPUT	WORD	Job status, error information with ERROR = "1"
RCVD_LEN	-	64	OUTPUT	INT	Number of bytes actually received
DATA	63	64	IN_OUT	ANY	Send or receive mailbox

In the initial state, the parameters REQ, BUSY, DONE and ERROR have the signal state “0”. You start the data transmission with a rising edge at parameter REQ. When called for the first time with “1”, the data is fetched from the area specified by the parameter DATA. The number of bytes specified at parameter LEN are sent, with their maximum size depending on the type of connection:

Connection type	Number of bytes
B#16#01	1 to 1460
B#16#11	1 to 8192
B#16#12	1 to 1452 (with CP) 1 to 8192 (without CP)

As long as the job is being executed, BUSY = “1”. The job has been completed successfully if BUSY = “0”, DONE = “1” and ERROR = “0”. If the job is executed with an error, BUSY = “0”, DONE = “0” and ERROR = “1”. The error is then specified at the parameter STATUS. BUSY, DONE and ERROR are set to “0” again if REQ is returned to “0”.

The data in the send area can then be changed again if either DONE or ERROR has the signal state “1”.

FB 64 TRCV
Receive data with logic connection

Function block FB 64 TRCV receives data with the TCP native or ISO on TCP protocol over an existing communication connection.

You identify the communication connection by assigning the parameter ID. The data must correspond to the variable *id* in the connection data. You specify the receive mailbox using the pointer at parameter DATA.

If the *parameter LEN* is “0”, the length data in the parameter DATA are used. Once a data block has been received, the number of received bytes is made available in the parameter RCVD_LEN, and NDR is set to “1”.

In the case of the *TCP native protocol*, neither the length of the message nor the start or end is sent during data transmission. To ensure that the sent number of bytes is received correctly, the parameter LEN on the receive block must be assigned the same value as the parameter LEN on the send block.

If the value of LEN on the receive block is selected larger, part of the subsequent message (from the next job) is also received. NDR is only set to “1” when the parameterized length has been received.

If the value of LEN is selected smaller, NDR is set to “1” when the parameterized length is reached, and the parameter RCVD_LEN contains the number of received bytes. A further data block is received with each further execution.

In the case of the *ISO-on-TCP protocol*, information on the length and the end of a message is transmitted. If LEN on the receive block is larger than on the send block, the sent data is received, NDR is set to “1”, and the number of received bytes written in RCVD_LEN. If LEN is selected smaller, an error message is output: ERROR = “1”, STATUS = W#16#8088.

The FB 64 TRCV only receives data if parameter EN_R has the signal state “1”.

As long as the job is running, BUSY = “1”. The job has been successfully completed if BUSY = “0”, NDR = “1” and ERROR = “0”. If an error occurs when executing a job, BUSY = “0”, NDR = “0” and ERROR = “1”. The error is then specified in the parameter STATUS. BUSY, NDR and ERROR are set to “0” again when EN_R is returned to “0”.

The data in the receive mailbox is consistent if NDR has the signal state “1”.

20.8.4 Data Transmission with UDP

The following blocks are available for data transmission with the connectionless protocol UDP:

▷ FB 67 TUSEND
Send data with UDP

▷ FB 68 TURCV
Receive data with UDP

▷ UDT 66 TADD_PAR
Data structure for the partner address

The parameters of the function blocks are listed in Table 20.30, the structure of the UDT in Table 20.31.

Table 20.30 Parameters for FB 67 TUSEND and FB 68 TURCV

Parameter	With FB		Declaration	Data type	Contents, Description
REQ	67	-	INPUT	BOOL	Start sending data (with rising edge)
EN_R	-	68	INPUT	BOOL	FB ready to receive (with "1")
ID	67	68	INPUT	WORD	Reference to communication connection
LEN	67	68	INPUT	INT	Number of bytes to be sent or received
DONE	67	-	OUTPUT	BOOL	Job running ("0") or executed without error ("1")
NDR	-	68	OUTPUT	BOOL	Job running ("1") or completed ("1")
BUSY	67	-	OUTPUT	BOOL	Job being processed ("1") or is finished ("0")
ERROR	67	68	OUTPUT	BOOL	Error occurred (with "1")
STATUS	67	68	OUTPUT	WORD	Job status, error information with ERROR = "1"
RCVD_LEN	-	68	OUTPUT	INT	Number of bytes actually received
DATA	67	68	IN_OUT	ANY	Send or receive mailbox
ADDR	67	68	IN_OUT	ANY	Pointer to address of sender or receiver

Tabelle 20.31 Structure of the Partner Address UDT 66 TADD_PAR

Byte	Parameter	Data type	Default value	Contents, Description
0 to 3	rem_ip_addr	ARRAY [1..4] OF BYTE	4(B#16#00)	IP address of partner
4 to 5	rem_port_nr	ARRAY [1..2] OF BYTE	2(B#16#00)	Port number of partner
6 to 7	spare	ARRAY [1..2] OF BYTE	2(B#16#00)	Must be occupied by 0000

Prior to transmission of the data, it is necessary to establish a connection to the communication layer of the operating system with the FB 65 TCON (see Chapter 20.8.2 "Establishment and Cancellation of Connections"). The address of the communication partner is in a data area which has the structure of the UDT 66 TADD_PAR hat.

FB 67 TUSEND
Send data with UDP

The function block FB 67 TUSEND sends data with the UDP protocol.

By assigning the parameter ID, you identify the connection between user program and communication layer of the operating system. The value must agree with the variable *id* in the connection data. You specify the send mailbox by the pointer on parameter DATA.

The data on the communication partner is present in a data area indicated by the pointer on parameter ADDR. The address, and thus the partner, can be changed with each new send job, without the communication access point having to be redefined with FB 65 TCON.

In the initial state, the parameters REQ, BUSY, DONE and ERROR have the signal state "0". You start the data transmission with a rising edge at parameter REQ. When called for the first time with "1", the data is fetched from the area specified by the parameter DATA. The number of bytes specified at parameter LEN are sent (1 to max. 1460).

As long as the job is running, BUSY = "1". The job has been successfully completed if BUSY = "0", DONE = "1" and ERROR = "0". If an error occurs when executing a job, BUSY = "0", DONE = "0" and ERROR = "1". The error is then specified in the parameter STATUS. BUSY, DONE and ERROR are set to "0" again when REQ is returned to "0".

The data in the send area can then be changed again if either DONE or ERROR has the signal state "1".

FB 68 TURCV
Receive data with UDP

The function block FB receives data with the UDP protocol.

By assigning the parameter ID, you identify the connection between user program and communication layer of the operating system. The value must agree with the variable *id* in the connection data. You specify the receive mailbox by the pointer on parameter DATA.

The data on the communication partner is present in a data area indicated by the pointer on parameter ADDR.

The number of bytes to be received is set in the parameter LEN (1 to max. 1460). Once a data block has been received, the number of received bytes is made available in the parameter RCVD_LEN, and NDR is set to "1".

Data is only received if the parameter EN_R has the signal state "1".

As long as the job is running, BUSY = "1". The job has been successfully completed if BUSY = "0", NDR = "1" and ERROR = "0". If an error occurs when executing a job, BUSY = "0", NDR = "0" and ERROR = "1". The error is then specified in the parameter STATUS. BUSY, NDR and ERROR are set to "0" again when EN_R is returned to "0".

The data in the receive area is consistent if NDR has the signal state "1".

UDP (User Data Protocol)

In the case of the UDP, a connection is not established. The communication partner is specified in the parameter ADDR of the send block (IP address and port number). The receive block then provides the IP address and the port number of the sender in the parameter ADDR.

The user-defined data type UDT 66 TADD_PAR contains the structure of the address information. The pointer on ADDR refers to a data area with this structure.

In the case of the UDP, information on the length and the end of a message is transmitted. If LEN on the receive block is larger, the sent data is copied into the receive mailbox, NDR is set to "1", and the number of received bytes written in RCVD_LEN. If LEN has been selected smaller, an error message is output: ERROR = "1", STATUS = W#16#8088.

UDT 66 TADD_PAR
Data structure of partner address

The UDT 66 contains the structure of the partner address when transmitting with the UDP protocol. The parameter ADDR on function blocks FB 67 TUSEND and FB 68 TURCV refers to a data area with this structure.

20.9 PtP Communication with S7-300C

20.9.1 Fundamentals

Using point-to-point communication (PtP), you transmit data via a serial interface to a communications partner, e.g. a printer or a SIMATIC S5 station. An RS 422/485 interface (X.27) is already integrated in a number of S7-300 compact CPUs.

The communications connections are specified in the interface properties by the Hardware Configuration when parameterizing the CPU. ASCII mode, the 3964(R) procedure and RK512 computer link are available as transmission protocols.

The communication functions are system function blocks (SFB), which are integrated in the S7-300C CPU operating system. The instance data blocks for these SFBs are in user memory. The SFBs do not perform any parameter check. Incorrect parameterization can cause the CPU to STOP. If you use the transmission protocol of the computer link, a synchronization data block is additionally used (once for all computer link SFBs in the user memory, Figure 20.35).

Configuring PtP communication

You use the Hardware Configuration to set the transmission protocol in the properties window of the point-to-point interface:

▷ ASCII mode
The data are transmitted as ASCII characters. Transmitted data are not acknowl-

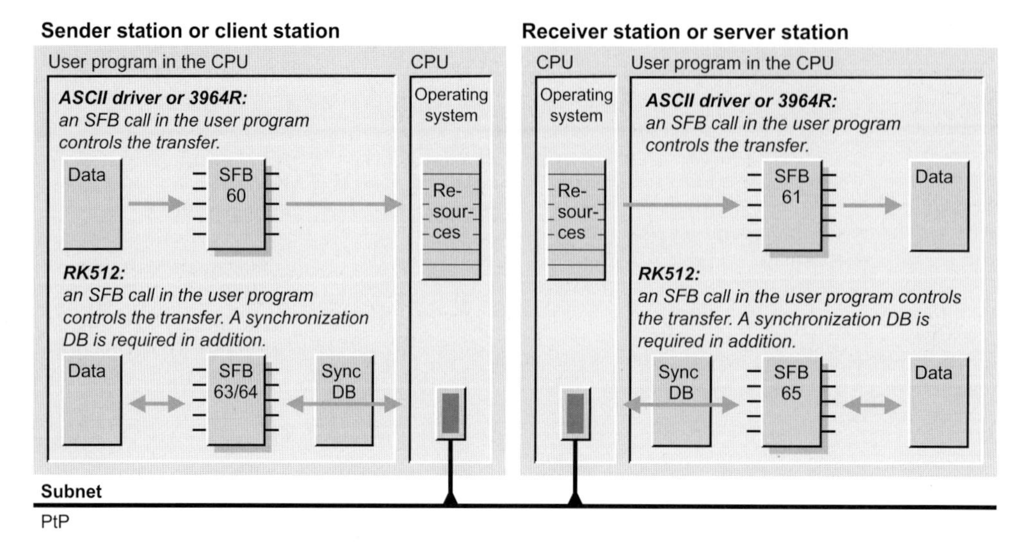

Figure 20.35 Point-to-point communication with S7-300C

edged. Setting of signal assignments and transmission parameters, such as baud rate, parity, end-of-text character.

▷ 3964(R) procedure
The data are sent to the communications partner, and positively acknowledged if received correctly. Setting of signal assignments and transmission parameters, such as baud rate, parity, block check.

▷ RK512 computer link
The data transmission can be coordinated using so-called interprocessor communication flags. Data reception and the fetching of data are acknowledged. Setting of signal assignments and transmission parameters, such as baud rate, parity, block check.

With the 3964(R) and RK512 transmission protocols, the communications partners must have different priorities in order to define the sequence if a simultaneous request to send is present.

20.9.2 ASCII Driver and 3964(R) Procedure

Using the point-to-point connection, you can send and receive data with application of the ASCII driver or the 3964(R) procedure. The system function blocks required for this are:

▷ SFB 60 SEND_PTP
Send data with ASCII driver or 3964(R) procedure

▷ SFB 61 RCV_PTP
Receive data with ASCII driver or 3964(R) procedure

▷ SFB 62 RES_RCVB
Delete receive buffer with ASCII driver or 3964(R) procedure

Table 20.32 shows the parameters of these system function blocks.

SFB 60 SEND_PTP Send data with ASCII driver or 3964R procedure

The system function block SFB 60 SEND_PTP is used to send a data area to a communications partner. You set the transmission protocol and the transmission parameters using the Hardware Configuration. You specify the area of data to be sent at parameter SD_1. The length of the sent data area depends on the parameterization of the interface, e.g. transmit up to length specified at parameter LEN or transmit up to an end-of-text character.

With the ASCII driver you can send frames up to a length of 1024 bytes. The SFB 60 SEND_PTP transmits the data in consistent blocks of

Table 20.32 SFB Parameters for Sending and Receiving Data with ASCII Driver or 3964(R) Procedure

Parameter	Present in SFB			Declaration	Data Type	Contents, Description
REQ	60	-	62	INPUT	BOOL	Trigger requests with signal status "1"
EN_R	-	61	-	INPUT	BOOL	Enable receive
R	60	61	62	INPUT	BOOL	With "1", the request is aborted
LADDR	60	61	62	INPUT	WORD	Submodule address of interface
DONE	60	-	62	OUTPUT	BOOL	With "1", the request is still busy
NDR	-	61	-	OUTPUT	BOOL	With "1", the request has been finished without fault
ERROR	60	61	62	OUTPUT	BOOL	With "1", a fault has occurred
STATUS	60	61	62	OUTPUT	WORD	Error information
SD_1	60	-	-	IN_OUT	ANY	Send mailbox
RD_1	-	61	-	IN_OUT	ANY	Receive mailbox
LEN	60	61	-	IN_OUT	INT	Number of transmitted bytes

206 bytes. You must not modify the data in the send area while the transmission is running.

Sending is triggered by the rising signal edge at parameter REQ. With a signal status "1" at parameter DONE, the SFB signals that a request has been completed successfully. In the event of an error, the parameter ERROR is set to "1" and the error information output at parameter STATUS.

With a signal status "1" at parameter R you abort a current send request and reset the call instance to the basic state.

SFB 61 RCV_PTP
Receive data with ASCII driver or 3964R procedure

The system function block SFB 61 RCV_PTP is used to receive a data area from a communications partner. You set the transmission protocol and the transmission parameters using the Hardware Configuration. The received data are entered in the area specified by the parameter RD_1. The number of received bytes is in the parameter LEN.

The SFB 61 RCV_PTP receives the data in consistent blocks of 206 bytes. You must not access the data in the receive area while the transmission is running.

The CPU-internal receive buffer has a size of 2048 bytes. During parameterization of the interface, you also define whether you use the complete length of the receive buffer for receiving data or limit the number of received frames.

You can enable data reception by a signal status "1" at parameter EN_R. The parameter NDR has the signal status "1" if new data have been received successfully. In the event of a fault, the parameter ERROR is set to "1", and the error information output at parameter STATUS.

With a signal status "1" at parameter R you abort a current receive request and reset the call instance to the basic state.

SFB 62 RES_RCVB
Delete receive buffer with ASCII driver or 3964R procedure

The system function block SFB 62 RES_RCVB deletes the receive buffer of the point-to-point interface. A frame received during the delete operation is not deleted.

Deleting is triggered by a rising signal edge at parameter REQ. With a signal status "1" at parameter DONE, the SFB signals that deleting has been completed successfully. In the event of a fault, the parameter ERROR is set to "1" and the error information output at parameter STATUS. DONE, ERROR and STATUS are only set for the duration of one call.

With a signal status "1" at parameter R you abort deleting and reset the call instance to the basic state.

20.9.3 RK512 Computer Link

Using the point-to-point connection, you can also send and receive data using the RK512 computer link. The system function blocks required are:

▷ SFB 63 SEND_RK
Send data with the RK512 computer link

▷ SFB 64 FETCH_RK
Fetch data with the RK512 computer link

▷ SFB 65 SERVE_RK
Receive and serve data with the RK512 computer link

Table 20.33 shows the parameters of these system function blocks.

Specification of the transmission area

The SFB 63 SEND_RK sends from the area specified by SD_1 and addresses a data block area in the partner device. The SFB 64 FETCH_RK can address all address areas in the partner device, and stores the fetched data in the data block area specified by the parameter RD_1. The SFB 65 SERVE_RK can store received data in a data block, and serve data from all address areas. Refer to Table 20.34 for the permissible assignment of the parameters. Certain values are only meaningful when using a SIMATIC S5 station as the partner device.

Synchronization data block

In addition to the instance data block, the system function blocks of the computer link interact

Table 20.33 SFB Parameters for Transmitting and Receiving Data with RK512 Computer Link

Parameter	Present in SFB			Declaration	Data Type	Contents, Description
SYNC_DB	63	64	65	INPUT	INT	Number of synchronization data block
REQ	63	64	-	INPUT	BOOL	Trigger request with “1”
EN_R	-	-	65	INPUT	BOOL	Enable receive with “1”
R	63	64	65	INPUT	BOOL	Abort request with “1”
LADDR	63	64	65	INPUT	WORD	Logical basic address of interface
R_CPU	63	64	-	INPUT	INT	CPU number of partner station
R_TYPE	63	64	-	INPUT	CHAR	Type of data block in the partner CPU
R_DBNO	63	64	-	INPUT	INT	Number of data block in the partner CPU
R_OFFSET	63	64	-	INPUT	INT	Number of start byte in the partner CPU
R_CF_BYT	63	64	-	INPUT	INT	Number of interprocessor communication flag byte in the partner CPU
R_CF_BIT	63	64	-	INPUT	INT	Number of interprocessor communication flag bit in the partner CPU
DONE	63	64	-	OUTPUT	BOOL	With “1”: request completed without error
NDR	-	-	65	OUTPUT	BOOL	With “1”: request completed without error
ERROR	63	64	65	OUTPUT	BOOL	With “1”: request completed with error
STATUS	63	64	65	OUTPUT	WORD	Error information
L_TYPE	-	-	65	OUTPUT	CHAR	Type of data area on local CPU
L_DBNO	-	-	65	OUTPUT	INT	Number of data block in local CPU
L_OFFSET	-	-	65	OUTPUT	INT	Number of start byte in local CPU
L_CF_BYT	-	-	65	OUTPUT	INT	Number of interprocessor communication flag byte in local CPU
L_CF_BIT	-	-	65	OUTPUT	INT	Number of interprocessor communication flag bit in local CPU
SD_1	63	-	-	IN_OUT	ANY	Send mailbox
RD_1	-	64	-	IN_OUT	ANY	Receive mailbox
LEN	63	64	65	IN_OUT	INT	Number of data bytes

Table 20.34 Specification of the Transmission Area

Parameter	Type	SFB 63 Send data	SFB 64 Fetch data	SFB 65 Receive data	SFB 65 Provide data	Description
R_CPU	INT	0 to 4	0 to 4	-	-	0 = single-processor operation 1..4 = number of CPU in multi-processor operation
R_TYPE	CHAR	D, X	D, X, M, E, A, T, Z	-	-	D = data block DB X = expanded data block DX M = flag memory area E = process input image A = process output image T = timer values Z = counter values
L_TYPE	CHAR	-	-	D	D, M, E, A, T, Z	
R_DBNO	INT	0 to 255	0 to 255	-		Number of data block (irrelevant with M, E, A, T and Z)
L_DBNO	INT	-	-	1 to n [1)]	1 to n [1)]	
R_OFFSET	INT	0 to 510	0 to 510	-	-	First byte with data blocks (must be an even address)
		-	0 to 255	-	-	First byte with M, E, A, T and Z
L_OFFSET	INT	-	-	0 to 1024	0 to 1024	Frame length

[1)] CPU-specific

with a synchronization data block which synchronizes and controls the activities of all computer link instances. The data block is present once in the user memory. You initialize it as a global data block with a minimum length of 240 bytes. You specify the number of the data block at parameter SYNC_DB.

Coordination with interprocessor communication flags

Data reception via the computer link can be coordinated with interprocessor communication flags. A interprocessor communication flag is a bit from the address area for bit memories M. Use a interprocessor communication flag for each transmission request whose address you specify at parameters R_CF_BYT and R_CF_BIT or L_CF_BYT and L_CF_BIT.

If the local CPU is the client, the system function blocks SFB 63 SEND_RK are used for transmitting data and SFB 64 FETCH_RK for fetching data. When sending and fetching data from the partner CPU, the address of the interprocessor communication flag is also specified. If this interprocessor communication flag has the signal status "0" in the partner CPU, the latter permits importing of the data packet into the user memory when sending, and reading of the data packet from the user memory when fetching. The interprocessor communication flag is then set by the communications function to indicate that data transmission has taken place. The data can then be processed or edited by the user program. If the interprocessor communication flag is then reset by the user program, the data transmission is enabled again. The interprocessor communication flag in the partner CPU thus permits control of the data transmission.

If the local CPU is the server, the data are received by the system function block SFB 65 SERVE_RK if the client sends data, or made available if the client fetches data. On the SFB you parameterize the local interprocessor communication flag (in the server) with which you then receive or provide the data in the user program. The SFB indicates at parameters L_CF_BYT and L_CF_BIT which interprocessor communication flag is being used for the currently executed request and has been set to "1". Following processing of the data (fetching or providing again) you reset the interprocessor communication flag by the program and then enable processing of the next transmission request.

SFB 63 SEND_RK
Send data with RK512 computer link

The system function block SFB 63 SEND_RK is used to send a data area to a communications partner. You set the transmission protocol and the transmission parameters using the Hardware Configuration. You specify the area of data to be sent at parameter SD_1. You specify the length of the sent data area at parameter LEN. Note that the number of bytes must be even.

The SFB 63 SEND_RK only sends the data if the interprocessor communication flag in the communications partner has the signal status "0". A frame can have a length of up to 1024 bytes. The data are transmitted in consistent blocks of 128 bytes. You must not modify the data in the send area while the transmission is running.

Sending is triggered by the rising signal edge at parameter REQ. With a signal status "1" at parameter DONE, the SFB signals that a request has been completed successfully. In the event of a fault, the parameter ERROR is set to "1" and the error information output at parameter STATUS.

With a signal status "1" at parameter R you abort a current send request and reset the call instance to the basic state.

SFB 64 FETCH_RK
Fetch data with the RK512 computer link

The system function block SFB 64 FETCH_RK is used to fetch a data area from a communications partner. You set the transmission protocol and the transmission parameters using the Hardware Configuration. The fetched data are entered in the area specified by the parameter RD_1. The number of received bytes is in the parameter LEN.

The SFB 64 FETCH_RK only fetches the data if the interprocessor communication flag in the communications partner has the signal status "0". A frame can have a length of up to 1024 bytes. The data are transmitted in consistent blocks of 128 bytes. You must not access the data in the receive area while the transmission is running.

You can enable data fetching by a signal status "1" at parameter EN_R. The parameter NDR has the signal status "1" if new data have been fetched successfully. In the event of a fault, the parameter ERROR is set to "1", and the error information output at parameter STATUS.

With a signal status "1" at parameter R you abort a current fetch request and reset the call instance to the basic state.

SFB 65 SERVE_RK
Receive and provide data with the RK512 computer link

The system function block SFB 65 SERVE_RK handles the server functionality for the RK512 computer link. It accepts a data area which has been sent by a communications partner, and provides a data area which is fetched by a communications partner. The received or provided data are entered in the area specified by the parameters L_TYPE, L_DBNO and L_OFFSET. The number of transmitted bytes is present in the parameter LEN.

The SFB 65 SERVE_RK transmits the data in consistent blocks of 128 bytes. You must not access the data in the transmission area while the transmission is running. You control coordination of the data transmission in the user program by means of a interprocessor communication flag.

You can enable request processing by a signal status "1" at parameter EN_R. The parameter NDR has the signal status "1" if new data have been received or fetched successfully. In the event of an error, the parameter ERROR is set to "1", and the error information output at parameter STATUS.

With a signal status "1" at parameter R you abort a current request and reset the call instance to the basic state.

20.10 Configuration in RUN

Configuration in RUN (CiR) means system modification in running operation. This functionality permits you to change the configuration of the distributed I/O of an S7 station without the CPU entering STOP or having to be set to STOP.

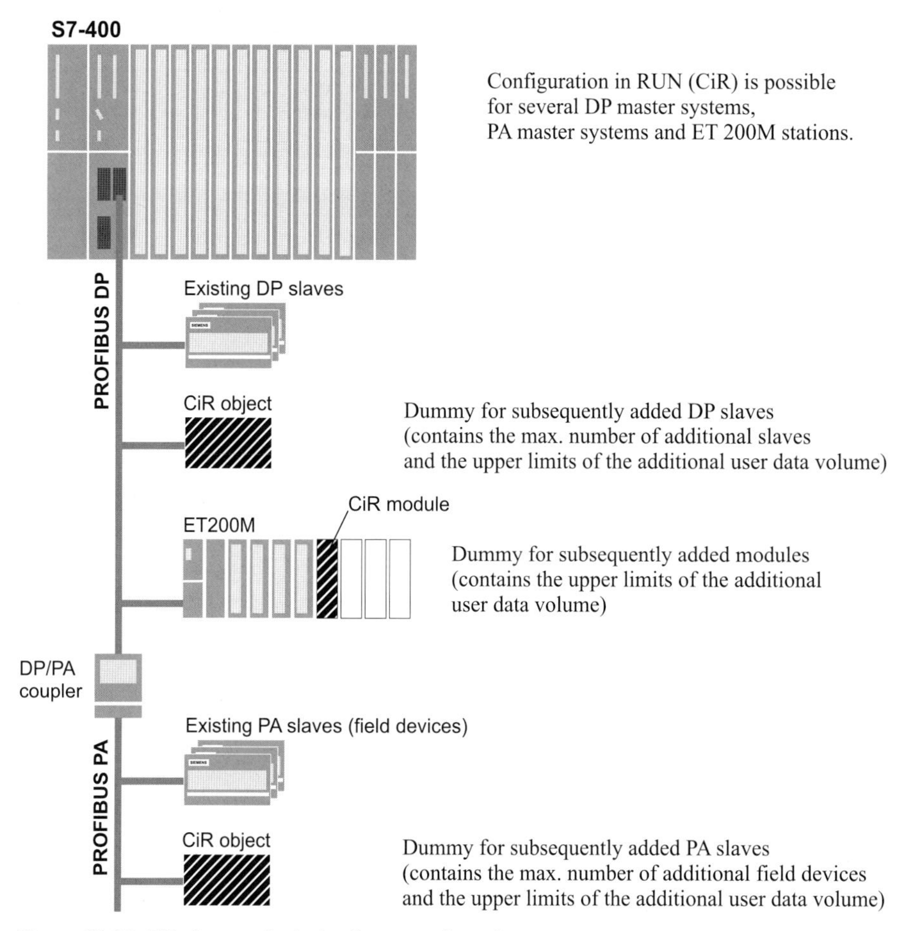

Figure 20.36 CiR elements in the hardware configuration

The changes comprise adding compact DP slaves, ET 200M stations, and PA master systems to an existing DP master system; adding modules to ET 200M stations; and adding PA slaves (field devices) to existing PA master systems. All objects added during running operation can also be removed during running operation (Figure 20.36).

In addition to the fact that all involved devices must be able to handle the CiR functionality, there are further prerequisites and limitations. For example, the PROFIBUS DP master system must be a mono master system and must not have any equidistant bus cycles, use of intelligent DP slaves is not permissible in the associated station components, and the module parameters must be saved on the CPU.

Components with and without CiR can be mixed; however, modifications are only possible on components with CiR capability.

During reconfiguration, process operation is stopped for a brief period (typically 1 s, can be parameterized). The time can be kept short by only carrying out a few modifications at any one time.

20.10.1 Preparation of Modifications to Configuration

With the Hardware Configuration, you can configure, for example, an S7-400 station (CPU with firmware release V3.1 or higher) with at least one PROFIBUS DP master system. Now add the DP slaves and – if envisaged – dummies for subsequent plant expansion (CiR object under PROFIBUS DP in the hardware catalog). Set the subsequent maximum configuration in the properties of this dummy. The Hardware Configuration calculates 244 input bytes and 244 output bytes for each additional DP slave. In reality, far fewer user data are usually required. You can modify the total of all required input and output bytes if you click the check box “Expanded settings”.

You can also reserve space in an ET 200M station for later expansions. Add a station with an interface module from IM 153-2BA00-0XB0 to the DP master system and check the "Module replacement during operation" checkbox in the station properties in the "Special" tab. In the lower window area, the necessary active bus modules are shown so that you can perform the module replacement (these are necessary for the mechanical design, but are not configured). It is recommended that the ET 200M station be equipped with active bus modules up to the planned expansion, because they must not be removed or inserted during operation.

You can now insert modules into the ET 200M station and – if you want to expand later – provide a dummy immediately after the last configured module (the *CiR Module* object under the IM 153 interface module used in the hardware catalog). Set the required number of additional input and output bytes in the properties of the CiR module.

If you wish to expand a PA master system connected to the DP master system, use an interface module IM 157-0AA82-0XA0 or better as the DP/PA link. A dummy for subsequent expansion with field devices is also provided here (*CiR object* under the used IM 157 interface module in the hardware catalog). Set the required number of additional input and output bytes in the properties of the CiR object.

With a CiR-capable DP master system selected and EDIT → MASTER SYSTEM → ENABLE CIR CAPABILITY, a CiR object is created on the DP master system and at each subordinate CiR-capable PA master system. In each CiR-capable ET 200M station, a CiR module is added. With EDIT → MASTER SYSTEM → DISABLE CIR CAPABILITY, all CiR objects and CiR modules are deleted.

20.10.2 Changing the Configuration

You can now change the configuration within the limits defined in the CiR elements and load them again in RUN. Possible changes include:

- ▷ Adding of compact and modular DP slaves to an existing DP master system (the added slaves must have a higher PROFIBUS address than the largest previously used address)
- ▷ Modification of partial process image assignment with existing DP slaves
- ▷ Adding of PA slaves (field devices) to an existing PA master system
- ▷ Adding of DP/PA couplers following an IM 157 interface module
- ▷ Adding of DP/PA links including PA master system to an existing DP master system
- ▷ Adding modules to an ET 200M station
- ▷ Reconfiguration of modules in an ET 200M station (e.g. new or changed assignment to a process image partition, activation of previously unused channels)
- ▷ Cancellation of above-mentioned modifications (starting from the highest addresses for modules and slaves)

The total of the configured addresses (real, immediately used) and the addresses for future use must not be larger than the quantity framework of the DP master (is checked during configuring), but can be larger than the quantity framework of the CPU (is only checked when “converting” into specific slaves or modules).

During a CiR procedure, the configuration can be modified on max. 4 DP master systems. In certain cases it is recommendable or even essential to repeatedly execute the CiR procedure. For example, if modules or slaves are to be replaced by others, the corresponding component must first be removed and the replacement added in a second CiR procedure.

20.10.3 Loading the Configuration

The (initial) loading of a configuration with CiR elements or with a modified CiR configuration is carried out with the CPU in the STOP status. In order to check whether the CiR capability is also present, you should subsequently load the CiR configuration again in RUN. Checking of the CiR capability using STATION → CHECK CIR COMPATIBILITY is not 100% possible offline. For example, the CiR synchronization time could be limited by the SFC 104 CiR.

To guarantee that the CPU remains in the RUN status during the CiR procedure, you must make sure that interrupts from unknown components are ignored. A corresponding program must be present in the following organization blocks:

- ▷ Hardware interrupts OB 40 to OB 47
- ▷ Timeout OB 80
- ▷ Diagnostics interrupt OB 82
- ▷ Hot swapping interrupt OB 83
- ▷ Program execution error OB 85
- ▷ Rack failure OB 86
- ▷ I/O access error OB 122

When adding modules or slaves, you should first load the configuration and then the matching user program. When removing modules and slaves, first load the matched user program and then the modified configuration. The adding or removal of the real modules or slaves must only be carried out following loading of the modified configuration (once the INTF LED on the CPU has gone out).

When reparameterizing modules, you must first load a user program which no longer addresses the associated modules or no longer evaluates their interrupts. Then load the modified configuration, change the hardware if necessary, and subsequently load the user program matched to the modification.

20.10.4 CiR Synchronization Time

Following loading of the new configuration into the CPU, the new data are checked and – if the check is positive – imported into the current configuration. This importing requires a certain time, the so-called CiR synchronization time. Process execution is stopped during this period.

The CiR synchronization time is calculated from the total of the CiR synchronization times of all involved DP and PA master systems. The synchronization time for a master system depends on the CPU used and on the real and planned I/O volumes in this master system. This time is shown by the Hardware Configuration in the properties of the CiR object in the master system. The worst case is always calculated, so that the actual CiR synchronization time is shorter. If modules are only reparameterized when changing the configuration, the synchronization time is 100 ms.

The CPU compares the calculated CiR synchronization time with the permissible upper limit whose default setting is 1000 ms. You can modify this upper limit using the SFC 104 CIR. If the calculated CiR synchronization time is greater than this upper limit, the change in the configuration is not carried out.

20.10.5 Effects on Program Execution

Execution of the user program is stopped during the CiR synchronization time. All process images retain their last value. The SIMATIC timers and the CPU clock continue. Any interrupts which occur are only processed following expiry of the CiR synchronization time. Communication with a connected programming device is limited; only the STOP command is accepted.

At the end of synchronization, the CPU starts the organization block OB 80 "Time error" with the value W#16#350A in the first word of the start information (variables OB80_EV_CLASS and OB80_FLT_ID). The required CiR synchronization time in ms is present in variable OB80_ERROR_INFO.

If modules are to be reparameterized, the CPU starts the organization block OB 83 "Insert/ Remove module interrupt" with the value W#16#3367 in the first word of the start information (variables OB83_EV_CLASS and OB83_FLT_ID). The modules are subsequently reparameterized. It may occur that the associated modules no longer deliver valid values.

Following reparameterization, the CPU starts the OB 83 again, this time with W#16#3267 in the first word of the start information. Faulty reparameterization is signaled by W#16#3968. The associated modules are then considered as not available. The described procedure takes place in every affected master system.

20.10.6 Controlling the CiR Procedure

Using the **SFC 104 CIR** you can disable the CiR procedure in the user program, limit it for a certain time, or enable it. The SFC parameters are shown in Table 20.35.

MODE = B#16#00 delivers the currently valid upper limit of the CiR synchronization time. Using MODE = B#16#01 you can set the CiR synchronization time to the default value 1000 ms and enable processing of the CiR procedure. MODE = B#16#02 always disables the CiR procedure, MODE = B#16#03 only if the CiR synchronization time calculated in the CPU is larger than that specified at parameter FRZ_TIME.

Table 20.35 Parameters of the SFC 104 CIR

Parameter	Declaration	Data Type	Contents, Description
MODE	INPUT	BYTE	Request ID B#16#00: information function B#16#01: enable CiR procedure B#16#02: disable CiR procedure B#16#03: conditionally disable CiR procedure
FRZ_TIME	INPUT	TIME	Upper limit of CiR synchronization time Default setting: T#1000 ms Permissible from T#200ms to T#2500ms
RET_VAL	RETURN	INT	Error information
A_FT	OUTPUT	TIME	Currently valid upper limit of CiR synchronization time

21 Interrupt Handling

Interrupt handling is always event-driven. When such an event occurs, the operating system interrupts scanning of the main program and calls the routine allocated to this particular event. When this routine has executed, the operating system resumes scanning of the main program at the point of interruption. Such an interruption can take place after every operation (statement).

Applicable events may be interrupts and errors. The order in which virtually simultaneous interrupt events are handled is regulated by a priority scheduler. Each event has a particular servicing priority. Several interrupt events can be combined into priority classes.

Every routine associated with an interrupt event is written in an organization block in which additional blocks can be called. A higher-priority event interrupts execution of the routine in an organization block with a lower priority. You can affect the interruption of a program by high-priority events using system functions.

21.1 General Remarks

SIMATIC S7 provides the following interrupt events (interrupts):

▷ Time-of-day interrupt
An interrupt generated by the operating system at a specific time of day, either once only or periodically

▷ Time-delay interrupt
An interrupt generated after a specific amount of time has passed; a system function call determines the instant at which this time period begins

▷ Watchdog interrupt
An interrupt generated by the operating system at periodic intervals

▷ Hardware interrupt
An Interrupt from a module, either via an input derived from a process signal or generated on the module itself

▷ DPV1 interrupt
An Interrupt from a PROFIBUS DPV1 slave

▷ Multiprocessor interrupt
An interrupt generated by another CPU in a multiprocessor network

▷ Synchronous cycle interrupt
An interrupt from the PROFIBUS DP master during the DP cycle

Other interrupt events are the synchronous errors which may occur in conjunction with program scanning and the asynchronous errors, such as diagnostic interrupts. The handling of these events is discussed in Chapter 23 "Error Handling".

Priorities

An event with a higher priority interrupts a program being processed with lower priority because of another event. The main program has the lowest priority (priority class 1), asynchronous errors the highest (priority class 26), apart from the start-up routine. All other events are in the intervening priority classes. In S7-300 systems the priorities are fixed; in S7-400 systems, you can change the priorities by parameterizing the CPU accordingly.

An overview of all priority classes, together with the default organization blocks for each, is presented in Chapter 3.1.2 "Priority Classes".

Disabling interrupts

The organization blocks for event-driven program scanning can be disabled and enabled with system functions SFC 39 DIS_IRT and SFC 40 EN_IRT and delayed and enabled with SFC 41 DIS_AIRT and SFC 42 EN_AIRT (see Chapter 21.9 "Handling Interrupts").

Current signal states

In an interrupt handling routine, one of the requirements is that you work with the current signal states of the I/O modules (and not with the signal states of the inputs that were updated at the start of the main program) and write the fetched signal states direct to the I/O (not waiting until the process-image output table is updated at the end of the main program).

In the case of a few inputs and outputs for the interrupt handling routine, it is enough to access the I/O modules direct with load and transfer operations (STL) or with the MOVE box (LAD, FBD). You are recommended here to maintain a strict separation between the main program and the interrupt handling routine with regard to the I/O signals.

If you want to process many input and output signals in the interrupt handling routine, the solution on the S7-400 CPUs is to use partial process images. When assigning addresses, you assign each module to a partial process image. With SFC 26 UPDAT_PI and SFC 27 UPDAT_PO, you update the partial process images in the user program (see also Chapter 20.2.1 "Process Image Updating").

On new S7-400 CPUs, you can assign an input and an output partial process image to each interrupt organization block (each interrupt priority class) and so cause the process images to be updated automatically when the interrupt occurs.

Start information, temporary local data

Each organization block delivers the start information in the first 20 bytes of its temporary local data. You can create the declaration of the start information yourself using own data, or you use the templates from the *Standard Library* under *Organization Blocks*.

In S7-300 systems, the available temporary local data have a fixed length of 256 bytes. In S7-400 systems, you can specify the length per priority class by parameterizing the CPU accordingly (parameter block "local data"), whereby the total may not exceed a CPU-specific maximum. Note that the minimum number of bytes for temporary local data for the priority class used must be 20 bytes so as to be able to accommodate the start information. Specify zero for unused priority classes.

Note that you can only directly read the start information of an organization block in the block itself since it is temporary local data. If you also require values from the start information in blocks which are present in lower call levels, call the system function SFC 6 RD_SINFO at the corresponding position in the program (see Chapter 20.2.5 "Start Information").

Current interrupt information

The interrupt organization block contains the specific information for the triggering interrupt in bytes 4 to 11 of the start information. In many cases, the interrupt-triggering component provides additional information which you can then read in the interrupt organization block using the system function block SFB 54 RALRM (see Chapter 21.9.3 "Reading Additional Interrupt Information").

21.2 Time-of-Day Interrupts

Time-of-day interrupts are used when you want to run a program at a particular time, either once only or periodically, for instance daily. In STEP 7, organization blocks OB 10 to OB 17 are provided for servicing time-of-day interrupts; which of these eight organization blocks are actually available depends on the CPU used.

You can configure the time-of-day interrupts in the Hardware Configuration data or control them at runtime via the program using system functions. The prerequisite for proper handling of the time-of-day interrupts is a correctly set real-time clock on the CPU.

Table 21.1 shows you the start information for the time-of-day interrupts. The dummy value xx represents the number of the associated interrupt organization block 10 to 17.

Table 21.1 Start Information for Time-of-Day Interrupts

Byte	Variable Name	Data Type	Description	Contents
0	OBxx_EV_CLASS	BYTE	Event class	B#16#11 = Incoming event
1	OBxx_STRT_INF	BYTE	Start request for the interrupt OB	B#16#11 = OB 10
2	OBxx_PRIORITY	BYTE	Priority class	Default value 2 for all time-of-day interrupts
3	OBxx_OB_NUMBR	BYTE	OB number	B#16#xx
4	OBxx_RESERVED_1	BYTE	Spare	-
5	OBxx_RESERVED_2	BYTE	Spare	-
6..7	OBxx_PERIOD_EXE	WORD	Interval with periodically called OBs	See description of SFC 28 SET_TINT
8..9	OBxx_RESERVED_3	INT	Spare	-
10..11	OBxx_RESERVED_4	INT	Sparc	-
12..19	OBxx_DATE_TIME	DATE_AND_TIME	Start of event	Call time of OB

xx represents the OB numbers 10 to 17

21.2.1 Handling Time-of-Day Interrupts

General remarks

To start a time-of-day interrupt, you must first set the start time, then activate the interrupt. You can perform the two activities separately via the Hardware Configuration data or using SFCs. Note that when activated via the Hardware Configuration data, the time-of-day interrupt is started automatically following parameterization of the CPU.

You can start a time-of-day interrupt in two ways:

▷ Single-shot: the relevant OB is called once only at the specified time, or

▷ Periodically: depending on the parameter assignments, the relevant OB is started every minute, hourly, daily, weekly, monthly or yearly.

Following a single-shot time-of-day interrupt OB call, the time-of-day interrupt is canceled. You can also cancel a time-of-day interrupt with SFC 29 CAN_TINT.

If you want to once again use a canceled time-of-day interrupt, you must set the start time again, then reactivate the interrupt.

You can query the status of a time-of-day interrupt with SFC 31 QRY_TINT.

Performance characteristics during startup

During a cold or warm restart, the operating system clears all settings made with SFCs. Settings made via the Hardware Configuration data are retained. On a hot restart, the CPU resumes servicing of the time-of-day interrupts in the first complete scan cycle of the main program.

You can query the status of the time-of-day interrupts in the start-up OB by calling SFC 31, and subsequently cancel or re-set and reactivate the interrupts. The time-of-day interrupts are serviced only in RUN mode.

Performance characteristics on error

If a time-of-day interrupt OB is called but was not programmed, the operating system calls OB 85 (program execution error). If OB 85 was not programmed, the CPU goes to STOP.

Time-of-day interrupts that were deselected when the CPU was parameterized cannot be serviced, even when the relevant OB is available. The CPU goes to STOP.

If you activate a time-of-day interrupt on a single-shot basis, and if the start time has already passed (from the real-time clock's point of view), the operating system calls OB 80 (timing error). If OB 80 is not available, the CPU goes to STOP.

If you activate a time-of-day interrupt on a periodic basis, and if the start time has already passed (from the real-time clock's point of view), the time-of-day interrupt OB is executed the next time that time period comes due.

If you set the real-time clock ahead by more than approx. 20 s, whether for the purpose of correction or synchronization, thus skipping over the start time for the time-of-day interrupt, the operating system calls OB 80 (timing error). The time-of-day interrupt OB is then executed precisely once.

If you set the real-time clock back by more than approx. 20 s, whether for the purpose of correction or synchronization, an activated time-of-day interrupt OB will no longer be executed at the instants which are already past.

If a time-of-day interrupt OB is still executing when the next (periodic) call occurs, the operating system invokes OB 80 (timing error). When OB 80 and the time-of-day interrupt OB have executed, the time-of-day interrupt OB is restarted.

Disabling, delaying and enabling

Time-of-day interrupt OB calls can be disabled and enabled with SFC 39 DIS_IRT and SFC 40 EN_IRT, and delayed and enabled with SFC 41 DIS_AIRT and SFC 42 EN_AIRT.

21.2.2 Configuring Time-of-Day Interrupts with STEP 7

The time-of-day interrupts are configured via the Hardware Configuration data. Open the selected CPU with EDIT → OBJECT PROPERTIES and choose the "Time-of-Day" tab from the dialog box.

In S7-300 controllers, the processing priority is permanently set to 2. In S7-400 controllers, you can set a priority between 2 and 24, depending on the CPU, for each possible OB; priority 0 deselects an OB. You should not assign a priority more than once, as interrupts might be lost when more than 12 interrupt events with the same priority occur simultaneously.

The "Active" option activates automatic starting of the time-of-day interrupt. The "Execution" option screens a list which allows you to choose whether you want the OB to execute on a single-shot basis or at specific intervals. The final parameter is the start time (date and time).

When it saves the Hardware Configuration, STEP 7 writes the compiled data to the *System Data* object in the offline user program *Blocks.* From here, you can load the parameter assignment data into the CPU while the CPU is at STOP; these data then go into force immediately.

21.2.3 System Functions for Time-of-Day Interrupts

The following system functions can be used for time-of-day interrupt control:

▷ SFC 28 SET_TINT
Set time-of-day interrupt

▷ SFC 29 CAN_TINT
Cancel time-of-day interrupt

▷ SFC 30 ACT_TINT
Activate time-of-day interrupt

▷ SFC 31 QRY_TINT
Query time-of-day interrupt

The parameters for these system functions are listed in Table 21.2.

SFC 28 SET_TINT
Set time-of-day interrupt

You determine the start time for a time-of-day interrupt by calling system function SFC 28 SET_TINT. SFC 28 sets only the start time; to start the time-of-day interrupt OB, you must activate the time-of-day interrupt with SFC 30 ACT_TINT. Specify the start time in the SDT parameter in the format DATE_AND_TIME, for instance DT#1997-06-30-08:30. The operating system ignores seconds and milliseconds and sets these values to zero. Setting the start time will overwrite the old start time value, if any. An active time-of-day interrupt is canceled, that is, it must be reactivated.

Table 21.2 SFC Parameters for Time-of-Day Interrupts

SFC	Parameter	Declaration	Data Type	Contents, Description
28	OB_NR	INPUT	INT	Number of the OB to be called at the specified time on a single-shot basis or periodically
	SDT	INPUT	DT	Start date and start time in the format DATE_AND_TIME
	PERIOD	INPUT	WORD	Period on which start time is based: W#16#0000 = Single-shot W#16#0201 = Every minute W#16#0401 = Hourly W#16#1001 = Daily W#16#1201 = Weekly W#16#1401 = Monthly W#16#2001 = Last in the month W#16#1801 = Yearly
	RET_VAL	RETURN	INT	Error information
29	OB_NR	INPUT	INT	Number of the OB whose start time is to be deleted
	RET_VAL	RETURN	INT	Error information
30	OB_NR	INPUT	INT	Number of the OB to be activated
	RET_VAL	RETURN	INT	Error information
31	OB_NR	INPUT	INT	Number of the OB whose status is to be queried
	RET_VAL	RETURN	INT	Error information
	STATUS	OUTPUT	WORD	Status of the time-of-day interrupt

SFC 30 ACT_TINT
Activate time-of-day interrupt

A time-of-day interrupt is activated by calling system function SFC 30 ACT_TINT. When a TOD interrupt is activated, it is assumed that a time has been set for the interrupt. If, in the case of a single-shot interrupt, the start time is already past, SFC 30 reports an error. In the case of a periodic start, the operating system calls the relevant OB at the next applicable time. Once a single-shot time-of-day interrupt has been serviced, it is, for all practical purposes, canceled. You can re-set and reactivate it (for a different start time) if desired.

SFC 29 CAN_TINT
Cancel time-of-day interrupt

You can delete a start time, thus deactivating the time-of-day interrupt, with system function SFC 29 CAN_TINT. The respective OB is no longer called. If you want to use this same time-of-day interrupt again, you must first set the start time, then activate the interrupt.

SFC 31 QRY_TINT
Query time-of-day interrupt

You can query the status of a time-of-day interrupt by calling system function SFC 31 QRY_TINT. The required information is returned in the STATUS parameter.

When the bits have signal state "1", they have the following meanings:

0 CPU is starting up

1 The interrupt has been disabled by the call of SFC 39 DIS_IRT

2 Time-of-day interrupt is activated and has not elapsed

3 (always "0")

4 An organization block with the number of OB_NR is loaded

5 (and following: always "0")

21.3 Time-Delay Interrupts

A time-delay interrupt allows you to implement a delay timer independently of the standard timers. In STEP 7, organization blocks OB 20 to OB 23 are set aside for time-delay interrupts; which of these four organization blocks are actually available depends on the CPU used.

The priorities for time-delay interrupt OBs are programmed in the Hardware Configuration data; system functions are used for control purposes.

Table 21.3 shows the start information for the time-delay interrupts. The dummy value xx represents the number of the associated interrupt organization block 20 to 23.

21.3.1 Handling Time-Delay Interrupts

General remarks

A time-delay interrupt is started by calling SFC 32 SRT_DINT; this system function also passes the delay interval and the number of the selected organization block to the operating system. When the delay interval has expired, the OB is called.

You can cancel servicing of a time-delay interrupt, in which case the associated OB will no longer be called.

You can query the status of a time-delay interrupt with SFC 34 QRY_DINT.

Performance characteristics during startup

On a cold or warm restart, the operating system deletes all programmed settings for time-delay interrupts. On a hot restart, the settings are retained until processed in RUN mode, whereby the "residual cycle" is counted as part of the start-up routine.

You can start a time-delay interrupt in the start-up routine by calling SFC 32. When the delay interval has expired, the CPU must be in RUN mode in order to be able to execute the relevant organization block. If this is not the case, the CPU waits to call the organization block until the start-up routine has terminated, then calls the time-delay interrupt OB before the first network in the main program.

Performance characteristics on error

If no time-delay interrupt OB has been programmed, the operating system calls OB 85 (program execution error). If there is no OB 85 in the user program, the CPU goes to STOP.

Table 21.3 Start Information for Time-Delay Interrupts

Byte	Variable Name	Data Type	Description	Contents
0	OBxx_EV_CLASS	BYTE	Event class	B#16#11 = Incoming event
1	OBxx_STRT_INF	BYTE	Start request for the interrupt OB	B#16#21 = OB 20
2	OBxx_PRIORITY	BYTE	Priority class	Default values 3 to 6 (OB 20 to OB 23)
3	OBxx_OB_NUMBR	BYTE	OB number	B#16#xx
4	OBxx_RESERVED_1	BYTE	Spare	-
5	OBxx_RESERVED_2	BYTE	Spare	-
6..7	OBxx_SIGN	WORD	Request ID	See description of SFC 32 SRT_DINT
8..11	OBxx_DTIME	TIME	Expired delay time	See description of SFC 32 SRT_DINT
12..19	OBxx_DATE_TIME	DATE_ AND_TIME	Start of event	Call time of OB

xx represents the OB numbers 20 to 23

If the delay interval has expired and the associated OB is still executing, the operating system calls OB 80 (timing error) or goes to STOP if there is no OB 80 in the user program.

Time-delay interrupts which were deselected during CPU parameterization cannot be serviced, even when the respective OB has been programmed. The CPU goes to STOP.

Disabling, delaying and enabling

The time-delay interrupt OBs can be disabled and enabled with system functions SFC 39 DIS_IRT and SFC 40 EN_IRT, and delayed and enabled with SFC 41 DIS_AIRT and SFC 42 EN_AIRT.

21.3.2 Configuring Time-Delay Interrupts with STEP 7

Time-delay interrupts are configured in the Hardware Configuration data. Simply open the selected CPU with EDIT → OBJECT PROPERTIES and choose the "Interrupts" tab from the dialog box.

In S7-300 controllers, the priority is permanently preset to 3. In S7-400 controllers, you can choose a priority between 2 and 24, depending on the CPU, for each possible OB; choose priority 0 to deselect an OB. You should not assign a priority more than once, as interrupts could be lost if more than 12 interrupt events with the same priority occur simultaneously.

When it saves the Hardware Configuration, STEP 7 writes the compiled data to the *System Data* object in the offline user program *Blocks*. From here, you can transfer the parameter assignment data while the CPU is at STOP; the data take effect immediately.

21.3.3 System Functions for Time-Delay Interrupts

A time-delay interrupt can be controlled with the following system functions:

- SFC 32 SRT_DINT
  Start time-delay interrupt
- SFC 33 CAN_DINT
  Cancel time-delay interrupt
- SFC 34 QRY_DINT
  Query time-delay interrupt

The parameters for these system functions are listed in Table 21.4.

SFC 32 SRT_DINT Start time-delay interrupt

A time-delay interrupt is started by calling system function SFC 32 SRT_DINT. The SFC call is also the start time for the programmed delay interval. When the delay interval has expired, the CPU calls the programmed OB and passes the time delay value and a job identifier in the start information for this OB. The job identifier

Table 21.4 SFC Parameters for Time-Delay Interrupts

SFC	Parameter	Declaration	Data Type	Contents, Description
32	OB_NR	INPUT	INT	Number of the OB to be called when the delay interval has expired
	DTIME	INPUT	TIME	Delay interval; permissible: T#1ms to T#1m
	SIGN	INPUT	WORD	Job identification in the respective OB's start information when the OB is called (arbitrary characters)
	RET_VAL	RETURN	INT	Error information
33	OB_NR	INPUT	INT	Number of the OB to be canceled
	RET_VAL	RETURN	INT	Error information
34	OB_NR	INPUT	INT	Number of the OB whose status is to be queried
	RET_VAL	RETURN	INT	Error information
	STATUS	OUTPUT	WORD	Status of the time-delay interrupt

is specified in the SIGN parameter for SFC 32; you can read the same value in bytes 6 and 7 of the start information for the associated time-delay interrupt OB. The time delay is set in increments of 1 ms. The accuracy of the time delay is also 1 ms. Note that execution of the time-delay interrupt OB may itself be delayed when organization blocks with higher priorities are being processed when the time-delay interrupt OB is called. You can overwrite a time delay with a new value by recalling SFC 32. The new time delay goes into force with the SFC call.

SFC 33 CAN_DINT
Cancel time-delay interrupt

You can call system function SFC 33 CAN_DINT to cancel a time-delay interrupt, in which case the programmed organization block is not called.

SFC 34 QRY_DINT
Query time-delay interrupt

System function SFC 34 QRY_DINT informs you about the status of a time-delay interrupt. You select the time-delay interrupt via the OB number, and the status information is returned in the STATUS parameter.

When the bits have signal state "1", they have the following meanings:

0 CPU is starting up
1 The interrupt has been disabled by the call of SFC 39 DIS_IRT
2 The time-delay interrupt is activated and has not elapsed
3 (always "0")
4 An organization block with the number of OB_NR is loaded
5 (and following: always "0")

21.4 Watchdog Interrupts

A watchdog interrupt is an interrupt which is generated at periodic intervals and which initiates execution of a watchdog interrupt OB. A watchdog interrupt allows you to execute a particular program periodically, independently of the processing time of the cyclic program.

In STEP 7, organization blocks OB 30 to OB 38 have been set aside for watchdog interrupts; which of these nine organization blocks are actually available depends on the CPU used.

Watchdog interrupt handling is set in the Hardware Configuration data when the CPU is parameterized.

Table 21.5 Start Information for Watchdog Interrupts

Byte	Tag name	Data type	Description	Assignment
0	OBxx_EV_CLASS	BYTE	Event class	B#16#11 = incoming event
1	OBxx_STRT_INF	BYTE	Start request for the interrupt OB	B#16#31 = OB 30
2	OBxx_PRIORITY	BYTE	Priority class	Default values 7 to 15 (OB 30 to 38)
3	OBxx_OB_NUMBR	BYTE	OB number	B#16#xx
4	OBxx_RESERVED_1	BYTE	Reserve	-
5	OBxx_RESERVED_2	BYTE	Reserve	-
6..7	OBxx_PHS_OFFSET	INT	Phase offset	ms, refer to Table 21.6
8..9	OBxx_RESERVED_3	INT	Reserve	-
10..11	OBxx_EXC_FREQ	INT	Interval	ms, refer to Table 21.6
12..19	OBxx_DATE_TIME	DATE_AND_TIME	Event occurrence	Call time of the OB

xx stands for the OB numbers 30 to 38

Table 21.5 shows the start information for the watchdog interrupts. The dummy value xx represents the number of the associated interrupt organization block 30 to 38.

21.4.1 Handling Watchdog Interrupts

Triggering watchdog interrupts in an S7-300

In an S7-300, there is a limited selection of watchdog interrupts with a fixed priority depending on the CPU. You can set the interval in the range from 1 millisecond to 1 minute, in 1-millisecond increments, by parameterizing the CPU accordingly (in increments of 500 µs for the CPU 319 from firmware release V2.6 onwards).

Triggering watchdog interrupts in an S7-400

You define a watchdog interrupt when you parameterize the CPU. A watchdog interrupt has three parameters: the interval, the phase offset, and the priority. You can set all three. Specifiable values for interval and phase offset are from 1 millisecond to 1 minute, in 1-millisecond increments; the priority may be set to a value between 2 and 24 or to zero, depending on the CPU (zero means the watchdog interrupt is not active).

STEP 7 provides the organization blocks listed in Table 21.6, in their maximum configurations.

Table 21.6 Defaults for Watchdog Interrupts

OB	Time Interval	Phase	Priority
30	5 s	0 ms	7
31	2 s	0 ms	8
32	1 s	0 ms	9
33	500 ms	0 ms	10
34	200 ms	0 ms	11
35	100 ms	0 ms	12
36	50 ms	0 ms	13
37	20 ms	0 ms	14
38	10 ms	0 ms	15

Phase offset

The phase offset can be used to execute watchdog interrupt handling routines which have a common time interval or a common multiple thereof at an exact interval. This permits a higher processing interval accuracy.

The start time of the time interval and the phase offset is the instant of transition from STARTUP to RUN. The call instant for a watchdog interrupt OB is thus the time interval plus the phase offset. Figure 21.1 shows an example of this. No phase offset is set in the left-hand part, therefore the start of processing of the lower-priority organization block is shifted by the current processing time of the higher-priority organization block in each case.

If, on the other hand, a phase offset is configured which is greater than the maximum processing time of the high-priority organization block, the lower-priority organization block is processed at the exact interval.

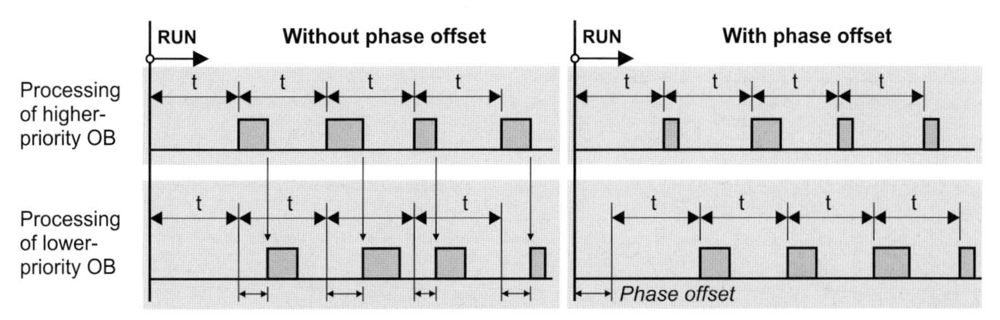

Figure 21.1 Example of Phase Offset for Watchdog Interrupts

Performance characteristics during startup

Watchdog interrupts cannot be serviced in the start-up OB. The time intervals do not begin until a transition is made to RUN mode.

Performance characteristics on error

When the same watchdog interrupt is generated again while the associated watchdog interrupt handling OB is still executing, the operating system calls OB 80 (timing error). If OB 80 has not been programmed, the CPU goes to STOP.

The operating system saves the watchdog interrupt that was not serviced, servicing it at the next opportunity. Only one unserviced watchdog interrupt is saved per priority class, regardless of how many unserviced watchdog interrupts accumulate.

Watchdog interrupts that were deselected when the CPU was parameterized cannot be serviced, even when the corresponding OB is available. The CPU goes to STOP in this case.

Disabling, delaying and enabling

Calling of the watchdog interrupt OBs can be disabled and enabled with system functions SFC 39 DIS_IRT and SFC 40 EN_IRT and delayed and enabled with SFC 41 DIS_AIRT and SFC 42 EN_AIRT.

21.4.2 Configuring Watchdog Interrupts with STEP 7

Watchdog interrupts are configured via the Hardware Configuration data. Simply open the selected CPU with EDIT → OBJECT PROPERTIES and choose the "Cyclic Interrupt" tab from the dialog box.

In S7-300 controllers, the processing priority is permanently set. In S7-400 controllers, you may set a priority between 2 and 24 for each possible OB (CPU-specific); priority 0 deselects the OB to which it is assigned. You should not assign a priority more than once, as interrupts might be lost if more than 12 interrupt events with the same priority occur simultaneously.

The interval for each OB is selected under "Execution", the delayed call instant under "Phase Offset".

When it saves the Hardware Configuration, STEP 7 writes the compiled data to the *System Data* object in the offline user program *Blocks*. From here, you can load the parameter assignment data into the CPU while the CPU is at STOP; the data take effect immediately.

21.5 Hardware Interrupts

Hardware interrupts are used to enable the immediate detection in the user program of events in the controlled process, making it possible to respond with an appropriate interrupt handling routine. STEP 7 provides organization blocks OB 40 to OB 47 for servicing process interrupts; which of these eight organization blocks are actually available, however, depends on the CPU.

Hardware interrupt handling is programmed in the Hardware Configuration data. With system functions SFC 55 WR_PARM, SFC 56 WR_DPARM and SFC 57 PARM_MOD, you can (re)parameterize the modules with process interrupt capability even in RUN mode.

Table 21.7 shows the start information for the process interrupts. The dummy value xx represents the number of the associated interrupt organization block 40 to 47.

21.5.1 Generating a Hardware Interrupt

A process interrupt is generated on the modules with this capability. This could, for example, be a digital input module that detects a signal from the process or a function module that generates a process interrupt because of an activity taking place on the module.

By default, process interrupts are disabled. A parameter is used to enable servicing of a process interrupt (static parameter), and you can specify whether the process interrupt should be generated for a coming event, a leaving event, or both (dynamic parameter). Dynamic parameters are parameters which you can modify at runtime using SFCs.

In an intelligent DP slave equipped for this purpose, you can initiate a process interrupt in the master CPU with SFC 7 DP_PRAL.

Table 21.7 Start Information for Hardware Interrupts

Byte	Variable Name	Data Type	Description	Contents
0	OBxx_EV_CLASS	BYTE	Event class	B#16#11 = incoming event
1	OBxx_STRT_INF	BYTE	Start request for the interrupt OB	B#16#41 = OB 40
2	OBxx_PRIORITY	BYTE	Priority class	Default values 16 to 23 (OB 40 to 47)
3	OBxx_OB_NUMBR	BYTE	OB number	B#16#xx
4	OBxx_RESERVED_1	BYTE	Spare	-
5	OBxx_IO_FLAG	BYTE	I/O ID	B#16#54 = input module, input submodule B#16#55 = output module, output submodule
6..7	OBxx_MDL_ADDR	WORD	Module starting address of component triggering the interrupt	
8..11	OBxx_POINT_ADDR	DWORD	Interrupt information	
12..19	OBxx_DATE_TIME	DATE_AND_TIME	Start of event	Call time of OB

xx represents the OB numbers 40 to 47

The process interrupt is acknowledged on the module when the organization block containing the service routine for that interrupt has finished executing.

Resolution on the S7-300

If an event occurs during execution of a process interrupt OB which itself would trigger generation of the same process interrupt, that process interrupt will be lost when the event that triggered it is no longer present following acknowledgment. It makes no difference whether the event comes from the module whose process interrupt is currently being serviced or from another module.

A diagnostic interrupt can be generated while a process interrupt is being serviced. If another process interrupt occurs on the same channel between the time the first process interrupt was generated and the time that interrupt was acknowledged, the loss of the latter interrupt is reported via a diagnostic interrupt to system diagnostics.

Resolution on the S7-400

If during execution of a process interrupt OB an event occurs on the same channel on the same module which would trigger the same process interrupt, that interrupt is lost. If the event occurs on another channel on the same module or on another module, the operating system restarts the OB as soon as it has finished executing.

21.5.2 Servicing Hardware Interrupts

Querying interrupt information

The starting address of the module that triggered the process interrupt is in bytes 6 and 7 of the process interrupt OB's start information. If this address is an input address, byte 5 of the start information contains B#16#54; otherwise it contains B#16#55. If the module in question is a digital input module, bytes 8 to 11 contain the status of the inputs; for any other type of module, these bytes contain the interrupt status of the module.

Interrupt handling in the start-up routine

In the start-up routine, the modules do not generate process interrupts. Interrupt handling begins with the transition to RUN mode. Any process interrupts pending at the time of the transition are lost.

Error handling

If a process interrupt is generated for which there is no process interrupt OB in the user program, the operating system calls OB 85 (program execution error). The process interrupt is acknowledged. If OB 85 has not been programmed, the CPU goes to STOP.

Hardware interrupts deselected when the CPU was parameterized cannot be serviced, even when the OBs for these interrupts have been programmed. The CPU goes to STOP.

Disabling, delaying and enabling

Calling of the process interrupt OBs can be disabled and enabled with system functions SFC 39 DIS_IRT and SFC 40 EN_IRT, and delayed and enabled with SFC 41 DIS_AIRT and SFC 42 EN_AIRT.

21.5.3 Configuring Hardware Interrupts with STEP 7

Hardware interrupts are programmed in the Hardware Configuration data. Open the selected CPU with EDIT → OBJECT PROPERTIES and choose the "Interrupts" tab in the dialog box.

In S7-300 systems, the priority for OB 40 is fixed and cannot be changed. In S7-400 systems, you can choose a priority between 2 and 24 for every possible OB (on a CPU-specific basis); priority 0 deselects execution of an OB. You should never assign the same priority twice because interrupts can be lost when more than 12 interrupt events with the same priority occur simultaneously.

You must also enable the triggering of process interrupts on the respective modules. To this purpose, these modules are parameterized much the same as the CPU.

When it saves the Hardware Configuration, STEP 7 writes the compiled data to the *System Data* object in offline user program *Blocks;* from here, you can load the parameterization data into the CPU while the CPU is in STOP mode. The parameterization data for the CPU go into force immediately following loading; the parameter assignment data for the modules take effect after the next start-up.

21.6 DPV1 Interrupts

PROFIBUS DPV1 slaves can trigger the following interrupts in addition to the types previously known with SIMATIC S7:

- ▷ Status interrupt if e.g. the DPV1 slave changes its operating status; the interrupt organization block OB 55 is called.
- ▷ Update interrupt if e.g. the DPV1 slave was reconfigured via the PROFIBUS or directly; the interrupt organization block OB 56 is called.
- ▷ Manufacturer specific interrupt if an associated event envisaged by the vendor occurs in the DPV1 slave; the interrupt organization block OB 57 is called. The events triggering the interrupt are defined by the vendor of the DPV1 slave.

The start information of the DPV1 interrupt organization blocks includes the origin of the interrupt, the interrupt specifier, and the length of additionally available interrupt information (Table 21.8). You can read the additional interrupt information using the system function block SFB 54 RALRM (Chapter 21.9.3 "Reading Additional Interrupt Information").

Response during startup

PROFIBUS DPV1 slaves can also generate interrupts if the master CPU is at STOP. At STOP, the master CPU cannot call an interrupt organization block; processing of the interrupts is not carried out subsequently either when the CPU enters RUN mode.

However, the received interrupt events are entered into the diagnostics buffer and into the module status data. You can read the module status data using the system function SFC 51 RDSYSST.

Table 21.8 Start Information for DPV1 Interrupts

Byte	Variable Name	Data Type	Description	Contents
0	OBxx_EV_CLASS	BYTE	Event class	B#16#11 = Incoming event
1	OBxx_STRT_INF	BYTE	Start request for the OB xx	B#16#xx
2	OBxx_PRIORITY	BYTE	Priority class	B#16#02 = default value
3	OBxx_OB_NUMBR	BYTE	OB number	B#16#xx
4	OBxx_RESERVED_1	BYTE	Spare	-
5	OBxx_IO_FLAG	BYTE	I/O ID	B#16#54 = input module, input submodule B#16#55 = output module, output submodule
6..7	OBxx_MDL_ADDR	WORD	Module starting address of component triggering the interrupt	-
8	OBxx_LEN	BYTE	Length of interrupt data record	-
9	OBxx_TYPE	BYTE	ID for type of interrupt	B#16#00 = spare B#16#01 = diagnostics interrupt B#16#02 = process interrupt B#16#03 = removal interrupt B#16#04 = insertion interrupt B#16#05 = status interrupt B#16#06 = update interrupt B#16#07..1F = spare B#16#20..7E = manufacturer specific interrupt B#16#7F = spare
10	OBxx_SLOT	BYTE	Slot number of component triggering the interrupt	-
11	OBxx_SPEC	BYTE	Specifier	Bits 1 and 0: 0 0 spare 0 1 Incoming event 1 0 Outgoing event with error 1 1 Outgoing event with further errors Bit 2: 0 no additional acknowledgment required 1 additional acknowledgment required Bits 3 to 7: spare
12..19	OBxx_DATE_TIME	DATE_AND_TIME	Start of event	Call time of OB

xx represents the OB number 55, 56 or 57

Error handling

If a DPV1 interrupt is generated for which there is no DPV1 interrupt OB in the user program, the operating system calls OB 85 (priority class error). The DPV1 interrupt is acknowledged. If OB 85 has not been programmed, the CPU enters the STOP mode.

Disabling, delaying and enabling

Calling of the DPV1 interrupt OBs can be disabled and enabled with system functions

SFC 39 DIS_IRT and SFC 40 EN_IRT, and delayed and enabled with SFC 41 DIS_AIRT and SFC 42 EN_AIRT.

Configuring DPV1 interrupts with STEP 7

DPV1 interrupts are programmed in the Hardware Configuration data. Open the selected CPU with EDIT → OBJECT PROPERTIES and choose the "Interrupts" tab in the properties window.

The default priority is 2. You can set the priority in the range from 2 to 24. Priority 0 deselects the interrupt. DPV1 interrupts which had been deselected cannot be executed, even if the corresponding OB exists. The CPU then goes to STOP.

In addition, you must program the interrupt triggering on the corresponding DPV1 slaves.

When saving the Hardware Configuration, STEP 7 writes the compiled data into the system data object in the offline user program blocks; from here, you can load the programming data into the CPU in the STOP status. The programming data for the CPU become immediately effective following loading, those for the DPV1 slaves following the next startup.

21.7 Multiprocessor Interrupt

The multiprocessor interrupt allows a synchronous response to an event in all CPUs in multiprocessor mode. A multiprocessor interrupt is triggered using SFC 35 MP_ALM. Organization block OB 60, which has a fixed priority of 25, is the OB used to service a multiprocessor interrupt.

Table 21.9 shows you the assignment of the start information for the multiprocessor interrupt.

General remarks

An SFC 35 MP_ALM call initiates execution of the multiprocessor interrupt OB. If the CPU is in single-processor mode, OB 60 is started immediately. In multiprocessor mode, OB 60 is started simultaneously on all participating CPUs, that is to say, even the CPU in which SFC 35 was called waits before calling OB 60 until all the other CPUs have indicated that they are ready.

The multiprocessor interrupt is not programmed in the Hardware Configuration data; it is already present in every CPU with multicomputing capability. Despite this fact, however, a sufficient number of local data bytes (at

Table 21.9 Start Information for the Multiprocessor Interrupt

Byte	Variable Name	Data Type	Description	Contents
0	OB60_EV_CLASS	BYTE	Event class	B#16#11 = Incoming event
1	OB60_STRT_INF	BYTE	Start request for the OB 60	B#16#61: multiprocessor interrupt triggered by own CPU B#16#62: multiprocessor interrupt triggered by a different CPU
2	OB60_PRIORITY	BYTE	Priority class	B#16#19 = default value (25dec)
3	OB60_OB_NUMBR	BYTE	OB number	B#16#3C (60dec)
4	OB60_RESERVED_1	BYTE	Spare	-
5	OB60_RESERVED_2	BYTE	Spare	-
6..7	OB60_JOB	INT	Job ID	Input variable JOB of SFC 35 MP_ALM
8..9	OB60_RESERVED_3	INT	Spare	-
10..11	OB60_RESERVED_4	INT	Spare	-
12..19	OB60_DATE_TIME	DATE_ AND_TIME	Start of event	Call time of OB

Table 21.10 Parameters for SFC 35 MP_ALM

Parameter	Declaration	Data Type	Contents, Description
JOB	INPUT	BYTE	Job identification in the range B#16#00 to B#16#0F
RET_VAL	RETURN	INT	Error information

least 20) must still be reserved in the CPU's "Local Data" tab under priority class 25.

Performance characteristics during startup

The multiprocessor interrupt is triggered only in RUN mode. An SFC 35 call in the start-up routine terminates after returning error 32 929 (W#16#80A1) as function value.

Performance characteristics on error

If OB 60 is still in progress when SFC 35 is recalled, the system function returns error code 32 928 (W#16#80A0) as function value. OB 60 is not started in any of the CPUs.

The unavailability of OB 60 in one of the CPUs at the time it is called or the disabling or delaying of its execution by system functions has no effect, nor does SFC 35 report an error.

Disabling, delaying and enabling

The multiprocessor OB can be disabled and enabled with system functions SFC 39 DIS_IRT and SFC 40 EN_IRT, and delayed and enabled with SFC 41 DIS_AIRT and SFC 42 EN_AIRT.

SFC 35 MP_ALM Multiprocessor interrupt

A multiprocessor interrupt is triggered with system function SFC 35 MP_ALM. Its parameters are listed in Table 21.10.

The JOB parameter allows you to forward a job identifier. The same value can be read in bytes 6 and 7 of OB 60's start information in all CPUs.

21.8 Synchronous Cycle Interrupts

Isochronous mode is when the reading, processing, and output of I/O signals occurs synchronously and at fixed (equidistant) time intervals. For PROFIBUS DP the time interval is the DP cycle, with which the I/O signals of the DP slaves are updated; for PROFINET IO it is the data cycle with which the signals of the I/O devices are updated, multiplied by a whole-number factor.

The user program executed in isochronous mode is present in organization blocks OB 61 to OB 64. For isochronous process image updating, there are the system functions SFC 126 SYNC_PI and SFC 127 SYNC_PO. Table 21.11 shows the start information for the synchronous cycle interrupts. The placeholder xx stands for the number of the affected interrupt organization block, 61 to 64.

21.8.1 Processing Synchronous Cycle Interrupts

Synchronous cycle interrupts are only processed in the RUN status. An synchronous cycle interrupt in the STARTUP, STOP or HOLD mode is rejected. The start information of the isochronous mode OB called for the first time during RUN contains the number of OB calls which have not been executed.

Error handling

If an synchronous cycle interrupt arrives before the associated synchronous cycle interrupt OB has been completed, a timing error is signaled. This can occur if the user program takes too long in an synchronous cycle interrupt OB or if the processing has been interrupted for too long because of program components of higher priority. The OB called by the "too early" interrupt is rejected, and the OB 80 "Timing error" is called. It is possible here to react to the timing error. In the next synchronous cycle interrupt

Table 21.11 Start information for synchronous cycle interrupts

Byte	Tag name	Data type	Description, remark	
0	OBxx_EV_CLASS	BYTE	Event class	B#16#11 = incoming event
1	OBxx_STRT_INF	BYTE	Start request for the OB xx	B#16#36: OB 61, … , B#16#67: OB 64
2	OBxx_PRIORITY	BYTE	Priority class	B#16#19 = default value (25dez)
3	OBxx_OB_NUMBR	BYTE	OB number	B#16#xx
4	OBxx_RESERVED_1	BYTE	Reserve	-
5	OBxx_RESERVED_2	BYTE	Reserve	-
6.0	OBxx_GC_VIOL	BOOL	GC violation	With PROFIBUS DP
6.1	OBxx_FIRST	BOOL	First execution following STARTUP or HOLD	For "1"
7	OBxx_MISSED_EXEC	BYTE	Number of discarded OB calls	since the last OB xx call
8	OBxx_DP_ID	BYTE	DP master system ID of the synchronous DP master system	is configured in the hardware configuration
9	OBxx_RESERVED_3	BYTE	Reserved	-
10..11	OBxx_RESERVED_4	WORD	Reserved	-
12..19	OBxx_DATE_TIME	DT	Event occurrence	Call time of the OB

OB processed, the start information contains the number of omitted synchronous cycle interrupts.

In the event of an error, the DP master can omit the Global_Control (GC) command or send it offset. This "GC violation" is shown in the start information of the next synchronous cycle interrupt OB which is called correctly.

Disabling, delaying and enabling

Calling of the synchronous cycle interrupt OB can be disabled and enabled with system functions SFC 39 DIS_IRT and SFC 40 EN_IRT, and delayed and enabled with SFC 41 DIS_AIRT and SFC 42 EN_AIRT.

21.8.2 Isochronous Updating of Process Image

Hardware Configuration can be used to assign process image partitions to a synchronous cycle interrupt OB. These are not updated automatically. The system function **SFC 126 SYNC_PI** should be used to update inputs and **SFC 127 SYNC_PO** should be used to update outputs. Updating is carried out isochronously and data-consistent. The two SFCs must only be called in an synchronous cycle interrupt OB. Direct access to the peripheral inputs and outputs of these process image partitions should be avoided.

Table 21.12 shows the parameters of the SFC 126 SYNC_PI and SFC 127 SYNC_PO.

If an error is detected, the partial process images are not updated. Exceptions:

- ▷ If an access error occurs during updating of the input partial process image, the inputs of faulty modules are set to signal status “0”; the OB 85 “Priority class error” is not called.
- ▷ If the complete data could not be transmitted consistently to the outputs, a consistency warning is generated. However, the data of individual slaves are consistent.
- ▷ If an access error occurs during updating of the output partial process image, the data of faulty modules are not transmitted; they remain unchanged in the partial process

Table 21.12 Parameters of the SFCs for isochronous Mode Updating of the Process Image

With SFC		Parameter Name	Declaration	Data Type	Contents, Description
126	127	PART	INPUT	BYTE	Number of partial process image B#16#01 to B#16#1E
126	127	RET_VAL	RETURN	INT	Error information
126	127	FLADDR	OUTPUT	WORD	In the case of an access error, the address of the first byte causing the error

image. Updating of non-affected modules is divided between two DP cycles (consistency warning).

21.8.3 Programming of Synchronous Cycle Interrupts with STEP 7

Programming of the synchronous cycle interrupts is carried out using the Hardware Configuration. Open the selected CPU with EDIT → OBJECT PROPERTIES, and select the "Synchronous cycle interrupts" in the properties window.

The default priority is 25. You can set the priority in the range from 2 to 26. Priority 0 deselects the interrupt. Synchronous cycle interrupts which have been deselected are not executed, even if the corresponding OB is present. Furthermore, you assign the isochronous DP master system and the involved partial process images to the interrupt OB.

In addition, you assign the isochronous DP master system or the isochronous PROFINET IO system and the involved process image partitions to the interrupt OB (see sections "Configuring constant bus cycle time and isochrone mode" in Chapter 20.4.3 "Special Functions for PROFIBUS DP" or "Isochronous mode" in Chapter 20.4.6 "Special functions for PROFINET IO").

By saving the Hardware Configuration, STEP 7 writes the compiled data into the object *System data* in the offline user program *Blocks*; you can load the parameterization data to the CPU from here in the STOP status. The parameterization data for the CPU become effective immediately following loading, and those for the DP components following the next startup.

21.9 Handling Interrupts

The system functions for disabling, delaying and enabling influence all interrupts and all asynchronous errors. The system functions SFC 36 to SFC 38 are available for handling synchronous errors.

21.9.1 Disabling and Enabling Interrupts

The following system functions are available for disabling and enabling interrupts and asynchronous errors:

▷ SFC 39 DIS_IRT
Disable interrupts

▷ SFC 40 EN_IRT
Enable disabled interrupts

Table 21.13 lists the parameters for these system functions.

SFC 39 DIS_IRT
Disabling interrupts

System function SFC 39 DIS_IRT disables servicing of new interrupts and asynchronous errors. All new interrupts and asynchronous errors are rejected. If an interrupt or asynchronous error occurs following a Disable, the organization block is not executed; if the OB does not exist, the CPU does not go to STOP.

The Disable remains in force for all priority classes until it is revoked with SFC 40 EN_IRT. After a cold or warm restart, all interrupts and asynchronous errors are enabled.

The MODE and OB_NR parameters are used to specify which interrupts and asynchronous errors are to be disabled. MODE = B#16#00 disables all interrupts and asynchronous errors.

Table 21.13 Parameters of the System Functions for Interrupt Handling

SFC	Parameter Name	Declaration	Data Type	Contents, Description
39	MODE	INPUT	BYTE	Disable mode (see text)
	OB_NR	INPUT	INT	OB number (see text)
	RET_VAL	RETURN	INT	Error information
40	MODE	INPUT	BYTE	Enable mode (see text)
	OB_NR	INPUT	INT	OB number (see text)
	RET_VAL	RETURN	INT	Error information
41	RET_VAL	RETURN	INT	(New) number of delays
42	RET_VAL	RETURN	INT	Number of delays remaining

MODE = B#16#01 disables an interrupt class whose first OB number is specified in the OB_NR parameter.

For example, MODE = B#16#01 and OB_NR = 40 disables all process interrupts; OB = 80 would disable all asynchronous errors. MODE = B#16#02 disables the interrupt or asynchronous error whose OB number you entered in the OB_NR parameter.

Regardless of a Disable, the operating system enters each new interrupt or asynchronous error in the diagnostic buffer.

SFC 40 EN_IRT
Enabling disabled interrupts

System function SFC 40 EN_IRT enables the interrupts and asynchronous errors disabled with SFC 39 DIS_IRT. An interrupt or asynchronous error occurring after the Enable will be serviced by the associated organization block; if that organization block is not in the user program, the CPU goes to STOP (except in the case of OB 81, the organization block for power supply errors).

The MODE and OB_NR parameters specify which interrupts and asynchronous errors are to be enabled. MODE = B#16#00 enables all interrupts and asynchronous errors. MODE = B#16#01 enables an interrupt class whose first OB number is specified in the OB_NR parameter. MODE = B#16#02 enables the interrupt or asynchronous error whose OB number you entered in the OB_NR parameter.

21.9.2 Delaying and Enabling Delayed Interrupts

The following system functions are available for delaying and enabling interrupts and asynchronous errors:

- SFC 41 DIS_AIRT
  Delay interrupts
- SFC 42 EN_AIRT
  Enable delayed interrupts

Table 21.13 lists the parameters for these system functions.

SFC 41 DIS_AIRT
Delaying interrupts

System function SFC 41 DIS_AIRT delays the servicing of higher-priority new interrupts and asynchronous errors. Delay means that the operating system saves the interrupts and asynchronous errors which occurred during the delay and services them when the delay interval has expired. Once SFC 41 has been called, the program in the current organization block (in the current priority class) will not be interrupted by a higher-priority interrupt; no interrupts or asynchronous errors are lost.

A delay remains in force until the current OB has terminated its execution or until SFC 42 EN_AIRT is called.

You can call SFC 41 several times in succession. The RET_VAL parameter shows the number of calls. You must call SFC 42 precisely the same number of times as SFC 41 in order to reenable the interrupts and asynchronous errors.

SFC 42 EN_AIRT
Enabling delayed interrupts

System function SFC 42 EN_AIRT reenables the interrupts and asynchronous errors delayed with SFC 41. You must call SFC 42 precisely the same number of times as you called SFC 41 (in the current OB). The RET_VAL parameter shows the number of delays still in force; if RET_VAL is = 0, the interrupts and asynchronous errors have been reenabled.

If you call SFC 42 without having first called SFC 41, RET_VAL contains the value 32896 (W#16#8080).

21.9.3 Reading Additional Interrupt Information

The system function block **SFB 54 RALRM** reads additional interrupt information – if present – from the components (modules or submodules) triggering the interrupt. It is called in an interrupt organization block or in a block called within this. Processing of the SFB 54 RALRM is synchronous, i.e. the requested data are available at the output parameters immediately following the call. Table 21.14 shows the block parameters of the SFB 54 RALRM.

The SFB 54 RALRM can be basically called in all organization blocks or execution levels for all events. If you call it in an organization block whose start event is not an interrupt from the I/O, correspondingly less information is available. Depending on the respective organization block and the components triggering the interrupt, different information is entered in the target areas specified by the parameters TINFO and AINFO (Table 21.15).

The target area TINFO (task information) contains the complete status information in bytes 0 to 19 of the organization blocks in which the SFB 54 RALRM was called, independent of the nesting depth in which it was called. The SFB 54 RALRM therefore partially replaces the system function SFC 6 RD_SINFO. Bytes 20 to 27 contain administration information, e.g. from which component the interrupt has been triggered.

The target area AINFO (alarm information) contains the header information in bytes 0 to 3 (e.g. number of received bytes of additional interrupt information or interrupt type) and in bytes 4 to 223 the component-specific additional interrupt information itself.

The assignment of the MODE parameter determines the operating mode of SFB 54 RALRM. With MODE = 0, the SFB shows you the component triggering of the interrupt in the ID parameter; NEW is occupied by TRUE. With MODE = 1, all output parameters are written. With MODE = 2, you can check whether the

Table 21.14 Parameters of the System Function Block SFB 54 RALRM

Parameter	Declaration	Data type	Assignment, description
MODE	INPUT	INT	Mode: 0 = Shows the component triggering the interrupt 1 = Describes all output parameters 2 = Checks whether the selected component triggered the interrupt
F_ID	INPUT	DWORD	Module starting address of the components to be queried
MLEN	INPUT	INT	Maximum number of bytes of the additional interrupt information to be requested
NEW	INPUT	BOOL	TRUE = A new interrupt was received
STATUS	OUTPUT	DWORD	Error ID
ID	OUTPUT	DWORD	Module starting address of interrupt-triggering component
LEN	OUTPUT	INT	Number of bytes of additional interrupt information received
TINFO	IN_OUT	ANY	Destination area for OB start and management information
AINFO	IN_OUT	ANY	Destination area for header information and additional interrupt information

component specified by the F_ID parameter triggered the interrupt. If this is the case, the NEW parameter is TRUE, and all other output parameters are written.

In order to work correctly, the SFB 54 RALRM requires separate instance data for each call in the various organization blocks, e.g. a separate instance data block in each case.

Table 21.15 Assignment of Parameters TINFO and AINFO

Interrupt Type	OB No.	TINFO		AIFO	
		OB start information	Administration information	Header information	Additional interrupt information
		Bytes 0 to 19	Bytes 20 to 27	Bytes 0 to 3 [1)]	Bytes 4 to 223 [2)]
Central process interrupt	40 to 47	yes	yes	yes	no
Decentral process interrupt	40 to 47	yes	yes	yes	As delivered by the station
Status interrupt	55	yes	yes	yes	yes
Update interrupt	56	yes	yes	yes	yes
Vendor interrupt	57	yes	yes	yes	yes
I/O redundancy error	70	yes	yes	no	no
Central diagnostics interrupt	82	yes	yes	yes	Diagnostics data record 1
Decentral diagnostics interrupt	82	yes	yes	yes	As delivered by DP slave
Central hot swapping interrupt	83	yes	yes	yes	no
Decentral hot swapping interrupt	83	yes	yes	yes	As delivered by DP slave
Rack, station failure	86	yes	yes	no	no
All other events		yes	no	no	no

with PROFINET IO:

1) 0 to 25

2) 26 to 1431

22 Restart Characteristics

22.1 General Remarks

22.1.1 Operating Modes

Before the CPU begins processing the main program following power-up, it executes a restart routine. START-UP is one of the CPU's operating modes, as is STOP or RUN. This chapter describes the CPU's activities on a transition from and to START-UP and in the restart routine itself.

Following power-up ①, the CPU is in the STOP mode (Figure 22.1). If the keyswitch on the CPU's front panel is at RUN or RUN-P, the CPU switches to START-UP mode ②, then to RUN mode ③. If an "unrecoverable" error occurs while the CPU is in START-UP or RUN mode or if you position the keyswitch to STOP, the CPU returns to the STOP mode ④ ⑤.

The user program is tested with breakpoints in single-step operation in the HOLD mode. You can switch to this mode from both RUN and START-UP, and return to the original mode when you abort the test ⑥ ⑦. You can also set the CPU to the STOP mode from the HOLD mode ⑧.

When you parameterize the CPU, you can define restart characteristics with the "Restart" tab such as the maximum permissible amount of time for the Ready signals from the modules following power-up or whether the CPU is to start up when the configuration data do not coincide with the actual configuration or in what mode the CPU restart is to be in.

SIMATIC S7 has three restart modes, namely *cold restart*, *warm restart* and *hot restart.* On a cold restart or warm restart, the main program is always processed from the beginning. A hot restart resumes the main program at the point of interruption, and "finishes" the cycle.

S7 CPUs supplied before 10/98 have warm restart and hot restart. The warm restart corresponds in functionality to the hot restart.

You can scan a program on a single-shot basis in START-UP mode. STEP 7 provides organization blocks OB 102 (cold restart), OB 100 (warm restart) and OB 101 (hot restart) expressly for this purpose. Sample applications are the parameterization of modules unless this was already taken care of by the CPU, and the programming of defaults for your main program.

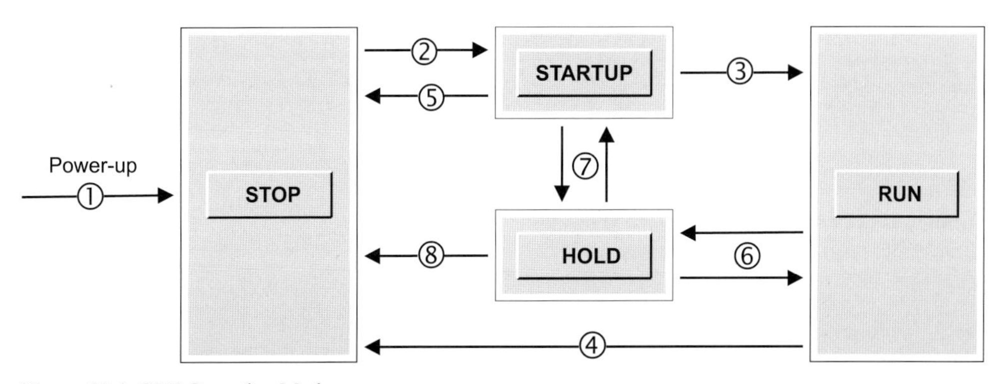

Figure 22.1 CPU Operating Modes

22.1.2 HOLD Mode

The CPU changes to the HOLD mode when you test the program with breakpoints (in "single-step mode"). The STOP LED then lights up and the RUN LED blinks.

In HOLD mode, the output modules are disabled. Writing to the modules affects the module memory, but does not switch the signal states "out" to the module outputs. The modules are not reenabled until you exit the HOLD mode.

In HOLD mode, everything having to do with timing is discontinued. This includes, for example, the processing of timers, clock memory and run-time meters, cycle time monitoring and minimum scan cycle time, and the servicing of time-of-day and time-delay interrupts. Exception: the real-time clock continues to function normally.

Every time the progression is made to the next statement in test mode, the timers for the duration of the single step run a little further, thus simulating a dynamic behavior similar to "normal" program scanning.

In HOLD mode, the CPU is capable of passive communication, that is, it can receive global data or take part in the unilateral exchange of data.

If the power fails while the CPU is in HOLD mode, battery-backed CPUs go to STOP on power recovery. CPUs without backup batteries execute an automatic warm restart.

22.1.3 Disabling the Output Modules

In the STOP and HOLD modes, modules are disabled (output disable, OD signal). Disabled output modules output a zero signal or, if they have the capability, the replacement value. Via a variable table, you can control outputs on the modules with the "Enable Peripheral Outputs" function, even in STOP mode.

During restart, the output modules remain disabled. Only when the cyclic scan begins are the output modules enabled.

On a cold restart (OB 102) and warm restart (OB 100), the process images and the module memory are cleared. If you want to scan inputs in OB 102 or in OB 100, you must load the signal states from the module using direct access. You can then set the inputs (transfer them, for instance, with load statements or with the MOVE box from address area PI to address area I), then work with the inputs.

On a hot restart, the "old" process-image input and process-image output tables, which were valid prior to power-down or STOP, are used in OB 101 and in the remainder of the cycle. At the end of that cycle, the process-image output table is transferred to module memory (but not yet switched through to the external outputs, since the output modules are still disabled).

You now have the option of parameterizing the CPU to clear the process-image output table and the module memory at the end of the hot restart. Before switching to OB 1, the CPU revokes the Disable signal so that the signal states in the module memory are applied to the external outputs.

22.1.4 Restart Organization Blocks

On a cold restart, the CPU calls organization block OB 102; on a warm restart, it calls organization block OB 100. In the absence of OB 100 or OB 102, the CPU begins cyclic program execution immediately.

On a hot restart, the CPU calls organization block OB 101 on a single-shot basis before processing the main program. If there is no OB 101, the CPU begins scanning at the point of interruption.

The start information in the temporary local data has the same format for the restart organization blocks; Table 22.1 shows the start information for OB 100. The reason for the restart is shown in the restart request (Byte 1):

B#16#81 Manual warm restart (OB 100)

B#16#82 Automatic warm restart (OB 100)

B#16#83 Manual hot restart (OB 101)

B#16#84 Automatic hot restart (OB 101)

B#16#85 Manual cold restart (OB 102)

B#16#86 Automatic cold restart (OB 102)

Tabelle 22.1 Start Information for the Restart OBs (example of OB 100)

Byte	Variable Name	Data Type	Description	Contents
0	OB100_EV_CLASS	BYTE	Event class	B#16#13
1	OB100_STRTUP	BYTE	Restart request	B#16#8x (see text)
2	OB100_PRIORITY	BYTE	Priority class	Default value 27
3	OB100_OB_NUMBR	BYTE	OB number	100, 101 or 102
4	OB100_RESERVED_1	BYTE	Spare	-
5	OB100_RESERVED_2	BYTE	Spare	-
6..7	OB100_STOP	WORD	Number of the stop event	(see Instruction Manual)
8..11	OB100_STRT_INFO	DWORD	Additional information on the current restart	(see Instruction Manual)
12..19	OB100_DATE_TIME	DT	Occurrence of event	Call time of OB

The number of the stop event and the additional information define the restart more precisely (tells you, for example, whether a manual warm restart was initiated via the mode selector). With this information, you can develop an appropriate event-related restart routine.

Note that no asynchronous system blocks can be executed in the startup program of an S7-300 CPU. You can set and reset outputs in the process image in the startup program, but transmission to the output modules only takes place when transferring to RUN mode.

22.2 Power-Up

22.2.1 STOP Mode

The CPU goes to STOP in the following instances

- ▷ When the CPU is switched on
- ▷ When the mode selector is set from RUN to STOP
- ▷ When an "unrecoverable" error occurs during program scanning
- ▷ When system function SFC 46 STP is executed
- ▷ When requested by a communication function (stop request from the programming device or via communication function blocks from another CPU)

The CPU enters the reason for the STOP in the diagnostic buffer. In this mode, you can also read the CPU information with a programming device in order to localize the problem.

In STOP mode, the user program is not scanned. The CPU retrieves the settings – either the values which you entered in the Hardware Configuration data when you parameterized the CPU or the defaults – and sets the modules to the specified initial state.

In STOP mode, the CPU can receive global data via GD communication and carry out passive unilateral communication functions. The real-time clock keeps running.

You can parameterize the CPU in STOP mode, for instance you can also set the MPI address, transfer or modify the user program, and execute a CPU memory reset.

22.2.2 Memory Reset

A memory reset sets the CPU to the "initial state". You can initiate a memory reset with a programming device only in STOP mode or with the mode selector: hold the switch in the MRES MRES position for at least 3 seconds then release, and after a maximum of 3 seconds hold it the MRES position again for at least 3 seconds.

The CPU erases the entire user program both in work memory and in RAM load memory. System memory (for instance bit memory, timers and counters) is also erased, regardless of retentivity settings. With a micro memory card, the contents of the load memory are retained following an overall reset.

The CPU sets the parameters for all modules, including its own, to their default values. The MPI parameters are an exception. They are not changed so that a CPU whose memory has been reset can still be addressed on the MPI bus. A memory reset also does not affect the diagnostic buffer, the real-time clock, or the run-time meters.

If a micro memory card or a memory card with Flash EPROM is inserted, the CPU copies the user program from the memory card to work memory. The CPU also copies any configuration data it finds on the memory card.

22.2.3 Restoration of Delivery State

In the case of newer CPUs, you can restore the factory settings with "Reset to factory settings". Proceed as follows:

▷ Switch the power supply off, and remove the memory card or micro memory card.

▷ Hold the mode selector in the MRES position, and switch the power supply on again.

▷ When the SF (S7-300) or INTF (S7-400), FRCE, RUN and STOP LEDs flash slowly, release the mode selector, return it to MRES again within 3 s, and hold in this position.

▷ Wait until only the SF or INTF LED flashes. During this time (approx. 5 s), you can abort the reset process by releasing the mode selector.

▷ When the SF or INTF LED lights up continuously, release the mode selector.

The CPU starts up without backup, and all LEDs light up. It carries out an overall reset and subsequently sets the MPI address to 2 and the MPI baud rate to 187.5 Kbit/s. In addition to the overall reset, the real-time clock is set to the starting date, and the runtime meter and the diagnostics buffer are deleted. The CPU then enters the event “Reset to factory settings” into the diagnostics buffer, and enters the STOP state.

22.2.4 Retentivity

A memory area is retentive when its contents are retained even when the mains power is switched off as well as on a transition from STOP to RUN following power-up. With the current S7-300 CPUs, the retentivity is achieved using a micro memory card. With the S7-400 CPUs, a battery backup is a prerequisite for retentivity.

Retentive memory areas may be those for bit memories, timers, counters and also data areas. The number of data in which areas can be made retentive depends on the CPU. You can specify the number of retentive memory bytes, timers and counters via the “Retentivity” tab when you parameterize the CPU.

The contents of data blocks in the work memory can also be retentive. The retentive area available is specific to the CPU. You define the retentivity of a data block using the block property *Non-Retain* (see Chapter 3.2.3 “Block Properties”).

In the case of the S7-300 with micro memory card, the bit memories, timers and counters set as retentive as well as the user program and user data are saved on the micro memory card where they are retentive even without a battery backup. When a warm restart is carried out, the non-retentive bit memories, timers and counters are deleted. The contents of the data blocks declared as “non-retentive” are initialized during a warm restart (loaded with the initial values from the load memory) or set to zero if a load memory object is not present.

With S7-400, a battery backup is required for retentivity. A cold restart deletes all address areas, and loads the user program and the (configured) user data from the load memory into the work memory. With a hot restart or warm restart, the values of the bit memories, timers and counters set as retentive are retained; the user program and the user data are not changed.

22.2.5 Restart Parameterization

On the “Startup” tab of the CPUs, you can affect a restart with the following settings:

▷ Restart when the set configuration is not the same as the actual configuration
A restart is executed even if the parameterized hardware configuration does not agree with the actual configuration. Exception:

the configured PROFIBUS DP interface modules must always be present and ready for operation.

- Reset outputs with hot restart
  The S7-400 CPUs delete all process output images and all peripheral outputs with a hot restart.

- Disable hot restart at manual restart
  Manual hot restart not permissible through manual operation or communications request.

- Restart following POWER UP
  Definition of the type of restart following power up

- Monitoring time for ready signal of the modules
  If the monitoring time for a module is exceeded, it is considered as non-existent. The CPU response is then determined by the setting "Startup with preset configuration not equal to actual configuration". The result is entered in the diagnostics buffer. This timeout is important for switching on the power on expansion racks or distributed I/Os.

- Monitoring time for transferring the parameters to the modules
  When the monitoring time has elapsed, it is considered as non-existent. The CPU response is then determined by the setting "Startup with preset configuration not equal to actual configuration". The event is entered in the diagnostics buffer. (In the event of this error, you can only parameterize the CPU with a higher monitoring time – without memory reset – if you transfer the system data of an "empty" project in which the new value of the monitoring time is entered, so that the module parameterization is completed within the "old" monitoring time.)

- Monitoring time for hot restart
  If the time between power off and power on or the time between STOP and RUN is greater than the monitoring time, no restart is carried out. The specification 0 ms switches the monitor off.

22.3 Types of Restart

22.3.1 START-UP Mode

The CPU executes a restart in the following cases

- When the mains power is switched on
- When switching on with the mode selector (key switch: rotate the mode selector from STOP to RUN or RUN-P, or set the toggle switch from STOP to RUN)
- On the request from a communication function (initiated from a programming device or via communication function blocks from another CPU)

A *manual* restart is initiated via the keyswitch or a communication function, an *automatic* restart by switching on the mains power.

The restart routine may be as long as required, and there is no time limit on its execution; the scan cycle monitor is not active.

During the execution of the restart routine, no interrupts will be serviced. Exceptions are errors that are handled as in RUN (call of the relevant error organization blocks).

In the restart routine, the CPU updates the timers, the run-time meters and the real-time clock.

During restart, the output modules are disabled, i.e., output signals cannot be transmitted. The output disable is only revoked at the end of the restart and prior to starting the cyclic program.

A restart routine can be aborted, for instance when the mode selector is actuated or when there is a power failure. The aborted restart routine is then executed from the beginning when the power is switched on. If a cold or warm restart is aborted, it must be executed again. If a hot restart is aborted, all restart types are possible.

Figure 22.2 shows the activities carried out by the S7-400 CPU during a restart.

22.3.2 Cold Restart

On a cold restart, the CPU sets both itself and the modules to the programmed initial state, deletes all data in the system memory (including the

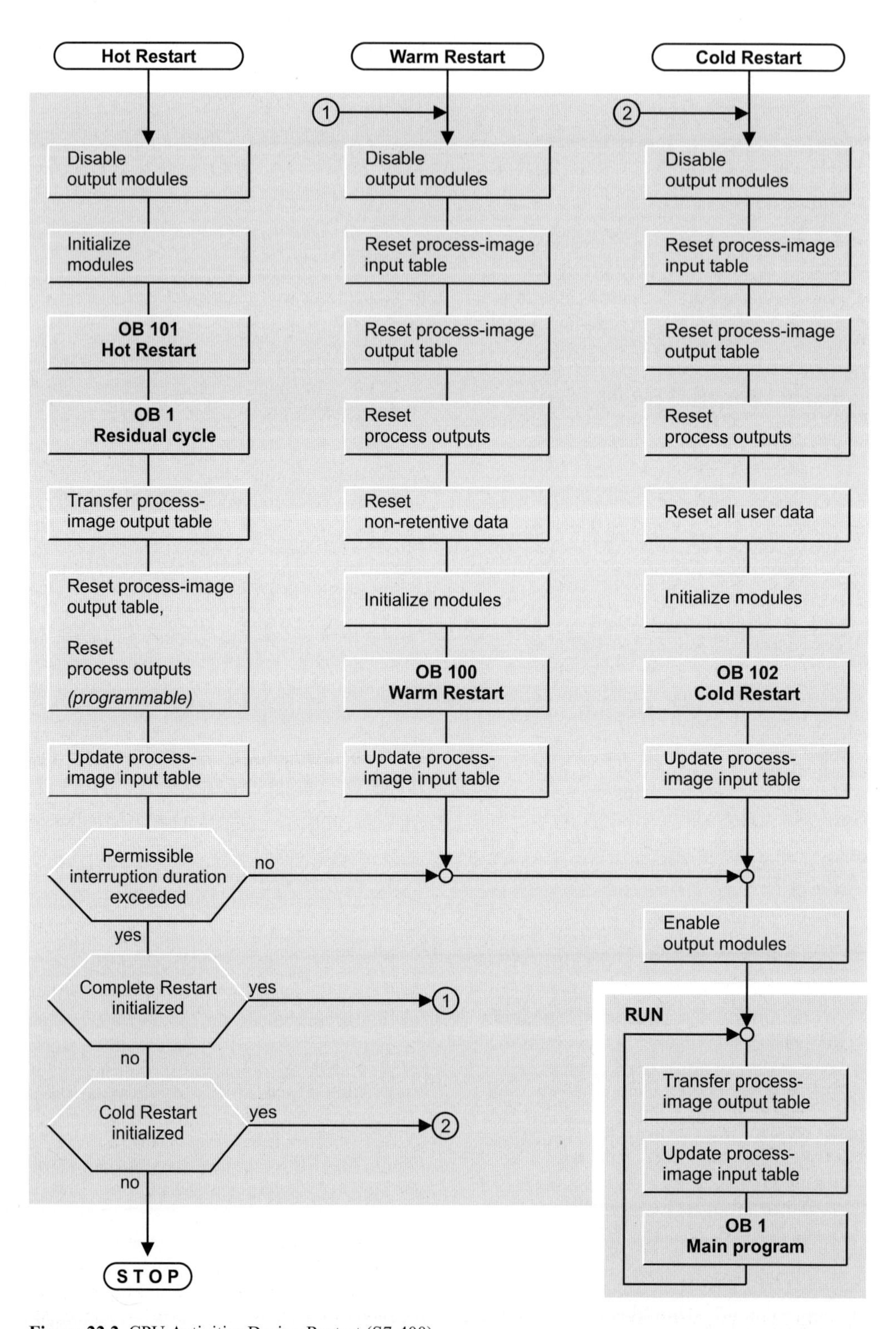

Figure 22.2 CPU Activities During Restart (S7-400)

retentive data), calls OB 102, and then executes the main program in OB 1 from the beginning.

The current program and the current data in work memory are deleted and with them also the data blocks generated by a system function; the program from load memory is reloaded. (In contrast to memory reset, a RAM load memory is not deleted.)

Manual cold restart

With newer CPUs, a cold restart can no longer be triggered manually using the mode selector. With older CPUs, a manual cold restart is triggered using the mode selector if the switch was held in the MRES position for at least 3 seconds on the transition from STOP to RUN or RUN-P.

A manual cold restart can also be triggered by a communication function from a PG or with a system function block (SFB) from another CPU. The mode selector must be in the RUN or RUN-P position.

A manual cold restart can always be initiated unless the CPU requests a memory reset.

Automatic cold restart

An automatic cold restart is initiated by switching on the mains power. The cold restart is executed if

- ▷ the CPU was not at STOP when the power was switched off
- ▷ the mode selector is at RUN or RUN-P
- ▷ the CPU was interrupted by a power outage while executing a cold restart
- ▷ "Cold restart" is parameterized as "Restart on POWER UP"

When operated without a backup battery, the CPU executes an automatic non-retentive warm restart. The CPU starts the memory reset automatically, then copies the user program from the memory card to work memory. The memory card must be a Flash EPROM.

22.3.3 Warm Restart

On a warm restart, the CPU sets both itself and the modules to the programmed initial state, erases the non-retentive data in the system memory, calls OB 100, and then executes the main program in OB 1 from the beginning.

The current program and the data set as retentive in work memory are retained, as are the data blocks created by SFC.

Manual warm restart

A manual warm restart is initiated in the following instances

- ▷ Via the mode selector on the CPU on a transition from STOP to RUN or RUN-P (on S7-400 CPUs with restart type switch, this is in the CRST position)
- ▷ Via a communication function from a PG or with a system function block (SFB) from another CPU; the mode selector must be in the RUN or RUN-P position.

A manual warm restart can always be initiated unless the CPU requests a memory reset.

Automatic warm restart

An automatic warm restart is initiated by switching on the mains power. The restart is executed if

- ▷ the CPU was not at STOP when the power was switched off
- ▷ the mode selector is at RUN or RUN-P
- ▷ the CPU was interrupted by a power outage while executing a warm restart
- ▷ "Warm restart" is parameterized as "Restart on POWER UP"

If there is a restart type switch, it remains without effect in the case of automatic warm restart.

If the CPU contains a micro memory card, it responds exactly like a CPU with backup battery. When operated without a micro memory card and without a backup battery, the CPU executes an automatic non-retentive warm restart. The CPU starts the memory reset automatically, then copies the user program from the memory card to work memory. The memory card must be a flash EPROM.

22.3.4 Hot Restart

A hot restart is possible only on an S7-400.

On a STOP or power outage, the CPU saves all interrupts as well as the internal CPU registers that are important to the processing of the user program. On a hot restart, it can therefore resume at the location in the program at which the interruption occurred. This may be the main program, or it may be an interrupt or error handling routine. All ("old") interrupts are saved and will be serviced.

The so-called "residual cycle", which extends from the point at which the CPU resumes the program following a hot restart to the end of the main program, counts as part of the restart. No (new) interrupts are serviced. The output modules are disabled, and are in their initial state.

A hot restart is permitted only when there have been no changes in the user program while the CPU was at STOP, such as modification of a block.

By parameterizing the CPU accordingly, you can specify how long the interruption may be for the CPU to still be able to execute a hot restart (from 100 milliseconds to 1 hour). If the interruption is longer, only a cold or warm restart is allowed. The length of the interruption is the amount of time between exiting of the RUN mode (STOP or power-down) and reentry into the RUN mode (following execution of OB 101 and the residual cycle).

Manual hot restart

A manual hot restart is initiated

▷ If the mode selector was at RUN or RUN-P when the CPU was switched on By moving the mode selector from STOP to RUN or RUN-P when the restart switch is at WRST (only on CPUs with restart type switch)

▷ Via a communication function from a PG or with a system function block (SFB) from another CPU; the mode selector must be at RUN or RUN-P.

A manual hot restart is possible only when the hot restart disable was revoked in the "Restart" tab when the CPU was parameterized. The cause of the STOP must have been a manual activity, either via the mode selector or through a communication function; only then can a manual hot restart be executed while the CPU is at STOP.

Automatic hot restart

An automatic hot restart is initiated by switching on the mains power. The CPU executes an automatic hot restart only in the following instances,

▷ If it was not at STOP when switched off

▷ If the mode selector was at RUN or RUN-P when the CPU was switched on

▷ "Hot restart" is parameterized as "Restart following POWER UP"

▷ If the backup battery is inserted and in working order

The position of the restart switch is irrelevant to an automatic hot restart.

22.4 Ascertaining a Module Address

Signal modules – or to be more exact, the user data on input/output modules – are addressed in two manners: you use the *logical address* in the user program in order to address the inputs and outputs. This corresponds to the absolute address, and can be made easier to read by using symbols. The smallest logical address is the base address or the module starting address. The CPU addresses the modules using the *geographical address*. You need the geographical address if you wish to ascertain the module slot. The same applies to the user data on stations of the distributed I/O.

With the following system blocks you can ascertain the geographical address from the logical address and vice versa:

▷ SFC 70 GEO_LOG
Ascertain the logical base address

▷ SFC 5 GADR_LGC
Ascertain the logical address of a module channel

- SFC 50 RD_LGADR
  Ascertain all logical addresses of a module
- SFC 71 LOG_GEO
  Ascertain the geographical address
- SFC 49 LGC_GADR
  Ascertain the slot address of a module

Table 22.2 shows the parameters for these SFCs.

The SFCs 5 GADR_LGC, 49 LGC_GADR and 50 RD_LGADR have IOID and LADDR as common parameters for the logical address (= address in the I/O area). IOID is either B#16#54, which stands for the peripheral inputs (PIs) or B#16#55, which stands for the peripheral outputs (PQs). LADDR contains an I/O address in the PI or PQ area which corresponds to the specified channel. If the channel is 0, it is the module starting address.

With the SFCs 70 GEO_LOG and 71 LOG_GEO, the logical address is only in the LADDR parameter. Bit 15 is used to distinguish whether the address is assigned to an input (= 0) or an output (= 1).

The hardware configuration data must specify an allocation between logical address (module starting address) and slot address (location of the module in a rack or a station for distributed I/O) for the addresses ascertained with these system blocks.

SFC 70 LOG_GEO
Ascertain the logical base address

System function SFC 70 LOG_GEO returns the logical base address of a module or station. The value of the MASTER parameter indicates whether the station or module is inserted in a rack (central design) or whether the station is operated in a PROFIBUS or PROFINET system. Specify the slot number in the rack or station in the SLOT parameter, and the number of the submodule in the SUBSLOT parameter. The LADDR parameter then delivers the base address of the submodule. SUBSLOT = 0 delivers the diagnostics address of the module or station.

SFC 70 GEO_LOG replaces the SFC 5 GADR_LGC, and can also be used in association with PROFINET IO.

SFC 5 GADR_LGC
Ascertain the logical address of a module channel

System function SFC 5 GADR_LGC returns the logical address of a channel when you specify the slot address ("geographical" address). Enter the DP master system ID in the SUBNETID parameter if the module belongs to the distributed I/O or B#16#00 if the module is plugged into a controller rack or expansion rack. The RACK parameter specifies the number of the rack or, in the case of distributed I/O, the number of the station. If the module has no submodule slot, enter B#16#00 in the SUBSLOT parameter. SUBADDR contains the address offset in the module's user data (W#16#0000, for example, stands for the module starting address)

SFC 71 GEO_ LOG
Ascertain the geographical address

SFC 71 GEO_LOG returns the geographical address of a module or station when you specify the logical base address for it. The value of the AREA parameter indicates the system in which the module is used (Table 22.3).

SFC 71 GEO_LOG replaces the SFC 49 LGC_GADR and can also be used in association with PROFINET IO.

SFC 49 LGC_GADR
Ascertain the slot address of a module

SFC 49 LGC_GADR returns the slot address of a module when you specify an arbitrary logical module address. Subtracting the address offset (parameter SUBADDR) from the specified user data address gives you the module starting address. The value in the AREA parameter specifies the system in which the module is operated (Table 22.3).

SFC 50 RD_LGADR
Ascertain all logical addresses for a module

SFC 50 RD_LGADR returns all logical addresses for a module when you specify an arbitrary address from the user data area.

Use the PEADDR and PAADDR parameters to define an area of WORD components (a word-based ANY pointer, for example P#DBzDBXy.x WORD nnn).

Table 22.2 Parameters of the System Blocks Used to Ascertain the Module Address

SFC	Parameter	Declaration	Data Type	Contents, Description
5	SUBNETID	INPUT	BYTE	Area identifier
	RACK	INPUT	WORD	Number of the rack
	SLOT	INPUT	WORD	Number of the slot
	SUBSLOT	INPUT	BYTE	Number of the submodule
	SUBADDR	INPUT	WORD	Offset in the module's user data address area
	RET_VAL	RETURN	INT	Error information
	IOID	OUTPUT	BYTE	Area identifier
	LADDR	OUTPUT	WORD	Logical address of the channel
50	IOID	INPUT	BYTE	Area identifier
	LADDR	INPUT	WORD	A logical module address
	RET_VAL	RETURN	INT	Error information
	PEADDR	OUTPUT	ANY	WORD field for the PI addresses
	PECOUNT	OUTPUT	INT	Number of PI addresses returned
	PAADDR	OUTPUT	ANY	WORD field for the PQ addresses
	PACOUNT	OUTPUT	INT	Number of PQ addresses returned
49	IOID	INPUT	BYTE	Area identifier
	LADDR	INPUT	WORD	A logical module address
	RET_VAL	RETURN	INT	Error information
	AREA	OUTPUT	BYTE	Area identifier
	RACK	OUTPUT	WORD	Number of the rack
	SLOT	OUTPUT	WORD	Number of the slot
	SUBADDR	OUTPUT	WORD	Offset in the module's user data address area
70	MASTER	INPUT	INT	Master system ID 0 = central I/O 1 to 31 = PROFIBUS DP 100 to 115 = PROFINET IO
	STATION	INPUT	INT	Station number, with central I/O: number of the rack
	SLOT	INPUT	INT	Number of the slot
	SUBSLOT	INPUT	INT	Number of the submodule
	RET_VAL	RETURN	INT	Error information
	LADDR	OUTPUT	WORD	Base address of station or module
71	LADDR	INPUT	WORD	Base address of station or module
	RET_VAL	RETURN	INT	Error information
	AREA	OUTPUT	INT	Area identifier (see Table 22.3)
	MASTER	OUTPUT	INT	Master system ID 0 = central I/O 1 to 31 = PROFIBUS DP 100 to 115 = PROFINET IO
	STATION	OUTPUT	INT	Station number, with central I/O: number of the rack
	SLOT	OUTPUT	INT	Number of the slot
	SUBSLOT	OUTPUT	INT	Number of the submodule
	OFFSET	OUTPUT	INT	Offset in the module's/submodule's address area

Table 22.3 Description of the Output Parameters of SFCs 49 LGC_GADR and 71 LOG_GEO

AREA	System	Meaning of output parameters with SFC 49 LGC_GADR	Meaning of output parameters with SFC 71 LOG_GEO
0	S7-400	RACK = rack number SLOT = slot number SUBADDR = address offset from base address	MASTER = 0 STATION = rack number SLOT = slot number SUBSLOT = 0 OFFSET = address offset from base address
1	S7-300		
2	Distributed I/O	RACK Low Byte = station number High Byte = DP master system ID SLOT = slot number SUBADDR = address offset from base address	With PROFIBUS DP: MASTER = DP master system ID STATION = station number SLOT = slot number SUBSLOT = 0 OFFSET = address offset from base address With PROFINET IO: MASTER = PROFINET IO system ID STATION = station number SLOT = slot number SUBSLOT = submodule number OFFSET = address offset from base address
3	S5 P area	RACK = rack number SLOT = slot number of adapter casing SUBADDR = address in the S5 area	MASTER = 0 STATION = rack number SLOT = slot number of adapter casing SUBSLOT = 0 OFFSET = address in the S5 area
4	S5 Q area		
5	S5 IM3 area		
6	S5 IM4 area		

SFC 50 then shows the number of entries returned in these areas in the PECOUNT and PACOUNT parameters.

22.5 Parameterizing Modules

22.5.1 General Remarks on Parameterizing Modules

Most S7 modules can be parameterized, that is to say, values may be set on the module which deviate from the default. To specify parameters, open the module in the Hardware Configuration and fill in the tabs in the dialog box. When you transfer the *System Data* object in the *Blocks* container to the PLC, you are also transferring the module parameters.

The CPU transfers the module parameters to the module automatically in the following cases

- ▷ On restart
- ▷ When a module has been plugged into a configured slot (S7-400)
- ▷ Following the "return" of a rack or a distributed I/O station.

Static and dynamic module parameters

The module parameters are divided into static parameters and dynamic parameters. You can set both parameter types offline in the Hardware Configuration. You can also modify the dynamic parameters at runtime by calling a system block. In the restart routine, the parameters set on the modules using the system blocks are overwritten by the parameters set (and stored on the CPU) via the Hardware Configuration).

The parameters for the signal modules are in two data records: the static parameters in data record 0 and the dynamic parameters in data record 1. You can transfer both data records to the module with system function SFC 57 PARM_MOD, data record 0 or 1 with system function SFC 56 WR_DPARM, and only data record 1 with system function SFC 55 WR_PARM. The data records must be in the system data blocks on the CPU.

After parameterization of an S7-400 module, the specified values do not go into force until bit 2 ("Operating mode") in byte 2 of the diagnostic data record has assumed the value "RUN". The diagnostics data record can be read with system function SFC 59 RD_REC or system function block SFB 52 RDREC.

Asynchronous processing of system blocks

Apart from the system function SFC 54 RD_DPARM, the system blocks for module parameterization and data record transfer work asynchronously. Execution of the function extends over several calls, and is triggered by the block parameter REQ = "1". During processing of the job, the BUSY parameter has a signal status "1" and the error information the value W#16#7001 (job being processed). The error information for the system functions is in the RET_VAL parameter, and for the system function blocks in bytes 2 and 3 of the STATUS parameter.

A specific job for a module is specified by the module starting address and the data record number. As long as BUSY = "1", a new call for the same job with REQ = "1" has no effect; the error information is set to W#16#7002.

If an error occurs when triggering a job, this is signaled by the error information, and BUSY remains "0".

If the job has been completed, BUSY has the signal status "0". If the termination is faulty, the error information has the value W#16#0000; with the system function SFC 59 RD_REC, the number of transferred bytes is present in RET_VAL. In the event of an error, the error information contains the error code.

Module and data record addressing

As far as addressing for data transfer is concerned, use the module starting address. With mixed modules having input and output areas, use the lower area starting address. If the input and output areas have the same starting address, use the identifier for an input address. Use the I/O identifier regardless of whether you want to execute a Read or a Write operation.

The module starting address is parameterized either using the IOID and LADDR parameters or – with newer system blocks – using the LADDR parameter on its own. In this case, bit 15 then defines whether it is an input ("0") or output ("1"). (With the system function blocks SFB 52 RDREC and SFB 53 WRREC, this is the ID parameter.)

Specify the data record number using the RECNUM or INDEX parameter.

Use the RECORD parameter with the data type ANY to define an area of BYTE components. This may be a variable of type ARRAY, STRUCT or UDT, or an ANY pointer of type BYTE (for example P#DB*z*DBX*y.x* BYTE *nnn*). If you use a variable, it must be a "complete" variable; individual array or structure components are not permissible.

Permissible data record numbers

Data records with the numbers 1 to 240 are permissible for the system functions for module parameterization. With the system blocks SFC 54 RD_DPARM, SFC 56 WR_DPARM and SFB 81 RD_DPAR, the specified data records must be present in the system data.

The system function SFC 58 WR_REC can process data records in the range from 2 to 240, the SFC 59 RD_REC in the range from 0 to 240. The system function blocks SFB 52 RDREC and SFB 53 WRREC transfer data records with the numbers 0 to 255.

The data records 0 and 1 have a special significance with SIMATIC S7:

- ▷ Data record 0: read diagnostics data (4 bytes) and write static module parameters
- ▷ Data record 1: read diagnostics data (data record 0 and further data) and write dynamic module parameters

A data record can be up to 240 bytes long.

Module parameterization with PROFINET IO

Connection of distributed I/O over PROFINET IO requires an extended quantity framework for the module parameterization compared to PROFIBUS DP, and there are new system blocks for this. These new system blocks can replace the previous ones. Figure 22.3 provides an overview of the system blocks for module parameterization.

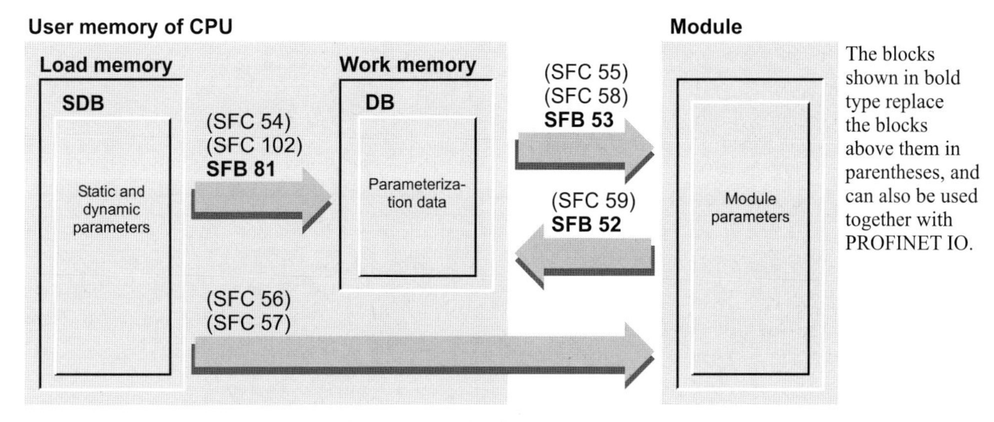

Figure 22.3 System Blocks for Module Parameterization

22.5.2 System Blocks for Module Parameterization

The following system blocks are available for parameterizing modules:

- ▷ SFB 81 RD_DPAR
  Read predefined parameters
- ▷ SFC 54 RD_DPARM
  Read predefined parameters
- ▷ SFC 55 WR_PARM
  Write dynamic parameters
- ▷ SFC 56 WR_DPARM
  Write predefined parameters
- ▷ SFC 57 PARM_MOD
  Parameterize module
- ▷ SFC 102 RD_DPARA
  Read predefined parameters

The parameters for these system functions are listed in Table 22.4.

SFB 81 RD_DPAR
Reading predefined parameters

System function block SFB 81 RD_DPAR transfers the data record with the number specified in the INDEX parameter from the corresponding SDB system data block to the target area specified in the RECORD parameter.

The transfer is asynchronous and can be distributed over several programs cycles; the BUSY parameter has a signal status "1" during the transfer. Following successful transfer, the VALID parameter has a signal status "1", and the number of data bytes transferred is present in the LEN parameter.

The read data record can then be e.g. evaluated or modified and written to the module with SFB 53 WRREC.

The SFB 81 RD_DPAR replaces the SFC 102 RD_DPARA and the SFC 54 RD_PARM.

SFC 102 RD_DPARA
Reading predefined parameters

System function SFC 102 RD_DPARA transfers the data record with the number specified in the RECNUM parameter from the relevant SDB system data block to the target area specified in the RECORD parameter.

The transmission is carried out asynchronously and may be distributed over several program cycles; the BUSY parameter is "1" during the transfer.

The SFC 102 replaces the synchronous SFC 54 RD_DPARM.

SFC 54 RD_DPARM
Reading predefined parameters

System function SFC 54 RD_DPARM transfers the data record with the number specified in the RECNUM parameter from the corresponding SDB system data block to the target area specified in the RECORD parameter.

The transfer is synchronous; the system function is executed until the data record has been

Table 22.4 Parameters of the System Blocks for Module Parameterization

Present in SFC				Parameter	Declaration	Data Type	Contents, Description
-	55	56	57	REQ	INPUT	BOOL	"1" = Write request
54	55	56	57	IOID	INPUT	BYTE	B#16#54 = Peripheral inputs (PIs) B#16#55 = Peripheral outputs (PQ)
54	55	56	57	LADDR	INPUT	WORD	Module starting address
54	55	56	-	RECNUM	INPUT	BYTE	Data record number
-	55	-	-	RECORD	INPUT	ANY	Source area for the data record
54	55	56	57	RET_VAL	RETURN	INT	Error information
-	55	56	57	BUSY	OUTPUT	BOOL	Transfer still in progress if "1"
54	-	-	-	RECORD	OUTPUT	ANY	Target area for data record

Present in		Parameter	Declaration	Data Type	Contents, Description
SFC 102	SFB 81	REQ	INPUT	BOOL	"1" = Write request
SFC 102	SFB 81	LADDR	INPUT	WORD	Module starting address
SFC 102	-	RECNUM	INPUT	BYTE	Data record number
-	SFB 81	INDEX	INPUT	INT	Data record number
SFC 102	-	RET_VAL	RETURN	INT	Error information
-	SFB 81	VALID	OUTPUT	BOOL	New data record received and is valid
SFC 102	SFB 81	BUSY	OUTPUT	BOOL	Transfer still in progress if "1"
-	SFB 81	ERROR	OUTPUT	BOOL	Error has occurred if "1"
-	SFB 81	STATUS	OUTPUT	DWORD	Call identifier or error information (bytes 2 and 3)
-	SFB 81	LEN	OUTPUT	INT	Length of read data
SFC 102	-	RECORD	OUTPUT	ANY	Target area for data record
-	SFB 81	RECORD	IN_OUT	ANY	Target area for data record

transferred. Since the load memory is read, the relatively long processing time with comprehensive data records may be disturbing, depending on the application. In this case, use the SFB 81 RD_DPAR or SFC 102 RD_DPARA which execute this function asynchronously.

The read data record can then be e.g. evaluated or modified and written to the module with SFB 53 WRREC or SFC 58 WR_REC.

SFC 55 WR_PARM
Writing dynamic parameters

System function SFC 55 WR_PARM transfers the data record addressed by RECORD to the module specified by the IOID and LADDR parameters. Specify the number of the data record in the RECNUM parameter. The data record must only contain the dynamic module parameters, and must not be data record 0. If the module parameters are present in the associated system data block SDB, they must not be identified as static.

When the job is initiated, the SFC reads the entire data record; the transfer may be distributed over several program scan cycles. The BUSY parameter is "1" during the transfer.

SFC 56 WR_DPARM
Writing predefined parameters

System function SFC 56 WR_DPARM transfers the data record with the number specified in the RECNUM parameter from the relevant SDB system data block to the module identified by the IOID and LADDR parameters.

The transfer may be distributed over several program scan cycles; the BUSY parameter is "1" during the transfer.

SFC 57 PARM_MOD
Parameterizing a module

System function SFC 57 PARM_MOD transfers all the data records programmed when the module was parameterized via the Hardware Configuration.

The transfer may be distributed over several program scan cycles; the BUSY parameter is "1" during the transfer.

22.5.3 Blocks for Data Record Transfer

The following system blocks are available for transferring data records:

▷ SFB 52 RDREC
Read data record

▷ SFC 59 RD_REC
Read data record

▷ SFB 53 WRREC
Write data record

▷ SFC 58 WR_REC
Write data record

The parameters of the listed system functions are described in Table 22.5, and those of the system function blocks in Table 22.6.

With a S7-300 CPU, you can process up to four write jobs and four read jobs simultaneously per DP line. With an S7-400 CPU, up to eight write jobs and eight read jobs can be simultaneously active per DP line. A maximum total of 32 write jobs and 32 read jobs may be processed simultaneously on external DP lines.

SFB 52 RDREC
Reading a data record

When the REQ parameter is "1", SFB 52 RDREC reads the data record INDEX from the module and places it in target area RECORD. The target area must be longer than or at least as long as the data record. Use the MLEN parameter to specify how many bytes you wish to read.

The transfer may be distributed over several program cycles; the BUSY parameter is "1" during the transfer.

A "1" at the VALID parameter signals that the data record has been read without errors. The LEN parameter then indicates the number of transferred bytes.

In the event of an error, ERROR is set to "1". The STATUS parameter then contains the error information.

The system function block SFB 52 RDREC contains the functionality of the system function SFC 59 RD_REC, and can replace the latter.

Table 22.5 Parameters for System Functions Used for Data Transfer

Present in		Parameter	Declaration	Data Type	Contents, Description
SFC 58	SFC 59	REQ	INPUT	BOOL	"1" = write request
SFC 58	SFC 59	IOID	INPUT	BYTE	B#16#54 = input module B#16#55 = output module
SFC 58	SFC 59	LADDR	INPUT	WORD	Module starting address
SFC 58	SFC 59	RECNUM	INPUT	BYTE	Data record number
SFC 58	-	RECORD	INPUT	ANY	Data record
SFC 58	SFC 59	RET_VAL	RETURN	INT	Error information
SFC 58	SFC 59	BUSY	OUTPUT	BOOL	Transfer still in progress if "1"
-	SFC 59	RECORD	OUTPUT	ANY	Data record

Table 22.6 Parameters of the System Function Blocks for Data Record Transfer

Present in		Parameter	Declaration	Data Type	Contents, Description
SFB 52	SFB 53	REQ	INPUT	BOOL	With "1", request to write
SFB 52	SFB 53	ID	INPUT	DWORD	Module starting address Bit 15 = "0": input address Bit 15 = "1": output address
SFB 52	SFB 53	INDEX	INPUT	INT	Data record number
SFB 52	-	MLEN	INPUT	INT	Maximum number of bytes of data record to be read
-	SFB 53	LEN	INPUT	INT	Maximum number of bytes of data record to be transferred
SFB 52	-	VALID	OUTPUT	BOOL	"1" = new data record was received and is valid
-	SFB 53	DONE	OUTPUT	BOOL	Data record was transferred
SFB 52	SFB 53	BUSY	OUTPUT	BOOL	With "1", transfer still in progress
SFB 52	SFB 53	ERROR	OUTPUT	BOOL	With "1", an error occurred
SFB 52	SFB 53	STATUS	OUTPUT	DWORD	Status codes
SFB 52	-	LEN	OUTPUT	INT	Number of bytes of read data
SFB 52	SFB 53	RECORD	OUTPUT	ANY	Data record

SFC 59 RD_REC
Reading a data record

When the REQ parameter is "1", SFC 59 RD_REC reads the data record addressed by the RECNUM parameter from the module and places it in target area RECORD. The target area must be longer than or at least as long as the data record. If the transfer is completed without error, the RET_VAL parameter contains the number of bytes transferred.

The transfer may be distributed over several program scan cycles; the BUSY parameter is "1" during the transfer.

S7-300s delivered prior to February 1997: the SFC reads as much data from the specified data record as the target area can accommodate. The size of the target area may not exceed that of the data record.

SFB 53 WRREC
Writing a data record

When the REQ parameter is "1", SFB 53 WRREC writes the data record INDEX from the source area RECORD to the module. Use the LEN parameter to specify how many bytes you wish to write.

The transfer may be distributed over several program cycles; the BUSY parameter is "1" during the transfer.

A "1" at the DONE parameter signals that the data record has been written without errors. In the event of an error, ERROR is set to "1". The STATUS parameter then contains the error information.

The system function block SFB 53 WRREC contains the functionality of the system function SFC 58 WR_REC, and can replace the latter.

SFC 58 WR_REC
Writing a data record

SFC 58 WR_REC transfers the data record addressed by the RECORD parameter and the number RECNUM to the module defined by the IOID and LADDR parameters. A "1" in the REQ parameter starts the transfer. When the job is initiated, the SFC reads the complete data record.

The transfer may be distributed over several program cycles; the BUSY parameter is "1" during the transfer.

23 Error Handling

The CPU reports errors or faults detected by the modules or by the CPU itself in different ways:

▷ Errors in arithmetic operations (overflow, invalid REAL number) by setting status bits (status bit OV, for example, for a numerical overflow)

▷ Errors detected while executing the user program (synchronous errors) by calling organization blocks OB 121 and OB 122

▷ Errors in the programmable controller which do not relate to program scanning (asynchronous errors) by calling organization blocks OB 80 to OB 87

The CPU signals the occurrence of an error or fault, and in some cases the cause, by setting error LEDs on the front panel. In the case of unrecoverable errors (such as invalid OP code), the CPU goes directly to STOP.

With the CPU in STOP mode, you can use a programming device and the CPU information functions to read out the contents of the block stack (B stack), the interrupt stack (I stack) and the local data stack (L stack) and then draw conclusions as to the cause of error.

The system diagnostics can detect errors/faults on the modules, and enters these errors in a diagnostic buffer. Information on CPU mode transitions (such as the reasons for a STOP) are also placed in the diagnostic buffer.

The contents of this buffer are retained on STOP, on a memory reset, and on power failure, and can be read out following power recovery and execution of a start-up routine using a programming device.

On the new CPUs, you can use CPU parameterization to set the number of entries the diagnostics buffer is to hold.

23.1 Synchronous Errors

The CPU's operating system generates a synchronous error when an error occurs in immediate conjunction with program scanning. A distinction is made between two error types:

A **programming error** is the case if program execution is faulty. Such errors include BCD conversion errors, errors with indirect addressing, addressing of missing timers, counters or blocks. In the event of a programming error, the organization block OB 121 is called.

An **I/O access error** is present if an attempt is made to access a faulty or non-existent module or an I/O address not known to the CPU. The operating system responds differently depending on the type of access:

▷ The I/O access is carried out by the user program. In this case, the I/O access error organization block OB 122 is called.

▷ The I/O access error occurs during automatic updating of a (sub)process image. The default response is that there is no entry in the diagnostics buffer and no OB is called in the case of S7-300 CPUs; S7-400 CPUs enter each access error into the diagnostics buffer and start the organization block OB 85. The response to a access error can be parameterized with newer CPUs (see "Program execution errors OB 85" in Chapter 23.3 "Asynchronous Errors").

▷ The I/O access error occurs if a partial process image is updated by a system function. In this case, the error and the address of the first byte signaling the error are returned by their parameters (system functions SFC 26 UPDAT_PI, SFC 27 UPDAT_PO, SFC 126 SYNC_PI and SFC 127 SYNC_PO).

If the corresponding organization block OB 121 or OB 122 is not present when a synchronous error occurs, the CPU enters the STOP status.

Table 23.1 shows the start information for both synchronous error organization blocks.

Table 23.1 Start Information for the Synchronous Error OBs 121 and 122

Byte	Variable Name	Data Type	Description, Contents
0	OB12x_EV_CLASS	BYTE	B#16#25 = Call programming error OB 121 B#16#29 = Call access error OB 122
1	OB12x_SW_FLT	BYTE	Error code (see Chapter 23.2.1 "Error Filters")
2	OB12x_PRIORITY	BYTE	Priority class in which the error occurred
3	OB12x_OB_NUMBR	BYTE	OB number (B#16#79 or B#16#80)
4	OB12x_BLK_TYPE	BYTE	Type of block interrupted (S7-400 only) OB: B#16#88, DB: B#16#8A, FB: B#16#8E, FC: B#16#8C
5	OB121_RESERVED_1 OB122_MEM_AREA	BYTE	Byte assignments (B#16#xy): 7... (x) ... 4: 1 Bit access, 2 Byte access, 3 Word access, 4 Doubleword access 3 ... (y) ... 0: 0 I/O area PI or PQ, 1 Process-image input table I, 2 Process-image output table Q
6..7	OB121_FLT_REG OB122_MEM_ADDR	WORD	OB 121: Error source: ▷ Errored address (at read/write access) ▷ Errored area (in the case of area error) ▷ Incorrect number of the block, timer/counter function OB 122: Address at which the error occurred
8..9	OB12x_BLK_NUM	WORD	Number of the block in which the error occurred (S7-400 only)
10..11	OB12x_PRG_ADDR	WORD	Error address in the block that caused the error (S7-400 only)
12..19	OB12x_DATE_TIME	DT	Time at which programming error was detected

The S7-400 CPUs differentiate between two types of I/O access errors: access to a non-existent module, and faulty access to a module entered as being present (acknowledgment delay QVZ). If a module fails during operation, this module is entered as "non-existent" after an access time of approx. 150 ms, so that the I/O access error is signaled with each further access. The CPU also signals access error if a non-existent module is accessed, either directly via the I/O area or indirectly via the process image.

If an I/O access error occurs during a write access to the I/O outputs, an S7-400 CPU updates the process output image, an S7-300 CPU does not.

A synchronous error OB has the same priority (class) as the block in which the error was caused. The values present in the block causing the error at the time of the abort are present in the synchronous error OB in the accumulators and address registers. The data block registers are deleted; the condition code word has an undefined assignment.

Note that when a synchronous error OB is called, its 20 bytes of start information are also pushed onto the L stack for the priority class that caused the error, as are the other temporary local data for the synchronous error OB and for all blocks called in this OB. The area reserved for the temporary local data must be designed for this in every affected priority class (program execution level) (fixed definition with S7-300 CPUs, adjustable during parameterization of the CPU in the "Memory" tab for S7-400 CPUs).

This applies similarly to the block nesting depth. The nesting depth permissible for a CPU depending on the priority class is the total of the nesting depth of the "normal" processing and

Table 23.2 SFC Parameters for Synchronous Error Handling

SFC	Parameter	Declaration	Data type	Assignment, description
36	PRGFLT_SET_MASK	INPUT	DWORD	New (additional) programming error mask
	ACCFLT_SET_MASK	INPUT	DWORD	New (additional) access error mask
	RET_VAL	RETURN	INT	W#16#0001 = The new mask overlaps with the existing mask
	PRGFLT_MASKED	OUTPUT	DWORD	Complete programming error mask
	ACCFLT_MASKED	OUTPUT	DWORD	Complete access error mask
37	PRGFLT_RESET_MASK	INPUT	DWORD	Programming error mask to be reset
	ACCFLT_RESET_MASK	INPUT	DWORD	Access error mask for resetting
	RET_VAL	RETURN	INT	W#16#0001 = The new mask contains bits that are not set (in the saved mask)
	PRGFLT_MASKED	OUTPUT	DWORD	Remaining programming error mask
	ACCFLT_MASKED	OUTPUT	DWORD	Remaining access error mask
38	PRGFLT_QUERY	INPUT	DWORD	Programming error mask for scanning
	ACCFLT_QUERY	INPUT	DWORD	Access error mask for scanning
	RET_VAL	RETURN	INT	W#16#0001 = The query mask contains bits that are not set (in the saved mask)
	PRGFLT_CLR	OUTPUT	DWORD	Programming error mask with error messages
	ACCFLT_CLR	OUTPUT	DWORD	Access error mask with error messages

the nesting depth of the synchronous error processing.

In the case of S7-400, another synchronous error OB can be called in an error OB. The block nesting depth for a synchronous error OB is 3 for S7-400 CPUs and 4 for S7-300 CPUs.

You can disable and enable a synchronous error OB call with system functions SFC 36 MSK_FLT, SFC 37 DMSK_FLT and SFC 38 READ_ERR.

23.2 Synchronous Error Handling

The following system functions are provided for handling synchronous errors:

- SFC 36 MSK_FLT
  Mask synchronous errors (disable OB call)
- SFC 37 DMSK_FLT
  Unmask synchronous error (re-enable OB call)
- SFC 38 READ_ERR
  Read error register

Independent of the use of the system functions SFC 36 to SFC 38, the operating system enters the synchronous error event in the diagnostics buffer. The parameters for these system functions are listed in Table 23.2.

23.2.1 Error Filters

The error filters are used to control the system functions for synchronous error handling. In the programming error filter, one bit stands for each programming error detected; in the access error filter, one bit stands for each access error detected. When you define an error filter, you set the bit that stands for the synchronous error you want to mask, unmask or query. The error filters returned by the system functions show a “1” for synchronous errors that are still masked or which have occurred.

The access error filter is shown in Table 23.3; the Error Code column shows the contents of

Table 23.3 Assignment of access error mask

Bit	Error code	Assignment
2	B#16#42	I/O access error when reading S7-300 and CPU 417: The module is not present or does not acknowledge S7-400 (except CPU 417): An existing module does not acknowledge on I/O access (time-out)
3	B#16#43	I/O access error when writing S7-300 and CPU 417: The module is not present or does not acknowledge S7-400 (except CPU 417): An existing module does not acknowledge on I/O access (time-out)

variable OB122_SW_FLT in the start information for OB 122.

The programming error filter is shown in Table 23.4; the Error Code column shows the contents of variable OB121_SW_FLT in the start information for OB 121.

The error filter bits not listed in the tables are not relevant to the handling of synchronous errors.

23.2.2 Masking Synchronous Errors

System function **SFC 36 MSK_FLT** disables synchronous error OB calls via the error filters. A "1" in the error filters indicates the synchronous errors for which the OBs are not to be called (the synchronous errors are "masked"). The masking of synchronous errors in the error filters is in addition to the masking stored in the operating system's memory. SFC 36 returns a function value indicating whether a (stored) masking already exists on at least one bit for the masking specified at the input parameters (W#16#0001).

SFC 36 returns a "1" in the output parameters for all currently masked errors.

If a masked synchronous error event occurs, the respective OB is not called and the error is entered in the error register. The Disable applies to the current priority class (priority level). For example, if you were to disable a synchronous error OB call in the main program, the synchronous error OB would still be called if the error were to occur in an interrupt service routine.

23.2.3 Unmasking Synchronous Errors

System function **SFC 37 DMSK_FLT** enables the synchronous error OB calls via the error filters. You enter a "1" in the filters to indicate the synchronous errors for which the OBs are once again to be called (the synchronous errors are "unmasked"). The entries corresponding to the specified bits are deleted in the error register. SFC 37 returns W#16#0001 as function value if no (stored) masking already exists on at least one bit for the unmasking specified at the input parameters.

SFC 37 returns a "1" in the output parameters for all currently masked errors.

If an unmasked synchronous error occurs, the respective OB is called and the event entered in the error register. The Enable applies to the current priority class (priority level).

23.2.4 Reading the Error tab

System function **SFC 38 READ_ERR** reads the error register. You must enter a "1" in the error filters to indicate the synchronous errors whose entries you want to read. SFC 38 returns W#16#0001 as function value when the selection specified in the input parameters included at least one bit for which no (stored) masking exists.

SFC 38 returns a "1" in the output parameters for the selected errors when these errors occurred, and deletes these errors in the error register when they are queried. The synchronous errors that are reported are those in the current priority class (priority level).

Table 23.4 Programming Error Filter

Bit	Error Code	Contents
1	B#16#21	BCD conversion error (pseudo-tetrad detected during conversion)
2	B#16#22	Area length error on read (address not within area limits)
3	B#16#23	Area length error on write (address not within area limits)
4	B#16#24	Area length error on read (wrong area in area pointer)
5	B#16#25	Area length error on write (wrong area in area pointer)
6	B#16#26	Invalid timer number
7	B#16#27	Invalid counter number
8	B#16#28	Address error on read (bit address <> 0 in conjunction with byte, word or doubleword access and indirect addressing)
9	B#16#29	Address area on write (bit address <> 0 in conjunction with byte, word or doubleword access and indirect addressing)
16	B#16#30	Write error, global data block (write-protected block)
17	B#16#31	Write error, instance data block (write-protected block)
18	B#16#32	Invalid number of a global data block (DB register)
19	B#16#33	Invalid number of an instance data block (DI register)
20	B#16#34	Invalid number of a function (FC)
21	B#16#35	Invalid number of a function block (FB)
26	B#16#3A	Called data block (DB) does not exist
28	B#16#3C	Called function (FC) does not exist
30	B#16#3E	Called function block (FB) does not exist

23.2.5 Entering a Substitute Value

SFC 44 REPL_VAL allows you to enter a substitute value in accumulator 1 from within a synchronous error OB. Use SFC 44 when you can no longer read any values from a module (for instance when a module is defective). When you program SFC 44, OB 122 ("access error") is called every time an attempt is made to access the module in question. When you call SFC 44, you can load a substitute value into the accumulator; the program scan is then resumed with the substitute value. Table 23.5 lists the parameters for SFC 44.

You may call SFC 44 in only one synchronous error OB (OB 121 or OB 122).

Table 23.5 Parameters of SFC 44 REPL_VAL

SFC	Parameter name	Declaration	Data type	Assignment, description
44	VAL	INPUT	DWORD	Substitute value
	RET_VAL	RETURN	INT	Error information

23.3 Asynchronous Errors

Asynchronous errors are errors which can occur independently of the program scan. When an asynchronous error occurs, the operating system calls one of the organization blocks listed below:

OB 80 Timing error

OB 81 Power supply error

OB 82 Diagnostic interrupt

OB 83 Insert/remove module interrupt

OB 84 CPU hardware fault

OB 85 Program execution error

OB 86 Rack failure

OB 87 Communication error

OB 88 Processing abort

The OB 82 call (diagnostic interrupt) is described in detail in Chapter 23.4 "System Diagnostics".

On the S7-400H, there are three additional asynchronous error OBs:

OB 70 I/O redundancy errors

OB 72 CPU redundancy errors

OB 73 Communications redundancy errors

The call of these asynchronous error organization blocks can be disabled and enabled with system functions SFC 39 DIS_IRT and SFC 40 EN_IRT, and delayed and enabled with system functions SFC 41 DIS_AIRT and SFC 42 EN_AIRT.

Timing errors OB 80

The operating system calls organization block OB 80 when one of the following errors occurs:

- ▷ Cycle monitoring time exceeded,
- ▷ OB request error (the requested OB is still being processed or an OB is requested too frequently within a priority class),
- ▷ Time-of-day error interrupt (time-of-day interrupt expired through setting ahead of time or following transition to RUN).

If OB 80 is not present, the CPU switches to STOP in the event of a time error. The CPU also goes to STOP if the OB is called a second time in the same program scan cycle due to a cycle time violation.

Power supply errors OB 81

The operating system calls organization block OB 81 if one of the following errors occurs:

- ▷ At least one backup battery in the central rack or in an expansion unit is empty,
- ▷ No backup voltage in the central rack or an expansion unit,
- ▷ Failure of the 24 V supply in the central rack or in an expansion unit.

OB 81 is called for incoming and outgoing events. If there is no OB 81, the CPU continues functioning when a power supply error occurs.

Insert/remove module interrupt OB 83

The operating system monitors the module configuration once per second. An entry is made in the diagnostic buffer and in the system status list each time a module is inserted or removed in RUN, STOP or START-UP mode.

In addition, the operating system calls the operation block OB 83 in RUN. If OB 83 is not present, the CPU switches to STOP in the event of an insert/remove module interrupt.

As much as a second can pass before the insert/remove module interrupt is generated. As a result, it is possible that an access error or an error relating to the updating of the process image could be reported in the interim between removal of a module and generation of the interrupt.

If a suitable module is inserted into a configured slot, the CPU automatically parameterizes that module, using data records already stored on that CPU. Only then is OB 83 called in order to signal that the connected module is ready for operation.

CPU hardware faults OB 84

The operating system calls organization block OB 84 when an interface error (e.g. MPI network, PROFIBUS DP) occurs or disappears. If there is no OB 84, CPUs with older operating systems go to STOP on a CPU hardware fault.

Program execution errors OB 85

The operating system calls organization block OB 85 when one of the following errors occurs:

▷ Start request for an organization block which has not been loaded

▷ Error occurred while the operating system was accessing a block (for instance no instance data block when a system function block (SFB) was called)

▷ I/O access error while executing (automatic) updating of the process image on the system side

On the S7-400 CPUs, OB 85 is called at every I/O access error (on the system side), i.e. when updating the process image in each cycle. The substitute value or zero is then entered in the relevant byte in the process-image input table at every update.

On the S7-300 CPUs, OB 85 is not called in the event of an I/O access error during automatic updating of the process image. At the first errored access, the substitute value or zero is entered in the relevant byte; it is then no longer updated.

With appropriately equipped CPUs, you can use CPU parameterization to influence the call mode of OB 85 in the event of an I/O access error on the system side:

▷ OB 85 is called every time. The affected input byte is overwritten with the substitute value or with zero each time.

▷ OB 85 is called in the event of the first error with the attribute "incoming". An affected input byte is only overwritten with the substitute value or with zero the first time; following this it is no longer updated. If the error is then corrected, OB 85 is called with the attribute "outgoing"; following this, a corresponding input byte is updated "normally".

▷ OB 85 is not called in the event of an access error. Affected input bytes are overwritten once with the substitute value or zero, and then no longer updated.

If there is no OB 85, the CPU goes to STOP on a program execution error.

Rack failure OB 86

The operating system calls organization block OB 86 if it detects the failure of an expansion unit (power failure, line break, defective IM; not with S7-300), a DP master system, or a distributed I/O station (PROFIBUS DP or PROFINET IO). OB 86 is called for both incoming and leaving errors.

In multiprocessor mode, OB 86 is called in all CPUs if a rack fails.

If there is no OB 86, the CPU goes to STOP if a rack failure occurs.

Communication error OB 87

The operating system calls organization block OB 87 when a communication error occurs. Some examples of communication errors are

▷ Invalid frame identification or frame length detected during global data communication

▷ Sending of diagnostic entries not possible

▷ Clock synchronization error

▷ GD status cannot be entered in a data block

If there is no OB 87, the CPU goes to STOP when a communication error occurs.

Processing abort OB 88

The operating system calls the organization block OB 88 if the processing of a block is aborted in the user program. Possible causes of the abort are:

▷ With a synchronous error, the permissible block nesting depth has been exceeded.

▷ With a block call, the permissible nesting depth has been exceeded.

▷ A fault has occurred when allocating the local data of a block.

If the OB 88 is not present, the CPU goes to STOP if a processing abort occurs. The CPU also goes to STOP if the OB is called in priority class 28.

I/O redundancy error OB 70

The operating system of an H CPU calls organization block OB 70 if a redundancy loss occurs on PROFIBUS DP, e.g. in the event of a bus failure on the active DP master or in the event of a fault in the interface of a DP slave.

If OB 70 does not exist, the CPU continues to operate in the event of an I/O redundancy error.

CPU redundancy error OB 72

The operating system of an H CPU calls organization block OB 72 if one of the following events occurs:

- ▷ Redundancy loss of the CPU
- ▷ Comparison error (e.g. in RAM, in the PIQ)
- ▷ Standby-master changeover
- ▷ Synchronization error
- ▷ Error in a SYNC submodule
- ▷ Update abort

If OB 72 does not exist, the CPU continues to operate in the event of a CPU redundancy error.

Communication redundancy error OB 73

The operating system of a H-CPU calls the organization block OB 73 when a fault-tolerant S7 connection loses redundancy for the first time. As long as at least one fault-tolerant S7 connection signals a loss of redundancy, the OB 73 is not called again if there is an additional loss of redundancy.

If the OB 73 does not exist, the CPU continues despite a communication redundancy error.

23.4 System Diagnostics

23.4.1 Diagnostic Events and Diagnostic Buffer

System diagnostics is the detection, evaluation and reporting of errors occurring in programmable controllers. Examples are errors in the user program, module failures or wirebreaks on signaling modules. These *diagnostic events* may be:

- ▷ Diagnostic interrupts from modules with this capability
- ▷ System errors and CPU mode transitions or
- ▷ User messages via system functions.

Modules with diagnostic capabilities distinguish between programmable and non-programmable diagnostic events. Programmable diagnostic events are reported only when you have set the parameters necessary to enable diagnostics. Non-programmable diagnostic events are always reported, regardless of whether or not diagnostics have been enabled. In the event of a reportable diagnostic event,

- ▷ The fault LED on the CPU goes on
- ▷ The diagnostic event is passed on to the CPU's operating system and
- ▷ A diagnostic interrupt is generated if you have set the parameters enabling such interrupts (by default, diagnostic interrupts are disabled).

All diagnostic events reported to the CPU operating system are entered in a *diagnostic buffer* in the order in which they occurred, and with date and time stamp. The diagnostic buffer is a battery-backed memory area on the CPU which retains its contents even in the event of a memory reset. The diagnostic buffer is a ring buffer whose size depends on the CPU. When the diagnostic buffer is full, the oldest entry is overwritten by the newest.

You can read out the diagnostic buffer with a programming device at any time. In the CPU's *System Diagnostics* parameter block you can specify whether you want expanded diagnostic entries (all OB calls). You may also specify whether the last diagnostic entry made before the CPU goes to STOP should be sent to a specific node on the MPI bus.

23.4.2 Writing User Entries in the Diagnostic Buffer

System function **SFC 52 WR_USMSG** writes an entry in the diagnostic buffer which may be sent to all nodes on the MPI bus. Table 23.6 lists the parameters for SFC 52.

The entry in the diagnostic buffer corresponds in format to that of a system event, for instance the start information for an organization block. Within the permissible boundaries, you may choose your own event ID (EVENTN parameter) and additional information (INFO1 and INFO2 parameters).

The event ID is identical to the first two bytes of the buffer entry (Figure 23.1). Permissible

Table 23.6 Parameters for SFC 52 WR_USMSG

SFC	Parameter name	Declaration	Data type	Assignment, description
52	SEND	INPUT	BOOL	For "1": Sending is enabled
	EVENTN	INPUT	WORD	Event ID
	INFO1	INPUT	ANY	Additional information 1 (one word)
	INFO2	INPUT	ANY	Additional information 2 (one double-word)
	RET_VAL	RETURN	INT	Error information

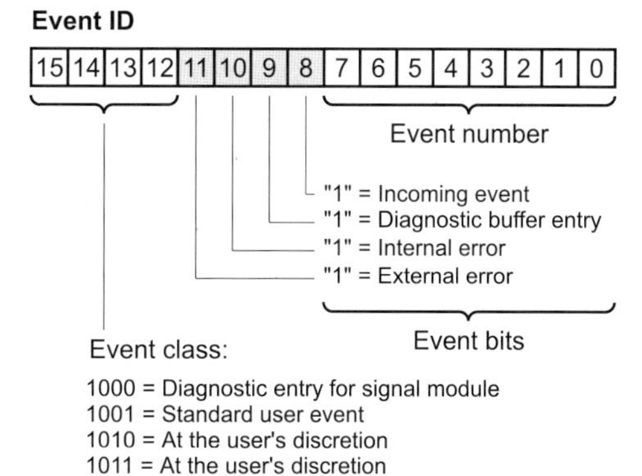

Figure 23.1 Event ID for Diagnostic Buffer Entries

for a user entry are the event classes 8 (diagnostic entries for signal modules), 9 (standard user events), A and B (arbitrary user events).

Additional information (INFO1) corresponds to bytes 7 and 8 of the buffer entry (one word) and additional information 2 (INFO2) to bytes 9 to 12 (one doubleword). The contents of both variables may be of the user's own choice.

Set SEND to "1" to send the diagnostic entry to the relevant node. Even if sending is not possible (because no node is logged in or because the Send buffer is full, for example), the entry is still made in the diagnostic buffer (when bit 9 of the event ID is set).

23.4.3 Evaluating Diagnostic Interrupts

In the event of a diagnostic interrupt that comes and goes, the operating system interrupts execution of the user program and calls the **organization block OB 82**. If OB 82 is not programmed, the CPU switches to STOP on a diagnostic interrupt. You can disable or enable the processing of OB 82 with the system functions SFC 39 DIS_IRT and SFC 40 EN_IRT, and delay or enable it with the system functions SFC 41 DIS_AIRT and SFC 42 EN_AIRT.

PROFIBUS DPV1 slaves can also generate a diagnostic interrupt if the master CPU is in STOP. A diagnostic interrupt in STOP of the CPU is acknowledged but not processed. Calling the organization block OB 82 is also not executed if the CPU goes to RUN.

Table 23.7 shows the startup information of the diagnostics interrupt OB 82. In the first byte of the start information, B#16#39 stands for an incoming diagnostic interrupt and B#16#38 for a leaving diagnostic interrupt. The sixth byte gives the address identifier (B#16#54 stands for an input, B#16#55 for an output); the subsequent INT variable contains the address of the module that generated the diagnostic interrupt. The next four bytes contain the diagnostic information provided by that module.

You can use system function SFC 59 RD_REC (read data record) in OB 82 to obtain detailed error information. The diagnostic information are consistent until OB 82 is exited, that is, they remain "frozen". Exiting of OB 82 acknowledges the diagnostic interrupt on the module.

A module's diagnostic data are in data records DS 0 and DS 1. Data record DS 0 contains four bytes of diagnostic data describing the current status of the module. The contents of these four bytes are identical to the contents of bytes 8 to 11 of the OB 82 start information. Data record DS 1 contains the four bytes from data record DS 0 and, in addition, the module-specific diagnostic data.

When using a CPU with DPV1 capability and a corresponding slave, you can obtain further information on the diagnostics interrupt by

Table 23.7 Start Information of the Organization Block OB 82 (Diagnostics Interrupt)

Byte	Variable Name	Data Type	Contents, Description
0	OB82_EV_CLASS	BYTE	B#16#38 = Outgoing event B#16#39 = Incoming event
1	OB82_FLT_ID	BYTE	Error code (B#16#42)
2	OB82_PRIORITY	BYTE	Priority class for the diagnostics interrupt OB
3	OB82_OB_NUMBR	BYTE	OB number (B#16#52)
4	OB82_RESERVED_1	BYTE	Spare
5	OB82_IO_FLAG	BYTE	I/O ID (B#16#54 = input, B#16#55 = output)
6..7	OB82_MDL_ADDR	WORD	Module starting address of module generating interrupt
8.0	OB82_MDL_DEFECT	BOOL	Module defect
8.1	OB82_INT_FAULT	BOOL	Internal fault
8.2	OB82_EXT_FAULT	BOOL	External fault
8.3	OB82_PNT_INFO	BOOL	Channel fault present
8.4	OB82_EXT-VOLTAGE	BOOL	External supply voltage missing
8.5	OB82_FLD_CONNCTR	BOOL	Front connector missing
8.6	OB82_NO_CONFIG	BOOL	Module not parameterized
8.7	OB82_CONFIG_ERR	BOOL	Incorrect parameters in the module
9	OB82_MDL_TYPE	BYTE	Bits 0 to 3: module class Bit 4: channel information present Bit 5: user information present Bit 6: diagnostics interrupt from proxy Bit 7: spare
10.0	OB82_SUB_MDL_ERR	BOOL	Incorrect or missing user module
10.1	OB82_COMM_FAULT	BOOL	Communications fault
10.2	OB82_MDL_STOP	BOOL	Operating status ("0" = RUN, "1" = STOP)
10.3	OB82_WTCH_DOG_FLT	BOOL	Timeout triggered
10.4	OB82_INT_PS_FLT	BOOL	Internal module voltage failed
10.5	OB82_PRIM_BATT_FLT	BOOL	Battery flat
10.6	OB82_BCKUP_BATT_ FLT	BOOL	Complete backup failed
10.7	OB82_RESERVED_2	BOOL	Spare
11.0	OB82_RACK_FLT	BOOL	Expansion unit failed
11.1	OB82_PROC_FLT	BOOL	Processor failure
11.2	OB82_EPROM_FLT	BOOL	EPROM fault
11.3	OB82_RAM_FLT	BOOL	RAM fault
11.4	OB82_ADU_FLT	BOOL	ADC/DAC fault
11.5	OB82_FUSE_FLT	BOOL	Blown fuse
11.6	OB82_HW_INTR_FLT	BOOL	Hardware interrupt lost
11.7	OB82_RESERVED_3	BOOL	Spare
12..19	OB82_DATE_TIME	DT	Recording time of diagnostics event

means of the system function block SFB 54 RALRM.

23.4.4 Reading the System Status List

The system status list (SZL) describes the current status of the programmable controller. Using information functions, the list can be read but not modified. Since the complete system status list is extremely extensive, reading is carried out in sublists and sublist extracts. Sublists are virtual lists, which means that they are made available by the CPU operating system only on request.

The SZL ID is available to identify a sublist. This contains the module type class to which the list applies, the number of the sublist extract, and the actual SZL sublist number (Figure 23.2). Together with the index, which specifies an object of a sublist, you are provided with the desired information. As standard, the CPU basically provides information on the automation system, but FM and CP modules can also use this service to provide information (see module documentation). You can find the possible system status lists of a CPU in the description of operations.

Reading the header information

With the SZL ID W#16#0Fxx you can read the header information of an SZL sublist, without the associated data record (xx = SZL sublist number). The parameter SZL_HEADER. N_ DR (number of data records) then returns the maximum possible data record number of the sublist extract which the module can deliver with an SZL job. With dynamic sublists, the

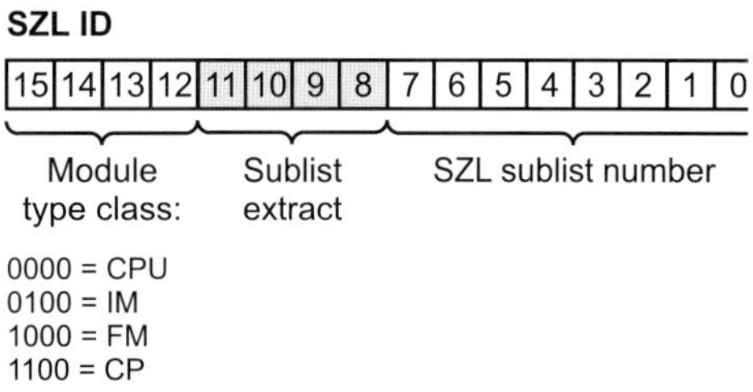

SZL ID

15	14	13	12	11	10	9	8	7	6	5	4	3	2	1	0

Module type class: (bits 15–12)
Sublist extract (bits 11–8)
SZL sublist number (bits 7–0)

0000 = CPU
0100 = IM
1000 = FM
1100 = CP

Figure 23.2 Structure of the SZL_ID

value can be larger than the current number that can be read. The length of a data record is present in SZL_HEADER. LENGTHDR. With this data in the header information, it is possible e.g. to initialize in the startup a sufficiently large data buffer for the associated SZL sublist.

SFC 51 RDSYSST Reading SZL sublist

With the system function SFC 51 RDSYSST you can read a sublist or a sublist extract of the system status list (SZL). The SFC 51 parameters are explained in Table 23.8.

REQ = “1” initiates the read operation, and BUSY = “0” tells you when it has been completed. The operating system can execute several asynchronous read operations quasi simultaneously; how many depends on the CPU being used. If SFC 51 reports a lack of resources via the function value (W#16#8085), you must resubmit your read request.

The assignment of the parameters SZL_ID and INDEX is CPU-dependent. If the INDEX parameter is not required for infor-

Table 23.8 Parameters of the SFC 51 RDSYSST

SFC	Parameter	Declaration	Data type	Assignment, description
51	REQ	INPUT	BOOL	For "1": Starts processing
	SZL_ID	INPUT	WORD	SZL_ID of partial list
	INDEX	INPUT	WORD	Type or number of the partial list object
	RET_VAL	RETURN	INT	Error information
	BUSY	OUTPUT	BOOL	For "1": Reading not yet completed
	SZL_HEADER	OUTPUT	STRUCT	Length and number of data records read
	DR	OUTPUT	ANY	Field for the read data records

mation, its assignment is irrelevant. The parameter SZL_HEADER has data type STRUCT with the variables LENGTHDR (data type WORD) and N_DR (WORD) as components. LENGTHDR contains the length of a data record, N_DR the number of data records read.

Use the DR parameter to specify the variable or data area in which SFC 51 is to enter the data records. For example, P#DB200.DBX0.0 WORD 256 would provide an area of 256 data words in data block DB 200, beginning with DBB 0. If the area provided is of insufficient capacity, as many data records as possible will be entered. Only complete data records are transferred. The specified area must be able to accommodate at least one data record.

23.5 Web Server

CPUs with Ethernet interface may have a Web server which provides information from the CPU. To read the information on the company-internal intranet or the Internet, you require a Web browser, e.g. Internet Explorer Version 6.0 or later, which displays the information on HTML pages.

23.5.1 Activate Web Server

You can activate the Web server when parameterizing the CPU with the Hardware Configuration. When a CPU is selected, EDIT → OBJECT PROPERTIES selects the "Web" tab on which you activate the Web server and select the language for the message texts and the entries in the diagnostics buffer. It is also possible to select several languages, depending on the memory capacity of the CPU.

The languages installed with STEP 7 are available. You can establish agreement with the languages installed in the project in the SIMATIC Manager: Select TOOLS → LANGUAGE FOR DISPLAY DEVICES and define the languages. You can define the access privileges to the Web server site by means of a user list. The Web server is ready for use after the configuration data has been loaded onto the CPU.

23.5.2 Reading Web Information

To select the CPU in the Web browser, enter the IP address of the CPU in the form *http://aaa.bbb.ccc.ddd* in the "Address" box. You can obtain the IP address of the CPU from the object properties of the PROFINET interface in the "General" tab.

The Web server also supports the Terminal Service of Windows so that thin client solutions can be used with mobile devices or HMI stations with the thin client option under Windows CE. In this case, specify the address in the form *http://aaa.bbb.ccc.ddd/basic*.

You can navigate to a further range of information from the start page of the CPU. Note that the information offered is static, and you must update the screen contents yourself. The most recent information from the CPU is always used for printouts independent of the display.

Note: provide a firewall to protect the Web server from unauthorized access.

23.5.3 Web Information

The Web server can provide the following information in an appropriately equipped CPU:

- ▷ Start page with general CPU information
- ▷ Identification information
- ▷ Diagnostics buffer
- ▷ Messages (without facility for acknowledgement)
- ▷ PROFINET interface
- ▷ Status of variables
- ▷ Tables of variables

The first page provided by the Web server is the welcome page. From here, you can select the start page by clicking on ENTER. If you wish to bypass this introductory page in the future, activate the option "Skip Intro".

Start and identification

The start page shows you general information and the status of the CPU at the time of scanning. The identification page contains the CPU's characteristic data, e.g. plant identifier, location identifier and Order No.

Diagnostics buffer

On this page you see the contents of the diagnostics buffer. Select the number of diagnostics buffer entries per display interval. Detailed information is displayed on the selected event.

You can select the display language in the window at the top right. If the selected language is not configured, hexadecimal code will be displayed.

Messages

Messages are displayed in chronological order with date and time. The messages cannot be acknowledged using the Web browser.

You can search for specific information using filter settings. Using the sorting function, you can sort the messages e.g. according to message number or status. Detailed information is displayed on the selected message.

You can select the display language in the window at the top right. If the selected language is not configured, hexadecimal code will be displayed.

PROFINET interface

The information on the PROFINET interface is present on the pages “Parameters” and “Statistics”. For example, MAC and IP addresses are displayed, as well as statistical evaluations of sent and received data packets.

Status of variables

On this page, you can monitor the status of up to 50 variables. Enter the address of the variable and the display format to obtain the value of the variable.

You can select the display language in the window at the top right. Note when entering the address that the mnemonics differ for the address input (e.g. I for input in English, E for input in German). A faulty syntax is displayed in red.

Table of variables

Using the Web server you can monitor up to 50 tables with up to 200 variables each. It could be the case that the memory space available in the CPU is too small to utilize all possibilities. If tables of variables are displayed incompletely, reduce the memory requirements for the messages and symbol comments as far as possible, use only one language, and keep the number of variables per table low.

Select one of the configured tables to display the variables. You must previously prepare the table for use by the Web server. If you select EDIT → OBJECT PROPERTIES with a table selected or create a new table of variables, the properties window is opened. Enter *VATtoWEB* as the family on the “General – Part 2” tab or alternatively activate the checkbox “Web server”.

SFC 99 WWW
Synchronize user websites

Using the configuration tool S7-Web2PLC, you can integrate self-generated websites into the CPU's Web server. The websites can show CPU data, controlled either by direct access or by the user program.

These websites are saved in special data blocks – the "Fragment DBs". One data block – the "Web control DB" – contains the structure information required to edit the websites.

The system function SFC 99 WWW makes the user websites known to the CPU's operating system. It must be called once for this, e.g. during startup.

In addition, the SFC 99 synchronizes the user program and the Web server. It must be called cyclically for this purpose, e.g. in the main program.

Table 23.9 shows the parameters of the SFC 99 WWW.

Table 23.9 Parameters of the SFC 99 WWW

SFC	Parameter name	Declaration	Data type	Assignment, description
99	CTRL_DB	INPUT	BLOCK_DB	Web control DB
	RET_VAL	RETURN	INT	Error information

Variable Handling

This section provides information on handling complex variables. Knowledge of the structure of data types, mastery of indirect addressing and the ability to determine the addresses of the variables at runtime are all requirements here.

Variables if elementary **data types** can be accessed direct with STL statements, whether you are dealing with binary logic operations, memory functions or load and transfer operations. With complex data types and user-defined data types, only the individual components can currently be accessed direct. If you still want to access variables of these data types, you must know the inner structure of the variables.

Indirect addressing allows you to access addresses whose addresses are not known until runtime. You can choose between memory-indirect and register-indirect addressing. You can even wait until runtime to use the address area. Indirect addressing allows you to access variables of complex and user-defined data types using absolute addressing.

Direct variable access loads the current address of a local variable. When you have determined the address, you can process local variables (and so also block parameters) of any data types. The two preceding chapters contain the information required for this purpose.

Several extensive examples – collected in Chapter 26.4 "Brief Description of the Message Frame Example" – explain the handling of complex variables. The examples "Message Frame Data", "Preparing a Message Frame" and "Clock Check" deal with handling user-defined data types and the use of variables of complex data types in conjunction with system functions and standard functions. The examples "Checksum" and "Data Item Conversion" describe how to access parameters of complex data types with the help of indirect addressing. The example "Save Message Frame" shows how to use the system function SFC 20 BLKMOV to transfer data areas whose addresses are not known until runtime.

24 Data Types
Elementary, complex and user-defined data types; declaration and structure of the data types

25 Indirect Addressing
Area pointers, DB pointers and ANY pointers; memory-indirect and register-indirect addressing, area-internal and area-crossing; working with address registers

26 Direct Variable Access
Addresses of local variables; data storage of variables; data storage with parameter transfer; 'Variable' ANY pointer; Message Frame example

24 Data Types

Data types determine the properties and characteristics of data, essentially the representation of the content of one or more related addresses and the permissible areas. STEP 7 provides predefined data types that you can compile in addition to user-defined data types. The data types are globally available; they can be used in any block.

Chapter 3.7 "Variables and Constants" gives an overview of all data types and the corresponding constant representation.

This chapter gives detailed information on elementary data types and complex data types and shows the structure of the relevant variables. You will learn how user-defined data types are created and used.

Examples of the data types can be found in the download files (download address: see pages 8-9) in the STL-Book library under the "Variable Handling" program in function blocks FB 101, FB 102 and FB 103 or source file Chap_24.

24.1 Elementary Data Types

Variables of elementary data types have a maximum length of one doubleword; they can therefore be processed with load and transfer functions or with binary logic operations.

24.1.1 Declaration of Elementary Data Types

Elementary data types can occupy one bit, one byte, one word or one doubleword.

Declaration

varname : *datatype* := *pre-assignment*;

varname is the name of the variable
datatype is an elementary data type
pre-assignment is a fixed value

The identifiers of the data types (for example, BOOL, REAL) are keywords; they can also be written in lower case. A variable of elementary data type can be declared globally in the symbol table or locally in the declaration section.

Data type CHAR

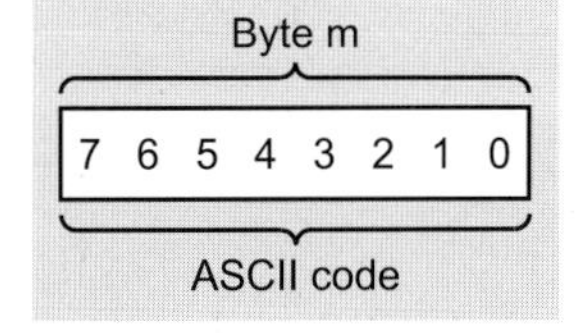

BCD number, 3 Decades

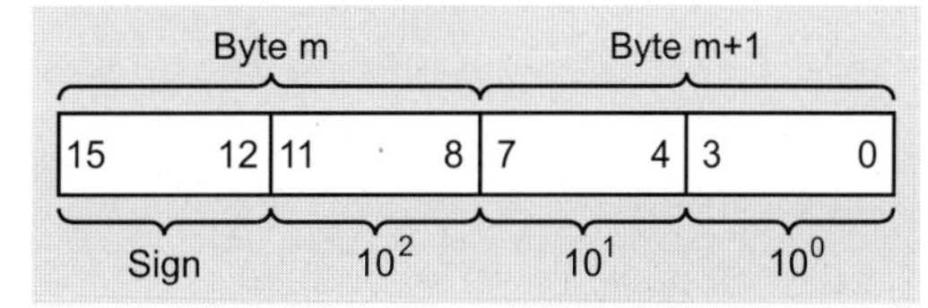

BCD numbers, 7 Decades

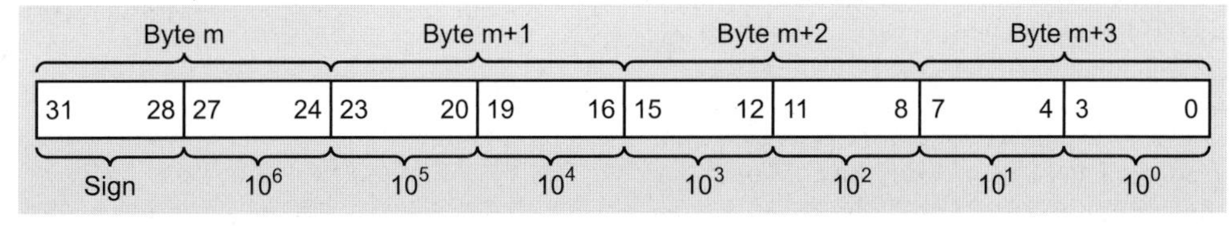

Figure 24.1 Representation of BCD Numbers and CHAR

Pre-assignment

The variable can be pre-assigned when it is declared (not as a block parameter in a function or as a temporary variable). The pre-assignment must be of the same data type as the variable.

Application

You can apply variables of elementary data type at the correspondingly declared block parameter (of the same data type POINTER or ANY) or you can access them with "normal" STL statements (for example, binary checks, load functions).

Storing the variables

A variable of elementary data type is stored in the same way as the relevant address. All address areas including block parameters are permissible.

24.1.2 BOOL, BYTE, WORD, DWORD, CHAR

A variable of data type BOOL represents a bit value (for example, input I 1.0). Variables of data types BYTE, WORD and DWORD are bit strings of 8, 16 or 32 bits. The individual bits are not evaluated. Chapter 3 "SIMATIC S7 Program" shows possible representations as constants.

Special forms of these data types include the BCD numbers and the counter value as used in conjunction with counter functions, as well as the data type CHAR that represents a character in ASCII code (Figure 24.1).

BCD numbers

BCD numbers have no special identifier in STL. You enter a BCD number with data type 16# (hexadecimal) and use only digits 0 to 9.

BCD numbers occur in coded loading of timer and counter values and in conjunction with conversion functions. Data type S5TIME# is available for specifying a timer value when starting a timer function (see below), and for specifying a counter value there is data type 16# or C#. A counter value C# is a BCD number between 000 and 999, where the sign is always 0.

In general, BCD numbers are unsigned numbers. In conjunction with conversion functions, the sign of a BCD number is accommodated in the extreme-left (highest) decade. This results in the loss of one decade in the number range.

In the case of a BCD number stored in a 16-bit word, the sign is in the upper decade with only bit position 15 being relevant. Signal state "0" signifies that the number is positive and signal state "1" represents a negative number. The sign does not affect the assignment of the individual decades. An equivalent assignment applies for a 32-bit word.

The number range available is 0 to ±999 for 16-bit BCD numbers and 0 to ±9 999 999 for 32-bit BCD numbers.

CHAR

A variable of data type CHAR (character) occupies one byte. The data type CHAR represents a single character stored in ASCII format. Example: 'A'. You can use every printable character in single inverted commas.

In conjunction with STL load statements, some special characters take the notation shown in Table 24.1. Example: L '$$' loads a dollar sign in ASCII code.

In addition, you can use other special forms of the data type CHAR when loading ASCII-coded characters into the accumulator. L 'a' loads one character (in this case, an a) right-justified into the accumulator, L 'aa' loads two characters and L 'aaaa' loads 4 characters.

Table 24.1 Special Characters for CHAR

CHAR	Hex	Meaning
$$	24_{hex}	Dollar sign
$'	27_{hex}	Single inverted comma
$L or $l	$0A_{hex}$	Line feed (LF)
$P or $p	$0C_{hex}$	Page break (FF)
$R or $r	$0D_{hex}$	Carriage return (CR)
$T or $t	09_{hex}	Tabulator

24.1.3 Number Representations

The data types INT, DINT and REAL are summarized in this section. Figure 24.2 shows the bit assignments of these data types.

INT

A variable of data type INT represents an integer (whole number) that is stored as a 16-bit fixed-point number. The data type INT has no special identifier.

A variable of data type INT occupies one word. The signal states of bits 0 to 14 represent the positional weight of the number; the signal state of bit 15 represents the sign (S). Signal state "0" means that the number is positive. Signal state "1" represents a negative number. Negative numbers are represented in two's complement.

The number range is

from +32,767 ($7FFF_{hex}$)

to –32,768 (8000_{hex}).

DINT

A variable of data type DINT represents an integer that is stored as a 32-bit fixed-point number. An integer is stored as a DINT variable if it is greater than +32,767 or less than –32,768 or if an L# precedes the number as the type identifier.

A variable of data type DINT occupies a doubleword. The signal states of bits 0 to 30 represent the positional weights of the number; the sign is stored in bit 31. This bit contains "0" for a positive number and "1" for a negative number. Negative numbers are stored in two's complement.

The number range is

from +2,147,483,647 (7FFF $FFFF_{hex}$)

to –2,147,483,648 (8000 0000_{hex}).

Example for STL: with L –100, you load an INT number into the accumulator, and with L# –100 you load a DINT number. The difference is in the assignment of the left word in the accumulator: in the example of INT number –100, this contains the value 0000_{hex}, and in the example of the DINT number –100, it contains the sign $FFFF_{hex}$.

Example for SCL: if you specify the constant value –100, the editor automatically converts the value to a DINT number when it is combined with a DINT variable ("implicit" data type conversion).

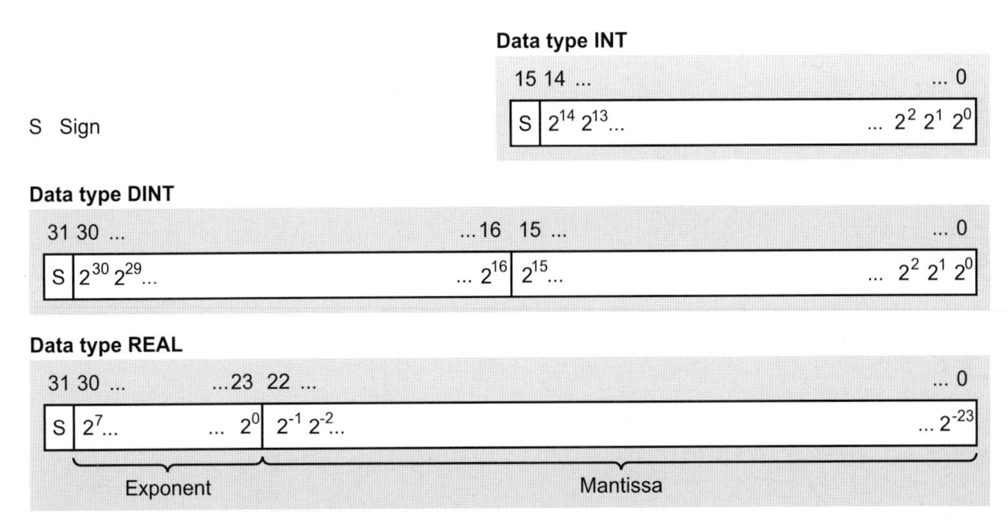

Figure 24.2 Bit Assignments of the Data Types INT, DINT and REAL

REAL

A variable of data type REAL represents a fraction that is stored as a 32-bit floating-point number. An integer is stored as a REAL variable if the decimal point is followed by a zero.

Example for STL: while100 or L#100 indicates the positive number 100 in INT or DINT format, you specify 100 in REAL format with 100.0 or 1.0e+2 (specification with decimal point with or without exponent).

Example for SCL: in conjunction with REAL variables, you can specify constant values in every numerical representation. The value 100, for example, is automatically converted by the editor to a REAL number when it is combined with a corresponding variable ("implicit" data type conversion).

In exponent representation, you can specify an integer or a fraction with 7 significant digits with sign before the "e" or "E". The specification following the "e" or "E" is the exponent to base 10. Conversion of the REAL variable into the internal representation of a floating-point number is handled by STEP 7.

With REAL numbers, a distinction is made between numbers that can be represented with total accuracy ("normalized" floating-point numbers) and numbers with restricted accuracy ("denormalized" floating-point numbers). The value range of a normalized floating-point number lies between the limits:

$-3.402\,823 \times 10^{+38}$ to $-1.175\,494 \times 10^{-38}$
± 0
$+1.175\,494 \times 10^{-38}$ to $+3.402\,823 \times 10^{+38}$

A denormalized floating-point number can lie within the following limits:

$-1.175\,494 \times 10^{-38}$ to $-1.401\,298 \times 10^{-45}$
and
$+1.401\,298 \times 10^{-45}$ to $+1.175\,494 \times 10^{-38}$

The S7-300 CPUs cannot perform calculations with denormalized floating-point numbers. The bit pattern of a denormalized number is interpreted as a zero. If the result of a calculation falls within this range, it is represented as a zero, with the status bits OV and OS being set (number range violation).

The CPUs calculate with the full accuracy of the floating-point numbers. Due to rounding errors in the conversion, the results displayed on the programming device may deviate from the theoretically accurate representation.

A variable of data type REAL consists internally of three components: the sign, the 8-bit exponent to base 2 and the 32-bit mantissa. The sign can assume the values "0" (positive) or "1" (negative). The exponent is stored incremented by one constant (bias, +127), so that it has a value range of 0 to 255. The mantissa represents the fraction component. The integer component of the mantissa is not stored since it is either always 1 (in the case of normalized floating-point numbers) or always 0 (in the case of denormalized floating-point numbers). Table 24.2 shows the internal range limits of a floating-point number.

Table 24.2 Range Limits of a Floating-Point Number

Sign	Exponent	Mantissa	Meaning
0	255	not equal to 0	Not a valid floating-point number (not a number)
0	255	0	+ infinite
0	1 ... 254	any	Positive normalized floating-point number
0	0	not equal to 0	Positive denormalized floating-point number
0	0	0	+ zero
1	0	0	– zero
1	0	not equal to 0	Negative denormalized floating-point number
1	1 ... 254	any	Negative normalized floating-point number
1	255	0	– infinite
1	255	not equal to 0	Not a valid floating-point number (not a number)

24.1.4 Time Representations

The data types S5TIME, DATE, TIME and TIME_OF_DAY are summarized in this section. Figure 24.3 shows the bit assignments of these data types.

A data type that fits into this category (DATE_AND_TIME) belongs to the complex data types since it occupies 8 bytes.

S5TIME

A variable of data type S5TIME is used for initializing the SIMATIC timer functions in the basic languages STL, LAD and FBD (SCL uses the representation of the data type TIME for this purpose). The data type S5TIME occupies a 16-bit word with 1 + 3 decades.

The time is specified in hours, minutes, seconds and milliseconds. Conversion to the internal representation is handled by STEP 7. The number is represented internally as a BCD number from 000 to 999. The time base can assume the following values: 10 ms (0000), 100 ms (0001), 1 s (0010) and 10 s (0011). The time is the product of the time base and the time value.

Examples:

S5TIME#500ms (= 0050_{hex})

S5T#2h46m30s (= 3999_{hex})

DATE

A variable of data type DATE is stored in a word as a un-signed fixed-point number. The contents of the variable correspond to the number of days since 01.01.1990. The representation contains the year, the month and the day, each separated by a hyphen. Examples:

DATE#1990-01-01 (= 0000_{hex})

D#2168-12-31 (= $FF62_{hex}$)

TIME

A variable of data type TIME occupies one doubleword. The representation contains the specifications for days (d), hours (h), minutes (m), seconds (s) and milliseconds (ms); individual specifications can be omitted. The contents of the variable are interpreted as milliseconds (ms) and stored as a 32-bit fixed-point number with sign.

S5TIME data type

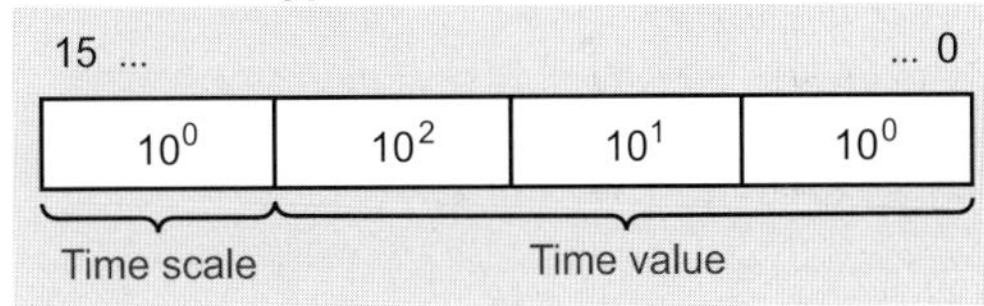

DATE data type

15 14 0

$2^{15} 2^{14} 2^{13}$... ... $2^2\ 2^1\ 2^0$

V = sign

TIME data type

31 30 ... ...16 15 0

V $2^{30} 2^{29}$... ... 2^{16} 2^{15}... ... $2^2\ 2^1\ 2^0$

TIME_OF_DAY data type

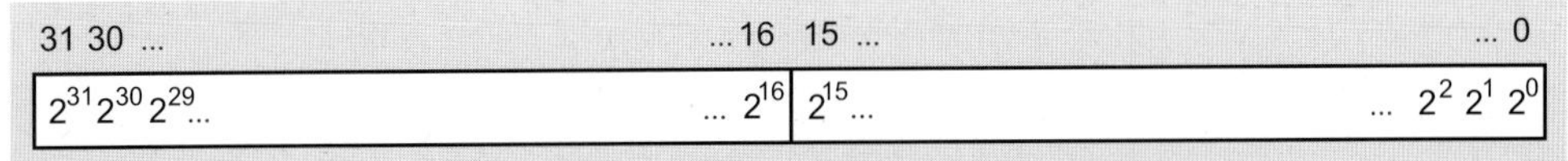

Figure 24.3 Bit Assignments of the Data Types S5TIME, DATE, TIME and TIME_OF_DAY

Examples:

TIME#24d20h31m23s647ms (= 7FFF_FFFF$_{hex}$)

TIME#0ms (= 0000_0000$_{hex}$)

T#–24d20h31m23s648ms (= 8000_0000$_{hex}$)

SCL uses this representation for the duration of SIMATIC timer functions (S5TIME). The editor then converts the specified TIME value to an S5TIME representation (1 + 3 decades) and rounds down where necessary.

A "decimal representation" is also possible for TIME, e.g. TIME#2.25h or T#2.25h. This representation is only permissible for positive values in SCL.

Examples:

TIME#0.0h (= 0000_0000$_{hex}$)

TIME#24.855134d (= 7FFF_FFFF$_{hex}$)

TIME_OF_DAY

A variable of data type TIME_OF_DAY occupies one doubleword. It contains the number of milliseconds since the start of the day (0:00 hours) as an unsigned fixed-point number. The representation contains the specifications for hours, minutes and seconds, each separated by a colon. Specification of the milliseconds, following the seconds and separated by a dot, can be omitted.

Examples:

TIME_OF_DAY#00:00:00 (= 0000_0000$_{hex}$)

TOD#23:59:59.999 (= 0526_5BFF$_{hex}$)

24.2 Complex Data Types

Complex data types are data types which (in their totality) cannot be processed direct by STL statements but are permissible in SCL expressions. STEP 7 defines the following four complex data types:

- DATE_AND_TIME
  date and time of day (BCD-coded)
- STRING
  character string with up to 254 characters
- ARRAY
  field (combination of variables of the same type)
- STRUCT
  structure (combination of variables of different types)

The data types are pre-defined, with the length of the data type STRING (character string) and the combination and size of the data types ARRAY and STRUCT (structure) being defined by the user.

You can declare variables of complex data types only in global data blocks, in instance data blocks, as temporary local data or as block parameters.

Table 24.3 Examples of the Declaration of DT Variables and STRING Variables

Name	Type	Initial Value	Comment
Date1	DT	DT#1990-01-01-00:00:00	DT variable minimum value
Date2	DATE_AND_TIME	DATE_AND_TIME# 2089-12-31-23:59:59.999	DT variable maximum value
First_name	STRING[10]	'Jack'	STRING variable, 4 characters out of 10 occupied
Last_name	STRING[14]	'Daniels'	STRING variable, 7 characters out of 14 occupied
NewLine	STRING[2]	'RL'	STRING variable, occupied by special characters
EmptyString	STRING[16]	''	STRING variable without entry

24.2.1 DATE_AND_TIME

The data type DATE_AND_TIME represents a time consisting of the date and the time of day. You can also use the abbreviation DT in place of DATE_AND_TIME.

Declaration

varname : DATE_AND_TIME := *Pre-assignment*;
varname : DT := *Pre-assignment*;

DATE_AND_TIME or DT are keywords; they can also be written in lower case.

Pre-assignment

At the declaration stage, the variable can be pre-assigned (not as a block parameter in a function, as an in/out parameter in a function block or as a temporary variable). The pre-assignment must be of the type DATE_AND_TIME or DT and must have the following appearance:

Keyword#Year-Month-Day-Hours:Minutes:Seconds.Milliseconds

Specification of the milliseconds can be omitted (Table 24.3).

Application

Variables of data type DT can be applied at block parameters of data type DT or ANY; for example, they can be copied with the system function SFC 20 BLKMOV. There are standard function blocks available for processing these variables ("IEC functions").

Structure of the variables

A variable of data type DATE_AND_TIME occupies 8 bytes (Figure 24.4). The variable begins at a word boundary (at a byte with an even address). All specifications are available in BCD format.

24.2.2 STRING

The data type STRING represents a character string consisting of up to 254 characters.

Declaration

varname : STRING[*maxNumber*] := *Pre-assignment*;

STRING is a keyword and can also be written in lower case.

maxNumber specifies the number of characters that a variable declared in this way can have (from 0 to 254). This specification can also be omitted; the Editor then uses a length of 254

Data format DT

Byte n	Year	0 bis 99
Byte n+1	Month	1 bis 12
Byte n+2	Day	1 bis 31
Byte n+3	Hour	0 bis 23
Byte n+4	Minute	0 bis 59
Byte n+5	Second	0 bis 59
Byte n+6	ms	0 bis 999
Byte n+7		Week-day

Week-day
from 1 = Sunday
to 7 = Saturday

Data format STRING

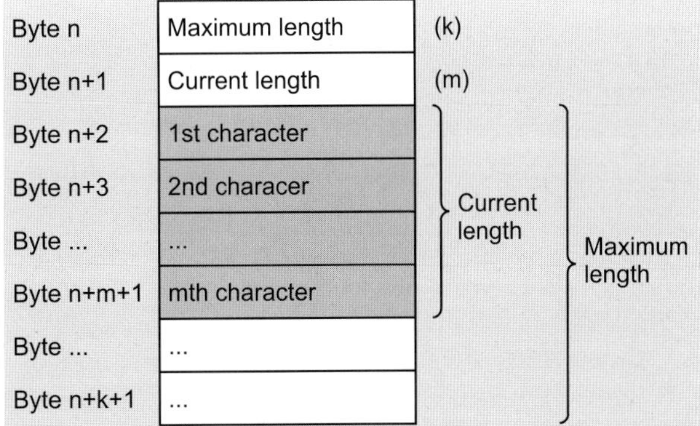

Figure 24.4 Structure of a DT and a STRING Variable

Table 24.4 Examples of field declarations

Name	Type	Initial value	Comment
Measured value	ARRAY[1..24]	0.4, 1.5, 11 (2.6, 3.0)	Array tag with 24 REAL components
	REAL		
Time of day	ARRAY[-10..10]	21 (TOD#08:30:00)	Time of day field with 21 components
	TIME_OF_DAY		
Result	ARRAY[1..24,1..4]	96 (L#0)	Two-dimensional array with 96 components
	DINT		
Character	ARRAY[1..2,3..4]	2 ('a'), 2 ('b')	Two-dimensional array with 4 components
	CHAR		

bytes. With functions FCs, the Editor does not permit length specifications or it demands the standard length of 254.

Pre-assignment

At the declaration stage, the variable can be pre-assigned at the declaration stage (not as a block parameter in a function, as an in/out parameter in a function block or as a temporary variable). The pre-assignment is made with ASCII-coded characters enclosed in single inverted commas or with a preceding dollar sign in the case of certain characters (see data type CHAR).

If the pre-assignment value is shorter than the declared maximum length, the remaining character positions are not occupied. When further processing a variable of data type STRING, only the currently occupied character positions are taken into account. Pre-assignment as "EmptyString" is possible.

Application

Variables of data type STRING can be applied at block parameters of data type STRING or ANY; for example, they can be copied with the system function SFC 20 BLKMOV. There are standard function blocks available for processing these variables ("IEC functions"). See Chapter 27.5.2 "Assignment of DT and STRING Variables" for special points regarding use in SCL.

Structure of the variables

A variable of data type STRING (character string) has a maximum length of 256 characters with 254 bytes of net data. It starts at a word boundary (at a byte with an even address).

When the variables are applied, their maximum length is defined. The current length (the actual used length of the character string = Number of valid characters) is entered when pre-assigning or when processing a character string. The first byte of the character string contains the maximum length and the second byte contains the current length; these are followed by the characters in ASCII format (Figure 24.4).

24.2.3 ARRAY

The data type ARRAY represents a field consisting of a fixed number of components of the same data type.

Declaration

fieldname : ARRAY [*minIndex..maxIndex*]
OF *datatype* := *pre-assignment*;
fieldname : ARRAY [$minIndex_1..maxIndex_1$,..,
$minIndex_6..maxIndex_6$]
OF *datatype* := *pre-assignment*;

ARRAY and OF are keywords and can also be written in lower case.

fieldname is the name of the field

minIndex is the lower limit of the field and *maxIndex* is the upper limit. Both limits are INT numbers in the range –32768 to +32767; *max-*

Index must be greater than or equal to *minIndex*. A field can have up to 6 dimensions whose limits can be specified separated by a comma.

data_type is every data type except for ARRAY itself (exception in SCL), including user-defined data types.

Pre-assignment

At the declaration stage, you can pre-assign values to individual field components (not as a block parameter in a function, as an in/out parameter in a function block or as a temporary variable). The data type of the pre-assignment value must match the data type of the field.

You do not require to pre-assign all field components; if the number of pre-assignment values is less than the number of field components, only the first components are pre-assigned. The number of pre-assignment values must not be greater than the number of field components.

The pre-assignment values are each separated by a comma. Multiple pre-assignment with the same values is specified within round brackets with a preceding repetition factor. With SCL, the pre-assigned values can be enclosed in square brackets.

Application

You can apply a field as a complete variable at block parameters of data type ARRAY with the same structure or at a block parameter of data type ANY. For example, you can copy the contents of a field variable using the system function SFC 20 BLKMOV. You can also specify individual field components at a block parameter if the block parameter is of the same data type as the components.

If the individual field components are of elementary data types, you can process them with "normal" STL statements.

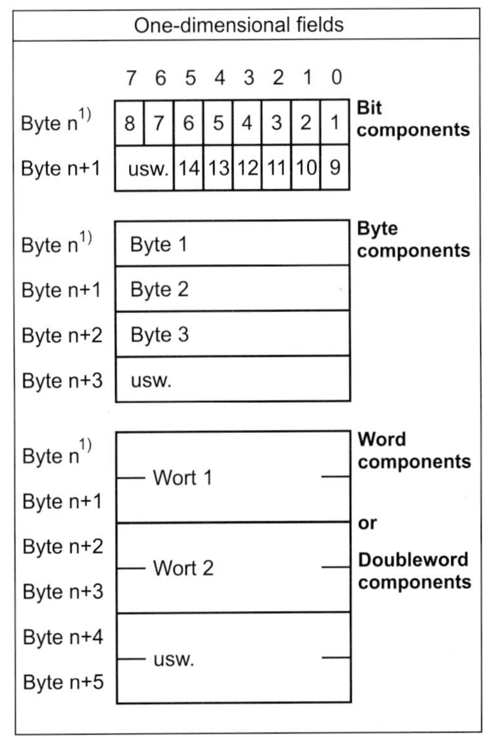

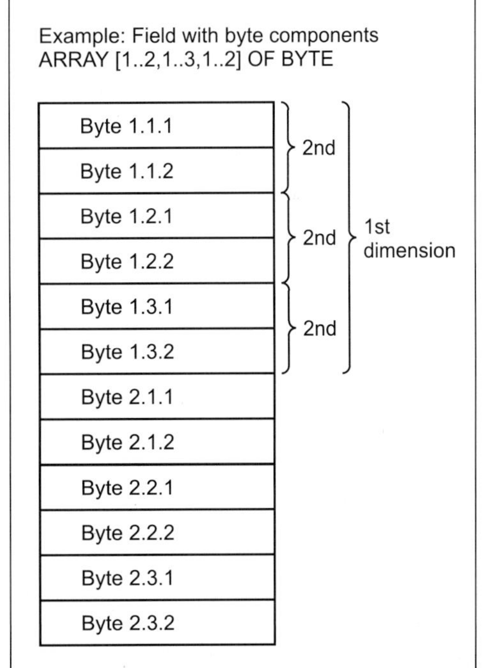

Figure 24.5 Structure of an ARRAY Variable

A field component is accessed with the field name and an index in square brackets. The index is a fixed value in STL and cannot be modified at runtime (no variable indexing possible).

In SCL, the index can also be a variable or an expression of data type INT whose value can be modified at runtime.

Multi-dimensional fields

Fields can have up to 6 dimensions. Multi-dimensional fields are analogous to one-dimensional fields. At the declaration stage, the ranges of the dimensions are written in square brackets, each separated by a comma.

With SCL, you can also declare a multi-dimensional array by specifying another array as an array component, for example:

Array: ARRAY [1..4] of ARRAY [1..4] OF INT;

When accessing the field components of multi-dimensional fields, you must always specify the indices of all dimensions in STL. In SCL, it is possible to address part fields (see Chapter 27.5.4 "Assigning Fields").

Structure of the variables

An ARRAY variable always begins at a word boundary, that is, at a byte with an even address. ARRAY variables occupy the memory up to the next word boundary.

Components of data type BOOL begin in the least significant bit; components of data type BYTE and CHAR begin in the right-hand byte (Figure 24.5 left). The individual components are listed in order.

In multi-dimensional fields, the components are stored line-wise (dimension-wise) starting with the first dimension (Figure 24.5 right). With bit and byte components, a new dimension always starts in the next byte, and with components of other data types a new dimension always starts in the next word (in the next even byte).

24.2.4 STRUCT

The data type STRUCT represents a data structure consisting of a fixed number of components that can each be of a different data type.

Declaration

```
structname : STRUCT
    komp1name : datatype := pre-assignment;
    komp2name : datatype := pre-assignment;
    ...
    END_STRUCT;
```

STRUCT and END_STRUCT are keywords that can also be written in lower case.

structname is the name of the structure.

komp1name, *komp2name* etc. are the names of the individual structure components.

datatype is the data type of the individual components. All data types can be used, including further structures.

Pre-assignment

In the declaration, the individual structure components can be preassigned with values (not as block parameters on a function, in-out parameters on a function block, or temporary tags). The data type of the default values must correspond to the data type of the component.

Table 24.5 Example of the declaration of a structure

Name	Type	Initial value	Comment
MotCont	STRUCT		Simple structure tag with 4 components
On	BOOL	FALSE	MotCont.On tag of type BOOL
Off	BOOL	TRUE	MotCont.Off tag of type BOOL
Delay	S5TIME	S5TIME#5s	MotCont.Delay tag of type S5TIME
maxSpeed	INT	5000	MotCont.maxSpeed tag of type INT
	END_STRUCT		

Application

You can apply a complete variable at block parameters of data type STRUCT with the same structure or at a block parameter of data type ANY. For example, you can copy the contents of a STRUCT variable with the system function SFC 20 BLKMOV. You can also specify an individual structure component at a block parameter if the block parameter is of the same data type as the component.

If the individual structure components are of elementary data types, you can process them with "normal" STL statements.

A structure component is accessed with the structure name and the component name separated by a dot.

Structure of the variables

A STRUCT variable always begins at a word boundary, that is, at a byte with an even address; following this, the individual components are located in the memory in the order of their declaration. STRUCT variables occupy the memory up to the next word boundary.

Components with data type BOOL commence in the least significant bit of the following byte; components with data type BYTE and CHAR in the following byte (Figure 24.6). Components with other data types commence at a word limit.

A nested structure is a structure as a component of another structure. A nesting depth of maximum 8 structures is possible. All components can be addressed individually with"normal" STL statements as long as they are elementary data types. The individual names are each separated by a dot.

Structure with elementary data types

	7	6	5	4	3	2	1	0	
Byte n[1)]	7	6	5	4	3	3	2	1	**Bit components**
Byte n+1						10	9	8	
Byte n+2	Byte 1								**Byte components**
Byte n+3	Byte 2								
Byte n+4	Byte 3								
Byte n+5	(Filler byte)								
Byte n+6	Word 1								**Word components**
Byte n+7									
Byte n+8	Word 2								
Byte n+9									
Byte n+10						3	2	1	**Bit components**
Byte n+11	Byte 4								**Byte component**
Byte n+12	etc.								
Byte ...									

[1)] n = even

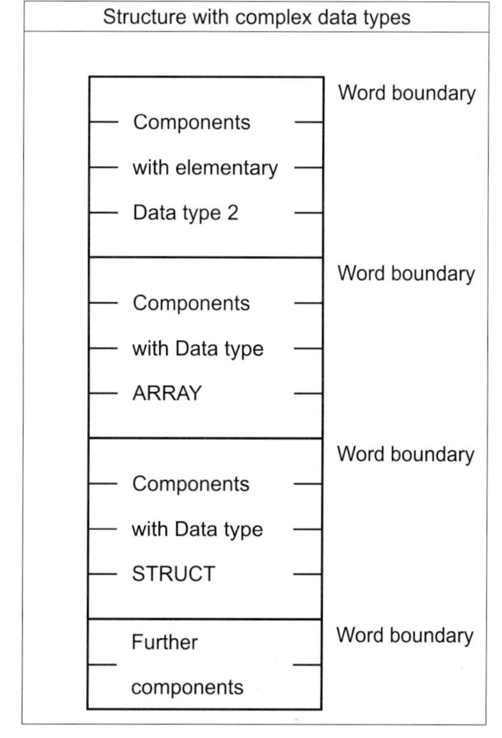

Figure 24.6 Structure of a STRUCT tag

24.3 User-Defined Data Types

A user-defined data type (UDT) corresponds to a structure (combination of components of any data type) with global validity. You can use a user-defined data type if a data structure occurs frequently in your program or if you want to assign a name to a data structure.

You create UDTs with the incremental editor or with the text editor as a source file. They are programmed and used in the STL and SCL programming languages in the same way (you can also use incrementally programmed UDTs in SCL if they are located in the *Blocks* object).

UDTs have global validity; i.e., once declared, they can be used in all blocks. UDTs can be addressed symbolically; you assign the absolute address in the symbol table. The data type of a UDT (in the symbol table) is identical with the absolute address.

If you want to give a variable the data structure defined in the UDT, assign the UDT to it at declaration like a "normal" data type. The UDT can be absolutely addressed (UDT 0 to UDT 65,535) or symbolically addressed.

You can also define a UDT for an entire data type. When programming the data block, you assign this UDT to the block as a data structure.

The example "Message Frame Data" in Chapter 26.4 "Brief Description of the Message Frame Example" shows you how to work with user-defined data types.

24.3.1 Programming UDTs Incrementally

You create a user-defined data type either in the SIMATIC Manager by selecting the *Blocks* object and then INSERT → S7 BLOCK → DATA TYPE, or in the editor by selecting FILE → NEW and entering "UDTn" in the "Object name" line.

A double-click on the *UDT* object in the program window opens a declaration table that looks exactly like the declaration table of a data block. A UDT is programmed in exactly the same way as a data block, with individual lines for Name, Type, Initial value and Comments. The only difference is that switching to the data view is not possible. (With a UDT, you do not create any variables but only a collection of data types; for this reason, there can be no actual values here).

The initial values you program in UDT are transferred to the tags in the declaration.

24.3.2 Source-File-Oriented Programming of UDTs

The source-oriented entry of a UDT corresponds to that of a STRUCT tag, "framed" by the keywords TYPE and END_TYPE.

Table 24.6 Example of a user-defined UDT data type

Name	Type	Initial value	Comment
	STRUCT		
Identifier	WORD	W#16#F200	UDT component identifier of type WORD
Number	INT	0	UDT component number of type INT
Arrival	TIME_OF_DAY	TOD#0:0:0.0	UDT component arrival of type TOD
	END_STRUCT		

Declaration

```
TYPE udtname
    STRUCT
    komp1name : datatype := pre-assignment;
    komp2name : datatype := pre-assignment;
    ...
    END_STRUCT
END_TYPE
```

TYPE, END_TYPE, STRUCT and END_STRUCT are keywords that can also be written in lower case.

udtname is the name of the user-defined data type. In place of *udtname*, you can also use the absolute address UDTn.

komp1name, *komp2name* etc. are the names of individual structure components.

datatype is the data type of the individual components. All data types can be used except POINTER and ANY (not even as components of a field or a structure).

User-defined data types are pre-assigned and used like structures; the structure is the same as for structures.

When pre-assigning a user-defined data type UDT, the method of writing constants in STL also applies in SCL (see the overview in Chapter 3.7.3 "Elementary Data Types").

Block properties

User-defined data types have a block header which contains the block properties. With the menu command FILE → PROPERTIES in the editor, you can view and change the properties of the currently opened block.

With source-oriented programming, use the keywords provided for this purpose (Table 24.7).

Table 24.7 Keywords for Programming User-Defined Data Types UDT

Block type	TYPE "*Symbol*" or *UDTn*
Header	TITLE = *Block title*
	//Block comment
	KNOW_HOW_PROTECT
	NAME : *Block name*
	FAMILY : *Block family*
	AUTHOR : *Author*
	VERSION : *Version*
Declaration	STRUCT
	name : Type := Pre-assignment;
	END_STRUCT
Block end	END_TYPE

25 Indirect Addressing

Indirect addressing gives you the ability to assign addresses that are not known until runtime. With indirect addressing, you can also effect multiple processing of program sections, for example, in a loop, and with each pass, you can assign a different address to the addresses used. This chapter shows how the STL programming language supports you here. Indirect addressing for the SCL programming language is described in the Chapter 27.2.3 “Indirect Addressing in SCL”.

Since the addresses are not calculated until runtime in the case of indirect addressing, there is a danger that memory areas could be inadvertently overwritten. *The programmable controller might then respond unpredictably! Please exercise the utmost caution when using indirect addressing!*

The examples in this chapter can be found in the download files (download address: see pages 8-9) under the “Variable Handling” program in function block FB 125 or source file Chap_25.

25.1 Pointers

The address for indirect addressing must be structured in such a way that it contains the bit address, the byte address and, if applicable, also the address area. It therefore has a special format, called Pointer. A pointer is used for “pointing” to an address.

STEP 7 recognizes three types of pointers:

▷ Area pointers; these are 32 bits long and contain a specific address

▷ DB pointers; these are 48 bits long and also contain the number of the data block in addition to the area pointer

▷ ANY pointer; these are 80 bits long and contain further specifications such as the data type of the address in addition to the DB pointer

Only the area pointer is significant for indirect addressing, the DB pointer and the ANY pointer are used when transferring block parameters. Since these pointer types contain the area pointer, this chapter also describes the structure of the DB pointer and the ANY pointer.

25.1.1 Area Pointer

The area pointer contains the address and possibly also the address area. Without the address area, it is an *area-internal* pointer; if the pointer also contains the address area, it is referred to as an *area-crossing* pointer.

You can address an area pointer direct and load it into the accumulator or into the address register, since it is 32 bits long. The notation for constant representation is as follows:

P#y.x for an area-internal pointer (for example P#22.0) and

P#Zy.x for an area-crossing pointer (for example P#M22.0)

where x = bit address, y = byte address and Z = area. As the area, you specify the address identifier. The assignment of bit 31 differentiates between the two pointer types (Figure 25.1).

The area pointer has a bit address that must always be specified even for digital addresses; with digital addresses, the bit address is 0 (zero). With the area pointer P#M22.0, you can address memory bit M 22.0 but also memory byte MB 22, memory word MW 22 or memory doubleword MD 22.

25.1.2 DB Pointer

A DB pointer also contains a data block number as a positive INT number in addition to the area pointer. It specifies the data block if the area pointer contains the address areas global data or instance data. In all other cases, the first two bytes contain zero.

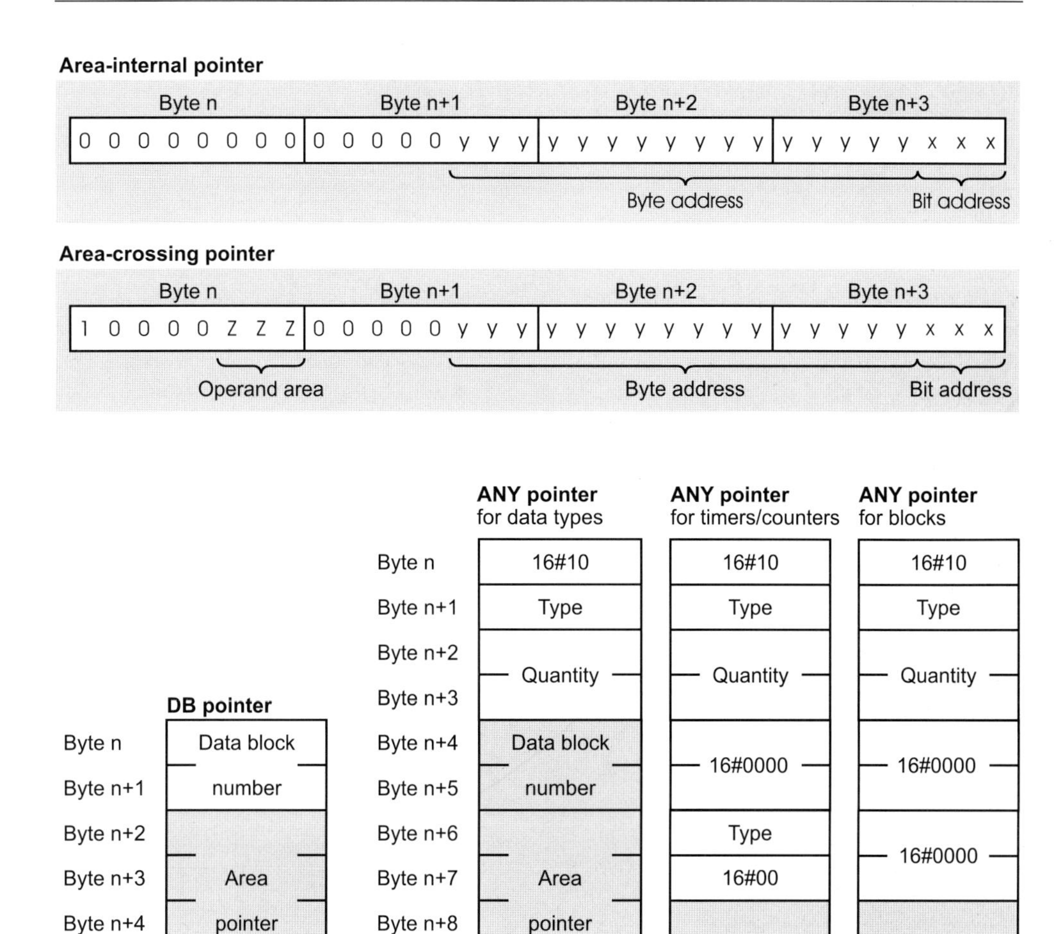

Address area:

0	0	0	Peripheral I/O (P)
0	0	1	Inputs (I)
0	1	0	Outputs (Q)
0	1	1	Memory bits (M)
1	0	0	Global data (DBX)
1	0	1	Instance data (DIX)
1	1	0	Temporary local data (L) [1]
1	1	1	Temporary local data of the predecessor block (V) [2]

[1] Not with area-crossing addressing

[2] Only with block parameter transfer

Type in the ANY pointer:

Elementary data types

01	BOOL
02	BYTE
03	CHAR
04	WORD
05	INT
06	DWORD
07	DINT
08	REAL
09	DATE
0A	TOD
0B	TIME
0C	S5TIME

Complex data types

0E	DT
13	STRING

Parameter types

17	BLOCK_FB
18	BLOCK_FC
19	BLOCK_DB
1A	BLOCK_SDB
1C	COUNTER
1D	TIMER

Zero pointer

00	NIL

Figure 25.1 Structure of the Pointer for Indirect Addressing

You are familiar with the notation of the pointer from the full addressing of data addresses. Here too, the data block and the data address are specified separated by a dot:

P#DataBlock.DataAddress

Example: P#DB 10.DBX 20.5

You cannot load this pointer; however, you can apply it at a block parameter of parameter type POINTER in order to point to a data address (not in SCL). STEP 7 uses this pointer type internally in order to transfer actual parameters.

25.1.3 ANY Pointer

The ANY pointer also contains the data type and a repetition factor in addition to the DB pointer. This makes it possible to also point to a data area.

The ANY pointer is available in two versions: for variables with data types and for variables with parameter types. If you point to a variable with a data type, the ANY pointer contains a DB pointer, the type and a repetition factor. If the ANY pointer points to a variable with parameter type, it contains only the number instead of the DB pointer, in addition to the type. With a timer or counter function, the type is repeated in the byte (n+6); byte (n+7) contains B#16#00. In all other cases, these two bytes contain the value W#16#0000.

The first byte of the ANY pointer contains the syntax ID; in STEP 7 it is always 10_{hex}. The type specifies the data type of the variables for which the ANY pointer applies. Variables of elementary data types, DT and STRING receive the type shown in Figure 25.1 and the quantity 1.

If you apply a variable of data type ARRAY or STRUCT (also UDT) to an ANY parameter, the Editor generates an ANY pointer to the field or the structure. This ANY pointer contains the identifier for BYTE (02_{hex}) as the type and the number of bytes making up the length of the variable as the quantity. The data type of the individual field or structure components are insignificant here. An ANY pointer thus points to a WORD field with double the quantity of bytes. Exception: A pointer to a field consisting of components of data type CHAR is also applied with CHAR type (03_{hex}).

You can apply an ANY pointer at a block parameter of parameter type ANY if you want to point to a variable or an address area (not in SCL).

The constant representation for *data types* is as follows:

P#[DataBlock.]Address Type Quantity

Examples:

▷ P#DB 11.DBX 30.0 INT 12
  Area with 12 words in DB 11 from DBB 30

▷ P#M 16.0 BYTE 8
  Area with 8 bytes from MB 16

▷ P#E 18.0 WORD 1
  Input word IW 18

▷ P#E 1.0 BOOL 1
  Input I 1.0

With *parameter types*, you write the pointer as follows:

L# Number Type Quantity

Examples:

▷ L# 10 TIMER 1 Timer function T10
  L# 2 COUNTER 1 Counter function Z2

The Editor then applies an ANY pointer that agrees in type and in quantity with the specifications in the constant representation. Please note that the memory location in the ANY pointer for data types must always be a bit address.

Specification of a constant ANY pointer is meaningful if you want to access a data area for which you have not declared a variable. In principle, you can also apply variables or addresses at an ANY parameter. For example, the representation "P#I 1.0 BOOL 1" is identical to "I 1.0" or the corresponding symbol address.

With parameter type ANY, you can also declare variables in the temporary local data. You use these variables to create an ANY pointer that can be modified at runtime (see Chapter 26.3.3 ""Variable" ANY Pointer").

If you do not specify any pre-assignment when declaring an ANY parameter in a function block, the Editor assigns 10_{hex} to the syntax ID and 00_{hex} to the remaining bytes. It then represents these (empty) ANY pointers (in the data view) thus: P#P0.0 VOID 0.

25.2 Types of Indirect Addressing in STL

This section describes indirect addressing for the STL programming language; for SCL, please refer to Chapter 27.2.3 "Indirect Addressing in SCL".

25.2.1 General

Indirect addressing is only possible with absolute addressing. You cannot indirectly address variables with symbolic addresses (you must also access the components of a field individually and directly in STL). If you want to access a variable indirectly, you must know the absolute address of the variable. STL supports you here with direct variable access (see next chapter). Absolute addressing recognizes the following

▷ immediate addressing,

▷ direct addressing and

▷ indirect addressing.

Addressing via block parameters is a special form of indirect addressing: By specifying the actual parameter at the block parameter, you determine the address to be processed at runtime.

We refer to *immediate addressing* when the number value is specified together with the operation. Examples of immediate addressing include loading a constant value into the accumulator, shifting by a fixed value and also setting and resetting the result of the logic operation with SET or CLR.

With *direct addressing*, you access the address direct, for example A I 1.2 or L MW 122. The value you want to combine or load into the accumulator is located in an address, that is, in a memory cell. You address this memory cell by specifying the address direct in the STL statement.

With *indirect addressing*, the STL statement indicates where the address is to be found instead of containing the address itself. We distinguish between two types of indirect addressing depending on the type of the indicator:

Memory-indirect addressing uses an address from the system memory to accommodate the address. Example: In the statement T QW [MD 220], the address of the output word to which the transfer is to be made, is located in the memory doubleword MD 220.

Register-indirect addressing uses an address register to determine the address of the address. Example: With the statement T QW [AR1,P#2.0] a transfer is made to the output word whose address is 2 (bytes) higher than the address located in the address register AR1.

You can use register-indirect addressing in two variants: With *area-internal* register-indirect addressing, you program in the statement the address area for which the address in the address register is to apply. The address in the address register therefore moves within an address area (example: L MW[AR1,P#0.0], load the memory word whose address is located in AR1). With *area-crossing* register-indirect addressing, you specify only the address width (bit, byte, word or doubleword) in the statement; the address area is located in the address register and can be modified dynamically (example: L W[AR1,P#0.0], load the word whose address area and address are located in the AR1).

25.2.2 Indirect Addresses

Addresses that can be specified indirectly can be divided into two categories:

▷ Addresses that can be assigned an elementary data type, and

▷ Addresses that can be assigned a parameter type.

You can use memory-indirect and register-indirect addressing with the former, but only memory-indirect addressing with the latter (Table

Table 25.1 Indirect Addresses

Addresses that can be specified indirectly	Addressing	Pointer
Peripheral I/O, inputs, outputs, memory bits, global data, instance data, temporary local data	Memory-indirect and register-indirect	Area pointers, either area-internal or area-crossing
Timers, counters, functions, function blocks, data blocks	Memory-indirect	16-bit number

25.1). The addresses that cannot have a bit address also require no bit address in the pointer, so that a 16-bit wide number is sufficient as the address (unsigned INT number).

The areas of the pointers have a theoretical size of 0 to 65535 (byte address or number). In practice, the addresses are restricted by the address volume of the CPU in each case. The bit address lies in the range from 0 to 7.

25.2.3 Memory-Indirect Addressing

With memory-indirect addressing, the address is located in an address. This address has doubleword width if an area pointer is required, or word width, if indirect addressing via a number is used.

The address can be within one of the following address areas:

▷ Bit memory
as absolute address or as symbolically addressed variable

▷ L stack (temporary local data)
as absolute address or as symbolically addressed variable

▷ Global data block
as absolute address
When using global data addresses, please ensure that the 'correct' data block is opened via the DB register. If, for example, you address a global data address indirect via a global data doubleword, both operations must be located in the same data block.

▷ Instance data block
as absolute address or as symbolically addressed variable
There are restrictions to the use of instance data as addresses; see below.

If you use instance data as addresses in functions, treat them in exactly the same way as global data addresses; you use only the DI register in place of the DB register. Symbolic addressing is not permissible in this case. You can use instance data as addresses in function blocks only if you compile the blocks as CODE_VERSION1 block (no multi-instance capability).

Indirect addressing with an area pointer

The area pointer required for memory-indirect addressing is always an area-internal pointer; that is, it consists of byte and bit address. If you want to address digital addresses, you must always specify 0 as the bit address.

Example: Memory doubleword MD 10 contains the pointer P#30.0. The statement A M [MD 10] accesses the memory bit whose address is located in memory doubleword MD 10; memory bit M 30.0 is therefore checked (Figure 25.2). With the statement L MW [MD 10], you load memory word MW 30 into the accumulator.

You can use memory-indirect addressing for all binary addresses in conjunction with the binary logic operations and the memory functions and for all digital addresses in conjunction with the load and transfer functions.

Indirect addressing with a number

The number for indirect addressing of timers, counters and blocks is 16 bits wide. An address of word width is sufficient for saving.

Example: Memory word MW 20 contains the number 133. The statement OPN DB [MW 20] opens the global data block whose number is located in memory word MW 20. With the statement SP T [MW 20] you start timer T 133 as a pulse.

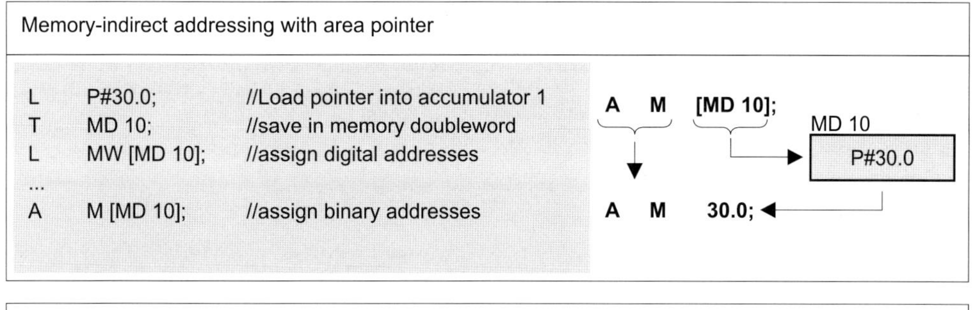

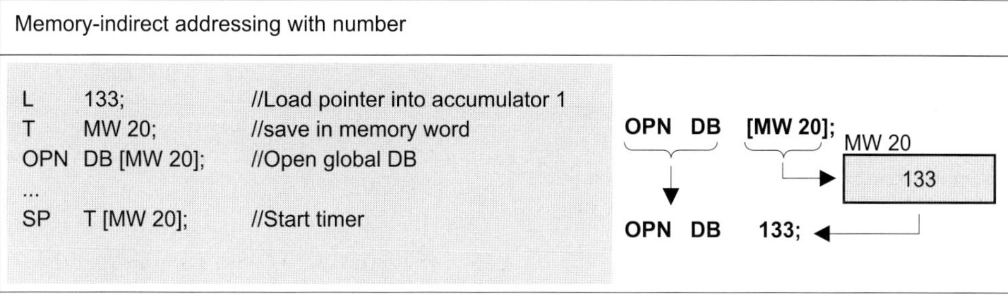

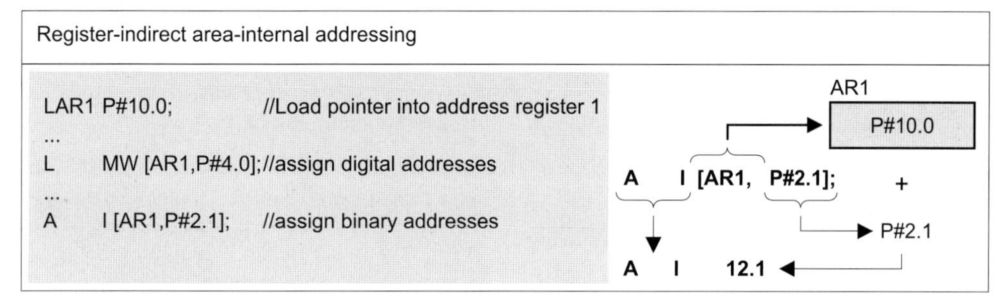

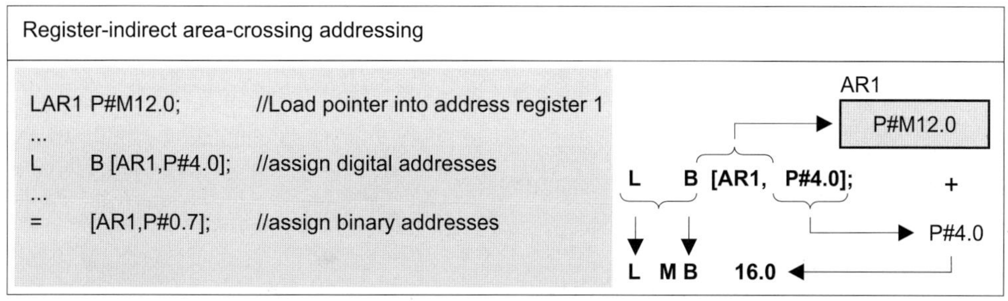

Figure 25.2 Types of Indirect Addressing

You can use all timer and counter operations together with indirect addressing. You can open a data block either via the DB register (OPN DB [..]) or via the DI register (OPN DI [..]). If the address word contains zero, the CPU executes a NOP operation and "closes" the DB.

You can address the call of code blocks indirectly with UC FC [..] and CC FC [..] or UC FB [..] and CC FB [..]. Calling with UC or CC simply changes to another block; the transfer of block parameters or the opening of an instance data block does not take place.

25.2.4 Register-Indirect Area-Internal Addressing

With register-indirect area-internal addressing, the address is located in one of the two address register. The contents of the address register is an area-internal pointer.

With register-indirect addressing, an offset is specified in addition to the address register. This offset is added to the contents of the address register when the operation is executed (without changing the contents of the address register). This offset has the format of an area-internal pointer. You must always specify it and you can only specify it as a constant. With indirect addressed digital addresses, this offset must have the bit address 0. The maximum value is P#8191.7.

Example: Address register AR1 contains the area pointer P#10.0 (with LAR1, you can load the pointer direct into address register AR1, see below). The statement A I [AR1,P#2.1] adds the pointer P#2.1 to address register AR1 and so forms the address of the input to be checked. With the statement L MW [AR1,P#4.0], you load memory word MW 14 into the accumulator.

Area-internal addressing with area-crossing pointers

If the address register contains an area-crossing pointer and if you use this address register in conjunction with area-internal operations, the address area in the address register is ignored.

Example: The following statements load an area-crossing pointer to the global data bit DBX 20.0 into address register AR1 and then execute area-internal addressing via AR1 on a memory doubleword. When the load statement is executed, memory doubleword MD 20 is loaded.

```
LAR1  P#DBX20.0;
L     MD[AR1,P#0.0];
```

25.2.5 Register-Indirect Area-Crossing Addressing

With register-indirect area-crossing addressing, the address is located in one of the two address registers. The contents of the address register is an area-crossing pointer.

With area-crossing addressing, you write the address area in conjunction with the area pointer into the address register. If you use indirect addressing you only specify an ID for the address width as the address: no specification for a bit, "B" for a byte, "W" for a word and "D" for a doubleword.

As with area-internal addressing, you work here with an offset that you specify with as a fixed value with bit address. The contents of the address register are not changed by the offset.

Example: Address register AR1 contains the area pointer P#M12.0 (with LAR1 P#M12.0, you can load the pointer direct into address register AR1, see below). The statement

```
L B [AR1,P#4.0]
```

adds the pointer P#4.0 to address register AR1 and so forms the address of the memory byte to be loaded (MB 16 in this case). With the statement

```
= [AR1,P#0.7]
```

you assign the result of the logic operation (RLO) to memory bit M 12.7.

25.2.6 Summary

When do you use which type of addressing? If possible, use register-indirect area-internal addressing. STL supports this type of addressing best. You see the accessed address area in the operation and the CPU processes register-indirect area-internal addressing fastest.

Memory-indirect addressing offers advantages if more than two pointers are currently involved in program execution. However, please note the "validity period" of a pointer: A pointer in the bit memory area is available without restriction during the entire program even across several program cycles. A pointer in a data block remains valid as long as the data block is open. A pointer in the temporary local data area remains valid only during the runtime of the block.

If address areas are also to be accessible with variable addressing during runtime, register-indirect area-crossing addressing is the right choice.

Table 25.2 shows a comparison of indirect addressing types. All statement sequences shown

Table 25.2 Comparison of Indirect Addressing Types

Memory-Indirect		Register-Indirect Area-Internal		Register-Indirect Area-Crossing	
L	P#4.7	LAR1	P#4.7	LAR1	P#Q4.7
T	MD 24				
S	Q [MD 24]	S	Q [AR1,P#0.0]	S	[AR1,P#0.0]

lead to the same result, the setting of output Q 4.7.

25.3 Working with Address Registers

The Figure 25.3 shows you the statements possible in combination with the address registers in the STL programming language, as a list above and as a diagram below.

All statements are executed without regard to any conditions and they do not affect the status bits.

25.3.1 Loading into an Address Register

The statement LAR*n* loads an area pointer into address register AR*n*. As the source, you can select an area-internal or area-crossing pointer or a doubleword from the address areas bit memory, temporary local data, global data and instance data. The contents of the doubleword must correspond to the format of an area pointer.

If you do not specify an address, LAR*n* loads the contents of accumulator 1 into address register AR*n*.

With the statement LAR1 AR2, you copy the contents of address register AR2 into address register AR1.

Examples:

```
LAR2  P#20.0;  //Load AR2 with P#20.0
L     P#24.0;
LAR1  ;        //Load AR1 with <Accum 1>
LAR1  MD 120;  //Load AR1 with <MD 120>
LAR1  AR2;     //Load AR1 with <AR2>
```

25.3.2 Transferring from an Address Register

The statement TAR*n* transfers the complete area pointer from address register AR*n*. As the destination, you can specify a doubleword from the address areas bit memory, temporary local data, global data and instance data.

If you do not specify an address, TAR*n* transfers the contents of address register AR*n* into accumulator 1. In so doing, the previous contents of accumulator 1 are shifted into accumulator 2; the previous contents of accumulator 2 are lost. Accumulators 3 and 4 remain unaffected.

With the statement TAR1 AR2, you copy the contents of address register AR1 into address register AR2.

Examples:

```
TAR2   MD 140;  //Transfer <AR2>
                  to MD 140
TAR1   ;        //Transfer <AR1>
                  to Accum 1
TAR1   AR2;     //Transfer <AR1>
                  to AR2
```

25.3.3 Swap Address Registers

The statement CAR swaps the contents of address registers AR1 and AR2.

Example: 8 Bytes of data are transferred between the bit memory area from MB 100 and the data area from DB 20.DBB 200. The direction of transfer is determined by memory bit M 126.6. If M 126.6 has signal state "0", the contents of the address registers are swapped. If you want to transfer data between two data blocks in this way, load the two data block registers (with OPN DB and OPN DI) together with the address registers and swap with the statement TDB.

```
     LAR1    P#M100.0;
     LAR2    P#DBX200.0;
     OPN     DB 20;
     A       M 126.6;
     JC      OV;
     CAR     ;
OV:  L       D[AR1,P#0.0];
     T       D[AR2,P#0.0];
     L       D[AR1,P#4.0];
     T       D[AR2,P#4.0];
```

LAR1	-	Load address register AR1
LAR2	-	Load address register AR2
	P#Zy.x	with an area-crossing pointer
	P#y.x	with an area-internal pointer
LAR1	-	Load address register AR1 with the content of
LAR2	-	Load address register AR2 with the contents of
	MD y	a memory doubleword
	LD y	a local data doubleword
	DBD y	a global data doubleword
	DID y	an instance data doubleword [1)]
LAR1		Load address register AR1 with the contents of accumulator 1
LAR2		Load address register AR2 with the contents of accumulator 1
LAR1	AR2	Load address register AR1 with the contents of address register AR 2
TAR1	-	Transfer the contents of address register AR1 to
TAR2	-	Transfer the contents of address register AR2 to
	MD y	a memory doubleword
	LD y	a local data doubleword
	DBD y	a global data doubleword
	DID y	an instance data doubleword [1)]
TAR1		Transfer the contents of address register AR1 to accumulator 1
TAR2		Transfer the contents of address register AR2 to accumulator 1
TAR1	AR2	Transfer the contents of address register AR1 to accumulator AR 2
CAR		Swap the contents of the address registers
+AR1		Add the contents of accumulator 1 to address register AR 1
+AR2		Add the contents of accumulator 1 to address register AR 2
+AR1	P#y.x	Add a pointer to the contents of address register AR1
+AR2	P#y.x	Add a pointer to the contents of address register AR2

[1)] There are restrictions to the use of these addresses (see "Special Features of Indirect Addressing" below).

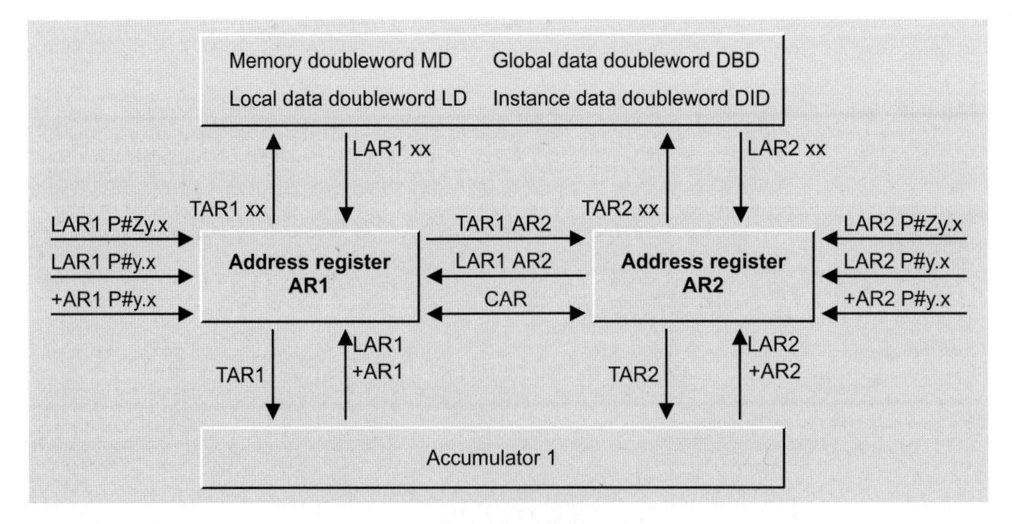

Figure 25.3 STL Statements in Conjunction with Address Registers

Note: System function SFC 20 BLKMOV is available for transferring larger data areas.

25.3.4 Adding to the Address Register

You can add a value to the address registers in order, for example, to increment an address at each loop pass in program loops. You either specify the value as a constant (as an area-internal pointer) in the statement, or the value is located in the right-hand word of accumulator 1. The type of pointer in the address register (area-internal or area-crossing) and the address area are retained.

Adding with pointers

The statements +AR1 P#y.x and +AR2 P#y.x add a pointer to the address register specified. Please note, that with these statements, the area pointer has a maximum size of P#4095.7. If the accumulator contains a value greater than P#4095.7, the number is interpreted as a fixed-point number in two's complement and subtracted (see below).

Example: A data area is to be compared word-wise with a value. If the comparison value is greater than the value in the data field, a memory bit is to be set to "1", otherwise it is to be set to "0".

```
      OPN     DB 14;
      LAR1    P#DBX20.0;
      LAR2    P#M10.0;
      L       Quantity_Data;
Loop: T       LoopCounter;
      L       ComparisonVal;
      L       W[AR1,P#0.0];
      >I      ;
      =       [AR2,P#0.0];
      +AR1    P#2.0;
      +AR2    P#0.1;
      L       LoopCounter;
      LOOP    Loop;
```

Adding with value in the accumulator

The statements +AR1 and +AR2 interpret the value in accumulator 1 as a number in INT format, extend it with the correct sign to 24 bits and add it to the contents of the address register. In this way, a pointer can also be reduced. Violation of the maximum range of the byte address (0 to 65535) has no further effects: The uppermost bits are "cut" (Figure 25.4).

Please note, that the bit address is located in bits 0 to 2. If you want to increment the byte address already in accumulator 1, you must add from bit 3 (shift the value by 3 to the left).

Example: In data block DB 14, the 16 bytes whose addresses are calculated from the pointer in memory doubleword MD 220 and a (byte) offset in memory byte MB 18 are to be deleted. Before adding to AR1, the contents of MB 18 must be adjusted (SLW 3).

Accumulator 1

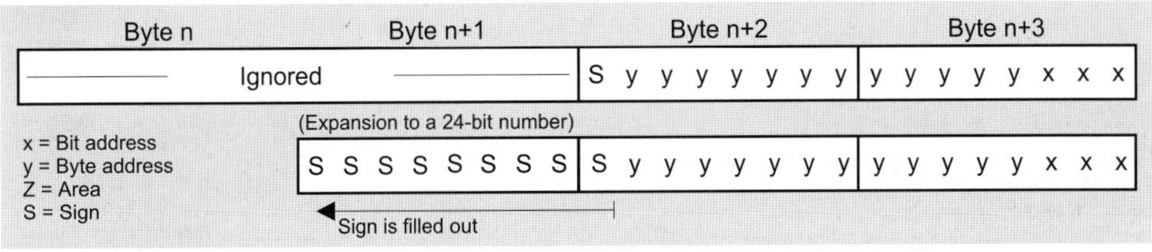

Address register

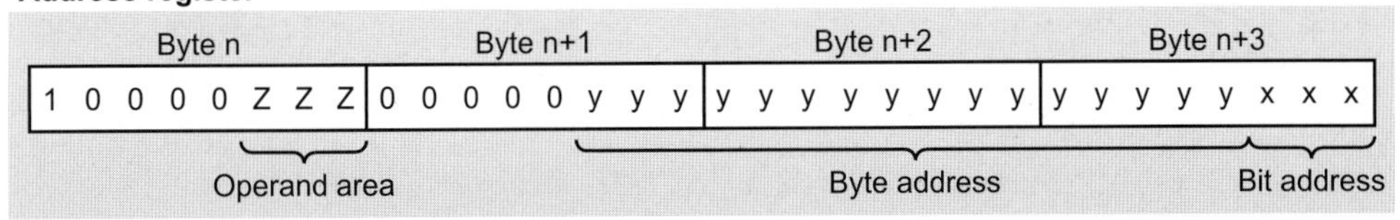

Figure 25.4 Adding to the Address Register

```
OPN   DB 14;
LAR1  MD 220;
L     MB 18;
SLW   3;
+AR1  ;
L     0;
T     DBD[AR1,P#0.0];
T     DBD[AR1,P#4.0];
T     DBD[AR1,P#8.0];
T     DBD[AR1,P#12.0];
```

Note: System function SFC 21 FILL is available for filling larger data areas with bit patterns.

25.4 Special Features of Indirect Addressing

25.4.1 Using Address Register AR1

STL uses address register AR1 to access block parameters that are transferred as DB pointers. In the case of functions, these include all block parameters of complex data type and in the case of function blocks, it means in/out parameters of complex data type.

When you access a block of this kind, in order, for example, to check a bit component of a structure or to write an INT value to a field component, the contents of address register AR1 are changed and so, incidentally, are the contents of the DB register. This also applies when you 'pass on' block parameters of this data type to called blocks.

If you use address register AR1, there must be no block parameter access as described above between loading the address register and indirect addressing. Otherwise, you must save the contents of AR1 before the access, and load them again following the access.

Example: You load a pointer into AR1 and use this address register for indirect addressing. In the meantime, you want to load the value of the structure component *Motor.Act*. Before loading *Motor.Act*, you save the contents of the DB register and address register AR1; after loading, you restore the contents of the registers (Figure 25.5 top).

25.4.2 Using Address Register AR2

With "multi-instance-capable" function blocks (block version 2), STEP 7 uses address register AR2 as the "Base address register" for instance data. When an instance is called, AR2 contains P#DBX0.0 and all accesses to block parameters or static local data in the FB use register-indirect area-internal addressing with the address area DI via this register. A call of a local instance increments the "base address" with +AR2 P#y.x, so that access can be made relative to this address within the called function block that uses the instance data block of the calling function block. In this way, function blocks can be called both as autonomous instances and as local instances (and here at any point in a function block, even several times).

If you program a function block with block version 1 (no "multi-instance capability"), STEP 7 does not use address register AR2.

So if you want to use address register AR2 in a function block with multi-instance capability, you must first save the contents and then restore them after use. You must not program any block parameter or static local data accesses in the area in which you work with address register AR2.

Within functions, there are no restrictions on working with address register AR2.

Example: You want to perform indirect addressing in a function block with AR2 and the DI register. First, you save their contents. You must not access block parameters or static local data again until you have restored the contents of AR2 and the DI register (Figure 25.5 bottom).

25.4.3 Restrictions with Static Local Data

With function blocks compiled with CODE_VERSION1 (no "multi-instance capability"), you can use all statements described in this chapter without restriction.

In the case of function blocks with "multi-instance capability", the Editor accesses instance data via address register AR2; that is, all accesses are indirect. This also applies in conjunction with indirect addressing or when handling address registers. If you use absolute ad-

```
//******************* Save address register AR1 ************************
...
VAR_TEMP
  AR1_Memory    : DWORD;
  DB_Memory     : WORD;
END_VAR
...
//Indirect addressing with AR1 and DB register
LAR1    P#y.x;
OPN     DB z;
...
//Save the register contents
L       DBNO;
T       DB_Memory;
TAR1    AR1_Memory;
//Access block parameters of complex data types
L       Motor.Act;
//Restore the register contents
OPN     DB [DBMemory];
LAR1    AR1_Memory;
T       DBW[AR1,P#0.0];// store loaded value

//******************* Save address register AR2 ************************
...
VAR_TEMP
  AR2_Memory : DWORD;
  DI_Memory  : WORD;
END_VAR
...
//Save the register contents
L       DINO;
T       DI_Memory;
TAR2    AR2_Memory;
//Indirect addressing with AR2 and the DI register
LAR2    P#y.x;
OPN     DI z;
...
L       DIW[AR2, P#0.0];
...
//Restore the register contents
OPN     DI [DI_Memory];
LAR2    AR2_Memory;
```

Figure 25.5 Examples: Saving Address Registers AR1 and AR2

dressing for the instance data in which you store area pointers, the Editor adopts the absolute address. However, as soon as you use symbolic addressing, the Editor rejects this programming as "double indirect addressing".

Table 25.3 gives two examples of this: If you are using memory-indirect addressing in the case of function blocks with "multi-instance capability", you cannot use direct a pointer that you want to store in the static local data. You copy the pointer into a temporary local data item and then you can work with it. You cannot load the pointer in the static local data direct into an address register and you cannot transfer the contents of an address register direct to the pointer (second example).

Table 25.3 Different Programming in the Case of Static Local Data

In the FB with CODE_VERSION1 (no "multi-instance capability")	In function block with "multi-instance capability"

`VAR`
`      sPointer : DWORD;`
`END_VAR`
	`VAR`
	`      sPointer : DWORD;`
	`END_VAR`
	`VAR_TEMP`
	`      tPointer : DWORD;`
	`END_VAR`

`L     MW[sPointer];`	`L     sPointer;`
	`T     tPointer;`
	`L     MW[tPointer];`

`LAR1 sPointer;`	`L     sPointer;` `LAR1;`
`TAR1 sPointer;`	`TAR1;` `T     sPointer;`

26 Direct Variable Access

This chapter shows you how to access the absolute addresses of local variables direct. The "normal" STL statements are available for local variables of elementary data types. Local variables of complex data types or block parameters of type POINTER or ANY cannot be handled "as a whole". To process these variables, you first calculate the starting address at which the variable is stored and you then process parts of the variable with indirect addressing. In this way, you can also process block parameters of complex data types.

The examples in this chapter can be found in the download files (download address: see pages 8-9) in the STL_Book library under the "Variable Handling" program in function block FB 126 or source file Chap_26.

26.1 Loading the Variable Address

The following statements give the starting address of a local variable

```
L     P#name;
LAR1  P#name;
LAR2  P#name;
```

with *name* as the name of the local variable. These statements load an area-crossing pointer into accumulator 1 or into address register AR1 or AR2. The area pointer contains the address of the first byte of the variable. If *name* cannot be identified uniquely as the local variable, insert a "#" before the name so that the statement becomes, for example: L P##*name*. Depending on the block, the variable areas listed in Table 26.1 are permissible for *name*.

With functions, the address of a block parameter cannot be loaded direct into an address register. You can take the route via accumulator 1 here (for example: L P#name; LAR1;).

In function blocks compiled with the keyword CODE_VERSION1 (no "multi-instance capability"), the absolute address of the instance variable is loaded.

In "multi-instance capable" functions blocks, the absolute address *relative to address register AR2* is loaded in the case of the static local data and the block parameters. If you want to calculate the absolute address of the variable in the instance data block, you must add the *area-internal pointer* (address only) of AR2 to the loaded variable address.

Example 1:
Load variable address into address register AR1

```
TAR2  ;
UD    DW#16#00FF_FFFF;
LAR1  P#name;
+AR1  ;
```

Table 26.1 Loading Permissible Addresses for Variable Addresses

Operation	*name* is a	OB	FC	FB V1	FB V2
L P#*name*	temporary local datum	x	x	x	x
	static local datum	-	-	x	x [1)]
	block parameter	-	x	x	x [1)]
LAR*n* P#*name*	temporary local datum	x	x	x	x
	static local datum	-	-	x	x [1)]
	block parameter	-	-	x	x [1)]

[1)] Variable address relative to the address register AR2

With the first two statements, the address in the AR2 is loaded into the accumulator and masked. The content of the AR1 is then added. As a result, the address of the tags *#name* is located in AR1.

Example 2:
Load variable address into accumulator 1

```
TAR2  ;
UD    DW#16#00FF_FFFF;
L     P#name;
+D    ;
```

Similarly to example 1, the result of this is that accumulator 1 then contains the address of the variable *#name*.

The addition of the area-internal pointer can be omitted if it has the value P#0.0. This is the case if you do not use the function block as a local instance.

Please note that "LAR2 P#name" overwrites address register AR2 that is used in the case of "multi-instance capable" function blocks as the "base address register" for addressing the instance data!

You can only access one overall variable with these load statements and not individual components of fields, structures or local instances. You cannot reach variables in global data blocks or in the address areas inputs, outputs, peripheral I/O and bit memory with these load statements.

Table 26.2 shows you how to calculate the address of an INT and a STRING variables in the static local data and how to work with this address. If you use the example program in a function block that you call as a local instance, you must add the base address to the variable address as shown above.

26.2 Data Storage of Variables

26.2.1 Storage in Global Data Blocks

The Editor stores the individual variables in the data block in the order of their declaration. Essentially, the following rules apply here:

▷ The first bit variable of an uninterrupted declaration sequence is located in bit 0 of

Table 26.2 Load Variable Address (Examples)

```
//Variable declaration (function block is not local instance!)
//Variable assignment begins at address P#0.0
VAR
   Field      : ARRAY [1..22] OF BYTE;  //ARRAY variable, occupies 22 bytes
   Number     : INT := 123;             //INT variable, occupies 2 bytes
   FirstName  : STRING[12] := 'Joane';  //STRING variable, occupies 5 bytes
END_VAR
```

`LAR1  P#Number;`	Loads the starting address of *Number* into AR1 AR1 now contains P#DIX22.0
`L     W[AR1,P#0.0];`	Corresponds to the statement L DIW 22 or L *Number*
`LAR1  P#FirstName;`	Loads the starting address of *FirstName* into AR1 AR1 now contains P#DIX24.0
`L     B[AR1,P#0.0];`	Loads the first byte (maximum length of the character string) into accumulator 1
`L     B[AR1,P#2.0];`	Loads the third byte (first relevant byte) into accumulator 1
`L     'John';` `T     D[AR1,P#2.0];`	Writes 'John' into the character string
`L     4;` `T     B[AR1,P#1.0];`	Corrects the current length of the character string to 4 The variable *FirstName* now contains 'John'

the next byte followed by the next bit variables.

- ▷ Byte variables are stored in the next byte.
- ▷ Word and doubleword variables always start at a word boundary, that is, at a byte with an even address.
- ▷ DT and STRING variables start at a word boundary.
- ▷ ARRAY variables start at a word boundary and are "filled" up to the next word boundary. This applies also for bit and byte fields. Field components of elementary data types are stored as described above. Field components of higher data types start at word boundaries. Each dimension of a field is aligned like an autonomous field.
- ▷ STRUCT variables begin at a word boundary and are "filled" up to the next word boundary. This applies also for purely bit and byte structures. Structure components of elementary data types are stored as described above. Structure components of higher data types begin at word boundaries.

By combining bit variables and arranging byte variables in pairs, you can optimize data storage in a data block.

In Figure 26.1, you can see one example each of non-optimized and optimized data storage. Please note that the Editor always "fills" ARRAY and STRUCT variables up to the next word, that is, no bit or byte variables can be stored in byte gaps. However, you can have optimized arrangement of the variables within the structure.

26.2.2 Storage in Instance Data Blocks

The Editor stores the variables in an instance data block in the following order:

- ▷ Input parameters
- ▷ Output parameters
- ▷ In/out parameters
- ▷ Local variables (including local instances)

Each variable is stored in the order of its declaration. The declaration areas each begin at a word boundary, that is, at a byte with an even address. Within the declaration areas, the individual variables are arranged as described in the previous chapter (as in a global data block). Figure 26.2 shows an example for the assignment of an instance data block.

26.2.3 Storage in the Temporary Local Data

Storage of the variables in the temporary local data (L stack) corresponds to storage in a global data block. The assignment always begins at (relative) byte 0. Please note, that in organization blocks, the first 20 bytes are occupied by the start information. Even if you do not use the start information, the first 20 bytes must be declared (even if only with a field of 20 bytes).

The Editor itself also uses local data, for example when transferring parameters in a block call. The Editor applies the symbolically declared temporary local data and the temporary local data it uses itself in the order of their declaration or their use. The absolutely addressed temporary local data are not taken into account here, so that overlaps can occur if you do not know which local data the Editor is applying. If you want or have to access local data with absolute addressing, you can, for example, declare at the first location of the temporary local data declaration a field that keeps free the required number of bytes (words, doublewords). You can then make absolute accesses in this field area. With organization blocks, you define the field after the 20 bytes for the start information.

The example in Figure 26.3 shows the assignment of the temporary local data of an organization block. The field *Ldata* starts immediately following the start information, at byte LB 20 and stretches in this example to byte LB 35. The Editor does not occupy this area with its own temporary data so you can use this area for absolute addressing.

The start information is omitted in functions and function blocks. If you require the temporary local data for absolute addressing, apply the field as the first variable in these blocks; it then begins at byte LB 0.

```
DATA_BLOCK StorageNotOptimized
STRUCT
  Bit1       : BOOL ;
  Bit2       : BOOL ;
  Bit3       : BOOL ;
  Real1      : REAL;
  Byte1      : BYTE;
  Bit array  : ARRAY [1..3] OF BOOL;
  Structure  : STRUCT
    S_Bit1   : BOOL ;
    S_Bit2   : BOOL ;
    S_Bit3   : BOOL ;
    S_Int1   : INT ;
    S_Byte   : BYTE;
    END_STRUCT;
  Characters: STRING[3];
  Date       : DATE ;
  Byte2      : BYTE;
END_STRUCT
BEGIN
END_DATA_BLOCK
```

```
DATA_BLOCK StorageOptimized
STRUCT
  Bit1       : BOOL ;
  Bit2       : BOOL;
  Bit3       : BOOL;
  Byte1      : BYTE;
  Real1      : REAL;
  Bit array  : ARRAY [1..3] OF BOOL;
  Structure  : STRUCT
    S_Bit1   : BOOL ;
    S_Bit2   : BOOL ;
    S_Bit3   : BOOL ;
    S_Byte   : BYTE;
    S_Int1   : INT ;
    END_STRUCT;
  Characters: STRING[3];
  Byte2      : BYTE;
  Date       : DATE ;
END_STRUCT
BEGIN
END_DATA_BLOCK
```

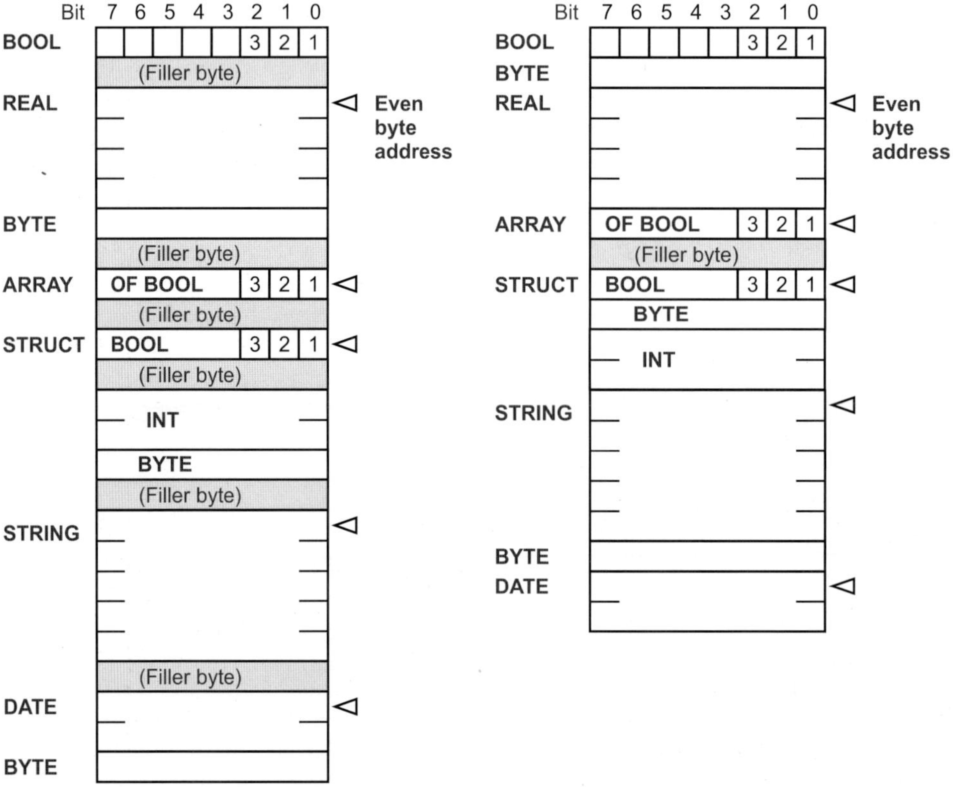

Figure 26.1 Example of assignment of a data block

```
FUNCTION_BLOCK StorageExample
VAR_INPUT
  E_Bit1      : BOOL;
  E_Bit2      : BOOL;
  E_Bit3      : BOOL;
  E_Real1     : REAL;
END_VAR
VAR_OUTPUT
  A_Byte1     : BYTE;
  A_BYTE2     : BYTE;
  A_BYTE3     : BYTE;
END_VAR
VAR_IN_OUT
  D_BYTE1     : BYTE;
  D_Bit1      : BOOL;
  D_Bit2      : BOOL;
  D_Bit3      : BOOL;
END_VAR
VAR
  Date1       : DATE;
  Character   : STRING[3];
  Bit_field   : ARRAY [1..3] OF BYTE;
END_VAR
BEGIN
//...
END_FUNCTION_BLOCK
```

```
ORGANIZATION_BLOCK Cycle
VAR_TEMP
  SInfo     : ARRAY [1..20] OF BYTE;
  LData     : ARRAY [1..16] OF BYTE;
  Temp1     : STRING [36];
  Temp2     : BOOL;
  Temp3     : BOOL;
  Temp4     : BOOL;
  Temp5     : BYTE;
  Temp6     : INT;
END_VAR
BEGIN
//Access via absolute addresses
  ...
  T     LW 20;
  ...
  =     L  22.2;
//Access symbolic
  T     Temp6;
  =     Temp3;
  T     LData[16];
//Load variable address
  L     P#Temp1;
  LAR1  P#Temp2;
//...
END_ORGANIZATION_BLOCK
```

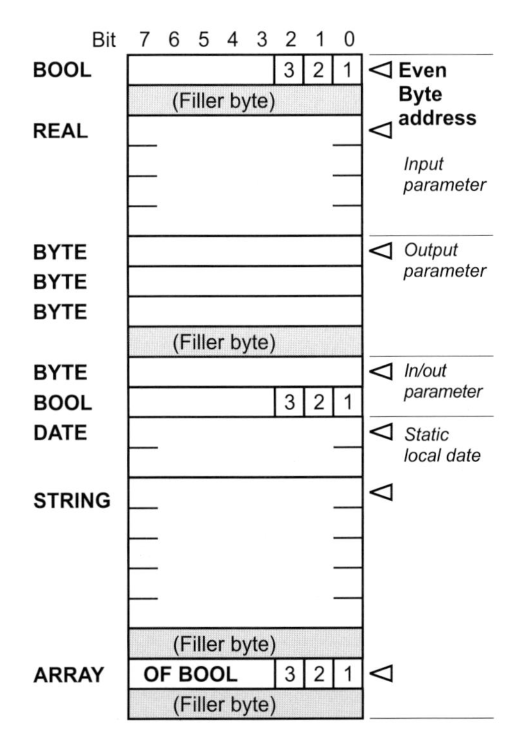

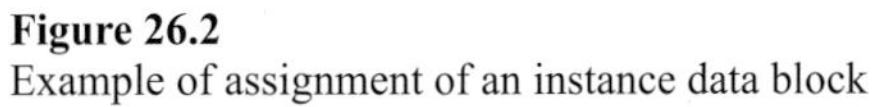

Figure 26.2
Example of assignment of an instance data block

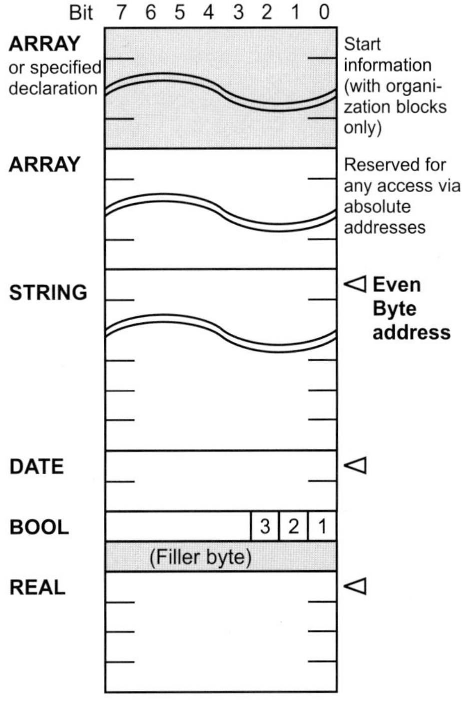

Figure 26.3
Example of assignment of the L stack for organization blocks

26.3 Data Storage when Transferring Parameters

The block parameters are stored differently in functions and function blocks. You as the user need not be concerned with this; you program the block parameters for both block types in the same way. However, this difference is extremely important for direct block parameter access.

26.3.1 Parameter Storage in Functions

The Editor stores a block parameter of a function as area-crossing area pointer in block code in accordance with its own call statement so that each block parameter requires one doubleword of memory. Depending on data type and declaration type, the pointer points to the actual parameter itself, to a copy of the actual parameter in the temporary local data of the calling block (set up by the Editor), or to a pointer in the temporary local data of the calling block that in turn points to the actual parameter (Table 26.3). Exception: With the parameter types TIMER, COUNTER and BLOCK_xx the pointer is a 16-bit number located in the left-hand word of the block parameter.

With elementary data types, the block parameter points direct to the actual parameter (Figure 26.4). However, with the area-pointer as block parameter, you cannot reach any constants or addresses located in data blocks. For this reason, at the compiling stage, the Editor copies a constant or a (fully-addressed) actual parameter located in a data block into the temporary local data of the calling block and directs the area pointer to this. This parameter area is called V (temporary local data of the preceding block, V area).

Copying to the V area takes place before the actual FC call with input parameters and with in/out parameters and after the call with in/out and output parameters and therefore also with the function value. For this reason, the rule that you can only check input parameters and write to output parameters also applies. If, for example, you transfer a value to an input parameter with a fully-addressed data address, the value will be stored in the temporary local data of the preceding block and it will be forgotten because no more copying take place to the "actual" variable in the data block.

It is a similar story for loading a corresponding output parameter: Since no copying takes place from the "actual" variable from the data block to the V area, you load an (indeterminate) value from the V area in this case.

Because of the copy operation, you have to overwrite an output parameter, and therefore also a function value, of elementary data type in the block with a defined value if a fully addressed data address is envisaged or could be envisaged as the actual parameter. If you do not assign a value to the output parameter (e.g. by previously exiting the block or by jumping the program location) the local datum will also not be initialized. It is then at the value it "happened" to have prior to the block call. The output parameter is then overwritten with this "undefined" value.

With complex data types (DT, STRING, ARRAY, STRUCT as well as UDT), the actual parameters are located either in a data block or in the V area. Since an area pointer cannot reach an actual parameter in a data block, the Editor creates a DB pointer in the V area at the compiling stage. Since pointer then points to the actual parameter in the data block (DB No. <> 0)

Table 26.3 Parameter Storage in Functions

Data Type	INPUT	IN_OUT	OUTPUT
	The parameter is an area pointer to		
Elementary	a value	a value	a value
Complex	a DB pointer	a DB pointer	a DB pointer
TIMER, COUNTER, BLOCK	a number	not possible	not possible
POINTER	a DB pointer	a DB pointer	a DB pointer
ANY	an ANY pointer	an ANY pointer	an ANY pointer

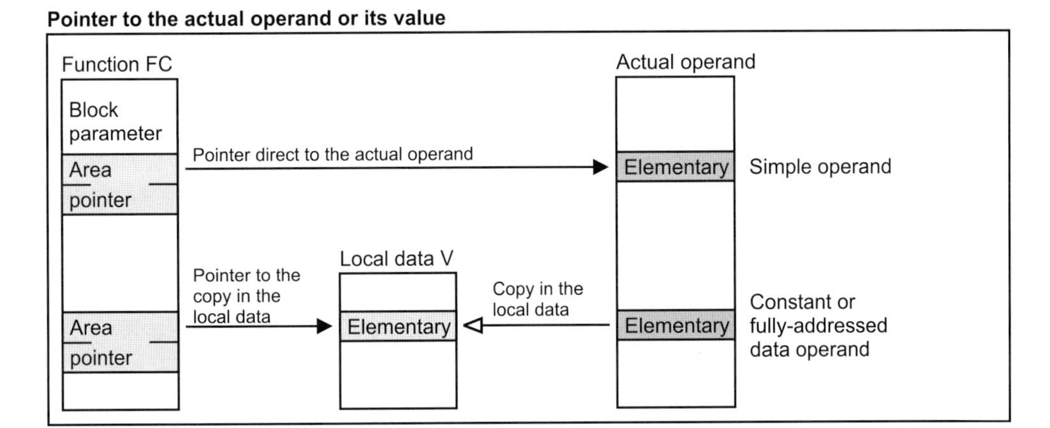

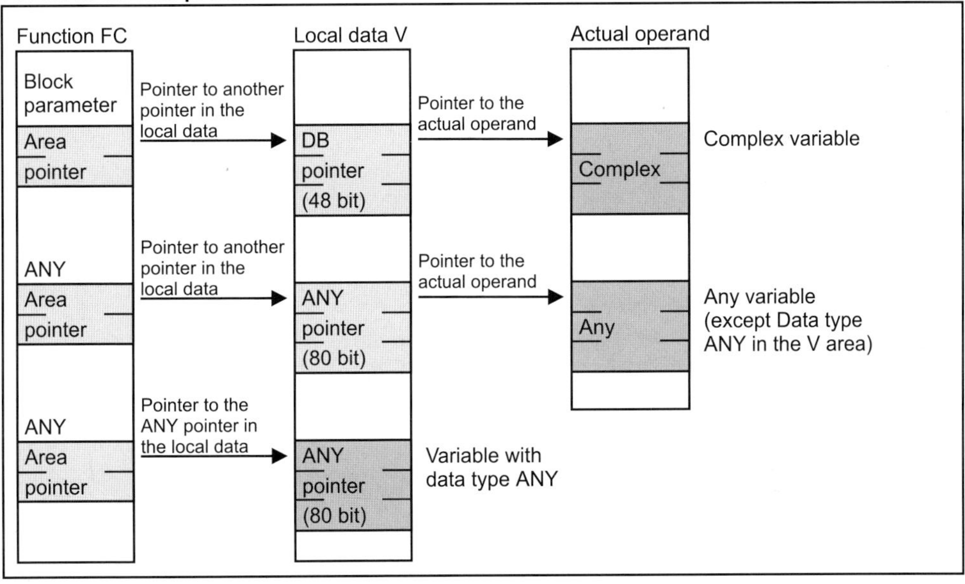

Figure 26.4 Transferring Parameters in Functions

or to the V area (DB No. = 0). The DB pointers for all declaration types in the V area are created before the 'actual' FC call.

With the parameter types TIMER, COUNTER and BLOCK_xx the block parameter contains a number (16 bits left-justified in the 32-bit parameter) instead of the area pointer.

The parameter type POINTER is handled in exactly the same way as a complex data type.

With the parameter type ANY, the editor creates a 10-byte ANY pointer in the V area which may point to any variable. The same principle applies as with complex data types.

The Editor makes an exception if you apply at a block parameter of type ANY an actual parameter that is located in the temporary local data and is of data type ANY. The Editor then does not create any more ANY pointers, but instead directs the area pointer (the block parameter) straight to the actual parameter (in this case, the ANY pointer can be modified at runtime, see Chapter 26.3.3 "“Variable” ANY Pointer").

26.3.2 Storing Parameters in Function Blocks

The Editor stores the block parameters of a function block in the function block's instance data block. At the function block call, the Editor generates statement sequences that copy the values of the actual parameters into the instance data block before the actual call and then copy them back from the instance data block to the actual parameters following the call. You do not see these statement sequences when you look at the compiled block. You only notice it indirectly in the occupied memory space.

In the instance data block, the block parameters are stored either as a value, as a 16-bit number or as a pointer to the actual parameter (Table 26.4). When stored as a value, the memory required depends on the data type of the block parameter; the number occupies 2 bytes, the pointers occupy 6 bytes (DB pointers) or 10 bytes (ANY pointers).

The relationships between block parameters, instance data assignments and actual parameters are shown in Figure 26.5. When copying actual parameters of complex data types into the instance data block (input parameters) or back to the actual parameter (output parameter), the Editor uses the system function SFC 20 BLKMOV, whose parameters it builds up in the temporary local data area of the calling block.

Copying block parameters that are stored as a value in the instance data block is carried out using statement sequences before the "actual" FB call in the case of input parameters and in/out parameters, and after the call in the case of in/out and output parameters. For this reason, the rule that you can only check input parameters and write to output parameters also applies. If, for example, you transfer a (new) value to an input parameter, the current value of the actual parameter is lost. If you load an output parameter, you load the (old) value in the instance data block and not that of the actual parameter.

Because the block parameters are stored in the instance data block, they need not be initialized every time the function block is called. When no initialization is made, the program works with the "old" value of the input or in/out parameter or you fetch the value of the output parameter from another subsequent position in the program. Outside the function block, you can access variables in the instance data block in the same way as you do variables in a global data block (with the symbolic name of the data block and the name of the block parameter). The same applies also for the static local data.

If you apply a temporary local variable of data type ANY at an ANY parameter, the Editor copies the contents of this variable into the ANY pointer (into the block parameter) in the instance data block.

26.3.3 "Variable" ANY Pointer

ANY parameters can only be parameterized with data areas or variables that must be defined already at the compiling stage. Example: Copying a variable into a data area with SFC 20 BLKMOV

```
CALL SFC20 (
  SRCBLK  := "ReceiveMailbox".Data,
  RET_VAL := SFC20Error,
  DSTBLK  := P#DB63.DBX0.0 BYTE 8);
```

It is possible to modify or re-define the variable or the data area at runtime because the Editor

Table 26.4 Storing Parameters in the Case of Function Blocks

Data Type	INPUT	IN_OUT	OUTPUT
Elementary	Value	Value	Value
Complex	Value	DB pointer	Value
TIMER, COUNTER, BLOCK	Number	not possible	not possible
POINTER	DB pointer	DB pointer	not possible
ANY	ANY pointer	ANY pointer	not possible

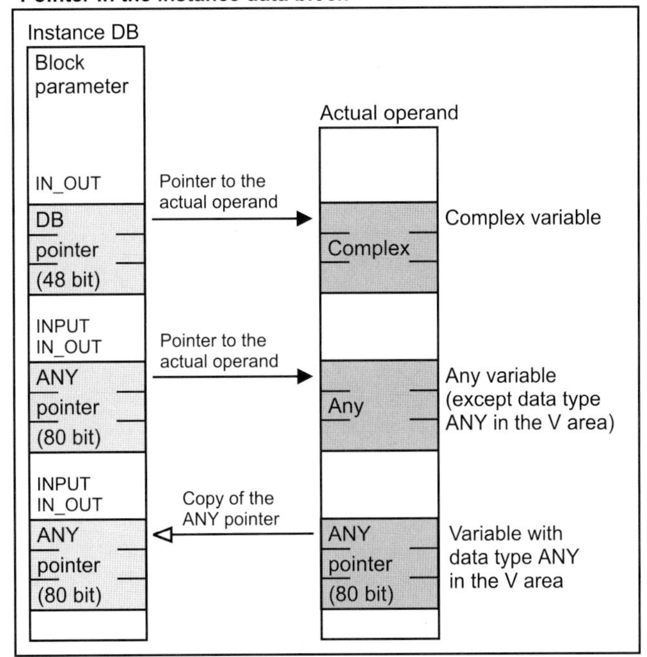

Figure 26.5 Transferring Parameters in the Case of Function Blocks

applies a fixed ANY pointer to the actual parameter in the temporary local data (see below in this chapter).

The Editor makes an exception to this if the actual parameter itself is in the temporary local data and the is of data type ANY. Then no further ANY pointer is set up, instead the Editor interprets these ANY variables as ANY pointers to the actual parameter. This means that the ANY variable must have the same structure as an ANY pointer.

You can now modify these ANY variables in the temporary local data at runtime and so specify another actual parameter for an ANY parameter. To apply this "variable" ANY pointer, proceed as follows:

▷ Applying a temporary local variable of data type ANY (The name of the ANY variable can be selected freely within the permissible framework for block-local variables.):

```
VAR_TEMP
  ANY_POINTER : ANY;
END_VAR
```

▷ Providing the ANY tag with values:

The address is known at runtime (e.g. from 0)	The address is not known at runtime
	LAR1 P#ANY_POINTER;
L W#16#1002;	L W#16#1002;
T LW 0;	T LW[AR1,P#0.0];
L 16;	L 16;
T LW 2;	T LW[AR1,P#2.0];
L 63;	L 63;
T LW4;	T LW[AR1,P#4.0];
L P#DBX0.0;	L P#DBX0.0;
T LD 6;	T LD[AR1,P#6.0];

▷ Initialize the ANY parameter, for example, at an SFC 20

```
CALL SFC20 (
  SRCBLK  := "ReceiveMailbox".Data,
  RET_VAL := SFC20Error,
  DSTBLK  := ANY_POINTER);
```

This procedure is not restricted to SFC 20 BLKMOV; you can use it on all ANY parameters of any blocks.

Example: We want to write a copy block that is to copy data areas between data blocks. The source and target area is to be parameterizable. We use SFC 20 BLKMOV for copying. The block – a function FC – has the following parameters:

```
VAR_INPUT
  QDB  : INT; //Source data block
  SSTA : INT; //Source starting
               address
  NUMB : INT; //Number of bytes
  DDB  : INT; //Destination data block
  DSTA : INT; //Destination starting
               address
END_VAR
```

The function value is to contain the error message of SFC 20 and can then also be evaluated as if we were using SFC 20 direct. In addition, the status bit BR is set to "0" in the event of an error.

Two ANY variables, one as a pointer for the source area and one as a pointer for the target area, are sufficient for the block-local data:

```
VAR_TEMP
  SANY : ANY; //ANY pointer source
  DANY : ANY; //ANY pointer
               destination
END_VAR
```

Since we know the addresses of the ANY pointers in the temporary local data, we can program them with absolute addresses, for example, the preparation of a source pointer:

```
L      W#16#1002;//Type BYTE
T      LW 0;
L      NUMB;     //Number of bytes
T      LW 2;
L      QDB;      //Source DB
T      LW 4;
L      SSTA;     //Start of the source
SLD 3;
OD     DW#16#8400_0000;
T      LD 6;
```

The destination pointer starting at address LB 10 is prepared in the same way.

It only remains to initialize SFC 20:

```
CALL SFC20 (
  SRCBLK  := SANY,
  RET_VAL := RET_VAL,
  DSTBLK  := DANY);
```

The function value RET_VAL of SFC 20 is initialized with the function value RET_VAL of our function FC.

This little example in full can be found in the download files (download address: see pages 8-9) (function FC 47 in the program "General Examples").

In this way, an ANY pointer can be assigned any value, for example, you can vary the type in word 2 or the area pointer so that in principle, you can address any variables and data areas, for example also the bit memory area.

Note: if the ANY pointer located in the temporary local data points to a variable that is also located in the temporary local data of the calling block, V must be entered as the address area because from the viewpoint of the called block, this variable is located in the temporary local data of the predecessor block.

26.4 Brief Description of the Message Frame Example

The following examples will deepen your understanding of how to handle complex variables. the program of the different blocks each emphasizes a specific aspect of this topic. The declared technological function of the examples such as "Generate_ Frame" and "Checksum", are intended only to make things clearer and, where necessary, are dealt with only briefly.

At this point, the examples are described with text and figures. The program can be found in the download files (download address: see pages 8-9) in the STL-Book library under the program "Message Frame Example".

This example consists of the following sections:

▷ Message frame data
(UDT 51, UDT 52, DB 61, DB 62, DB 63)
shows how to handle self-defined data structures

▷ Clock check (FC 61)
shows how to handle system blocks and standard blocks

▷ Checksum (FC 62)
shows how to use direct variable access

- ▷ Generate frame (FB 51)
  shows how to use SFC 29 BLKMOV with fixed addresses
- ▷ Store frame (FB 52)
  Shows how to use the "variable" ANY pointer
- ▷ Date conversion (FC 63)
  shows the processing of variables of complex data types

Message frame data example

The example shows how you can define frequently occurring data structures as your own data type and how to use this data type in variable declaration and parameter declaration.

We establish a database for incoming and outgoing frames: A send mailbox with the structure of a message frame, a receive mailbox with the same structure and a (receive) ring buffer for intermediate storage of the incoming message frames (Figure 26.6).

Since the data structure of the message frame occurs frequently, we want to make it a user-defined data type (UDT) Frame. The frame contains a frame header; we also want to give a name to the structure of the frame header. The send mailbox and the receive mailbox are to be data blocks each containing a variable with the structure of Frame. Finally, there is the ring buffer, a data block with a field of eight components that also have the data structure of Frame.

First, we define the UDT Header, then the UDT Frame. Frame consists of a structure Header, a field *Measured_values* with 4 components and a variable *Check*. All components are initialized with zero. In the data blocks "Send_mailb" and "Rec_mailb", a variable *Data* with the structure *Frame* is defined in each case.

The variables can now be individually initialized in the initialization section of the data block. In the example, the component *ID* receives a value in each case that deviates from the initialization in the UDT. The data block "Buffer" contains the variable *Entry* as a field with 8 components of the structure *Frame*.

Here too, the individual components can be initialized with different values in the initialization section (for example:
Entry[1].Header.Numb := 1).

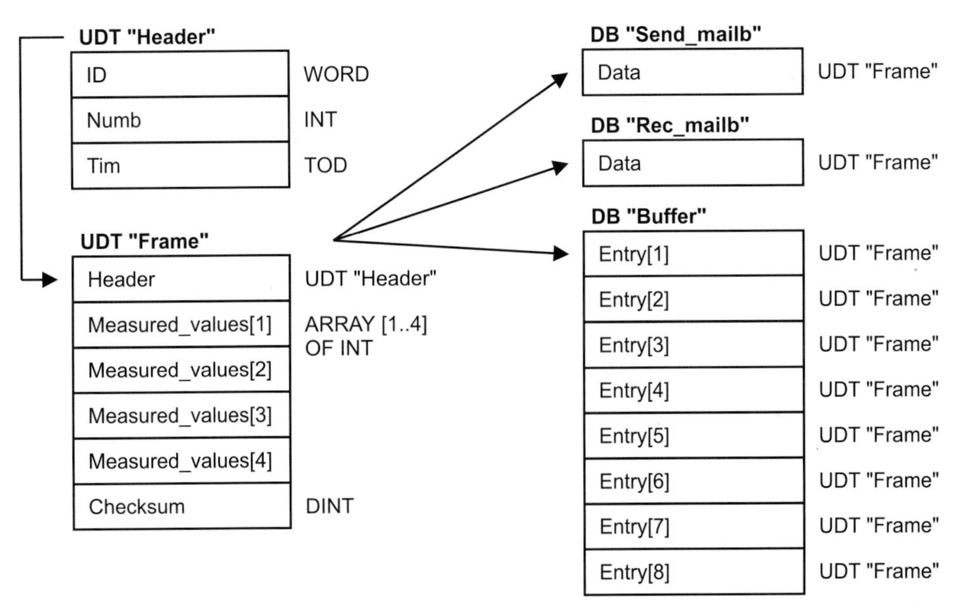

Figure 26.6 Data Structure for the Message Frame Data Example

This example contains the following objects that are used in the examples that follow:

UDT 51 User-defined data type Header

UDT 52 User-defined data type Frame

DB 61 Send mailbox (Send_mailb)

DB 62 Receive mailbox (Rec_mailb)

DB 63 Buffer

Clock check example

The example shows how to handle system blocks and standard blocks (error evaluation, copying from the library, renaming).

The function "Clock_check" is to output the time-of-day in the CPU-integrated real-time clock as a function value. For this purpose, we require the system function SFC 1 READ_CLK that reads the date and time-of-day of the real-time clock in data format DATE_AND_TIME or DT. Since we only want to read the time-of-day, we also require the IEC function FC 8 DT_TOD. This function fetches the time-of-day in format TIME_OF_DAY or TOD from the data format DT (Figure 26.7)

The time specification of the real-time clock is stored in data block "Data66" since we still require this information for the "Date conversion" example. Without this additional use, we could have also declared a temporary local variable instead of the variable *CPU_Tim*.

Error evaluation

The system functions signal an error via the binary result BR and the via the function value RET_VAL. An error exists if the binary result BR = "0"; the function value is then also negative (bit 15 is set). The IEC standard functions signal an error only via the binary result.

Both types of error evaluation are shown in the example. In the "Clock_Check" function, the binary result is first set to "1"; if an error exists, the binary result is set to "0" by the relevant block. Then an invalid value is output for the time-of-day. After the "Clock_check" function has been called, you can also check for an error via the binary result.

Offline programming of system functions

Before compiling the example program or before calls in incremental program input, the offline user program must contain the system function SFC 1 and the standard function FC 8.

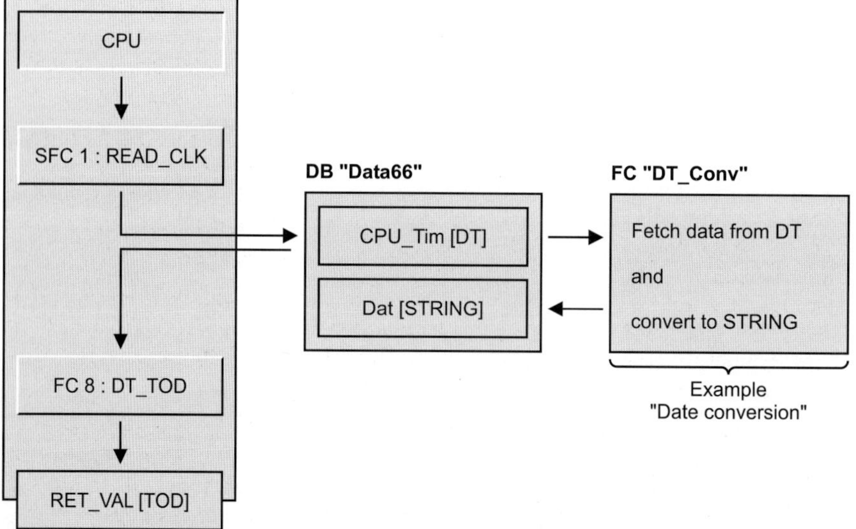

Figure 26.7 Clock Check Example

Both functions are included in the scope of supply of STEP 7. You can find these functions in the block libraries supplied. (For the system functions integrated in the CPU, the library contains an interface description instead of the program of the system functions. The function can be called offline via this interface description; the interface description is not transferred to the CPU. The loadable functions such as the IEC functions are stored in the library as executable programs.)

With FILE → OPEN in the SIMATIC Manager, you select the library *Standard Library* and open the library *System Function Blocks*. Under *Blocks* here you will find all interface descriptions for the system functions. If you have not opened the project window of your project, you can arrange both windows next to each other with WINDOW → ARRANGE → VERTICALLY and drag the selected system functions into your program with the mouse (mark the SFC with the mouse, "hold" it, "drag" to *Blocks* or to its open window and "drop"). You can copy the standard function FC 8 in the same way. You will find it in the library *IEC Function Blocks*. FC 8 is a loadable function; it therefore occupies user memory, in contrast to SFC 1.

If a standard block is called from the Editor's Program Element Catalog under "Libraries" during incremental programming, it is automatically copied to *Blocks* and entered in the symbol table.

Renaming standard functions

You can rename a loadable standard function. You mark the standard function (for example, FC 8) in the project window and click (again) on the identifier. A frame appears around the name and you can specify a new address (for example, FC 98). If you now press F1 while the standard function (renamed to FC 98) is marked, you will still nevertheless receive the online Help function for the original standard function (FC 8).

If an identically addressed block exists when copying is performed, a dialog box appears to allow to choose between overwriting and renaming.

Symbol address

In the symbol table, you can assign names to the system functions and the standard functions, so that you can also access these functions symbolically. You have a free choice of names within the framework of the permissible definitions for block names. In the example, a symbolic name has been selected for each block name (for improved identification).

Checksum example

This example shows direct access to a block parameter of type ANY with calculation of the variable address and use of indirect addressing.

A checksum is to be generated from a data structure by simply adding all bytes with no account being taken of any carry (overflow, number range violation for DINT).

All data structures (STRUCT and UDT) are treated by the Editor like a field with byte components if they are applied to a block parameter of parameter type ANY. With this program, therefore, you can generate the checksum not only from a field with byte components (ARRAY OF BYTE), but also from structure variables. If you also want to use the program on variables of other data types, you must modify the relevant check (type ID in the ANY pointer).

The checksum function uses direct variable access to get the absolute address of the block parameter (more precisely: the address at which the Editor has stored the ANY pointer).

First, a check is made to ensure that the type ID "Byte" and a repetition factor >1 has been entered. In the event of an error, the binary result is set to "0" and the function is exited with a function value equal to zero.

The starting address of the actual parameter (at runtime) is in the ANY pointer. It is loaded into address register AR1. If the variable is located in a data block, this data block is also opened.

The next network adds the values of all bytes making up the actual parameter. The program loop runs until the variable *Quantity* has the value zero (LOOP decrements this value).

Then, the total is transferred to the function value.

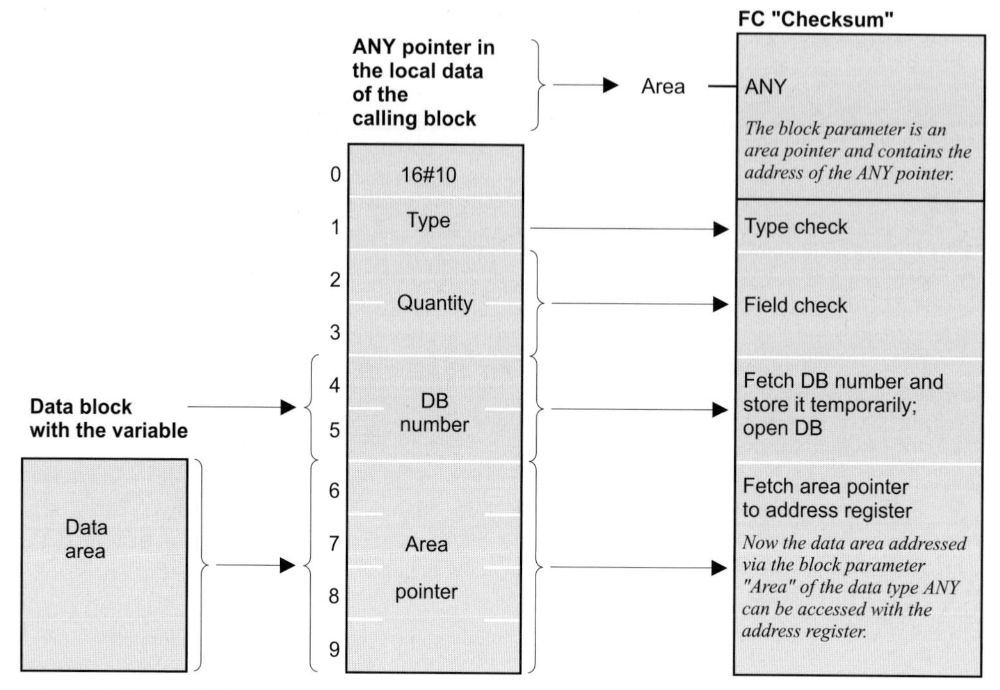

Figure 26.8 Example: Checksum

Generate frame example

The example shows you how to copy complex variables with the function SFC 20 BLKMOV.

The data block "Send_mailb" is to be filled with the data of a message frame. We use a function block that has stored the ID and the consecutive number in its instance data block. The net data are located in a global data block; they are copied to the send mailbox with the system function BLKMOV.

We get the time-of-day from the real-time clock in the CPU with the help of the function "Clock_check" (see previous example) and we generate the checksum by simply adding all bytes in the message frame header and the data (see "Checksum" examples). Figure 26.9 shows the program and the data structures.

The first network in the function block FB "Generate_Frame" transfers the ID stored in the instance data block to the frame header. The consecutive number is incremented by +1 and is also entered in the frame header.

The second network contains the call of the function "Clock_check" that fetches the time-of-day from the real-time clock and enters it in the format TIME_OF_DAY in the frame header.

In the third network, you can see a method of using system function SFC 20 BLKMOV to copy variables selected at runtime without using indirect addressing. It is therefore also not necessary to know the absolute address and the structure of the variable.

The principle is extremely simple: The desired copy function is selected with the jump distributor. The numbers 1 to 4 are permissible as selection criteria. The example "Buffer entry" shows the same functionality this time with a variable destination range using a pointer calculated at runtime.

The next network generates the checksum via the frame header and the frame data. Since the function "Checksum" generates the checksum over a single data area, the frame header and the data are first combined in the temporary variable *Block*. The contents of *Block* are then added

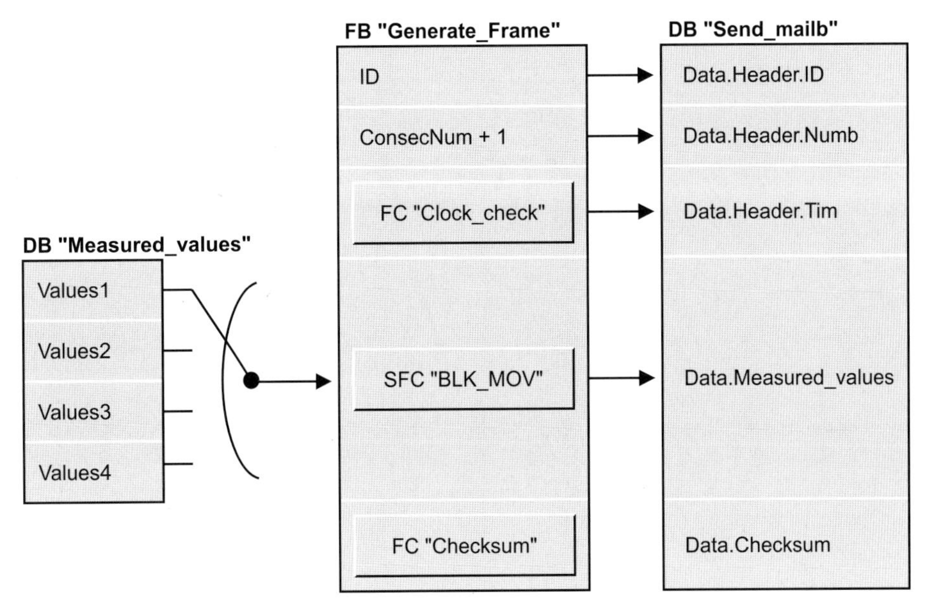

Figure 26.9 Example: Generate Frame

bytewise and stored in the checksum in the send frame.

The FB "Generate_Frame" is programmed in such a way that it can be called via a signal edge for the purpose of generating the frame.

Store frame example

This example concentrates on showing you how to use an "variable" ANY pointer.

A frame in the data block "Rec_mailb" is to be entered at the next location in the data block "Buffer". The block-local variable Entry determines the location in the ring buffer; the address of the ring buffer is calculated from the value in this location (Figure 26.10).

If the number of the frame in the receive mailbox has changed, the frame is to be written to a buffer at the next location. The buffer is to be a data block that can accommodate 8 frames. After the eighth frame has been entered, the next frame is to b entered at the first location again.

The function block "Store_Frame" compares the entered frame number with the stored number in the data block "Rec_mailb". If the frame numbers are different, the stored number is corrected and the frame in the receive mailbox is copied to the data block "Buffer" in the next entry. The system function SFC 20 BLKMOV handles the copying. Since the destination can be different depending on the value of *Entry*, we calculate the absolute address of the target area, generate an ANY pointer from this in the variable *ANY_Pointer* and transfer it to the SFC at the parameter DSTBLK. Please note that you only use area-internal addressing for indirect addressing of a temporary local variable.

The data structure *Frame* has a length of 20 bytes (header: 8 bytes, Meas: 8 bytes, Check: 4 bytes). The variable *Receive* in the data block "Rec_mailb" is therefore 20 bytes long, just as every component of the field *Entry* in the data block *Buffer* is also 20 bytes long. Consequently, the individual components *Entry[n]* begin at byte address n × 20, where n corresponds to the variable *Entry*.

Date conversion example

The example concentrates on showing the processing of variables of complex data types using direct variable access and indirect addressing with both address registers.

The global data block "Data66" contains the variables *CPU_Tim* (data type DATE_AND_

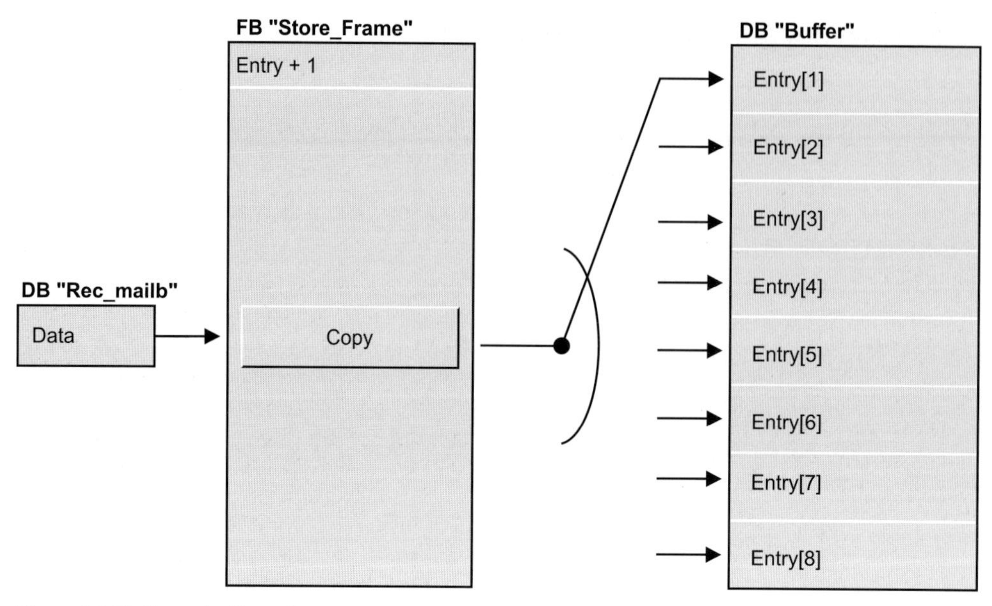

Figure 26.10 Example: Store Frame

TIME) and *Dat* (data type STRING). The date is to be fetched from the variable *CPU_Tim* and stored as a character string with the format "YYMMTT" in the variable *Dat*.

The subsequent program in the function "DT_Conv" uses address register AR1 and the DB register for the pointer to the input parameter *Tim* and address register AR2 and the DI register for the pointer to the function value (corresponding to the STRING variable *Dat* in the data block "Data66"). The program is located in a function so that both data block registers and both address registers are available without restriction.

The program in the first network calculates the address for the actual parameter at the block parameter *Tim*, valid at runtime, and stores the address in the DB register in AR1. An actual parameter of complex data type can only be located in a data block (global or instance data) or in the temporary local data of the calling block (in the V area). If the actual parameter is in a data block, the data block number would be loaded into the DB register and the area pointer in AR1 would contain the address area DB. If the actual parameter is in the V area, zero would be loaded into the DB register and the are pointer in AR1 would contain the address area V.

The second network contains the equivalent program for the function value whose address is then located in address register AR2 and in the DI register. In order to be able to address indirect via the DI register as well, the address area DI must be entered in AR2. However, depending on the memory occupied by the actual parameter, either DB for data block or V for V area would be found here. By setting bit 24 in AR2 to "1", we change the address area from DB to DI but we do not change any address area V.

Prepared in this way, the maximum length fixed for the actual parameter at the function value can be checked in the next network. The length must be at least 6 characters. If it is less than 6, "0" in is entered in the binary result BR (otherwise "1") and block processing is terminated. In this way, you can check for processing errors via the binary result after calling the function "DT_Conv".

The next network fetches the year and the month from *Tim* (in BCD form), converts the values into ASCII characters (precedes them with a 3) and writes them back to the function value. The same happens with the days.

The program is ended with the correction of the current length in the function value.

Structured Control Language (SCL)

Structured Control Language (SCL) is a high-level programming language for SIMATIC S7. The language is based on the "Structured Text" part of the DIN EN 61131-3 standard and has PLC-Open certification Base Level from V4.01 if international mnemonics are used (the German mnemonics are used in this book). SCL is optimized for the programming of programmable controllers and contains PASCAL language elements as well as typical PLC elements like inputs and outputs.

SCL is especially suitable for programming complex algorithms or for tasks from the data management area. SCL supports the STEP 7 block structure and so enables creation of an S7 program in conjunction with the STL, LAD and CSF programming languages.

S7-SCL is available as optional software for STEP 7 Basis. The description in this book is based on the SCL version 5.3 SP5.

Following installation, SCL is fully integrated in the SIMATIC Manager and is then used in the same way as one of the basic programming languages (e.g. STL). Using the SCL program editor, you create within an S7 project the program sources that you then compile with the SCL compiler. The user program contains the compiled SCL blocks; it can also contain compiled blocks created in other languages. You can test the blocks created with SCL online in the CPU using the SCL Debugger.

The **language elements** of SCL differ in instruction notation from the language elements of the basic programming languages (operators, expressions, value assignments). They all share data types, address areas, symbolic programming and block structure.

Using the **control statements**, you can execute program branches (alternatives), repeat program sections (program loops) or exit linear program execution and then continue it at another location in the block (jumps).

You program **blocks** with SCL and you can call blocks (integrate them into your program so to speak) that have been created with SCL or with another S7 programming language. With SCL, you have access to all system functions.

The standard functions like conversion functions are available as **SCL functions**, or you can program your own functions with SCL or STL. In addition, the **IEC functions** supplied with the STEP 7 Basic Package enable the handling of variables with complex data types.

27 Introduction, Language Elements

This chapter gives an overview of the requirements for programming with SCL. Chapters 2 and 3 contain the detailed description; reference is made to these chapters where appropriate.

Chapter 2 “STEP 7 Programming Software” introduces the “programming tools”: symbol editor, SCL program editor, compiler and debugger. The chapter also indicates the environment in which you can write an SCL program.

Chapter 3 “SIMATIC S7 Program”, shows you how a user program is structured. It describes the different types of program execution, the program block structure and it lists all the required keywords for block programming. There is an introduction to variable addressing and to the STEP 7 data types.

The examples in this chapter can be found in the download files (download address: see pages 8-9) in the SCL_Book library under the program “27 Language Elements”.

27.1 Integration in SIMATIC

27.1.1 Installation

Installation of SCL requires a SIMATIC Manager of a suitable version. SCL is installed with the SETUP program; memory requirement is approximately 50 MB if all languages and examples are installed. You require the relevant authorization for SCL, supplied on its own diskette or on a USB flash drive.

27.1.2 Setting Up a Project

The SIMATIC Manager is the central tool for SCL too. To program with SCL, you must start it and, in the same way as for a standard programming language, set up a project (Chapter 2.1 “STEP 7 Basic Package”). You can either use the project wizard or create the project “manually”.

When configuring the station, it is sufficient to assign a CPU so that the SIMATIC Manager sets up the containers for the associated S7 program. You can also set up an S7 program direct under a project and assign it later to a CPU.

You can also use an existing project. The containers *S7 Program*, *Sources* and *Blocks* must be available, as well as *Symbols* for the symbol table. If an object does not exist, you generate it by marking the (higher-level) containers and then selecting the menu command INSERT.

If you use an existing project, then STL sources or compiled blocks, written say in CSF, may already exist. This does not upset the SCL editor. You can even call previously compiled blocks in the SCL program regardless of the language used to write them.

27.1.3 Editing the SCL Source

Mark the *Sources* container and select INSERT → SCL SOURCE. This menu command is only available if you have installed SCL. You can now rename the inserted object *SCL Source(1)*. Double click on the SCL source to call the SCL program editor which then displays an empty source file. Now you can enter the SCL program.

Use of the SCL program editor is described in Chapter 2.5.4 “SCL Program Editor”. You begin program entry by editing a block. Chapter 3.5 “Programming Code Blocks with SCL” describes the structure of a block and gives the keywords.

Here is a simple example to get you started: we’ll program a “Delimiter” function that limits an input value between an upper and lower value and we’ll call this function in the organization block OB 1 (Figure 27.1).

```
FUNCTION Limiter : INT
VAR_INPUT
   MAXI : INT;                                  //Maximum value
   IN   : INT;                                  //Input value
   MINI : INT;                                  //Minimum value
END_VAR
BEGIN
IF IN > MAXI THEN Limiter := MAXI;              //Restricted to upper limit
   ELSIF IN < MINI THEN Limiter := MINI;        //Restricted to lower limit
   ELSE Limiter := IN;                          //Value lies between the limits
END_IF;
END_FUNCTION

ORGANIZATION_BLOCK Main
VAR_TEMP
   SINFO : ARRAY [1..20] OF BYTE;
END_VAR
BEGIN
Result := Limiter (MAXI := Maximum, IN := Input_value, MINI := Minimum);
END_ORGANIZATION_BLOCK
```

Figure 27.1 Example: "Limiter"

The example program begins with the definition of the block type for the delimiter (a function FC) and defines the data type of the function value (INT). This is followed by the declaration of the block parameters: an INT input for the maximum value, the minimum value and the input value. The program itself follows the declaration section. If the input value IN is greater than the maximum value, the function value is to assume the maximum value. If this is not the case, and the input value is less than the minimum value, the minimum value is assigned to the function value. If neither applies, the function value assumes the contents of the input value.

Then we call the "Delimiter" function in the organization block "Main Program". In SCL you must also reserve 20 bytes of temporary local data as start information in an organization block whether you use them or not.

In contrast to the standard programming languages, a function FC with function value is a "real" function in SCL that you can incorporate in an expression in place of an address, provided the data types are compatible. When calling the "Delimiter" function in the organization block "Main Program", its value is assigned to a global variable "Result"; this variable now contains the "Input_value" limited between the limits "Maximum" and "Minimum".

An SCL source can contain one or more blocks. You can also create several sources that are then compiled with a compiler control file in the order specified.

Save the source file with FILE → SAVE. Because we use symbols instead of operands in the program, we need to fill out the symbol table before compiling.

27.1.4 Completing the Symbol Table

The symbol table is completed in SCL in the same way as in the standard programming languages (see Chapter 2.5.2 "Symbol Table"). You can also supplement an existing, partly completed, symbol table with the desired entries (there can only be one single symbol table in any given S7 program). The symbol table is represented by the object *Symbols* in the *S7 Program* container.

You can call the symbol editor either by selecting OPTIONS → SYMBOL TABLE in the SCL program editor or by double clicking on *Symbols* in the SIMATIC Manager. You enter your

Table 27.1 Symbol Table for the "Delimiter" Example

Symbol	Address	Data Type	Comment
Main	OB 1	OB 1	Cyclic program execution block
Limiter	FC 271	FC 271	Function for limiting an INT variable
Input_value	MW 10	INT	Specified value
Maximum	MW 12	INT	Upper limit
Minimum	MW 14	INT	Lower limit
Result	MW 16	INT	Limited value

symbols in an empty symbol table or supplement existing entries (Table 27.1).

You can, of course, select other addresses to try them out. We then save the symbol table.

27.1.5 Compiling the SCL Program

To compile, open the SCL source, if it is not already open. You can find the options required for compiling under OPTIONS → CUSTOMIZE on the "Compiler". (If blocks are to be generated, for example, select the "Blocks generated" option).

The source is compiled with FILE → COMPILE; the compiled blocks are stored in the *Blocks* container. Chapter 2.5.4 "SCL Program Editor" contains more details on compiling.

With a compiler control file, you can also batch compile several sources in any order. Please note that called blocks or functions must be available at compiling, either as compiled blocks in the Blocks container, as (error-free) program sources before being called in the source file or as standard functions in the standard library.

27.1.6 Loading SCL Blocks

If the programming device is connected to a CPU, the compiled blocks are loaded into the CPU's user memory with PLC → LOAD. The CPU must be in the STOP mode because the sequential order on loading may differ from the sequential call order. Please refer to Chapter 2.6 "Online Mode" for details of any other points you must take into account.

You can also handle the blocks with the SIMATIC Manager in the offline or online window.

27.1.7 Testing SCL Blocks

The SCL debugger can test individual blocks in Program Status ("continuous monitoring") or in single-step mode. In Program Status, you can see the assignment of the variables during continuous program execution. In single-step mode, you can stop the program at a breakpoint and execute it statement by statement while monitoring the variable values (Chapter 2.7 "Testing the Program").

The variable table can also be used to test the SCL program. Here you can specify variable values while the program is running and monitor the results.

27.1.8 Addresses and Data Types

Address areas

The addresses and variables in SCL correspond to those of the standard programming languages (see Chapter 1.5 "Address Areas").

- ▷ Inputs I, outputs Q, memory bits M
- ▷ Peripheral inputs PI, Peripheral outputs PQ
- ▷ Global data addresses D
- ▷ Temporary and static local data (symbolic addressing only)
- ▷ Organization blocks OBs, function blocks FBs, functions FCs with and without function value, data blocks DBs

Timer functions T and counter functions C are handled in SCL as "standard functions" (see Chapters 30.1 "Timer Functions" and 30.2 "Counter Functions").

Note: The global data addresses have a different address identifier to the standard programming languages. Please refer to Chapter 27.2.1 "Absolute Addressing" for details of the address identifiers used in SCL.

In SCL, function calls that return a function value can also be used as addresses in expressions.

Data types

The definition of a data type contains:

- ▷ Type and meaning of the data elements (e.g. integer, character string)
- ▷ Permissible ranges (number range, length of a character string)
- ▷ Permissible operations that can be executed with a data type
- ▷ Method of writing the constants

The data types in SCL are the same as in the standard programming languages: Chapter 3.7 "Variables and Constants" gives an overview in tabular form; Chapter 24 "Data Types" contains the detailed descriptions.

Numerical values can be represented as decimal numbers, hexadecimal numbers, octal numbers (8#17 corresponds to 16#F or 15dec) and binary numbers.

Data type classes

In conjunction with the combination of values, SCL defines classes of data types that show the same behavior in the relevant combination:

- ▷ ANY_INT comprises the data types INT and DINT
- ▷ ANY_NUM comprises the data types INT, DINT and REAL
- ▷ ANY_BIT comprises the data types BOOL, BYTE, WORD and DWORD

These data type classes have been introduced in order to make description of the operations and operators clearer; variables cannot be declared with these data type classes.

Constant notation

A constant is a fixed value that generally does not change with program execution. Constants are used to pre-assign initial values to variables at variable declaration or to combine them with other variables in the program, e.g. as limit values.

In SCL, a constant does not receive the data type until it is combined arithmetically. The constant 1234, for example, can have data type INT or data type REAL, depending on the application:

```
int1  := int2 + 1234;
      //INT constant
real1 := real2 + 1234;
      //REAL constant
```

In SCL, you can also assign a data type to a constant ("type-defined" constant notation). With a suitable prefix, you can, for example, pre-assign a WORD variable in the declaration with a decimal, hexadecimal, octal or binary number. The example below shows the contents of the variable with the same value in each case but with different representations:

```
w1 : WORD := W#1234;   //decimal
w2 : WORD := W#16#04D2;//hexadecimal
w3 : WORD := W#8#2322;//octal
w4 : WORD := W#2#0000_0100_1101_0010;
                        //binary
```

Data type in the case of absolute addresses

An absolute Address is always of the data type ANY_BIT (e.g. memory doubleword MD10 has the data type DWORD). Only if the address is symbolic ("when it has been turned into a variable") or after a data type conversion, can the address be used with a data type, e.g. DINT or REAL.

```
MW14 := SHL(IN := MW12, N := 2);
real1 := real2 + DWORD_TO_REAL(MD10);
```

Data type STRING

A character string is represented in single inverted commas; non-printable control characters can be entered with $hh (hh represents the value of the ASCII character in hexadecimal form).

```
string1 := '$0A$0D'; //new line
```

The characters ‘$>’ and ‘<$’ are available for interrupting a character string, e.g. at the end of a line or for comments that are not to be printed or displayed.

```
string2 := 'ABCDEFGHIJKLMNOP$>
      <$QRSTUVWXYZ';
```

27.1.9 Data Type Views

In SCL, you can assign additional data types to an already declared variable (more precisely: you can assign additional data type views). It is then possible to address the contents of the variable in whole or in part areas with different data types.

Example: you declare an input parameter with the name *Station* and the data type STRING. You transfer the variable *Station* to a called block in order to further process it, e.g. supplementing it with a number. In addition, you want to calculate the current length of *Station*. For this purpose, you apply an additional data type view in the form, for example, of a structure of two bytes via the variable *Station*. The second byte then contains the current string length. The additional data type view is to receive the name *Len* and the components are to be called *max* and *cur*.

```
VAR_INPUT
  Station : STRING[24] := ' ';
  Len AT Station : STRUCT
      max : BYTE; //Maximum length
      cur : BYTE; //Current length
      END_STRUCT;
END_VAR
....
IF WORD_TO_INT(Len.cur) > 12
      THEN ...
END_IF;
...
```

First, you declare the variable with the “original” data type and with any pre-assignments. Then, you can assign an additional data type view to this variable with the keyword AT:

```
View AT Variable : Data_type; //Comment
```

You can apply several data type views to a variable, distinguishing between them by name. Pre-assignment with fixed values (initialization) is not possible.

The memory requirement of the data type view must not be greater than the variable to which the view has been assigned (the new data type must “fit” into the variable).

Table 27.2 Permissible Data Type Views

Block	The variable is declared in the block	The variable is of data type			
		Elementary	Complex	POINTER	ANY
FC	VAR_INPUT	E	C		
	VAR_OUTPUT	E	C		
	VAR_IN_OUT	E	C		
	VAR [1)]	E C	E C A		C
	VAR_TEMP	E C	E C A		C
FB	VAR_INPUT	E C	E C P A	C	C
	VAR_OUTPUT	E C	E C		
	VAR_IN_OUT	E	C		
	VAR	E C	E C		
	VAR_TEMP	E C	E C A		C

[1)] corresponds to temporary local data
Data type of the view:
E elementary (BOOL, CHAR, BYTE, WORD, DWORD, INT, DINT, REAL, S5TIME, TIME, DATE, TIME_OF_DAY)
C complex (DATE_AND_TIME, STRING, ARRAY, STRUCT) and UDT
P POINTER
A ANY

You use a data type view like any other variable but only locally in the block. In the example above, the calling block initializes the input parameter Station with a character string: the data type view is not accessible to it as a byte structure.

A data type view can be applied via block parameters and via temporary and static local data. The data type view must be declared in the same declaration block as the variables.

Table 27.2 shows which data type views you can apply to a variable of a specific data type. If, for example, the variable is located in the temporary local data of an FC and if it is of a complex data type, the data type views applied to it can be of the elementary, complex, POINTER or ANY data types.

Variables of the type TIMER, COUNTER and BLOCK_xx cannot have data type views applied to them.

27.2 Addressing

27.2.1 Absolute Addressing

Absolute addressing assigns addresses in relation to the start of the address area; e.g. I1.0 (input bit 0 in byte 1). Absolute addressing in SCL corresponds to that of the standard programming languages (Chapter 3.3 "Addressing Variables") with the exception that the address identifiers of global data addresses are different (Table 27.3).

In SCL, access to global data addresses is only possible with complete addresses. The data block can also be a block parameter of the type BLOCK_DB (see also Chapter 27.2.3 "Indirect Addressing in SCL").

Note: Please note that in SCL there must be no separator (space or tab) between the address ID and the address.

Differences to the standard programming languages: no absolute addressing of temporary and static local data; calling a data block with subsequent part address access is not possible; calculation of the number and length of the current global instance data block is not possible.

27.2.2 Symbolic Addressing

Symbolic addressing assigns names to addresses and variables. For global data, the name is assigned in the symbol table; for local data it is assigned in the declaration section of the block.

Symbolic addressing in SCL corresponds to symbolic addressing in the standard programming languages (Chapter 3.3 "Addressing Variables"). Mixed absolute/symbolic identifiers such as:

```
DB10.Setpoint
"Motor1Data".DW12.
```

are permissible for full address access to global data addresses.

In SCL, you can assign names to constants in the declaration section of a block and use these names as symbols in the program.

Table 27.3 Address Identifiers for Absolute Addressing

Address area	Bit	Byte	Word	Doubleword
Inputs	Iy.x	IBy	IWy	IDy
Outputs	Qy.x	QBy	QWy	QDy
Peripheral inputs	-	PIBy	PIWy	PIDy
Peripheral outputs	-	PQBy	PQWy	PQDy
Memory bits	My.x	MBy	MWy	MDy
Global data addresses	DBz.DXy.x DBz.Dy.x	DBz.DBy	DBz.DWy	DBz.DDy

x = bit address, y = byte address, z = data block number

27.2.3 Indirect Addressing in SCL

Indirect assignment of global addresses

Indirect assignment of global addresses is based on absolute addressing. Instead of the memory location, an INT variable is specified in square brackets; in the case of bit addresses, two INT variables are used:

▷ I[*byteindex*, *bitindex*]

▷ MB[*byteindex*]

byteindex and *bitindex* are constants, variables that can be modified at runtime, or expressions of data type INT. You can address the following areas in this way:

▷ Peripheral inputs PI, peripheral outputs PQ (no bit addressing in either case)

▷ Inputs I, outputs Q, and memory bits M

▷ Global data addresses D (data block and data address)

▷ Timer functions T and counter functions C (no bit addressing for neither of the functions)

Indirect assignment of global data addresses

Indirect assignment of global data addresses is based on absolute addressing but the data address as well as the data block address can be modified at runtime.

You can assign either an absolute address or a symbolic address to the data block:

▷ DB10.DX[*byteindex*, *bitindex*]
▷ MotorDaten.DW[*byteindex*]

byteindex and *bitindex* are constants, variables that can be modified at runtime or expressions of data type INT.

With the WORD_TO_BLOCK conversion function, you can assign an indirect address to a data block. The DB number is specified as a variable or as an expression of data type WORD (see Figure 27.2 for examples).

▷ WORD_TO_BLOCK_DB(*dbindex*).DW0

dbindex is a variable that can be modified at runtime or an expression of data type WORD.

If the data block is addressed indirectly, the data address cannot be accessed symbolically.

```
//*******************************************************************
//Example of indirect assignment of global addresses
k := 120; FOR i := 48 TO 62 BY 2 DO
MW[k] := PIW[i]; k := k + 2; END_FOR;

//
*********************************************************************
//Indirect addressing of data blocks
//The DB index is available in data type WORD
M0.0 := WORD_TO_BLOCK_DB(dbindex_w).DX0.0;
M0.0 := WORD_TO_BLOCK_DB(dbindex_w).DX[byteindex,bitindex];

//The DB index is available in data type INT
M0.0 := WORD_TO_BLOCK_DB(INT_TO_WORD(dbindex_i)).DX0.0;
M0.0 := WORD_TO_BLOCK_DB(INT_TO_WORD(dbindex_i)).DX[byteindex,bitindex];

//
*********************************************************************
//Indirect addressing via a block parameter
//With the name "Data" and the parameter type BLOCK_DB
M0.0 := Data.DX0.0;//absolute addressing
M0.0 := Data.DX[byteindex,bitindex];//indirect addressing
```

Figure 27.2 Examples of Indirect Addressing of Global Addresses

Assignment of data addresses via a block parameter BLOCK_DB

If the data block is available as a block parameter, its data addresses can be assigned absolutely and indirectly (Figure 27.2). Example: The input parameter *Data* is of the type BLOCK_DB:

▷ Data.DW0
▷ Data.DX2.0

▷ Data.DW[*byteindex*]
▷ Data.DX[*byteindex.bitindex*]

byteindex and *bitindex* are constants, variables that can be modified at runtime or expressions of data type INT.

If the data block is addressed via a block parameter, the data address cannot be accessed symbolically.

Addressing fields

In SCL, you can use either a constant or a variable or an expression of data type INT as a field index and so modify it at runtime. You can also address part fields as variables (Chapter 27.5.4 "Assigning Fields").

At pre-assignment, repetition factors can be assigned for the individual field dimensions.

27.3 Operators

An expression represents a value. It can consist of a single address (a single variable) or several addresses (variables) combined using operators.

Example: $a + b$;
a and b are addresses, + is the operator.

The order of combinations is specified by the priority of the operators and can be controlled with parentheses. Expressions can be mixed provided the data types created in the calculation of the expression permit it.

SCL provides the operators listed in Table 27.4. Operators of the same priority are executed from left to right.

Table 27.4 Operators in SCL

Combination	Name	Operator	Priority
Parentheses	(*Expression*)	(,)	1
Arithmetic	Power	**	2
	Unary plus, unary minus (sign)	+, -	3
	Multiplication, division	*, /, DIV, MOD	4
	Addition, subtraction	+, -	5
Comparison	Less than, less than/equal to, greater than, greater than/equal to	<, <=, >, >=	6
	equal to, not equal to	=, <>	7
Binary combination	Negation (unary)	NOT	3
	AND logic operation	AND, &	8
	Exclusive OR	XOR	9
	OR logic operation	OR	10
Assignment	Assignment	:=	11

"unary" means this operator is assigned to an address

27.4 Expressions

An expression is a formula for calculating a value and consists of addresses (variables) and operators. In the simplest case, an expression is an address, a variable or a constant. A sign or negation can be included here.

An expression can consist of addresses combined using operators. Expressions themselves can be combined with operators so that an expression can have an extremely complex structure. Parentheses can be used to control the order of execution of an expression.

The result of an expression can be assigned to a variable or a block parameter or it can be used as a condition in a control instruction.

Expressions are divided, according to the type of combination, into arithmetic expressions, comparison expressions and logical expressions.

Table 27.5 Data types and operators in SCL expressions

Operation	Operator	1st operand	2nd operand	Result
Arithmetic expressions				
Power	**	ANY_NUM	INT	REAL
Multiplication	*	ANY_NUM	ANY_NUM	ANY_NUM
		TIME	ANY_INT	TIME
Division	/	ANY_NUM	ANY_NUM	ANY_NUM
Integer division	DIV	ANY_INT	ANY_INT	ANY_INT
		TIME	ANY_INT	TIME
Division with remainder as result	MOD	ANY_INT	ANY_INT	ANY_INT
Addition	+	ANY_NUM	ANY_NUM	ANY_NUM
		TIME	TIME	TIME
		TOD	TIME	TOD
		DT	TIME	DT
Subtraction	–	ANY_NUM	ANY_NUM	ANY_NUM
		TIME	TIME	TIME
		TOD	TIME	TOD
		DATE	DATE	TIME
		TOD	TOD	TIME
		DT	TIME	DT
Comparison expressions				
Comparison as equal, unequal, lesser than, lesser than or equal to, greater than, greater than or equal to	=, <>, <, <=, >, >=,	ANY_NUM	ANY_NUM	BOOL
		CHAR or STRING	CHAR or STRING	BOOL
		TIME	TIME	BOOL
		DATE	DATE	BOOL
		TIME_OF_DAY	TIME_OF_DAY	BOOL
Comparison as equal and unequal	=, <>	ANY_BIT	ANY_BIT	BOOL
Logical expressions				
Negation	NOT	ANY_BIT	-	ANY_BIT
AND logic operation (conjunction)	AND, &	ANY_BIT	ANY_BIT	ANY_BIT
Exclusive OR (exclusive disjunction)	XOR	ANY_BIT	ANY_BIT	ANY_BIT
OR logic operation (disjunction)	OR	ANY_BIT	ANY_BIT	ANY_BIT

27.4.1 Arithmetic Expressions

An arithmetic expression either consists of a numerical value or it combines two values or expressions with arithmetic operators. Example:

```
Voltage * Current
```

Table 27.5 lists the permissible data types for arithmetic expressions and the data type of the result.

Specification of the data type class ANY_NUM means that the data type of the first and second operands can be INT, DINT, or REAL. If you link an INT operand and a DINT operand, the result is of data type DINT; if you link an INT or DINT operand with a REAL operand, the result is of data type REAL. The program editor carries out a corresponding data type conversion (not visible to the user) prior to the arithmetic operation (see also Table 30.4 "Implicit conversion functions").

In the case of a division, the second address must not be equal to zero. Figure 27.3 gives an example for arithmetic expressions in conjunction with value assignments.

27.4.2 Comparison Expressions

A comparison expression compares the values of two addresses and yields a Boolean value; if the comparison is met, it yields the result TRUE, otherwise it yields the result FALSE. Example:

```
Voltage1 > Voltage2
```

The addresses compared must be of the same data type or the same data class (ANY_INT, ANY_NUM, ANY_BIT) (Table 27.5). To enhance clarity, the use of parentheses is recommended in comparison expressions.

Comparison expressions can be combined with logical operators such as:

```
(Value1 > 40) AND NOT (Value2 = 20)
```

Comparison of variables of data type CHAR is carried out according to the ASCII character code.

```
(****************************** Assignment *********************************)
Automatic := TRUE;//Assignment of a constant value
Setpoint  := StartSetpoint;//Assignment of a variable
Deviation := ActualValue - Setpoint;//Assignment of an expression
Display   := INT_TO_WORD(Deviation);//Assignment of a function value

(************************** Arithmetic expressions ***************************)
Power      := Voltage * Current;
Volume     := 4/3 * PI * Radius**3;
Solution1  := -P/2 + SQRT(SQR(P/2)-Q);
MeanValue  := (Motor[1].Power + Motor[2].Power)/2;

(************************** Comparison expressions ***************************)
TooLarge   := Voltage_Act > Voltage_Set;
Warning    := (Voltage * Current) >= 20_000;
M101.0     := Setpoint = ActualValue;
IF Deviation > 2_000 THEN Display := 16#F002; END_IF;

(************************** Logical expressions ******************************)
Q4.0       := I1.0 & I1.1;
ON         := (Manual_on OR Auto_on) AND NOT Fault;
MW30       := MW32 AND Mask;
Pulses     := (Edge_mem_bits XOR ID16) AND ID16; Edge_mem_bits := ID16;
```

Figure 27.3 Operators, Expressions and Value Assignment

The IEC functions are available for comparing variables of data type STRING and DT: IEC functions are loadable FC blocks in the *Standard Library* in the *IEC Function Blocks* program.

Figure 27.3 gives some examples of comparison expression in conjunction with value assignments.

27.4.3 Logical Expressions

A logical expression combines addresses and expressions of data type ANY_BIT according to AND, OR and exclusive OR.

Example:

```
Automatic AND NOT Manual_on
```

The logical expressions also include (Boolean) negation; it is handled similarly to the sign of a number.

A logical expression yields a value of data type class ANY_BIT. The result of a logical expression is of data type BOOL if both addresses are also of data type BOOL. If one or both of the addresses are a bit pattern of data type BYTE, WORD or DWORD, the result will be of the "more powerful" of the data types involved.

Figure 27.3 gives some examples of logical expressions in conjunction with value assignments.

27.5 Value Assignments

With a value assignment, a variable receives the value of another variable or of an expression. On the left of the assignment operator := is the variable that is to assume the value of the address or expression on the right.

The data types of both sides of the assignment must be identical. An exception to this is "Implicit data type conversion": if the data type of the variable has at least the same bit width as, or a greater bit width than, the data type of the expression, the data type of the expression is implicitly converted (the value of the expression is automatically converted in the data type and assigned to the variable). Otherwise, explicit conversion (with conversion functions) is necessary.

27.5.1 Assignment for Elementary Data Types

A constant value, another variable, an address or an expression can be assigned to a variable or an address (Figure 27.3).

Absolute addresses (e.g. MW 10) are of data type ANY_BIT; i.e. depending on "data width" BOOL, BYTE, WORD or DWORD. If you want to assign a value of a different data type to an absolute address, use data type conversion or assign a name and the desired data type to the address in the symbol table.

27.5.2 Assignment of DT and STRING Variables

Every DT variable can be assigned the value of another DT variable or a DT constant.

Every STRING variable can be assigned the value of another STRING variable or a character STRING. If the assigned character string is longer than the variable on the left of the assignment operator, a warning is issued at the compiling stage.

No pre-assignment is possible at the declaration stage in the temporary local data. If you use STRING processing functions, such as the IEC functions, that check the STRING variable (as well as the output parameter) for valid assignment, you must program the pre-assignment out.

27.5.3 Assignment of Structures

A STRUCT variable can only be assigned to another STRUCT variable if

- ▷ the data structures agree,
- ▷ the structure components agree in data type,
- ▷ the structure components agree in name.

Individual structure components can be handled like variables of the same data type, for example, a structure component *Motor1.Setpoint* of data type INT can be assigned to another INT variable, or an INT value can be assigned to this structure component.

27.5.4 Assigning Fields

An ARRAY variable can only be assigned to another ARRAY variable if the data types of the field components as well as the field limits with the smallest and largest field index agree.

Individual field components can be handled in the same way as variables of the corresponding data type.

In the case of multi-dimensional fields, you can handle **part fields** in the same way as correspondingly dimensioned variables: starting from the right, leave out field indices to get a lower dimensioned part area of the original field. Example:

`Field1 : ARRAY [1..8,1..16] OF INT` represents a two-dimensional field; you can now address the entire field with `Field1`, a part field with `Field1[i]` (corresponds to the lines of the matrix) and a field component with `Field1[i,j]`.

You can assign part field `Field1[i]` to a correspondingly dimensioned field, e.g. `Field2 := Field1[i]`, where i = 1 to 8 and `Field2 : ARRAY [1..16] OF INT`.

28 Control Statements

With the control statements, you can execute program branchings, repeat program sections or jump to another point in the program of the block. SCL provides the following control statements:

IF	Program branch dependent on a Boolean value
CASE	Program branch dependent on an INT value
FOR	Program loop with a run variable
WHILE	Program loop with an execution condition
REPEAT	Program loop with an cancel condition
CONTINUE	Cancellation of the current loop pass
EXIT	Exit the program loop
GOTO	Jump to a jump label
RETURN	Exit the block

Note: Please ensure that the cycle monitoring time is not exceeded when using program loops.

The examples in this chapter can be found in the download files (download address: see pages 8-9) in the SCL_Book library under the "28 Control Statements" program.

28.1 IF Statement

The IF statement controls program flow dependent on a Boolean value. You can program different types of IF statement, depending on the type of branching.

```
IF condition
   THEN statements;
END_IF;
```

Condition is an address or an expression with a Boolean value. If *condition* has the value TRUE, the statements following THEN are executed. If *condition* has the value FALSE, program execution is continued with the statement following END_IF. END_IF terminates an IF statement.

```
IF condition
   THEN statements1;
   ELSE statements0;
END_IF;
```

As in the previous example, *condition* here either has the value TRUE or FALSE. If TRUE, the statements following THEN are executed, if FALSE, the statements following ELSE are executed.

```
IF condition1
   THEN statements1;
   ELSIF condition2
      THEN statements2;
   ELSE statements0;
END_IF;
```

IF statements can be parenthesized. If *condition1* is met (TRUE), *statements1* are executed and then program execution is continued following END_IF. If *condition1* has the value FALSE, *condition2* is tested; if the value is TRUE, *statements2* are executed and program execution is continued following END_IF.

You can insert any number of ELSIF... THEN... combinations between IF...THEN... and ELSE. If no condition is met, the statements following ELSE are executed. ELSE and the subsequent statements are not mandatory.

Example: If the variable *Actual_value* is greater than the variable *Setpoint*, the statements following THEN are executed. Otherwise, if *Actual_value* is found to be less than *Setpoint*, the statements following ELSIF are executed. If neither of the two comparisons is fulfilled, the statements following ELSE are executed.

```
IF Actual_value > Setpoint
   THEN greater_than := TRUE;
      less_than := FALSE;
      equal_to := FALSE;
   ELSIF Actual_value < Setpoint
      THEN less_than := TRUE;
      greater_than := FALSE;
      equal_to := FALSE;
   ELSE equal_to:= TRUE;
      greater_than := FALSE;
      less_than := FALSE;
END_IF;
```

28.2 CASE Statement

With the CASE statement, you can process one of several statement sequences dependent on an INT value.

The general structure of a CASE statement takes the following form:

```
CASE Selection OF
   Const1 : Statements1;
   Const2 : Statements2;
   ...
   Constx : StatementsX;
   ELSE     Statements0;
END_CASE;
```

Selection is an address or an expression of data type INT. If *Selection* has the value of *Const1*, *Statements1* are executed and program execution is then continued following END_CASE. If *Selection* has the value of *Const2*, *Statements2* are executed, etc.

If there is no value in the value list corresponding to *Selection*, the statements following ELSE are executed. The ELSE branch is not mandatory.

The value list with *Const1, Const2*, etc. consists of INT constants. Various expressions are possible for a component in the value list:

▷ a single INT number,

▷ a range of INT numbers (e.g. 15..20) or

▷ a list of INT numbers and INT number ranges (e.g. 21,25,30..33).

Each value must occur only once in the value list.

CASE statements can be parenthesized. Instead of a statement block, another CASE statement can stand in the selection table of a CASE statement.

Example: A value is assigned to the *Error_number* variable dependent on the assignment of *ID* variable.

```
CASE ID OF
0      : Error_number := 0;
1,3,5  : Error_number := ID + 128;
6..10  : Error_number := ID;
ELSE     Error_number := 16#7F;
END_CASE;
```

28.3 FOR Statement

With the FOR statement, a program loop is repeated for as long as a run variable remains within a value range.

The general representation of a FOR statement is as follows:

```
FOR Runtime_variable := Starting_value
    TO End_value
    BY Step_width
    DO Statements;
END_FOR;
```

In the start statement, a starting value is assigned to a run variable. You define the run variable yourself; it must be a variable of data type INT or DINT. *Starting_value* is any INT or DINT expression, as are *End_value* and *Step_width*.

At the start of loop execution, the run variable is set to the starting value. At the same time, *End_value* and *Step_width* are calculated and "frozen" (modification of these values during loop execution has no effect on execution of the loop). Then the terminating condition is scanned and – if is not fulfilled – the program loop is executed.

Following each loop pass, the run variable is increased by the step width (in the case of a positive step width) or decreased by the step width (in the case of a negative step width). The specification 'BY step width' is not mandatory; +1 is then used as the step width. If the run variable is outside the range between starting value and end value, program execution is continued following END_FOR.

The last loop pass is made with the end value, or the value *End_value* minus *Step_width* if the end value has not been reached exactly. After exiting of a fully passed program loop, the run variable has the value of the last loop pass plus *Step_width*.

FOR loops can be parenthesized: within the FOR loop, further FOR loops can be programmed with other runtime variables.

In the FOR loop, the current program pass can be aborted with CONTINUE; EXIT terminates the entire FOR loop.

Example: The peripheral I/O words PIW 128 to PIW 142 are read into the memory words MW 128 to MW 142.

```
FOR i := 128 TO 142 BY 2 DO
  MW[i] := PEW[i];
END_FOR;
```

28.4 WHILE Statement

With the WHILE statement, a program loop is repeated as long as an execution condition is met.

The general representation of a WHILE statement is as follows:

```
WHILE Condition DO
   Statements;
END_WHILE;
```

Condition is an address or an expression of data type BOOL. The statements following DO are repeated for as long as *Condition* has the value TRUE.

Condition is scanned before every pass. If the value is FALSE, program execution is continued following END_WHILE. This can be the case even before the first loop pass (the statements in the program loop are then not executed).

WHILE loops can be parenthesized: further WHILE loops can be programmed within one WHILE loop.

In the WHILE loop, the current program pass can be aborted with CONTINUE; EXIT terminates the entire WHILE loop.

Example: data block DB10 is searched for the bit pattern 16#FFFF: data word DW0 contains either 16#FFFF or the interval to the next data word which contains either 16#FFFF or the interval to the next data word again.

```
i := 0;
WHILE DB10.DB[i] = 16#FFFF DO
  i := i + WORD_TO_INT(DB10.DB[i]);
END_WHILE;
```

28.5 REPEAT Statement

With the REPEAT statement, a program loop is repeated as long as a terminating condition is not met.

The general representation of a REPEAT statement is as follows:

```
REPEAT
   Statements;
UNTIL Condition
END_REPEAT;
```

Condition is an address or an expression of data type BOOL. The statements following REPEAT are repeated as long as *Condition* has the value FALSE. *Condition* is scanned following every loop. If the value is TRUE, program execution is continued following END_REPEAT. The program loop is executed at least once even if the terminating condition is met from the start.

REPEAT loops can be parenthesized: other REPEAT loops can be programmed within the REPEAT loop.

In the REPEAT loop, the current program pass can be aborted with CONTINUE; EXIT terminates the entire REPEAT loop.

Example: SFC25 COMPRESS is invoked in the restart program until it has completed compression of the user memory.

```
REPEAT
  SFC_ERROR := COMPRESS(
       BUSY := busy,
       DONE := done);
  UNTIL done
END_REPEAT;
```

28.6 CONTINUE Statement

CONTINUE terminates the current program pass in a FOR, WHILE or REPEAT loop.

After CONTINUE has been executed, the conditions for continuing the program loop are scanned (in the case of WHILE and REPEAT), or the run variable is changed by the step width and tested to see if it is still within the run range. If the conditions are met, the next loop pass starts after CONTINUE.

CONTINUE terminates the program pass of the loop immediately surrounding the CONTINUE statement.

Example: memory bits are set with two parenthesized FOR loops. If the byte address (i) is equal to zero, and the bit address (k) is less than 2, the subsequent statements of the inner FOR loop are not executed (setting starts at memory bit M0.3).

```
FOR i := 0 TO 2 DO
 FOR k := 0 TO 7 DO
  IF (k<2 & i=0) THEN CONTINUE; END_IF;
  M[i,k] := TRUE;
 END_FOR;
END_FOR;
```

28.7 EXIT Statement

With EXIT, you exit a FOR, WHILE or REPEAT loop at any position regardless of conditions. The loop pass is aborted immediately and the program is executed following END_FOR, END_WHILE or END_REPEAT.

EXIT exits the loop immediately surrounding the EXIT statement.

Example: memory bits are set with two parenthesized FOR loops. If the byte address (i) equals 2, and the bit address (k) is greater than 5, execution of the inner FOR loop is aborted (setting ends with memory bit M2.5).

```
FOR i := 0 TO 2 DO
 FOR k := 0 TO 7 DO
  IF (i=2 & k>5) THEN EXIT; END_IF;
  M[i,k] := TRUE;
 END_FOR;
END_FOR;
```

In the example, execution of the FOR loop is aborted with run variable *k* at EXIT. Execution of the outer FOR loop with run variable *i* is not affected by this. However, the example is designed in such a way that the EXIT statement becomes effective in the last pass of the "i loop".

28.8 RETURN Statement

The RETURN statement exits the currently executing block without conditions. Program execution is continued in the invoking block or in the operating system if an organization block is exited.

RETURN is not mandatory at block end.

RETURN transfers the signal state of the OK variable to the ENO output of the exited block.

Example: conditional block end

```
IF Error <> 0 THEN RETURN; END_IF;
```

28.9 GOTO Statement

With GOTO, you can continue program execution at another point.

Example:

```
      GOTO M1;
      ...;     //jumped
      ...;     //statements
M1:   ...;     //jump destination
```

The connection between the GOTO statement and the jump destination is represented by the jump label. You must declare jump labels in the declaration section of the block between the keywords LABEL and END_LABEL. The name of a jump label has the same structure as the name of a block-local variable.

A jump label must be unique; it must be assigned only once in the block. You can jump from several GOTO statements to one jump label.

Following execution of the GOTO statement, program execution is continued at the statement with the jump label. Jump label and statement are separated by a colon.

A jump label must always be followed by a statement. An "empty statement" is also permissible:

```
Label1: ;
```

The jump destination must be within a block. If statements form a defined block, e.g. a program rump within a program loop,

- ▷ the jump destination must be within this statement block if the GOTO statement is also within the statement block,
- ▷ you cannot jump from "outside" into this statement block.

Example:

```
...
LABEL
M1, M2, M3, END;
END_LABEL;
...
GOTO CASE Selection DO;
1 : GOTO M1;
2 : GOTO M2;
3 : GOTO M3;
ELSE GOTO End;
END_CASE;
M1: ...statements1...;
GOTO End;
M2: ...statements2...;
GOTO End;
M3: ...statements3...;
End: ;
```

Note: GOTO is not defined in the standard. SCL provides all the statements and functions required for structured programming, so that GOTO can be dispensed with.

29 SCL Blocks

29.1 SCL Blocks – General

SCL uses the block structure in exactly the same way as the standard programming languages. You can program individual blocks with SCL that you then invoke in, say, an FBD block, or you invoke blocks in SCL that you have created in STL, for example.

To be able to use blocks in the user program that have been created with different languages, the block interface must have a "standardized" structure. This essentially means initialization of the EN input and the ENO output (see Chapter 29.4 "EN/ENO Mechanism").

Programming examples in this chapter can be found in the download files (download address: see pages 8-9) in the SCL_Book library in the "29 Block Calls" program.

User program structure

The organization blocks represent the interface between the operating system and the user program. Organization blocks are called by the operating system of the CPU when certain events occur, such as interrupts. "Normal" program execution for a programmable controller is cyclic execution; the assigned organization block is OB 1 (Chapter 3.1 "Program Processing").

You can subdivide the user program in OB 1 into individual subroutines ("blocks") to suit your requirements. The user program is located in code blocks and user data in data blocks. Code blocks are subroutines that you must invoke to execute (Chapter 20.1 "Program Organization").

Blocks

STEP 7 provides functions FCs and function blocks FBs as code blocks. Function blocks FBs are invoked in conjunction with a data block in which the block-local variables are stored ("memory" of the block). This data block, assigned to an FB call, is called an *instance data block*; it can be a data block in itself or it can be part of a "higher-level" data block. Functions FCs have no data block but they can have a *function value*. This function value makes it possible, for example, to combine a function FC (or more precisely, its function value) with another variable in an arithmetic expression (Chapter 3.2 "Blocks").

Both block types can have block parameters. Block parameters make it possible to parameterize the execution rule (the block function). You declare the block parameters when programming the block: as *input parameter* (VAR_INPUT) if you only scan or read its value in the block program, as *output parameter* (VAR_OUTPUT) if you only write to it, or as *in-out parameter* (VAR_IN_OUT) if it is to be read and written to.

If you address a block parameter in the block program, use a *formal parameter* with the name of the block parameter. The formal parameter serves as a dummy for the *actual parameter* used by the CPU during program execution. You assign the actual parameters to the block parameters when you call the block; they represent the values with which the block is to work, or the values the block is to yield.

29.2 Programming SCL Blocks

The tools for programming SCL blocks are described in Chapter 2; the relevant keywords can be found in Chapter 3.5 "Programming Code Blocks with SCL". Data blocks and user-defined data types are generally programmed in the same way as in STL (Chapters 3.6 "Programming Data Blocks" and 24.3 "User-Defined Data Types").

To highlight the programming differences between the different code blocks we will implement the "Delimiter" function from the introduction to Chapter 27 "Introduction, Language Elements" as the following

- ▷ Function FC 291 without function value
- ▷ Function FC 292 with function value
- ▷ Function block FB 291 with its own data block DB 291
- ▷ Function block FB 291 as local instance in function block FB 290

Then we will call all blocks in a function block (FB 290 with DB 290 as instance data block). The program is always the same; only the declaration and the initialization of the parameters change.

Note: Since the "Delimiter" program does not store local data and it returns a value, a function FC with function value is the optimal block type.

29.2.1 Function FC without a Function Value

A function FC without a function value is of data type VOID. In our example, function FC 291 has the input parameters MAXI, IN, MINI and the output parameter OUT.

```
FUNCTION FC291 : VOID
VAR_INPUT
  MAXI : INT;
  IN   : INT;
  MINI : INT;
END_VAR
VAR_OUTPUT
  OUT : INT;
END_VAR
BEGIN
IF IN > MAXI THEN OUT := MAXI;
  ELSIF IN < MINI THEN OUT := MINI;
  ELSE OUT := IN;
END_IF;
END_FUNCTION
```

All output parameters of elementary data type in a function must be set in a defined manner on execution and must also be executed at runtime. Input parameters may only be read and output parameters may only be written to.

29.2.2 Function FC with Function Value

A function FC with function value has the data type of the function value (return value). In our example, function FC 292 has the input parameters MAXI, IN, MINI and a function value that has the address (name) of the function either in absolute or symbolic form. The data type of the function value is specified after the block name, separated by a colon.

```
FUNCTION FC292 : INT
VAR_INPUT
  MAXI : INT;
  IN   : INT;
  MINI : INT;
END_VAR
BEGIN
IF IN > MAXI THEN FC292 := MAXI;
  ELSIF IN < MINI THEN FC292 := MINI;
  ELSE FC292 := IN;
END_IF;
END_FUNCTION
```

You can use all elementary data types as the data type of the function value, as well as the data types DATE_AND_TIME, STRING and user-defined data types UDT. ARRAY, STRUCT, POINTER and ANY are not permissible.

If the function value is of data type STRING, the reserved length is determined by the compiler setting (and not the maximum length given in square brackets).

All output parameters of elementary data type in a function must be set in a defined manner on execution and must also be executed at runtime. Input parameters may only be read and output parameters may only be written to.

In the FC program, a value must be assigned to the function value, for example, with an expression of the same data type. This assignment must also be executed at runtime.

29.2.3 Function Block FB

A function block has an instance data block in which it can store its variables (the function block is either called with its own data block or it uses the data block of the called function block). We want to make use of this and declare the limit values as static local variables. The input value IN and the result OUT remain as block parameters.

```
FUNCTION_BLOCK FB291
VAR_INPUT
  IN  : INT;
END_VAR
VAR_OUTPUT
  OUT : INT;
END_VAR
VAR
  MAXI : INT := 10_000;
  MINI : INT := -5_000;
END_VAR
BEGIN
IF IN > MAXI THEN OUT := MAXI;
  ELSIF IN < MINI THEN OUT := MINI;
  ELSE OUT := IN;
END_IF;
END_FUNCTION_BLOCK
```

Input parameters may only be read and output parameters may only be written to.

There are two variants for the call: call with own data block or call as local instance. The type of the subsequent block call need not be taken into account when programming the function block. However, please ensure that when used as a local instance, at least one block parameter or one static local data item is available: the instance length must not be zero.

Note: Input and output parameters of complex data types are stored as a value in the instance data block, in-out parameters are stored as pointers to the actual parameters (see Chapter 26.3.2 "Storing Parameters in Function Blocks").

29.2.4 Temporary Local Data

All code blocks have temporary local data that you can use as intermediate storage in the block. You use the temporary local data in SCL in the same way as in the standard programming languages. Please refer to Chapter 18.1.5 "Temporary Local Data" for more detailed information.

You declare the temporary local data in the declaration section of the block under VAR_TEMP. All elementary, complex and user data types are permissible, as well as the data types POINTER and ANY. Special rules apply for ANY (see below).

Temporary local data cannot be pre-assigned at the declaration stage. This is why, when assigning the L stack, the Editor reserves the length entered on the "Compiler" tab under OPTIONS → CUSTOMIZE for STRING variables.

If temporary local data are to be assigned meaningful values, they must first be written to. This also applies to (temporary) STRING variables created at an output parameter, e.g. in the case of IEC functions. When writing, the IEC function checks that a meaningful (valid) value has been entered in the length information of the STRING variable. You achieve this by assigning a value (any value) to the variable in the program before it is used.

In SCL, you can declare variables of the same data type as a list:

```
VAR_TEMP
Value1, Value2, Value3 : INT;
...
END_VAR
```

Please note that with SCL the temporary local data are only addressed symbolically.

Data type ANY

Temporary local data of data type ANY can store the address of an instruction or of a global or block-local variable:

```
any_var := MW10;
any_var := Setpoint;
any_var := DB10.Field1;
```

You can also pre-assign a temporary local variable of data type ANY with NIL, a pointer "to zero".

```
any_var := NIL;
```

Example: Various data records are to be copied to a send mailbox with SFC 20 BLKMOV dependent on an identifier:

```
...
VAR_TEMP
Address := ANY;
...
END_VAR
...
CASE Identifier OF
1: Address := DataRecord1;
2: Address := DataRecord2;
...
ELSE Address := NIL;
END_CASE;
SFC_ERROR := BLKMOV(
    SRCBLK := Address,
    DSTBLK := SendMailBox);
```

You can edit the individual components of an ANY pointer, such as the DB number or the address, direct with the help of a data type view (see Chapter 27.1.9 "Data Type Views").

29.2.5 Static Local Data

The static local data are the "memory" of a function block. They are stored in the instance data block and retain their value until changed by the program just like data addresses in a global data block.

In the static local data, you also declare the local instances of function blocks and system function blocks. Please refer to Chapter 18.1.6 "Static Local Data" for more detailed information.

You declare the static local data with the keywords VAR and END_VAR. All elementary, complex and user data types are permissible, as well as the data types POINTER and ANY.

In SCL you can declare variables of the same data type as a list. Variables declared in this way cannot be pre-assigned:

```
VAR
Value1, Value2, Value3 : INT;
...
END_VAR
```

Please note that with SCL, the static local data in the function block are only addressed symbolically.

Since the static local data are located in a data block, they can also be accessed in the same way as global data addresses. They are accessed with full addressing specifying the data block and the data address.

29.2.6 Block Parameters

The block parameters constitute the interface between the calling block and the called block. They are declared as input, in/out and output parameters (Chapter 19.1.3 "Declaration of the Block Parameters").

You may only scan input parameters and you may only write to output parameters. If you want to read, modify and write back to a block parameter, use an in/out parameter.

In the case of functions FCs, the block parameters are pointers to the actual parameters or to another pointer. In the case of function blocks FBs, the block parameters are stored in the instance data block (Chapter 26.3 "Data Storage when Transferring Parameters").

In SCL, you can declare block parameters of the same data type as a list. Variables declared in this way cannot be pre-assigned. Example:

```
VAR_INPUT
Value1, Value2, Value3 : INT;
...
END_VAR
```

Since the block parameters are located in a data block, they can also be accessed in the same way as global data addresses. They are accessed with full addressing specifying the data block and the data address.

```
Result := DB279.DW20;
Result := DB279.Total;
Result := Totalizer.Total;
Result := Totalizer.DW20;
```

In the case of output parameters, it is in fact, the only possibility of further processing their values (see block calls in Chapters 29.3.3 "Function Block with its Own Data Block" and 29.3.4 "Function Block as Local Instance").

Pre-assignment of block parameters

Pre-assignment of block parameters is optional and only permissible in the case of function blocks if the block parameter is stored as a value. This applies to all block parameters of elementary data type and to input and output parameters of complex data type.

If you make no initialization, the editor uses zero, the least value, or a space as the initialization value depending on the data type. The default initialization in the case of parameters of type BLOCK_DB is DB1 (DB0 is not permissible since it does not exist).

If you do not specify length information for STRING variables, the Compiler sets 254 as the maximum length and 0 as the current length or it takes the setting on the "Compiler" tab under OPTIONS → CUSTOMIZE.

29.2.7 Formal Parameters

You use formal parameters to address the block parameters in the block program. The formal parameters have the same name as the block parameters and are used in the statements in place of an address.

Formal parameters of elementary data type

You can use formal parameters of elementary data type instead of addresses of the same data type in any expression, and you can "pass them on" to block parameters of called blocks.

You can assign several data type views to block parameters of elementary data type and so access them with different formal parameters.

Formal parameters of complex data type and UDTs

You can use formal parameters of complex data type and of user-defined data types instead of addresses of the same data type in an assignment, and you can "pass them on" to block parameters of called blocks. You can treat individual components of data types ARRAY, STRUCT and UDT in the same way.

You can assign several data type views to block parameters of complex data type and so access them with different formal parameters. This is especially useful with the data types DT and STRING whose individual bytes you otherwise cannot process.

Formal parameters of parameter types TIMER and COUNTER

Formal parameters of the parameter types TIMER and COUNTER can be processed with the SIMATIC timer functions or SIMATIC counter functions (Chapters 30.1 "Timer Functions" and 30.2 "Counter Functions"). Formal parameters of these data types can also be passed on to the parameters of called blocks.

Formal parameters of parameter types BLOCK_xx

With a formal parameter of the type BLOCK_DB, you can access a data address in a data block (see Chapter 27.2.3 "Indirect Addressing in SCL"). You can also pass a formal parameter of this type to a parameter of the called block.

Formal parameters of parameter types BLOCK_FB and BLOCK_FC can only be passed on to called blocks in SCL (no processing of the formal parameter in the block).

Formal parameters of data types POINTER and ANY

Formal parameters of data types POINTER and ANY can be passed on to called blocks in SCL as entire units. Exception: if the actual parameter is located in the temporary local data, passing on is not permissible.

You can assign several data type views to block parameters of data types POINTER and ANY and so access them with different formal parameters. This is especially useful in the case of the data type ANY since you can, for example, modify an ANY pointer at runtime in this way.

29.3 Calling SCL Blocks

When calling blocks, SCL differentiates between blocks with and without function value.

Function blocks FBs and functions FCs without function value are simply program branchings in the sense of subroutines; this also includes system function blocks SFBs and system functions SFCs without function value.

Functions FCs with function value can be used in value assignments and expressions in place of variables. Table 29.1 gives an overview of the block calls.

You call system function blocks SFBs in exactly the same way as function blocks FBs and you call system functions SFCs in exactly the same way as functions FCs. If you call system function blocks SFBs with a data block, the data block is located in the user program.

Table 29.1 SCL Block Calls

Calling a function	
with function value	without function value
Variable := FCx(...); Variable := FC_name(...);	FCx(...); FC_name(...);
Calling a function block	
with data block	as local instance
FBx.DBx(...); FB_name.DB_name(...);	local_name(...);

When calling a block with block parameters, the block parameters are initialized with *actual parameters*. These are the values (constants, variables or expressions) with which the block works at runtime and in which it stores its results

All block parameters must be initialized when calling functions FCs and system function SFCs.

When calling function blocks FBs and system function blocks SFBs, the block parameters can be freely initialized. Output parameters in the case of FBs and SFBs are initialized with direct access to the instance data instead of with actual parameters when called.

29.3.1 Function FC without Function Value

```
FC291(MAXI:= Maximum,
      IN  := InputValue,
      MINI:= Minimum,
      OUT := Result);
```

The call is made under specification of the block address (absolute or symbolic) followed by the parameter list in parentheses.

All parameters must be initialized and the order is optional. The parentheses must be written even if a function FC has no parameters.

If a function has an input parameter as its single parameter, the parameter name can be omitted at initialization.

Example: conversion of the INT variable Speed to a STRING variable Display:

```
Display := I_STRNG(Speed);
```

29.3.2 Function FC with Function Value

```
Result := FC292(
      MAXI := Maximum,
      IN   := InputValue,
      MINI  := Minimum);
```

A function FC with function value can be used in any expression in place of a variable of the same data type; in a value assignment in the example. The global variable *Result* is assigned to the function value of function FC 292.

The call is made under specification of the block address (absolute or symbolic) followed by the parameter list in parentheses.

All parameters must be initialized and the order is optional. The parentheses must be written even if a function FC has no parameters.

If a function has an input parameter as its single parameter, the parameter name can be omitted at initialization.

If you use the EN input at block call, and if this input has the value FALSE, the function value is undefined (assigned any value).

29.3.3 Function Block with its Own Data Block

The instance data block is specified when calling the function block. It can either be programmed in the program source (following the function block and before its invocation), or SCL generates the data block specified in the invocation after checking, provided it does not already exist. The instance data block can also be programmed incrementally in SCL without source (Chapter 3.6.1 “Programming Data Blocks Incrementally”).

Any free data block can be used as the instance data block. The symbolic name can be selected freely within the permissible framework.

```
DATA_BLOCK DB291
  FB291
BEGIN
END_DATA_BLOCK
```

Call with instance data block:

```
FB291.DB291(IN := InputValue);
Result := DB291.OUT;
```

The call is made under specification of the function block followed by the instance data block, separated by a colon, and the parameter list in parentheses. The addresses (names) can be specified either absolutely or symbolically.

Initialization of function block parameters is free. Since in/out parameters of complex data types are stored as pointers, they should be initialized the first time the function block is called so that a meaningful value is entered. If a block parameter is not initialized, it retains its last set value. The parentheses must be written even if no parameters are initialized.

All parameters can also be addressed as global data addresses with specification of the instance data block and the parameter name. In the example, the limit values are assigned constants. They can also be initialized prior to the function block call with

```
DB291.MAXI := Maximum;
DB291.MINI := Minimum;
```

Output parameters cannot be initialized at a block call. If required, their values are read direct from the instance data block and further processed without intermediate storage:

```
IF DB291.OUT > 10_000 THEN ... END_IF;
```

29.3.4 Function Block as Local Instance

Other function blocks can be declared as local instances and called in a function block. The function blocks to be called then store their local data in the instance data block of the calling function block.

```
FUNCTION_BLOCK FB290
...
VAR
  Delimiter : FB291;
END_VAR
...
BEGIN
...
Delimiter(IN := InputValue);
Result := Delimiter.OUT;
...
END_FUNCTION_BLOCK
```

You make the declaration as local instance in the static local data; you assign a name (e.g. Delimiter) and assign the function block (FB291 or its symbolic name) as the data type. When compiling, the function block to be called must exist, either as a compiled block in the container *Blocks* or as an (error-free) program source that is compiled prior to being called.

You choose the same procedure when you call a system function block SFB as a local instance.

The call as local instance is made under specification of the variable name followed by the parameter list in parentheses. Initialization of function block parameters is free. Since in/out parameters of complex data types are stored as pointers, they should be initialized the first time the function block is called so that a meaningful value is entered. If a block parameter is not initialized, it retains its last set value. The parentheses must be written even if no parameters are initialized.

You can create several local instances with different names for the same function block.

All parameters of a local instance can also be addressed as components of a structure variable under specification of the local instance name and the parameter name. In the example, the limit values are assigned constants. They can also be initialized prior to calling the local instance with

```
Delimiter.MAXI := Maximum;
Delimiter.MINI := Minimum;
```

Output parameters cannot be initialized at a function block call (also applies to local instances). If required, their values are read as components of the local instance:

```
Result := Delimiter.OUT;
```

You can also access the parameters of a local instance from "outside" the calling function block. Access takes place like access to global data addresses under specification of the data block (DB 290), the local instance (Delimiter) and the parameter name:

```
DB290.Delimiter.MAXI := Maximum;
DB290.Delimiter.MINI := Minimum;
Result := DB290.Delimiter.OUT;
```

29.3.5 Actual Parameters

When a block is called, you initialize the block parameters with the current values ("actual parameters") by making an assignment (see previous section). The same statements apply to actual parameters in SCL as in STL (see Chapter 19.3 "Actual Parameters") with the following exceptions:

- ▷ Block parameters of the complex data types DT and STRING can be initialized with constant values in SCL.
- ▷ Block parameters of data type POINTER cannot be initialized with constants, or with a pointer of the form P#Operand. Exception: pre-assignment with a zero pointer NIL is permissible.
- ▷ Block parameters of data type ANY cannot be initialized with constants or with an ANY pointer of the form P#[Data

block.]Operand Type Quantity. Exception: pre-assignment with a zero pointer NIL is permissible.

▷ You can initialize block parameters with expressions that supply a value of the same data type as the block parameter. For example, a function FC with function value can also be an actual parameter.

Note: If you initialize a formal parameter of type POINTER or ANY with a temporary variable when calling an FB or an FC, you cannot pass this parameter in the called block on to another block. The addresses of the temporary variables lose their values when passed on.

29.4 EN/ENO Mechanism

In SCL, you can check certain expressions for correct execution, for example, you can check if the result of a computational function is still within the permissible number range. The result of this scan is stored in the OK variable. You can also communicate the assignment of OK to the calling block via the ENO output of the block. Finally, you can execute the block call with EN, dependent on the conditions.

You can use the pre-defined variables EN and ENO for all blocks (FCs, SFCs, FBs, SFBs and also IEC functions), for all standard functions (e.g. shift and conversion functions) except the timer and counter functions.

Chapter 15 "Status Bits", in particular Chapter 15.4 "Using the Binary Result", describes how the EN/ENO mechanism is handled in the standard programming languages.

29.4.1 OK Variable

SCL provides an initialized variable with the name "OK" and the data type BOOL. This variable indicates errors in program execution in an SCL block but only if you have selected the option "Set OK flag" in the "Compiler" tab under OPTIONS → CUSTOMIZE in the SCL Program Editor.

The editor or the compiler to do not check whether this option is set or not when you use the OK variable in the program.

At the start of the block, the OK variable has the value TRUE. In the event of a program error, OK is set to FALSE. You can scan the OK variable with SCL statements or assign a value to the OK variable at any time.

```
SUM := SUM + IN;
IF OK
   THEN (* no error occurred *);
   ELSE (* errored addition *);
END_IF;
```

The OK variable is affected by arithmetic expressions and by some conversion functions (Chapter 30.5.2 "Explicit Conversion Functions"). If an error occurs when executing standard functions, such as math functions, it is reported via the ENO output (see below).

When exiting the block, the value of the OK variable is assigned to the ENO output.

29.4.2 ENO Output

The called block stores the result of the OK variable in the ENO output (enable output). ENO is of the data type BOOL. Following the block call, ENO can be used to see if the block was executed properly (ENO = TRUE) or if an error occurred (ENO = FALSE).

```
FC15 (In1 := ..., In2 := ...);
IF ENO
   THEN (* all in order *);
   ELSE (* error occurred *);
END_IF;
```

If you want to "pass on" a group error reported with ENO to the calling block following the block call, you must set the OK variable accordingly:

```
FC15 (In1 := ..., In2 := ...);
OK := ENO;
```

You an also assign a value to the ENO output in the block by setting the OK variable accordingly.

```
IF (* error occurred *)
   THEN OK := FALSE; RETURN;
END_IF;
```

ENO is not a block parameter but a sequence of statements generated by the program editor when you use ENO. ENO is not declared. You scan ENO immediately after calling the block.

If you control the block call with the EN input (see next chapter), and EN has the value FALSE, so that the block call is not executed, the ENO output will also have the value FALSE.

Note: If a block written with the standard programming languages uses the binary result BR as an error message, you can scan the error message in SCL with the ENO output following this block call (see also Chapter 15.4 “Using the Binary Result”).

29.4.3 EN Input

You control a block call with the Boolean EN input. If EN is initialized with TRUE, the called block will be executed. If EN is initialized with FALSE, the called block is not executed. A jump is then made beyond the block call to the next statement.

```
FC15 (EN := I1.0,
      In1 := ...,
      In2 := ...);
(*FC15 is only executed if I1.0 = "1"*)
```

If you do not use EN, the block will always be executed.

EN is not a block parameter but a sequence of statements generated by the program editor when you use EN. EN is not declared. You use EN in the parameter list in the same way as an input parameter.

You can initialize EN with ENO, in which case the called block is only executed if the previously called block has been properly executed. Example:

FC16 is only called if FC15 has been executed and no errors have occurred.

```
FC15 (EN := I1.0,
      In1 := ...,
      In2 := ...);
FC16 (EN := ENO,
      In1 := ...,
      In2 := ...);
```

If no block has been called previously on the same call level, ENO has the value TRUE.

Please note that a function FC or a system function SFC yields an undefined function value (any assignment) if you control its execution with EN and EN has the value FALSE.

30 SCL Functions

30.1 Timer Functions

The timers in the system memory of the CPU are addressed in SCL as functions with a function value. The function names for the different behaviors of the timer functions are as follows:

- ▷ S_PULSE (pulse time)
- ▷ S_PEXT (extended pulse)
- ▷ S_ODT (ON delay)
- ▷ S_ODTS (latching ON delay)
- ▷ S_OFFDT (OFF delay)

All timer functions have the parameters shown in Table 30.1. Timer function call example:

```
Time_BCD := S_PULSE(
        T_NO := Timer_address,
        S    := Start_input,
        TV   := Timer_duration,
        R    := Reset,
        Q    := Timer_status,
        BI   := Binary_time);
```

The behavior of the timer functions with pulse diagrams is described in detail in Chapter 7 “Timer Functions”. Please note that enabling of a timer function is not available in SCL.

The following rules apply for initializing the parameters of a timer function:

- ▷ T_NO must always be initialized.
- ▷ S and TV can be omitted in pairs.
- ▷ Q can be omitted.
- ▷ BI can be omitted.

In addition to the SIMATIC timer functions, the correspondingly set up CPUs are provided with “IEC timer functions” as system function blocks SFBs:

- ▷ SFB 3 TP
  Pulse generation
- ▷ SFB 4 TON
  ON delay
- ▷ SFB 5 TOF
  OFF delay

These functions are described in Chapter 7.7 “IEC Timer Functions”. The block shells are stored in the *Standard Library* in the *System Function Blocks* program.

Examples of the SIMATIC timer functions and the IEC timer functions can be found in the download files (download address: see pages 8-9) in the SCL_Book library in the “Timer Functions” source file of the “30 SCL Functions” program.

Table 30.1 Parameters for the SIMATIC Timer Functions

Parameter	Declaration	Data type	Meaning
T_NO	INPUT	TIMER	Timer address
S	INPUT	BOOL	Start timer
TV	INPUT	S5TIME	Timer value to be set
R	INPUT	BOOL	Reset timer
Function value	OUTPUT	S5TIME	Current time BCD coded
Q	OUTPUT	BOOL	Timer status
BI	OUTPUT	WORD	Current time binary coded

30.2 Counter Functions

The counters in the system memory of the CPU are addressed by SCL as functions with a function value. The function names for the different behaviors of the counter functions are as follows:

- S_CU (up counter)
- S_CD (down counter)
- S_CUD (up-down counter)

The counter functions have the parameters shown in Table 30.2. Counter function call example:

```
BCD_count_value := S_CU(
      C_NO := Counter_address,
      CU   := Count_up,
      S    := Set_input,
      PV   := Count_value,
      R    := Reset,
      Q    := Counter_status,
      CV   := Binary_count_value);
```

The behavior of the counter functions is described in detail in Chapter 8 "Counter Functions". Please note that enabling of counter functions is not available in SCL.

The following rules apply for initializing the parameters of a counter function:

- The CD parameter is not available in the S_CU counter function.
- The CU parameter is not available in the S_CD counter function.
- C_NO must always be initialized.
- CU and CD must be initialized depending on the counter function
- S and PV can be omitted in pairs
- Q can be omitted
- CV can be omitted

An INT number in the range 0 to 999 or a hex number in the range 16#000 to 16#3E7 can be applied as a constant to the PV counter value to be set.

In addition to the SIMATIC counter functions, the correspondingly set up CPUs are provided with "IEC counter functions" as system function blocks SFBs:

- SFB 0 CTU
  Up counter
- SFB 1 CTD
  Down counter
- SFB 2 CTUD
  Up-down counter

These functions are described in Chapter 8.6 "IEC Counter Functions". The block shells are stored in the *Standard Library* in the *System Function Blocks* program.

Examples of the SIMATIC counter functions and the IEC counter functions can be found in the download files (download address: see pages 8-9) in the SCL_Book library in the "Counter Functions" source file of the "30 SCL Functions" program.

Table 30.2 Parameters for the SIMATIC Counter Functions

Parameter	Declaration	Data type	Meaning
C_NO	INPUT	COUNTER	Counter address
CU	INPUT	BOOL	Count up
CD	INPUT	BOOL	Count down
S	INPUT	BOOL	Set counter
PV	INPUT	WORD	Count to be set
R	INPUT	BOOL	Reset counter
Function value	OUTPUT	WORD	Current count value BCD coded
Q	OUTPUT	BOOL	Counter status
CV	OUTPUT	WORD	Current count value binary coded

30.3 Math Functions

SCL provides the following math functions:

▷ Trigonometric functions:
SIN Sine
COS Cosine
TAN Tangent

▷ Arc functions:
ASIN Arc sine
ACOS Arc cosine
ATAN Arc tangent

▷ Logarithmic functions:
EXP Exponentiation to the base e
EXPD Exponentiation to the base 10
LN Natural logarithm
LOG Decimal logarithm

▷ Other math functions:
ABS Generate absolute value
SQR Generate square
SQRT Generate square root

A math function processes INT, DINT and REAL numbers. If you enter an INT or a DINT number as the input parameter, it is automatically converted to a REAL number.

The math functions operate with REAL numbers internally and yield a REAL number as the result. Exception: ABS yields the data type available at the input as the result.

A trigonometric function expects as an input value an angle in radian measure in the range 0 to 2π (where π = +3.141593e+00) corresponding to 0° to 360°. The arc functions are inverse trigonometric functions; they yield an angle in radian measure. The permissible value ranges for the arc functions are:

Function	Permissible range	Returned value
ASIN	–1 to +1	$-\pi/2$ to $+\pi/2$
ACOS	–1 to +1	0 to π
ATAN	Entire range	$-\pi/2$ to $+\pi/2$

Examples:

```
Reactive_power :=
      Voltage * Current * SIN(phi);
Volume := SQR(Radius) * Level * PI;
c := SQRT(SQR(a) + SQR(b));
```

30.4 Shifting and Rotating

The general function call for the shift and rotate functions are:

```
Result := Function(
      IN := Input_value,
      N  := Shift_number);
```

The shift and rotate functions have two input parameters: parameter N of data type INT indicates the number of positions by which the shift or rotation is to be made. Parameter IN indicates the variable to be shifted in data type ANY_BIT (BOOL, BYTE, WORD, DWORD). The function value is of the same data type as the input value.

Examples:

```
MW14 := SHL(IN := MW12, N := 2);
res_dword := ROR(
      IN := in_dword,
      N  := shift_int);
```

Table 30.3 Shift and Rotate Functions

SHL	Shift left	Input value IN is shifted to the left by N positions; the vacated positions are filled with zeros.
SHR	Shift right	Input value IN is shifted to the right by N positions; the vacated positions are filled with zeros.
ROL	Rotate left	Input value IN is rotated to the left by N positions; the vacated positions are filled with the shifted positions.
ROR	Rotate right	Input value IN is rotated to the right by N positions; the vacated positions are filled with the shifted positions.

30.5 Conversion Functions

When you combine variables, the variables must be of the same data type. This also applies when you make value assignments or when you initialize function parameters or block parameters. If a variable is not available in the required data type, the data type must be changed. This is the purpose of the conversion functions.

SCL provides two types of conversion function. "Class A" conversions can be executed automatically ("implicitly") in SCL since they are not associated with information loss (e.g. conversion from BYTE to WORD). You must specify "Class B" conversions explicitly (e.g. conversion from REAL to INT). Any threat of information loss can be anticipated and avoided with an upstream check, or you can scan the OK variable in such cases (must be set in the Compiler Properties).

You can convert and process variables of data type DATE_AND_TIME and STRING with the IEC functions (*Standard Library* and *IEC Function Blocks*, described in Chapter 31 "IEC Functions").

30.5.1 Implicit Conversion Functions

Implicit conversion functions are executed "automatically" by SCL. You can also program them, e.g. when you want to enhance the clarity or readability of the program.

Table 30.4 shows the implicit conversion functions available in SCL.

When converting from CHAR_TO_STRING, a STRING variable with the length 1 is generated and the OK variable is set to FALSE.

Examples:

```
MB10 := M7.0;
real_var := int_var;
string_var := char_var;
```

In the example, memory bit M10.0 receives the signal state of memory bit M7.0. The remaining bits are set to signal state "0".

30.5.2 Explicit Conversion Functions

You must specify explicit conversion functions in the program; nevertheless, with some of these functions, conversion does not take place and no code is executed (indicated in Table 30.5 with "Accepted without change"). The OK variable is affected by some conversion functions.

Examples:

```
MB10 := CHAR_TO_BYTE(char_var);
int_var := WORD_TO_INT(MW20);
real_var := DWORD_TO_REAL(MD30);
```

Please note that no number conversion takes place in the last example. The bit pattern of the memory doubleword is accepted unchanged into the REAL variable.

Table 30.4 Implicit Conversion Functions

Function	OK	Conversion
BOOL_TO_BYTE	N	Filled with leading zeros
BOOL_TO_WORD	N	
BOOL_TO_DWORD	N	
BYTE_TO_WORD	N	
BYTE_TO_DWORD	N	
WORD_TO_DWORD	N	
INT_TO_DINT	N	Leading positions filled with the sign
INT_TO_REAL	N	-
DINT_TO_REAL	N	At conversion, accuracy, among other things, is reduced
CHAR_TO_STRING	Y	Conversion to a character string with one character

Table 30.5 Explicit Conversion Functions (Part 1)

Function	OK	Conversion	Remarks
BYTE_TO_BOOL	Y	Least significant bit is accepted	The OK variable has TRUE if a bit is set to "1" in the unaccepted part of the variable
WORD_TO_BOOL	Y		
DWORD_TO_BOOL	Y		
WORD_TO_BYTE	Y	Least significant byte is accepted	
DWORD_TO_BYTE	Y		
DWORD_TO_WORD	Y	Least significant word is accepted	
CHAR_TO_BYTE	N	Without change to the assignment	
BYTE_TO_CHAR	N	Without change to the assignment	
CHAR_TO_INT	N	Most significant byte filled with zeros	
INT_TO_CHAR	Y	Acceptance of the least significant byte without change	OK = TRUE if a bit is set in the left byte
STRING_TO_CHAR	Y	Acceptance of the first character	OK = FALSE if the STRING length is not equal to 1
WORD_TO_INT	N	Acceptance without change	
DWORD_TO_DINT	N		
INT_TO_WORD	N		
DINT_TO_DWORD	N		
REAL_TO_DWORD	N		No conversion !
DWORD_TO_REAL	N		No conversion !
DINT_TO_INT	Y	Bits for the sign copied	OK = FALSE if the number range is violated
REAL_TO_INT	Y	Rounded to INT	
REAL_TO_DINT	Y	Rounded to DINT	
ROUND	Y	Conversion of REAL to DINT with rounding	As for REAL_TO_DINT
TRUNC	Y	Conversion of REAL to DINT without rounding ("truncation" of the fractional component)	OK = FALSE if the number range is violated
DINT_TO_TIME	N	Acceptance without change	
DINT_TO_TOD	Y		OK = FALSE if the range for TOD is violated
DINT_TO_DATE	Y		OK = FALSE if the left word is assigned
DATE_TO_DINT	N		
TIME_TO_DINT	N		
TOD_TO_DINT	N		
WORD_TO_BLOCK_DB	N	Acceptance without change	
BLOCK_DB_TO_WORD	N	Acceptance without change	

Table 30.6 Explicit Conversion Functions (Part 2)

Function	OK	Conversion
BOOL_TO_INT	N	WORD_TO_INT(BOOL_TO_WORD(x))
BOOL_TO_DINT	N	DWORD_TO_DINT(BOOL_TO_DWORD(x))
BYTE_TO_INT	N	WORD_TO_INT(BYTE_TO_WORD(x))
BYTE_TO_DINT	N	DWORD_TO_DINT(BYTE_TO_DWORD(x))
WORD_TO_DINT	N	INT_TO_DINT(WORD_TO_INT(x))
DWORD_TO_INT	Y	DINT_TO_INT(DWORD_TO_DINT(x))
INT_TO_BOOL	Y	WORD_TO_BOOL(INT_TO_WORD(x))
INT_TO_BYTE	Y	WORD_TO_BYTE(INT_TO_WORD(x))
INT_TO_DWORD	N	WORD_TO_DWORD(INT_TO_WORD(x))
DINT_TO_BOOL	Y	DWORD_TO_BOOL(DINT_TO_DWORD(x))
DINT_TO_BYTE	Y	DWORD_TO_BYTE(DINT_TO_DWORD(x))
DINT_TO_WORD	Y	DWORD_TO_WORD(DINT_TO_DWORD(x))
INT_TO_STRING	N	Like loadable IEC function FC 16 I_STRNG
DINT_TO_STRING	N	Like loadable IEC function FC 5 DI_STRNG
REAL_TO_STRING	N	Like loadable IEC function FC 30 R_STRNG
STRING_TO_INT	N	Like loadable IEC function FC 38 STRNG_I
STRING_TO_DINT	N	Like loadable IEC function FC 37 STRNG_DI
STRING_TO_REAL	N	Like loadable IEC function FC 39 STRNG_R
BCD_TO_INT(x)	N	x with the data type WORD or DWORD is interpreted as a BCD-coded number between ±999 and ±9 999 999. If x contains a pseudo tetrad (numerical value 10 to 15 or A to F in hexadecimal representation), the organization block OB 121 (programming error) is called. If it is not present, the CPU got to STOP.
WORD_BCD_TO_INT(x)	N	
BCD_TO_DINT(x)	N	
DWORD_BCD_TO_DINT(x)	N	
INT_TO_BCD(x)	N	x with the data type INT or DINT is interpreted as an integer between ±999 and ±9 999 999. The result is a BCD-coded number with data type WORD or DWORD.
INT_TO_BCD_WORD(x)	N	
DINT_TO_BCD(x)	N	
DINT_TO_BCD_DWORD(x)	N	

Examples:

```
MW10 := BOOL_TO_WORD(M20.3)
MW10 := M20.3
```

The conversion BOOL_TO_WORD is an implicit function, and need not be specified. If M20.3 has a signal status "1", MW10 has the value W#16#0001, otherwise W#16#0000.

```
M20.3 := WORD_TO_BOOL(MW10)
```

This conversion must be programmed. The signal status of the least significant bit (M11.0 in the example) is applied. If one of the other bits not involved in the conversion has a signal status "1", the OK variable is set to TRUE.

```
int_var := BOOL_TO_INT(M20.3)
```

The signal status of the bit is applied in the least significant digit of the INT variable, so that the value of *int_var* is either 0 or 1.

```
M20.3 := INT_TO_BOOL(int_var)
```

The signal status of the least significant bit is applied, i.e. M20.3 is set to "1" in the case of an odd value of the INT variable. If one of the remaining bits has a signal status "1", the OK variable is set to TRUE.

30.6 Numerical Functions

SCL provides the following functions for selecting values:

- ▷ SEL Binary selection
- ▷ MUX Multiple selection
- ▷ MAX Maximum selection
- ▷ MIN Minimum selection
- ▷ LIMIT Limiter

Apart from MUX, these functions are also available as loadable IEC functions provided in the *Standard Library* in the *IEC Function Blocks* program with STEP 7. The functions integrated in SCL exactly correspond – partially differing from the loadable functions – to the IEC 61131-3 standard. Table 30.7 lists the parameters of the numerical functions.

SEL Binary selection

Call: *any* :=
SEL (G := *bool*, IN0 := *any*, IN1 := *any*);

The SEL function selects one of two variable values (IN0 and IN1) depending on a switch (parameter G). Variables of all data types are permissible as input values for the IN0 and IN1 parameters, with the exception of S5TIME, ARRAY, STRUCT and the parameter types. The two input variables (current parameters) and the function value must be of the same class of data type.

Table 30.7 Parameters of the Numerical SCL Functions

Function	Parameter	Declaration	Data Type	Contents, Description
SEL	G	INPUT	BOOL	Selection criterion (G = “0” or “1”)
	IN0	INPUT	ANY 1)	First input value
	IN1	INPUT	ANY 1)	Second input value
	Function value	RETURN	ANY 1)	Selected input value
MUX	K	INPUT	INT	Selection criterion (K = 0 to 31)
	IN0	INPUT	ANY 1)	First input value
	IN1	INPUT	ANY 1)	Second input value
	IN*n*	INPUT	ANY 1)	*n* = 2 to 31 (optional input values)
	INELSE	INPUT	ANY 1)	Alternative input value (optional)
	Function value	RETURN	ANY 1)	Selected input value
MAX	IN1	INPUT	ANY_NUM 2)	First input value
	IN2	INPUT	ANY_NUM 2)	Second input value
	IN*n*	INPUT	ANY_NUM 2)	*n* = 3 to 32 (optional input values)
	Function value	RETURN	ANY_NUM 2)	Largest input value
MIN	IN1	INPUT	ANY_NUM 2)	First input value
	IN2	INPUT	ANY_NUM 2)	Second input value
	IN*n*	INPUT	ANY_NUM 2)	*n* = 3 to 32 (optional input values)
	Function value	RETURN	ANY_NUM 2)	Smallest input value
LIMIT	MN	INPUT	ANY_NUM 2)	Lower limit
	IN	INPUT	ANY_NUM 2)	Input value
	MX	INPUT	ANY_NUM 2)	Upper limit
	Function value	RETURN	ANY_NUM 2)	Limited input value

1) Except ARRAY, STRUCT and parameter types
2) Plus the time data types, except S5TIME

MUX Multiple selection

Call: *any* :=
MUX (K := *int*, IN0 := *any*, IN1 := *any*, ..., IN31 := *any*, INELSE := *any*);

From 2 to 32 numerical variable values, the MUX function selects the value whose number is specified in the K parameter. If the value of K is outside the number of input parameters, the alternative value from the INELSE parameter is output. If INELSE is missing, the value at IN0 is output.

Variables of all data types except S5TIME, ARRAY, STRUCT and the parameter types are permissible as input values. All input values (current parameters) and the function value must be of the same class of data type. The function value accepts the data type of highest significance.

Example:

```
selection:= MUX(
   K   := int0,
   IN0 := int1,
   IN1 := int2,
   IN2 := dint1,
   IN3 := dint2,
   IN4 := real1,
   INELSE:= real2);
```

If the variables to be selected are of data types INT, DINT and REAL, the *selection* variable accepts the data type of highest significance, i.e. REAL. Depending on the value of the *int0* variable, the values of the selected variables are imported into *selection*, if necessary with the implicit data type conversion INT_TO_REAL or DINT_TO_REAL.

MAX Maximum selection

Call: *any_num* :=
MAX (IN1 := *any_num*, IN2 := *any_num*, ..., IN32 := *any_num*);

From 2 to 32 numerical variable values, the MAX function selects the largest one. Variables of the data type class ANY_NUM and the time data types except S5TIME are permissible as input values. All input values (current parameters) and the function value must be of the same class of data type. The function value accepts the data type of highest significance.

MIN Minimum selection

Call: *any_num* :=
MIN (IN1 := *any_num*, IN2 := *any_num*, ..., IN32 := *any_num*);

From 2 to 32 numerical variable values, the MIN function selects the smallest one. Variables of the data type class ANY_NUM and the time data types except S5TIME are permissible as input values. All input values (current parameters) and the function value must be of the same class of data type. The function value accepts the data type of highest significance.

LIMIT Limiter

Call: *any_num* :=
LIMIT (MN := *any_num*, IN := *any_num*; MX := *any_num*);

The LIMIT function limits the numerical value of the IN variable to the limits specified in the MN and MX parameters. Variables of the data type class ANY_NUM and the time data types except S5TIME are permissible as input values. All input values (current parameters) and the function value must be of the same class of data type. The function value accepts the data type of highest significance. The lower limit (MN parameter) must be less than the upper limit (MX parameter).

30.7 Programming Your Own Functions with SCL

If you do not find any suitable functions among the SCL standard functions and the IEC functions, SCL allows you to write your own functions that you can then adapt to your own requirements.

The correct block type for this purpose is the function FC with function value. Programming and calling a function FC with function value are described in Chapter 29.2.2 "Function FC with Function Value" and in Chapter 29.3.2 "Function FC with Function Value", respectively.

In many cases, the language resources of SCL are insufficient for programming the desired function. In such cases there is still the possibility of implementing the function with STL (see

Chapter 30.8 “Programming Your Own Functions with STL”). But with the principle of data type views, SCL also makes it possible to process complex variables. Chapter 27.1.9 “Data Type Views” shows you which data type views you can assign to which variables.

Bit-wise processing of variables of elementary data types

Example: you want to process individual bits in a doubleword variable in some way, e.g. scan them or logically combine them and write the result to another bit. For this purpose, you apply a data type view to the variable in the form of a bit field and you can now address the individual bits as field components.

```
VAR_TEMP
DW_Var : DWORD;
Pattern AT DW_Var : ARRAY [0..31] OF BOOL;
END_VAR
...
Pattern[1] := Pattern[10] & Pattern[11];
...
```

In the small example, bits 10 and 11 of the variable DW_VAR are combined for logic AND and the result is assigned to bit 1.

Processing of variables of data types DT and STRING

Variables of data types DT and STRING are usually handled as “whole” variables by SCL, for example when initializing function inputs or when passing on from one block parameter to another. The IEC functions are available to you from the STEP 7 Standard Library for processing variables of data type DT or STRING.

If you want to process parts of a variable of data types DT and STRING with SCL statements, apply a data type view to the variable whose components can be processed with SCL. BYTE fields are suitable for representing DT and STRING variables (Table 30.8).

Table 30.8 Frequently Used Data Type Views

Data type of variables	Data type view	Declaration example for a variable with the name TEMPVAR and a data type view with the name VIEW
Elementary	Bit field	`TEMPVAR : DWORD;` `VIEW AT TEMPVAR : ARRAY[0..31] OF BOOL;`
DT	BYTE field	`TEMPVAR : DT;` `VIEW AT TEMPVAR : ARRAY [1..8] OF BYTE;`
STRING	CHAR field	`TEMPVAR : STRING[max];` `VIEW AT TEMPVAR : ARRAY [1..max] OF CHAR;`

ARRAY	STRUCT	`TEMPVAR : ARRAY[0..255] OF BYTE;`
		`VIEW1 AT TEMPVAR : STRUCT`
		`name : data_type;`
		`.... : ...`
		`END_STRUCT;`
		`VIEW2 AT TEMPVAR : STRUCT`
		`name : data_type;`
		`.... : ...`
		`END_STRUCT;`

ANY	STRUCT	`TEMPVAR : ANY;`
		`VIEW AT TEMPVAR : STRUCT`
		`ID : BYTE;`
		`TYP : BYTE;`
		`NUM : INT;`
		`DBN : INT;`
		`PTR : DWORD;`
		`END_STRUCT;`

Example of an SCL function

The function "Hour" is to extract the hour information from the data format DT and supply it with the data type INT.

```
FUNCTION Hour : INT
VAR_INPUT
  DAT : DT;
  TMP AT DAT : ARRAY [1..8] OF BYTE;
END_VAR
Hour :=
  WORD_TO_INT(SHR(IN:=TMP[4],N:=4))*10 +
  WORD_TO_INT(TMP[4] AND 16#0F);
END_FUNCTION
(* READ THE CPU-TIME AND
CALL THE FUNCTION "Hour" *)
SFC_ERROR := READ_CLK(DATE_TIME);
IF Hour(DATE_TIME) >= 18
  THEN FINISH_WORK := TRUE;
END_IF;
```

Different views of fields and structures

Variables of data types ARRAY and STRUCT can be assigned data type views that are themselves of data types ARRAY and STRUCT. One application of this is setting up a data area for a send or receive mailbox for message frames.

You set up the maximum length of the mailbox using a byte field, for example. For each message frame that you want to process in the mailbox, you can apply to the mailbox a data type view that has the structure of the message frame. The data type view is specially adapted to the relevant message frame: it can therefore also be shorter than the mailbox.

Manipulation of the ANY pointer

If you create a variable of data type ANY in the temporary local data, the Compiler interprets this variable as a pointer and passes it on direct to an ANY input parameter, for example, of a called block (see Chapter 29.2.4 "Temporary Local Data").

You can manipulate this ANY pointer at runtime with the help of a data type view, enabling you, for example, to dynamically specify different source data areas for copy blocks.

Example: copying from a data area that has been specified with the variables *DataBlock*, *DataStart* and *Num_of_Bytes*, to a variable called *Send_MailBox*.

```
FUNCTION_BLOCK COPY
VAR_INPUT
  AREA            : ANY;
  DATABLOCK       : INT;
  DATASTART       : INT;
  NUM_OF_BYTES    : INT;
END_VAR
VAR_TEMP
  SFC_ERROR       : INT;
  SEND_MAILBOX    : ANY;
  VIEW AT SEND_MAILBOX : STRUCT
    ID  : WORD;
    TYP : BYTE;
    NUM : INT;
    DBN : INT;
    PTR : DWORD;
    END_STRUCT;
END_VAR
BEGIN
VIEW.ID  := 16#10;
VIEW.TYP := 16#02;
VIEW.NUM := NUM_OF_BYTE;
VIEW.DBN := DATABLOCK;
VIEW.PTR := INT_TO_WORD(8*DATASTART);
SFC_ERROR := BLKMOV(
   SRCBLK := AREA,
   DSTBLK := SEND_MAILBOX);
END_FUNCTION_BLOCK

(* Call of the FB *)
COPY.COPYDATA(
  AREA           := SEND_MAILBOX,
  DATABLOCK      := 309,
  DATASTART      := 32,
  NUM_OF_BYTE    := 32);
```

More examples on this topic can be found in the download files (download address: see pages 8-9) in the SCL_Book library under the "General Examples" program.

30.8 Programming Your Own Functions with STL

The function FC with function value allows you to program your own functions but still with the possibilities of the SCL programming language. However, since you can mix blocks created with various languages in your program, it is also possible to program functions FCs with STL and then invoke them in SCL. This gives you access to the more extensive function range of STL, such as direct access to variable addresses or addressing via the address register.

You can program STL blocks in two different ways: incrementally or source-file-oriented (Chapter 3.4 "Programming Code Blocks with STL"). If you select source-file-oriented programming, the procedure is identical to SCL blocks:

1) Create an STL source in the *Source Files* container.
2) Open the STL source with a double-click.
3) Program the source program with the STL programming language (see notes below)
4) If you have chosen symbolic names for the functions, update the symbol table.
5) Compile the STL program in order to have the compiled functions available in the *Blocks* container.
6) You can call the new functions in the same way as, say, the standard functions in the SCL program.

Source-file-oriented STL programming uses almost the same keywords for block programming as SCL (see Table 3.3 in Chapter 3.4.2 "Programming STL Code Blocks Incrementally"). The main difference regarding functions with function value is that the function value in the program has the name RET_VAL (or ret_val). You then assign the value of the function to the RET_VAL variable in the program.

For our little example, we select functions for scanning, starting and resetting a timer function in order to achieve simpler handling of the timer functions. Chapter 7 "Timer Functions" shows how timer functions are programmed in STL.

The function T_SCAN yields the status of the parameterized timer address:

```
FUNCTION T_SCAN : BOOL
VAR_INPUT
  T_NO : TIMER;
END_VAR
BEGIN
  U T_NO; = RET_VAL;
END_FUNCTION
```

The function T_PULSE starts a timer address as a pulse via an input:

```
FUNCTION T_PULSE : VOID
VAR_INPUT
  T_NO : TIMER; Start : BOOL;
  Time_value : S5TIME;
END_VAR
BEGIN
  U Start; L Time_value; SI T_NO;
END_FUNCTION
```

The function T_RESET resets a timer address at every call:

```
FUNCTION T_RESET : VOID
VAR_INPUT
  T_NO : TIMER;
END_VAR
BEGIN
  SET; R T_NO;
END_FUNCTION
```

Following compiling, these functions could be used in an SCL program as follows:

```
IF NOT T_SCAN(T1)
   THEN T_PULSE(T_NO := T2,
                Start := E1.0,
                Time_value := S5T#5s);
   ELSE T_RESET(T3);
END_IF;
```

These examples can be found in the download files (download address: see pages 8-9) in the SCL_Book library under the "General Examples" program.

30.9 Brief Description of the SCL Examples

30.9.1 Conveyor Example

The "Conveyor" example shows the application of binary logic operations, set/reset functions and block calls. It is designed for the STL programming language. If you have knowledge of STL and you want to learn SCL, you will find suggestions here for how to convert typical STL functions into SCL.

Figure 30.1 shows the program and data structure of this example. Please refer to the following sections for the detailed description

- ▷ 5.5 "Example of a Conveyor Belt Control System" (FC 11)
- ▷ 8.7 "Parts Counter Example" (FC 12)
- ▷ 19.5.1 "Conveyor Belt Example" (FB 21)

Conveyor example

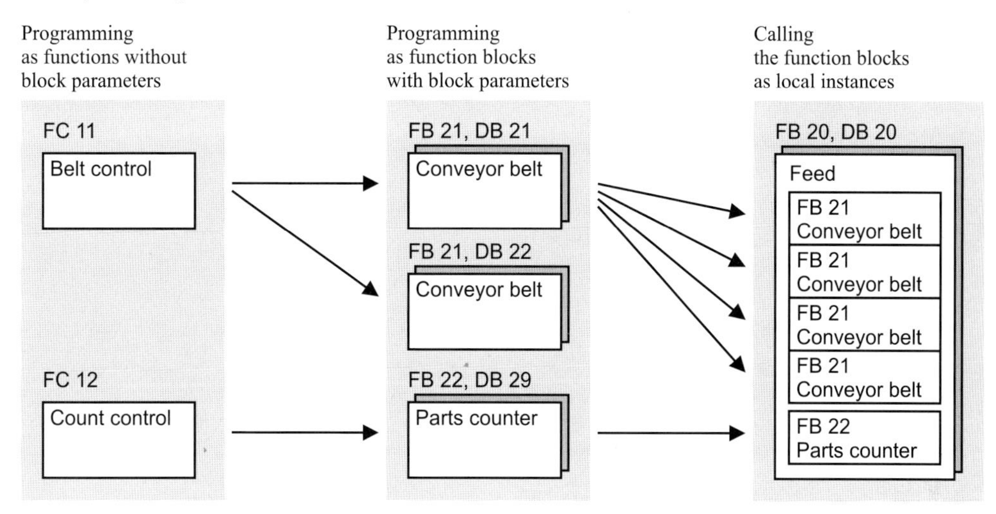

Figure 30.1 Data and Program Structure for the Conveyor Example

- ▷ 19.5.2 “Parts Counter Example” (FB 22)
- ▷ 19.5.3 “Feed Example” (FB 20)

The program of this example can be found in the download files (download address: see pages 8-9) in the SCL_Book library under the “Conveyor Example” program.

30.9.2 Message Frame Example

The “Message frame” example shows how to handle user-defined data types and how to copy data areas. In the STL programming language, the use of indirect addressing with the address registers and the manipulation of the ANY pointer is represented in this context (Chapter 26.4 “Brief Description of the Message Frame Example”).

The same functionality can be implemented with the SCL programming language – more elegantly, to a certain extent. You can get specific support here from the possibility of processing individual field components at runtime with indexing (Figure 30.2).

SCL is also better suited to the formulation of this task (clearer and therefore easier to use and less prone to errors). However, with direct access to variables, STL offers a functionality that is missing in SCL so that the solution was implemented to a certain extent somewhat differently than in STL.

The program of this example can be found in the download files (download address: see pages 8-9) in the SCL_Book library under the “Message Frame Example” program.

30.9.3 General Examples

The general examples focus on the processing of variables of complex data types and the manipulation of the ANY pointer with the help of data type views.

The following functions execute data type conversion with the SCL language resources:

- ▷ FC 61 DT_TO_STRING
  Extracts the date and converts to a STRING variable
- ▷ FC 62 DT_TO_DATE
  Extracts the date and converts to a DATE variable
- ▷ FC 63 DT_TO_TOD
  Extracts the time-of-day and converts to a TOD variable

The following function blocks provide access to variables of complex data types as well as data management in a ring buffer and in a FIFO register

▷ FB 61 Variable_length

▷ FB 62 Checksum

▷ FB 63 Ring_buffer

▷ FB 64 FIFO_register

The source programs "STL Functions" and "Call STL Functions" show you how to write simple functions in STL and integrate these functions in your SCL program. You will learn how to use SIMATIC timers and counters in SCL in the same way as in the standard programming languages.

These examples can be found in the download files (download address: see pages 8-9) in the SCL_Book library under the "General Examples" program.

Example Message Frame

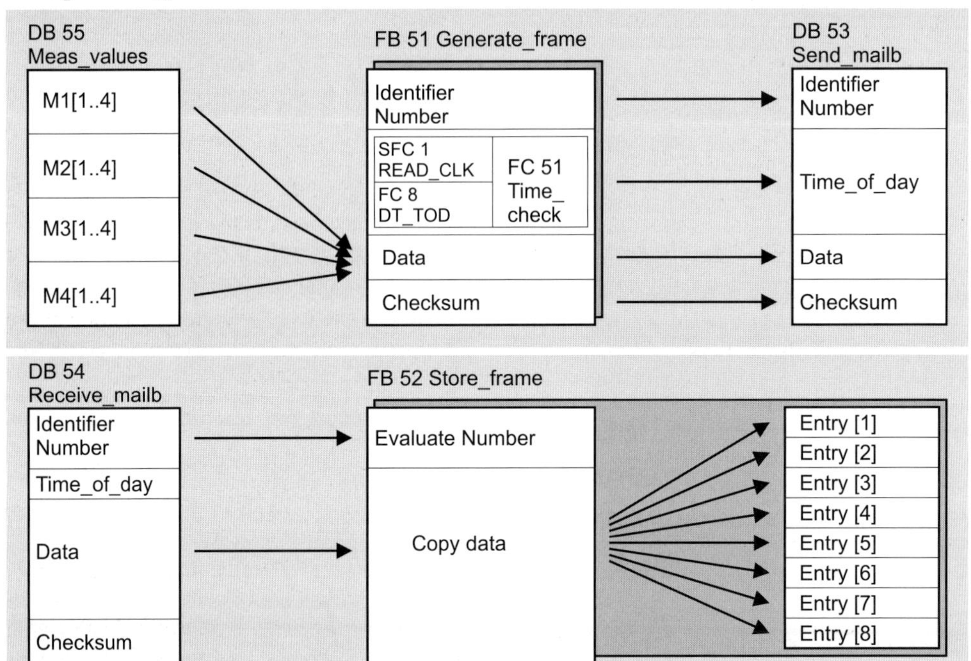

Figure 30.2 Data and Program Structure for the Message Frame Example

31 IEC functions

The IEC functions are loadable functions FCs supplied with STEP 7. They are located in the *Standard Library* in the *IEC Function Blocks* program. They supplement the standard functions of SCL and can also be used by other languages such as STL. You can find the overview of all IEC functions in the Appendix. At this point, they are arranged according to the following function groups:

▷ Conversion functions

▷ Comparison functions for DATE_AND_TIME

▷ Comparison functions for STRING

▷ STRING functions

▷ Date/time-of-day functions

▷ Numerical functions

The call is represented in SCL notation. If you use the IEC functions in STL, the function value has the name RET_VAL and represents the first output parameter. Example:

Call in SCL:

```
CompResult := EQ_STRNG(
      S1 := string1,
      S2 := string2);
```

Call in STL:

```
CALL EQ_STRNG(
      S1 := string1,
      S2 := string2,
      RET_VAL := CompResult);
```

Some IEC functions set the binary result BR as a group error message. BR can be scanned in SCL via the ENO output and in STL direct via binary scanning or jump functions.

31.1 Conversion Functions

General

The conversion functions convert the data type of a variable. The value to be converted is at the function input and the function value has the new data type.

General call:

```
var_aus :=
      Conversion_function(var_in);
```

Some conversion functions set the binary result BR or the ENO output to FALSE if an error occurs during conversion. In such a case, conversion does not take place.

Example: The INT value in the variable *Speed* is to be converted to a character string located in the variable *Display*.

```
Display := I_STRNG(Speed);
IF ENO
   THEN (*Conversion in order*);
   ELSE (*Error occurred*);
END_IF;
```

If you assign a STRING function value to a STRING variable located in the temporary local data, you must assign a defined value with the required length to this variable in the program (pre-assignment per declaration is not possible in the temporary local data).

A specific space (number of bytes) is reserved for the STRING variable declared in the temporary local data. You can set this length in the Compiler Properties. If you make no setting, 254 (+2) bytes are assigned.

FC 33 S5TI_TIM
Data type conversion S5TIME to TIME

The function FC 33 S5TI_TIM converts the data format S5TIME to the format TIME.

The function does not report errors.

FC 40 TIM_S5TI
Data type conversion TIME to S5TIME

The function FC 40 TIM_S5TI converts the data format TIME to the format S5TIME. The conversion rounds down.

If the input parameter is greater than the representable S5TIME format (greater than TIME#02:46:30.000), S5TIME#999.3 is output as the result and the binary result or the ENO output is set to FALSE.

FC 16 I_STRNG
Data type conversion INT to STRING

The function FC 16 I_STRING converts a variable in the INT format to a character string. The character string is represented with a leading sign (number of digits plus sign).

If the variable specified at the function value is too short, conversion does not take place and the binary result BR or the ENO output is set to FALSE.

FC 5 DI_STRNG
Data type conversion DINT to STRING

The function FC 5 DI_STRING converts a variable in DINT format to a character string. The character string is represented with a leading sign (number of digits plus sign).

If the variable specified at the function value is too short, conversion does not take place and the binary result BR or the ENO output is set to FALSE.

FC 30 R_STRNG
Data type conversion REAL to STRING

The function FC 30 R_STRNG converts a variable in REAL format to a character string. The character string is represented with 14 digits:

±v.nnnnnnnE±xx
± Sign
v 1 place before the point
n 7 decimal places
x 2 exponent places

If the variable specified at the function value is too short, or if there is no valid floating-point number at the input parameter, conversion does not place and the binary result or the ENO output is set to FALSE.

FC 38 STRNG_I
Data type conversion STRING to INT

The function FC 38 STRNG_I converts a character string to a variable in INT format. The first character of the string can be a sign or a digit and the subsequent characters must be digits.

If the length of the character string is zero or greater than 6, or if there are illegal characters in the string, or if the converted value exceeds the INT number range, conversion does not take place and the binary result BR or the ENO output is set to FALSE.

FC 37 STRNG_DI
Data type conversion STRING to DINT

The function FC 37 STRNG_DI converts a character string to a variable in DINT format. The first character of the string can be a sign or a digit and the subsequent characters must be digits.

If the length of the character string is zero or greater than 11, or if there are illegal characters in the string, or if the converted value exceeds the DINT number range, conversion does not take place and the binary result BR or the ENO output is set to FALSE.

FC 39 STRNG_R
Data type conversion STRING to REAL

The function FC 39 STRNG_R converts a character string to a variable in REAL format. The character string must exist in the following format:

±v.nnnnnnnE±xx
± Sign
v 1 place before the point
n 7 decimal places
x 2 exponent places

If the length of the character string is less 14, or if it is not structured as shown above, or if the converted value exceeds the REAL number range, conversion does not take place and the binary result BR or the ENO output is set to FALSE.

31.2 Comparison Functions

The comparison functions compare the values of two variables and report the comparison result via the function value. The function value is TRUE if the comparison is met, otherwise it is FALSE. A comparison function does not report errors. There are comparison functions for DT variables and for STRING variables.

General call:

```
Result :=
      Comparison_function_DT(
      DT1 := DT_var1,
      DT2 := DT_var2);


Result :=
      Comparison_function_STRNG(
      S1 := STRING_var1,
      S2 := STRING_var2);
```

FC 9 EQ_DT
Comparison of DT for equal to

The function FC 9 EQ_DT compares the contents of two variables in the DATE_AND_TIME format for equal to, and only outputs TRUE as the function value if the time at parameter DT1 is equal to the time at parameter DT2.

FC 28 NE_DT
Comparison of DT for not equal to

The function FC 28 NE_DT compares the contents of two variables in the DATE_AND_TIME format for not equal to, and only outputs TRUE as the function value if the time at parameter DT1 is not equal to the time at parameter DT2.

FC 14 GT_DT
Comparison of DT for greater than

The function FC 14 GT_DT compares the contents of two variables in the DATE_AND_TIME format for greater than, and only outputs TRUE as the function value if the time at parameter DT1 is greater (later) than the time at parameter DT2.

FC 12 GE_DT
Comparison of DT for greater than or equal to

The function FC 12 GE_DT compares the contents of two variables in the DATE_AND_TIME format for greater than or equal to, and only outputs TRUE as the function value if the time at parameter DT1 is greater (later) than the time at parameter DT2, or if both times are equal.

FC 23 LT_DT
Comparison of DT for less than

The function FC 23 LT_DT compares the contents of two variables in the DATE_AND_TIME format for less than, and only outputs TRUE as the function value if the time at parameter DT1 is less (earlier) than the time at parameter DT2.

FC 18 LE_DT
Comparison of DT for less than or equal to

The function FC 18 LE_DT compares the contents of two variables in the DATE_AND_TIME format for less than or equal to, and only outputs TRUE as the function value if the time at parameter DT1 is less (earlier) than the time at parameter DT2, or if both times are equal.

FC 10 EQ_STRNG
Comparison of STRING for equal to

The function FC 10 EQ_STRNG compares the contents of two variables in the STRING format for equal to, and only outputs TRUE as the function value if the character string at parameter S1 is equal to the character string at parameter S2.

FC 29 NE_STRNG
Comparison of STRING for not equal to

The function FC 29 NE_STRNG compares the contents of two variables in the STRING format for not equal to, and only outputs TRUE as the function value if the character string at parameter S1 is not equal to the character string at parameter S2.

FC 15 GT_STRNG
Comparison of STRING for greater than

The function FC 15 GT_STRNG compares the contents of two variables in the STRING format for greater than, and only outputs TRUE as the function value if the character string at parameter S1 is greater than the character string at parameter S2. The characters are compared starting from the left via their ASCII codes (e.g. is 'a' greater than 'A'). The first variant character decides the comparison result. If the first characters are equal, the longer string is taken as the greater.

FC 13 GE_STRNG
Comparison of STRING for greater than or equal to

The function FC 13 GE_STRNG compares the contents of two variables in the STRING format for greater than or equal to, and only outputs TRUE as the function value if the character string at parameter S1 is greater than the character string at parameter S2, or if both strings are equal. The characters are compared starting from the left via their ASCII codes (e.g. is 'A' greater than 'a'). The first variant character decides the comparison result. If the first characters are equal, the longer string is taken as the greater.

FC 24 LT_STRNG
Comparison of STRING for less than

The function FC 24 LT_STRNG compares the contents of two variables in the STRING format for less than and only outputs TRUE as the function value if the character string at parameter S1 is less than the character string at parameter S2. The characters are compared starting from the left via their ASCII codes (e.g. is 'A' less than 'a'). The first variant character decides the comparison result. If the first characters are equal, the shorter string is taken as being "less than".

FC 19 LE_STRNG
Comparison of STRING for less than or equal to

The function FC 19 LE_STRNG compares the contents of two variables in the STRING format for less than or equal to and only outputs TRUE as the function value if the character string at parameter S1 is less than the character string at parameter S2, or if both strings are equal. The characters are compared starting from the left via their ASCII codes (e.g. is 'A' less than 'a'). The first variant character decides the comparison result. If the first characters are equal, the shorter string is taken as being "less than".

31.3 STRING Functions

The STRING functions allow you to use character strings. Some STRING functions set the binary result BR or the ENO output to FALSE if an error occurs during execution of the STRING function.

The STRING functions check the actual parameter for validity (e.g. is a block parameter applied to the STRING variable long enough). If you declare a STRING variable in the temporary local data and then use it as an actual parameter, you must first assign (any) character string of the required length to this variable. The reason is that variables in the temporary local data cannot be pre-assigned by the compiler. That is, their values are semi random, and in the case of STRING variables, so are the bytes for the maximum and current length. These bytes receive meaningful values when the string is assigned.

A specific space (number of bytes) is reserved for a STRING variable declared in the temporary local data. You can set this length in the Compiler Properties. If you make no setting, 254 (+2) bytes are assigned.

FC 21 LEN
Length of a STRING variable

Call: *int* := LEN (*string*);

The function FC 21 LEN outputs the current length of a character string (number of valid characters) as the function value. An empty string has the length zero. The maximum length is 254.

The function does not report errors.

FC 11 FIND
Searching in a STRING variable

Call: *int* := FIND (IN1 := *string*, IN2 := *string*);

The function FC 11 FIND yields the position of the second character string (IN2) within the first character string (IN1). The search begins on the left; the first occurrence of a character string is reported. If the second character string is not contained in the first, zero is returned.

The function does not report errors.

FC 20 LEFT
Left section of a STRING variable

Call: *string* := LEFT (IN := *string*, L := *int*);

The function FC 20 LEFT yields the first L character of a string. If L is greater than the current length of the STRING variable, the input value is returned. If L = 0 and if the input value is an empty string, an empty string is returned.

If L is negative, an empty string is output and the binary result BR or the ENO output is set to FALSE.

FC 32 RIGHT
Right section of a STRING variable

Call: *string* := RIGHT (IN := *string*, L := *int*);

The function FC 32 RIGHT yields the last L character of a string. If L is greater than the current length of the STRING variable, the input value is returned. If L = 0 and if the input value is an empty string, an empty string is returned.

If L is negative, an empty string is output and the binary result BR or the ENO output is set to FALSE.

FC 26 MID
Mid section of a STRING variable

Call: *string* := MID (IN := *string*, L := *int*, P := *int*);

The function FC 26 MID yields the mid section of a string (L characters from the P character inclusive). If the sum of L and P exceeds the current length of the STRING variable, a string from the P character is yielded to the end of the input value.

In all other cases (P outside the current length, P and/or L equal to zero or negative), an empty string is output and the binary result BR or the ENO output is set to FALSE.

FC 2 CONCAT
Concatenation of two STRING variables

Call: *string* := CONCAT (IN1 := *string*, IN2 := *string*);

The function FC 2 CONCAT joins two STRING variables together to form one.

If the resulting string is longer than the variable applied at the output parameter, the resulting string is limited to the maximum set length and the binary result or the ENO output is set to FALSE.

FC 17 INSERT
Insertion into a STRING variable

Call: *string* := INSERT (IN1 := *string*, IN2 := *string*, P := *int*);

The function FC 17 INSERT inserts the character string at parameter IN2 into the string at parameter IN1 following the P character. If P equals zero, the second string is inserted in front of the first string. If P is greater than the current length of the first string, the second string is appended to the first.

If P is negative, an empty string is output and the binary result BR or the ENO output is set to FALSE. The binary result or the ENO output are also set to FALSE if the resulting character string is longer than the variable specified at the output parameter; in this case, the resulting character string is limited to the maximum set length.

FC 4 DELETE
Deletion of a STRING variable

Call: *string* := DELETE (IN := *string*, L := *int*, P := *int*);

The function FC 4 DELETE deletes L characters from the P character (inclusive) in a character string. If L and/or P are equal to zero or if P is greater than the current length of the input string, the input string is returned. If the sum of L and P is greater than the input string, the characters up to the end of the string are deleted.

If L and/or P are negative, an empty string is output and the binary result BR or the ENO output is set to FALSE.

FC 31 REPLACE
Replacement in a STRING variable

Call: *string* := REPLACE (IN1 := *string*, IN2 := *string*, L := *int*, P := *int*);

The function FC 31 REPLACE replaces L characters of the first character string (IN1) from the P character (inclusive) with the second character string (IN2). If L is equal to zero, the first string is returned. If P is equal to zero or one, replacement is made from the 1st character (inclusive). If P is outside the first character string, the second string is appended to the first string.

If L and/or P are negative, an empty string is output and the binary result BR or the ENO output is set to FALSE. The binary result or the ENO output are also set to FALSE if the resulting character string is longer than the variable specified at the output parameter; in this case, the resulting character string is limited to the maximum set length.

31.4 Date/Time-of-Day Functions

With the date/time-of-day functions, you handle variables of data types DATE, TIME_OF_DAY and DATE_AND_TIME.

Some date/time-of-day functions set the binary result BR or the ENO output to FALSE if an error has occurred during execution of the function.

FC 3 D_TOD_DT
Combining DATE and TIME_OF_DAY to DT

Call: *date_and_time* := D_TOD_DT (IN1 := *date*; IN2 := *time_of_day*);

The function FC 3 D_TOD_DT combines the data formats DATE (D#) and TIME_OF_DAY (TOD#) and converts these formats to the DATE_AND_TIME (DT#) format. The input value IN1 must be between the limits DATE#1990-01-01 and DATE#2089-12-31.

The function does not report errors.

FC 6 DT_DATE
Extraction of DATE from DT

Call: *date* := DT_DATE (*date_and_time*);

The function FC 6 DT_DATE extracts the data format DATE (D#) from the DATE_AND_TIME (DT#) format. DATE is between the limits DATE#1990-01-01 and DATE#2089-12-31.

The function does not report errors.

FC 7 DT_DAY
Extraction of the day of the week from DT

Call: *int* := DT_DAY (*date_and_time*);

The function FC 7 DT_DAY extracts the day of the week from the DATE_AND_TIME (DT#) format. The day of the week is available in the INT data format:

1 Sunday
2 Monday
3 Tuesday
4 Wednesday
5 Thursday
6 Friday
7 Saturday

The function does not report errors.

FC 8 DT_TOD
Extraction of TIME_OF_DAY from DT

Call:
time_of_day := DT_DAY (*date_and_time*);

The function FC 8 DT_TOD extracts the data format TIME_OF_DAY (TOD#) from the DATE_AND_TIME (DT#) format.

The function does not report errors.

FC 1 AD_DT_TM
Adding a time period to a time

Call: *date_and_time* := AD_DT_TM (T := *date_and_time*, D := *time*);

The function FC 1 AD_DT_TM adds a time period in the TIME (T#) format to a time in the DATE_AND_TIME (DT#) format and yields a new time in the DATE_AND_TIME (DT#) format. The time (parameter T) must be within the range DT#1990-01-01-00:00:00.000 and

DT#2089-12-31-59:59:59.999. The function does not execute an input check.

If the result of the addition is not within the range given above, the result is limited to the relevant value and the binary result BR or the ENO output is set to FALSE.

FC 35 SB_DT_TM
Subtracting a time period from a time

Call: *date_and_time* := SB_DT_TM (T := *date_and_time*, D := *time*);

The function FC 35 SB_DT_TM subtracts a time period in the TIME (T#) format from a time in the DATE_AND_TIME (DT#) format and yields a new time in the DATE_AND_TIME (DT#) format. The time (parameter T) must be within the range DT#1990-01-01-00:00:00.000 and DT#2089-12-31-59:59:59.999. The function does not execute an input check.

If the result of the subtraction is not within the range given above, the result is limited to the relevant value and the binary result BR or the ENO output is set to FALSE.

FC 34 SB_DT_DT
Subtracting two times

Call: *time* := SB_DT_DT (T1 := *date_and_time*, T2 := *date_and_time*);

The function FC 34 SB_DT_DT subtracts two times in the DATE_AND_TIME (DT#) format and yields a time period in the TIME (T#) format. The times must be within the range

DT#1990-01-01-00:00:00.000 and

DT#2089-12-31-59:59:59.999.

The function does not execute an input check. If the first time (parameter T1) is greater (later) than the second (parameter T2), the result is positive; if the first time is less (earlier) than the second, the result is negative.

If the result of the subtraction is outside the TIME number range, the result is limited to the relevant value and the binary result BR or the ENO output is set to FALSE.

31.5 Numerical Functions

The numerical functions leave the function value unchanged and set the binary result or the ENO output to FALSE if

▷ a parameterized variable is of an impermissible data type,

▷ all parameterized variables do not share the same data type,

▷ a REAL variable does not represent a valid floating-point number.

FC 22 LIMIT
Delimiter

Call: *any_num* := LIMIT (MN := *any_num*, IN := *any_num*; MX := *any_num*);

The function FC 22 limits the numerical value of the variable IN to the limit values specified at the parameters MN and MX. Variables of data type INT, DINT and REL are permissible as input values. All input values (actual parameters) must be of the same data type. The lower limit value (parameter MN) must be less than the upper limit value (parameter MX).

The function reports an error if, in addition to the errors listed above, the lower limit value MN is not less than the upper limit value MX.

FC 25 MAX
Selecting the maximum

Call: *any_num* := MAX(IN1 := *any_num*, IN2 := *any_num*, IN3 := *any_num*);

The function FC 25 MAX selects the highest of three numerical variable values. Variables of data type INT, DINT and REL are permissible as input values. All input values (actual parameters) must be of the same data type.

FC 27 MIN
Selecting the minimum

Call: *any_num* := MIN(IN1 := *any_num*, IN2 := *any_num*, IN3 := *any_num*);

The function FC 27 MIN selects the lowest of three numerical values. Variables of data type INT, DINT and REL are permissible as input values. All input values (actual parameters) must be of the same data type.

FC 36 SEL
Binary selection

Call: *any* := SEL (G := *bool*, IN0 := *any*, IN1 := *any*);

The function FC 36 SEL selects one of two variable values (IN0 and IN1) dependent on a switch (parameter G). Variables of all elementary data types except BOOL are permissible as input values at the parameters IN0 and IN1. Both input variables and the function value must be of the same data type.

Appendix

This section of the book contains instructions for converting a STEP 5 program into a STEP 7 program, an overview of the contents of the STEP 7 block libraries and an overview of all STL and SCL statements and functions.

With the optional package **S5/S7-Converter**, you can convert an existing STEP 5 program into a (STEP 7) STL program as a source file.

The scope of supply of STEP 7 includes **Block Libraries** with loadable functions and function blocks and with block headers and interface descriptions of system functions SFCs and system function blocks SFBs.

The loadable functions FCs and function block SFBs are compiled blocks that you copy to your user program (or more precisely, to the offline container *Blocks*) and then call. These blocks occupy memory space like entirely "normal" user blocks and they are also loaded into the CPU.

You can rename loadable functions and function blocks, for example, if you have already assigned your own blocks to their numbers. However, you still get the correct online Help (function key F1 with the block selected) since the help function is oriented around the block properties FAMILY and NAME.

The system functions SFCs and system functions blocks SFBs are blocks in the operating system of the CPU. In order to call these blocks offline, the standard library contains the block header and the interface description of these blocks (the program is, of course, located in the CPU). You can copy the interface descriptions like a compiled block into the offline container *Blocks* and then call the system block. The program editor learns from the interface description how many block parameters the system block has and the data type and name of a block parameter.

In the case of incremental programming, you drag the library blocks from the program elements catalog to the program window and thus call them. The program editor then copies these blocks automatically into your program.

If you call the library blocks with symbolic names from the library's symbol table in the case of source-file-oriented programming, the standard block will also be automatically copied to your program at the compilation stage.

The book ends with an **STL Operation View** and an **SCL Statement Overview**.

32 S5/S7 Converter

With the S5/S7 Converter, you can convert a STEP 5 program into a STEP 7 STL source file. The Converter turns all directly convertible statements into the corresponding STEP 7 statements. STEP 5 statements that cannot be converted to STEP 7 statements, are commented out. The Converter takes over all comments. As an option, the assignment list can also be converted to an importable symbol table.

To convert a sequential control with GRAPH 5 to a STEP 7 program, you must create the program again with S7-GRAPH.

The S5/S7 Converter is included in the scope of supply of the STEP 7 Standard Package. You do not require authorization to use the Converter.

In the electronic catalog CA01 (CD), you will find support for the hardware conversion of a SIMATIC S5 configuration to a SIMATIC S7 configuration under the menu point SELECTION AIDS → SIMATIC. After selecting the S5 configuration with EDIT → GENERATE SIGNAL LIST and EDIT → GENERATE CONFIGURATION, you generate an S7 station from the specifications for the S5 configuration.

32.1 General

To convert a STEP 5 program, you require the program file *name*ST.S5D and the cross-reference list *name*XR.INI as well as the assignment list *name*Z0.SEQ if available. In addition, you can create a macro file. It contains statement sequences that the Converter can use in place of certain STEP 5 statements. From these files, the Converter generates a STEP 7 source file and, if required, a symbol table. All generated files are stored in the same directory as the STEP 5 files.

The Converter transfers organization blocks with user program into the corresponding STEP 7 organization blocks and all other code blocks into functions FCs. The block numbers of the FC blocks start at zero and are numbered consecutively; you can change the suggested block numbers in a dialog window.

The Converter detects standard blocks in the following Siemens block packages:

- ▷ Floating-point arithmetic
- ▷ Signal functions
- ▷ Basic functions with analog functions
- ▷ Math functions

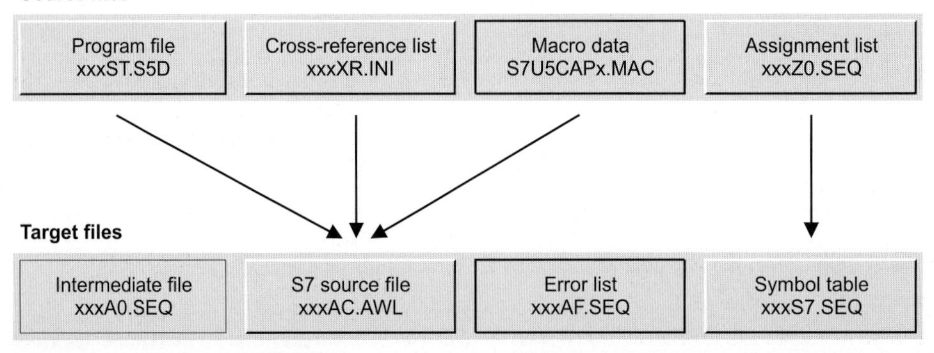

Figure 32.1 Files for the Converter

The library *S5-S7 Converting Blocks*, included in the scope of supply of STEP 7, contains replacement blocks for the standard blocks from these packages. You can also find standard blocks ("integral functions") in this library, that replace some of the function blocks integrated into the S5-115U CPUs.

If the STEP 5 program contains blocks from these packages, the Converter converts the call and signals which blocks occur in the program. You must then copy the relevant blocks from the library to your user program before compiling the converted program.

You can follow the procedure below when converting a STEP 5 program:

▷ Program executability check in the destination environment

▷ If necessary, preparation of the STEP 5 program (removal of non-convertible sections that are replaced, for example, by CPU parameterization)

▷ If necessary, creation of macros (replacement of STEP 5 statements with self-selected STEP 7 statement sequences when converting)

▷ Conversion (generation of a STEP 7 source program)

▷ Setting up of a STEP 7 project with importing of the source program and the symbol table into the STEP 7 project, if necessary, copying of the standard function blocks used

▷ If post editing is required, correct or supplement the STEP 7 source program

▷ Compilation

The conversion sequence is not fixed. You can, for example, convert a STEP 5 program without preparation and then make all the corrections in the STEP 7 source program.

32.2 Preparation

32.2.1 Checking Executability on the Target System (PLC)

If you want to use an existing STEP 5 program in a SIMATIC S7, you must first check that the program can execute on the target system (PLC). For example:

▷ Does the destination CPU have the required properties? Do the required program execution characteristics exist?

▷ Which modules has the STEP 5 program worked with? Which modules are accessed in the STEP 7 program?

▷ Does the destination CPU have the required number of addresses (for example, inputs, outputs, blocks)?

You can operate an S5 expansion unit via the IM 463-2 interface module or you can operate certain S5 modules in an adapter casing in an S7-400. It is also possible to connect SIMATIC S5 modules to SIMATIC S7 as distributed I/O via PROFIBUS DP.

32.2.2 Checking Program Execution Characteristics

The program execution levels familiar to you from SIMATIC S5 generally correspond to the program execution levels in SIMATIC S7, now called priority classes. You replace the settings you have made in data block DB 1 or DX 0 or perhaps in the system data with the parameterization of the S7-CPU (for example, restart characteristics, watchdog interrupt handling).

The integral organization blocks and the integral function blocks in S5 correspond to the S7 system blocks. If you have used integral functions in S5, you must imitate this functionality in S7 with system blocks or with CPU parameterization.

Data block DB 1

On the S5-115U, the program execution characteristics are set in data block DB 1 or in the system data RS. The top of Table 32.1 shows how these characteristics can be implemented with SIMATIC S7.

System utilities

The CPUs of the S5-115U provide system utilities that you can use with organization block OB 250 (CPU 945) or via system datum RS 125 (CPU 941 to CPU 944). The middle section of

Table 32.1 contains suggestions for converting these system utilities to SIMATIC S7.

Data block DX 0

With the CPUs of the upper performance range, the entries in data block DX 0 determine the program execution characteristics. The bottom section of Table 32.1 shows the conversion to SIMATIC S7.

32.2.3 Checking the Modules

I/O modules

Compare the technical specifications of the I/O modules used with those of the SM modules in S7. Are there analog modules available with the required area? When you access analog modules direct, please note the differences in data format to S5.

Intelligent I/O modules IPs

You can also use some IP modules in the S7-400 in conjunction with the adapter casing:

- ▷ IP 240 Positioning, module, position decoding module and counter module
- ▷ IP 242B Counter module
- ▷ IP 244 Temperature control module
- ▷ IP 246/247 Positioning modules
- ▷ WF 721/723 Positioning modules
- ▷ WF 705 Position decoding module

There are standard blocks for these modules that are supplied together with the module and adapter casing. If you have been using these modules, you must exchange the S5 standard blocks for the S7 standard blocks and adapt your program in accordance with the new parameter initialization. For the remaining IP modules, you use comparable FM modules.

Communications processors CPs

The communications processors used in S5 are replaced with CP modules with corresponding functionality. The CP modules are accessed in S7 via SFB communications replacing the S5 data handling blocks. The functionality is similar but implemented with STEP 7 language resources. You must adapt a corresponding S5 program with data handling functions to the SFB blocks.

S5 modules in S7-400

You connect S5 expansion units to an S7-400 using an *IM 463-2 interface module*. You can connect up to four S5 expansion units to each of the two interfaces; a maximum of four IM 463-2s can be used in one central mounting rack. In the S5 expansion unit, an IM 314 interface module handles the connection. Only digital and analog modules are permissible. Process interrupts cannot be transferred. You set the I/O areas of the S5 modules on the IM 314 S5 interface module (as usual with S5). The I/O areas P, Q, IM3 and IM4 are available.

You can operate some IP and WF modules in an S7-400 using an *adapter casing* (see above). You set the S5 addresses in the familiar way on the modules.

You parameterize the assignment of the S5 addresses to the S7 addresses in the Hardware Configuration. You can find the IM 463-2 interface module and the adapter casing in the module catalog under SIMATIC 400 Æ IM-400 Æ S5 ADAPTER. After arranging the modules in the rack, you address these modules, like S7 signal modules, in the peripheral I/O area, separated into input and output address. Please ensure that one the one hand, the S7 address areas, and on the other hand, S5 address areas do not overlap.

32.2.4 Checking the Addresses

Check the number of available addresses in the selected destination CPU. Are there sufficient inputs, outputs, memory bits, timers and counters available? The Converter converts the memory bits from the extended area (S memory bits) to memory bits from M 256.0.

In S7 there is one single peripheral I/O area. All modules addressed in the S5 I/O areas P, Q, IM3 and IM4 and in the global area are now addressed in the S7 peripheral I/O area P (you must take careful note of this if you have addressed a large number of modules in the extended I/O areas and you are connecting these

Table 32.1 Comparison of Program Execution Characteristics

Data Block in DB 1 and System Data (S5-115U)			
Function	941 - 944	945	Replaced in S7 with
Restart delay	x	x	CPU parameter "Restart"
Retentive feature	x	x	CPU parameter "Retentivity"
Cycle time monitoring	x	x	CPU parameter "Cycle / Clock memory"
Time interval for watchdog interrupts	x	x	CPU parameter "Watchdog interrupt"
Software protection	x	x	CPU parameter "Protection"
Output disable process images	x	x	Handling via partial process images: SFC 26 UPDAT_PI, SFC 27 UPDAT_PO
Integral clock	x	x	CPU parameter "Diagnostics / Clock" SFC 0 SET_CLK, SFC 1 READ_CLK
Delay interrupt OB 6			
Time duration	x	x	CPU parameter "Interrupts"
Execution priority	x	-	CPU parameter "Interrupts"
Sequential process image transfer	-	x	- omitted -
Reduced PIQ transfer	-	x	- omitted -
System utilities OB 250 and BS 125			
Function	OB 250	BS 125	Replaced in S7 with
Time intervals for watchdog interrupts	x	-	CPU parameter "Watchdog interrupt"
Time duration of delay interrupt	x	-	CPU parameter "Interrupts"
Reduced PIQ transfer	x	-	- omitted -
Read/write DBA/DBL register	x	-	Read with, e.g., L DBNO, L DBLG direct writing omitted
Call DX/FX blocks indirectly	x	-	Indirect block call
Change block ID	x	-	- omitted -
Update configuration image	x	x	- omitted -
Set up block address list	x	x	- omitted -
Create data block	x	x	SFC 22 CREAT_DB
I/O accesses without QVZ	x	x	Handle synchronization error events: SFC 36 MSK_FLT, SFC 37 DMSK_FLT SFC 38 READ_ERR
Disable/enable digital outputs	x	x	Master Control Relay MCR
Delete block	x	x	Delete data block: SFC 23 DEL_DB
Update process image	-	x	Handling via partial process images: SFC 26 UPDAT_PI, SFC 27 UPDAT_PO
Interpret data block DB 1	-	x	- omitted -
Data block DX 0			
Function	135U	155U	Replaced in S7 with
Restart characteristics	x	x	CPU parameter "Restart"
Number of processed timer cells	x	x	- omitted - (fixed)
Cycle time monitoring	x	x	CPU parameter "Cycle / clock memory"
Multiprocessor restart, interprocessor communication flags	x	x	- omitted -
Accuracy of floating-point arithmetic	x	-	- omitted -
Timed interrupt handling	-	x	CPU parameter "Watchdog interrupt"
Process interrupt handling, interrupt	x	x	CPU parameter "Interrupts"
Process interrupts level/level-triggered	x	-	Module parameterization
Addressing error monitoring	x	-	OB 122 (I/O access errors)
Error handling (system stop)	x	-	replaced by handling of error OBs

modules via, for example, the IM 463-2 to an S7-400). The page memory area is omitted without replacement.

The Converter converts all blocks with user program (except organization blocks) into functions, that is, the total number of all program blocks (PBs), step blocks without sequencer program (SBs) and functions blocks (FBs and FXs) must not exceed the permissible number of functions (FCs). Similarly, the total number of data blocks (DBs and DXs) must not exceed the permissible number of S7 data blocks. These restrictions are only relevant in practice if you are using the S7-300 as the target system (PLC).

The system data areas RI, RJ, RS and RT are omitted without replacement in S7. Any information you have buffered in these areas is stored in S7 in global data blocks or in memory bits. You now get system information from the RS area via system functions; you implement functions initiated via this area via system functions or CPU parameterization.

Preparing the STEP 5 Program

Before conversion, you can prepare your STEP 5 program for its future use as a STEP 7 program (but you do not have to do this; you can also carry out all corrections after conversion to the STEP 7 source file). With this adaptation, you can reduce the number of error messages and warnings. For example, you can make the following adaptations before conversion:

- ▷ Deletion of the data blocks with program characteristics DB 1 or DX 0
- ▷ Removal of all calls of integral blocks or accesses to the system data area RS whose functionality can be reached via the parameterization of the S7-CPU
- ▷ Adaptation of the address areas inputs, outputs, peripheral I/O to the (new) module addresses (you should ensure here that the STEP 5 address range is not exceeded, otherwise an error will be signaled already in the first conversion run; these statements are then not converted)
- ▷ With non-convertible program sections that occur repeatedly, you can delete the sections down to one "unique" STEP 5 statement per program section. You assign a macro (a STEP 7 statement sequence) to this "unique" statement that is to replace the program section.
- ▷ If your program contains many (long) data blocks that have no data structure (those used, for example, as a data buffer), you can significantly reduce the number of statements to be compiled and therefore the source code if you delete all but one of the data words in this data block. After conversion (and before compiling) program the contents of these data blocks in the source file with a field declaration, for example Buffer : ARRAY [1..256] OF WORD.

You can use the Converter to convert not only whole programs but also individual blocks.

32.3 Converting

32.3.1 Creating Macros

You can create macros before conversion for replacing non-convertible STEP 5 statements or for making a change different to the standard conversion. You create conversion macros with the Converter. If a macro is defined twice, the first definition is used. Macros with the SIMATIC instruction set (German) are stored in the file S7U5CAPA.MAC; macros with the international instruction set (English) are stored in the file S7U5CAPB.MAC. The Converter distinguishes between instruction macros and OB macros. You can create 256 instruction macros and 256 OB macros.

Instruction macros replace a STEP 5 statement with a sequence of specified STEP 7 statements.

General structure of an instruction macro:

```
$MACRO: <STEP 5 statement>
<STEP 7 statement sequence>
$ENDMACRO
```

The STEP 5 statement must be specified in full (with complete addresses). The Converter then inserts the specified STEP 7 statement sequence in their place.

Example: You use a delay interrupt (organization block OB 6) in the STEP 5 program for the

CPU 945. You have started this interrupt by calling special function OB 250:

```
L   KF  +200
L   KB     1
JU  OB   250
```

The first load statement contains the number of milliseconds by which the call of OB 6 is to be delayed. This statement can remain and you replace the remaining two statements with a STEP 5 statement that otherwise does not occur in your program, for example, TB RT 200.0, so that your STEP 5 program appears as follows prior to conversion:

```
L   KF  +200
TB  RT 200.0
```

You now write the following instruction macro:

```
$MACRO: TB RT 200.0
T MD 250;
CALL SFC 32 (
  OB_NO    := 20,
  DTIME    := MD 250,
  SIGN     := W#16#0000,
  RET_VAL := MW 254);
$ENDMACRO
```

The STEP 5 statement TB RT 200.0 is replaced at conversion with the specified STEP 7 statement sequence. The delay time in ms is loaded into the (scratchpad) memory word MW 250 and then SFC 32 is called. In the dialog window before starting, the Converter suggests the number 20 instead of the number 6 for the delay OB.

OB macros replace an OB call (JU OB or JC OB) with the specified STEP 7 statement sequence.

The general structure of an OB macro is as follows:

```
$OBCALL: <Number of the OB>
<STEP 7 statement sequence>
$ENDMACRO
```

Example: In the STEP 5 program for the CPU 945, you use organization block OB 160 to start a waiting time. In STEP 7, a waiting time is implemented by system function SFC 47 WAIT. If you enter the macro

```
$OBCALL: 160
T MW 250;
CALL SFC 47 (WT := MW 250);
$ENDMACRO
```

the Converter replaces every OB 160 call (even a conditional call) with the specified statement sequence.

Input of the macros begins with EDIT → REPLACE MACRO. You enter the macros in the opened file S7U5CAPA.MAC and save the file with FILE → SAVE. Terminate macro input with FILE → EXIT.

32.3.2 Preparing the Conversion

If there is still no cross-reference list *name*XR.INI for your STEP 5 program, you must create one for conversion (under STEP 5 with MANAGE → CREATE XREF).

You can now

- ▷ create your own working directory for the conversion and copy the required data into this directory or
- ▷ execute the conversion in the directory (folder) containing the STEP 5 files (if you have worked with the same programming device under STEP 5) or
- ▷ execute the conversion on diskette (if you have generated the STEP 5 files on another programming device).

The directory for the conversion must contain the files *name*ST.S5D and *name*XR.INI as well as *name*Z0.SEQ if appropriate. The Converter also puts the destination files *name*AC.AWL as well as *name*A0.SEQ and, if appropriate, *name*AF.SEQ and *name*S7.SEQ into this directory.

The file S7S5CAPx.MAC is stored in the Windows directory.

32.3.3 Starting the Converter

You call the S5/S7 Converter via the Windows 95/NT taskbar: START → SIMATIC → STEP 7 → S5 CONVERT FILE. With FILE → OPEN, you select the S5 program file you want to convert. If you click on “OK”, the Converter displays the source and destination files as well as the assignment of the old blocks to the new. If necessary, you can change the names of the destination files in the text field. To change the suggested block numbers, double click on the line and enter the new block number in the dialog field. The converter identifies standard blocks with a star (you must then copy these blocks from the block library into your offline user program before compiling the S7 source file).

You start the Converter with the "Start" button. In the first run, it compiles the S5 program into an S5-ASCII text file (*name*A0.seq) and in the second run it compiles this into the S7 source file. The assignment list is compiled into the symbol table. The conversion is completed with the display of error messages and warnings. All errors and warnings are contained in the error file *name*AF.SEQ.

Error messages are output if parts of the S5 program are not convertible and can only be accommodated in the S7 program as comments. Warnings contain information on possible problems; they are output if the converted statements require to be checked again. The messages refer in part to the S5 program (for example, if an illegal MC 5 code is found) or to the S7 program (for example, if a non-convertible statement is found). If you click on a message, the Converter displays the environment of the message in a window.

It is advisable to print out the error list in order to process the error messages.

32.3.4 Convertible Functions

Table 32.2 lists the statements that are converted essentially unchanged. These also include statements with addresses that are replaced in STEP 7 with others, such as the extended S memory bits that are replaced by the M memory bits from 256. Syntax changes can also occur (for example, +G becomes +R). You will normally not have to correct these statements.

The substitution statements (accesses to block parameters) are largely converted. Some editing is required with statements that access both timer and counter functions (for example SEC =*parname*) as well as in the processing of block parameters (DO =*parname*). Here, either code blocks or data blocks can be used as actual addresses and (important!): the block number can change as a result of the conversion.

Organization blocks contain the numbers used in STEP 7. All other blocks with user program become functions FC. The Converter converts data blocks DB to global data blocks with the same number. Data blocks DX are converted to data blocks DB from number 256 (DX 1 becomes DB 257, etc.). The Converter suggests numbers; you can change all the suggested block numbers in a window prior to the conversion run.

Table 32.2 Conversion of the Operations

Functions with STEP 5	Functions with STEP 7
Binary logic operations, memory functions	Binary logic operations, memory functions
Timer and counter functions	Timer and counter functions
Bit test functions	replaced with SET followed by check or with double negation Set/Reset
Load and transfer functions (without system data and absolute address)	Load and transfer functions
Comparison functions	Comparison functions
Calculation functions	Calculation functions
Digital logic operations	Word logic operations
Shift functions	Shift functions
Jump functions	Jump functions
Conversion functions	Conversion functions
Disable/enable interrupts	Replaced with SFC 41, SFC 42
Stop functions	Replaced with SFC 46
Null operations (NOP, ***, blank line)	NOP, NETWORK, // (blank line comment)

The converter takes over the library number of the blocks as AUTHOR in the block header. The name of a function block is taken over as NAME provided it does not contain any special characters (otherwise it is taken over without special characters with the original name as comment).

Special function calls are not converted (they must be replaced with system functions, for example).

The addresses of the inputs and outputs are taken over unchanged. In the case of load and transfer statements with addresses from the P area, the Converter uses peripheral inputs PI and peripheral outputs PQ with unchanged addresses. Addresses from the Q area are mapped to the peripheral I/O area (P) from address 256 (L OB 0 becomes L PIB 256, T OB 1 becomes T PQB 257, etc.).

The addresses of the memory bits F are taken over unchanged. This also applies for the memory bits used as 'scratchpad memory' from memory byte FY 200 to FY 255. If you convert your STEP 5 program largely unchanged, you can retain the scratchpad memory bits as usual. If you want to continue to use the STEP 5 program or parts of it in a STEP 7 environment, I recommend that you store the 'scratchpad memory' blockwise in the temporary local data. This applies especially if you want to transfer your own program standards from STEP 5 to STEP 7. The extended S memory bits are mapped to the memory bits from address 256 (A S 0.0 becomes A M 256.0, L SY 2 becomes L MB 258, etc.)

Timer and counter functions are converted unchanged. Direct access to the individual bits of the timer or counter word is not longer possible under STEP 7. Influencing of the edge memory bits in these words with the bit test statements can be replaced with SET and CLR in conjunction with the relevant timer and counter operation.

Please note that in STEP 7 the data are address bytewise (in STEP 5, by contrast, wordwise). Thus, DL 0 becomes DBB 0, DR 0 becomes DBB 1; you can see the conversion for any addresses in Table 32.3. With direct and indirect addresses, the Converter uses the correct S7 address; with data addressed via block parameters, you must make the conversion to bytewise addressing yourself.

Table 32.3
Address Conversion for Data Addresses

STEP 5	STEP 7
DL [n]	DBB [2n]
DR [n]	DBB [2n+1]
DW [n]	DBW [2n]
DD [n]	DBD [2n]
D [(n).0..7]	DBX [(2n+1).0..7]
D [(n).8..15]	DBX [(2n).0..7]

Floating-point numbers are taken over unchanged, provided they are specified as constants in load statements or they have been used as actual parameters, and they are treated at conversion like STEP 7 floating-point numbers. The standard blocks supplied as a replacement for the STEP 5 standard function blocks also process floating-point numbers in the STEP 7 format (data type REAL). If you have put together floating-point yourself in your STEP 5 program or if you have taken them over from other devices via, for example, a computer link, you must adapt the STEP 5 representation of these floating-point numbers to the data type REAL. A conversion example can be found in the library STL_Book under the program "General examples" (FC 45 GP_TO_REAL). You can download the library from the publisher's Web site (see page 8).

32.4 Post-Editing

32.4.1 Creating the STEP 7 Project

To complete the conversion, you create a STEP 7 project that corresponds in structure to your target system (PLC) (if you have not already created it in order to learn the S7 module addresses). If you want to change module addresses, parameterize modules or change the execution properties of the CPU, you require a hardware configuration (that is, a fully set up project). If the default settings of the module characteristics cannot be changed, it is enough to set up a module-independent program.

- ▷ You create a station (S7-300 or S7-400), open the object *Hardware* and configure the station. You also set the properties of the CPU with the Hardware Configuration (for example, numbers of the interrupt OBs). Together with the CPU, the SIMATIC Manager also sets up the lower-level object containers.
- ▷ With the object *Sources* marked, you fetch the generated file *name*AC.AWL into the source program container with INSERT → EXTERNAL SOURCE FILE.
- ▷ If your program uses S5 standard blocks, open the library *S5/S7 Converting Blocks* under *Standard Library* and copy the S7 standard blocks, indicated by the Converter in the block list with a star, into the offline user program *Blocks* of your project. If you use S7 system blocks in the converted program (for example, SFC 20 BLKMOV), open the library *System Function Blocks* and copy the system blocks used into the offline user program *Blocks*.
- ▷ If you have been working with symbolic programming, open the (empty) symbol table *Symbols* and fetch the converted symbols *name*S7.SEQ with SYMBOLTABLE → IMPORT.

Following these preparations, you can now process the source file with the Editor before compiling it (you can reduce the number of error messages, if you carry out all corrections before compiling).

32.4.2 Non-convertible Functions

After conversion, you usually have to post-edit the source file. This affects all the statements listed in Table 32.4.

32.4.3 Address Changes

The address changes affect essentially the input and output modules. Under certain circumstances, you must adapt the accesses to the inputs and outputs as well as the direct peripheral I/O accesses to the (new) module addresses. You can carry out this adaptation before conversion in the STEP 5 file (if the address volume suitable for STEP 5) or you can swap the absolute addresses in the S7 source file with the help of the 'Replace' function of the Editor used (use caution if the old and new address areas overlap).

In the case of programming with symbolic addressing, you can also generate a source with symbol addresses, change the absolute addresses in the symbol table and then re-compile. Proceed as follows here:

- ▷ A requirement is that you have a symbol table with symbols for all the absolute addresses to be changed and a program compiled free of errors (the blocks in which the absolute addresses occur must be available in compiled form).

Table 32.4 Unconvertible Functions

Functions in STEP 5	Remarks
Load and transfer functions	
with system data	Replaced, for example, with system functions
with absolute addresses	Must be replaced with a new program
Register functions (LIR, TIR, LDI, TDI, MBA, MAB, MSA, MAS, MBA, MSB, MBR, ABR, ACR)	Must be replaced with a new program
Block transfer (TNB, TNW, TXB, TXW)	Replaced with SFC 20 BLKMOV
DO functions	
DO DW, DO FW	Converted
DO RS	Must be replaced with a new program
Calling special functions	Replace special functions with SFCs
LIM, SIM, IAE, RAE	Can be replaced with SFC 39 .. SFC 42
Semaphore functions (SED, SEE, TSC, TSG)	No replacement
Other (IAI, RAI, ASM, UBE)	No replacement

▷ You set the Editor to symbolic addressing: OPTIONS → CUSTOMIZE displays a dialog field; select the option SYMBOLIC REPRESENTATION in the 'Editor' tab.

▷ You generate a new source file using the Editor with FILE → GENERATE SOURCE FILE. After entering the file name, you select all blocks in the dialog window shown that you want as a source file with symbolic addressing. The new source file now contains the statements with symbolic addressing.

▷ Next, correct all absolute addresses in the symbol table from the (old) S5 version to the (new) S7 version.

▷ If you now compile the new source file, the new absolute addresses will be contained in the compiled blocks.

32.4.4 Indirect Addressing

The Converter can also understand indirect addressing with DO MW and DO DW with STEP 7 statements. However, it is necessary here to convert the pointer to the STEP 7 format, which, in conjunction with the buffering of accumulator contents and the status word, leads to an increased memory requirement.

With suitable programming you can usually execute indirect addressing, whether it is memory-indirect or register-indirect, with fewer statements and a clearer program structure.

If indirect addressing occurs frequently, STEP 7-adapted programming is certainly of advantage.

▷ Indirect addressing of timers, counters and blocks
This is converted into memory-indirect addressing using a temporary local data word.

▷ Indirect addressing of blocks
Allocation of the new block numbers cannot be taken into account (manual correction)

▷ Indirect addressing
Converted bitwise and wordwise using AR1, buffering of STW, Accum 1 and 2 in temporary local data (see below)

▷ Indirect addressing via the BR register
No conversion possible, change manually via address registers

▷ Other indirect addressing
Must be changed manually

Jump functions	Replaced with jump distributor SPL
Shift functions	Replaced with shift functions with number of positions in accum 2
TNB, TNW	Replaced with SFC 20 BLKMOV with "variable" ANY pointer
LIR, TIR	No direct replacement available
Decrementing/ incrementing	No direct replacement available

The Converter changes indirect addressing with DO MW and DO DW of binary logic operations, memory functions, and load and transfer functions to a STEP 7 program. The STEP 5 pointer must be changed to the format of an area-internal STEP 7 pointer (with buffering of the accumulator contents and the status word). The result is a long sequence of statements (see example).

If you have used a large number of indirect addresses in your program, manual conversion could be of advantage. As index register, you have unrestricted access to the two address registers AR1 and AR2 (in functions FCs). You can also address memory bits or data memory-indirect as in STEP 5, but you then require one doubleword per index register instead of one word.

The example in Table 32.5 shows in the first column a STEP 5 program which is compared with a data field with the bit pattern of an input word; if they are identical, a memory bit is set in each case. The second column contains the converted program. Using both address registers, you can write a directly comparable program requiring significantly fewer statements.

First, the address registers are loaded with the pointers (take account of bytewise addressing of the data!). Access to the data words and the memory bits is then register-indirect. After every comparison, address register AR1 is incremented by 2 bytes and address register AR2 is incremented by one bit (conversion to the byte address is omitted). In the example, the pointer to the data words is used as the break criterion,

Table 32.5 Converting Indirect Addressing

STEP 5 program	Converted program	Optimized program
FB 174 Name : COMP	FUNCTION FC 4 : VOID NAME: COMP VAR_TEMP conv_accum1 :dword; conv_accum2 :dword; conv_stw :word; END_VAR BEGIN NETWORK	FUNCTION FC 4 : VOID NAME: COMP BEGIN
:L KB 20 :T DW 2 :L KB 50 :T DW 3 LOOP :L IW 10	L 20; T DBW 4; L 50; T DBW 6; LOOP: L IW 10;	LAR1 P#40.0; LAR2 P#50.0; LOOP: L IW 10;
	T conv_accum1; L STW; T conv_stw; L DBB 5; SLW 4; LAR1; L conv_stw; T STW; L conv_accum1;	
:DO DW 2 :L DW 0	L DBW[AR1,P#0.0];	L DBW[AR1,P#0.0];
:>F	>I;	>I;
	T conv_accum1; TAK; T conv_accum2; L STW; T conv_stw; L DBB 6; SLW 5; SRW 5; L DBB 7; SLW 3; OW; LAR1; L conv_stw; T STW; L conv_accum2; L conv_accum1;	
:DO DW 3 := F 0.0	= M[AR1,P#0.0];	= M[AR2,P#0.0];
:L DW 2 :I 1 :T DW 2 :L KB 100 :>F :JC =END :L DL 3 :I 1 :T DL 3 :L KB 8 :<F :JC =LOOP :L DR 3 :I 1 :T DW 3 :JU =LOOP END :NOP 0 :BE	L DBW 4; INC 1; T DBW 4; L 100; >I; JC END; L DBB 6; INC 1; T DBB 6; L 8; <I; JC LOOP; L DBB 7; INC 1; T DBW 6; JU LOOP; END: NOP 0; END_FUNCTION	+AR1 P#2.0; CAR1; L P#200.0; >D; JC END; +AR2 P#0.1; JU LOOP; END: NOP 0; END_FUNCTION

just as in STEP 5; at this point STEP 7 provides use of the loop jump LOOP.

32.4.5 Access to "Excessively Long" Data Blocks

Access to "excessively long" data blocks, that is access to data addresses that had a byte address > 255, was carried out under STEP 5 with absolute addressing. The data block start address was calculated, the address offset was added and the data address was accessed either direct with LIR/TIR or via the BR register with LRW/TRW.

With STEP 7, you can assign data addresses direct up to the permissible limit (8095 on the S7-300, 32767 on the S7-400). You can therefore replace access via the absolute address with a "normal" STL statement.

32.4.6 Working with Absolute Addresses

If is necessary to handle absolute memory addresses in STEP 5 if you assign data addresses in "excessively long" data blocks, or if you address indirectly with the BR register, or if you use the block transfer. Access to absolute memory addresses in no longer possible with STEP 7; the STEP address counter (with the associated operations) has been removed without replacement.

Access to data addresses in "excessively long" data blocks is carried out direct in STEP 7 with "normal" statements. In this regard, calculation of the data block is also omitted. The obvious solution for indirect addressing via the BR register is register-indirect addressing, if necessary also area-crossing.

The system function SFC 20 BLKMOV replaces block transfer. You specify the variables or memory areas to be copied direct as parameters. If you want to change the source or target area at runtime, use a "variable" ANY pointer as actual parameter.

32.4.7 Parameter Initialization

The converter takes over the actual parameters at block calls without change. If you have specified addresses with an actual parameter, you check and, if necessary, modify this address specification.

Examples:

- ▷ Specifying a data word number:
  must be converted to bytewise addressing
- ▷ Specifying an I/O address:
  the new module address must be used
- ▷ Transferring a block:
  must be provided with the new block number

32.4.8 Special Function Organization Blocks

In STEP 7, you can use system functions or STL statements to replace the organization blocks with special functions (Table 32.6). Some functions are omitted completely (for example, page addressing, system program accesses).

32.4.9 Error Handling

Signaling a range violation via the status bits OV and OS in STEP 7 is similar to STEP 5, but there are some minor deviations. If you check OV and OS, make sure you know the precise functionality in conjunction with the associated statement (for example, arithmetic function).

Almost all system functions SFCs signal any errors via the function value RET_VAL. You can evaluate this value in the program.

STEP 7 has error organization blocks for synchronous errors (OB 121, OB 122) and asynchronous errors (OB 80 to OB 87). Table 32.7 shows you how you can replace the STEP 5 error organization blocks in STEP 7.

Table 32.6 Converting the Special Function Organization Blocks

Function	115U	135U	155U	S7 Replacement
Process condition code byte	-	110	-	Statement sequence
Process accumulators	-	111 - 113	131 - 133	Statement sequence
Handle interrupts	-	120 - 123	122 141 - 143	SFC 39 DIS_IRT, SFC 40 EN_IRT, SFC 41 DIS_AIRT, SFC 42 EN_AIRT
Activate a timer job	-	151	151	SFC 28 SET_TINT, SFC 29 CAN_TINT, SFC 30 ACT_TINT,.SFC 31 QRY_TINT
Handle a delay interrupt	-	153	153	SFC 32 SRT_DINT, SFC 33 QRY_DINT, SFC 34 CAN_DINT
Variable waiting time	160	-	-	SFC 43 WAIT
Delete block	-	-	124	Data block: SFC 23 DEL_DB
Create block	125	-	125	Data block: SFC 22 CREAT_DB
Read block stack	-	170	-	- omitted -
Test data block	-	181	-	SFC 24 TEST_DB
Data block access	-	180	-	- omitted -
Copy data blocks	183, 184	254, 255	254, 255	SFC 20 BLKMOV (data areas)
Copy memory areas	182 190 - 193	182 190 - 193	-	SFC 20 BLKMOV
Set and read time-of-day	-	150	121, 150	SFC 0 SET_CLK, SFC 1 READ_CLK
Cycle statistics	-	152	-	Start information OB 1, SFC 6 RD_SINFO
Read status information	-	228	-	Start information, SFC 6 RD_SINFO
Multiprocessor communi-cations	-	200 - 205	200 - 205	Replacement: GD communication
Compare restart types	-	223	223	- omitted -
Transfer interprocessor communication flags	-	224	-	GD communication
Set cycle time	-	221	-	CPU parameterization
Cycle time triggering	-	222	31, 222	SFC 43 RE_TRIGR
Transfer process images	254, 255	-	126	SFC 26 UPDAT_PI, SFC 27 UPDAT_PO
Counter loop	-	160 - 163	-	Statement sequence
Sign extension	220	220	-	Statement sequence
Page accesses	-	216 - 218		- omitted -
System program access	-	226, 227	-	- omitted -
Process shift register	-	240 - 248	-	- omitted -
Handling blocks	-	230 - 237	-	SFB blocks for communications
PID algorithm	251	250 - 251	-	Standard blocks for PID control
Execute system service	250	-	-	(see above under 'Checking Program Ex-ecution Characteristics')

Table 32.7 Converting the Error Organization Blocks

Function	S5-115	S5-135	S5-155	S7 Replacement
Calling an unloaded block	19	19	19	OB 121
Acknowledgment delay in the case of direct access to I/O modules	23	23	23	OB 122
Acknowledgment in the case of updating the process image	24	24	24	OB 122
Addressing errors	-	25	25	OB 122
Cycle time exceeded	26	26	26	OB 80
Substitution errors	27	27	27	-
Conditional Stop	-	28	-	-
Acknowledgment delay in the case of input byte IB 0	-	-	28	OB 85
Illegal operation code	-	29	-	STOP
Acknowledgment delay in the case of direct access in the extended I/O area	-	-	29	OB 122
Illegal parameters	-	30	-	-
Parity errors or acknowledgment in the case of access to the user memory	-	-	30	OB 122
Special function group errors	-	31	-	-
Transfer errors in data blocks	32	32	32	OB 121
Watchdog errors in the case of time-controlled execution	33	33	33	OB 80
Battery failure	34	-	-	OB 81
Controller errors	-	34	-	-
Error in creating a data block	-	-	34	(SFC)
I/O errors	35	-	-	OB 86
Interface errors	-	35	-	OB 84
Self-test errors	-	-	36	-

33 Block Libraries

The STEP 7 Basic software includes the *Standard Library* which contains the following library programs:

- ▷ Organization Blocks
- ▷ System Function Blocks
- ▷ IEC Function Blocks
- ▷ S5-S7 Converting Blocks
- ▷ TI-S7 Converting Blocks
- ▷ PID Control Blocks
- ▷ Communication Blocks
- ▷ Miscellaneous Blocks
  Time synchronization and time tagging

Further supplied libraries are **SIMATIC_NET_CP,** which contains the communications blocks for the CP modules in the *CP 300* and *CP 400* library programs, **Redundant IO MGP** with blocks for module redundancy, and **Redundant IO CGP** with blocks for the redundancy of individual module channels.

You can copy blocks or interface descriptions from the above library programs into your own projects or libraries.

33.1 Organization Blocks

(Prio = Default priority class)

OB	Prio	Designation
1	1	Main program
10	2	Time-of-day interrupt 0
11	2	Time-of-day interrupt 1
12	2	Time-of-day interrupt 2
13	2	Time-of-day interrupt 3
14	2	Time-of-day interrupt 4
15	2	Time-of-day interrupt 5
16	2	Time-of-day interrupt 6
17	2	Time-of-day interrupt 7
20	3	Time-delay interrupt 0
21	4	Time-delay interrupt 1
22	5	Time-delay interrupt 2
23	6	Time-delay interrupt 3
30	7	Watchdog interrupt 0 (5 s)
31	8	Watchdog interrupt 1 (2 s)
32	9	Watchdog interrupt 2 (1 s)
33	10	Watchdog interrupt 3 (500 ms)
34	11	Watchdog interrupt 4 (200 ms)
35	12	Watchdog interrupt 5 (100 ms)
36	13	Watchdog interrupt 6 (50 ms)
37	14	Watchdog interrupt 7 (20 ms)
38	15	Watchdog interrupt 8 (10 ms)
40	16	Hardware interrupt 0
41	17	Hardware interrupt 1
42	18	Hardware interrupt 2
43	19	Hardware interrupt 3
44	20	Hardware interrupt 4
45	21	Hardware interrupt 5
46	22	Hardware interrupt 6
47	23	Hardware interrupt 7
55	2	DPV1 status interrupt
56	2	DPV1 update interrupt
57	2	DPV1 manufacturer specific interrupt
60	25	Multiprocessor interrupt

OB	Prio	Designation
61	25	Synchronous cycle interrupt 0
62	25	Synchronous cycle interrupt 1
63	25	Synchronous cycle interrupt 2
64	25	Synchronous cycle interrupt 3
65	25	Technology synchronous interrupt
70	25	I/O redundancy error
72	28	CPU redundancy error
73	25	Communication redundancy error
80	26	Time error [1)]
81	26	Power supply fault [1)]
82	26	Diagnostics interrupt [1)]
83	26	Insert/remove-module interrupt [1)]
84	26	CPU hardware fault [1)]
85	26	Priority class error [1)]
86	26	DP error [1)]
87	26	Communications error [1)]
88	28	Processing abort
90	29	Background processing
100	27	Warm restart
101	27	Hot restart
102	27	Cold restart
121	-	Programming error
122	-	I/O access error

[1)] Prio = 28 at restart

33.2 System Function Blocks

CPU clock and run-time meter

SFC	Name	Designation
0	SET_CLK	Set clock
1	READ_CLK	Read clock
2	SET_RTM	Set run-time meter
3	CTRL_RTM	Modify run-time meter
4	READ_RTM	Read run-time meter
48	SNC_RTCB	Synchronize slave clocks
64	TIME_TCK	Read system time
100	SET_CLKS	Set clock and clock status
101	RTM	Set operating hours counter

IEC timers and IEC counters

SFB	Name	Designation
0	CTU	Up counter
1	CTD	Down counter
2	CTUD	Up/down counter
3	TP	Pulse
4	TON	On delay
5	TOF	Off delay

S7 communication

SFB	Name	Designation
8	USEND	Uncoordinated send
9	URVC	Uncoordinated receive
12	BSEND	Block-oriented send
13	BRCV	Block-oriented receive
14	GET	Read data from partner
15	PUT	Write data to partner
16	PRINT	Write data to printer
19	START	Initiate cold or warm restart in the partner
20	STOP	Set partner to STOP
21	RESUME	Initiate restart in the partner
22	STATUS	Check status of partner
23	USTATUS	Receive status of partner

SFC	Name	Designation
62	CONTROL	Check communications status
87	C_DIAG	Determine connection status

S7 basic communication

SFC	Name	Designation
65	X_SEND	Send data externally
66	X_RCV	Receive data externally
67	X_GET	Read data externally
68	X_PUT	Write data externally
69	X_ABORT	Abort external connection
72	I_GET	Read data internally
73	I_PUT	Write data internally
74	I_ABORT	Abort internal connection

Global data communications

SFC	Name	Designation
60	GD_SND	Send GD packet
61	GD_RCV	Receive GD packet

Point-to-point coupling S7-300C

SFB	Name	Designation
60	SEND_PTP	Send data (ASCII, 3964 (R))
61	RCV_PTP	Receive data (ASCII, 3964(R))
62	RES_RCVB	Delete receive buffer (ASCII, 3964(R))
63	SEND_RK	Send data (RK 512)
64	FETCH_RK	Fetch data (RK 512)
65	SERVE_RK	Receive and serve data (RK 512)

Integrated functions S7-300C

SFB	Name	Designation
44	ANALOG	Positioning with analog output
46	DIGITAL	Positioning with digital output
47	COUNT	Control counter
48	FREQUENC	Control frequency measurement
49	PULSE	Control pulse-width modulation

Integrated functions CPU 312/314/614

SFB	Name	Designation
29	HS_COUNT	High-speed counter
30	FREQ_MES	Frequency meter
38	HSC_A_B	Control "Counter A/B"
39	POS	Control "Positioning"
41	CONT_C	Continuous closed-loop control
42	CONT_S	Step-action control
43	PULSEGEN	Generate pulse

SFC	Name	Designation
63	AB_CALL	Call assembler block

Drum

SFB	Name	Designation
32	DRUM	Drum

H-CPU

SFC	Name	Designation
90	H_CTRL	Control operating modes on H-CPU

Interrupt events

SFC	Name	Designation
28	SET_TINT	Set time-of-day interrupt
29	CAN_TINT	Cancel time-of-day interrupt
30	ACT_TINT	Activate time-of-day interrupt
31	QRY_TINT	Query time-of-day interrupt
32	SRT_DINT	Start time-delay interrupt
33	CAN_DINT	Cancel time-delay interrupt
34	QRY_DINT	Query time-delay interrupt
35	MP_ALM	Trigger multiprocessor alarm
36	MSK_FLT	Mask synchronous errors
37	DMSK_FLT	Unmask synchronous errors
38	READ_ERR	Read event status register
39	DIS_IRT	Disable asynchronous errors
40	EN_IRT	Enable asynchronous errors
41	DIS_AIRT	Delay asynchronous errors
42	EN_AIRT	Enable asynchronous errors

Address modules

SFC	Name	Designation
5	GADR_LGC	Determine logical address
49	LGC_GADR	Determine slot
50	RD_LGADR	Determine all logical addresses
70	GEO_LOG	Determine logical address
71	LOG_GEO	Determine slot

Data record transfer

SFB	Name	Designation
52	RDREC	Read data record from a DP slave
53	WRREC	Write data record to a DP slave
81	RD_DPAR	Read predefined parameters

SFC	Name	Designation
54	RD_DPARM	Read predefined parameters
55	WR_PARM	Write dynamic parameters
56	WR_DPARM	Write predefined parameters
57	PARM_MOD	Parameterize module
58	WR_REC	Write data record
59	RD_REC	Read data record
102	RD_DPARA	Read predefined parameters

Process image updating

SFC	Name	Designation
26	UPDAT_PI	Update process image input
27	UPDAT_PO	Update process image output
79	SET	Set I/O bit field
80	RSET	Reset I/O bit field
126	SYNC_PI	Update partial process image input in isochronous mode
127	SYNC_PO	Update partial process image output in isochronous mode

Diagnostics

SFC	Name	Designation
6	RD_SINFO	Read start information
51	RDSYSST	Read partial system status list
52	WR_USMSG	Entry in the diagnostic buffer
99	WWW	Synchronize websites

Distributed I/O

SFB	Name	Designation
54	RALRM	Receive alarm
73	RCVREC	Receive data record
74	PRVREC	Provide data record
75	SALRM	Trigger alarm
104	IP_CONF	Set IP config.

SFC	Name	Designation
7	DP_PRAL	Trigger hardware interrupt
11	DPSYC_FR	SYNC/FREEZE
12	D_ACT_DP	Activate or deactivate DP slave
13	DPNRM_DG	Read diagnostic data
14	DPRD_DAT	Read slave data
15	DPWR_DAT	Write slave data
103	DP_TOPOL	Determine bus topology

Create block-related messages

SFB	Name	Designation
31	NOTIFY_8P	Messages without acknowledgment display
33	ALARM	Messages with acknowledgment display
34	ALARM_8	Messages without accompanying values
35	ALARM_8P	Messages with accompanying values
36	NOTIFY	Messages without acknowledgment display
37	AR_SEND	Send archive data

SFC	Name	Designation
9	EN_MSG	Enable messages
10	DIS_MSG	Disable messages
17	ALARM_SQ	Messages that can be acknowledged
18	ALARM_S	Messages that are always acknowledged
19	ALARM_SC	Determine acknowledgment status
105	READ_SI	Read dynamic system resources
106	DEL_SI	Delete dynamic system resources

107	ALARM_DQ	Alarms for acknowledgment
108	ALARM_D	Alarms always acknowledged

Copy and block functions

SFC	Name	Designation
20	BLKMOV	Copy memory area
21	FILL	Pre-assign data area
22	CREAT_DB	Create data block in work memory
23	DEL_DB	Delete data block
24	TEST_DB	Test data block
25	COMPRESS	Compress memory
44	REPL_VAL	Enter substitute value
81	UBLKMOV	Copy data area without gaps
82	CREA_DBL	Create data block in load memory
83	READ_DBL	Read load memory
84	WRIT_DBL	Write load memory
85	CREA_DB	Create data block in work memory

Program control

SFC	Name	Designation
43	RE_TRIGR	Retrigger cycle time monitor
46	STP	Change to STOP state
47	WAIT	Wait for delay time
78	OB_RT	Determine OB runtime
104	CIR	Plant modification during runtime
109	PROTECT	Change protection level

Blocks for PROFINET CbA

SFC	Name	Designation
112	PN_IN	Update inputs
113	PN_OUT	Update outputs
114	PN_DP	Update DP connection

33.3 IEC Function Blocks

String functions

FC	Name	Designation
21	LEN	Length of a STRING
20	LEFT	Left section of a STRING
32	RIGHT	Right section of a STRING
26	MID	Middle section of a STRING
2	CONCAT	Concatenate STRINGs
17	INSERT	Insert STRING
4	DELETE	Delete STRING
31	REPLACE	Replace STRING
11	FIND	Find STRING
16	I_STRNG	Convert INT to STRING
5	DI_STRNG	Convert DINT to STRING
30	R_STRNG	Convert REAL to STRING
38	STRNG_I	Convert STRING to INT
37	STRNG_DI	Convert STRING to DINT
39	STRNG_R	Convert STRING to REAL

Date and time functions

FC	Name	Designation
3	D_TOD_DT	Combine DATE and TOD to DT
6	DT_DATE	Extract DATE from DT
7	DT_DAY	Extract day-of-the-week from DT
8	DT_TOD	Extract TOD from DT
33	S5TI_TIM	Convert S5TIME to TIME
40	TIM_S5TI	Convert TIME to S5TIME
1	AD_DT_TM	Add TIME to DT
35	SB_DT_TM	Subtract TIME from DT
34	SB_DT_DT	Subtract DT from DT

Comparisons

FC	Name	Designation
9	EQ_DT	Compare DT for equal to
28	NE_DT	Compare DT for not equal to
14	GT_DT	Compare DT for greater than

FC	Name	Designation
12	GE_DT	Compare DT for greater than or equal to
23	LT_DT	Compare DT for less than
18	LE_DT	Compare DT for less than or equal to
10	EQ_STRNG	Compare STRING for equal to
29	NE_STRNG	Compare STRING for not equal to
15	GT_STRNG	Compare STRING for greater than
13	GE_STRNG	Compare STRING for greater than or equal to
24	LT_STRNG	Compare STRING for less than
19	LE_STRNG	Compare STRING for less than or equal to

Math functions

FC	Name	Designation
22	LIMIT	Limiter
25	MAX	Maximum selection
27	MIN	Minimum selection
36	SEL	Binary selection

33.4 S5-S7 Converting Blocks

Floating-point arithmetic

FC	Name	Designation
61	GP_FPGP	Convert fixed-point to floating-point
62	GP_GPFP	Convert floating-point to fixed-point
63	GP_ADD	Add floating-point numbers
64	GP_SUB	Subtract floating-point numbers
65	GP_MUL	Multiply floating-point numbers
66	GP_DIV	Divide floating-point numbers
67	GP_VGL	Compare floating-point numbers
68	GP_RAD	Find the square root of a floating-point number

Signal functions

FC	Name	Designation
69	MLD_TG	Clock pulse generator
70	MLD_TGZ	Clock pulse generator with timer function
71	MLD_EZW	Initial value single blinking wordwise
72	MLD_EDW	Initial value double blinking wordwise
73	MLD_SAMW	Group signal wordwise
74	MLD_SAM	Group signal
75	MLD_EZ	Initial value single blinking
76	MLD_ED	Initial value double blinking
77	MLD_EZWK	Initial value single blinking (wordwise) memory bit
78	MLD_EZDK	Initial value double blinking (wordwise) memory bit
79	MLD_EZK	Initial value single blinking memory bit
80	MLD_EDK	Initial value double blinking memory bit

Integrated functions

FC	Name	Designation
81	COD_B4	BCD-binary conversion 4 decades
82	COD_16	Binary-BCD conversion 4 decades
83	MUL_16	16-bit fixed-point multiplier
84	DIV_16	16-bit fixed-point divider

Basic functions

FC	Name	Designation
85	ADD_32	32-bit fixed-point adder
86	SUB_32	32-bit fixed-point subtractor
87	MUL_32	32-bit fixed-point multiplier
88	DIV_32	32-bit fixed-point divider

89	RAD_16	16-bit fixed-point square root extractor
90	REG_SCHB	Bitwise shift register
91	REG_SCHW	Wordwise shift register
92	REG_FIFO	Buffer (FIFO)
93	REG_LIFO	Stack (LIFO)
94	DB_COPY1	Copy data area (direct)
95	DB_COPY2	Copy data area (indirect)
96	RETTEN	Save scratchpad memory (AG 155U)
97	LADEN	Load scratchpad memory (AG 155U)
98	COD_B8	BCD-binary conversion 8 decades
99	COD_32	Binary-BCD conversion 8 decades

Analog functions

FC	Name	Designation
100	AE_460_1	Analog input module 460
101	AE_460_2	Analog input module 460
102	AE_463_1	Analog input module 463
103	AE_463_2	Analog input module 463
104	AE_464_1	Analog input module 464
105	AE_464_2	Analog input module 464
106	AE_466_1	Analog input module 466
107	AE_466_2	Analog input module 466
108	RLG_AA1	Analog output module
109	RLG_AA2	Analog output module
110	PER_ET1	ET 100 distributed I/O
111	PER_ET2	ET 100 distributed I/O

Math functions

FC	Name	Designation
112	SINUS	Sine
113	COSINUS	Cosine
114	TANGENS	Tangent
115	COTANG	Cotangent
116	ARCSIN	Arc sine
117	ARCCOS	Arc cosine
118	ARCTAN	Arc tangent
119	ARCCOT	Arc cotangent
120	LN_X	Natural logarithm
121	LG_X	Logarithm to base 10
122	B_LOG_X	Logarithm to any base
123	E_H_N	Exponential function with base e
124	ZEHN_H_N	Exponential function with base 10
125	A2_H_A1	Exponential function with any base

33.5 TI-S7 Converting Blocks

FB	Name	Designation
80	LEAD_LAG	Lead/lag algorithm
81	DCAT	Discrete control time interrupt
82	MCAT	Motor control time interrupt
83	IMC	Index matrix comparison
84	SMC	Matrix scanner
85	DRUM	Event maskable drum
86	PACK	Collect/distribute table data

FC	Name	Designation
80	TONR	Latching ON delay
81	IBLKMOV	Transfer data area indirectly
82	RSET	Reset process image bit by bit
83	SET	Set process image bit by bit
84	ATT	Enter value in table
85	FIFO	Output first value in table
86	TBL_FIND	Find value in table
87	LIFO	Output last value
88	TBL	Execute table operation
89	TBL_WRD	Copy value from the table
90	WSR	Save datum
91	WRD_TBL	Combine table element
92	SHRB	Shift bit in bit shift register
93	SEG	Bit pattern for 7-segment display
94	ATH	ASCII-hexadecimal conversion
95	HTA	Hexadecimal-ASCII conversion
96	ENCO	Least significant set bit
97	DECO	Set bit in word
98	BCDCPL	Generate ten's complement
99	BITSUM	Count set bits

100	RSETI	Reset PQ byte by byte
101	SETI	Set PQ byte by byte
102	DEV	Calculate standard deviation
103	CDT	Correlated data tables
104	TBL_TBL	Table combination
105	SCALE	Scale values
106	UNSCALE	Unscale values

33.6 PID Control Blocks

FB	Name	Designation
41	CONT_C	Continuous control
42	CONT_S	Step control
43	PULSGEN	Generate pulse
58	TCONT_CP	Continuous temperature control
59	TCONT_S	Step temperature control

33.7 Communication Blocks

FB	Name	Designation
8	USEND	Uncoordinated send
9	URCV	Uncoordinated receive
12	BSEND	Block-oriented send
13	BRCV	Block-oriented receive
14	GET	Read data from partner
15	PUT	Write data to partner
28	USEND_E	Uncoordinated send
29	URCV_E	Uncoordinated receive
34	GET_E	Read data from partner
35	PUT_E	Write data to partner

FC	Name	Designation
1	DP_SEND	Send data
2	DP_RECV	Receive data
3	DP_DIAG	Diagnostics
4	DP_CTRL	Control
62	C_CNTR	Scan connection status

For DP standard slaves and PROFINET IO devices

FB	Name	Designation
20	GETIO	Read inputs
21	SETIO	Set outputs
22	GETIO_PA	Read inputs consistent
23	SETIO_PA	Set outputs consistent

IE communication

FB	Name	Designation
63	TSEND	Send data
64	TRCV	Receive data
65	TCON	Establish connection
66	TDISCON	Cancel connection
67	TUSEND	Send data over UDP
68	TURCV	Receive data over UDP

UDT	Name	Designation
65	TCON_PAR	Data structure for configuration of connection
651	TCON_PAR	TCP_conn_active
652	TCON_PAR	TCP_conn_passive
653	TCON_PAR	ISOonTCP_conn_active
654	TCON_PAR	ISOonTCP_conn_passive
655	TCON_PAR	ISOonTCP_conn_CP_active
656	TCON_PAR	ISOonTCP_conn_CP_passive
657	TCPN_PAR	UDP_local_open
66	TADD_PAR	Address structure of communication partner
661	TADD_PAR	UDP_rem_address and port

FB	Name	Designation
210	FW_TCP	TCP server for FETCH/WRITE
220	FW_IOT	ISO-on-TCP server for FETCH/WRITE

33.8 Miscellaneous Blocks

FC	Name	Designation
60	LOC_TIME	Read local time and summer ID
61	BT_LT	Convert module time

		into local time
62	LT_BT	Convert local time into module time
63	S_LTINT	Set time interrupt according to local time

FB	Name	Designation
60	SET_SW	Summer/winter time switchover
61	SET_SW_S	Summer/winter time switchover with time status
62	TIMESTMP	Transmit messages with time stamp

UDT	Name	Designation
60	WS_RULES	Rules for summer/winter time switchover

33.9 SIMATIC_NET_CP

CP 300 library program

FB	Name	Designation
		FMS communication:
2	IDENT	Identify partner
3	READ	Read data from partner
4	REPORT	Transmit variables
5	STATUS	Request status information from partner
6	WRITE	Write data to partner
8	USEND	Uncoordinated send
9	URCV	Uncoordinated receive
12	BSEND	Block-oriented send
13	BRCV	Block-oriented receive
14	GET	Read data from partner
15	PUT	Write data to partner
40	FTP_CMD	FTP commands (replaces FC 40 to 44)
52	PNIO_REC	Transmit data record
54	PNIOALRM	Receive alarm
55	IP_CONF	Transmit configuration
56	LOG_TRIG	Trigger ERPC communication

FC	Name	Designation
1	DP_SEND	Send data
2	DP_RECV	Receive data
3	DP_DIAG	Diagnostics
4	DP_CTRL	Control
5	AG_SEND	Send data (PB FDL and Ethernet)
6	AG_RECV	Receive data (PB FDL and Ethernet)
7	AG_LOCK	Block data exchange (Industrial Ethernet)
8	AG_UNLOC	Enable data exchange (Industrial Ethernet)
10	AG_CNTRL	Diagnose and initialize connections
11	PNIO_SND	Data transfer on PROFINET
12	PNIO_RCV	Data receipt on PROFINET
40	FTP_CONN	Establish connection to server
41	FTP_STOR	Transfer data block to server
42	FTP_RETR	Transfer file to client
43	FTP_DELE	Delete file on server
44	FTP_QUIT	Terminate connection
62	C_CNTRL	Scan connection status

UDT	Name	Designation
1	-	FILE_DB_HEADER

CP 400 library program

FB	Name	Designation
		FMS communication:
2	IDENT	Identify partner
3	READ	Read data from partner
4	REPORT	Transmit variables
5	STATUS	Request status information from partner
6	WRITE	Write data to partner
40	FTP_CMD	FTP commands (replaces FC 40 to 44)
55	IP_CONF	Transmit connection configuration

FC	Name	Designation
5	AG_SEND	Send data (PROFIBUS FDL and

		Industrial Ethernet)
6	AG_RECV	Receive data (PROFIBUS FDL and Industrial Ethernet)
7	AG_LOCK	Disable data exchange (Industrial Ethernet)
8	AG_UNLOC	Enable data exchange (Industrial Ethernet)
10	AG_CNTRL	Diagnose and initialize connections
40	FTP_CONN	Establish connection to server
41	FTP_STOR	Transmit data block to server
42	FTP_RETR	Transmit file to client
43	FTP_DELE	Delete file on server
44	FTP_QUIT	Cancel connection
50	AG_LSEND	Send data to PROFIBUS CP
53	AG_SSEND	Send data to Ethernet CP
60	AG_LRECV	Receive data from PROFIBUS CP
63	AG_SRECV	Receive data from Ethernet CP

UDT	Name	Designation
1	-	FILE_DB_HEADER

33.10 Redundant IO MGP V31

Support of redundancy for modules

***Red_IO* library program**

FB	Name	Designation
450	RED_IN	Read redundant I/O signals
451	RED_OUT	Output redundant I/O signals
452	RED_DIAG	Diagnose redundant I/O
453	RED_STAT	Read status of redundant I/O

FC	Name	Designation
450	RED_INIT	Initialize I/O redundancy
451	RED_DEPA	Trigger depassivation

33.11 Redundant IO CGP V40

Support of redundancy for individual module channels

***Red_IO* library program**

FB	Name	Designation
450	RED_IN	Read redundant I/O signals
451	RED_OUT	Output redundant I/O signals
452	RED_DIAG	Diagnose redundant I/O
453	RED_STAT	Read status of redundant I/O

FC	Name	Designation
450	RED_INIT	Initialize I/O redundancy
451	RED_DEPA	Trigger depassivation

33.12 Redundant IO CGP V51

Support of redundancy for individual module channels

***Red_IO* library program**

FB	Name	Designation
450	RED_IN	Read redundant I/O signals
451	RED_OUT	Output redundant I/O signals
452	RED_DIAG	Diagnose redundant I/O
453	RED_STAT	Read status of redundant I/O

FC	Name	Designation
450	RED_INIT	Initialize I/O redundancy
451	RED_DEPA	Trigger depassivation

34 STL Operation Overview

The overview below lists the operations with absolute addresses.

The following are also possible with the addressing types:

A I [doubleword	memory-indirect with the doublewords MD Memory doubleword LD Local data doubleword DBD Global data doubleword DID Instance data doubleword	all addresses
A I [AR1, P#offset] A I [AR2, P#offset] A [AR1, P#offset] A [AR2, P#offset]	register-indirect area-internal with AR1 register-indirect area-internal with AR2 register-indirect area-crossing with AR1 register-indirect area-crossing with AR2	no timer functions, no counter functions and no blocks
A #name	parameter-indirect	all addresses

34.1 Basic Functions

34.1.1 Binary Logic Operations

A	-	AND with check for "1"
AN	-	AND with check for "0"
O	-	OR with check for "1"
ON	-	OR with check for "0"
X	-	Exclusive OR with check for "1"
XN	-	Exclusive OR with check for "0"
-	I	an input
-	Q	an output
-	M	a memory bit
-	L	a local data bit
-	T	a timer function
-	C	a counter function
-	DBX	a global data bit
-	DIX	an instance data bit
-	==0	Result equal to zero
-	<>0	Result not equal to zero
-	>0	Result greater than zero
-	>=0	Result greater than or equal to zero
-	<0	Result less than zero
-	<=0	Result less than or equal to zero
-	UO	Result invalid
-	OV	Overflow
-	OS	Stored overflow
-	BR	Binary result

A(	AND open bracket
AN(	AND NOT open bracket
O(	OR open bracket
ON(	OR NOT open bracket
X(	Exclusive OR open bracket
XN(	Exclusive OR NOT open bracket
)	Close bracket
O	OR combination of AND
NOT	Negate RLO
SET	Set RLO
CLR	Reset RLO
SAVE	Save RLO to BR

34.1.2 Memory Functions

=	-	Assign
S	-	Set
R	-	Reset
FP	-	Positive edge
FN	-	Negative edge

-	I	an input
-	Q	an output
-	M	a memory bit
-	L	a local data bit
-	DBX	a global data bit
-	DIX	an instance data bit

34.1.3 Transfer Functions

L	-	Load
T	-	Transfer

-	IB	an input byte
-	IW	an input word
-	ID	an input doubleword
-	QB	an output byte
-	QW	an output word
-	QD	an output doubleword
-	MB	a memory byte
-	MW	a memory word
-	MD	a memory doubleword
-	LB	a local data byte
-	LW	a local data word
-	LD	a local data doubleword
-	DBB	a global data byte
-	DBW	a global data word
-	DBD	a global data doubleword
-	DIB	an instance data byte
-	DIW	an instance data word
-	DID	an instance data doubleword
-	STW	the status word

L	PIB	Load peripheral input byte
L	PIW	Load peripheral input word
L	PID	Load peripheral input doubleword
T	PQB	Transfer peripheral output byte
T	PQW	Transfer peripheral output word
T	PQD	Transfer peripheral output doubleword

L	T	Direct loading of a timer value
LC	T	Coded loading of a timer value
L	C	Direct loading of a counter value
LC	C	Coded loading of a counter value
L	const	Load a constant
L	P#..	Load a pointer
L	P#var	Load a variable start address

Accumulator functions

PUSH	Shift accums "forward"
POP	Shift accums "back"
ENT	Shift accums (without CC1)
LEAVE	Shift accums (without CC1)
TAK	Swap accum 1 and accum 2
CAW	Swap bytes 0 and 1 in accum 1
CAD	Swap all bytes in accum 1

34.1.4 Timer Functions

SP	T	Start timer as pulse
SE	T	Start as extended pulse
SD	T	Start as ON delay
SS	T	Start as retentive ON delay
SF	T	Start as OFF delay
R	T	Reset timer function
FR	T	Enable timer function

34.1.5 Counter Functions

CU	C	Count up
CD	C	Count down
S	C	Set counter function
R	C	Reset counter function
FR	C	Enable counter function

34.2 Digital Functions

34.2.1 Comparison Functions

==I	INT comparison for equal to
<>I	INT comparison for not equal to
>I	INT comparison for greater than
>=I	INT comparison for greater than or equal to
<I	INT comparison for less than
<=I	INT comparison for less than or equal to

==D	DINT comparison for equal to
<>D	DINT comparison for not equal to
>D	DINT comparison for greater than
>=D	DINT comparison for greater than or equal
<D	DINT comparison for less than
<=D	DINT comparison for less than or equal to
==R	REAL comparison for equal to
<>R	REAL comparison for not equal to
>R	REAL comparison for greater than
>=R	REAL comparison for greater than or equal to
<R	REAL comparison for less than
<=R	REAL comparison for less than or equal to

34.2.2 Math Functions

SIN	Sine
COS	Cosine
TAN	Tangent
ASIN	Arc sine
ACOS	Arc cosine
ATAN	Arc tangent
SQR	Finding the square
SQRT	Finding the square root
EXP	Exponent to base e
LN	Natural logarithm

34.2.3 Arithmetic Functions

+I	INT addition
-I	INT subtraction
*I	INT multiplication
/I	INT division
+D	DINT addition
-D	DINT subtraction
*D	DINT multiplication
/D	DINT division (integer)
MOD	DINT division (remainder)
+R	REAL addition
-R	REAL subtraction
*R	REAL multiplication
/R	REAL division
+ const	Adding a constant
+ P#..	Adding a pointer
DEC n	Decrementing
INC n	Incrementing

34.2.4 Conversion Functions

ITD	Conversion of INT to DINT
ITB	Conversion of INT to BCD
DTB	Conversion of DINT to BCD
DTR	Conversion of DINT to REAL
BTI	Conversion of BCD to INT
BTD	Conversion of BCD to DINT
	Conversion of REAL to DINT with
RND+	Rounding to next higher number
RND-	Rounding to next lower number
RND	Rounding to next integer
TRUNC	Without rounding
INVI	INT one's complement
INVD	DINT one's complement
NEGI	INT negation
NEGD	DINT negation
NEGR	REAL negation
ABS	REAL absolute-value generation

34.2.5 Shift Functions

SLW -	Shift left wordwise
SLD -	Shift left doublewordwise
SRW -	Shift right wordwise
SRD -	Shift right doublewordwise
SSI -	Shift with sign wordwise
SSD -	Shift with sign doublewordwise
RLD -	Rotate left doublewordwise
RRD -	Rotate right doublewordwise
- n	by n positions
-	with number of positions in accum 2
RLDA	Rotate left through CC1
RRDA	Rotate right through CC1

34.2.6 Word Logic Operations

AW -	AND wordwise
AD -	AND doublewordwise
OW -	OR wordwise
OD -	OR doublewordwise
XOW-	Exclusive OR wordwise
XOD -	Exclusive OR doublewordwise
- *const*	with a word/doubleword constant
-	with the contents of accum 2

34.3 Program Flow Control

34.3.1 Jump Functions

JU	*label*	Unconditional jump
		Jump if
JC	*label*	RLO = "1"
JCB	*label*	RLO = "1" store with RLO
JCN	*label*	RLO = "0"
JNB	*label*	RLO = "0" store with RLO
JBI	*label*	BR = "1"
JNBI	*label*	BR = "0"
		Jump if result
JZ	*label*	zero
JN	*label*	not zero
JP	*label*	greater than zero
JPZ	*label*	greater than or equal to zero
JM	*label*	less than zero
JMZ	*label*	less than or equal to zero
JUO	*label*	invalid
JO	*label*	Jump on overflow
JOS	*label*	Jump on stored overflow
JL	*label*	Jump distributor
LOOP	*label*	Jump loop

34.3.2 Master Control Relay

MCRA	Activate MCR area
MCRD	Deactivate MCR area
MCR(	Open MCR zone
)MCR	Close MCR zone

34.3.3 Block Functions

CALL	FB	Call function block
CALL	FC	Call function
CALL	SFB	Call system function block
CALL	SFC	Call system function
UC	FB	Call function block unconditionally
CC	FB	Call function block conditionally
UC	FC	Call function unconditionally
CC	FC	Call function conditionally
BEU		Unconditional block end
BEC		Conditional block end
BE		Block end
OPN	DB	Call global data block
OPN	DI	Call instance data block
CDB		Swap data block registers
L	DBNO	Load global data block number
L	DINO	Load instance data block number
L	DBLG	Load global data block length
L	DILG	Load instance data block length
NOP	0	Null operation
NOP	1	Null operation
BLD	n	Program display instruction

34.4 Indirect Addressing

LAR1	-	Load AR1 with
LAR2	-	Load AR2 with
-	MD	a memory doubleword
-	LD	a local data doubleword
-	DBD	a global data doubleword
-	DID	an instance data doubleword
LAR1		Load AR1 with accum 1
LAR2		Load AR2 with accum 1
LAR1	AR2	Load AR1 with AR2
LAR1	P#..	Load AR1 with a pointer
LAR2	P#..	Load AR2 with a pointer
LAR1	P#*var*	Load AR1 with a variable start address
LAR2	P#*var*	Load AR2 with a variable start address
TAR1	-	Transfer AR1 to
TAR2	-	Transfer AR2 to
-	MD	a memory doubleword
-	LD	a local data doubleword
-	DBD	a global data doubleword
-	DID	an instance data doubleword
TAR1		Transfer AR1 to accum 1
TAR2		Transfer AR2 to accum 1
TAR1	AR2	Transfer AR1 to AR2
CAR		Swap AR1 and AR2
+AR1		Add accum 1 to AR1
+AR2		Add accum 1 to AR2
+AR1	P#..	Add pointer to AR1
+AR2	P#..	Add pointer to AR2

35 SCL Statement and Function Overview

35.1 Operators

Combination	Name	Operator	Priority
Parenthesis	(*Expression*)	(,)	1
Arithmetic	Exponentiation	**	2
	Unary plus, unary minus (sign)	+, -	3
	Multiplication, division	*, /, DIV, MOD	4
	Addition, subtraction	+, -	5
Comparison	Less than, less than or equal to, greater than, greater than or equal to	<, <=, >, >=	6
	Equal to, not equal to	=, <>	7
Binary logic	Negation (unary)	NOT	3
	AND logic operation	AND, &	8
	Exclusive OR	XOR	9
	OR logic operation	OR	10
Assignment	Assignment	:=	11

35.2 Control Statements

IF	Program branching with BOOLean value
CASE	Program branching with INT value
FOR	Program loop with run variable
WHILE	Program loop with execution condition
REPEAT	Program loop with abort condition
CONTINUE	Abort current loop pass
EXIT	Exit program loop
GOTO	Jump to a jump label
RETURN	Exit the block

35.3 Block Calls

Functions FCs with function value	*Variable* := FC*x*(...); *Variable* := *FCname*(...);
System functions SFCs with function value	*Variable* := SFC*x*(...); *Variable* := *SFCname*(...);
Functions FCs without function value	FC*x*(...); *FCname*(...);
Function blocks FBs with data block	FB*x*.DB*x*(...); *FBname.DBname*(...);
System function blocks SFBs with data block	SFB*x*.DB*x*(...); *SFBname.DBname*(...);
Function blocks FBs and system function blocks SFBs as local instance	*localname*(...);

Initialization of the block parameters is mandatory with FC blocks and SFC blocks, and it is optional with SFBs.

35.4 SCL Standard Functions

35.4.1 Timer Functions

Call	Data type
`Time_BCD :=`	WORD
`  Timer_function(`	(see below)
`  T_NO := Timer_address,`	TIMER
`  S    := Start_input,`	BOOL
`  TV   := Timer_duration,`	S5TIME
`  R    := Reset,`	BOOL
`  Q    := Timer_status,`	BOOL
`  BI   := Binary_time);`	WORD

with timer function

S_PULSE	Pulse time
S_PEXT	Extended pulse
S_ODT	ON delay
S_ODTS	Latching OFF delay
S_OFFDT	Off delay

35.4.2 Counter Functions

Up counter call	Data type
`BCD_count_value :=`	WORD
`  S_CU(`	
`  C_NO := Count_address,`	COUNTER
`  CU   := Count_up,`	BOOL
`  S    := Set_input,`	BOOL
`  PV   := Count_value,`	WORD
`  R    := Reset,`	BOOL
`  Q    := Counter_status,`	BOOL
`  CV   := Bin_count_val);`	WORD

Down counter call	Data type
`BCD_count_value :=`	WORD
`  S_CD(`	
`  C_NO := Count_address,`	COUNTER
`  CD   := Count_down,`	BOOL
`  S    := Set_input,`	BOOL
`  PV   := Count_value,`	WORD
`  R    := Reset,`	BOOL
`  Q    := Counter_status,`	BOOL
`  CV   := Bin_count_val);`	WORD

Up Down counter call	Data type
`Count value_BCD :=`	WORD
`  S_CUD(`	
`  C_NO := Count_operand,`	COUNTER
`  CU   := Count_up,`	BOOL
`  CD   := Count_down,`	BOOL
`  S    := Set_input`	BOOL
`  PV   := Count_value,`	WORD
`  R    := Reset,`	BOOL
`  Q    := Counter_status,`	BOOL
`  CV   := Count_value_dual);`	WORD

35.4.3 Conversion Functions

Implicit conversion functions

Function	Description
BOOL_TO_BYTE BOOL_TO_WORD BOOL_TO_DWORD BYTE_TO_WORD BYTE_TO_DWORD WORD_TO_DWORD	Supplement with leading zeros
INT_TO_DINT INT_TO_REAL DINT_TO_REAL	With sign extension
CHAR_TO_STRING	

Explicit conversion functions

Function	Description
BYTE_TO_BOOL WORD_TO_BOOL DWORD_TO_BOOL WORD_TO_BYTE DWORD_TO_BYTE DWORD_TO_WORD	Least significant bit/ byte/word is applied
CHAR_TO_BYTE BYTE_TO_CHAR CHAR_TO_INT INT_TO_CHAR	Without changing the bit assignment
STRING_TO_CHAR	
WORD_TO_INT DWORD_TO_DINT INT_TO_WORD DINT_TO_DWORD REAL_TO_DWORD DWORD_TO_REAL	Without changing the bit assignment (no conversion!)
DINT_TO_INT REAL_TO_DINT REAL_TO_INT	With rounding to INT or DINT
TRUNC ROUND	Conversion from REAL to DINT
DINT_TO_TIME DINT_TO_TOD DINT_TO_DATE DATE_TO_DINT TIME_TO_DINT TOD_TO_DINT	Without changing the bit assignment
BLOCK_DB_TO_WORD WORD_TO_BLOCK_DB	Without changing the bit assignment
BOOL_TO_INT BOOL_TO_DINT BYTE_TO_INT BYTE_TO_DINT WORD_TO_DINT DWORD_TO_DINT	Supplement with leading zeros

Continued on next page

Explicit conversion functions (continued)

INT_TO_BOOL INT_TO_BYTE INT_TO_DWORD DINT_TO_BOOL DINT_TO_BYTE DINT_TO_WORD	Least significant bit/ byte/word is applied
INT_TO_STRING DINT_TO_STRING REAL_TO_STRING STRING_TO_INT STRING_TO_DINT STRING_TO_REAL	Corresponding to the loadable IEC functions (Chapter 31)
BCD_TO_INT WORD_BCD_TO_INT BCD_TO_DINT WORD_BCD_TO_DINT	Conversion of BCD to INT or DINT
INT_TO_BCD INT_TO_WORD_BCD DINT_TO_BCD DINT_TO_WORD_BCD	Conversion of INT or DINT to BCD

35.4.4 Math functions

Call ABS	Data type
`Result :=` `ABS(input value);`	ANY_NUM ANY_NUM

Call	Data type

`Result :=`
`MathFunction(`
`Input_value);`
	REAL
	(see below)
	ANY_NUM

with math_function:

SIN Sine
COS Cosine
TAN Tangent

ASIN Arc sine
ACOS Arc cosine
ATAN Arc tangent

EXP Exponentiation to base e
EXPD Exponentiation to base 10

LN Natural logarithm
LOG Decade logarithm

SQR Generate square
SQRT Generate square root

Call ABS	Data type
`Result :=` `ABS(Input_value);`	ANY_NUM ANY_NUM

35.4.5 Shift and Rotate

Call	Data type

`Result :=`
`Shift_function(`
`IN := Input_value,`
`N := Num_of_places);`
	ANY_BIT
	(see below)
	ANY_BIT
	INT

with the shift function:

SHL Shift to left
SHR Shift to right
ROL Rotate to left
ROR Rotate to right

Index

D

S

Abbreviations

AI	Analog Input
AO	Analog Output
AS	Automation System
ASI	Actuator-Sensor-Interface
BR	Binary Result
CFC	Continuous Function Chart
CP	Communication Processor
CPU	Central Processing Unit
DB	Data Block
DI	Digital Input
DO	Digital Output
DP	Distributed I/O
DS	Data set (record)
EPROM	Erasable Programmable Read Only Memory
FB	Function Block
FBD	Function Block Diagram
FC	Function Call
FEPROM	Flash Erasable Programmable Read Only Memory
FM	Function Module
IM	Interface Module
LAD	Ladder Diagram
MC	Memory Card
MCR	Master Control Relay
MMC	Micro Memory Card
MPI	Multi Point Interface,
OB	Organization Block
OP	Operator Panel
PG	Programming Device
PS	Power Supply
RAM	Random Access Memory
RLO	Result of Logic Operation
SCL	Structured Control Language
SDB	System Data Block
SFB	System Function Block
SFC	System Function Call
SM	Signal Module
STL	Statement List
SZL	System Status List
UDT	User Data Type
VAT	Variable Table

Raimond Pigan, Mark Metter

Automating with PROFINET

Industrial Communication based on Industrial Ethernet

2nd edition, 2008, 462 pages,
271 illustrations, 237 tables, hardcover
ISBN 978-3-89578-294-7, € 59.90

This book serves as an introduction to PROFINET technology. Engineers, technicians and students are given an overview of the concept and the fundamentals for solving automation tasks. Technical relationships and practical applications are described using SIMATIC products as example.

Siemens AG (Eds.)

Fundamentals of Motion Control

October 2012, ca. 224 pages,
ca. 100 illustrations, softcover
ISBN 978-3-89578-423-1, € 19.90

The book addresses apprentices or students of engineering occupations and, moreover, everybody requiring basic information on motion control and related topics. It presents basic principles of electromagnetism and the functionality of motion control systems, followed by a closer look on the different types of electrical motors and feedback components. Then it explains operation principles of speed control units on the basis of the Sinamics family, and an overview of the motion control system Simotion allows deeper insights into programming and commands. The book concludes with an example application, a glossary and a list of resources for further studies.

Nicolai Andler

Tools for Project Management, Workshops and Consulting

A Must-Have Compendium of Essential Tools and Techniques

2nd revised and enlarged edition, 2011,
382 pages, 136 illustrations, 55 tables, hardcover
ISBN 978-3-89578-370-8, € 39.90

This well accepted standard work is a unique reference work and guide for those wanting to learn about or who are active in the fields of consulting, project management and problem solving in general. It presents cookbook-style access to more than 120 most important tools, including a rating of each tool in terms of applicability, ease of use and effectiveness.

Elisabeth Bittner, Walter Gregorc

Experiencing Project Management

Projects, Challenges & Lessons Learned

2010, 231 pages, 88 colored
photos and illustrations, hardcover
ISBN 978-3-89578-378-4, € 24.90

In this book, experienced project managers tell about exciting tasks in different countries, the personal approach to handle a task, about daily life as a project manager and about what they learned during their work in projects. That way, readers experience the fascination and specific features of a project manager's career, including the ability to always be prepared for new challenges, and they will get an idea of how personal commitment and professional, cultural and social capabilities may perfectly act together in this unique profession.

Ulf Pillkahn

Using Trends and Scenarios as Tools for Strategy Development

Shaping the Future of Your Enterprise

2008, 452 pages,
167 colored illustrations, hardcover
ISBN 978-3-89578-304-3, € 47.90

The book presents the two most powerful tools for future planning: environmental analysis, based on the use of trends, as well as the development of visions of the future through the use of scenarios. While scenarios are generally regarded as a classical management tool, it is expected that the importance of trends will gain tremendously in the coming years. Pillkahn demonstrates how to build robust strategies by aligning the results of environmental and enterprise scenarios, thereby offering entirely new insights.

Siemens AG

Arenas for the Future

Smart Stadium Solutions

November 2012, ca. 480 pages,
ca. 400 colored illustrations,
27,7 cm × 23 cm, hardcover with book jacket
ISBN 978-3-89578-406-4, ca. € 59.90

“Arenas for the Future” is based on the experience gained in about 150 stadiums worldwide during the last few years. It provides a wealth of information from all disciplines of stadium technology, supports managers and technical experts in planning and provides investors, managers and city planners with a comprehensive view of how modern technologies help to create smart solutions for stadiums and their environment.

The book also shows how to master the extreme situations of days with and without event or match economically reasonable and in a sustainable way. It helps to translate the demands by FIFA and UEFA into practicable solutions and to create a pleasant place to stay for fans and their communities.